Yorikiyo Nagashima

Elementary Particle Physics

Related Titles

Nagashima, Y.

Elementary Particle Physics
Volume 1: Quantum Field Theory and Particles

2010
ISBN: 978-3-527-40962-4

Brock, I., Schörner-Sadenius, T. (eds.)

Physics at the Terascale

2011
ISBN: 978-3-527-41001-9

Hauptman, J.

Particle Physics Experiments at High Energy Colliders

2011
ISBN: 978-3-527-40825-2

Mandl, F., Shaw, G.

Quantum Field Theory

2010
ISBN: 978-0-471-49684-7

Russenschuck, S.

Field Computation for Accelerator Magnets
Analytical and Numerical Methods for Electromagnetic Design and Optimization

2010
ISBN: 978-3-527-40769-9

Stock, R. (ed.)

Encyclopedia of Applied High Energy and Particle Physics

2009
ISBN: 978-3-527-40691-3

Griffiths, D.

Introduction to Elementary Particles

2008
ISBN: 978-3-527-40601-2

Belusevic, R.

Relativity, Astrophysics and Cosmology

2008
ISBN: 978-3-527-40764-4

Lilley, J. S.

Nuclear Physics
Principles and Applications

2001
ISBN: 978-0-471-97935-7

Thomas, A. W., Weise, W.

The Structure of the Nucleon

2001
ISBN: 978-3-527-40297-7

Yorikiyo Nagashima

Elementary Particle Physics

Volume 2: Foundations of the Standard Model

WILEY-VCH

WILEY-VCH Verlag GmbH & Co. KGaA

The Author

Yorikiyo Nagashima
Osaka University
Japan
nagashimayori@ybb.ne.jp

Cover Image
Japanese symbol that denotes "void" or "nothing"; also symbolizes "supreme state of matter" or "spirit".

Library of Congress Card No.:
applied for

British Library Cataloguing-in-Publication Data:
A catalogue record for this book is available from the British Library.

Bibliographic information published by the Deutsche Nationalbibliothek
The Deutsche Nationalbibliothek lists this publication in the Deutsche Nationalbibliografie; detailed bibliographic data are available on the Internet at http://dnb.d-nb.de.

Print ISBN 978-3-527-40966-2
ePDF ISBN 978-3-527-64891-7
ePub ISBN 978-3-527-64890-0
mobi ISBN 978-3-527-64889-4
oBook ISBN 978-3-527-64888-7

Cover Design Adam-Design, Weinheim
Composition le-tex publishing services GmbH, Leipzig
Printing and Binding Markono Print Media Pte Ltd, Singapore

Printed in Singapore
Printed on acid-free paper

Contents

Preface

The unified theory of the electroweak interaction and QCD (Quantum chromodynamics) referred to as the Standard Model proposed in the late 1960s has been very successful and explains all the known phenomena at least in principle. The fundamental concept goes beyond particle physics and is at the core of cosmology, nuclear theory and even in some part of condensed matter physics. This book (*Elementary Particle Physics Volume II*) explains basic concepts of the Standard Model (SM) and its experimental foundations. Volume I offered an introduction to the field theory and explained how the Standard Model was historically formulated. Although this book can be considered as its follow up, it is self-contained and stands on its own, except occasional references to Volume 1 for basic formulas.

Target readers are graduate students concentrating on careers as particle physicists, though the author also has physicists who are not specialized in experimental particle physics in mind. Emphasis is placed on experimental tests of main themes of unified gauge theories rather than the detailed explanation of data. The author tried to collect some examples of initial trial, error and later correction on the experimental side because he believes that finding a problem is more important than solving a given exercise, and there is a lesson to be learned. All basic, (i.e., tree level) formulas of important concept and cross sections, are derived from the first principle in a self-contained way. Readers should be able to derive all the formulas given in this volume by themselves, except those containing higher order corrections if they have basic knowledge of the field theory[1)] as given in Vol. 1 of this textbook. In this way, the author wishes to bridge between two extreme tendencies of books in particle physics, one too experimental and the other too theoretical.

For higher order calculations, only qualitative discussions are given, that is, except those that are necessary to reproduce the precise Z resonance data. Some theoretical formulas and tools required to understand equations for specific processes are supplemented in appendixes. However, the reader may just as well accept given formulas in the main text and use them to interpret the given data. For those who want to elaborate on details, an extensive list of references are provided. Phenomena explained in this volume are new ones discovered after proposition, and were instrumental in establishing the validity of the Standard Model (SM). Choice of

1) Knowledge to carry out Dirac's γ matrix algebra and to quantize fields.

experiments are exemplary, but to some extent arbitrary. They are not necessarily the first nor the most advanced. They inevitably reflect the author's prejudice. Those that existed in the pre-SM era or were discovered after the SM but could be explained by pre-SM theories were already discussed in Vol 1.

At the Lagrangian level, the electroweak theory is conceptually more complicated due to mixing of the two symmetries $SU(2)$, $U(1)$ and the spontaneous symmetry breakdown. However, their coupling strength is weak as exemplified by the fine structure constant $\alpha = 1/137$. In general, a tree approximation (lowest order perturbation) calculation is sufficient to reproduce experimental data to a reasonable accuracy. In some cases, such as data at the Z resonance, higher order calculations are necessary to meet precision experiments quantitatively, but no new concepts are needed to obtain desired accuracies in order to match experimental precision.

Applications of the strong interaction theory, however, despite its relative simplicity of the Lagrangian, must cater to specific aspects of phenomena under different circumstances. For a certain class of reactions, theorems or laws at intermediary levels are necessary to handle the QCD phenomenology. Mathematical tools have been accumulated and the skills required to maneuver them have formed a sort of QCD engineering.

The reason is two-fold: Firstly, because of the confinement, accurate correspondence between theory and experiment is hard to realize. The partons (quarks and gluons) are typically observed as jets. Clear connections between the partons at the theoretical level and hadronic jets at the experimental level are yet to be established. Hadrons are composites of the quarks, and the observed phenomena are end results of multibody interactions. Extraction of the interested process requires complicated processing, that is, experimentally as well as theoretically. Secondly, though coupling strength of the gluon (the QCD gauge particles) is small enough so that one can use perturbation expansions, it is not small enough for tree approximation formulae to be applied directly to measurements. In almost all cases, higher order calculations are needed to obtain reasonable agreement with experiments as well as to establish theoretical stability. Phenomenological treatments using effective theories, albeit based on the QCD axiom, have to be developed for a certain class of reactions. They act as intermediaries between the QCD axioms and experimental data. Sometimes, they establish themselves as a discipline of their own like lattice QCD and chiral perturbation theory. Selection of topics in QCD inevitably reflects the author's own prejudice.

Because of the reasons stated above, the author chose to start from the electroweak phenomena (Part One) and discuss the QCD later (Part Two). However, treatment of the electroweak and strong phenomena are reasonably self-contained and the reader may as well start from Part Two (Chapter 7 and thereafter) if he or she chooses to do so.

This book is structured as follows. Chapter 1 describes the basic concept of the electroweak gauge interaction, prepares tools to calculate cross sections necessary to interpret electroweak phenomena. The Feynman rules are given. A simple explanation on the roles of the gauge symmetry and the Higgs to maintain unitarity in the framework of spontaneously broken gauge theory is given. It is somewhat on the technical side because physical concept of the gauge theory (or geometrical interpretation to be exact) is already given in Vol. 1, Chapt. 18. Then, a description of how the GWS (Glashow–Weinberg–Salam) model was established as the electroweak theory, first via the discovery of the neutral current (Chapter 2) and W, Z bosons (Chapter 3), and then by precision tests using the Z resonance data (Chapter 4).

Chapter 5 is an exception to others in the sense that it mathematically goes one step beyond the tree level to handle loop integrals and discuss how to deal with ultraviolet divergences. The electroweak higher order corrections are explicitly derived and compared with experiments. Accurate predictions including higher order corrections and subsequent confirmation by precision Z data firmly established the validity of the electroweak theory beyond doubt. In this sense, the role of Z resonance precision data may be compared to the Lamb shift and $g-2$ in QED. In fact, predictions are so perfect that there is little space for new physics. However, the Standard Model is not a perfect theory. We need to clarify mixings among the quark as well as lepton flavors because the electroweak theory by its construction closes basically within a generation and the flavor mixing is a bridge to possible new physics. Chapter 6 describes the quark flavor mixing, that is, the CKM (Cabibbo–Kobayashi–Maskawa) matrix and clarifies the framework in which the CP violation manifests itself. Dynamics of the Higgs, the first and the last ingredient of the electroweak theory, is not covered in this book because it was only discovered at the time when this book was about to be published and its dynamic properties are yet to be established.

Chapter 7, the first chapter in Part Two, describes the basics of the QCD Lagrangian and introduces the asymptotic freedom and renormalization group equation which are building blocks of the QCD framework. Concepts of the LLA (leading logarithmic approximation), which is similar to but not identical to the conventional first order perturbation expansion, are given. Chapter 8 introduces the concept of factorization and an evolution equation, namely, the DGLAP equation. It can describe both deep inelastic scattering phenomena and fragmentations of the partons. Chapters 9 and 10 explain that jets are end products of partons, and construct jet phenomenology which connects partons at the theoretical level and hadrons that are actually observed. Soft gluons are emitted with the same mechanism as soft photons in QED, but because of the non-Abelian nature of the QCD and confinement, they behave very differently and appear in a variety of phenomena which are by no means exhausted. They constitute a field of active research and only a glimpse of it is touched in Chapter 10. Jet phenomena produced in high p_T hadronic reactions are interpreted as essentially simple elementary processes like elastic scatterings of two partons. However, the multiparton structure of hadrons complicates the situation. Chapter 11 shows that the factorization theorem provides a powerful tool for disentangling the seemingly complicated hadronic processes which are, in reality, simple elementary processes at the parton level. Part Two culminates in the discovery of the top quark, the last and heaviest quark in the Standard Model.

It is impossible to cover all the interesting phenomena of the Standard Model in a single book. This book did not discuss decays of heavy quarks, many of flavor physics, and notably the so-called soft processes. The asymptotic freedom cannot be applied to the soft processes and nonperturbative treatments like lattice QCD or chiral perturbation theory require. We only had a glimpse of chiral perturbation theory in extracting the quark masses.

The author apologizes to the reader for promising to, but unfortunately not covering physics beyond the Standard Model in Volume 2. The reason is two-fold: firstly, LHC started producing data just when the book was about to be published and secondly, my description of the SM grew in size more than expected in the beginning. Since the LHC results are most likely to change the future course of particle physics, topics that deal with physics beyond the SM faced a danger of being obsolete before its publication. Therefore, the author decided to separate the latter half of Volume 2 into Volume 3 and delay publication of the latter for a year. In this way, Volume 2 is not likely to be obsolete in the near future, at the same time Volume 3 will be more up-to-date by at least picking some new discoveries. Intended topics at this stage include the Higgs, neutrino, concept of GUTS (grand unified theories), supersymmetry, extra dimension, axion and cosmology.

Osaka, 15 April 2012 *Yorikiyo Nagashima*

Acknowledgments

The author would like to express his gratitude to the authors cited in the text and to the following publishers for permission to reproduce the various photographs, figures and tables:

- American Physical Society, publisher of the Physical Review, Physical Review Letters and the Review of Modern Physics, and for permission to reproduce figures: 2.10, 2.13, 2.23, 2.25, 3.10, 3.11, 3.14, 4.21, 4.28, 4.32, 5.11 6.1, 6.2, 6.6, 6.9, 6.16, 6.17, 6.22, 8.19, 9.6b, 9.8, 10.10, 11.4, 11.6, 11.7, 11.9b, 11.11, 11.13, 11.14, 11.15, 11.16, 11.17, 11,18, 11.31a, 11.32b, 11.34, 11.36.
- Annual Reviews, publisher of the Annual Review of Nuclear and Particle Science for permission to reproduce figures: 3.12, 6.4, 9.4b, 11.1a, 11.2c, 11.10.
- Elsevier Science Ltd., publisher of Nuclear Instruments and Methods, Nuclear Physics, Physics Letters, Physics Reports, Progress of Particle and Nuclear Physics for permission to reproduce figures: 2.2, 2.5, 2.7, 2.8, 2.11, 2.12, 2.15, 2.19, 2.21, 3.1, 3.2, 3.3, 3.4, 3.5, 3.9a, 3.13, 3.17, 3.18, 3.19, 4.2, 4.5, 4.6, 4.7, 4.8, 4.9, 4.10, 4.13, 4.14, 4.23, 4.24, 4.29, 4.31, 4.34, 4.35, 6.5, 6.20, 7.5, 8.10, 8.11, 8.12, 8.14, 8.15, 9.4a, 9.6a, 9.7, 9.9, 9.13a, 9.17a, 9.23, 9.27, 9.29, 9.31b, 10.3, 10.6b, 10.11, 10.13b, 10.18, 11.2b, 11.9a, 11.20, 11.22, 11.23, 11.25, 11.29a, 11.30a, 11.31b.
- Institute of Physics Publishing Ltd., publisher of the Journal of Physics, Physica Scripta, Report on Progress of Physics for permission to reproduce figures and tables: 3.6, 4.11, 4.12, 4.15, 4.16, 4.18, 4.22, 4.33, 9.22, 9.30, Table 1, Table 2, Table 9.3, in Chapter 9, 9.31a, 10.1a, 10.7, 10.12b, 10.19, 10.20a, 11.12.
- Japan Physical Society, publisher of Butsuri for permission to reproduce figures; 2.14, 4.1, 6.19.
- Nature Publishing Group, publisher of Nature for permission to reproduce Figure 6.13.
- Particle Data Group, publisher of Review of Particle Physics for permission to reproduce figures: 3.15, Table 4.3, 4.25, 5.7, Table 5.1, 5.8, 5.9, 5.10, 6.26, 8.16, 8.17, 9.17b, 9.18, 10.12a, 10.13a, 10.15, 11.1b, 11.32a.
- Springer Verlag, publisher of European Journal of Physics and Zeitschrift für Physik for permission to reproduce figures: 2.18, 2.22, 4.19, 4.27, 7.6, 8.13, 10.1b, 10.6a, 10.16, 10.20b, 11.2a.

Color Plates

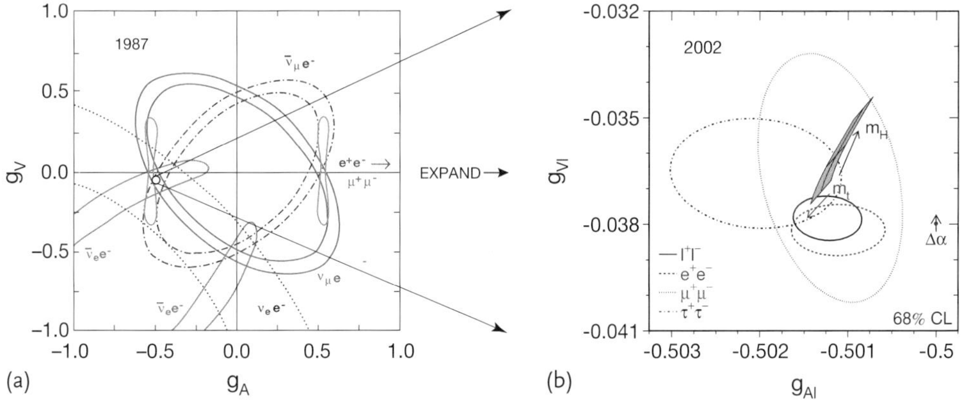

Figure 2.7 (a) The neutrino scattering and e^+e^- annihilation data circa 1987 constrained the values of g_V and g_A (v_e and a_e in the text) to lie within broad bands, whose intersections helped establish the validity of the Standard Model and were consistent with the hypothesis of the lepton universality. (b) shows the results of the LEP/SLD measurements in 2002 at a scale expanded by a factor of 65 (see Figure 4.34). The flavor-specific measurements demonstrate the universal nature of the lepton couplings unambiguously on a scale of approximately 0.001. The shaded region in the lepton plot shows predictions of the Standard Model for $m_t = 178.0 \pm 4.3$ GeV and $m_H = 300^{+700}_{-186}$ GeV. Figure (a) adapted from [1], (b) from [2, 7].

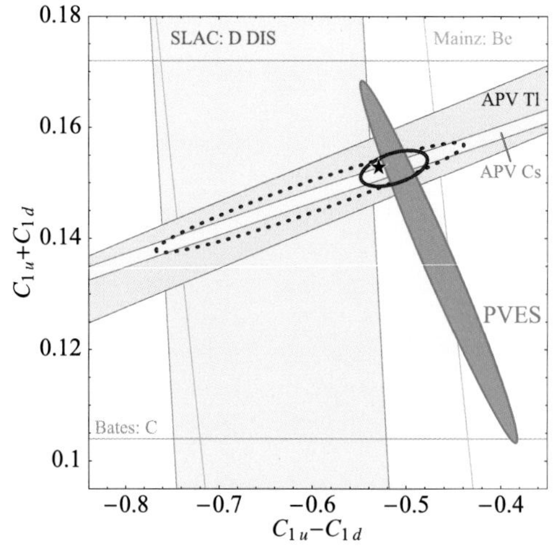

Figure 2.13 Allowed region determined by the polarized $e-D$ scattering, APV (atomic parity violation) experiments and recent PVES (parity violating electron scattering) experiments. The dotted contour displays the previous experimental limits (95% CL) reported in the PDG [3] together with the prediction of the standard model (black star). The filled ellipse denotes the new constraint provided by recent high precision PVES scattering measurements on hydrogen, deuterium, and helium targets (at 1 standard deviation), while the solid contour (95% CL) indicates the full constraint obtained by combining all results [4].

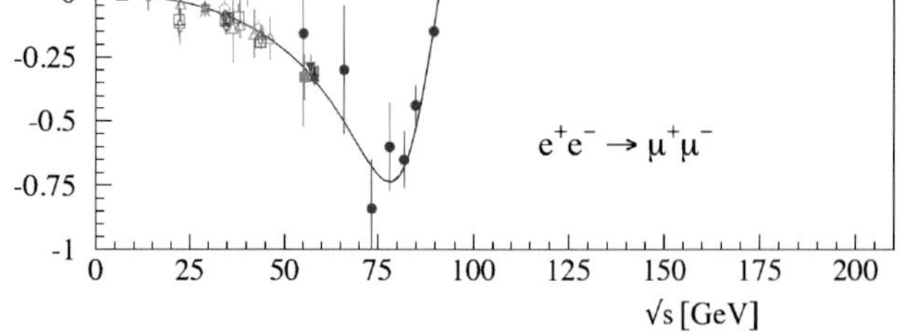

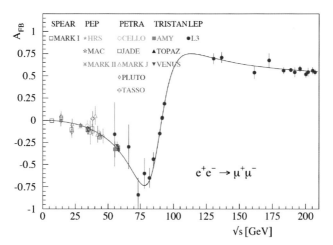

Figure 2.19 A compilation of the asymmetry data A^{μ}_{FB} compared to the Standard Model prediction [5, 9].

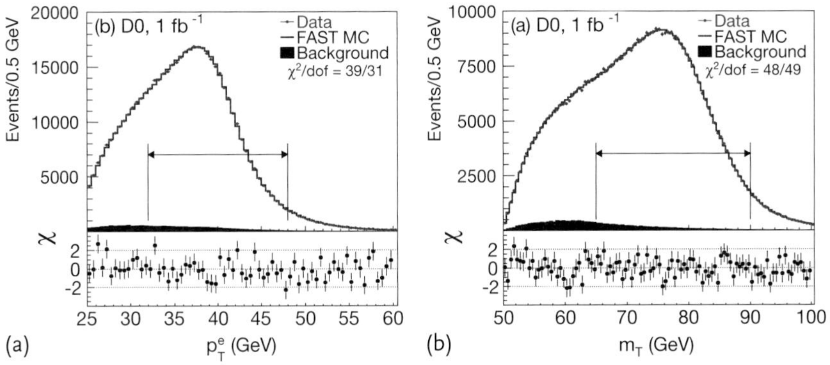

Figure 3.11 Distributions of (a) p_{eT} and (b) m_T, and FASTMC simulation with backgrounds. The χ values are shown below each distribution where $\chi_i = [N_i - (FASTMC_i)]/\sigma_i$ for each point in the distribution, N_i is the data yield in bin i, and only the statistical uncertainty is used. The fit ranges are indicated by the double-ended horizontal arrows [10].

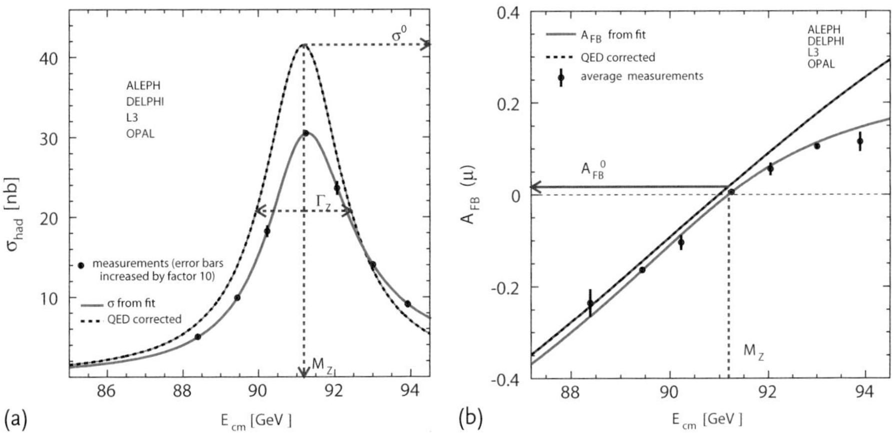

Figure 4.2 Average over measurements of the hadronic cross sections (a) and of the muon forward–backward asymmetry (b) by the four experiments at the LEP, as a function of center of mass energy. The full line represents the results of model-independent fits to the measurements. Correcting for QED photonic effects yields the dashed curves, which define the Z parameters described in the text.

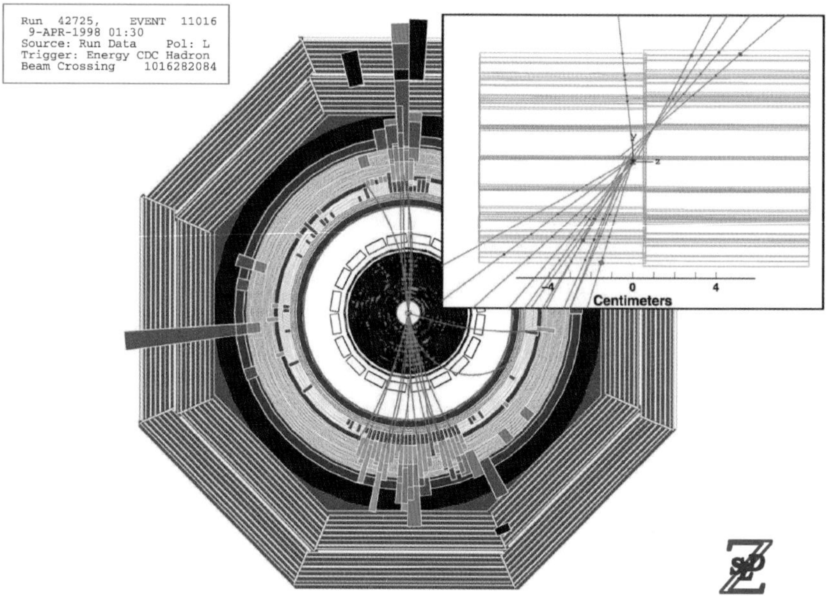

Figure 4.6 $e^-e^+ \rightarrow b\bar{b}$ event display by SLD. The inset is an enlarged *r–z* view of the vertex detector showing the secondary vertices [7].

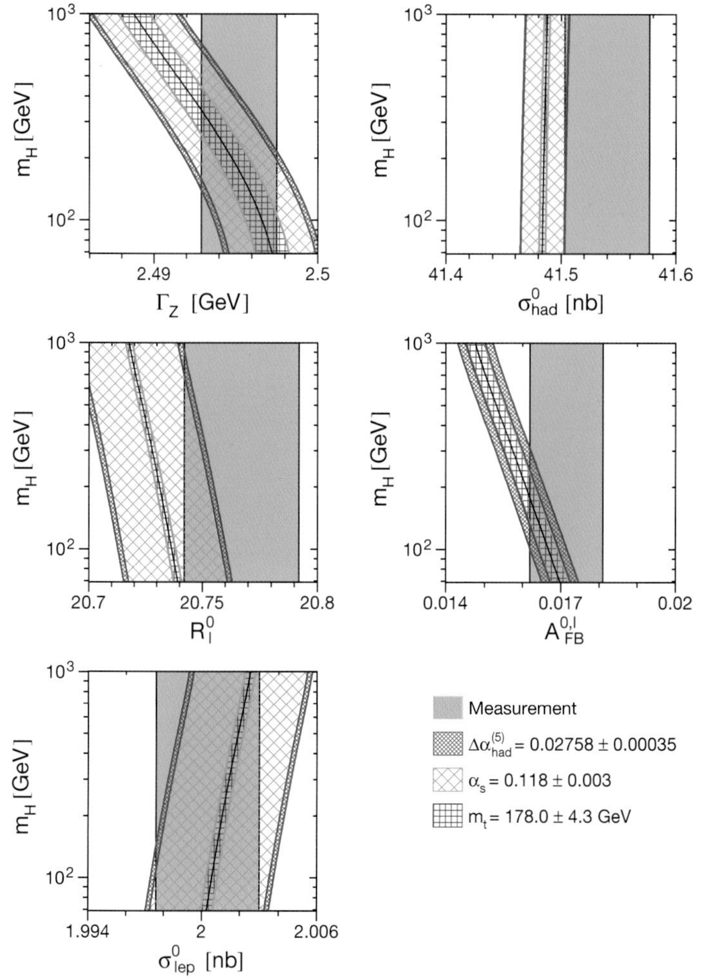

Figure 4.24 Comparisons of the LEP combined measurements of Γ_Z, σ^0_{had}, R^0_ℓ, $A^{0,\ell}_{FB}$, and σ^0_{lep} with the Standard Model prediction as a function of the Higgs boson mass. $\sigma^0_{lep} = (12\pi/m_Z^2)(\Gamma_l/\Gamma_Z)^2$ is the lepton analog of the hadronic cross section σ^0_{had}. The measured values with their uncertainty are shown as vertical bands. The width of the SM band arises due to uncertainties in $\Delta\alpha^{(5)}_{had}(m_Z^2)$, $\alpha_s(m_Z^2)$ and m_t in the ranges indicated. $\Delta\alpha^{(5)}_{had}(m_Z^2)$ is a radiative correction to the fine structure constant which contains contributions from five flavors of fermions [7].

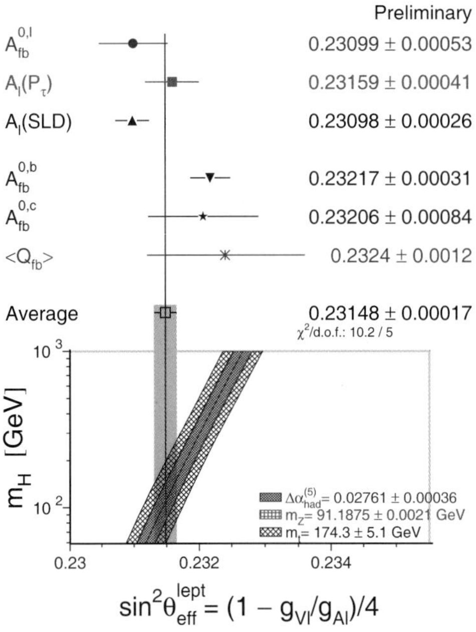

Figure 4.34 Comparison of the effective electroweak mixing angle $\sin^2\theta_{\text{eff}}$ derived from measurements depending on lepton couplings only (top) and also quark couplings (bottom). Also shown is the SM prediction for $\sin^2\theta_{\text{eff}}$ as a function of m_H. The additional uncertainty of the Standard Model prediction is parametric and dominated by the uncertainties in $\Delta\alpha^{(5)}_{\text{had}}(m_Z^2)$ and m_t, shown as the bands. The total width of the band is the linear sum of these effects [2, 7].

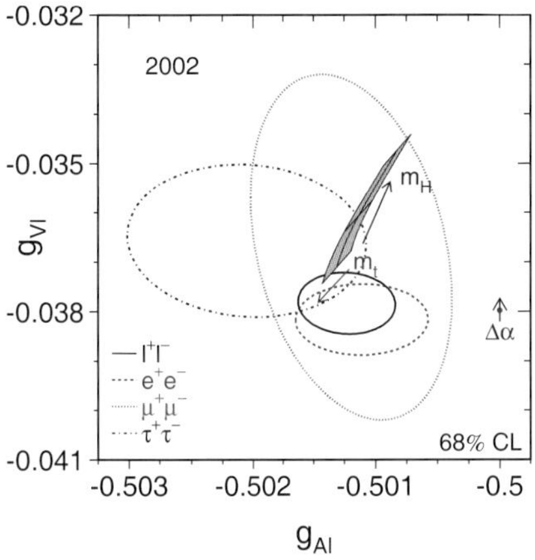

Figure 4.35 Comparison of the effective vector and axial-vector coupling constants for leptons. Here, a_l, v_l is denoted as g_{Al}, g_{Vl}. The shaded region in the lepton plot shows the predictions within the Standard Model for $m_t = 178.0 \pm 4.3$ GeV and $m_H = 300^{+700}_{-186}$ GeV. Varying the hadronic vacuum polarization by $\Delta\alpha^{(5)}_{\text{had}} = 0.02758 \pm 0.00035$ yields an additional uncertainty on the SMs prediction shown by the arrow labeled $\Delta\alpha$. This is the one shown in Figure 2.7 in Section 2.2 [2, 7].

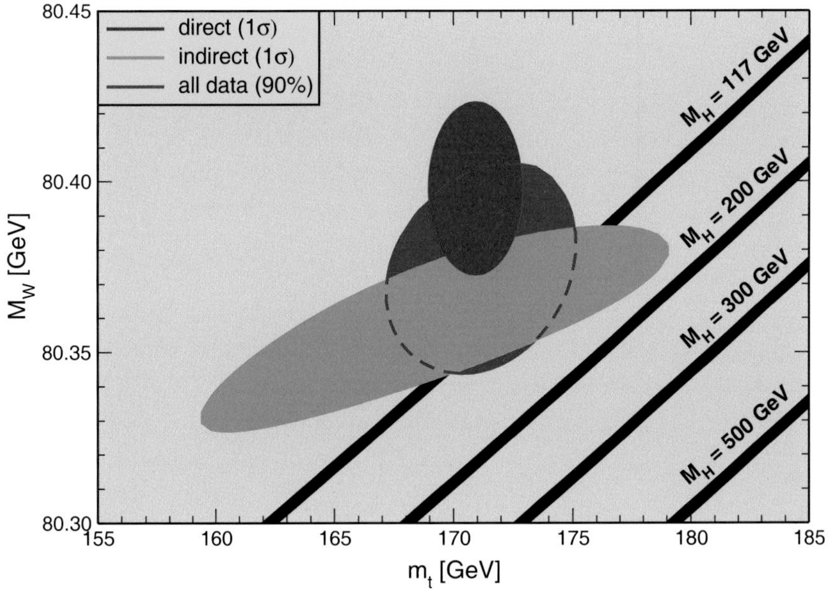

Figure 5.9 One standard deviation region in m_W as a function of m_t for the direct production and indirect (inferred from radiative correction) data and 90% region allowed by all data. The Standard Model prediction as a function of m_H is also indicated. Widths of the m_H lines reflect the theoretical uncertainty from $\alpha(m_Z)$ [11].

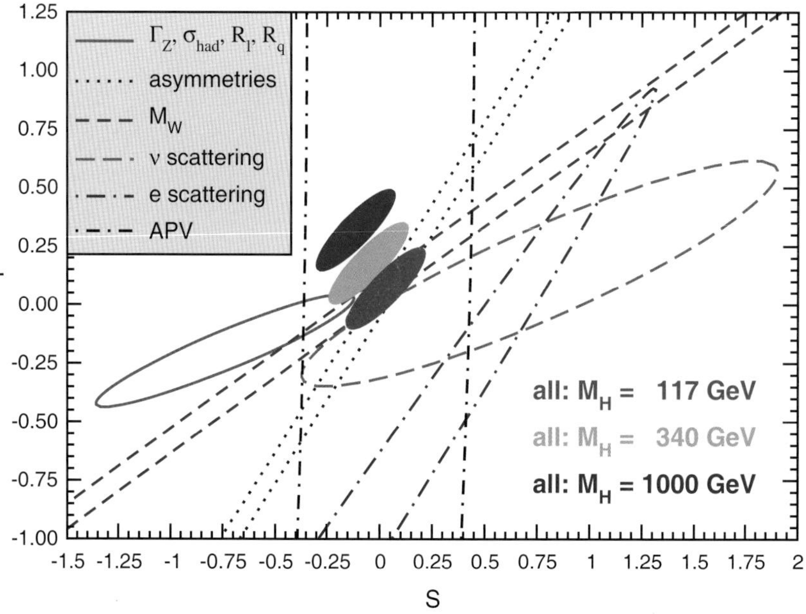

Figure 5.10 1σ constraints (39.35%) on S_{new} and T_{new} from various inputs combined with m_Z. S and T only represent contributions of new physics (The contour assumes $U = 0$). (Uncertainties from m_t are included in the errors.) The contours assume $m_h = 117$ GeV [3].

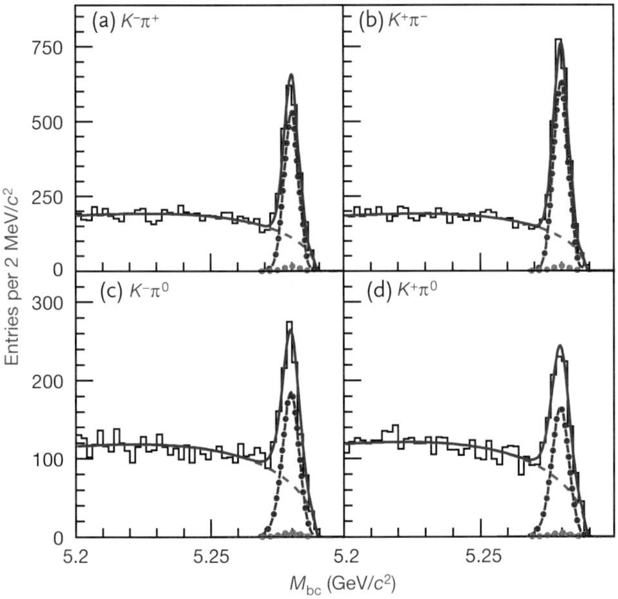

Figure 6.13 M_{bc} projections for $K^-\pi^+$ (a), $K^+\pi^-$ (b), $K^-\pi^0$ (c) and $K^+\pi^0$ (d). Histograms are data, solid blue lines are the fit projections, point-dashed lines are the signal components, dashed lines are the continuum background, and gray dotted lines are the $\pi^\pm\pi$ signals that are misidentified as $K^\pm\pi$. M_{bc} is a variable that should agree with the B meson mass ($=$ 5.279 GeV) if the observed $K\pi$ events are decay products of the B meson. It is defined by $\sqrt{E_{\text{beam}}^2 - (\boldsymbol{p}_K + \boldsymbol{p}_\pi)^2}$, $E_{\text{beam}} = m\{\Upsilon(4S)\}/2$ [12].

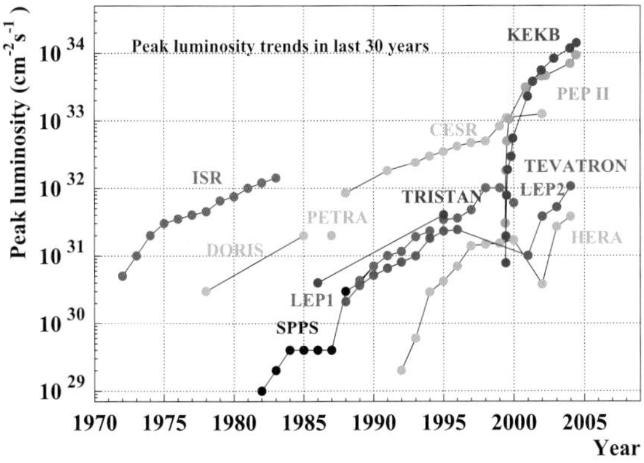

Figure 6.19 Peak luminosity trends achieved by the world collider accelerators which aimed for luminosity frontiers in the last 30 years. TRISTAN (e^-e^+), LEP (e^-e^+) aimed the energy frontier at the construction. No simple comparisons can be made with colliders of different particle species, but ISR ($p-p$), SppS ($p-p$), TEVATRON ($\overline{p}-p$), and HERA ($e-p$) are also listed for reference. One sees a quantum jump in the luminosity achieved by the B-factories (KEKB and PEPII) [6].

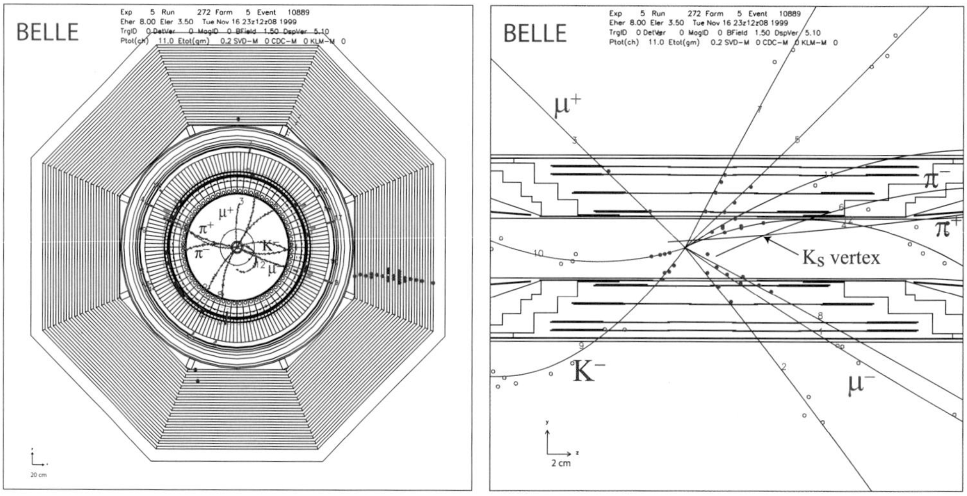

(a) (b)

Figure 6.21 A typical Belle event $B^0(\overline{B}^0) \rightarrow J/\psi + K_s$, $J/\psi \rightarrow \mu^-\mu^+$, $K_s \rightarrow \pi^+\pi^-$ [13]. (a) is $x-y$ view, (b) is expanded $y-z$ view of the vertex detector.

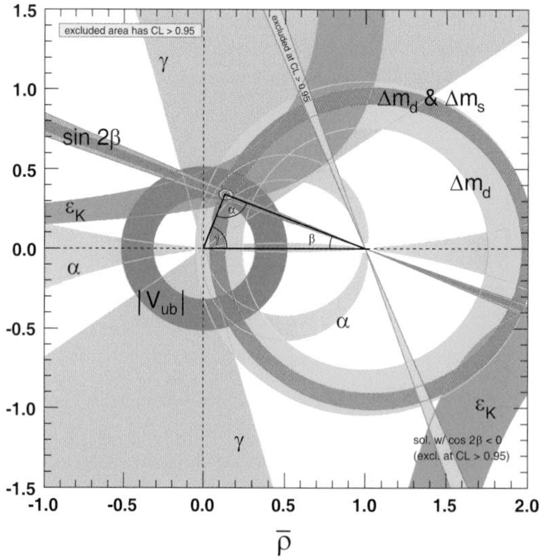

Figure 6.26 Constraints on the $\overline{\rho} - \overline{\eta}$ plane by various measurements. Note the angles are $\alpha = \phi_2, \beta = \phi_1, \gamma = \phi_3$. The shaded area is 95% CL bands [3].

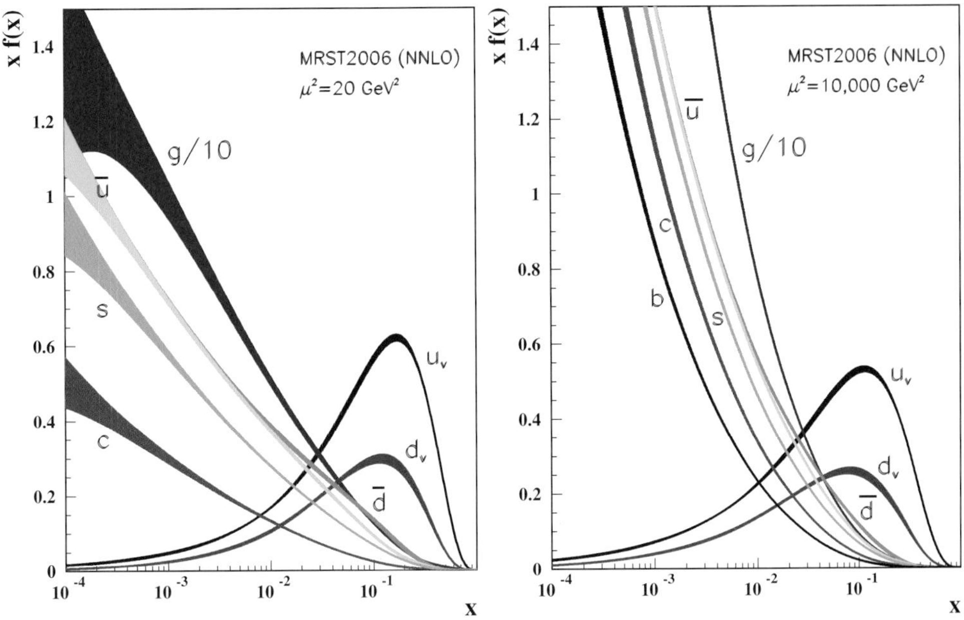

Figure 8.16 Parton distribution functions by MRS. Distributions of x times the unpolarized parton distributions $f(x)$ (where $f = u_v, d_v, u, d, s, c, b, g$) and their associated uncertainties using the NNLO MSTW2008 parameterization [15] at a scale $\mu^2 = 20\,\text{GeV}^2$ and $\mu^2 = 10\,000\,\text{GeV}^2$ [3].

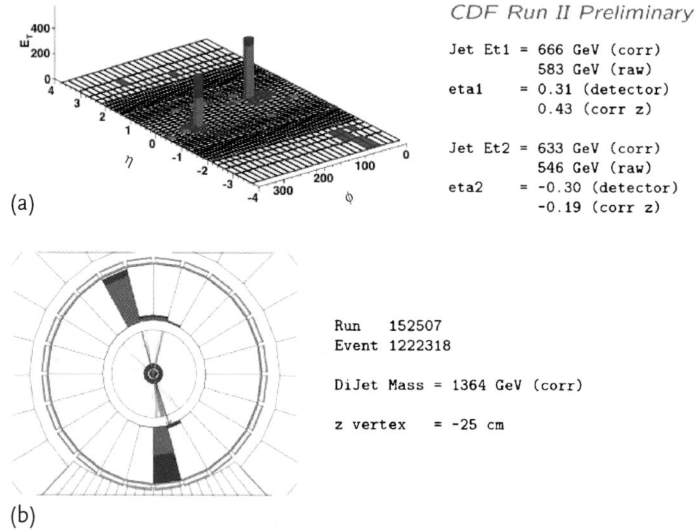

Figure 9.2 Event display of 2-jets in hadronic reactions with mass 1.36 TeV. Energies (666 and 633 GeV) of the two jets are shown as two columns (a) or two colored bars (b). The pink bars correspond to energy collected by the electromagnetic calorimeter and the blue bars to the hadronic calorimeter. Their angle positions ($\eta = \ln \cot \theta$ and ϕ clearly indicate that they are emitted in back-to-back configuration). Figure from CDF/FNAL [22].

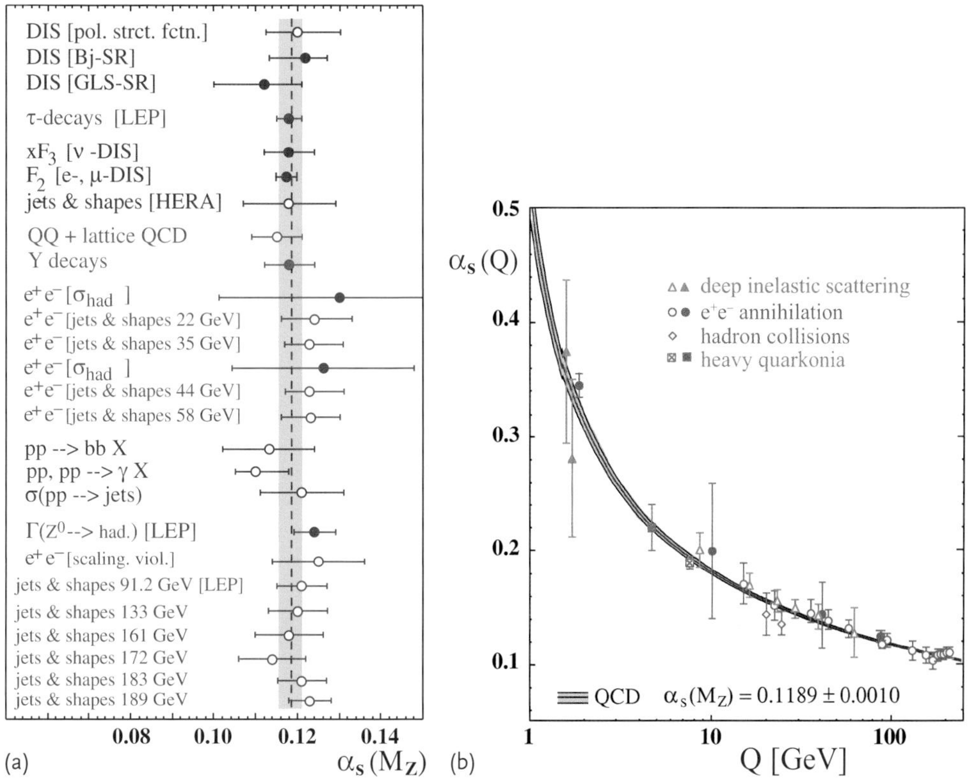

Figure 9.31 Asymptotic freedom and running coupling constant: (a) All $\alpha_s(m_Z^2)$s in the fourth column of Table 9.4 compared with a fourth order theoretical prediction (dashed line) [23, 24]. (b) $\alpha_s(Q^2)$s in the third column of Table 9.4 plotted as a function of Q [24, 25].

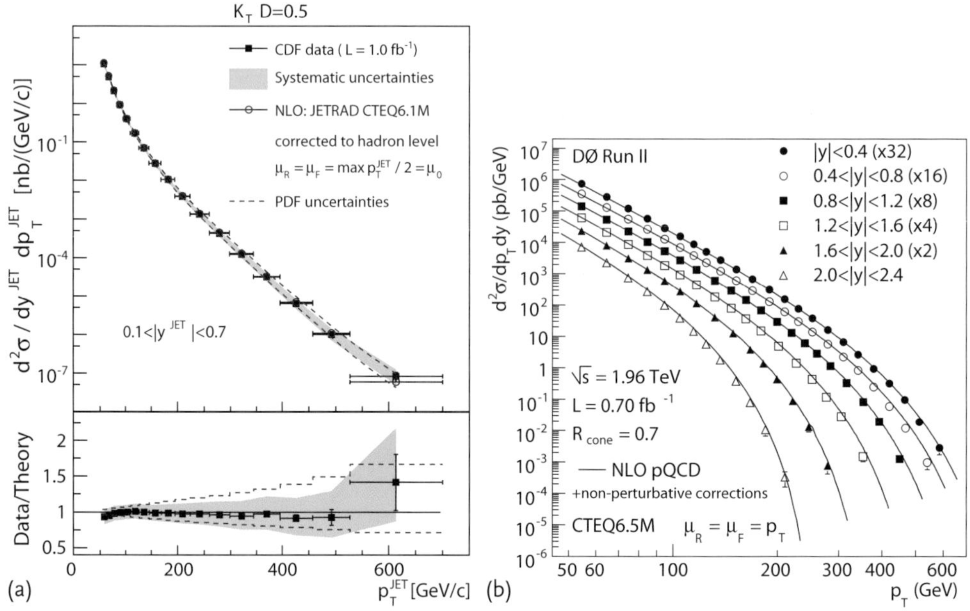

Figure 11.6 $d^2\sigma/dp_T dy$: (a) compares data with the hadronization program JETRAD [26]. The data used k_T algorithm (see Section 11.4.2 $D = 0.5, 1.0$) to define the jet energy. The lower part shows ratios of data/theory [27]; (b) compares data with the perturbative QCD (pQCD) at various values of jet rapidity y (η_c or η_d in the text). The data uses jet energy algorithm with $R = 0.7$ to define jets [28].

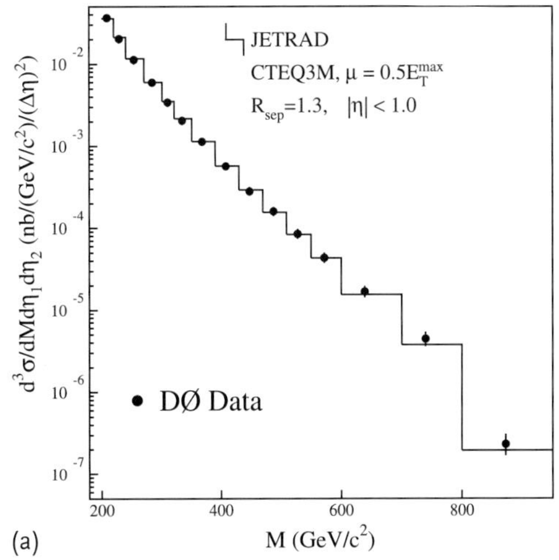

(a)

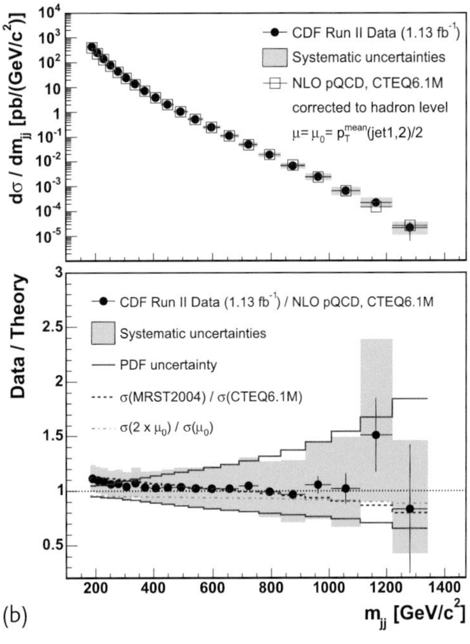

(b)

Figure 11.7 (a) D0 data for dijet production cross section $d^3\sigma/dM\,d\eta_1\,d\eta_2$ for $\eta_{\text{jet}} < 1.0$. The histogram represents the JETRAD prediction. [29, 30] CDF data for dijet mass cross section: (b) measured dijet mass spectrum for both jets to have $|y| < 1$ compared to the NLO pQCD prediction obtained using the CTEQ6.1 PDFs. (c) The ratio of the data to the NLO pQCD prediction [31, 32].

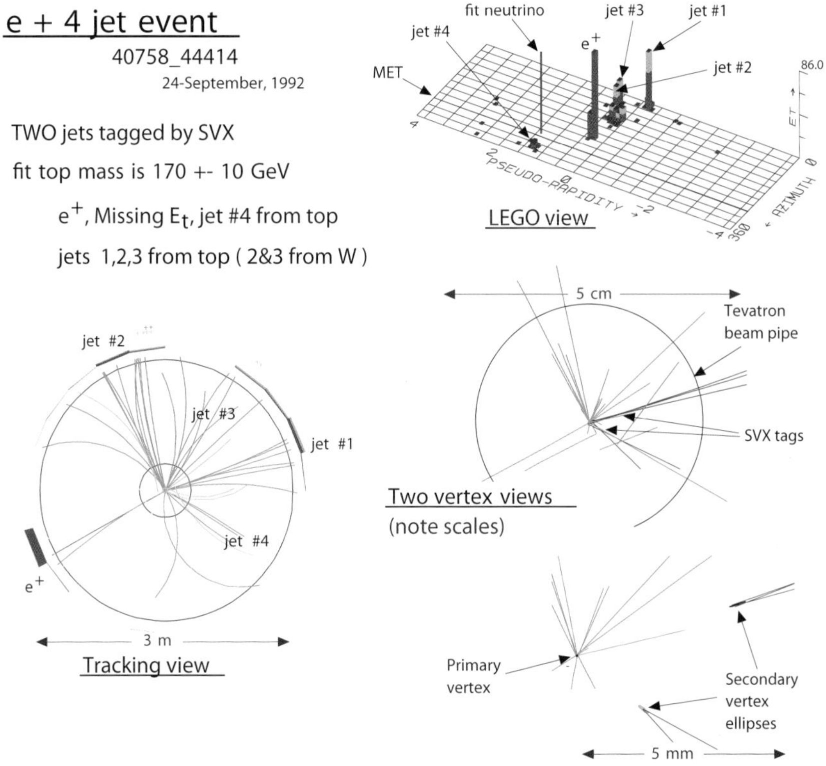

Figure 11.33 Top Events Display [33]. $q\overline{q} \rightarrow t\overline{t} \rightarrow e^{+} + 4$ jets. Two jets from secondary vertices are tagged as *b*-quarks.

Part One Electroweak Dynamics

1
The Standard Model

1.1
Introduction

The electroweak theory (also known as the Glashow–Weinberg–Salam theory) is a unified theory describing the weak and electromagnetic interaction of elementary particles. Its prototype was Glashow's [34] model to combine the weak and electromagnetic interaction in the framework of $SU(2) \times U(1)$ symmetry. Weinberg and Salam [35, 36] supplemented the Higgs mechanism [37] to generate masses of gauge particles and fermions, and succeeded in placing the model in the mathematical framework of gauge theories. Its renormalizability was proved by 't Hooft [38, 39], completing it as a self-consistent mathematical theory.

The quantum chromodynamics (QCD) is a gauge theory based on color $SU(3)$ symmetry. Its prototype was an idea that the color charge which comes in three kinds is the source of the strong interaction acting among the quarks [40, 41]. It has been elevated to the theory of strong interaction when the asymptotic freedom was discovered by Gross, Politzer, and Wilczek [42–44].

The essence of the Standard Model, a name to denote GWS theory and/or QCD, can be summarized in the following phrases.

1. Building blocks of matter are quarks and leptons.
2. Their interactions are described in the mathematical framework of the gauge field theory.
3. The vacuum is in a sort of superconducting phase.

This chapter gives a simple introduction to axioms of the electroweak theory and prepares tools for calculating cross sections at least at the tree level. For understanding the physical (or rather geometrical) picture of the gauge theory, we refer the reader to Vol. 1, Chapt. 18.
Starting from the fundamental Lagrangian based on a gauge symmetry and applying spontaneous symmetry breaking, one obtains a Lagrangian which is more relevant for describing observed phenomena and which can provide associated Feynman rules. A simple description on the role of gauge symmetry as well as the Higgs

Elementary Particle Physics, Volume 2: Foundations of the Standard Model, First Edition. Yorikiyo Nagashima.

field to maintain unitarity in the framework of the spontaneously broken gauge theory is also given.

1.2
Weak Charge and $SU(2) \times U(1)$ Symmetry

Based on the accumulated evidence described in Vol. 1 of this book, we choose to start from axioms of the electroweak theory and derive equations of motion. We will use the word “charge” for an object that is capable of producing a force field. This is in analogy to the electric charge which is the source of the electromagnetic force. Then, the source of the weak force may be called the weak charge. It has remarkable characteristics similar to the electric charge in the sense that the static weak charge produces Coulomb-like weak force potential and the moving charge, that is, the weak current, like the electric current produces “weak” magnetic fields whose dynamic characteristic is very similar to that in QED. However, there are two conspicuous differences between the two forces.

The weak charge appears in two kinds and the symmetry they obey is $SU(2)$, in contrast to $U(1)$ of the electric charge. Its value could be, and generally are, different for the left- and right-handed particles in contrast to the electric charge which does not differentiate between the two. This is referred to as the chiral symmetry. Understanding its concept is fundamental for clarification of the weak phenomena.

The electromagnetic force is the essential force at molecular and atomic levels. The electromagnetic phenomena at their fundamental level do not differentiate left from right unless we elaborate in order to closely examine it. We have been accustomed to this concept for over a hundred years. Besides, a massive particle can be in either state depending on an observer's position. Accordingly, we take it for granted that the left- and right-handed particles are the same particle, though just in different states. The discovery of the different weak charge carried by the left- and right-handed particles made us recognize that they could be different particles. Since the fundamental forces which we know respect the local gauge symmetry, we need to apply gauge transformations differently for the left- and right-handed particles, which is referred to as the chiral gauge transformation. The name is a misnomer because it connotes a different operation from the ordinary gauge transformation. Indeed, for the fermion, one can define the chiral gauge transformation by

$$\psi \to \psi' = e^{-i\alpha\gamma^5}\psi \tag{1.1}$$

Since $\gamma^5\psi_{\substack{L\\R}} = \gamma^5[(1 \mp \gamma^5)/2]\psi = \mp\psi_{\substack{L\\R}}$, it certainly produces different transformations for left- and right-handed fields. However, the weak charge carriers are not restricted to fermions. We had better avoid the use of γ^5 in the gauge operator. Instead, we shall consider that the difference lies in the operand field. We must consider that the operation is just the same old gauge transformation, but the operand acts differently depending on the weak charge it carries. All we need to recognize is that fermions of different chirality carry different charges.

As an example, let us consider the chiral gauge transformation which follows the $U(1)$ symmetry and denote the charge operator as "Y" (call it hypercharge). Then, the gauge transformation for the fermion field $f = \psi_{\rm f}$ or scalar ϕ is[1)]

$$\begin{aligned}
f_{\rm L} &\to f'_{\rm L} = e^{-i\alpha Y} f_{\rm L} = e^{-i\alpha Y(f_{\rm L})} f_{\rm L} \\
f_{\rm R} &\to f'_{\rm R} = e^{-i\alpha Y} f_{\rm R} = e^{-i\alpha Y(f_{\rm R})} f_{\rm R} \\
\phi &\to \phi' = e^{-i\alpha Y} \phi = e^{-\alpha Y(\phi)} \phi
\end{aligned} \tag{1.3}$$

where $Y(f_{\rm L})$, and so on, are the hypercharges that each field carries.

We said that the weak charge comes in two varieties and respects the $SU(2)$ symmetry. Actually, the true weak and the electromagnetic force are based on a mixture of $SU(2)$ and $U(1)$, and we need to be careful about the terminology. The weak force in its original form, that is, before mixing and spontaneous symmetry breakdown, has chiral $SU(2)$ symmetry. In $SU(2)$ terminology, all the weak force carriers constitute isospin multiplets. All the left-handed fermions constitute doublets. For instance, the electric-type neutrino ν_e and the electron e^- are members of a doublet. $\Psi_{\rm L}^T \equiv (\nu_{e\rm L}, e_{\rm L}^-)$. All the right-handed particles belong to $SU(2)$ singlets ($I = I_3 = 0$), namely, they do not carry weak charges. In the Standard Model, all the leptons can be classified by their isospin component as

$$\text{Leptons} \quad \begin{cases} I_3 = +1/2 \\ I_3 = -1/2 \end{cases} \quad \Psi_{\rm L} = \begin{pmatrix} \nu_e \\ e^- \end{pmatrix}_{\rm L}, \quad \begin{pmatrix} \nu_\mu \\ \mu^- \end{pmatrix}_{\rm L}, \quad \begin{pmatrix} \nu_\tau \\ \tau^- \end{pmatrix}_{\rm L} \tag{1.4}$$
$$I = I_3 = 0 \qquad e_{\rm R}^-, \mu_{\rm R}^-, \tau_{\rm R}^-$$

The leptons which have $I_3 = 1/2$, that is, the neutrinos are electrically neutral and those which have $I_3 = -1/2$ have electric charge $Q = -1$ in units of the positron charge. In the Standard Model, the right-handed neutrinos do not exist.[2)] For the quarks

$$\text{Quarks} \quad \begin{cases} I_3 = +1/2 \\ I_3 = -1/2 \end{cases} \quad \begin{pmatrix} u \\ d' \end{pmatrix}_{\rm L}, \quad \begin{pmatrix} c \\ s' \end{pmatrix}_{\rm L}, \quad \begin{pmatrix} t \\ b' \end{pmatrix}_{\rm L} \tag{1.5}$$
$$I = I_3 = 0 \qquad u_{\rm R}, d_{\rm R}, c_{\rm R}, s_{\rm R}, t_{\rm R}, b_{\rm R}$$

where $D'^T \equiv (d', s', b')$ are Cabibbo–Kobayashi–Maskawa rotated fields:

$$D' = V_{\rm CKM} D, \quad \to \quad \begin{bmatrix} d' \\ s' \\ b' \end{bmatrix} = \begin{bmatrix} V_{ud} & V_{us} & V_{ub} \\ V_{cd} & V_{cs} & V_{cb} \\ V_{td} & V_{ts} & V_{tb} \end{bmatrix} \begin{bmatrix} d \\ s \\ b \end{bmatrix} \tag{1.6}$$

1) The equation is an abbreviated version. In the quantum field theoretical treatment, the transformation equation should read

$$e^{i\alpha Y} \psi e^{-i\alpha Y} = e^{-i\alpha Y(\psi)} \psi \tag{1.2}$$

where Y is the generator of the gauge transformation (see Vol. 1, Eq. (9.159)) and $Y(\psi)$ is the eigenvalue of Y that the field carries.

2) In reality, they do exist as demonstrated by the discovery of the neutrino oscillation. In the context of this textbook, it is more simple and no inconvenience is encountered by assuming the massless neutrino. The neutrino oscillation phenomena will be treated separately in Vol. 3.

The quarks with $I_3 = 1/2$ have $Q = 2/3$ and those with $I_3 = -1/2$ have $Q = -1/3$. Each quark carries another degree of freedom, that is, three colors which are the source of the strong interaction. Its dynamics will be treated in detail later, as now we must put aside the strong interaction in the discussion of the electroweak force and simply consider that they have an extra three degrees of freedom.

The original Lagrangian for the weak force is invariant under $SU(2)$ gauge transformations which are generically denoted as U. Denoting the isospin operator as $\boldsymbol{t}$, we have

$$\begin{aligned}\Psi_{\rm L}(x) \to \Psi'_{\rm L}(x) &= U\Psi_{\rm L}(x) = \exp(-ig_{\rm W}\boldsymbol{\alpha}(x)\cdot\boldsymbol{t})\Psi_{\rm L}(x) = \exp[-ig_{\rm W}\boldsymbol{\alpha}\cdot\boldsymbol{\tau}/2]\Psi_{\rm L}(x)\\ \Psi_{\rm R}(x) \to \Psi'_{\rm R}(x) &= U\Psi_{\rm R}(x) = \exp(-ig_{\rm W}\boldsymbol{\alpha}(x)\cdot\boldsymbol{t})\Psi_{\rm R}(x) = \Psi_{\rm R}(x)\end{aligned} \tag{1.7}$$

where $\Psi_{\rm L}$ is any fermion doublet in Eqs. (1.4) and (1.5), $\boldsymbol{\alpha} = (\alpha_1(x), \alpha_2(x), \alpha_3(x))$ is a set of three independent continuous variables and x is a simplified notation for the Lorentz coordinate variables $x \equiv (x^0, \boldsymbol{x})$. $\boldsymbol{\tau}$ is the Pauli 2×2 matrix that operates on the isospin components of doublets.

$$\tau_1 = \begin{bmatrix} 0 & 1 \\ 1 & 0 \end{bmatrix}, \quad \tau_2 = \begin{bmatrix} 0 & -i \\ i & 0 \end{bmatrix}, \quad \tau_3 = \begin{bmatrix} 1 & 0 \\ 0 & 1 \end{bmatrix} \tag{1.8}$$

The field $\Psi_{\rm R}$ does not receive any change by $SU(2)$ transformation simply because $I(\Psi_{\rm R}) = 0$. If we denote the gauge bosons of $SU(2)$ as

$$\boldsymbol{W}(x) = (W_1(x), W_2(x), W_3(x)) \tag{1.9}$$

which constitute an isospin triplet ($I = 1$), the gauge transformation changes them to

$$\boldsymbol{W}_\mu \cdot \boldsymbol{t} \to \boldsymbol{W}'_\mu \cdot \boldsymbol{t} = U\boldsymbol{W}_\mu \cdot \boldsymbol{t}U^{-1} + \frac{i}{g_{\rm W}} U\partial_\mu U^{-1} \tag{1.10}$$

but keeps the covariant derivative

$$D_\mu = \mathbf{1}\partial_\mu + ig_{\rm W}\boldsymbol{W}_\mu \cdot \boldsymbol{t} \tag{1.11}$$

invariant, that is, $D' = \mathbf{1}\partial_\mu + ig_{\rm W}\boldsymbol{W}'_\mu \cdot \boldsymbol{t}$.

Now, we must treat the $U(1)$ part of $SU(2) \times U(1)$ in the GWS theory. It acts on all the leptons and quarks. We tentatively call it the B-force. It respects chiral $U(1)$ symmetry whose gauge transformation was given by Eq. (1.3). Each left- or right-handed fermion has its own hypercharge in addition to isospin component I_3 due to $SU(2)$. The weak isospin and the hypercharge satisfy Nishijima–Gell–Mann's law:

$$Q = I_3 + Y/2 \tag{1.12}$$

From this relation, one deduces that

$$\begin{aligned}&Y(\nu_{e\rm L}) = Y(e^-_{\rm L}) = -1\,, \quad Y(\nu_{e\rm R}) = 0\,, \quad Y(e^-_{\rm R}) = -2\\ &Y(u_{\rm L}) = Y(d_{\rm L}) = 1/3\,, \quad Y(u_{\rm R}) = 4/3\,, \quad Y(d_{\rm R}) = -2/3\end{aligned} \tag{1.13}$$

Assignment of Y to other fermions is obtained similarly.

We denote the gauge bosons of $U(1)$ as B_μ and the covariant derivative as $D_\mu = \partial_\mu + i(g_B/2)B_\mu$, where the factor 1/2 is introduced for later convenience. Then, the covariant derivative including both W and B is given by

$$\begin{aligned} D_\mu &= \partial_\mu + i g_W \mathbf{W}_\mu \cdot \mathbf{t} + \frac{i g_B}{2} B_\mu \\ &= \partial_\mu + \frac{i g_W}{2}(W_\mu^1 \tau_1 + W_\mu^2 \tau_2 + W_\mu^3 \tau_3) + \frac{i g_B}{2} B_\mu \end{aligned} \tag{1.14}$$

There is another important player of the electroweak force, namely, the Higgs field,

$$\Phi = \begin{bmatrix} \phi^+ \\ \phi^0 \end{bmatrix} \tag{1.15}$$

in the electroweak interaction. ϕ^+, ϕ^0 are two complex scalar fields. Together, they constitute an isospin doublet ($I = 1/2$, $I_3 = \pm 1/2$) and carry hypercharge $Y(\phi^+) = Y(\phi^0) = 1$. The self-interaction of the Higgs field is the cause of the spontaneously symmetry breaking of the $SU(2)_L \times U(1)$, giving mass to the gauge bosons as well as to the fermions. The original Lagrangian before mixing and spontaneous symmetry breaking is given by

$$\begin{aligned} \mathcal{L}_{EW} &= \overline{\Psi} i \gamma^\mu D_\mu \Psi - \frac{1}{4}\mathbf{F}_{\mu\nu} \cdot \mathbf{F}^{\mu\nu} - \frac{1}{4} B_{\mu\nu} B^{\mu\nu} \\ &\quad + (D_\mu \Phi)^\dagger (D^\mu \Phi) - V(\Phi) - G_f[\overline{e}_R(\Phi^\dagger \Psi_L) + (\overline{\Psi}_L \Phi) e_R] \end{aligned} \tag{1.16}$$

$$\mathbf{F}_{\mu\nu} = \partial_\mu \mathbf{W}_\nu - \partial_\nu \mathbf{W}_\mu - g_W \mathbf{W}_\mu \times \mathbf{W}_\nu \tag{1.17a}$$

$$B_{\mu\nu} = \partial_\mu B_\nu - \partial_\nu B_\mu \tag{1.17b}$$

$$D_\mu = \partial_\mu + i g_W \mathbf{W}_\mu \cdot \mathbf{t} + i\left(\frac{g_B}{2}\right) Y B_\mu \tag{1.17c}$$

$$V(\Phi) = \lambda \left(|\Phi|^2 + \frac{\mu^2}{2\lambda} \right)^2 \quad \lambda > 0 \tag{1.17d}$$

This is the master equation to calculate reaction rates of any electroweak processes. We shall use, for example, ν_e, e^- to denote quantized fields, that is, $\nu_e(x) = \psi_{\nu_e}(x)$, $e^-(x) = \psi_e(x)$, where there is no confusion. Here, we have only written down the Lagrangian of the $\Psi = (\nu_e, e^-)$ which will be needed in the following discussions. The Lagrangian for other fermions can be written down similarly. The first line of Eq. (1.16) is referred to as the gauge sector and the second line as the Higgs sector. $V(\Phi)$ is the self-interacting potential of the Higgs field. The whole expression satisfies the $SU(2) \times U(1)$ gauge symmetry manifestly. It is important to remember that both the gauge and the Higgs sectors are constructed to respect the gauge symmetry separately. The last term of Eq. (1.16) referred to as the Yukawa interaction was added to generate fermion masses. It can be written down as

$$\overline{e}_R(\Phi^\dagger \Psi_L) + (\overline{\Psi}_L \Phi) e_R = \overline{e}_R \nu_{eL} \phi^- + \overline{\nu}_{eL} e_R \phi^+ + \overline{e}_R e_L \phi^{0\dagger} + \overline{e}_L e_R \phi^0 \tag{1.18}$$

The case that this term is also $SU(2) \times U(1)$ invariant can be illustrated as follows. Because we constructed the Higgs field as belonging to the $I = 1/2$ doublet, and the product with another isospin doublet Ψ_L, then $(\Phi^\dagger \Psi_L)$ is therefore isospin rotation invariant. By multiplying e_R which is a scalar in $SU(2)$ transformation, the expression becomes Lorentz invariant because each term is of the form $\overline{\psi}_R \psi_R \phi$ or $\overline{\psi}_L \psi_R \phi$. One can show that each term is also chiral $U(1)$ invariant by referring to the hypercharges in Eq. (1.13). For instance, take a look at the first term on the right hand side of Eq. (1.18). By the chiral $U(1)$ gauge transformation

$$\begin{aligned} e_R &\to e'_R = e^{-iY(e_R)\beta} e_R \,, \quad \nu_{eL} \to \nu'_{eL} = e^{-iY(\nu_{eL})\beta} \nu_{eL} \,, \\ \phi^- &\to \phi^{-\prime} = e^{-iY(\phi^-)\beta} \phi^0 \end{aligned} \tag{1.19}$$

Considering $Y(e_R) = -2$, $Y(\nu_{eL}) = -1$, $Y(\phi^-) = -1$, one sees that the product $(\overline{e}_R \nu_{eL})\phi^-$ is $U(1)$ gauge invariant. Other terms can be proven similarly.

Mixing of *SU*(2) and *U*(1) The interaction of the gauge boson with the fermion is contained in the covariant derivative of Eq. (1.16). It is given by

$$-\mathcal{L}_{Wff} = g_W \overline{\Psi} \gamma^\mu \boldsymbol{W}_\mu \cdot \boldsymbol{t} \Psi + \frac{g_B}{2} \overline{\Psi} \gamma^\mu \Psi B_\mu \tag{1.20}$$

Since $\boldsymbol{t} = \boldsymbol{\tau}/2$ acts only on the left-handed fields and B_μ couples to the hypercharge Y which acts on the left- and right-handed fields differently, we have

$$\begin{aligned} -\mathcal{L}_{Wff} &= \frac{g_W}{2} \overline{\Psi}_L \gamma^\mu \left(W^1_\mu \tau_1 + W^2_\mu \tau_2 + W^3_\mu \tau_3 \right) \Psi_L \\ &+ \frac{g_B}{2} B_\mu \left(\overline{\Psi}_L \gamma^\mu Y \Psi_L + \overline{\Psi}_R \gamma^\mu Y \Psi_R \right) \end{aligned} \tag{1.21}$$

We wrote "Y" explicitly to remind the reader that B_μ is acting on the hypercharge.

Since both W_3 and B can couple to the same fermion, they mix and constitute the electromagnetic field (the photon) and the neutral gauge boson Z. To determine how they mix, we consider their coupling to the electron doublet.

$$\Psi_L(x) = \begin{pmatrix} \nu_e(x) \\ e^-(x) \end{pmatrix}_L \tag{1.22}$$

Then, the coupling to the neutral boson part (terms containing W^3_μ and B_μ in Eq. (1.21)) is expressed as

$$\frac{g_W}{2} (\overline{\nu}_{eL} \gamma^\mu \nu_{eL} - \overline{e}_L \gamma^\mu e_L) W^3_\mu + \frac{g_B}{2} (-\overline{\nu}_{eL} \gamma^\mu \nu_{eL} - \overline{e}_L \gamma^\mu e_L - 2\overline{e}_R \gamma^\mu e_R) B_\mu \tag{1.23a}$$

Rearranging the equation, we obtain

$$\frac{1}{2} \left[\overline{\nu}_{eL} \gamma^\mu \nu_{eL} \left(g_W W^3_\mu - g_B B_\mu \right) - \overline{e}_L \gamma^\mu e_L \left(g_W W^3_\mu + g_B B_\mu \right) - 2 g_B (\overline{e}_R \gamma^\mu e_R) B_\mu \right] \tag{1.23b}$$

Since the electromagnetic field should not couple to the neutrino, we define the weak neutral boson Z[3] by $Z \sim g_W W^3_\mu - g_B B_\mu$ and its orthogonal component as the electromagnetic field. With suitable normalization, they can be expressed as

$$A^\mu = \cos\theta_W B^\mu + \sin\theta_W W_3^\mu = \frac{1}{\sqrt{g_W^2 + g_B^2}} \left[g_W B^\mu + g_B W_3^\mu \right] \quad (1.24a)$$

$$Z^\mu = -\sin\theta_W B^\mu + \cos\theta_W W_3^\mu = \frac{1}{\sqrt{g_W^2 + g_B^2}} \left[-g_B B^\mu + g_W W_3^\mu \right] \quad (1.24b)$$

which also defines the Weinberg angle θ_W[4]. The Weinberg angle is related to the gauge coupling strength of the $SU(2)$ (g_W) and $U(1)$ (g_B) by

$$\tan\theta_W = \frac{g_B}{g_W} \quad (1.25)$$

As a result, the Z boson couples to the right-handed component of the fermion (via B^μ) as well as to their left-handed component (via W_3^μ). This is the reason that we have to be careful about saying that the weak force works only on the left-handed particles. The statement is true only for charged current reactions which couple to $W^\pm$. The neutral current which couples to Z contains the right-handed components. The photon couples to I_3 component and hypercharge of the particle which is the origin of the Nishijima–Gell–Mann's law.[5]

The kinetic energy part (derivatives of the fields) of the gauge fields can be rewritten as

$$\begin{aligned}
\mathcal{L}_{KE} &= -\frac{1}{4} \sum_{a=1}^{3} F^a_{\mu\nu} F^{a\mu\nu} - \frac{1}{4} B_{\mu\nu} B^{\mu\nu} \\
&= -\frac{1}{4} \left(2F^-_{\mu\nu} F^{+\mu\nu} + F^Z_{\mu\nu} F^{Z\mu\nu} + F^A_{\mu\nu} F^{A\mu\nu} \right) \\
F^\mp_{\mu\nu} &= \partial_\mu W^\mp_\nu - \partial_\nu W^\mp_\mu, \quad F^Z_{\mu\nu} = \partial_\mu Z_\nu - \partial_\nu Z^\mu, \quad F^A_{\mu\nu} = \partial_\mu A_\nu - \partial_\nu A^\mu
\end{aligned} \quad (1.26)$$

where $W^\mp_\mu = (W^1_\mu \pm i W^2_\mu)/\sqrt{2}$ are charged boson field operators.

3) The nomenclature "neutral boson" is used in general to mean any charge neutral member including the photon, but hereafter we use the word "neutral gauge boson" to specifically mean the Z unless otherwise noted.

4) In the following, $\sin\theta_W$ appears more often than θ_W itself. It will also be referred to as the Weinberg angle.

5) This is a circular logic. Historically, the hypercharge is assigned to satisfy the Nishijima–Gell–Mann law. Here, however, we are starting from an axiom that the hypercharge is the fundamental constants that all the particles possess transcendentally.

1.3
Spontaneous Symmetry Breaking

When the equation of motion satisfies a certain symmetry, its solution generally possesses the same symmetry. However, the solution is not necessarily stable. In such a case, the chosen state could break the symmetry. When the ground state (or vacuum in the field theory) does not respect the symmetry which the equation of motion has, we have the spontaneous symmetry breaking. Let us see how it happens. The Higgs potential Eq. (1.17d) contains a quartic term as well as the quadratic term.

$$V(\Phi) = V(0) + \mu^2|\Phi|^2 + \lambda|\Phi|^4\,, \quad |\Phi| = \sqrt{|\phi^+|^2 + |\phi^0|^2} \tag{1.27}$$

$V(0)$ is arbitrary and usually assumes a value to make the vacuum energy vanish. Only the energy difference from the vacuum matters unless we deal with the gravity.

1.3.1
An Intuitive Picture of Spontaneous Symmetry Breaking

Let us consider what happens to the field after the symmetry breakdown. For simplicity, let us consider a case $V(0) = 0$ and Φ is an isospin singlet complex field $\Phi = \phi = (\phi_1 + i\phi_2)/\sqrt{2}$. The potential is expressed as

$$V(\phi) = \mu^2\phi^\dagger\phi + \lambda(\phi^\dagger\phi)^2 \tag{1.28}$$

The symmetry which the field satisfies is the Abelian $U(1)$ gauge symmetry, namely, the Lagrangian is invariant under the transformation $\phi \to \phi' = e^{-i\alpha}\phi$. If $\lambda = 0$ and $\mu^2 > 0$, the potential represents a harmonic oscillator. Energy excited states appear as particles in the quantized field theory and μ represents the mass of the quanta. If $\lambda \neq 0$[6], the potential still has a minimum at $|\phi| = 0$ as described by the curve denoted $T > T_c$ in Figure 1.1a. In this case, the particle picture still holds, the quartic term represents self-interactions of the particles.

Let us consider μ^2 not as an a priori given mass squared, but some dynamical object that changes with temperature, that is, $\mu^2 = C(T - T_c)$, where T_c is a critical temperature whose meaning will soon be clarified. By doing so, we regard the vacuum not as an empty space, but some dynamical object which changes its characteristics with temperature just like any medium in condensed matter physics. As the temperature goes down, the μ^2 changes sign at $T = T_c$, and below T_c, the potential develops minima at $\phi \neq 0$. The μ^2 being negative cannot be interpreted as the mass squared, but should be regarded as a part of the potential. If the field ϕ is complex as we assumed, the potential shape is like a Mexican hat (Figure 1.1b). The old vacuum $|\phi| = 0$ is no longer a stable point and the vacuum moves to a stable point, the minimum at $|\phi| = \sqrt{-\mu^2/2\lambda} \equiv v/\sqrt{2}$. The field ϕ is not zero at the new vacuum, or the vacuum expectation value of the field is finite.

6) We do not consider the case $\lambda < 0$ because the vacuum becomes unstable.

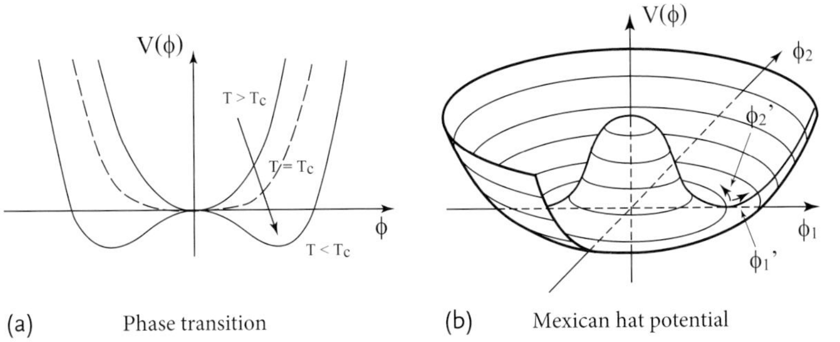

Figure 1.1 (a) Depending on the temperature, the potential changes its shape. (b) Illustration of a Mexican hat potential as required for the Higgs field. The ground state is degenerate and the choice of one specific value spontaneously breaks the symmetry.

For the complex ϕ, the new vacuum is infinitely degenerate because any point on the circle that satisfies the above condition is eligible as a vacuum. The vacuum the nature picks up can be anywhere on the circle. We conveniently set a new coordinate system such that the new vacuum is at $\phi_1 = v/\sqrt{2}, \phi_2 = 0$. As the experimentally observed energy is the excitation from the new vacuum, we are now dealing with new fields

$$\phi_1' = \phi_1 - v\,, \quad \phi_2' = \phi_2 \tag{1.29}$$

and the Higgs Lagrangian is rewritten as

$$\begin{aligned} \mathcal{L} &= \partial_\mu \phi^\dagger \partial^\mu \phi - V(\phi) \Rightarrow \partial_\mu \phi'^\dagger \partial^\mu \phi' - V(v + \phi') \\ &= \frac{1}{2}\left(\partial_\mu \phi_1' \partial^\mu \phi_1' - 2\lambda v^2 \phi_1' \phi_1'\right) + \frac{1}{2}\left(\partial_\mu \phi_2' \partial^\mu \phi_2'\right) \\ &\quad - \left[\lambda \phi_1' \left\{(\phi_1')^2 + (\phi_2')^2\right\} + \frac{\lambda}{4}\left\{(\phi_1')^2 + (\phi_2')^2\right\}^2\right] \end{aligned} \tag{1.30}$$

As one can see, the terms in the penultimate line show that the new field ϕ_1' has mass $m_H^2 = 2\lambda v^2$ and the ϕ_2' has vanishing mass. The last line describes the interaction between the two fields and among themselves. The different mass of the two fields can be pictorially understood from Figure 1.1b. ϕ_1' is in the bottom of the potential valley and it takes energy to excite the field, that is, to climb up the potential. However, for ϕ_2', there is no potential to hinder the motion, that is, it takes no inertia to move on the circle. The massless field ϕ_2' is referred as the Goldstone boson. They no longer satisfy the gauge symmetry. Remnants of the original symmetry can be seen in the fact that any point on the circle is equivalent and could have been chosen as the new vacuum.

The important point is that the new vacuum, once chosen, is fixed there. This is because of the infinite degrees of freedom that the field possesses. It may not be apparent intuitively, so perhaps an example in condensed matter physics that obeys the same mathematics may help to clarify it. ϕ is referred to as the order parameter there. Think of a ferromagnet which we discussed in Vol. 1, Chapt. 18. The order parameter in this case represents the magnetization. Above the critical temperature, the spin is not aligned because of thermal motion and there is no magnetization, namely, the medium is a paramagnet. Naturally, the order parameter vanishes in the ground state of the paramagnet. The medium respects the rotational symmetry because the spin has no preferred orientation. Below the Curie temperature, the spins begin to be aligned due to spin-spin force, and at absolute zero temperature, the whole medium is aligned. The magnetization in the ground state is finite and the medium is a ferromagnet. The spontaneous symmetry breaking is nothing but the phase transition in condensed matter physics and it is a metamorphose of the entire medium.

In a medium where all the spins are aligned in the same direction, it is possible to excite a spin or two by giving a small energy ΔE to each spin to change its direction. The turbulence propagates as a wave which, if quantized, is the Goldstone boson represented by ϕ_2'. However, changing the whole ground state, namely, changing the spin orientation of the entire medium, takes energy $\Delta E \times N_A$ where N_A is of the order of the Avogadro number which is large but finite. Macroscopically, it is possible to heat the medium above the critical temperature to transfer the ferromagnet back to paramagnet. However, the vacuum in the field theory is the vacuum of the universe. It has infinite degrees of freedom. The vacuum, once chosen, is impossible to change. We cannot heat up the whole universe!

1.3.2
Higgs Mechanism

Now, we go back to the original $SU(2)$ non-Abelian Higgs field Eq. (1.15) and we include the gauge field. The reason to consider a doublet Higgs field is that we want to use three Goldstone bosons later and an isospin doublet of the Higgs fields is the minimum requirement within the $SU(2)$ symmetry. The physics aspect of the theory changes from that of the ferromagnet, but the role of the Higgs field in inducing the spontaneous symmetry breaking is the same. Below the critical temperature, the Higgs field acquires the vacuum expectation value (vev)

$$|\Phi^2| = |\phi^+|^2 + |\phi^0|^2 = v^2/2 \tag{1.31}$$

Notice that there are three fields that can vanish. The vacuum is infinitely degenerate and we choose the new vacuum at $\mathrm{Re}[\phi^0] = v/\sqrt{2}$, $\mathrm{Im}[\phi^0] = \phi^+ = 0$, namely,

$$\Phi' = \begin{bmatrix} 0 \\ (v + H)/\sqrt{2} \end{bmatrix} \tag{1.32}$$

where we use nomenclature $H = \mathrm{Re}[\phi^0] - v\sqrt{2}$ to denote it as a real observable Higgs field. This is the spontaneous symmetry breakdown of the $SU(2)$ symmetry (or $SU(2) \times U(1)$ if we include Z instead of W_3). The Higgs field before the symmetry breaking can be parametrized as

$$\Phi(x) = \begin{bmatrix} \phi^+ \\ \phi^0 \end{bmatrix} = \frac{1}{\sqrt{2}} \begin{bmatrix} (\omega_2 + i\omega_1)/2 \\ v + H - i\omega_3/2 \end{bmatrix} \tag{1.33}$$

Without loss of generality, it can be rewritten as

$$\Phi = \exp\left(i\frac{\boldsymbol{\omega}}{v}\cdot\frac{\boldsymbol{\tau}}{2}\right)\begin{bmatrix} 0 \\ \frac{v+H}{\sqrt{2}} \end{bmatrix} \tag{1.34}$$

In fact, in the vicinity of the chosen vacuum where $v \gg |H|, |\boldsymbol{\omega}|$, Eq. (1.34) reduces to Eq. (1.33). The reason to place three of the fields in an exponent is because we can realize the new Higgs field Eq. (1.32) by a gauge transformation

$$\Phi \to \Phi' = U\Phi = \exp[-i\boldsymbol{\omega}\cdot\boldsymbol{\tau}/(2v)]\Phi \tag{1.35}$$

The $\boldsymbol{\omega}$ components disappear from the Higgs field. As the gauge transformation also changes the gauge field $\boldsymbol{W}_\mu$, they are transformed to $\boldsymbol{W}'_\mu$.

$$\begin{aligned} \boldsymbol{W}_\mu\cdot\boldsymbol{t} \to \boldsymbol{W}'_\mu\cdot\boldsymbol{t} &= U\boldsymbol{W}_\mu\cdot\boldsymbol{t}U^{-1} + \frac{i}{g_W}U\partial_\mu U^{-1} \\ &= U\boldsymbol{W}_\mu\cdot\boldsymbol{t}U^{-1} - \frac{1}{g_W v}\partial_\mu\boldsymbol{\omega}\cdot\boldsymbol{t} \end{aligned} \tag{1.36}$$

with $\boldsymbol{t} = \boldsymbol{\tau}/2$. The field $\boldsymbol{\omega}$ has reappeared as the third (longitudinal) component of the gauge bosons. Namely, three of the Higgs fields are absorbed by the gauge field and have become their third component.[7] We rename the new fields Φ' and $\boldsymbol{W}'_\mu$ again as Φ, $\boldsymbol{W}_\mu$. They are the physical observables because we live in a vacuum where the gauge symmetry has been spontaneously broken.

Now, we realize that when the broken symmetry is a local gauge symmetry, the Goldstone bosons do not appear. Three Goldstone bosons are produced by a doublet Higgs, but they are absorbed by the three gauge bosons ($W^\pm$, Z). The new gauge bosons acquire the third component which means they have become massive and the electroweak force has become short ranged. The appearance of the mass term will be explicitly shown later in Eq. (1.37). Analogous phenomenon actually happens in the superconductor below the critical temperature where the gauge symmetry is spontaneously broken and the electromagnetic force is converted to a short ranged force. The phenomenon that the magnetic force cannot penetrate into the superconducting medium and is wholly repelled is known as the Meissner effect. This is the reason that we said the physics outcome is different if the broken symmetry is the local gauge symmetry. The whole mechanism, the spontaneous breakdown of the gauge symmetry and associated mass generation of the gauge bosons, is referred to as the Higgs mechanism. The mathematics we developed is more appropriate to the superconductivity rather than the ferromagnetism. Indeed, the nonrelativistic Hamiltonian derived from the Abelian Lagrangian after spontaneous symmetry breakdown has identical form to the Ginzburg–Landau free energy of the superconductivity (see the boxed paragraph in Vol. 1, Sect. 18.5.2). In this case, the order parameter is the wave function of the Cooper pair whose condensation induces the superconductivity. This is the reason we said that the vacuum where we live is in a sort of superconducting phase.

7) They are sometimes referred as the would-be-Goldstone bosons because if the symmetry is not the gauge symmetry, they would have manifested themselves as physically observable Goldstone bosons.

1.3.3 Unitary Gauge

We have seen that the Higgs field after the spontaneous symmetry breaking ($= \Phi'$) is equivalent to that obtained as the result of the gauge transformation, that is, $\Phi' = \exp[-i\boldsymbol{\omega} \cdot \boldsymbol{t}/v]\Phi$. Since we chose the vacuum to fix the Higgs field in this form, this simply means a special gauge is chosen and fixed. This gauge is referred to as the unitary gauge or *U*-gauge. The unitary gauge is one where the physical meaning of the various fields is clearest.

Notice that the new fields do not respect the gauge symmetry, not because the symmetry was broken, but because we chose the new field (old Φ') as the excitation from a specific vacuum point to describe phenomena in our world. In doing so, we also fixed the gauge. Mathematically, the whole procedure is just choosing an appropriate gauge and fixing it. We can equally describe nature using old variables in the original gauge. From a theoretical point of view, it is much more convenient because it respects the gauge symmetry manifestly. However, then it is hard to make connections with observables and the physical interpretation of mathematics gets complicated. The symmetry was not really broken, rather it was hidden as a result of choosing new variables.

What is the physical meaning of the phase field?

In the *U*-gauge, the Higgs field takes the form given in Eq. (1.34). Mathematically, it removes extra Higgs components beautifully which are absorbed by the gauge bosons. It can also be cast in a form given by Eq. (1.33) which looks more familiar. What is the difference between the two choices? Let us remember that a particle picture of the quantized field is obtained by quantizing a harmonic potential which is quadratic in the field variables. The potential of the field in general, however, is not necessarily quadratic. It may have quite a complicated structure depending on the property of the field. In the framework of the quantum field theory, we refer to particles as excited states, or small perturbations around a stable environment which we call the vacuum. Mathematically, it may be a local minimum and we always expand the potential in a Taylor power series. If we use only up to quadratic terms, we get the particle picture of the quantized field.

Higher order terms are treated as the interaction among particles. The interaction is different depending on where and how we expand the potential. The expanded power series contains hints for the global structure of the whole potential. Inclusion of the higher order terms is to consider more global characteristics of the field which may not behave like particles. Consider, for instance, a superconducting object which we modeled in developing the Higgs formalism. It behaves as a macroscopic quantum fluid rather than as particles. The vacuum we developed as the result of the spontaneously broken symmetry is in a state of a superconducting phase. In the local vicinity of which we call the vacuum, the potential may well be approximated by the quadratic potential and the familiar particle picture is valid.

However, globally, it may be a part of circulating field or may be spatially expanding just as our universe. If it is happening in the cosmic scale, we will not notice it because we are only dealing with particles which are simply small excitations of a local minimum. The ω field in Eq. (1.34) is referred to as the phase filed. Locally, however, it is equivalent to linear fields in Eq. (1.33), that is, from a phenomenological point of view, there is no difference. The potential form given in Eq. (1.30) could be considered as representing the global circle structure of the vacuum, but for a description of phenomena we observe locally, it only appears as higher order corrections. However, a glimpse of the global structure may be obtained by investigating the particle interactions.

Our next task is to find what the Higgs sector of the Lagrangian have produced after the spontaneous symmetry breakdown.

1.3.4
Mass Generation

Mass of the Gauge Boson First, we take a look at the kinetic energy part of the Higgs sector. Substituting Eq. (1.32) in the first term of the second line of Eq. (1.16), we obtain

$$\begin{aligned}(D_\mu \Phi)^\dagger (D^\mu \Phi) &= (\partial_\mu \Phi)^\dagger (\partial^\mu \Phi) + \Phi^\dagger \left(g_W\, \boldsymbol{W}_\mu \cdot \boldsymbol{t} + (g_B/2) B_\mu \cdot Y\right)^2 \Phi \\ &= (\partial_\mu \Phi)^\dagger (\partial^\mu \Phi) + \frac{(v+H)^2}{8}\left[2 g_W^2\, W_\mu^- W^{+\mu} + (-g_W\, W_\mu^0 + g_B B_\mu)^2\right] \\ &= (\partial_\mu \Phi)^\dagger (\partial^\mu \Phi) + \left(\frac{g_W(v+H)}{2}\right)^2 W_\mu^- W^{+\mu} + \frac{1}{2}\left(\frac{g_Z(v+H)}{2}\right)^2 Z_\mu Z^\mu\end{aligned} \tag{1.37}$$

Linear terms in the derivative vanish and we used

$$g_W\, W_\mu^0 - g_B B_\mu = g_Z Z_\mu\,, \quad g_Z = \sqrt{g_W^2 + g_B^2} \tag{1.38}$$

In the absence of the interaction with the Higgs field (i.e., when $H = 0$), the second and third term in the last line of Eq. (1.37) gives the mass terms for W and Z. The H term gives the interaction of the Higgs field with W and Z. Considering that there is a factor 1/2 difference between the charged and neutral vector bosons, we have

$$m_W = \frac{g_W v}{2}\,, \quad m_Z = \frac{g_Z v}{2} \tag{1.39}$$

We have just explicitly shown that the gauge bosons have acquired mass after symmetry breaking. Notice that Eq. (1.37) does not contain the electromagnetic field. Therefore, the photon does not acquire mass. This is not accidental. Remember that the symmetry was broken by the vacuum. This means that the vacuum expectation value of both the isospin ($SU(2)$) operator $\boldsymbol{t}$ and the hypercharge operators Y

do not vanish.

$$t\langle\Phi\rangle = \frac{\boldsymbol{\tau}}{2}\begin{bmatrix} 0 \\ \frac{v}{\sqrt{2}} \end{bmatrix} \neq 0\,, \quad Y\langle\Phi\rangle = \langle\Phi\rangle \neq 0 \tag{1.40}$$

which is another statement of the symmetry breaking in the new vacuum. On the other hand, as $Q = I_3 + Y/2$ and $Y(\Phi) = 1$,

$$Q\langle\Phi\rangle = \frac{1}{2}(1+\tau_3)\begin{bmatrix} 0 \\ \frac{v}{\sqrt{2}} \end{bmatrix} = 0 \tag{1.41}$$

Therefore, the vacuum is not invariant under $SU(2)_{\mathrm{L}}$ or $U(1)_{\mathrm{Y}}$ gauge transformation separately, but is invariant under the combined gauge transformation $U = \exp[-Qe\alpha(x)]$. The photon field which is generated by the charge operator Q has kept its freedom of the gauge transformation which means the gauge field has vanishing mass.

Careful readers might have noticed that a massive vector has three degrees of freedom (two transverse polarizations and one longitudinal polarization), whereas the massless vector boson has only two degrees of freedom. Is it not a contradiction? As a matter of fact, the extra degree of freedom was provided by the Higgs components $(\omega_1, \omega_2, \omega_3)$ that had disappeared. This can be seen in the expression of Eq. (1.36) in which the second term provides the longitudinal component. Originally, there were four components in the Higgs field. Three of them are absorbed by the three gauge bosons, and the degree of freedom that the gauge boson had, increased from 2×3 to 3×3. The total number of the degrees of freedom is invariant.

The Vacuum Expectation Value We can determine the value of the vacuum expectation value by comparing the muon decay amplitude with the conventional four-Fermi interaction Lagrangian.

The fermion interaction with charged gauge bosons can be extracted by inspecting the covariant derivative of the fermion (the first term in the first line of Eq. (1.16)). By including the second fermion doublet with the muon-flavor, the interaction Lagrangian takes the form

$$\mathcal{L}_{Wff} = -\frac{g_W}{\sqrt{2}}(\overline{\nu}_{e\mathrm{L}}\gamma^\mu e_{\mathrm{L}} + \overline{\nu}_{\mu\mathrm{L}}\gamma^\mu \mu_{\mathrm{L}})\,W_\mu^- + h.c. \tag{1.42}$$

Here $h.c.$ means hermitian conjugate. From Eq. (1.42), one can extract the tree decay amplitude for the $\mu^-(p_1) \to e^-(p_4) + \nu_\mu(p_3) + \nu_e(p_2)$ (see Feynman rules in Section 1.6 or Vol. 1, Chapt. 6 for more details)

$$\begin{aligned} S_{fi} - \delta_{fi} &= -(2\pi)^4\delta^4(p_1 - p_2 - p_3 - p_4)\overline{u}(p_3)\left(\frac{ig_W}{\sqrt{2}}\gamma^\mu\right)\frac{(1-\gamma^5)}{2}u(p_1) \\ &\quad \times \frac{-i\left(g_{\mu\nu} - \frac{q^\mu q^\nu}{m_W^2}\right)}{q^2 - m_W^2 + i\epsilon}\overline{u}(p_4)\left(\frac{ig_W}{\sqrt{2}}\gamma^\nu\right)\frac{(1-\gamma^5)}{2}v(p_2) \\ &\xrightarrow{m_W^2 \gg q^2} \sim i\frac{g_W^2}{8m_W^2}\overline{u}(p_3)\gamma^\mu(1-\gamma^5)u(p_1)\overline{u}(p_4)\gamma_\mu(1-\gamma^5)v(p_2) \end{aligned} \tag{1.43}$$

The expression agrees with the transition amplitude obtained from the four-Fermi theory provided (compare this with Vol. 1, Eq. (15.75))

$$\frac{G_F}{\sqrt{2}} = \frac{g_W^2}{8m_W^2} = \frac{1}{2v^2} \tag{1.44}$$

Inserting the value of $G_F = 1.16637 \times 10^{-5}\ \text{GeV}^{-2}$, we obtain $v = 246$ GeV.

Mass of the Fermion Let us take a look at the Higgs-fermion interaction (the third term in the second line of Eq. (1.16)).

$$-\mathcal{L}_{Hee} \equiv G_e[\overline{e}_R(\Phi^\dagger \Psi_L) + (\overline{\Psi}_L \Phi)e_R] \tag{1.45}$$

After the symmetry breakdown, the Higgs-fermion interaction becomes

$$-\mathcal{L}_{Hee} \to \frac{G_e}{\sqrt{2}}(v + H)(\overline{e}_R e_L + \overline{e}_L e_R) = m_e \overline{e}e + \frac{m_e}{v}(\overline{e}e)H\,, \quad m_e = \frac{G_e v}{\sqrt{2}} \tag{1.46}$$

The first term gives the mass and the second, the Higgs interaction with the fermion. As a by-product of obtaining the fermion mass term, a new interaction of the Higgs with the fermion has appeared. Its interaction can be obtained by replacing the mass term v by $v + H$, just as the interaction with the gauge bosons in Eq. (1.37). Notice that the coupling strength is proportional to the mass of the fermion. Feynman diagrams for the Higgs-fermion interaction are given in Section 1.6.

One important side effect of the fermion mass generation is that the axial current no longer conserves.

$$\partial_\mu A^\mu(x) = \partial_\mu \overline{\psi}(x)\gamma^5\gamma^\mu\psi(x) = 2m\overline{\psi}(x)\psi(x) \neq 0 \tag{1.47}$$

This applies to the charged current as well as to the neutral current. Therefore, the chiral gauge symmetry is broken by the mass term and only the vector current conserves.[8] One refers to the phenomenon as the symmetry breakdown of $SU(2)_L \times SU(2)_R \to SU(2)_V$.

Mass of the Higgs The Lagrangian of the Higgs field can be extracted from the first and second term on the second line of Eq. (1.16). If we use the transformed Φ, we note that $\Phi^\dagger\Phi = (v + H)^2/2$ and $(\partial_\mu \Phi)^\dagger(\partial^\mu \Phi) = \partial_\mu H \partial^\mu H/2$. Putting aside

8) This is not to be confused with the chiral symmetry breaking in QCD. In QCD, the chiral symmetry breaking also occurs due to spontaneous breakdown of the QCD vacuum. More details will be given in Section 7.1.5. It is a global chiral symmetry breaking caused by the strong interaction among the quarks, and $q\overline{q}'$ plays the role of Higgs. Incidentally, this was the Nambu's original proposal of the spontaneous symmetry breaking [45, 46]. In QCD, additional explicit chiral symmetry breaking is induced by the fermion mass term. It is treated as an external perturbation because it is a product of the electroweak symmetry breaking.

the interactions with the gauge field and fermions, the Higgs part of the Lagrangian is expressed as

$$\begin{aligned}\mathcal{L}_H &= \frac{1}{2}\partial_\mu H \partial^\mu H - \lambda v^2 H^2 - \left(\lambda v H^3 + \frac{\lambda}{4} H^4\right) \\ &= \frac{1}{2}\left(\partial_\mu H \partial^\mu H - m_H^2 H^2\right) - \frac{g_W}{4 m_W} m_H^2 H^3 - \frac{1}{32}\frac{g_W^2}{m_W^2} m_H^2 H^4 \end{aligned} \tag{1.48a}$$

$$m_H^2 = 2\lambda v^2 \tag{1.48b}$$

where Eq. (1.39) was used to express the coupling constants in terms of masses and gauge couplings. The coupling strength, again, is proportional to the masses of the interacting particles. This is a conspicuous feature of the Standard Model. From this Lagrangian, one can construct the Higgs propagator and Feynman rules for the interactions of the Higgs field. They are listed in Section 1.6.

1.4 Gauge Interactions

Coupling with Fermions Interactions of the gauge boson with fermions are contained in the covariant derivative and are given by Eq. (1.21). As it is the starting point to derive the Feynman rules for the gauge interactions, we reproduce it here.

$$\begin{aligned}-\mathcal{L}_{Wff} &= g_W \overline{\Psi}\gamma^\mu \boldsymbol{W}_\mu \cdot \boldsymbol{t}\Psi + \frac{g_B}{2}\overline{\Psi}\gamma^\mu \Psi B_\mu \\ &= \frac{g_W}{2}\overline{\Psi}_L\gamma^\mu\left(W_\mu^1\tau_1 + W_\mu^2\tau_2 + W_\mu^3\tau_3\right)\Psi_L \\ &\quad + \frac{g_B}{2} B_\mu\left(\overline{\Psi}_L\gamma^\mu Y_L \Psi_L + \overline{\Psi}_R\gamma^\mu Y_R \Psi_R\right)\end{aligned} \tag{1.49a}$$

where $\boldsymbol{t} = \boldsymbol{\tau}/2$ was used. We removed the right-handed fields from the W interaction and added the hypercharge operator "Y" explicitly to remind the reader that both $\boldsymbol{W}$ and B_μ act differently on the left- and right-handed fields. Using Eq. (1.24) to rewrite W_μ^3 and B_μ in terms of A_μ and Z_μ, and denoting the charged W bosons as $W^\mp = (W_1 \pm W_2)/\sqrt{2}$, we have

$$\begin{aligned}-\mathcal{L}_{Wff} &= \frac{g_W}{\sqrt{2}}\overline{\Psi}_L\gamma^\mu\left(W_\mu^+\tau_+ + W_\mu^-\tau_-\right)\Psi_L \\ &\quad + \overline{\Psi}_L\gamma^\mu\left[g_W I_3(c_W Z_\mu + s_W A_\mu) + \frac{g_B}{2} Y_L(-s_W Z_\mu + c_W A_\mu)\right]\Psi_L \\ &\quad + \overline{\Psi}_R\gamma^\mu\left[\frac{g_B}{2} Y_R(-s_W Z_\mu + c_W A_\mu)\right]\Psi_R\end{aligned} \tag{1.49b}$$

where we abbreviated $s_W = \sin\theta_W$, $c_W = \cos\theta_W$. Using $Q = I_3 + Y/2$, $I_3\Psi_R = 0$, we finally obtain

$$\begin{aligned}
-\mathcal{L}_{Wff} &= \frac{g_W}{\sqrt{2}}\overline{\Psi}_L\gamma^\mu\left(W_\mu^+\tau_+ + W_\mu^-\tau_-\right)\Psi_L + g_Z\overline{\Psi}\gamma^\mu(I_{3L} - Q s_W^2)\Psi Z_\mu \\
&+ e\overline{\Psi}\gamma^\mu Q\Psi A_\mu \\
e &= g_W \sin\theta_W\,, \quad g_Z = g_W/\cos\theta_W = e/\sin\theta_W\cos\theta_W
\end{aligned} \tag{1.50}$$

which defines the electromagnetic coupling constant (i. e. the electric charge) in terms of g_W and the Weinberg angle. Here, L in I_{3L} is there to remind the reader that it only acts on the left-handed components.

For actual calculations, it is more convenient to separate couplings to the left and right or alternatively, to vector and axial vector parts. Coupling types which appear in the Feynman amplitude rule are given by the matrix element $i\mathcal{L}_{Wff}$. Therefore, omitting the field operators and attaching a suffix to differentiate the fermion flavor, they are expressed as

$$\gamma - ff: \quad -iQ_f e\gamma^\mu \tag{1.51a}$$

$$W^\pm - ff: \quad -i\frac{g_W}{2\sqrt{2}}\gamma^\mu(1-\gamma^5) \tag{1.51b}$$

$$\begin{aligned}
Z - ff: \quad &-i\frac{g_Z}{2}\gamma^\mu\left[\epsilon_L(f)(1-\gamma^5) + \epsilon_R(f)(1+\gamma^5)\right] \\
&= -i\frac{g_Z}{2}\gamma^\mu\left(v_f - a_f\gamma^5\right)
\end{aligned} \tag{1.51c}$$

$$\epsilon_L(f) = I_3 - Q_f s_W^2\,, \quad \epsilon_R(f) = -Q_f s_W^2 \tag{1.51d}$$

$$v_f = I_3 - 2Q_f s_W^2\,, \quad a_f = I_3 \tag{1.51e}$$

The Feynman rules are given in Section 1.6.

Self-interactions of the Gauge Boson The Lagrangian of the non-Abelian gauge field contains, in addition to the quadratic kinetic term, higher powers of the field's operator which represent self-interactions.

$$\mathcal{L}_{\text{gauge}} = -\frac{1}{4}\boldsymbol{F}_{\mu\nu}\cdot\boldsymbol{F}^{\mu\nu} - \frac{1}{4}F_{B\mu\nu}F_B^{\mu\nu} = -\frac{1}{4}\sum_A F_{A\mu\nu}F_A^{\mu\nu} - \frac{1}{4}F_{B\mu\nu}F_B^{\mu\nu} \tag{1.52a}$$

$$\begin{aligned}
\boldsymbol{F}_{A\mu\nu} &= \partial_\mu \boldsymbol{W}_{A\nu} - \partial_\nu \boldsymbol{W}_{A\mu} - g_W(\boldsymbol{W}_\mu \times \boldsymbol{W}_\nu)_A \\
(\boldsymbol{W}_\mu \times \boldsymbol{W}_\nu)_A &= \sum_{B,C}\epsilon_{ABC}W_{B\mu}W_{C\nu}
\end{aligned} \tag{1.52b}$$

$$F_{B\mu\nu} = \partial_\mu B_\nu - \partial_\nu B_\mu \tag{1.52c}$$

The self energy of the gauge boson is given by its triple and quartic terms in Eq. (1.52a). It can be unpacked into

$$\begin{aligned}\mathcal{L}_{GWS} &= \mathcal{L}_{KE} + \mathcal{L}_{SE}\\ &= -\frac{1}{2}(\partial_\mu \mathbf{W}_\nu - \partial_\nu \mathbf{W}_\mu)\cdot(\partial^\mu \mathbf{W}^\nu)\\ &+ g_W(\mathbf{W}_\mu \times \mathbf{W}_\nu \cdot \partial^\mu \mathbf{W}^\nu) - \frac{g_W^2}{4}\left[(\mathbf{W}_\mu \cdot \mathbf{W}^\mu)^2 - (\mathbf{W}_\mu \cdot \mathbf{W}_\nu)(\mathbf{W}^\mu \cdot \mathbf{W}^\nu)\right]\end{aligned} \tag{1.52d}$$

From this Lagrangian, one can derive Feynman amplitudes corresponding to 3-gauge and 4-gauge boson vertices.

3-W vertex: Let us first treat the 3-W vertex matrix elements. Rewriting

$$\begin{aligned}\mathbf{W}_\mu \times \mathbf{W}_\nu \cdot \partial^\mu \mathbf{W}^\nu &= (W_{1\mu} W_{2\nu} - W_{2\mu} W_{1\nu})\partial^\mu W_3^\nu + (\text{cyclic})\\ &= i(W_\mu^- W_\nu^+ - W_\mu^+ W_\nu^-)\partial^\mu (c_W Z + s_W A)^\nu\\ &+ (\text{cyclic})\end{aligned} \tag{1.53}$$

where we used $W^\pm = (W_1 \mp i W_2)/\sqrt{2}$ and Eq. (1.24). Feynman diagrams of the interaction can be obtained by calculating matrix elements of Eq. (1.53).

As an example, let us consider the $\gamma \to W^+ W^-$ vertex. For symmetry reasons, we treat all the momenta as incoming (see Figure 1.5). Denoting the polarization and momentum of the incoming photon as ϵ_3, k_3 and those of outgoing W's as $\epsilon_n, -k_n, n = 1, 2$, the scattering matrix element of the 3-W vertex is given by

$$\begin{aligned}S_{fi} - \delta_{fi} &= -(2\pi)^4\delta^4(k_1 + k_2 + k_3) i\mathcal{M}_{fi}\\ &= i\int d^4x \langle \epsilon_1(W^-, -k_1); \epsilon_2(W^+, -k_2)|[\cdots]|\epsilon_3(\gamma, k_3)\rangle\\ &= (2\pi)^4\delta^4(k_1 + k_2 + k_3)(i)(i g_W s_W)\\ &\times \left[\{(\epsilon_{1\mu}\epsilon_{2\nu}) - (\epsilon_{2\mu}\epsilon_{1\nu})\}(-i k_3^\mu)\epsilon_3^\nu + (\text{cyclic})\right]\end{aligned} \tag{1.54a}$$

$$\begin{aligned}[\cdots] = \Big[&(i g_W)\left\{W_\mu^-(x) W_\nu^+(x) - W_\mu^+(x) W_\nu^-(x)\right\}\\ &\times \partial^\mu\{c_W Z(x) + s_W A(x)\}^\nu\Big]\end{aligned} \tag{1.54b}$$

Omitting the ever present δ function, the matrix element becomes

$$\begin{aligned}-i\mathcal{M}_{fi} &= (-i g_W s_W)[(k_3\cdot\epsilon_1)(\epsilon_2\cdot\epsilon_3) - (k_3\cdot\epsilon_2)(\epsilon_3\cdot\epsilon_1) + (\text{cyclic})]\\ &= ie[\{(k_1 - k_2)\cdot\epsilon_3\}(\epsilon_1\cdot\epsilon_2) + \{(k_2 - k_3)\cdot\epsilon_1\}(\epsilon_2\cdot\epsilon_3)\\ &+ \{(k_3 - k_1)\cdot\epsilon_2\}(\epsilon_3\cdot\epsilon_1)]\\ &= \varepsilon^\mu(W^-)\varepsilon^\nu(W^+)\\ &\times [(ie)[g_{\mu\nu}(k_1 - k_2)_\lambda + g_{\nu\lambda}(k_2 - k_3)_\mu + g_{\lambda\mu}(k_3 - k_1)_\nu]]\varepsilon^\lambda(\gamma)\end{aligned} \tag{1.55}$$

The Feynman amplitude for the vertex is given by the content in the bracket.

$$ie[g_{\mu\nu}(k_1 - k_2)_\lambda + g_{\nu\lambda}(k_2 - k_3)_\mu + g_{\lambda\mu}(k_3 - k_1)_\nu] \tag{1.56}$$

4-W vertex: The Lagrangian of quartic interaction is given by

$$\begin{aligned}
\mathcal{L}_{4W} &= -\frac{g_W^2}{4}(\mathbf{W}_\mu \times \mathbf{W}_\nu)\cdot(\mathbf{W}^\mu \times \mathbf{W}^\nu) \\
&= -\frac{g_W^2}{4}\left[(\mathbf{W}_\mu \cdot \mathbf{W}^\mu)(\mathbf{W}_\nu \cdot \mathbf{W}^\nu) - (\mathbf{W}_\mu \cdot \mathbf{W}_\nu)(\mathbf{W}^\mu \cdot \mathbf{W}^\nu)\right] && (1.57a) \\
&= -\frac{g_W^2}{4}\Big[(W_1^2 + W_2^2 + W_3^2)^2 \\
&\quad - (W_{1\mu}W_{1\nu} + W_{2\mu}W_{2\nu} + W_{3\mu}W_{3\nu})\left(W_1^\mu W_1^\nu + W_2^\mu W_2^\nu + W_3^\mu W_3^\nu\right)\Big] \\
&= -\frac{g_W^2}{2}\Big[\{W_1^2 W_2^2 - (W_1 \cdot W_2)^2\} \\
&\quad + \{(W_1^2 + W_2^2)\, W_3^2 - (W_1 \cdot W_3)^2 - (W_2 \cdot W_3)^2\}\Big] \\
&= g_W^2\left[\frac{1}{2}\left\{(W^+)^2(W^-)^2 - (W^+ \cdot W^-)^2\right\}\right. \\
&\quad \left. - \left\{(W^+ \cdot W^-)X^2 - (W^+ \cdot X)(W^- \cdot X)\right\}\right] && (1.57b)
\end{aligned}$$

$$X = c_W Z + s_W A \tag{1.57c}$$

where $A^2, (A \cdot B)$ stand for the Lorentz scalar product $A_\mu A^\mu, (A_\mu B^\mu)$. The first bracket describes the $W^+W^- \leftrightarrow W^+W^-$ process and the second $A(Z)A(Z) \leftrightarrow W^+W^-$.

Next, we consider $A_\alpha Z_\beta \to W_\mu^+ W_\nu^-$. By taking the matrix element and removing the δ function, we can get the scattering matrix amplitude

$$\begin{aligned}
-i\mathcal{M} &\sim i\int d^4x \mathcal{L}_{4W}(x) \\
&\sim -ig_W^2 s_W c_W \langle\varepsilon(W^-)\varepsilon(W^+)|[(W^+ \cdot W^-)X^2 \\
&\quad - (W^+ \cdot X)(W^- \cdot X)]|\varepsilon(\gamma)\varepsilon(Z)\rangle \\
&= -ie^2\cot\theta_W[2\{\varepsilon(W^-)\cdot\varepsilon(W^+)\}\{\varepsilon(\gamma)\cdot\varepsilon(Z)\} \\
&\quad - \{\varepsilon(W^-)\cdot\varepsilon(Z)\}\{\varepsilon(W^+)\cdot\varepsilon(\gamma)\} - \{\varepsilon(W^-)\cdot\varepsilon(\gamma)\}\{\varepsilon(W^+)\cdot\varepsilon(Z)\}] \\
&= \varepsilon^\mu(W^+)\varepsilon^\nu(W^-)\left[-ie^2\cot\theta_W(2g_{\alpha\beta}g_{\mu\nu} - g_{\alpha\mu}g_{\beta\nu} - g_{\alpha\nu}g_{\beta\mu})\right] \\
&\quad \times \varepsilon^\alpha(\gamma)\varepsilon^\beta(Z) && (1.58)
\end{aligned}$$

Other processes can be calculated similarly. The Feynman diagrams for the triple and quartic coupling of the gauge bosons are summarized in Section 1.6.

1.5 Higgs Interactions

Coupling with Gauge Bosons We have seen that the mass is generated in the Higgs sector and that its value is proportional to the vacuum expectation value "v." As the physical Higgs always appears in combination with the vacuum expectation value, its coupling is closely related with the mass of particles which it couples. It is obtained by replacing v by $(v + H)$ as is seen in Eq. (1.37) and Eq. (1.46). The interaction Lagrangian for the Higgs-gauge coupling is obtained by expanding Eq. (1.37) and picking terms that contain the Higgs field H, that is,

$$\begin{aligned}
\mathcal{L}_{W-H} &= \left(\frac{g_W(v+H)}{2}\right)^2 W_\mu^- W^{+\mu} + \frac{1}{2}\left(\frac{g_Z(v+H)}{2}\right)^2 Z_\mu Z^\mu - \mathcal{L}_{\text{mass}} \\
&= \frac{g_W^2}{2} v H W_\mu^- W^{+\mu} + \frac{g_Z^2}{4} v H Z_\mu Z^\mu + \frac{g_W^2}{4} H^2 W_\mu^- W^{+\mu} \\
&\quad + \frac{g_Z^2}{8} H^2 Z_\mu Z^\mu \\
&= \left[g_W m_W W_\mu^- W^{+\mu} + \frac{1}{2} g_Z m_Z Z_\mu Z^\mu\right] H \\
&\quad + \left[\frac{g_W^2}{4} W_\mu^- W^{+\mu} + \frac{1}{2}\frac{g_Z^2}{4} Z_\mu Z^\mu\right] H^2
\end{aligned} \tag{1.59}$$

It contains interactions of the type $H W W$, $H Z Z$ and $H^2 W^2$, $H^2 Z^2$. Because of Eq. (1.39), the coupling strength can be expressed in terms of the mass of the particle.

$$\begin{aligned}
&H W W \;:\; \frac{g_W^2 v}{2} = g_W m_W\,, \qquad && H Z Z \;:\; \frac{g_Z^2 v}{2} = g_Z m_Z \\
&H H W W \;:\; \frac{g_W^2}{4} = \frac{m_W^2}{v^2}\,, \qquad && H H Z Z \;:\; \frac{g_Z^2}{4} = \frac{m_Z^2}{v^2}
\end{aligned} \tag{1.60}$$

Feynman diagrams for these interactions are given in Section 1.6.

Coupling with Fermions The coupling is again obtained from the fermion mass term by replacing $v \to v + H$ which is given by the second term of the last equality of Eq. (1.46).

$$\mathcal{L}_{Hff} = -\frac{m_f}{v}\overline{\psi}_f \psi H = -g_W \frac{m_f}{2 m_W}\overline{\psi}_f \psi_f H \tag{1.61}$$

Self Coupling The self-interaction of the Higgs is already given in Eq. (1.48).

$$\mathcal{L}_{\text{H-self}} = -\frac{g_W}{4 m_W} m_H^2 H^3 - \frac{1}{32}\frac{g_W^2}{m_W^2} m_H^2 H^4 \tag{1.62}$$

The expression is simple, but in extracting matrix elements, one needs to count the symmetry factor carefully, for instance, $\langle 0|H^3|h_1 h_2 h_3\rangle \to 3!$ and $\langle 0|H^4|h_1 h_2 h_3 h_4\rangle \to 4!$ where h_i's are the ith Higgs particles. Then, the Feynman rules are given by

$$-i\frac{3}{2}\frac{g_W}{m_W}m_H^2: \quad \text{triple-Higgs interaction} \tag{1.63a}$$

$$-i\frac{3}{4}\frac{g_W^2}{m_W^2}m_H^2: \quad \text{quartic-Higgs interaction} \tag{1.63b}$$

1.6 Feynman Rules of Electroweak Theory

Now that we have given all the rules to calculate transition matrix elements for the electroweak interaction, we summarize the Feynman rules in the unitary gauge below.

Feynman Rule 1: External Lines: We attach wave functions to fermions or polarizations to bosons for each incoming or outgoing particle (Fig. 1.2). Spinor indices for fermions are sometimes omitted.

Feynman Rule 2: Internal Lines To each internal line, we attach one of the propagators depicted in Figure 1.3, depending on the particle species. For fermions, the sign of momentum follows that of an arrow.

Feynman Rule 3: Fermion-Gauge Boson Vertices For vertices of fermions and gauge bosons, we attach coupling constants and appropriate γ factors (Fig. 1.4). The photon couples to the electromagnetic current with charge $Q_f e$ and is of the vector

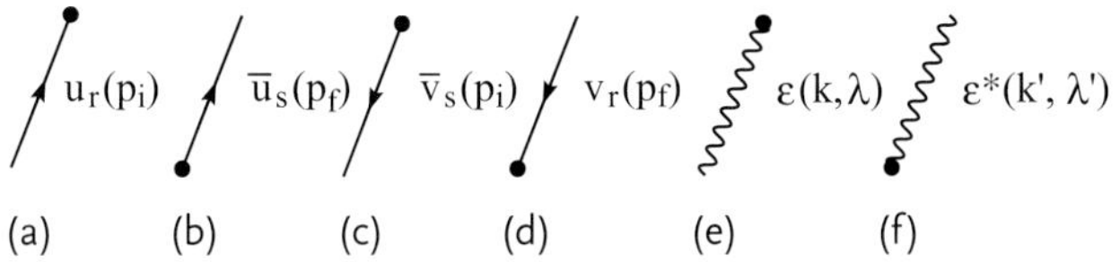

Figure 1.2 Wave functions to fermions and polarization vectors to bosons are to be attached to each external line.

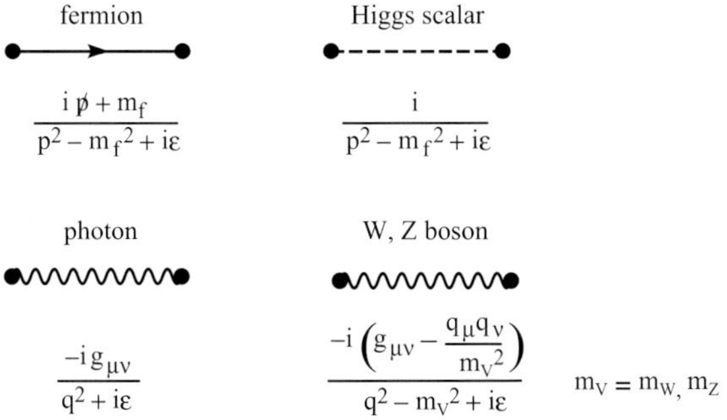

Figure 1.3 Propagators are to be attached to each internal line.

type. Q_fs are given by

$$\begin{aligned} &\text{Leptons}: \begin{cases} Q_{\nu_e} = Q_{\nu_\mu} = Q_{\nu_\tau} = 0 \\ Q_e = Q_\mu = Q_\tau = -1 \end{cases} \\ &\text{Quarks}: \begin{cases} Q_u = Q_c = Q_t = +2/3, \\ Q_d = Q_s = Q_b = -1/3 \end{cases} \end{aligned} \tag{1.64}$$

with opposite charge assigned to the antifermions. The neutral Z boson couples to the neutral current which is a mixture of the left- and right-handed fermions. Its coupling constant is a product of a common constant

$$g_Z = \frac{e}{\sin\theta_W \cos\theta_W} \tag{1.65}$$

and flavor dependent constants

$$\epsilon_L(f) = I_{3f} - Q_f \sin^2\theta_W, \quad \epsilon_R(f) = -Q_f \sin^2\theta_W \tag{1.66}$$

where $f = e, \mu, \tau$ or u, d, s, c, b, t. An alternative expression in terms of the vector and axial vector couplings is also used. Using

$$g_Z \gamma^\mu \left(\varepsilon_L(f) \frac{1-\gamma^5}{2} + \varepsilon_R(f) \frac{1+\gamma^5}{2} \right) = \frac{g_Z}{2} \gamma^\mu (v_f - a_f \gamma^5) \tag{1.67}$$

the vector and the axial vector couplings are expressed as

$$v_f = I_3 - 2Q_f \sin^2\theta_W, \quad a_f = I_3 \tag{1.68}$$

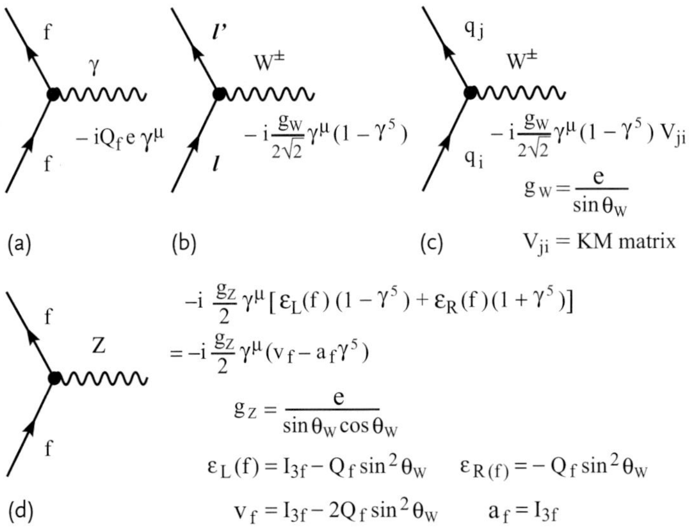

Figure 1.4 Vertices of fermions with gauge bosons.

They are mutually related by

$$v_f = \epsilon_{\rm L}(f) + \epsilon_{\rm R}(f) , \quad a_f = \epsilon_{\rm L}(f) - \epsilon_{\rm R}(f) \tag{1.69}$$

The charged W boson couples to left-handed fermions and its strength is given by

$$g_{\rm W} = \frac{e}{\sin\theta_{\rm W}} \tag{1.70}$$

Notice that the fields that appear in the original Lagrangian are so-called weak eigenstates. However, when we calculate cross sections, we use mass eigenstates. For the electromagnetic and neutral current interactions, we do not need to differentiate the mass eigenstates from the weak eigenstates, but for the charged current interactions they are different. They are related by the Cabibbo–Kobayashi–Maskawa (CKM) matrix V_{ji}[9]. Therefore, for the $W^{\pm}$-fermion interaction with up-quark j $(= u, c, t)$ and down-quark i $(= d, s, b)$, the CKM elements V_{ji} have to be attached.

Feynman Rule 4: Nonlinear Couplings of the Gauge Bosons: Because of the non-Abelian nature of the electroweak theory, the gauge bosons have self couplings which were given in Eqs. (1.56) and (1.58). Their Feynman graphs are shown in Figure 1.5. Note that there are no γZZ or ZZZ couplings. In the figure, all the momenta are taken to be inward going.

Feynman Rule 5: Higgs Couplings: In Figure 1.6 we list vertices which include at least one Higgs particle. Notice, the coupling strength is proportional to the mass of the connecting particles.

Feynman Rule 6: Momentum Assignment and Loops The momenta of external lines are fixed by experimental conditions. Then, at each vertex, the energy-momenta have to conserve. The energy-momentum conservation constrains that sum of all energy-momenta of external lines have to vanish assuming all the external momenta are inward going. It also fixes all the momenta for tree diagrams which do not contain loops. Each loop leaves one momentum unconstrained and has to be integrated, leading to divergent integrals. The integration includes a sum over spinor indices and polarizations, depending on the particle species that form the loop. For each closed fermion loop, an extra sign (–) has to be attached. It is a result of the anticommutativity of the fermion fields.

Amplitude for $e^-e^+ \to f\overline{f}$: Once all the Feynman diagrams are given, calculation of scattering amplitudes can be carried out in a straightforward way. As an example, we construct an amplitude for the reaction $e^-e^+ \to f\overline{f}$ in the $O(\alpha^2)$ process where f is any of the leptons or quarks.

9) Details of the CKM matrix elements are discussed in Chapter 6.

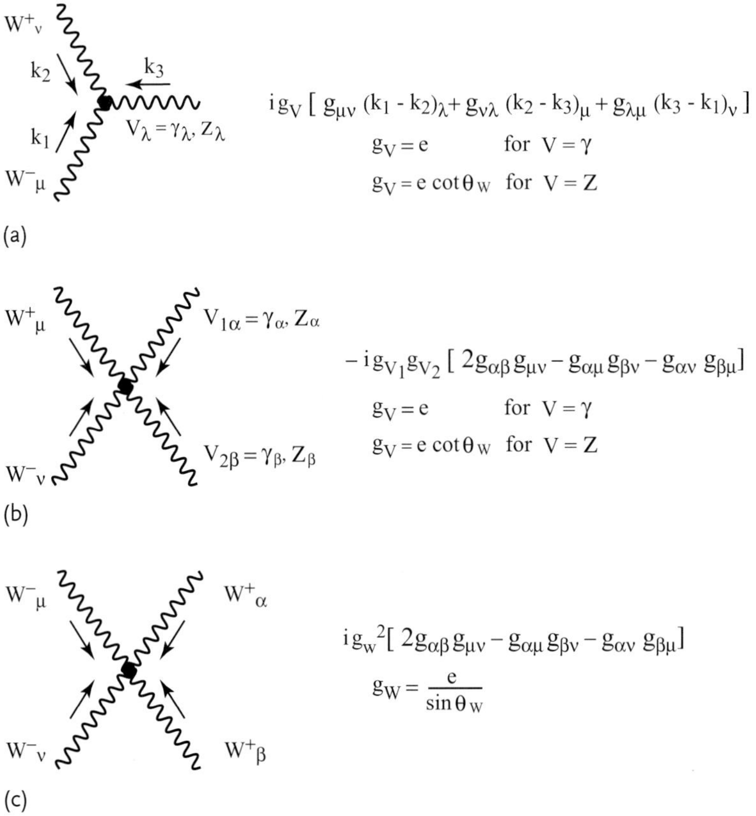

Figure 1.5 Nonlinear gauge boson couplings.

According to the Feynman rules we just described, we attach appropriate functions to every part of the Feynman diagram as shown in Figure 1.7. The S-matrix and the cross section is written as

$$S_{fi} = \delta_{fi} - (2\pi)^4 \delta^4(p_1 + p_2 - p_3 - p_4) i \mathcal{M} \tag{1.71a}$$

$$d\sigma = \frac{1}{F} \overline{\sum}_{\substack{\text{spin} \\ \text{pol}}} |\mathcal{M}|^2 dLIPS \tag{1.71b}$$

$$dLIPS = (2\pi)^4 \delta^4(p_1 + p_2 - p_3 - p_4) \frac{d^3 p_3}{(2\pi)^3 2E_3} \frac{d^3 p_4}{(2\pi)^3 2E_4} \tag{1.71c}$$

$$F = 4\left[(p_1 \cdot p_2)^2 - (m_1 m_2)^2\right] \simeq 2s \quad \text{for} \quad s = (p_1 + p_2)^2 \gg m_1^2, m_2^2 \tag{1.71d}$$

where F is the initial flux and $dLIPS$ is the Lorentz invariant phase space of the final state. $\overline{\sum}$ denotes the average of the initial state and sum over final state degrees of freedom which is valid when polarizations are not observed. Referring to the

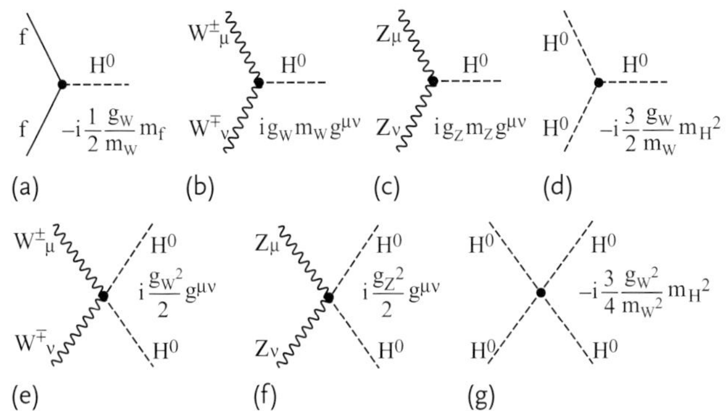

Figure 1.6 Vertices that include the Higgs boson.

Feynman diagram in Figure 1.7, the transition amplitude $\mathcal{M}$ can be written as

$$-i\mathcal{M} = \left[\overline{u}(p_3)\left\{-i\frac{g_Z}{2}\gamma^\mu(v_f - a_f\gamma^5)\right\}v(p_4)\right]\frac{-i\left(g_{\mu\nu} - \frac{q_\mu q_\nu}{m_Z^2}\right)}{q^2 - m_Z^2 + i\epsilon}$$
$$\times\left(-i\Sigma^{\nu\lambda}_{\gamma Z}(q^2)\right)\frac{-ig_{\lambda\rho}}{q^2 + i\epsilon}\left[\overline{v}(p_2)\left(-iQ_i e\gamma^\rho\right)u(p_1)\right] \qquad (1.72a)$$

$$-i\Sigma^{\nu\lambda}_{\gamma Z}(q^2) = -\int\frac{d^4p}{(2\pi)^4}\text{Tr}\left[\frac{i(\not{q} - \not{p}) + m_f}{(q-p)^2 - m_f^2 + i\epsilon}\left\{-i\frac{g_Z}{2}\gamma^\nu(v_f - a_f\gamma^5)\right\}\right.$$
$$\left.\times\frac{i\not{p} + m_f}{p^2 - m_f^2 + i\epsilon}(-iQ_f e\gamma^\lambda)\right] \qquad (1.72b)$$

where $\overline{u}(p_3)$, $u(p_1)$, ... are plane wave solutions of the Dirac equation [see Appendix A]. We have separated the fermion loop part of the Feynman diagram because it has to be integrated over the internal momentum and an extra (–) sign has been attached according to the rule (6). $\Sigma^{\nu\lambda}_{\gamma Z}(q^2)$ is a diverging integral and has to be treated with a renormalization prescription which will be discussed in detail in Chapter 5 and in Appendix C[10].

A Note on the Ghosts The loop Feynman diagram in Figure 1.7 was given just to illustrate how the Feynman rules work in the unitary gauge. In general, once we go to higher order diagrams which contain loops, things are more complicated and we need to consider ghost's contributions. We did not include the ghosts in our Lagrangian and their associated Feynman rules were not given either because technical details of higher order calculations are beyond the scope of this book. We only mention their role in the non-Abelian gauge theories.[11]

10) Generally, in this book, we derive cross sections only at the tree level and describe higher order corrections qualitatively. The only exception is the description of precision Z resonance data in Chapter 5 which are compared with theoretical calculations including radiative corrections.

11) Some notes are given in the Appendix D for settings of the R-gauge and Feynman rules for the ghosts.

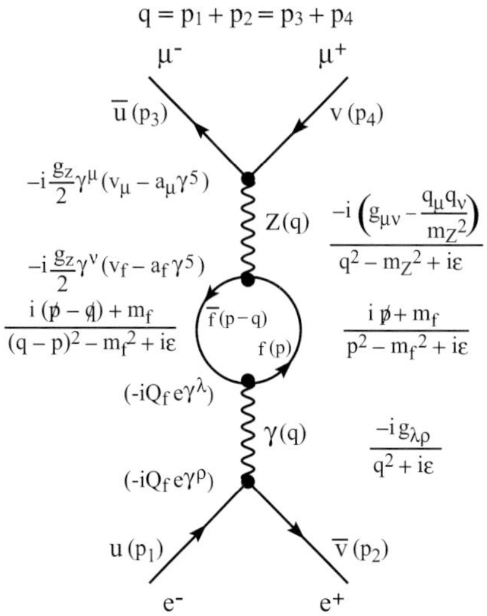

Figure 1.7 An example of the Feynman diagram in order $O(\alpha^2)$ for the process $e^-e^+ \to \mu^-\mu^+$. To every element of the Feynman diagram (wave functions, vertices and propagators), corresponding functions are attached.

Ghosts are fictitious scalar fields having the same isospin degrees of freedom as the gauge particles, but obeying the Fermi-Dirac statistics. They are mathematical artifacts that appear in the covariant gauge and only appear in the internal lines of the Feynman diagrams. They do not appear in the physical gauge[12]. They only couple to the gauge fields. Their sole role is to compensate the unphysical degrees of contributions in the loop generated by self-interactions of the non-Abelian gauge fields in the internal lines. Unphysical contributions are generated by unphysical components, that is, scalar components of the massive gauge particle. Therefore, whenever loop diagrams of the non-Abelian gauge fields like that in Figure 1.8a appear, the ghosts (Figure 1.8b) have to be included to compensate unphysical contributions.

1.7 Roles of the Higgs in Gauge Theory

Unitary Gauge and R-Gauge So far, we emphasized the role of the Higgs field in attaching mass to the gauge as well as matter particles. Here, we describe another role in maintaining renormalizability of spontaneously broken gauge theories.

A major difficulty in the theory of weak interaction is the existence of massive gauge bosons because they violate the gauge symmetry. In the GWS theory, it has been shown that the symmetry is not broken, but hidden. In the unitary gauge, the dynamical variables are chosen to match observed phenomena. However, in this

12) Coulomb gauge in QED and axial gauge in non-Abelian gauge. See Section 7.1.1.

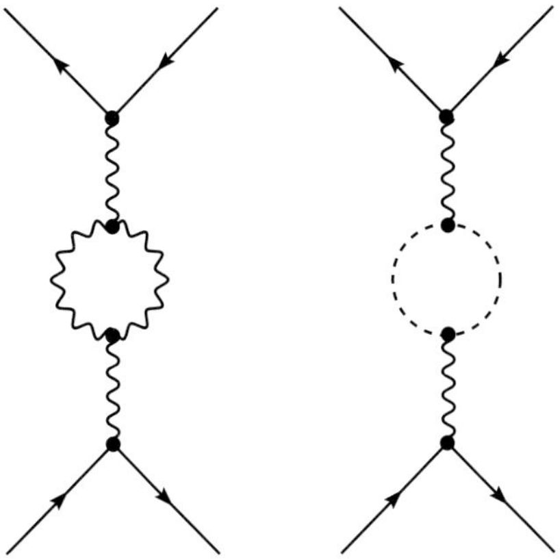

(a) (b)

Figure 1.8 The non-Abelian ghost field which only appears as internal lines. The propagation of the ghosts are denoted by dashed lines. Feynman diagrams containing a ghost loop (b) cancels unphysical contributions created by a self-interacting non-Abelian gauge loop (a).

gauge, the massive vector bosons have three degrees of freedom corresponding to three polarization states. The longitudinal polarization has components

$$\epsilon^{\mu}(3) = (0, 0, 0, 1) \tag{1.73}$$

in the particle's rest frame. When it is in motion, it is Lorentz boosted and has components

$$\epsilon_{\mathrm{L}} = \left(\frac{|\boldsymbol{k}|}{m}, 0, 0, \frac{\omega}{m}\right) = \frac{k^{\mu}}{m} + \frac{m}{\omega + |\boldsymbol{k}|}(-1, \hat{k}), \quad \hat{k} = \frac{\boldsymbol{k}}{|\boldsymbol{k}|} \tag{1.74}$$

in a coordinate system where the particle momentum is expressed as $k^{\mu} = (\omega, 0, 0, k), \omega = \sqrt{|\boldsymbol{k}|^2 + m^2}$. Accordingly, the propagator of a massive vector meson takes a form

$$i\Delta_F(k) = \frac{-i\left(g_{\mu\nu} - \frac{k_{\mu}k_{\nu}}{m_V^2}\right)}{k^2 - m_V^2} \tag{1.75}$$

At high energy as $k \to \infty$, the value of the propagator approaches a constant and is the cause of bad divergences in the loop integral (see Appendix C or Vol. 1, Sect. 15.8). In QED, we saw that the gauge invariance controlled the divergence in order to not grow faster than the logarithm of the momenta. There, the gauge propagator behaved like $\sim 1/k^2$ and the divergences were removed by introducing a few number of counter terms, in other words, the theory was renormalizable (see Vol. 1, Chapt. 8). However, presence of the longitudinal polarization adds an extra diverging contribution. This is why the unitarity of the process involving the massive gauge boson is broken. However, if the gauge symmetry is not really broken, but merely hidden, there must be a mechanism in the framework of spontaneous symmetry breaking to guarantee that the divergence created by the longitudinal polarization of the gauge boson is somehow canceled.

’t Hooft conceived of a clever gauge containing ξ as a parameter. Its formal setting is given in Appendix D. Here, we only mention its usefulness for higher order calculations and for the role of the longitudinal polarization. In this gauge (referred to as the R-gauge), the vector boson propagator is expressed as

$$i\Delta_F(k) = -\left[g^{\mu\nu} - (1-\xi)\frac{k^\mu k^\nu}{k^2 - \xi m^2}\right]\frac{i}{k^2 - m^2} \tag{1.76}$$

For $m \to 0$, it reproduces the photon propagator and by setting $\xi = 1$, it becomes the Feynman gauge in QED. The ordinary massive vector propagator can be reproduced by setting $\xi = \infty$. However, for such a setting, which is the case in the U-gauge, many divergent terms appear. If the theory is convergent as claimed, one has to carry out the algebra very carefully, otherwise one easily gets lost.

As long as ξ is kept finite, the propagator has a built-in cut-off and the longitudinal part can be calculated without difficulty. Setting $\xi = 1$, which is referred to as the ’t Hooft Feynman gauge, makes the calculation especially simple. Therefore, it is the preferred setting for most theoretical calculations. Only logarithmic divergences appear in the R-gauge and the calculation can be carried out in a straightforward way.

From a renormalization point of view, one sees that $\Delta_F \to 1/k^2$ as $k^2 \to \infty$ and is assured of the healthy theory applying the same logic to prove the renormalizability of the massless gauge boson, that is, the QED. Since it includes the U-gauge as a special choice of $\xi \to \infty$, the gauge invariance assures the renormalizability of the spontaneously broken gauge symmetry.

The price to pay is the appearance of the ξ dependent poles which has to be removed because it is not physical. Also, would-be-Goldstone bosons reappear. They vanish in the U-gauge because they are absorbed by the gauge bosons to become their longitudinal component. In the R-gauge, they appear as redundant degrees of freedom. However, it has been shown that the redundant would-be-Goldstone bosons exactly cancels the unwanted fictitious pole of the gauge bosons. The second price is that dynamical properties of each chosen variable in this gauge are not directly connected to observable quantities. Obtained mathematical results are hard to interpret in physical terms. Thus, the usual convention is to use the unitary gauge for physical interpretation, but rely on the R-gauge for actual calculation in order to address various theoretical technicalities.

Calculations are not difficult as long as one stays in the tree approximation. As we do not want to get involved in the higher order loop calculations too much, we will work in the unitary gauge in the following and restrict ourselves to qualitative discussions using a simple example. We calculate certain tree processes faithfully a’la Feynman rules and show how the renormalizability is restored in the spontaneously broken symmetry frame. Hopefully, we can obtain intuitive and clear insight by staying in the unitary gauge.

How is the Unitarity Maintained? Postulating the $SU(2)$ gauge symmetry for the weak interaction, the existence of the neutral vector boson W^0 was required. In the spontaneously broken symmetry, the existence of the Higgs field is also required. We shall see that they are necessary ingredients to keep the unitarity. Specifically, we show in qualitative arguments that their role is to eliminate badly diverging integrals induced by the massive gauge bosons. We take a simple example and see how the unitarity of the theory can be maintained.

$\boldsymbol{\nu\overline{\nu} \to W^+W^-}$: Let us consider the process of $\nu\overline{\nu} \to W^+W^-$ (Figure 1.9a). This is not a practically doable process, but serves as an illustration where the theoretical problem lies.

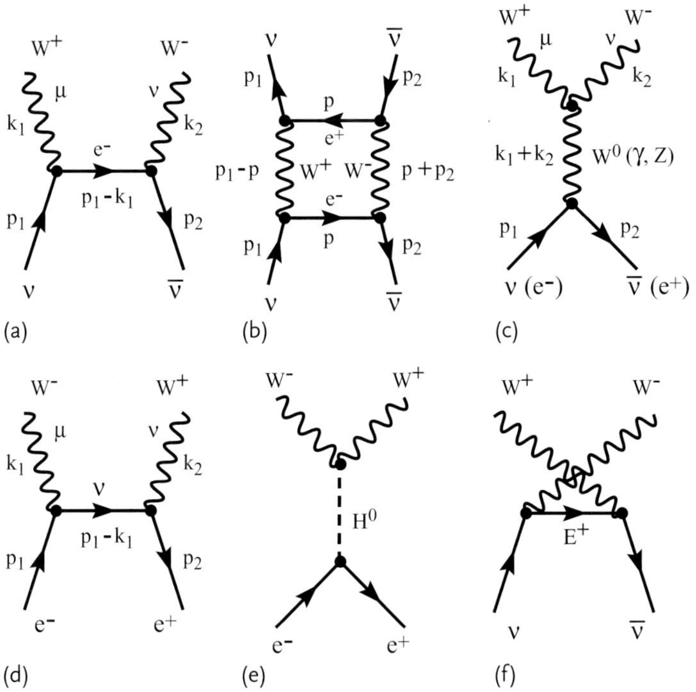

Figure 1.9 The role of various processes in the non-Abelian gauge theory. (a) Cross section for $\nu\overline{\nu} \to W^+W^-$ increases in proportion to the energy squared ($\sim s$). (b) Because of (a), integration over the W^+W^- intermediate state diverges badly. The remedy is two-fold: by introducing an additional contribution due to either a neutral W^0 in the s-channel (c) or a new charged E^+ in the u-channel (f), one possibly compensates the leading divergent term. (d) A massive fermion introduces another divergence. The wrong helicity component associated with massive fermions diverges. As it contributes to the $J = 0$ partial wave, it can be compensated by introducing a scalar particle (the Higgs) intermediate state (e).

Using the Feynman rules listed in Section 1.6, the scattering amplitude for the process is given by

$$
\begin{aligned}
-i\mathcal{M}_a &= \epsilon^*_{2\mu}(\lambda_2)\overline{v}(p_2)\gamma^\mu\left(-i\frac{g_W}{2\sqrt{2}}(1-\gamma^5)\right)\\
&\times\frac{i(\not{p}_1-\not{k}_1)+m_e}{(p_1-k_1)^2-m_e^2}\gamma^\nu\left(-i\frac{g_W}{2\sqrt{2}}(1-\gamma^5)\right)u(p_1)\epsilon_{1\nu}(\lambda_1)\\
&=-i\frac{g_W^2}{4}\overline{v}(p_2)\frac{\not{\epsilon}_2(\not{p}_1-\not{k}_1)\not{\epsilon}_1(1-\gamma^5)}{m_W^2-2(p_1\cdot k_1)}u(p_1)
\end{aligned}
\tag{1.77}
$$

The polarization vectors $\epsilon_\mu(\lambda), \lambda = 1-3$ satisfy $\epsilon(\lambda)\cdot\epsilon(\lambda') = -\delta_{\lambda\lambda'},\ \ k\cdot\epsilon(\lambda) = 0$. An expression for the longitudinal polarization was given in Eq. (1.74). It shows that at high energy ($|\boldsymbol{k}| \gg m_W$), where we are interested in, the longitudinal polarization can be approximated as $\epsilon^\mu \simeq k^\mu/m_W$. Replacing ϵ^μ with k^μ/m_W, we have

$$
\begin{aligned}
&\not{\epsilon}_2(\not{p}_1-\not{k}_1)\not{\epsilon}_1(1-\gamma^5)\sim\not{k}_2(\not{p}_1-\not{k}_1)\not{k}_1(1-\gamma^5)\\
&=\not{k}_2\left\{2(p_1\cdot k_1)-\not{k}_1\not{p}_1-m_W^2\right\}(1-\gamma^5)\\
&=-D\not{k}_2(1-\gamma^5)-m\not{k}_2\not{k}_1(1+\gamma^5)
\end{aligned}
\tag{1.78a}
$$

where $D = m_W^2 - 2(p_1\cdot k_1)$ and we used $\not{p}_1 u(p_1) = m u(p_1), \overline{v}(p_2)\not{p}_2 = -\overline{v}(p_2)m$. For $\nu\overline{\nu}$ reactions, $m = 0$, but we retain it for later discussions when the neutrino is replaced with the electron. Using $p_1 - k_1 = k_2 - p_2$, it can be rewritten as

$$
\begin{aligned}
&=\not{k}_2(\not{k}_2-\not{p}_2)\not{k}_1(1-\gamma^5)=\left\{m_W^2-2(p_2\cdot k_2)+\not{p}_2\not{k}_2\right\}\not{k}_1(1-\gamma^5)\\
&=D\not{k}_1(1-\gamma^5)-m\not{k}_2\not{k}_1(1-\gamma^5)
\end{aligned}
\tag{1.78b}
$$

$$
=\frac{1}{2}D(\not{k}_1-\not{k}_2)(1-\gamma^5)-m\not{k}_2\not{k}_1 \tag{1.78c}
$$

Equation (1.78c) is obtained by taking average of Eq. (1.78a) and Eq. (1.78b). The second term in Eq. (1.78c) can further be rewritten as

$$
-m\not{k}_2\not{k}_1 = mD + m^2(\not{k}_1-\not{k}_2) \tag{1.78d}
$$

The second term is $O(m^2/s)$ compared to the first and can be neglected. Substituting Eqs. (1.78c) and (1.78d) in Eq. (1.77), the amplitude for $\nu\overline{\nu} \to W^+W^-$ at high energy is given by

$$
-i\mathcal{M}_a = -i\frac{g_W^2}{8m_W^2}\overline{v}(p_2)[(\not{k}_1-\not{k}_2)(1-\gamma^5)+2m]u(p_1) \tag{1.79}
$$

It rises linearly with $|\boldsymbol{k}|$. As $\overline{v}(p)u(p) \sim E$ in the relativistic normalization, the cross section grows like $d\sigma \sim |\mathcal{M}|^2/s \sim s = (k_1+k_2)^2$.

Let us first consider the massless neutrino case. Then, the second term in the bracket is absent. We remark that the first term is a pure $J = 1$ amplitude. To prove it, we insert an explicit representation for the γ matrices and the plane wave solution for the Dirac particle (see Appendix A).

$$u(p) = \begin{bmatrix} \sqrt{E - \boldsymbol{p}\cdot\boldsymbol{\sigma}}\,\xi_r \\ \sqrt{E + \boldsymbol{p}\cdot\boldsymbol{\sigma}}\,\xi_r \end{bmatrix}, \quad v(p) = \begin{bmatrix} \sqrt{E - \boldsymbol{p}\cdot\boldsymbol{\sigma}}\,\eta_r \\ -\sqrt{E + \boldsymbol{p}\cdot\boldsymbol{\sigma}}\,\eta_r \end{bmatrix}, \quad \overline{v}(p) = v^\dagger(p)\gamma^0 \tag{1.80a}$$

$$\not{k} = \begin{bmatrix} 0 & \omega - \boldsymbol{k}\cdot\boldsymbol{\sigma} \\ \omega + \boldsymbol{k}\cdot\boldsymbol{\sigma} & 0 \end{bmatrix}, \quad \gamma^5 = \begin{bmatrix} -1 & 0 \\ 0 & 1 \end{bmatrix}, \quad \gamma^0 = \begin{bmatrix} 0 & 1 \\ 1 & 0 \end{bmatrix} \tag{1.80b}$$

$$\xi_+ = \begin{bmatrix} 1 \\ 0 \end{bmatrix}, \quad \xi_- = \begin{bmatrix} 0 \\ 1 \end{bmatrix}, \quad \eta = i\sigma_2\xi^* \tag{1.80c}$$

Choosing helicity eigenstates for ξ_r, η_r, and evaluating in the center of mass frame ($\boldsymbol{k}_1 = -\boldsymbol{k}_2 = \boldsymbol{k}, \boldsymbol{p}_1 = -\boldsymbol{p}_2 = \boldsymbol{p}$), we have for $\nu(h = -1)\overline{\nu}(h = +1) \to W^+W^-$,

$$\begin{aligned} \overline{v}(p_2)(\not{k}_1 - \not{k}_2)(1-\gamma^5)u(p_1) &= (\sqrt{E+p}\,\xi_+, -\sqrt{E-p}\,\xi_+) \\ &\times \begin{bmatrix} 0 & 1 \\ 1 & 0 \end{bmatrix} \begin{bmatrix} 0 & -2\boldsymbol{k}\cdot\boldsymbol{\sigma} \\ 2\boldsymbol{k}\cdot\boldsymbol{\sigma} & 0 \end{bmatrix} \begin{bmatrix} 2 & 0 \\ 0 & 0 \end{bmatrix} \begin{bmatrix} \sqrt{E+p}\,\xi_- \\ \sqrt{E-p}\,\xi_- \end{bmatrix} \\ &= 4(E+p)\xi_+(\boldsymbol{k}\cdot\boldsymbol{\sigma})\xi_- = 4(E+p)|\boldsymbol{k}|\sin\theta\, e^{-i\phi} \end{aligned} \tag{1.81}$$

If the scattering amplitude is expressed in terms of the Jacob–Wick's partial wave expansion (see Vol. 1, Eq. (9.47)),

$$\begin{aligned} \mathcal{M}_a &= 8\pi\sqrt{s}\, f_{\lambda_3\lambda_4,\lambda_1\lambda_1} \\ f_{\lambda_3\lambda_4,\lambda_1\lambda_1} &= \frac{1}{2i|\boldsymbol{p}|}\sum_J (2J+1)\langle\mu|(S_J - 1)|\lambda\rangle d^J_{\mu,\lambda}(\theta) \\ &\sim \frac{1}{|\boldsymbol{p}|}\sum_J (2J+1)e^{i\delta_J}\sin\delta_J\, d^J_{\mu,\lambda}(\theta) \end{aligned} \tag{1.82}$$

where $d^J_{\mu,\lambda}$ is the rotation matrix elements with angular momentum J. Setting the initial helicity $\lambda = 1/2 - (-1/2) = 1$ and the final helicity $\mu = 0 - 0 = 0$, $d^1_{0,1}(\theta) = \sin\theta$, it proves that the first term of Eq. (1.81) is a pure $J = 1$ contribution. Since the unitarity ($S_J = e^{2i\delta_J}$) constrains $|e^{i\delta_J}\sin\delta_J| \le 1$ which, in turn, means $|\mathcal{M}|$ should not grow more than a constant. Therefore, Eq. (1.81) violates the unitarity badly at high energy.

It is also the cause of the diverging loop integral. Consider Figure 1.9b. The diagram, if cut in half, is the scattering amplitude $\nu\overline{\nu} \to W^+W^-$ squared. Indeed, the unitarity of the scattering matrix dictates that the imaginary part of the forward scattering amplitude in Figure 1.9b is proportional to the total cross section of $\nu\overline{\nu} \to W^+W^-$ (see Eq. (I.36)). Since the intermediate state can have any momentum p as can be seen from Figure 1.9b, it has to be integrated over p which results in a bad divergence, the integrand growing $\sim s$.

To ameliorate the situation, we consider adding other diagrams just to cancel the bad divergence. Cancellations either in the u-channel or in the s-channel are possible. One in the t-channel does not help because it gives a similar amplitude with the same sign. Since Eq. (1.77) is in the pure $J = 1$ state, we consider adding another neutral vector boson V^0_μ in the s-channel.

The corresponding amplitude is depicted in Figure 1.9c. The amplitude can be calculated to give

$$-i\mathcal{M}_c = \overline{v}(p_2)\left(-i\gamma^{\delta}\frac{g_{V1}}{2}(1-\gamma^5)\right)u(p_1)\frac{-i\left(g_{\delta\lambda}-\frac{q_\delta q_\lambda}{m_V^2}\right)}{q^2-m_V^2+i\epsilon}$$
$$\times (ig_{V2})\epsilon_{1\mu}\epsilon_{2\nu}V^{\nu\mu\lambda}(-k_2,-k_1)$$
$$V^{\mu\nu\lambda}(k_1,k_2) = g^{\mu\nu}(k_1-k_2)_\lambda + g^{\nu\lambda}(k_2-k_3)^\mu + g^{\lambda\mu}(k_3-k_1)^\nu$$
$$q = k_1+k_2\,,\quad k_3 = -(k_1+k_2) \tag{1.83}$$

The trilinear coupling part can be simplified by using $(\epsilon_1\cdot k_1)=(\epsilon_2\cdot k_2)=0$ and $k_3=(k_1+k_2)$.

$$\epsilon_{1\mu}\epsilon_{2\nu}V^{\nu\mu\lambda}(-k_2,-k_1) = \left[(\epsilon_1\cdot\epsilon_2)(k_1-k_2)^\lambda - 2(\epsilon_2\cdot k_1)\epsilon_1^\lambda + 2(\epsilon_1\cdot k_2)\epsilon_2^\lambda\right] \tag{1.84a}$$

Inserting $\epsilon_i \sim k_i/m_W$, we have

$$\epsilon_{1\nu}\epsilon_{2\mu}V^{\mu\nu\lambda}(-k_2,-k_1) \simeq -\frac{1}{m_W^2}(k_1\cdot k_2)(k_1-k_2)^\lambda \tag{1.84b}$$

Substituting Eq. (1.84b) in Eq. (1.83), the matrix element becomes

$$-i\mathcal{M}_c = +i\frac{g_{V1}g_{V2}}{4m_W^2}\overline{v}(p_2)(\not{k}_1-\not{k}_2)(1-\gamma^5)u(p_1)\frac{(k_1\cdot k_2)}{(k_1\cdot k_2)+m_V^2/2} \tag{1.85}$$

If $g_{V1}g_{V2}=g_W^2/2$, Eq. (1.85) cancels the first term of Eq. (1.79) for $|\boldsymbol{k}| \gg m_V^2$. This is exactly the case for the V to be I_3 member of the gauge boson in $SU(2)$. This is also true for the case of $SU(2)\times U(1)$ where W^0 is replaced by γ and Z. For the γ, Eq. (1.51) and Figure 1.5 gives

$$\frac{g_{V1}}{2}(1-\gamma^5) \to Q_f e = Q_f g_W \sin\theta_W\,,\quad g_{V2} = e = g_W\sin\theta_W \tag{1.86a}$$

Note, we deliberately retained the charge Q_f to do similar arguments later for the electron. For the neutrino, $Q_\nu=0$ and the photon does not contribute. For the Z, Eq. (1.51e) and Figure 1.5 gives

$$\frac{g_{V1}}{2}(1-\gamma^5) \to \frac{g_Z}{2}(v_f - a_f\gamma^5)\,,\quad g_{V2} = e\cot\theta_W = g_W\cos\theta_W$$
$$g_Z = \frac{g_W}{\cos\theta_W}\,,\quad v_f = I_{3f} - 2Q_f\sin^2\theta_W\,,\quad a_f = I_{3f} \tag{1.86b}$$

Then,

$$\sum_{\gamma,Z} g_{V1}g_{V2} = 2Q_f g_W^2\sin^2\theta_W + g_W^2\left(I_{3f} - 2Q_f\sin^2\theta_W^2 - I_{3f}\gamma^5\right)$$
$$= g_W^2 I_{3f}(1-\gamma^5)$$
$$= \frac{g_W^2}{2}(1-\gamma^5)\quad\text{for}\quad \nu \tag{1.87}$$

Namely, the gauge invariance provides exactly the proper counter terms in order to not give bad high energy behavior.

In passing, we mention that the divergence can also be compensated by introducing a “heavy electron E^+” in the u-channel, provided it also has the coupling required by the gauge symmetry.[13] The model eliminates the neutral current. It was proposed as an alternative to the GWS theory before the discovery of the neutral current, and hence was ruled out by its discovery. However, the example illustrates the power of the gauge symmetry in controlling the divergence.

Problem 1.1

Prove that the diagram Figure 1.9f cancels the dominant contribution of Figure 1.9a provided the same coupling constant is used.

Higgs in the renormalization When the initial fermion pair is massive as is the case for the electron, the diverging term reappears. This is because the massive fermion can have opposite helicity for the same chirality. The left-handed neutrino is in a pure helicity ($h = -1$) state, but the electron can have a positive helicity component with amplitude proportional to its mass $\sim m_e/p$[14]. This induces an extra component, namely, the second term in Eq. (1.79). Note, the first term also changes sign because $W^\pm$ is interchanged in the transition $e^- e^+ \to W^- W^+$ as shown in Figure 1.9d. Then, the contribution of Figure 1.9c also changes sign because $I_3(e^-) = -1/2$, and thus the compensation mechanism is again valid here.

As is clear from the expression, this term has $J = 0$ and is proportional to the fermion mass. This is not the dominant divergence as that of the $J = 1$ component. Nevertheless, it is a divergence that grows faster than the logarithm. It gives the $O(s)$ term to the loop integral. A scalar meson is necessary to compensate it and this is the place where the Higgs comes in. The coupling of the Higgs field after spontaneous symmetry breakdown is proportional to the fermion's mass (Eq. (1.46)) just as required. It does not couple to the neutrino because it is not needed. The scattering amplitude corresponding to $e^- e^+ \to H^0 \to W^+ W^-$ (Figure 1.9e) is given by

$$-i\mathcal{M}_e = \overline{v}(p_2)\left(-\frac{i}{2}\frac{g_W}{m_W}m_e\right)u(p_1)\frac{i}{(k_1+k_2)^2 - m_H^2}(i g_W m_W g^{\mu\nu})\epsilon_{1\mu}\epsilon_{2\nu} \tag{1.88a}$$

13) This is equivalent to asking that E^+ be a member of the multiplet in which ν and e^- belong. In other words, the fermions constitute a triplet as opposed to a doublet in the Standard Model. The model is based on $SU(2)$ gauge theory and identifies I_3 as the electric charge operator [47].

14) See arguments in Vol. 1, Sect. 4.3.5. It is also apparent in Eq. (1.80a) because for $E > |\boldsymbol{p}|$, the opposite helicity component does not vanish.

Inserting $\epsilon_{i\mu} \to k_{i\mu}/m_W$ again and neglecting m_H^2 relative to $(k_1 \cdot k_2)$, we obtain

$$-i\mathcal{M}_e \simeq i\frac{g_W^2}{4m_W^2} m_e \overline{v}(p_2)u(p_1) \tag{1.88b}$$

which cancels the second term of Eq. (1.79).

$WW \to WW$ scattering The cancellation mechanism we just mentioned is at work for the vector boson scattering as well. The two diagrams with two triple boson vertices in Figure 1.10a,b produces a term $\sim (k_3 \cdot k_4)\{(k_3 - k_4) \cdot (k_1 - k_2)\}(k_1 \cdot k_2) \sim s^3$ which is reduced to $\sim s^2$ by the propagator. Therefore, the amplitude includes terms $O(s^2) + O(s) + O(\ln s)$. The term of $O(s^2)$ is compensated by the quartic coupling diagram of Figure 1.10c if the latter has the right coupling as is required by the gauge theory. The $O(s)$ term can be compensated by a term which includes a scalar field in the intermediate state if it has the right coupling as is given by the second and last term of Eq. (1.37). Thus, the remaining divergence is at most logarithmic and can be handled with the renormalization prescription.

In summary, the gauge theory which has broken spontaneously has a built-in mechanism to compensate all the annoying divergences and make the theory renormalizable [38, 39]. Conversely, if one tries to compensate diverging integrals by introducing additional particles and determines their particle species, coupling

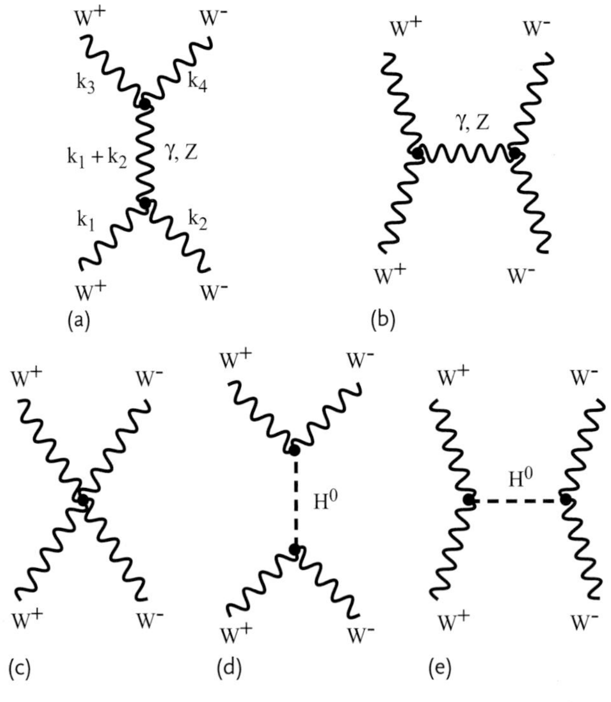

Figure 1.10 *W-W* scattering goes through triple vector coupling (a,b), quartic coupling (c) and Higgs coupling (d,e).

constants and masses, they always end up with the spontaneously broken gauge theory [48–51]. We conclude, therefore, that the spontaneously broken gauge theory is the only renormalizable theory that can handle massive vector bosons.

In retrospect, the role of the Higgs to rescue the divergence problems whenever they occur is clear if one accepts the fundamental role of the gauge symmetry to overcome the difficulties and looks at the original Lagrangian. The gauge sector (the first line of Eq. (1.16)) and the Higgs sector (the second line of Eq. (1.16)) both independently satisfy the $SU(2) \times U(1)$ gauge symmetry. The masses are generated in the Higgs sector. If the symmetry is broken spontaneously, which is equivalent to rewriting fields in a certain gauge, the gauge invariance as a whole is still maintained mathematically. The gauge invariance is broken when one separates the mass terms, add them alone to the gauge sector and discard the rest of the Higgs contributions. Since the whole Lagrangian which includes the Higgs part is gauge invariant, it is no wonder that the difficulty is solved by including the Higgs contributions.

2
Neutral Current

The Glashow–Weinberg–Salam theory (abbreviated as GWS), when it was proposed, was revolutionary because it unified the electromagnetic and weak interaction. However, in order to be a successful theory in physics, mathematical completeness is not enough. It has to describe phenomena that are actually happening. By its construction, it was made to reproduce the four-Fermi interaction in the low energy limit. Therefore, it covers at least experimental data which have been observed up to the time of its proposal. Its verification has to be done either by reproducing the yet to be carried out high energy processes correctly, or by predicting new phenomena which were not included in the Fermi theory. As was described in the previous chapter, the GWS theory required the existence of the neutral current and predicted the mass of the gauge bosons, both of which were not found when it was proposed. In this chapter, we derive basic formulas to describe neutral current processes and demonstrate that they behaved exactly as predicted. The treatment is extended to cover processes relevant to more recent developments and those in the hadronic reactions. Properties of the gauge bosons W and Z will be described in Chapters 3 and 4.

2.1
Discovery of the Neutral Current

The Fermi theory, originally introduced to explain nuclear beta-decays of nuclear reactions, had undergone a few modifications, but was very successful in explaining all the known weak interaction processes during the pre-Standard Model era. Its Lagrangian is of the current-current coupling type and the current is limited to the so-called "charged current."

$$\mathcal{L}_{\text{Fermi}}(x) = \frac{G_F}{\sqrt{2}} J_\mu^-(x) J^{+\mu}(x) \tag{2.1a}$$

$$J^{+\mu} = \sum_a \left[\overline{u}_a(x)\gamma^\mu(1-\gamma^5)d_a(x) + \overline{\nu}_a\gamma^\mu(1-\gamma^5)l_a\right] \tag{2.1b}$$

$$J^{-\mu} = (J^{+\mu})^\dagger \tag{2.1c}$$

$$u_a = u, c, t\,, \quad d_a = d', s', b'\,, \quad l_a = e, \mu, \tau \tag{2.1d}$$

Elementary Particle Physics, Volume 2: Foundations of the Standard Model, First Edition. Yorikiyo Nagashima.

where (d', s', b') are Cabibbo–Kobayashi–Maskawa (CKM) rotated (d, s, b). The quark current $\overline{u}\gamma^{\mu}(1-\gamma^5)d$ and the lepton current $\overline{\nu}_l\gamma^{\mu}(1-\gamma^5)l$ annihilate the d-quark and create $u, \overline{l}, \nu_l$. Unlike the electromagnetic current $(\overline{u}\gamma^{\mu}u, \ldots)$, the quarks and leptons change their identity and their electric charge is converted. Because of this, they are referred to as the charged currents. In the Standard Model, that is, the GWS model, the Lagrangian given in Eq. (2.1) is an effective one induced by $W^{\pm}$ boson exchange and valid only in the low energy region where its energy scale is much lower than the mass of the $W^{\pm}$ $(Q \ll m_W = 80.4\,\text{GeV})$. This condition is still satisfied in the majority of weak interaction phenomena covered in this book. Only when we discuss production of W, Z themselves or special high energy neutrino reactions, do explicit inclusion of the W, Z propagators become necessary.

Let us consider the simplest charged current reaction, $\nu_e e^-$ and $\overline{\nu}_e e^-$ scattering. It can happen through exchange of $W^{\pm}$ in the t- as well as s-channel as shown in Figure 2.1a,b. Note that the process was not observed at the time of GWS proposal. If the GWS theory is right, contributions from the neutral current Figure 2.1c,d should exist as well. Its existence can be proved, albeit indirectly, from the measured value of the cross section $\sigma(\overline{\nu}_e e^- \to \overline{\nu}_e e^-)$ by subtracting the contribution of Figure 2.1a,b. At that time, however, the neutrino experiment was in its infancy, and a number of events was too meager to make any quantitative evaluation. It is more true for the pure leptonic reactions. According to the Fermi theory, $\sigma \propto G_F^2 s \approx G_F^2(2E_\nu m_{\text{target}})$ and if the target is an electron, it has the reduction factor $m_e/m_p \sim 2000$ compared with a proton target. For the reactor neutrino which has $E_\nu \sim 1\,\text{MeV}$, the cross section is $\sim 10^{-44}\,\text{cm}^2$, a small number indeed compared with the strong interaction cross section $\sim 10^{-26}\,\text{cm}^2$. Even the electromagnetic interaction cross section is as large as $\sim 10^{-30}\,\text{cm}^2$[1].

Let us consider, then, $\nu_\mu e^-$ (or $\overline{\nu}_\mu e^-$) scattering. As ν_μ is not a member of the doublet (ν_e, e^-), it can not exchange $W^{\pm}$ with the electron because by doing so, it changes its identity (and the target electron's identity too) to become the other member of the doublet which is not the participating member of the reaction. In other words, the ν_μ can change itself to μ^- by exchanging $W^{\pm}$, but not to the electron. The "lepton flavor" conservation is at work here. For the case of Z exchange, its electric charge is not altered, the flavor is conserved and the ν_μ can stay as ν_μ. Namely, the $\nu_\mu e^-$ scattering is a pure neutral current phenomena which exists in the Standard Model, but not in the Fermi theory. To prove the existence of the neutral current by subtracting the $W^{\pm}$ contribution from the $\overline{\nu}_e e^-$ reaction is difficult to achieve, but with the ν_μ–e scattering, discovery of one event proves it.

A group at CERN [2] lead by Lagarrigue exposed a giant 10-ton bubble chamber Gargamelle to an intense ν_μ beam produced by PS, the 28 GeV proton synchrotron accelerator, and took about a million photographs. This was one of the first large, organized, team-efforts involving some fifty scientists and using the most advanced technology that has since become main stream in high energy experiments. Fig-

1) The proton Thomson scattering cross section $\sigma_T = (8\pi/3)\alpha^2/m_p^2$.
2) Conseil Europeen pour la Recherche Nucleaire (European Organization for Nuclear Research). It is located in Geneva, Switzerland.

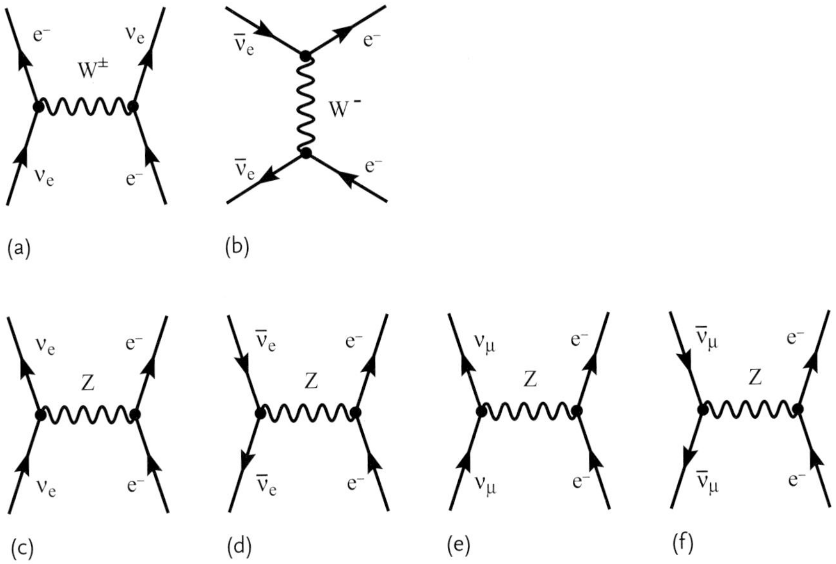

Figure 2.1 (a,b) Charged current reactions: $\nu_e(\overline{\nu}_e)e^- \rightarrow \nu_e(\overline{\nu}_e)e^-$ (c,d) Neutral current reactions: $\nu_e(\overline{\nu}_e)e^- \rightarrow \nu_e(\overline{\nu}_e)e^-$ (e,f) Neutral current reactions: $\nu_\mu(\overline{\nu}_\mu)e^- \rightarrow \nu_\mu(\overline{\nu}_\mu)e^-$.

ure 2.2 shows a photograph of the first neutral current event [53]. An antineutrino ($\overline{\nu}_\mu$) enters the bubble chamber from the right and kicks off an electron. The recoiled electron exhibits its characteristic behavior by a sequence of bremsstrahlung and pair creation (V-shape track topology). Absence of the muon (experimentally characterized as a long track because of its meagre interaction with matter except for its ionization) is the evidence of its being the neutral current event. This is a pure leptonic event, but quantitative data that can be compared with theoretical predictions were obtained from hadron production events on the nuclear target [54]. Figure 2.3 shows an example of hadron production by the neutral current interaction.

2.2 $\nu-e$ Scattering

In order to offer a quantitative discussion of the νe^- scatterings, we define the energy-momenta of the reaction as

$$\nu(p_1) + e^-(p_2) \rightarrow \nu(p_3) + e^-(p_4) \tag{2.2}$$

and set the kinematical variables as shown in Figure 2.4. The neutrino beam is obtained by bombarding high energy protons on matter (typically, a beryllium target),

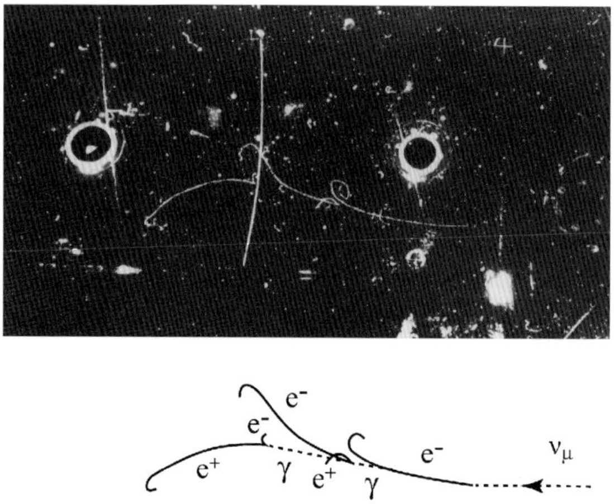

Figure 2.2 The first neutral current reaction event display [53]. A neutrino entering from the right kicks off an electron. The electron, in turn, emits a photon (invisible) and creates electron–positron pairs in cascade. Bright rings are reflections of lights. (Photo: CERN)

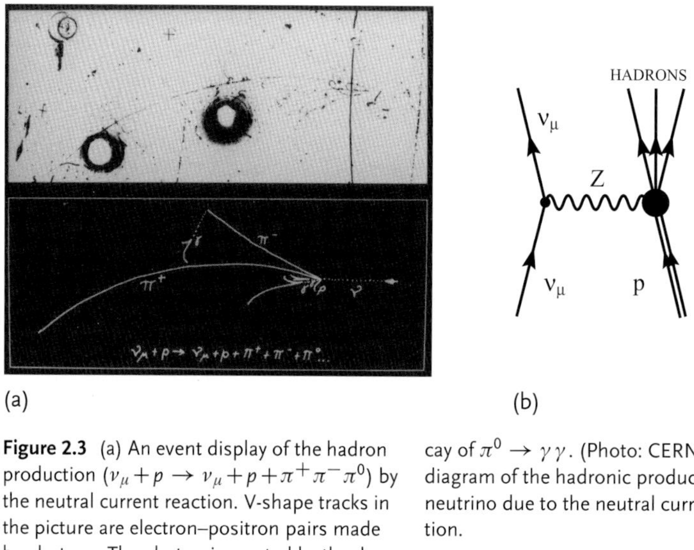

(a) (b)

Figure 2.3 (a) An event display of the hadron production ($\nu_\mu + p \to \nu_\mu + p + \pi^+ \pi^- \pi^0$) by the neutral current reaction. V-shape tracks in the picture are electron–positron pairs made by photons. The photon is created by the decay of $\pi^0 \to \gamma\gamma$. (Photo: CERN) (b) Feynman diagram of the hadronic production by the neutrino due to the neutral current interaction.

producing a secondary intense pion beam which, in turn, decay to neutrinos.

$$\begin{aligned} \pi^+ &\to \mu^+ + \nu_\mu \\ \pi^- &\to \mu^- + \overline{\nu}_\mu \end{aligned} \tag{2.3}$$

One can calculate other variables in the laboratory (LAB) frame by

$$s = (p_1 + p_2)^2 = 2(p_1 \cdot p_2) + m_e^2 = 2E_\nu m_e + m_e^2 \tag{2.4a}$$

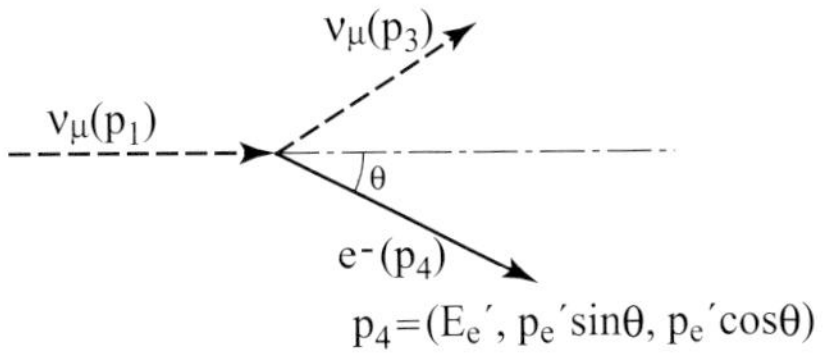

Figure 2.4 Variables of $\nu_\mu + e^- \to \nu_\mu + e^-$ in LAB frame.

$$E_\nu = (p_1 \cdot p_2)/m_e\,, \quad E'_e = (p_4 \cdot p_2)/m_e \tag{2.4b}$$

$$\cos\theta = \left(1 + \frac{m_e}{E_\nu}\right)\left(\frac{E'_e - m_e}{E'_e + m_e}\right)^{1/2} \tag{2.4c}$$

The parton variable y (see Eq. (8.5) or Vol. 1, Sect. 17.5.2 for details) is defined as

$$y = \begin{cases} = \dfrac{-q^2}{s - m_e^2} = -\dfrac{(p_2 - p_4)^2}{2m_e E_\nu} = \dfrac{E'_e - m_e}{E_\nu} = \dfrac{E_\nu - E'_\nu}{E_\nu} & \text{in LAB} \\ \simeq \dfrac{1 - \cos\theta^*}{2} & \text{in CM} \end{cases} \tag{2.5}$$

$$0 \le y \le \left(1 + \frac{m_e}{2E_\nu}\right)^{-1} \simeq 1$$

where variables in the center of mass frame (CM) are denoted with an asterisk. For $E_\nu, E'_e \gg m_e$, Eq. (2.4c) gives

$$\theta^2 \simeq 2m_e \frac{E_\nu - E'_e}{E_\nu E'_e} \simeq \frac{2m_e}{E'_e}(1 - y) \tag{2.6}$$

namely,

$$E'_e \theta^2 \lesssim 2m_e \tag{2.7}$$

Assuming $E'_e \sim 1$ GeV, we have $\theta^2 \lesssim 10^{-3}$, meaning the scattered electron is emitted in the forward direction.[3] Experimentally, this fact helps to eliminate background reactions

$$\begin{aligned} &\nu_\mu + N(\text{nucleon}) \to \nu_\mu + \pi^0 + X, \quad \pi^0 \to 2\gamma\,, \quad \gamma + e^- \to e^- + \gamma \\ &\nu_\mu + n(\text{neutron}) \to e^- + p(\text{proton}) \end{aligned} \tag{2.8}$$

Another background is produced by the ν_e components in the ν_μ beam through $\nu_e + e^- \to e^- + \nu_e$ scattering. The origin of the ν_e in the ν_μ beam is two-fold:

$$\pi^\pm \to \mu^\pm + \overset{\leftrightarrow}{\nu}_\mu \quad \text{dominant } \nu_\mu \text{ component} \tag{2.9a}$$

$$\mu^\pm \to \overset{\leftrightarrow}{\nu}_\mu + e^\pm + \overset{\leftrightarrow}{\nu}_e \quad \text{background } \nu_e \text{ component} \tag{2.9b}$$

3) This fact is used to find the incident neutrino direction, for instance, in the astronomical neutrino observation, like a solar neutrino ($= \nu_e$), despite the fact that the energy is much lower. Note, $\nu_e(\overline{\nu}_e)$ scattering by nuclei gives isotropic angular distribution for $E_\nu \ll m_N$ which was the case for the neutrino ($= \overline{\nu}_e$) observation from supernova SN1897A.

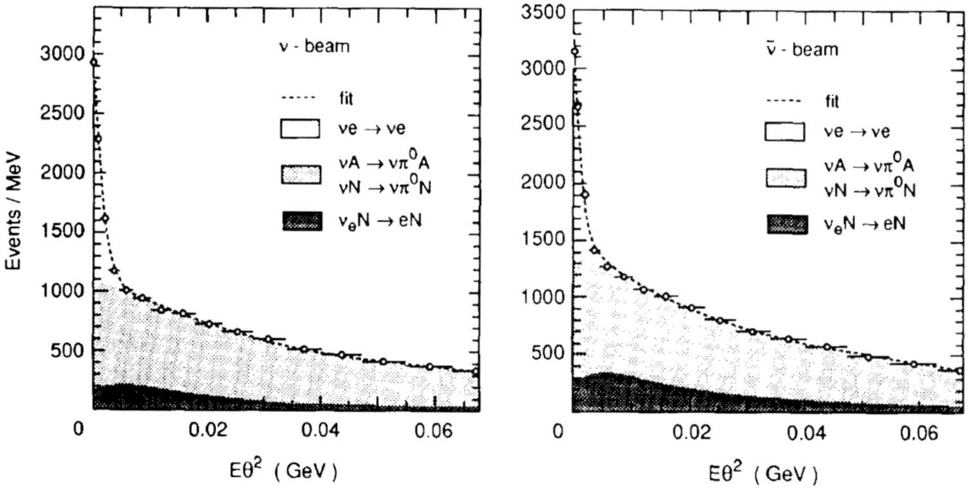

Figure 2.5 Experimental data are plotted as a function of $E_e'\theta^2$: data are shown as circles and the fit result is displayed as a dashed line. The different background components are added on top of each other [55].

$$K^{\pm} \to \pi^{\pm} + e^{\pm} + \nu_e \quad \text{another } \nu_e \text{ source} \tag{2.9c}$$

where $\overset{\leftrightarrow}{\nu}$ is either the neutrino or antineutrino depending on the parent's electric charge. Figure 2.5 shows angular distributions of the scattered electron measured by CHARM group at CERN [55]. One can see events are concentrated in the forward region.

Cross Section The interaction Lagrangian is given by Eq. (1.50). Using the Feynman rules in Section 1.6, the scattering amplitude is given by

$$\langle \nu_\mu(p_3)e(p_4)|S - \delta_{fi}|\nu_\mu(p_1)e(p_2)\rangle = -i(2\pi)^4\delta^4(p_1 + p_2 - p_3 - p_4)\mathcal{M}_{fi} \tag{2.10a}$$

$$\begin{aligned} -i\mathcal{M}(\nu_\mu e \to \nu_\mu e) = {} & \left[\overline{u}(p_3)\left(-i\frac{g_Z}{2}\gamma^\rho\frac{1}{2}(1-\gamma^5)\right)u(p_1)\right]\frac{-i\left(g_{\rho\sigma} - \frac{q_\rho q_\sigma}{m_Z^2}\right)}{q^2 - m_Z^2 + i m_Z\Gamma_Z} \\ & \times\left[\overline{u}(p_4)\left(-i\frac{g_Z}{2}\gamma^\sigma\right)\{\varepsilon_L(e)(1-\gamma^5) + \varepsilon_R(e)(1+\gamma^5)\}u(p_2)\right] \end{aligned} \tag{2.10b}$$

$$\begin{aligned} \epsilon_L(e) &= I_{3e} - Q_e\sin^2\theta_W = -\frac{1}{2} + x_W \\ \epsilon_R(e) &= -Q_e\sin^2\theta_W = x_W \end{aligned} \tag{2.10c}$$

where $\overline{u}(p_3), u(p_1), \ldots$ are plane wave solutions of the Dirac equation. We used $\epsilon_L(\nu) = \frac{1}{2}$, $\epsilon_R(\nu) = 0$ in the neutrino coupling. $x_W = \sin^2\theta_W$ will be used as a simple notation for the Weinberg angle. The second term in the numerator of the Z propagator vanishes if it acts on the conserved current which applies to the vector part of the neutral current. When it acts on the axial current which is not conserved, it gives a contribution proportional to the fermion mass. For the electron, we can neglect it as well. The width Γ_Z does not appear in the tree approximation (it appears as a result of the higher order calculation), but we included it for phenomenological treatment. In the Fermi theory, q^2 is neglected compared with the m_Z^2 and Eq. (2.10b) reduces to

$$\begin{aligned} -\mathcal{M}_Z &= \frac{G_N}{\sqrt{2}}[\overline{u}(p_3)\gamma_\mu(1-\gamma^5)u(p_1)] \\ &\quad \cdot \left[\overline{u}(p_4)\gamma^\mu\left\{\varepsilon_L(e)(1-\gamma^5)+\varepsilon_R(e)(1+\gamma^5)\right\}u(p_2)\right] \\ \frac{G_N}{\sqrt{2}} &= \frac{g_Z^2}{8m_Z^2} \equiv \rho\frac{G_F}{\sqrt{2}} \end{aligned} \tag{2.11}$$

which has the form given by the Fermi interaction, except inclusion of the right-handed current and replacement of the coupling constant ($G_F \to G_N$). G_N is the coupling strength of the neutral current. In the tree approximation, $\rho = 1$ and G_N agrees with the Fermi coupling constant G_F, but we keep the nomenclature G_N for the sake of later discussions. The cross section is given by (see Eq. (A.12) or Vol. 1, Eqs. (6.86), (6.90), (6.91))

$$d\sigma = \frac{1}{2s\lambda}|\overline{\mathcal{M}}|^2 dLIPS = \frac{1}{64\pi^2 s}\left(\frac{p_f}{p_i}\right)|\overline{\mathcal{M}}|^2 d\Omega_{CM} \tag{2.12a}$$

$$\lambda = \lambda(1, m_\nu^2/s, m_e^2/s) \simeq 1, \quad \lambda(x,y,z) = x^2+y^2+z^2-2xy-2yz-2zx \tag{2.12b}$$

$$dLIPS = (2\pi)^4\delta^4(p_1+p_2-p_3-p_4)\frac{d^3p_3}{(2\pi)^3 2E_3}\frac{d^3p_4}{(2\pi)^3 2E_4} \tag{2.12c}$$

where $|\overline{\mathcal{M}}|^2$ is the sum of final and average of initial spin states. It is warned, however, that no right-handed neutrinos are produced in any reaction, the neutrino beam contains only the left-handed components and hence no spin average of the neutrino is necessary (i.e., do not divide by two). Carrying out the standard γ trace algebra, we obtain

$$\frac{d\sigma}{dy}(\nu_\mu e) = \frac{G_N^2 s}{\pi}\left[\varepsilon_L^2(e) + \varepsilon_R^2(e)(1-y)^2 - \varepsilon_L(e)\varepsilon_R(e)y m_e/E_\nu\right] \tag{2.13a}$$

The third term in Eq. (2.13a) can be neglected for the high energy neutrino reactions, but not for the reactor neutrinos whose average energy is $\sim$ a few MeV. The cross section for $\overline{\nu}_\mu e^-$ can be obtained by simply interchanging L $\leftrightarrow$ R:

$$\frac{d\sigma}{dy}(\overline{\nu}_\mu e) = \frac{G_N^2 s}{\pi}\left[\varepsilon_L^2(e)(1-y)^2 + \varepsilon_R^2(e) - \varepsilon_L(e)\varepsilon_R(e)y m_e/E_\nu\right] \tag{2.13b}$$

Combining the two equations, we have

$$\frac{d\sigma}{dy}[\nu_\mu(\overline{\nu}_\mu)e] = \frac{G_N^2 s}{4\pi}\left[(v_e \pm a_e)^2 + (v_e \mp a_e)^2(1-y)^2 - (v_e^2 - a_e^2)y m_e/E_\nu\right] \tag{2.14a}$$

$$v_e = \varepsilon_L(e) + \varepsilon_R(e) = I_{3e} - 2Q_e e \sin^2\theta_W \equiv -\frac{1}{2} + 2x_W \tag{2.14b}$$

$$a_e = \varepsilon_L(e) - \varepsilon_R(e) = I_{3e} = -\frac{1}{2} \tag{2.14c}$$

As it was difficult to obtain enough statistics in the early experiments for the leptonic $\nu_\mu(\overline{\nu}_\mu)e$ scattering, we compare an integrated cross section with the experimental data. In the approximation to neglect the third term,

$$\sigma[\nu_\mu(\overline{\nu}_\mu)e] = \frac{G_N^2 s}{4\pi}\left[(v_e \pm a_e)^2 + \frac{1}{3}(v_e \mp a_e)^2\right] \tag{2.15}$$

Origin of $(1-y)^2$ in the cross section for νe scattering

As the cross section of the form given for the $\nu(\overline{\nu}) - e$ scattering in Eq. (2.13) is universal in the 2-body scattering for massless spin 1/2 particle and appear frequently in the following, it is illuminating to review its origin. We obtained cross sections in the CM frame for the chiral massless $e-\mu$ scattering in Vol. 1, Eq. (7.31) and (7.32).

$$\left.\frac{d\sigma}{d\Omega}\right|_{LL} = \left.\frac{d\sigma}{d\Omega}\right|_{RR} = \frac{\alpha^2}{t^2}s$$
$$\left.\frac{d\sigma}{d\Omega}\right|_{LR} = \left.\frac{d\sigma}{d\Omega}\right|_{RL} = \frac{\alpha^2}{t^2}\frac{u^2}{s} = \frac{\alpha^2}{t^2}s(1-y)^2$$
$$s = (p_1+p_2)^2\,, \quad t = (p_1-p_3)^2\,, \quad u = (p_1-p_4)^2 \tag{2.16}$$

where suffixes a, b $(=\text{L}, \text{R})$ denote chirality of the incoming and target particles. Replacing the photon propagator term by the Fermi constant $\frac{4\pi\alpha}{t} \to \frac{4G_F}{\sqrt{2}}$ with additional factors $\varepsilon_L^2(e)$, $\epsilon_R^2(e)$, we can convert the polarized $e\mu$ scattering cross section to that of the neutrino. Therefore, Eqs. (2.13a) and (2.13b) are proportional to $d\sigma_L = (d\sigma_{LL} + d\sigma_{LR})/2$, $d\sigma_R = (d\sigma_{RL} + d\sigma_{RR})/2$ and $d\sigma/dy = 4\pi d\sigma/d\Omega$.

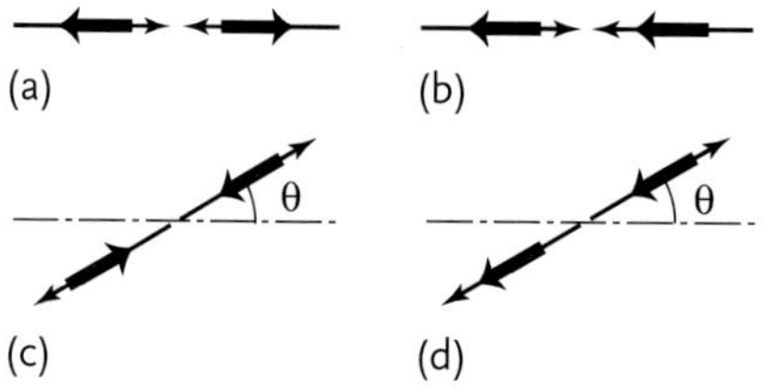

Figure 2.6 Helicity states for LL (a) and (c), and LR (b) and (d) before and after scattering. Thin arrows denote momenta and thick arrows denote spin.

The Fermi interaction is a contact interaction which means only the S-wave contributes. Then, the total angular momentum in their CM frame is either zero or one and the rotation matrix gives $d^0_{0,0} = 1$, $d^1_{1,1} = (1+\cos\theta^*)/2 = 1-y$. For LL (RR) scattering, the two particles have the same helicity and $J_Z = 0$ at any angle (Figure 2.6a,c). For LR (RL) scattering, they have opposite helicity with

$J_Z = \pm 1$ and only their projected helicity component along the incident direction is picked up (Figure 2.6b,d) which is the rotation matrix $d^1_{1,1}$. Namely, the factor 1 or $(1-y)^2$ simply reflects the angular momentum conservation.

At this stage, we also give the cross section formula for $\nu_e(\overline{\nu}_e)e$ scattering. It includes contribution from $W^\pm$ exchange diagrams (see Figure 2.1a,b) in addition to Z exchange diagrams (Figure 2.1c,d). The $W^\pm$ exchange Lagrangian is given by

$$\mathcal{L}_W = \frac{G_F}{\sqrt{2}}\left[\overline{\nu}_e\gamma_\mu(1-\gamma^5)e\right]\left[\overline{e}\gamma^\mu(1-\gamma^5)\nu_e\right] \tag{2.17}$$

The above equation can be rewritten in the form of Eq. (2.11) by the Fierz transformation (see Appendix E) with $G_N \to G_F$, $\varepsilon_L(e) \to 1$, $\varepsilon_R(e) \to 0$.

$$\mathcal{L}_W = \frac{G_F}{\sqrt{2}}\left[\overline{\nu}_e\gamma_\mu(1-\gamma^5)\nu_e\right]\left[\overline{e}\gamma^\mu(1-\gamma^5)e\right] \tag{2.18}$$

Therefore, assuming $G_N = G_F$, the cross section for $\nu_e(\overline{\nu}_e)e$ scattering which includes additional contribution from the charged current can be obtained from that of $\nu_\mu(\overline{\nu}_\mu)e$ by replacing $\varepsilon_L(e) \to \varepsilon_L(e) + 1$

ν_e, a_e, $\sin^2\theta_W$ A measure of the νe scattering cross section is given by

$$\sigma_0 = G_F^2 s/\pi = 1.73 \times 10^{-41}\,(E_\nu/\text{GeV})\,\text{cm}^2 \tag{2.19}$$

The measured total cross sections are given by [56]

$$\sigma(\nu_\mu e^-) = 1.80 \pm 0.20_{(\text{stat})} \pm 0.25_{(\text{sys})} \times 10^{-42}\,(E_\nu/\text{GeV})\,\text{cm}^2 \tag{2.20a}$$

$$\sigma(\overline{\nu}_\mu e^-) = 1.17 \pm 0.16_{(\text{stat})} \pm 0.13_{(\text{sys})} \times 10^{-42}(E_\nu/\text{GeV})\,\text{cm}^2 \tag{2.20b}$$

Thus, the existence of the neutral current reaction was confirmed. Once its existence is established, the next task is to determine the value of v_e and a_e or equivalently, that of the Weinberg angle in the Standard Model. Given two sets of data, Eq. (2.15) defines two ellipses on the a_e-v_e plane, giving four solutions (Figure 2.7a, in which a_e, v_e are denoted as g_A, g_V). This is because one cannot determine their sign. The data from the reactor neutrino event rate $(\overline{\nu}_e e^-)$ gives another ellipse with its center shifted and reduces the number of solutions to two. In order to remove the last ambiguity, one can use data on $e^-e^+ \to \mu^-\mu^+$ [57] which is depicted as two vertical strips in the figure and the solution is constrained to a line (reflecting the ambiguity of the unknown parameter, the Weinberg angle) within the white circle. Historically, however, the asymmetry in the scattering of the polarized electron $(e\uparrow)$ by deuterium (D) was used to remove the ambiguity [58, 59] which will be described in Section 2.4.2. The reason we are reluctant to use it at this stage is because it is not a pure leptonic reaction. It involves the $e-q$ (quark) interaction and its coupling could be different from that of the lepton. Of course, the unified theory relates them, but it is desirable to test the validity of the theory

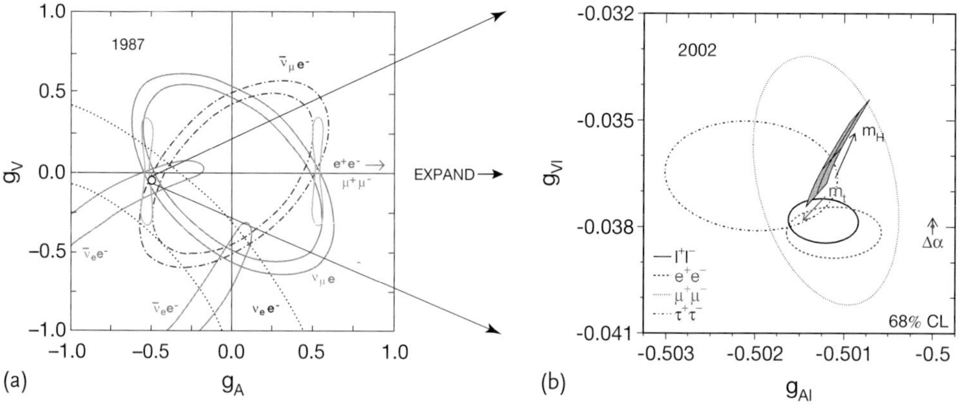

Figure 2.7 (a) The neutrino scattering and e^+e^- annihilation data circa 1987 constrained the values of g_V and g_A (v_e and a_e in the text) to lie within broad bands, whose intersections helped establish the validity of the Standard Model and were consistent with the hypothesis of the lepton universality. (b) shows the results of the LEP/SLD measurements in 2002 at a scale expanded by a factor of 65 (see Figure 4.34). The flavor-specific measurements demonstrate the universal nature of the lepton couplings unambiguously on a scale of approximately 0.001. The shaded region in the lepton plot shows predictions of the Standard Model for $m_t = 178.0 \pm 4.3$ GeV and $m_H = 300^{+700}_{-186}$ GeV. Figure (a) adapted from [1], (b) from [2, 7]. For a color version of this figure, please see the color plates at the beginning of the book.

independently and to see if the theory gives the right relation. Consequently, both v_e and a_e were determined as

$$\begin{aligned} v_e &= -0.035 \pm 0.017 \\ a_e^e &= -0.503 \pm 0.017 \end{aligned} \tag{2.21}$$

The value of a_e agrees with the theoretical value $-1/2$, but $v_e \approx 0$ means that the value of the Weinberg angle is very close to 1/4 ($\sin^2\theta_W \approx 1/4$). Its value can be determined, in principle, by measuring either one of the $\nu_\nu(\overline{\nu}_\mu)e$ cross sections alone, but experimentally (at least, in the early days), one can get more accurate value by using the ratio of the cross sections.

$$\begin{aligned} R &\equiv \frac{\sigma(\nu_\mu e)}{\sigma(\overline{\nu}_\mu e)} = 3\frac{1 - 4x_W + (16/3)x_W^2}{1 - 4x_W + 16x_W^2} \\ x_W &= \sin^2\theta_W \end{aligned} \tag{2.22}$$

The present world average gives [11]

$$\sin^2\theta_W(\nu_\mu e) = 0.2310 \pm 0.0077^{4)} \tag{2.23}$$

4) There are several different definitions of the Weinberg angle which are equivalent at the tree level, but obtain different radiative corrections. When quoting values of the Weinberg angle, one has to specify its exact definition. This one that used the $\overline{MS}$ scheme is ~ 0.008 larger than that of the on-shell scheme ($\sin^2\theta_W \equiv 1 - m_W^2/m_Z^2$). The latter definition will be used throughout this book unless otherwise noted. Detailed discussions on variants of the Weinberg angle will be given in Chapter 5.

If one uses absolute values of the cross sections for data fitting, the value of $\rho = G_N/G_F, \rho = \rho_0(1 + \Delta\rho)$ can be determined as well. Here, ρ_0 is the tree approximation value and $\Delta\rho$ is the radiative correction. The present value is given by [11]

$$\rho_0 = 1.0002^{+0.0007}_{-0.0004} \tag{2.24}$$

$\rho_0 = 1$ for the Standard Model, but could be different for other models.

In Figure 2.7b, we give values of a_e, v_e for each $\ell = e, \mu, \tau$ determined from LEP/SLAC data at the Z resonance. The experimental precision is drastically improved, but the data is still within the prediction of the Standard Model. Agreement of the data with predictions including radiative corrections are so accurate that they can be used to predict the mass value of the top quark or Higgs which appears in the higher order corrections of the SM. Details of the radiative corrections will be discussed in Chapter 5.

2.3 $\nu N \to \nu + X$

As the GWS model established its validity and universality in the lepton sector, that is, in $\nu - e, e^- e^+ \to \mu^- \mu^+$, the next task is to prove its validity in the quark/hadron sector. In terms of the quarks, the neutrino-quark cross sections have the same form as those of the neutrino-electron because all of them are point particles with spin 1/2, the only difference being their coupling strength associated with chirality. The cross sections of Eq. (2.13) for $\nu_\mu e$ and $\overline{\nu}_\mu e$ can be directly applied to $\nu_\mu q$ and $\overline{\nu}_\mu q$ scatterings. They are given by

$$\frac{d\sigma}{dy}(\nu_\mu q) = \frac{d\sigma}{dy}(\overline{\nu}_\mu \overline{q}) = \frac{G_N^2 \hat{s}}{\pi}\left[\epsilon_L(q)^2 + \epsilon_R(q)^2(1-y)^2\right] \tag{2.25a}$$

$$\frac{d\sigma}{dy}(\nu_\mu \overline{q}) = \frac{d\sigma}{dy}(\overline{\nu}_\mu q) = \frac{G_N^2 \hat{s}}{\pi}\left[\epsilon_L(q)^2(1-y)^2 + \epsilon_R(q)^2\right] \tag{2.25b}$$

$$\epsilon_L(q) = I_{3q} - Q_q x_W\,, \quad \epsilon_R(q) = -Q_q x_W \tag{2.25c}$$

where hats $\hat{}$ over the variable s denotes that it refers to partons inside hadrons. The cross section for the antineutrino can be obtained from that of the neutrino by crossing, namely, changing $p_1 \to -p_3, p_3 \to -p_1$ or $\hat{s} = (p_1 + p_2)^2 \leftrightarrow \hat{u} = (p_2 - p_3)^2$. It amounts to the exchange $\hat{s} \leftrightarrow \hat{s}(1-y)^2$ in Eqs. (2.25a) and (2.25b). $G_N = \rho G_F$ is the coupling constant for the neutral current. Strictly speaking, the ρ gets radiative corrections different from those of the leptonic reactions, but the difference is small and we neglect it at this stage.

In order to obtain cross sections for the neutrino-nucleon scattering, we note that the partons (quarks and gluons) in the nucleon carry fractional momenta of the parent nucleon

$$\hat{p}_q = x_q P \tag{2.26}$$

Then, for sufficiently high energy,

$$\hat{s} = (p_1 + xP)^2 = 2x(p_1 \cdot P) + m_N^2 \simeq 2x(p_1 \cdot P) \simeq x(p_1 + P)^2 = xs \quad (2.27)$$

If the quark or antiquark in the nucleon has momentum distribution $q(x), \overline{q}(x)$, each quark (or antiquark) has probability $q(x)dx$ or $\overline{q}(x)dx$ to carry momentum x. Then, the neutrino-nucleon (hereafter denoted as νN) cross section is expressed as

$$\frac{d\sigma}{dy}(\nu N) = \sum_{q,\overline{q}} \frac{d\sigma}{dy}(\nu q) q(x) dx = \frac{G_N^2 s}{\pi}\left[\sum_q \{\epsilon_L(q)^2 + \epsilon_R(q)^2(1-y)^2\}\, x q(x) + \sum_{\overline{q}} \{\epsilon_R(q)^2 + \epsilon_L(q)^2(1-y)^2\}\right] x\overline{q}(x) dx \quad (2.28)$$

where we used $\epsilon_R(\overline{q})^2 = (-\epsilon_R(q))^2 = \epsilon_R(q)^2$. Therefore,

$$\frac{d^2\sigma}{dxdy}(\nu_\mu N) = \frac{G_N^2 s}{\pi}\left[\{\epsilon_L^2 x q(x) + \epsilon_R^2 x \overline{q}(x)\} + \{\epsilon_R^2 x q(x) + \epsilon_L^2 x \overline{q}(x)\}(1-y)^2\right] \quad (2.29a)$$

$$\frac{d^2\sigma}{dxdy}(\overline{\nu}_\mu N) = \frac{G_N^2 s}{\pi}\left[\{\epsilon_R^2 x q(x) + \epsilon_L^2 x \overline{q}(x)\} + \{\epsilon_L^2 x q(x) + \epsilon_R^2 x \overline{q}(x)\}(1-y)^2\right] \quad (2.29b)$$

Integrating over x, y and putting

$$\sigma_{TOT}(\nu N) = \sigma_{NC}^{\nu}, \quad \sigma_0 = G_F^2 s/\pi$$
$$Q \equiv \int dx\, x q(x), \quad \overline{Q} \equiv \int dx\, x \overline{q}(x) \quad (2.30)$$

we obtain the total cross sections

$$\sigma_{NC}^{\nu} = \sigma_0 \rho^2 \left[\left(\epsilon_L^2 + \epsilon_R^2/3\right) Q + \left(\epsilon_L^2/3 + \epsilon_R^2\right) \overline{Q}\right] \quad (2.31a)$$

$$\sigma_{NC}^{\overline{\nu}} = \sigma_0 \rho^2 \left[\left(\epsilon_L^2/3 + \epsilon_R^2\right) Q + \left(\epsilon_L^2 + \epsilon_R^2/3\right) \overline{Q}\right] \quad (2.31b)$$

At this point, we express the u- and d-quark distributions in the proton as $u(x)$ and $d(x)$, neglect the strange and c-quark components in the nucleon. From isospin symmetry, u- and d-quark distributions in the neutron are given by $d(x)$ and $u(x)$ respectively. Assuming isoscalar target which has equal number of protons and neutrons, we take the average of u and d in the above expressions. Namely,

$$\epsilon_L^2 Q \to \left[\epsilon_L^2(u) U + \epsilon_L^2(d) D + (U \leftrightarrow D)\right]/2 \equiv G_L^2(U + D)/2$$
$$\epsilon_R^2 Q \to \left[\epsilon_R^2(u) U + \epsilon_R^2(d) D + (U \leftrightarrow D)\right]/2 \equiv G_R^2(U + D)/2$$
$$U = \int dx\, x u(x), \quad D = \int dx\, x d(x) \quad (2.32)$$

where

$$G_L^2 = \epsilon_L^2(u) + \epsilon_L^2(d) = \frac{1}{2} - x_W + \frac{5}{9}x_W^2 \tag{2.33a}$$

$$G_R^2 = \epsilon_R^2(u) + \epsilon_R^2(d) = \frac{5}{9}x_W^2 \tag{2.33b}$$

Rewriting $P = (U + D)/2, \overline{P} = (\overline{U} + \overline{D})/2$, Eq. (2.31) becomes

$$\sigma_{NC}^{\nu} = \sigma_0\rho^2 \left[G_L^2(P + \overline{P}/3) + G_R^2(P/3 + \overline{P})\right] \tag{2.34a}$$

$$\sigma_{NC}^{\overline{\nu}} = \sigma_0\rho^2 \left[G_L^2(P/3 + \overline{P}) + G_R^2(P + \overline{P}/3)\right] \tag{2.34b}$$

On the other hand, the total cross sections for the charged current reactions are obtained by replacing $\epsilon_L \to 1/\sqrt{2}, \epsilon_R \to 0$ (see Eq. (1.51a)) or equivalently $G_L^2 \to 1, G_R^2 \to 0$:

$$\sigma_{CC}^{\nu} = \sigma_0\rho_{CC}^2(P + \overline{P}/3) \tag{2.35a}$$

$$\sigma_{CC}^{\overline{\nu}} = \sigma_0\rho_{CC}^2(P/3 + \overline{P}) \tag{2.35b}$$

where $\rho_{CC} \simeq 1$ is the radiatively corrected ρ parameter for charged current interaction. Defining

$$R^{\nu} = \sigma_{NC}^{\nu}/\sigma_{CC}^{\nu}, \quad R^{\overline{\nu}} = \sigma_{NC}^{\overline{\nu}}/\sigma_{CC}^{\overline{\nu}}, \quad r = \sigma_{CC}^{\overline{\nu}}/\sigma_{CC}^{\nu} \tag{2.36}$$

and expressing them in terms of the coupling constants, we have

$$R^{\nu} = (\rho^{\nu h})^2 \left(G_L^2 + G_R^2 r\right) \tag{2.37a}$$

$$R^{\overline{\nu}} = (\rho^{\nu h})^2 \left(G_L^2 + \frac{G_R^2}{r}\right) \tag{2.37b}$$

$$\rho^{\nu h} = \frac{\rho}{\rho_{CC}} \tag{2.37c}$$

Combining all the expressions, we finally obtain [60]

$$R^{\nu} = (\rho^{\nu h})^2 \left[\frac{1}{2} - x_W + \frac{5}{9}x_W^2(1 + r)\right] \tag{2.38a}$$

$$R^{\overline{\nu}} = (\rho^{\nu h})^2 \left[\frac{1}{2} - x_W + \frac{5}{9}x_W^2\left(1 + \frac{1}{r}\right)\right] \tag{2.38b}$$

Experimentally, the neutral current and the charged current reactions can be measured simultaneously. As the neutrino deep inelastic scattering is the sum of the neutrino quark scatterings, one gets

$$\text{CC reactions} \quad \begin{cases} \nu_\mu + d \to \mu^- + u, & \overline{\nu}_\mu + u \to \mu^+ + d \\ \nu_\mu + \overline{u} \to \mu^- + \overline{d}, & \overline{\nu}_\mu + \overline{d} \to \mu^+ + \overline{u} \end{cases}$$

$$\text{NC reactions} \quad \begin{cases} \nu_\mu + q(= u, d) \to \nu_\mu + q, & \overline{\nu}_\mu + q \to \overline{\nu}_\mu + q \\ \nu_\mu + \overline{q}(= \overline{u}, \overline{d}) \to \nu_\mu + \overline{q}, & \overline{\nu}_\mu + \overline{q} \to \overline{\nu}_\mu + \overline{q} \end{cases} \tag{2.39}$$

Signals to identify the type of reactions are: (1) if there is a long noninteracting track which is most likely the muon, it is the charged current reaction, (2) if there is no muons in an event, it is most likely the neutral current reactions.[5)] We refer to Vol. 1, Sect. 17.7 for simple "how to do neutrino experiments" and reference [61] for detailed discussion of the experimental setups and recent data. From the measured ratios R^{ν}, $R^{\overline{\nu}}$, one can calculate $G_{\rm L}$, $G_{\rm R}$ (see Figure 2.8). Agreement with the GWS prediction is good.

The value of the Weinberg angle obtained in the semileptonic reactions agreed with that obtained from the pure leptonic reactions. Figure 2.9 shows a similar, but more recent model independent analysis using compilation of elastic as well as inelastic scattering data of the hadrons [9]. The data agrees with the GWS theory and the value of the Weinberg angle was determined. All the neutrino data demonstrated that the theory is equally applicable to both leptonic and semileptonic reactions with the common Weinberg angle. In summary, the Standard Model has been confirmed experimentally as the valid unified theory of the electromagnetic and weak interactions.

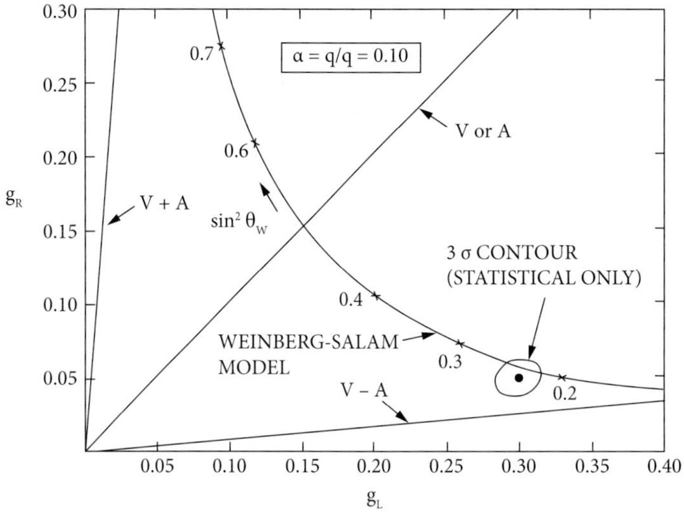

Figure 2.8 Graphical illustration of the left-handed versus right-handed NC coupling parameters $g_{\rm L}(e)$ and $g_{\rm R}(e)$ ($G_{\rm L}$ and $G_{\rm R}$ in the text). The relative antiquark-quark ratio is assumed to be $\alpha = 0.10$. The deviation of iron from an isoscalar target and the effect of the strange sea quarks are taken into account. The experimental point is shown together with the three standard deviation contour (statistical error). In addition, the predictions of $V-A$, pure V and A, $V+A$ and the GWS model (the curve punctuated with values of $\sin^2\theta_{\rm W}$) are given [62].

5) If the resolution of the calorimeter (energy measuring device) is not so good, the charged current reactions by $\nu_e N \to e^- + X$ are misidentified as the neutral current reactions. However, if the ν_e component in the ν_μ beam is small, they are generally small corrections.

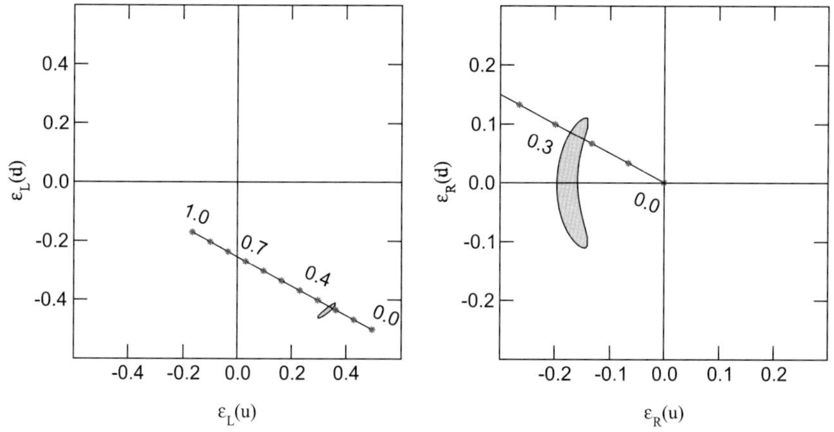

Figure 2.9 Model independent analysis of semileptonic neutrino reactions using a compilation of data for elastic as well as inelastic neutrino data. The line is the prediction of the GWS theory with numbers denoting the value of the Weinberg angle. The GWS theory with the universal value of the Weinberg angle reproduces high precision data well [9].

The Weinberg Angle The Weinberg angle can be determined from R^{ν}, $R^{\overline{\nu}}$ alone or both, but when a precise value is wanted, their accuracy needs to be enhanced. In such a case, various nonideal experimental conditions as well as theoretical uncertainties have to be corrected. They include:

1. the target (normally iron) is not exactly an isoscalar ($n_P \neq n_n$),
2. contribution of the strange quark in the nucleon,
3. production of the charm quark,
   Items (2, 3) include the momentum distribution in the nucleon, use of exact formula at the production threshold, the exact value of the mass and so on.
4. radiative corrections: mainly due to top and Higgs loops (see Chapter 5 for details)
5. W-Z propagator difference
6. finite muon mass

It is claimed best to use the following formula [63] which has small theoretical ambiguity (i.e., independent of hadronic models), that is,

$$R^{-} \equiv \frac{\sigma_{\rm NC}^{\nu} - \sigma_{\rm NC}^{\overline{\nu}}}{\sigma_{\rm CC}^{\nu} - \sigma_{\rm CC}^{\overline{\nu}}} = \frac{R^{\nu} - r R^{\overline{\nu}}}{1 - r} = (\rho^{\nu h})^2 (G_{\rm L}^2 - G_{\rm R}^2) = (\rho^{\nu h})^2 \left(\frac{1}{2} - x_{\rm W}\right) \tag{2.40}$$

Due to limited statistics of the antineutrino data, it was difficult to obtain a precise experimental value. A precision experiment of the NuTeV group using the Main

Injector beam at Fermilab gives a value [64]

$$\sin^2\theta_W(\text{on-shell}) = 0.2277 \pm 0.0013_{\text{stat}} \pm 0.0009_{\text{sys}}{}^{6)} \tag{2.41}$$

which is compared with world data including LEP Z resonance data [11].

$$0.22308 \pm 0.0003 \tag{2.42}$$

2.4
Parity Violation in the Electromagnetic Processes

2.4.1
Atomic Process

The Z boson couples to the neutral current which is expressed as

$$J_\mu^{\text{NC}} = g_Z \overline{\psi}\gamma^\mu \left[I_{3\text{L}} \frac{(1-\gamma^5)}{2} - Q\sin^2\theta_W \right] \psi \tag{2.43}$$

where ψ is any fermion field. The photon couples to the neutral current generated by charged particles. This means that the Z exchange contribution has to be included for processes hitherto considered as pure electromagnetic interaction. This applies to all phenomena in condensed matter, molecular and atomic physics. Inclusion of the Z exchange means that the process violates parity conservation. Under normal conditions, the contribution of the Z is suppressed by factor k^2/m_Z^2 where k is the typical energy of the process and for $k = 1\,\text{eV}$ (applicable to condensed matter physics), the suppression factor is $\sim 10^{-22}$ which can safely be neglected. Here, we pick up two examples to positively look into the parity violation effect as a test of the GWS theory. The first example exhibits the parity violation in atomic physics which provides the test of the Standard Model at an extremely different energy scale from ordinary high energy experiments. It is also a table-top experiment which is a rare species in high energy physics nowadays. The other is the polarized electron-deuteron ($e\uparrow - D$) scattering experiment which historically played a definitive role in establishing the validity of the GWS theory in 1978, excluding early erroneous atomic experiments which did not agree with the prediction of the GWS theory at the time. Today, very precise measurements have been carried out and they all agree with the Standard Model.

The dominant part of the parity violating Lagrangian in the electron-quark system arises from cross terms between γ and Z exchange, and is described as [3]

$$\mathcal{L}_{eq} = \frac{G_N}{\sqrt{2}} \sum_{i=u,d} \left[C_{1i}(\overline{e}\gamma^\mu\gamma^5 e)(\overline{q}_i\gamma_\mu q_i) + C_{2i}(\overline{e}\gamma^\mu e)(\overline{q}_i\gamma_\mu\gamma^5 q_i) \right] \tag{2.44a}$$

$$C_{1u} = -\rho'_{eq}\left(\frac{1}{2} - \frac{4}{3}\kappa'_{eq} x_W\right) + \lambda' \tag{2.44b}$$

6) Note that at the quoted precision level, there is 3σ discrepancy between the NuTeV data and the world average, which some people suspect as a possible sign of new physics. Here, they are treated as though they are in excellent agreement.

$$C_{1d} = -\rho'_{eq}\left(-\frac{1}{2} + \frac{2}{3}\kappa'_{eq} x_{\mathrm{W}}\right) - 2\lambda' \tag{2.44c}$$

$$C_{2u} = -\rho_{eq}\left(\frac{1}{2} - 2\kappa_{eq} x_{\mathrm{W}}\right) + \lambda_u \tag{2.44d}$$

$$C_{2d} = -\rho_{eq}\left(-\frac{1}{2} + 2\kappa_{eq} x_{\mathrm{W}}\right) + \lambda_d \tag{2.44e}$$

where $\rho'_{eq} = \rho_{eq} = \kappa'_{eq} = \kappa_{eq} = 1$ and $\lambda' = \lambda_u = \lambda_d = 0$ in the tree approximation. Radiative corrections change them at a few percent level, so they have to be considered in deriving accurate values, but can be neglected for qualitative discussions. The C_{1i} represent the axial electron and the vector quark currents while the C_{2i} represent the vector electron and the axial vector quark currents. In the following, we consider only the C_{1i} term because it is more easily determined compared to the C_{2i} term. This is because all the quarks in the first term contribute coherently in a heavy nucleus, whereas those in the second term couple to the spin of the nucleus due to $\gamma^0\gamma^5$, which becomes $\boldsymbol{\sigma}\cdot(\boldsymbol{p}+\boldsymbol{p}')/(2m)$ in the low energy limit (see Vol. 1, Table 4.2). Therefore, it is a spin flipping process. Since $\gamma^\mu \sim (1, \boldsymbol{v}/c) \to (1,0,0,0)$ at low energy, for an atom with Z[7] protons and N neutrons, the parity-nonconserving (referred as PNC in the following) amplitude that will be measured is [65, 66]

$$A_{\mathrm{PNC}} = \frac{G_{\mathrm{F}}}{2\sqrt{2}m_e c} Q_{\mathrm{W}} \int d^3x \rho_{\mathrm{nucl}} \langle n'L'|\overline{\psi}_e(\boldsymbol{\sigma}\cdot\boldsymbol{p}_e)\psi_e|nL\rangle \tag{2.45a}$$

$$\sim \frac{G_{\mathrm{F}}}{2\sqrt{2}m_e c} Q_{\mathrm{W}} \langle n'L'|\boldsymbol{\sigma}\cdot\nabla|nL\rangle|_{x=0} \tag{2.45b}$$

$$Q_{\mathrm{W}} = 2[(2Z+N)C_{1u} + (Z+2N)C_{1d}] = \rho'_{eq}\left[-N + Z(1 - 4\kappa'_{eq}\sin^2\theta_{\mathrm{W}})\right] \tag{2.45c}$$

where $|nL\rangle$ denote electron wave function states with principal quantum number n and angular momentum L. With $|nL\rangle \propto r^L Z^{L+1/2}$[8] as $r \to 0$, the matrix elements are nonvanishing only for $L = 1$, $L' = 0$, that is, S and P states are mixed. In the electromagnetic transitions, M1 (magnetic dipole) transition is the dominant process between equal parity states and E1 (electric dipole) transition is the dominant process between opposite parity states. The parity nonconserving interaction mixes states of different parity. Its intensity is $\sim 3\times10^{-22}$ which is to be contrasted with a typical allowed transition of order 1. Thus, direct observations are impossible. Observation of an interference term is the possibility. Still, it is reduced by $\sim 10^{-11}$ and the dominant process has to be suppressed to do a viable experiment. Some enhancement is obtained by choosing heavy elements because $\langle n'S| \sim Z^{1/2}, |nP\rangle \sim Z^{3/2}$, $Q_{\mathrm{W}} \sim Z$ and the transition rate increases as $\sim Z^3$.

7) Z is an atomic number, not to be confused with the gauge boson.
8) For a qualitative discussion of the angular momentum dependence, see Vol. 1, Sect. 13.5.3. The electron wave function is spread to atomic distance $\sim 10^{-8}$ cm, while the size of the nucleus is $\sim 10^{-(12-13)}$ cm. Therefore, $\int d^3x\rho_{\mathrm{nucl}} \sim \int d^3x\delta^3(\boldsymbol{r})$, meaning the electron wave function is evaluated at $\boldsymbol{x} = 0$.

The first experiments to produce results looked at interference with an allowed M1 amplitude. These experiments, for technical reasons, were designed to detect the optical rotation of the polarization of lights as it passes through a sample of atomic vapor like bismuth [67–69]. In these experiments, the interference between the parity conserving M1 amplitude and the parity nonconserving amplitude gives rise to a difference in refractive index for left and right circularly polarized light. A laser beam was prepared in a state of very clean linear polarization, then passing through a vapor cell that contains atoms of interest and then through a second nearly crossed polarizer. This polarizer blocks out almost all the laser light unless its polarization has been rotated in the vapor, in which case some light passes through.

In another experiments (so called Stark induced), E1 transition was induced by applying a dc electric field to an atomic sample. In the experiment [65, 66, 70, 71] which we describe below, the cesium atom ($Z = 55$, $J = 7/2$; J is the spin of the nucleus) was chosen. This has a very simple configuration, one S-state electron outside a tightly bound Xe core. M1 transition between $6S_{1/2}$ and $7S_{1/2}$ is highly suppressed. The PNC effect induces mixing of $6S_{1/2}$ state with other $|nP\rangle$ states. The experiment uses the Stark effect to induce mixing similar to the PNC by applying an external electric field. The reason is two-fold. It is parity changing and induces the same mixing as the PNC effect, but much stronger. Namely, it provides the main process with which the PNC amplitude can interfere. At the same time, it provides control over the interference amplitude to flip its sign. By clever arrangement of the apparatus, it can also reduce M1 transition further to a controllable level. Now that the $6S_{1/2}$ state has mixture of nP states, it can be excited to $7S_{1/2}$ by E1 transition. The idea is to excite the atom with a laser from $6S_{1/2} \to 7S_{1/2}$ and measure the transition rate by detecting its fluorescence from the second decay in the transition process $[7S_{1/2} \to (6P_{1/2}, 6P_{3/2}) \to 6S_{1/2}]$ (Figure 2.10a). Taking sum and difference of polarity flipped contributions, the parity violating effect can be extracted.

Three amplitudes contribute to the transition $6S_{1/2} \to 7S_{1/2}$.

1. A_{E}: Combination of the Stark effect induced ($\langle n''P| - e\boldsymbol{E} \cdot \boldsymbol{r}|nS\rangle$) and laser beam induced ($\langle n''P| - e\boldsymbol{\epsilon} \cdot \boldsymbol{r}|nS\rangle$) E1 excitation allow atoms to go through the intermediate states $|n''P\rangle$ to make $6S_{1/2} \to n''P \to 7S_{1/2}$ transitions happen. Here, $\boldsymbol{E}$ and $\boldsymbol{\epsilon}$ denote the applied external electric field and the polarization vector of the laser beam respectively. The amplitude is a product of external field ($\boldsymbol{E} \cdot \boldsymbol{r}$) and laser ($\boldsymbol{\epsilon} \cdot \boldsymbol{r}$) induced transition amplitudes. This is the main process.
2. A_{M}: Direct M1 transition $6S_{1/2} \to 7S_{1/2}$ which is highly suppressed to $\sim 10^{-7}$ by choosing Cs as the sample, but still $\sim$ 20 000 times larger than the parity violating process. It is further reduced by the experimental arrangement.
3. A_{PNC}: Combination of parity violating transition ($\langle n''P|H_{\text{PNC}}|nS\rangle$) and laser induced ($\langle n''P| - e\boldsymbol{\epsilon} \cdot \boldsymbol{r}|nS\rangle$) E1 transition to make a transition similar to the main process.

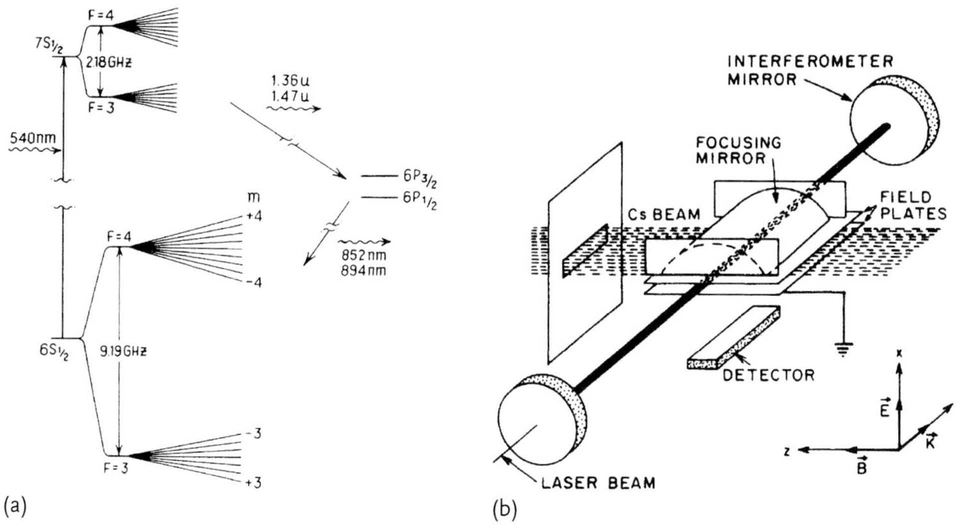

Figure 2.10 (a) cesium ($Z = 55, J = 7/2$) energy-level diagram showing hyperfine and weak-field Zeeman splitting of the 6S and 7S states. Each state $6S_{1/2}$, $7S_{1/2}$ couples with nuclear spin ($J = 7/2$) and splits, showing the hyperfine structure denoted by total spin $F = 7/2 \pm 1/2 = 3, 4$. Degenerate levels with differing m split due to the Zeeman effect by applying the external weak magnetic field. (b) The experimental apparatus. A dc laser beam was constructed between two mirrors which go through between two transparent electrodes. The atomic Cs beam goes through the laser beam in the transverse direction to the laser beam. Not shown is a Helmholtz coil that produces the weak magnetic field (75 Gauss). Directions of the external electric and magnetic field are shown. A detector which detects the fluorescence is placed below the plates and the upper focusing mirror focuses the emitted light from different locations to the detector [71, 72].

In order to understand the experimental strategy, we show the relevant level structure of the cesium in Figure 2.10a and write down processes that contribute to the $6S_{1/2} \to 7S_{1/2}$ transition.

The amplitudes can be written down as [65]

$$\begin{aligned} A &= A_{\mathrm{E}} + A_{\mathrm{M}} + A_{\mathrm{PNC}} \\ &\sim \underbrace{\alpha(\boldsymbol{E}\cdot\boldsymbol{\epsilon})\delta_{FF'}\delta_{mm'} + \Big(i\beta(\boldsymbol{E}\times\boldsymbol{\epsilon})}_{A_{\mathrm{E}}} + \mathrm{M1}(\hat{\boldsymbol{k}}\times\boldsymbol{\epsilon}) + E1_{\mathrm{PNC}}\boldsymbol{\epsilon}\Big)\cdot\langle F'm'|\boldsymbol{\sigma}|Fm\rangle \end{aligned} \tag{2.46}$$

where α and β are the scalar and tensor polarizabilities respectively, $\hat{\boldsymbol{k}}$ is the unit vector in the direction of the laser beam, and F, F' are the total spin and m, m' their components which are degenerate under normal circumstances.

The experimental apparatus is shown in Figure 2.10b. Due to two mirrors, the laser beam creates a standing wave between the two transparent electrodes that produces the external field $\boldsymbol{E}$ which directs in the x direction as shown in the figure. A weak external magnetic field (74 Gauss) is applied by a Helmholtz coil (not shown) to split the spin components m by the Zeeman effect. The reason will be described soon. The standing electromagnetic field means that it consists of two propagating waves in opposite directions with $\hat{\boldsymbol{k}} = \pm 1$ suppressing the M1 contribution as is shown by the term $\hat{\boldsymbol{k}} \times \boldsymbol{\epsilon}$. The polarity of $\boldsymbol{E}$, $\boldsymbol{B}$ and the laser handedness are used to invert the parity and to reduce the systematics. The reason the external magnetic field is applied is because otherwise, $\pm m$ components contribute equally with differing sign in the matrix element of $\boldsymbol{\sigma}$ and the net parity violating contribution vanishes in the total yield. By applying the external magnetic field, it is possible to select transitions between specific magnetic levels. By choosing the $\Delta F \neq 0$ transition, the main contribution (α term) in Eq. (2.46) is removed ($\beta \sim \alpha/10$).[9] The transition rate was monitored by observing the fluorescence which was provided by the transition $6P_{1/2} \rightarrow 6S_{1/2}$. The wavelength of the fluorescence is much longer and helps to reduce backgrounds from the scattering of the main beam. With a typical voltage of 2500 V/cm, the detector current of $\sim 3 \times 10^{-10}$ A and a parity nonconserving fraction of $\sim 1.3 \times 10^{-6}$ was obtained. The end result was

$$Q_W = -69.4 \pm 1.5_{\text{exp}} \pm 3.8_{\text{theor(ME)}} \tag{2.47a}$$

[10]

$$\sin^2\theta_W(\text{on-shell}) = 0.219 \pm 0.007 \pm 0.018 \tag{2.47b}$$

which agrees well with the Standard Model.

2.4.2 Polarized *e*–*D* Scattering

The experiment was carried out at SLAC (Stanford Linear Accelerator Center) using a polarized electron beam with energy 20 GeV. Unlike the atomic process, we need to consider the contribution of the Z-pole in the high energy scattering. As the beam is longitudinally polarized, we separate the beam into the left- and right-handed components and consider a fully polarized target for the sake of calculation. We define the kinematical variables as

$$e_a(p_1) + q_b(p_2) \rightarrow e_a(p_3) + q_b(p_4) \tag{2.48}$$

where subfix a, b = L, R denote the left- and right-handed particle. Considering the vector or axial vector interaction conserves the handedness of the particle (see Vol. 1, Sect. 4.3.5), the scattering amplitude due to polarized electron and a polar-

9) Note that early experiments with the Stark field [70, 73] used different target/configuration and did not apply the magnetic field. Therefore, polarization of the emitted light, not the total intensity, was measured and the small circular polarization was detected.

10) Subscript theor(ME) is the theoretical uncertainty in evaluating the transition matrix elements.

ized quark inside the nucleus can be described as follows.

$$\begin{aligned}\mathcal{M} &= i\left[\overline{u}_a(p_3)(-i\gamma^\mu)u_a(p_1)\right]\\
&\quad\times\left[\frac{Q_e e(-ig_{\mu\nu})Q_q e}{t}+\frac{g_Z\varepsilon_a(e)\left(-ig_{\mu\nu}+\frac{q_\mu q_\nu}{m_Z^2}\right)g_Z\epsilon_b(q)}{t-m_Z^2+im_Z\Gamma_Z}\right]\\
&\quad\times\left[\overline{u}_b(p_4)(-i\gamma^\nu)u_b(p_2)\right]\\
&\equiv -\left[\overline{u}_a(p_3)\gamma^\mu u_a(p_1)\right]\frac{e^2}{t}[\overline{u}_b(p_4)\gamma_\mu u_b(p_2)]F_{ab}(t)\end{aligned} \tag{2.49a}$$

$$u_L(p)=\frac{1-\gamma^5}{2}u(p)\,,\quad u_R(p)=\frac{1+\gamma^5}{2}u(p) \tag{2.49b}$$

$$F_{ab}(t)=Q_e Q_q+\frac{\sqrt{2}G_N m_Z^2}{\pi\alpha}\frac{t\epsilon_a(e)\epsilon_b(q)}{t-m_Z^2+im_Z\Gamma_Z} \tag{2.49c}$$

$$\varepsilon_L(e)=-\frac{1}{2}+x_W\,,\quad \varepsilon_R(e)=x_W\,,\quad x_W=\sin^2\theta_W \tag{2.49d}$$

$$\epsilon_L(q)=I_{3q}-Q_q x_W\,,\quad \epsilon_R(q)=-Q_q x_W \tag{2.49e}$$

From this amplitude, one can calculate the helicity specific "eq" scattering cross section [11]. The standard gamma trace calculation gives

$$\begin{aligned}&\sum_{\text{spin}}\left|[\overline{u}_a(p_3)\gamma^\mu u_a(p_1)][\overline{u}_b(p_4)\gamma_\mu u_b(p_2)]\right|^2\\
&\quad=\begin{cases}4s^2 & ;\quad ab=\text{LL, RR}\\ 4u^2=4s^2(1-y)^2 & ;\quad ab=\text{LR, RL}\end{cases}\end{aligned} \tag{2.50}$$

where y is defined in Eq. (2.5). In the massless limit, $y=(1-\cos\theta^*)/2$ in CM (center of mass frame) where θ^* is the angle between $\boldsymbol{p}_1$ and $\boldsymbol{p}_3$, and the energy loss $y=(E_1-E_3)/E_1$ in LAB (laboratory frame). Using

$$\left.\frac{d\sigma}{d\Omega}\right|_{\text{CM}}=\left(\frac{p_f}{p_i}\right)\frac{1}{64\pi^2 s}|\mathcal{M}|^2 \tag{2.51}$$

and neglecting $|t|$ compared to m_Z^2, we obtain

$$d\sigma(e_L q_L\to e_L q_L)=\sigma_0|F_{LL}(t)|^2 \tag{2.52a}$$

$$d\sigma(e_L q_R\to e_L q_R)=\sigma_0|F_{LR}(t)|^2(1-y)^2 \tag{2.52b}$$

$$d\sigma(e_R q_L\to e_R q_L)=\sigma_0|F_{RL}(t)|^2(1-y)^2 \tag{2.52c}$$

$$d\sigma(e_R q_R\to e_R q_R)=\sigma_0|F_{RR}(t)|^2 \tag{2.52d}$$

11) L, R is the chirality and not the helicity state, but in the massless limit, they are identical except for the helicity reversal for the antiparticle.

where

$$\sigma_0 = N_c \frac{\alpha^2}{t^2} s \tag{2.53}$$

and $N_c = 3$ is the color factor. As the value of the momentum squared is small ($|t| \ll m_Z^2$), the square of the Z-pole term can be neglected and $|F_{ab}(t)|^2$ are approximated by

$$|F_{ab}(t)|^2 \simeq Q_q^2 + \frac{2\sqrt{2} Q_q G_N t}{\pi\alpha} \varepsilon_a(e)\epsilon_b(q) \tag{2.54}$$

In the actual experiment, an unpolarized target was used. We assume that the $e-D$ cross section is given by sum of $e-q$ scattering as is done in the discussion of $\nu - N$ inelastic scattering in the previous section. We have to make the average over the target polarization and sum over the quark flavors to obtain

$$d\sigma_L = \frac{1}{2} \sum_q (d\sigma_{LL} + d\sigma_{LR}), \quad d\sigma_R = \frac{1}{2} \sum_q (d\sigma_{RL} + d\sigma_{RR}) \tag{2.55}$$

For the beam with polarization P, the cross section becomes

$$d\sigma = \frac{1-P}{2} d\sigma_L + \frac{1+P}{2} d\sigma_R \tag{2.56}$$

The asymmetry is defined by

$$A = \frac{d\sigma_R - d\sigma_L}{d\sigma_L + d\sigma_R} {}^{12)} = \frac{\sqrt{2} G_N(-t)}{2\pi\alpha} \frac{1}{\{1 + (1-y)^2\} \sum_q Q_q^2} \times \sum_q Q_q \left[a_e v_q \{1 + (1-y)^2\} + v_e a_q \{1 - (1-y)^2\}\right]$$

$$\begin{aligned} v_e &= \varepsilon_L(e) + \varepsilon_R(e) = -1/2 + 2x_W, \quad a_e = \varepsilon_L(e) - \varepsilon_R(e) = -1/2 \\ v_q &= \epsilon_L(q) + \epsilon_R(q) = I_{3q} - 2Q_q x_W, \quad a_q = \epsilon_L(q) - \epsilon_R(q) = I_{3q} \end{aligned} \tag{2.57}$$

We only used the dominant term and dropped the interference term in the denominator. As the target is a deuteron which has an equal number of protons and neutrons, that is, $I = 0$, the number of u quarks and d quarks are equal. Inserting $q = u, d$, $Q_u = 2/3$, $Q_d = -1/3$, $I_{3u} = 1/2$, $I_{3d} = -1/2$, the asymmetry can be expressed as

$$A = -\frac{G_N|t|}{2\sqrt{2}\pi\alpha} \frac{9}{10} \left[\left(1 - \frac{20}{9} x_W\right) + (1 - 4x_W) \frac{1 - (1-y)^2}{1 + (1-y)^2}\right] \tag{2.58}$$

Inserting $|t| \simeq 1\,\mathrm{GeV}^2$ and the standard value of the $x_W = \sin^2\theta_W$, we obtain

$$A \simeq 9 \times 10^{-9} \tag{2.59}$$

The asymmetry value does not depend on the beam energy. As it is very small, it is a difficult experiment. We show the experimental result carried out at SLAC in Figure 2.11 [58, 59].

12) The sign definition of the asymmetry is opposite to what is normally used in this book. It is adopted here to be consistent with the figures.

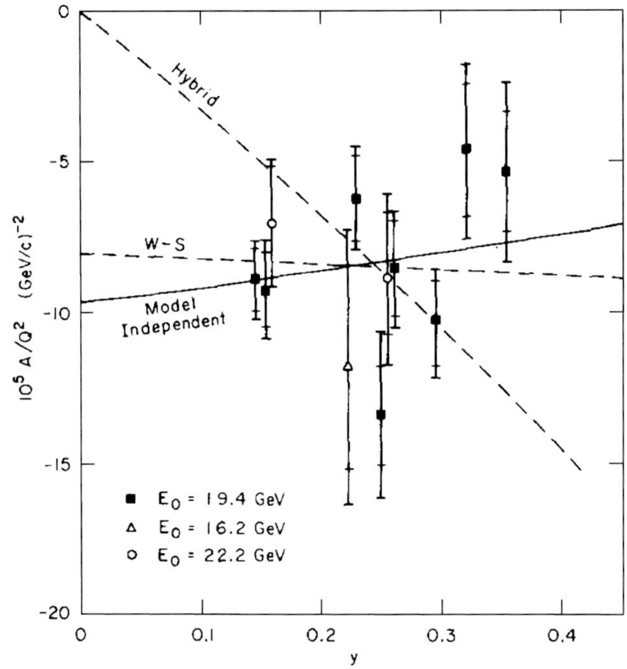

Figure 2.11 Asymmetries measured at three incident energies are plotted against $y = (E_0 - E')/E_0$. The total error bar gives the combined statistical and systematic error. The inner error corresponds to the statistical part only. The data are compared with two $SU(2) \times U(1)$ model predictions, the Weinberg–Salam model and the hybrid model. In each case, $\sin^2 \theta_W$ has been adjusted to minimize χ^2. A two parameter model-independent fit $(a_1 + a_2[1-(1-y)^2]/[1+(1-y)^2])$ based only on the simple parton model assumption fit to the data; the hybrid model appears to be ruled out [59].

In the experiment, the polarized beam was created by scattering electrons with a circularly polarized laser beam and obtaining the polarization $P_e = 0.37$. As the experiment did not measure A, but AP_e, a test was carried out by changing the polarization and determining if the asymmetry changes accordingly. The polarization P_e can be measured from the asymmetry of the Möller $(e^- e^-)$ scattering and was carried out separately. The polarization of the electron beam can also be varied. It is made from a linearly polarized beam by passing through a calcite prism and its polarization can be varied by rotating the calcite. Figure 2.12a,b shows the asymmetry as a function of the rotation angle of the calcite and the beam energy. One sees that it behaves just as it should. Note that the asymmetry itself does not change as a function of the beam energy, but the polarization does because of the spin precession in the magnetic field due to the anomalous magnetic moment $(g-2)$ (see Vol. 1, Eq. (8.94)).

The allowed region on the C_{1u}–C_{1d} plane determined by the atomic parity violation experiments and the polarized e–D experiments are shown in Figure 2.13.

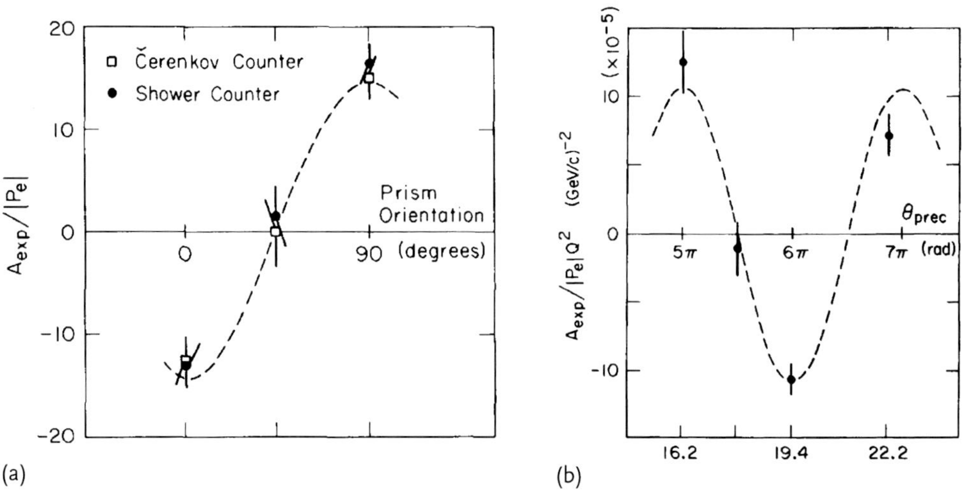

Figure 2.12 (a) The experimental asymmetry shows the expected variation (dashed lines) as the beam helicity changes due to the change in orientation of the calcite prism. The data are for 19.4 GeV and deuterium. Since the same scattered particles strike both counters, they are not statistically independent. No systematic errors are shown. No corrections have been made for helicity dependent differences in beam parameters. (b) The experimental asymmetry shows the expected variation (dashed lines) as the beam helicity changes due to the $g - 2$ precession in the beam transport system. The data are for the shower counter and the deuterium [58].

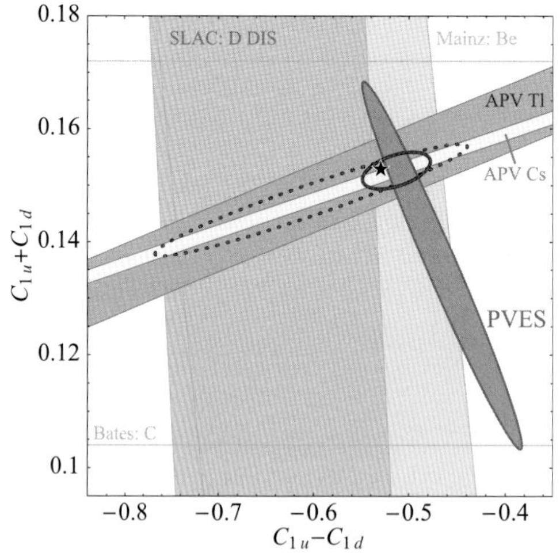

Figure 2.13 Allowed region determined by the polarized $e-D$ scattering, APV (atomic parity violation) experiments and recent PVES (parity violating electron scattering) experiments. The dotted contour displays the previous experimental limits (95% CL) reported in the PDG [3] together with the prediction of the standard model (black star). The filled ellipse denotes the new constraint provided by recent high precision PVES scattering measurements on hydrogen, deuterium, and helium targets (at 1 standard deviation), while the solid contour (95% CL) indicates the full constraint obtained by combining all results [4]. For a color version of this figure, please see the color plates at the beginning of the book.

Establishment of the GWS model as the electroweak theory

Looking at Figure 2.13, accuracy of the polarized electron-deuteron scattering does not look so impressive in determining the Weinberg angle. The atomic parity violation experiments are far better in this sense. From a historical point of view, however, it played a crucial role. In the late 1970s, roughly ten years after the GWS model was proposed, evidence was accumulating and many people already believed in it. However, there was one hitch before declaring that the GWS model was the right electroweak theory. An atomic parity violation experiment did not meet the GWS prediction. At that time, the neutral current data was like that shown in Figure 2.7a, minus the $e^+e^- \to \mu^+\mu^-$ data. Namely, there was two-fold ambiguity in the $v_e - a_e$ solution and the atomic parity violation experiment agreed with neither of them. The SLAC $e{-}D$ data, despite its large error, was good enough to choose one of the two solutions and also deny the erroneous atomic parity experiment. Namely, it singled out the GWS model as the only model which is compatible with the experimental data. At the same conference [ICHEP78 (International Conference on High Energy Physics in 1978)] that the $e\uparrow{-}D$ data was presented, a compilation of various neutral current data was presented and the value of the Weinberg angle derived from them agreed to each other very well (Figure 2.14). This is the incident which elevated the GWS model from a model to a theory.

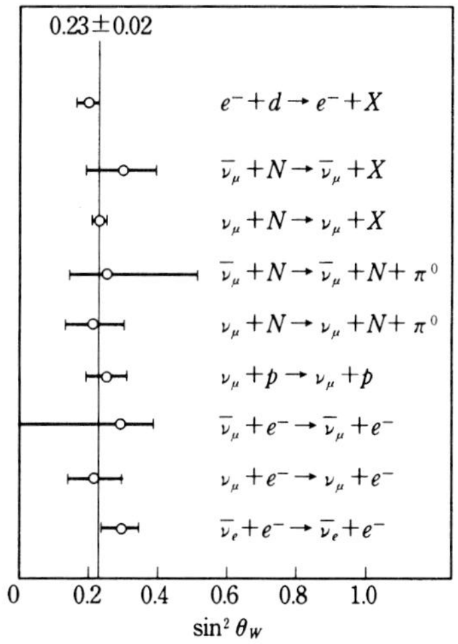

Figure 2.14 The Weinberg angles determined from various neutral current data showed good agreement with each other [74].

2.5
Electroweak Unification at High Q^2

The effect of the weak boson was small in the $e\uparrow -D$ scattering to say nothing of the atomic parity experiment. However, the smallness was a result of the energy scale Q being much smaller compared to the W, Z mass. At HERA,[13] where high energy $e-p$ scattering is possible, we might just see a comparable effect as the momentum transfer Q can go up to $\sim$ 200 GeV. The scattering cross section with quarks inside the proton $d\sigma(eq)$ including the Z-pole contribution are already given in Eq. (2.52). The difference is that, this time, the Z-pole term plays a crucial role.

For the partons having the fractional momentum $x p_2$ inside the proton, the variables are expressed as $\hat{s} = (p_1 + x p_2)^2 \simeq 2x(p_1 \cdot p_2) \simeq x(p_1 + p_2)^2 = xs$, $\hat{t} = (p_1 - p_3)^2 = t = -Q^2$ and to obtain the cross section for the nucleon target, we have to multiply the probability $q(x)dx$ of the momentum being at $x \sim x + dx$.

$$d\sigma(ep) = \sum_q \int dx d\sigma(eq) q(x)$$

$$\therefore \quad \frac{d\sigma}{dx dQ^2}(ep) = \frac{4\pi}{s} \sum_q q(x) \left.\frac{d\sigma}{d\Omega}\right|_{\text{CM}} (eq) \tag{2.60}$$

where we have integrated over the azimuthal angle. A slight complication arises when the sum includes the antiquark. Because of the helicity flip for the same chirality, the cross section for the antiquark target is obtained by changing the sign of the axial vector coupling $a_q \to -a_q$ and that for the positron beam by $a_e \to -a_e$.

In order to obtain a more conventional form for the cross section which is parametrized using generalized structure functions, we define

$$\begin{aligned}
&F_2^\gamma \equiv \textstyle\sum_q Q_q^2 x[q(x) + \overline{q}(x)]\,, && F_3^\gamma = 0 \\
&F_2^{\gamma Z} \equiv \textstyle\sum_q 2Q_q v_q x[q(x) + \overline{q}(x)]\,, && F_3^{\gamma Z} \equiv \textstyle\sum_q 2Q_q a_q[q(x) - \overline{q}(x)] \\
&F_2^Z \equiv \textstyle\sum_q \left(v_q^2 + a_q^2\right) x[q(x) + \overline{q}(x)]\,, && F_3^Z \equiv \textstyle\sum_q 2v_q a_q[q(x) - \overline{q}(x)]
\end{aligned} \tag{2.61}$$

We also use the approximation

$$\begin{aligned}
\chi &= \frac{g_Z^2}{4e^2} \text{Re}\left[\frac{t}{t - m_Z^2 + i m_Z \gamma_Z}\right] \simeq \kappa \frac{Q^2}{Q^2 + m_Z^2}\,, \qquad Q^2 = |t| \\
\kappa &= \frac{1}{4 \sin^2 \theta_W \cos^2 \theta_W}
\end{aligned} \tag{2.62}$$

13) It is an electron proton collider, a unique accelerator of that kind. The energy of the electron and the proton is 30 GeV and 820 GeV, respectively.

The final expression for the $e^{\pm}p$ scattering cross section becomes [3, 75, 76]

$$\left.\frac{d^2\sigma}{dx\,dQ^2}(e^{\pm}p)\right|_{\rm NC} = \frac{2\pi\alpha^2}{xQ^4}\left[Y_+\tilde{F}_2 \mp Y_- x\tilde{F}_3 - y^2\tilde{F}_L\right]^{14)} \qquad (2.63a)$$

$$Y_{\pm} = 1 \pm (1-y)^2 \qquad (2.63b)$$

$$\tilde{F}_2 = F_2^{\gamma} - (v_e \pm P a_e)\chi F_2^{\gamma Z} + \left[v_e^2 + a_e^2 \pm P(2v_e a_e)\right]\chi^2 F_2^Z \qquad (2.63c)$$

$$x\tilde{F}_3 = (a_e \pm P v_e)\chi x F_3^{\gamma Z} - \left[2v_e a_e \pm P\,(v_e^2 + a_e^2)\right]\chi^2 x F_3^Z \qquad (2.63d)$$

$$\tilde{F}_{\rm L} = \tilde{F}_2 - 2x\tilde{F}_1 \qquad (2.63e)$$

For comparison, we also give the charged current cross section ($e^- p \to \nu + X$):

$$\left.\frac{d^2\sigma}{dx\,dQ^2}(e^{\pm}p)\right|_{\rm CC} = \frac{G_{\rm F}^2}{4\pi x}\left(\frac{m_{\rm W}^2}{Q^2+m_{\rm W}^2}\right)^2\left[Y_+ W_2^{\pm} \mp Y_- x W_3^{\pm} - y^2 W_{\rm L}^{\pm}\right] \qquad (2.64)$$

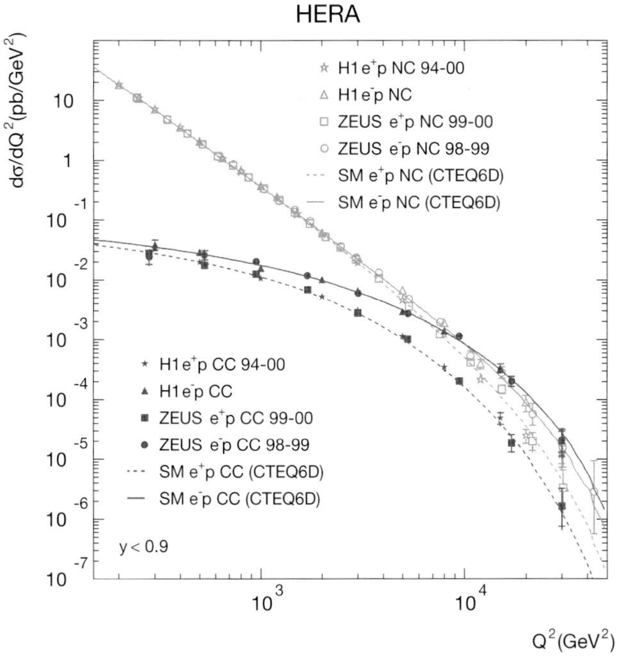

Figure 2.15 Electroweak unification. The neutral current cross section which is dominated by the photon exchange in the small Q^2 region becomes comparable around $Q^2 \gtrsim m_W^2$ to that of the charged current cross which is a pure weak interaction [76, 77].

14) The third term does not exist in the Callan–Gross approximation [78], that is, if the proton is made of only partons with spin 1/2. $\tilde{F}_{\rm L} = 0$ in the parton model and in the leading log approximation in QCD, too (see Chapter 8). It is important only at large y^2.

$$
\begin{aligned}
W_2^+ &= x[\overline{u}(x) + \overline{c}(x) + d(x) + s(x)] \\
x\,W_3^+ &= x[d(x) + s(x) - \overline{u}(x) - \overline{c}(x)] \\
W_2^- &= x[u(x) + c(x) + \overline{d}(x) + \overline{s}(x)] \\
x\,W_3^- &= x[u(x) + c(x) - \overline{d}(x) - \overline{s}(x)]^{15)}
\end{aligned}
\tag{2.65}
$$

Figure 2.15 shows the combined data of H1 and ZEUS [76, 77]. One can see that the neutral current cross section which is dominated by the QED photon exchange process at low Q^2 becomes comparable in the region $Q^2 \sim m_W^2$ with that of the charged cross section which is a pure weak interaction. Thus, the notion that at high energies where $Q^2 \gtrsim m_W^2$, the electromagnetic and the weak interaction are unified, has been beautifully confirmed.

2.6
Asymmetry in the $e^-e^+ \to l\bar{l}, c\bar{c}, b\bar{b}$

The parity violation is an outstanding feature of the weak interaction. Its appearance in the neutral current interactions was tested and proved in the atomic process as well as in the high energy e–D scattering. Its effect appears as an interference between the photon and Z exchange process and will also manifest itself as the asymmetry in the angular distribution of $e^-e^+ \to f\overline{f}$. It grows with the energy and will be maximal in the energy region $\sqrt{s} = 30$–80 GeV, decreasing again as the energy nears the Z-pole (see Figure 2.16). At the Z resonance peak ($\sqrt{s} = m_Z = 91$ GeV) where the photon exchange process is negligible, the asymmetry still exists because Z exchange process by itself contains both parity conserving amplitude $\sim v_f$ and parity violating amplitude $\sim a_f$.

However, the asymmetry is very small at the Z resonance peak, as shown in Figure 2.16. This is because of the accidental value of the Weinberg angle being

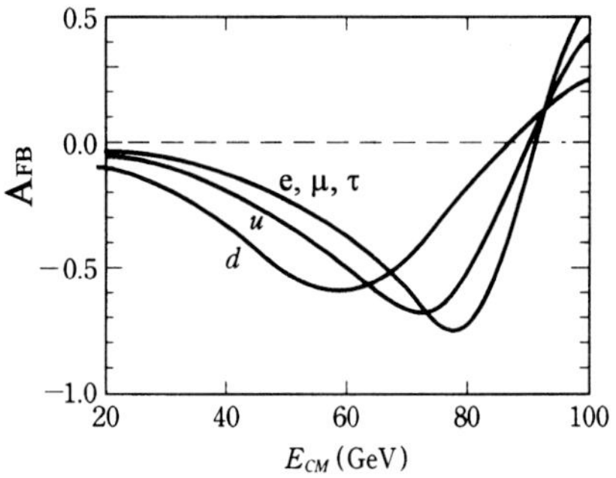

Figure 2.16 Forward-backward asymmetry $A_{FB}(f)$ in the $e^-e^+ \to f\overline{f}$ reaction predicted by the GWS theory as a function of the total energy ($\sqrt{s}$).

15) Contributions from the charm and strange quark components in the hadron are usually nelected in this book. Here, they are included for generality [76].

$x_W \sim 1/4$. Therefore, the asymmetry parameter is best measured in the energy region $\sqrt{s} = 30$–80 GeV, where e^-e^+ colliders PEP-I ($\sqrt{s} = 30$ GeV), PETRA ($\sqrt{s} = 34$ GeV), and TRISTAN ($\sqrt{s} = 58$ GeV) are located. The disadvantage of the small asymmetry at Z is more than compensated by its large production rate. The rich physics at the Z resonance is a subject by itself and will be discussed separately in Chapter 5.

Asymmetry formula The equation for $e^-e^+ \to f\bar{f}$ is obtained from that of $e^- + f \to e^- + f$ scattering by crossing (see Figure 2.17). Therefore, the scattering amplitude can be obtained from that of polarized e–D scattering by crossing, namely by exchanging $p_3 \to -p_3, p_2 \to -p_2$ and $t = (p_1 - p_3)^2$ to $s = (p_1 + p_3)^2$ inEq. (2.49a)

$$\mathcal{M}(e^-e^+ \to f\bar{f}) = -\left[\bar{v}_a(p_3)\gamma^\mu u_a(p_1)\right]\frac{e^2}{s}[\bar{u}_b(p_4)\gamma_\mu v_b(p_2)]F_{ab}(s) \quad (2.66a)$$

$$F_{ab}(s) = Q_e Q_f + \frac{\sqrt{2}G_N m_Z^2}{\pi\alpha}\frac{s\epsilon_a(e)\epsilon_b(f)}{s - m_Z^2 + i m_Z \Gamma_Z} \quad (2.66b)$$

$$(u_L(p), v_L(p)) = \left(\frac{1-\gamma^5}{2}u(p), \frac{1-\gamma^5}{2}v(p)\right),$$

$$(u_R(p), v_R(p)) = \left(\frac{1+\gamma^5}{2}u(p), \frac{1+\gamma^5}{2}v(p)\right) \quad (2.66c)$$

Then, the helicity cross section for $e_L\bar{e}_R \to f_{L(R)}\bar{f}_{R(L)}$ can be calculated to give

$$\frac{d\sigma}{d\Omega}(e_L\bar{e}_R \to f_L\bar{f}_R) \equiv d\sigma_{LL} = \sigma_0|F_{LL}(s)|^2\left(\frac{1+\cos\theta}{2}\right)^2 \quad (2.67a)$$

$$\frac{d\sigma}{d\Omega}(e_L\bar{e}_R \to f_R\bar{f}_L) \equiv d\sigma_{LR} = \sigma_0|F_{LR}(s)|^2\left(\frac{1-\cos\theta}{2}\right)^2 \quad (2.67b)$$

$$\frac{d\sigma}{d\Omega}(e_R\bar{e}_L \to f_L\bar{f}_R) \equiv d\sigma_{RL} = \sigma_0|F_{RL}(s)|^2\left(\frac{1-\cos\theta}{2}\right)^2 \quad (2.67c)$$

$$d\sigma(e_R\bar{e}_L \to f_R\bar{f}_L) \equiv d\sigma_{RR} = \sigma_0|F_{RR}(s)|^2\left(\frac{1+\cos\theta}{2}\right)^2 \quad (2.67d)$$

where

$$\sigma_0 = \frac{\alpha^2}{s}N_c^f \quad (2.68)$$

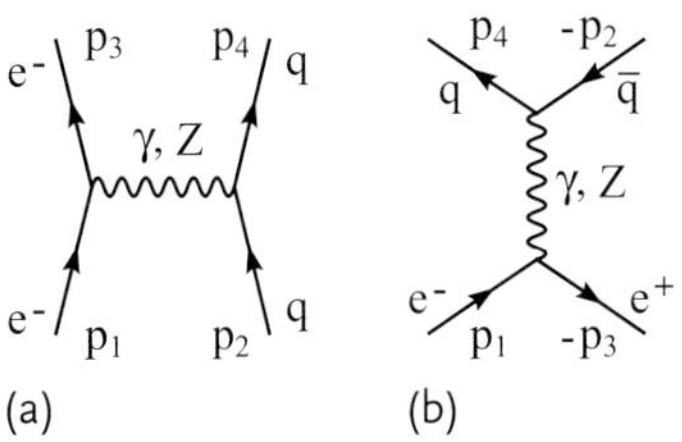

Figure 2.17 The process $e^-e^+ \to f\bar{f}$ is related to the elastic scattering $e^- + f \to e^- + f$ by crossing.

and N_c^f [$= 3(1+\delta_{\rm QCD})$ for $f = q$, $= 1$ for $f = \ell$] is the color factor including the QCD correction. $\delta_{\rm QCD} \sim 0.05$ at $\sqrt{s} \sim 60$ GeV and less in the lower energy region.

The cross section for the unpolarized beam is given by

$$\begin{aligned}\frac{d\sigma}{d\Omega} &= \frac{1}{4}(d\sigma_{\rm LL} + d\sigma_{\rm LR} + d\sigma_{\rm RL} + d\sigma_{\rm RR}) \\ &= \frac{\alpha^2}{4s} N_c^f \left[(1+\cos^2\theta)F_1 + 2\cos\theta\, F_2\right]\end{aligned} \tag{2.69}$$

where the factor 1/4 accounts for the average component of each polarization. Here, θ is the scattering angle of the outgoing fermion with respect to the direction of e^-.

$$\begin{aligned}F_1 &= \frac{1}{4}\left(|F_{\rm LL}|^2 + |F_{\rm RR}|^2 + |F_{\rm LR}|^2 + |F_{\rm RL}|^2\right) \\ &= \left[(Q_f Q_e)^2 - 2Q_f Q_e|\chi(s)|\cos\delta_{\rm R} v_e v_f + |\chi(s)|^2\left(v_e^2 + a_e^2\right)\left(v_f^2 + a_f^2\right)\right] \\ F_2 &= \frac{1}{4}\left(|F_{\rm LL}|^2 + |F_{\rm RR}|^2 - |F_{\rm LR}|^2 - |F_{\rm RL}|^2\right) \\ &= \left[-2Q_f Q_e|\chi(s)|\cos\delta_{\rm R} a_e a_f + 4|\chi(s)|^2 v_e a_e v_f a_f\right]\end{aligned} \tag{2.70a}$$

with

$$\chi(s) = \frac{G_{\rm N} m_Z^2}{2\sqrt{2}\pi\alpha}\frac{s}{s - m_Z^2 + i m_Z\Gamma_Z} \tag{2.70b}$$

$$\tan\delta_{\rm R} = \frac{m_Z\Gamma_Z}{m_Z^2 - s} \tag{2.70c}$$

$$v_e = \varepsilon_{\rm L}(e) + \varepsilon_{\rm R}(e) = -1/2 + 2x_{\rm W}\,, \quad a_e = \varepsilon_{\rm L}(e) - \varepsilon_{\rm R}(e) = -1/2 \tag{2.70d}$$

$$v_f = \varepsilon_{\rm L}(f) + \varepsilon_{\rm R}(f) = I_{3f} - 2Q_f x_{\rm W}\,, \quad a_f = \varepsilon_{\rm L}(f) - \varepsilon_{\rm R}(f) = I_{3f} \tag{2.70e}$$

The $\chi(s)$ is the Z propagator where we included the decay width for phenomenological treatment. $Q_e = -1$, $v_e = v_f(f = e)$, $a_e = a_f(f = e)$ are the electron part of the vector and axial vector coupling constants. Below, we give formulas for the R-value which is the ratio of the hadronic total cross section relative to that of muon pair production in QED and the asymmetry $A_{\rm FB}$ of the angular distribution of the final fermion pair.

$$R = \frac{\sigma(e^-e^+ \to \text{hadrons})}{\sigma(e^-e^+ \to \mu^-\mu^+)} = \sum_q R_q F_1 \tag{2.71a}$$

$$R_q = N_c^q \frac{\sigma(e^-e^+ \to q\bar{q})}{\sigma(e^-e^+ \to \mu^-\mu^+)} = 3(1+\delta_{\rm QCD})\sum_q |Q_q|^2 \tag{2.71b}$$

$$A_{\rm FB} \equiv \frac{\int_0^1 d\sigma - \int_{-1}^0 d\sigma}{\int_0^1 d\sigma + \int_{-1}^0 d\sigma} = \frac{3}{4}\frac{F_2}{F_1} \tag{2.71c}$$

where the integration is over $x = \cos\theta$.

Lepton asymmetry A^l_{FB} The Z-pole term is the dominant one at LEP ($\sqrt{s} = m_Z =$ 91 GeV), but can be neglected at PEP, PETRA and TRISTAN where $\sqrt{s} = 30 \sim 60$ GeV compared to the photon and interference terms. Note that $v_e = -1/2 + 2x_W$ nearly vanishes because $x_W = \sin^2\theta_W$ is accidentally very close to 1/4. Hence, $F_1 \approx 1$ and the weak interaction effect hardly appears in the total cross section or R-value below $\sqrt{s} \sim 60$ GeV. However, the asymmetry is large.

$$A_{FB} \sim -\frac{3}{2}a_e a_l |\chi(s)| \cos\delta(s) = -\frac{3}{8}|\chi(s)| \cos\delta(s) \tag{2.72}$$

Figure 2.18 shows the angular distributions for $e^-e^+ \to \mu^-\mu^+$ and $e^-e^+ \to \tau^-\tau^+$ obtained by the VENUS group at TRISTAN. Their distributions are asymmetric and the data agree with the prediction of the GWS theory. Figure 2.19 shows a compilation of all the asymmetry data in $e^-e^+ \to \mu^-\mu^+$ including them and those beyond the Z peak [5]. The line shows the prediction of the Standard Model which is given by Eq. (2.71c) and Figure 2.16. The data agree with the Standard Model.

Isospin of τ, c, b Having confirmed that the asymmetry in $e^-e^+ \to \mu^-\mu^+$ exactly follows the prediction of the Standard Model, we could apply the same argument to other particles, τ, c and b. Here, we choose to discuss the data from a somewhat different point of view and try to determine their isospin. At the time when the GWS theory was proposed, the known isodoublets were (u, d'), (ν_e, e^-), (ν_μ, μ^-), which were known to interact weakly via the exchange of $W^\pm$, but whether (c, s') really makes a doublet was not thoroughly tested. The top quark was not discovered yet and the possibility that the τ and/or b are singlet members was not out of the question.

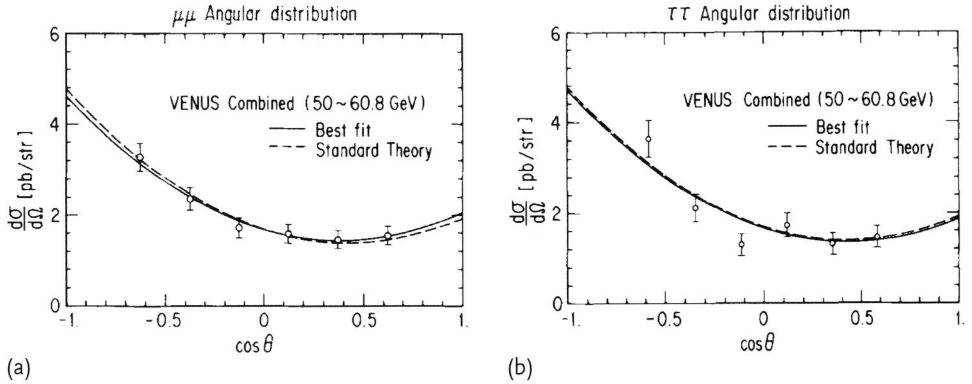

Figure 2.18 The differential cross section of $e^-e^+ \to \mu^-\mu^+$ (a) and $\tau^-\tau^+$ (b) obtained by the VENUS group at TRISTAN ($\sqrt{s} = 56.6$ GeV). The solid curve is a fit to the data and the dashed line is the prediction of the Standard Model [79]. For a color version of this figure, please see the color plates at the beginning of the book.

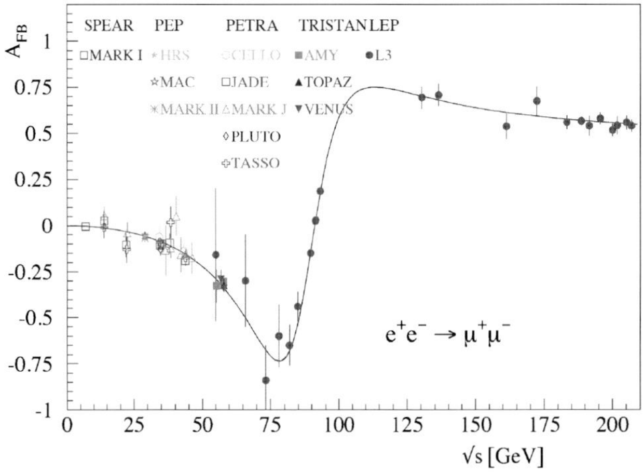

Figure 2.19 A compilation of the asymmetry data A^{μ}_{FB} compared to the Standard Model prediction [5, 9].

The τ signal

Experimentally, a clear τ signal can be obtained from an acoplanar $\ell\bar{\ell}'$ pair where $\ell, \ell' = e, \mu$. Production and decay of the τ pair go through reactions:

$$
\begin{aligned}
&e^+e^+ \to \tau^- + \tau^+ \\
&\qquad \tau^- \to \nu_\tau + e^-(\mu^-) + \overline{\nu}_e(\overline{\nu}_\mu) \quad \text{or} \to \nu_\tau + \text{hadrons} \\
&\qquad \tau^+ \to \overline{\nu}_\tau + e^+(\mu^+) + \nu_e(\nu_\mu) \quad \text{or} \to \overline{\nu}_\tau + \text{hadrons}
\end{aligned} \tag{2.73}
$$

The incoming electron pair and the outgoing τ pair define a plane. The decay leptons are not on the plane, and hence the lepton pair ($e\overline{e}$ or $\mu\overline{\mu}$, or their combination) makes an acoplanar lepton pair which is a clean τ pair production signal. However, the branching ratio is small. A majority of the decay is semileptonic and has a topology in which a single energetic track (lepton or pion with direction $\hat{s}$) is emitted in one hemisphere and several tracks (a jet) are emitted in the other hemisphere [see Figure 2.20]. The jet axis (direction $\hat{e}_{\mathrm{T}}$ referred as the thrust axis) can be defined.[16] The two hemispheres are separated by a plane perpendicular to the thrust axis (denoted by a dashed line). The energetic single track and the jet most likely retain their parent τs direction and are in back to back configuration. Therefore, their direction defines the emission angle of the τ pair.[17] An example of event display at LEP is given in Figure 4.5.

16) The thrust $\hat{e}_T$ is defined as the unit vector which maximize

$$T = \sum_I (\boldsymbol{p}_i \cdot \hat{e}_T) / \sum_i |\boldsymbol{p}_i| \tag{2.74}$$

where sum is over particle i in the jet. Its detailed treatment is given in Section 9.1.2.

17) To be more specific, $\cos\theta$ is defined by $(-q\hat{s}\cdot\hat{z})(\hat{e}_T\cdot\hat{z})/|\hat{s}\cdot\hat{e}_T|$ where q is the charge of the energetic single track and $\hat{z}$ is a unit vector in electron beam direction.

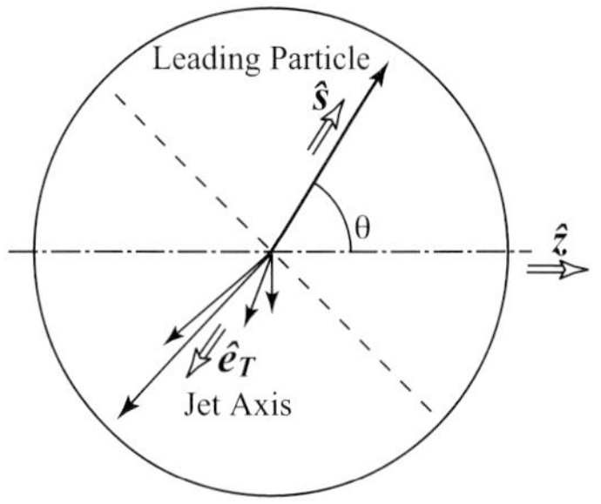

Figure 2.20 Jet axis.

As to the τ, we already saw that the asymmetry agrees well with the prediction of the Standard Model which means $I_3(\tau) = -1/2$ and is a member of an isodoublet. If it has other values like 0 or $+1/2$, the asymmetry would be quite different. As to c, b, there is additional difficulty because of the quark confinement, namely, selecting and identifying their clean samples.

c asymmetry: A^c_{FB} There was a strong indication that the charm quark is a member of an isodoublet because its existence was required by the GIM mechanism (see Vol. 1, Sect. 16.7.1) and was supposed to couple to s'. Still, it is worthwhile to check it by looking at the asymmetry. Measuring the asymmetry in $c\bar{c}$ and/or $b\bar{b}$ is not trivial, that is, because they can only be observed as jets due to confinement. The jets show their conspicuous characteristics at high energies which are addressed in Chapter 9. At low energies, they just began to show their jet-like behavior and it is rather difficult to apply conventional methods which are routinely used nowadays at higher energies.

Fortunately, in the case of the charm quark, there are characteristic decay modes that one can use. Consider a case where the production of an energetic D^* hadron, which is in a $J^P = 1^-$ state of $c\bar{u}$ or $c\bar{d}$ composite, was observed. It is likely that the primary c-quark which is produced in $e^- e^+ \to c\bar{c}$ has picked up $\bar{u}$ or $\bar{d}$ from the vacuum and hadronized without loosing much of its energy. Then, the D^* angular distribution can be considered as representing that of the charm quark. The D^* can be identified by its decay mode $D^* \to D^0 \pi_s$, $D^0 \to K^- \pi^+, K^- \pi^+ (\pi^0), K^- \pi^+ \pi^- \pi^+$. Figure 2.21 shows an angular distribution in which the final decay products are fully reconstructed.

However, full reconstruction of the decay products suffers from poor statistics. Since the D^* decay mode has very low Q-value ($Q = m(D^*) - m(D^0) - m_{\pi_s} = 5.8$ MeV), the pion with suffix "s" is produced almost at rest in its parent's rest system. Therefore, it flies with the D^0 side by side, that is, the π_s's direction can be identified with that of D^*.

Figure 2.22 shows a p_T distribution of all the particles with respect to the jet axis. Jet reconstruction algorithm will be discussed in Chapter 9. Here, suffice it to say, a jet is a bunch of particles flying more or less in the same direction (see inset

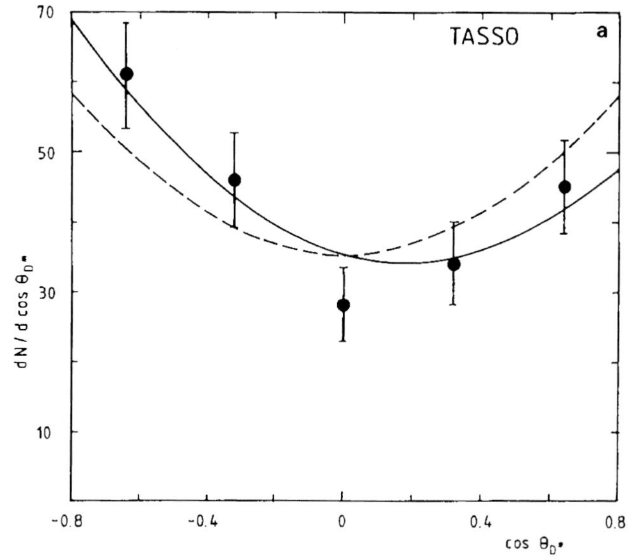

Figure 2.21 The D^{*+} production angular distribution for reactions $D^{*+} \to D^0\pi_s$, $D^0 \to K^-\pi^+, K^-\pi^+\pi^-\pi^+$. θ is the angle between the directions of the incoming e^- and the outgoing $D^0\pi^+$ system. The solid curve represents the result of the fit. The dashed curve shows a line of $1 + \cos^2\theta$ [80, 83].

in Figure 2.22). One can see a conspicuous forward peak showing the π_s from D^* decays. Figure 2.22 is the angular distribution of the jet axes which includes a particle with $p_T^2 < 0.0075\,(\text{GeV}/c)^2$.

From these angular distributions, the TASSO group derived [80]

$$A_{FB}^c = -0.16 \pm 0.072_{\text{stat}} \pm 0.011_{\text{sys}} \quad \sqrt{s} = 36.2\,\text{GeV} \tag{2.75a}$$

$$A_{FB}^c = -0.168 \pm 0.047_{\text{stat}} \pm 0.027_{\text{sys}} \quad \sqrt{s} = 35.0\,\text{GeV} \tag{2.75b}$$

Since the two methods are statistically independent, they are combined. Then, the value of $a_e a_c$ can be extracted which gives

$$a_e a_c = -0.276 \pm 0.073 \tag{2.76}$$

This agrees well with the Standard Model value of $-1/4$ and has confirmed that indeed, the c-quark has $I_3 = +1/2$.

b-asymmetry: A_{FB}^b Identification of the b-jet is easier compared to the c-quark because of its large mass ($\sim 5\,\text{GeV}$) as compared to the c-quark ($\lesssim 2\,\text{GeV}$). It is much heavier than u, d, s whose mass values are less than 0.2 GeV (see Table 7.1). As a result, decay leptons from $b \to c l \nu_l$ have maximum momentum $\sim 5/2 = 2.5\,\text{GeV}/c$ in the b quark rest frame. In the laboratory frame, the transverse component is conserved if boosted along the b flight direction and has large transverse momentum $p_T \approx 1 \sim 2\,\text{GeV}/c$. In contrast to this, decay leptons from the light quarks have

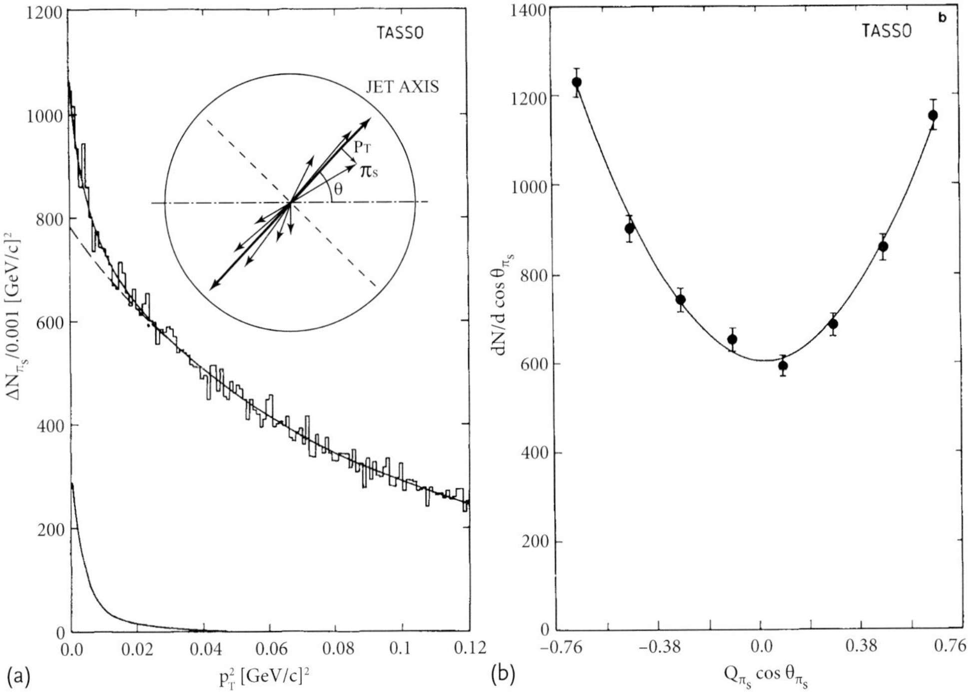

Figure 2.22 (a) Distribution of the transverse momentum squared of all particles with respect to the thrust axis. Forward peak corresponds to D^*. The contributions of the signal and the background components are also shown separately. The inset illustrates the jet axis and p_T. (b) Jet axis distribution of samples from (a) accompanied by a π_s with $p_T^2 < 0.0075\ (\text{GeV}/c)^2$. The solid curve is a fit [80].

$p_T \ll 1\,\text{GeV}/c$. The c-quark is in between and its value marginal, but the method still works, though with less efficiency and more background. Notice that the sign of the leptons from the b-decay is opposite to that from the c-decay which helps to discriminate b from c. The cascade decay $b \to c \to l$ gives the wrong sign and can be misidentified as that of genuine c, but the low p_T value helps here again.

Therefore, if there are leptons which have large transverse momentum with respect to the jet axis, probability for b being its origin is high. The AMY and VENUS group used this fact to select b quarks and obtained

$$A_{\text{FB}}^b = -0.72 \pm 0.28_{\text{stat}} \pm 0.13_{\text{sys}} \quad \text{AMY [81]} \tag{2.77a}$$

$$= -0.64 \pm 0.26_{\text{stat}} \pm 0.07_{\text{sys}} \quad \text{VENUS [82]} \tag{2.77b}$$

These values are consistent with the Standard Model value −0.56. Note, however, the above observed values have to be corrected for the B^0–$\overline{B}^0$ mixing effect. To explain it simply, there is $B^0(\bar{b}q) - \overline{B}^0(b\bar{q})(q = d, s)$ mixing and some of the b and $\bar{b}$ quarks were its products[18]. It is similar to $K^0 - \overline{K}^0$ oscillation (see Vol. 1,

18) The phenomena will be discussed in detail in Chapter 6.

Sect. 16.1.3), but the mixing rate is much faster. Therefore, the observed asymmetry has to be corrected to give the true asymmetry by the relation

$$A^b_{\mathrm{FB}}(\text{observed}) = (1 - 2\chi) A^b_{\mathrm{FB}}(\text{no mixing}) \tag{2.78a}$$

$$\chi = 0 \cdot f_u + \chi_d f_d + \chi_s f_s \tag{2.78b}$$

where $\chi_q(q = d, s)$ are mixing rates of $B^0_d(= \overline{b}d)$ and $B^0_s(= \overline{b}s)$. f_u, f_d, f_s are the probability to pick up u, d, s quarks from the vacuum which is the probability that the b quarks become B^+, B_d, B_s and are estimated to be 0.4,0,4,0,1. Inserting observed values for $\chi_d = 0.188$, $\chi_s = 0.5$ [11], we have $(1 - 2\chi) = 0.75 \pm 0.01$, which slightly improves the value of the asymmetry. The fact that the asymmetry values agreed with the Standard Model was used to assign $I_3(b) = -1/2$ to the b-quark.

However, this logic assumes that the right-handed b-quark is an isospin singlet which was poorly tested. If one makes both $I_{3\mathrm{L}}(b)$, $I_{3\mathrm{R}}(b)$ as free parameters, the Z-coupling in the Standard Model is modified to

$$v_b = I_{3\mathrm{L}b} + I_{3\mathrm{R}b} - 2Q_b \sin^2\theta_{\mathrm{W}}\,, \quad a_b = I_{3\mathrm{L}b} - I_{3\mathrm{R}b} \tag{2.79}$$

Since the width and asymmetry at the Z resonance are given by (see Eqs. (4.11c) and (4.45c))

$$\Gamma(Z \to b\overline{b}) = \frac{G_{\mathrm{F}} m_Z^3}{2\sqrt{2}\pi}\left(v_b^2 + a_b^2\right) \tag{2.80a}$$

$$A_{\mathrm{FB}}(\sqrt{s} = m_Z) = \frac{3}{4}\frac{2v_e a_e}{\left(v_e^2 + a_e^2\right)}\frac{2v_b a_b}{\left(v_b^2 + a_b^2\right)} \tag{2.80b}$$

the observed asymmetry only defines a band in the $I_{3\mathrm{L}} - I_{3\mathrm{R}}$ plane. Although there had been mounting circumstantial evidence for the existence of the top quark,[19] it was not discovered by direct production until 1994. Thus, there was a considerable interest to prove if the left-handed b-quark really has $I_3 = -1/2$. Figure 2.23 shows one such attempt using the early LEP data and the A^b_{FB}, which restricts the allowed region to a narrow circle on the $I_{3\mathrm{L}} - I_{3\mathrm{R}}$ plane [87]. The asymmetry at LEP restricts the region between two lines which crosses near the center. Since $(I_{3\mathrm{L}}, I_{3\mathrm{R}})$ has to be located at discrete points of half integers, this leaves two-fold ambiguity for $(I_{3\mathrm{L}}, I_{3\mathrm{R}}) = (-1/2, 0)$ or $(1/2, 0)$. The asymmetry at the lower energy region was good enough to resolve it. From the experimental data [7] at LEP as well as PEP, PETRA, and TRISTAN [88–90],

$$\begin{aligned} I_{3\mathrm{L}}(b) &= -0.490^{+0.015}_{-0.012} \Rightarrow I_{3\mathrm{L}}(b) = -1/2 \\ I_{3\mathrm{R}}(b) &= -0.028 \pm 0.056 \Rightarrow I_{3\mathrm{R}}(b) = \;\; 0 \end{aligned} \tag{2.81}$$

19) For instance, a singlet b breaks the GIM mechanism, allowing the flavor changing neutral current. The ratio $\mathrm{BR}(B \to l^+ l^- X)/BR(B \to l\nu_l X)$ could be as large as 0.3 which was denied by the CLEO and UA1 experiment [84–86].

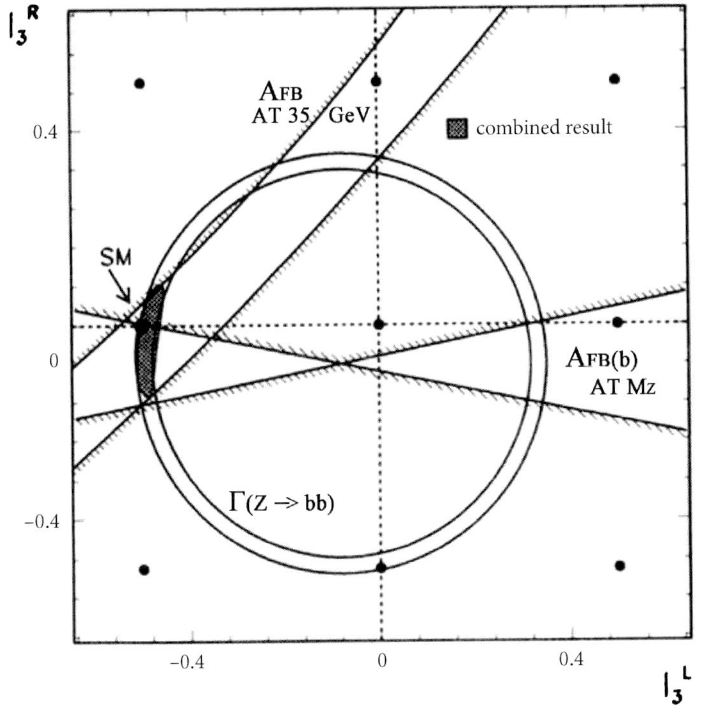

Figure 2.23 Experimental constraints for the isospin of the b-quark in the $(I_{3L} - I_{3R})$ plane. The circle and the pair of straight lines centered at $\sin^2\theta_W = -0.08$ represent 1σ bounds from $\Gamma(Z \to b\bar{b})$ and $A_{FB}(b)$ at the Z resonance. The grid points mark the allowed half integer values of I_{3L}, I_{3R}. Experimentally, the $b\bar{b}$ asymmetry at PETRA, PEP, and TRISTAN (marked as A_{FB} at 35 GeV) resolved the ambiguity between the two solutions at $I_{3L} = \pm\frac{1}{2}$ [87].

This implies that the b-quark must have a weak isospin partner, that is, the top quark with $I_{3L}(t) = +1/2$ has to exist. However, the mass could not be predicted.

In summary, the existence of the top quark within the Standard Model had firm foundations experimentally before its discovery.

A^c_{FB}, A^b_{FB} at LEP At Z resonance, the asymmetry measurement was drastically improved for two reasons. The energy is much higher and the cross section is much larger ($\sim 500\times$). Being at high energies, two more methods, lifetime tagging and jet tagging, became available to identify c and b in addition to D^* tagging and lepton-tagging which we mentioned. They are subjects of the Z physics at the resonance. The heavy quark tagging will be discussed in Section 4.3.4 and the asymmetry in Section 4.4.4 in more detail.

2.7
Asymmetry in $q + \overline{q} \to l + \overline{l}$

We can measure the asymmetry of the reaction

$$q + \overline{q} \to \gamma^*/Z \to e^- + e^+ \tag{2.82a}$$

$$\to \mu^- + \mu^+ \tag{2.82b}$$

using hadron colliders. Equation (2.82b) is the familiar Drell–Yan process which will be discussed in more detail in Section 8.6. It is the reverse reaction of $e^- + e^+$(or $\mu^- + \mu^+$) $\to q + \overline{q}$ and its asymmetry expression is given exactly by the same formula. The quark flavor in the hadronic reaction consists mainly of u and d, while in the electron collider, the extracted quark flavor is mainly c, b because of the difficulty in separating the light quarks.[20] Therefore, the asymmetry measurements in the hadron collider provide complementary information to that of $e^-e^+ \to q\overline{q}$. Its advantage lies in its ability to determine the asymmetry (and the differential cross section) at the highest energies in the wide range not reachable by electron colliders. For the same reason, it is a powerful tool for probing possible new physics effects beyond the Standard Model [91–95].

The Drell–Yan process within the parton model was discussed in Vol. 1, Sect. 17.9. Setting the energy-momenta of the participating particles as

P_A, P_B	Incoming proton and antiproton momenta
$p_a = x_a P_A, p_b = x_b P_B$	Incoming quark–antiquark momenta
k_1, k_2	Produced leptons' momenta

$$Q = (Q_0, \boldsymbol{Q}_T, Q_z) = k_1 + k_2$$
$$s = (P_A + P_B)^2, \quad \hat{s} = (p_a + p_b)^2 \simeq x_a x_b s \tag{2.83}$$

where $s \gg Q^2 \gg Q_T^2$ are assumed and all the particles are treated massless. $Q_T = |\boldsymbol{Q}_T|$ is the transverse momentum of the lepton pair relative to the beam (z) axis. The cross section for producing the lepton pair is given by

$$\frac{d\sigma}{dQ^2 d\Omega} = \sum_{a,b} \int dx_a dx_b \left[f_A^a(x_a) f_B^b(x_b) + (a \leftrightarrow b) \right]$$
$$\times \delta\left(Q^2 - x_a x_b s\right) \frac{d\sigma}{d\Omega}(ab \to l\overline{l}; \hat{s} = Q^2) \tag{2.84}$$

Here, a quark in the proton A having fractional momentum $p_a = x_a P_A$ of the parent proton interacts with an antiquark in the antiproton B which has fractional momentum $p_b = x_b P_B$ of the parent antiproton. $f_A^a(x_a)$, $f_B^b(x_b)$ are momentum distribution functions of the quarks with flavor 'a', 'b' inside hadrons and are known functions in QCD referred to as the parton distribution function (pdf). Details of

20) Note that the Weinberg angle obtained from the deep inelastic scattering [for instance, that of NuTeV in Eq. (2.41) [64]] also determines the coupling of the Z boson to the u and d quarks.

the parton distribution function are discussed in Chapter 8, but here, suffice it to say, higher order effects like soft gluon emissions by the initial partons are already taken into account in the QCD corrected parton distribution function. The hard gluon emission is separated (factorization) and treated differently in QCD.

Actually, the lepton pairs are produced with some finite transverse momentum (Q_{T}) relative to the beam direction due to associated gluon emissions. However, as long as $Q_{\mathrm{T}}^2 \ll Q^2$, the transverse momentum is a result of recoils to soft gluon emission whose effect is in the distribution function. Therefore, we may proceed treating the lepton pair as products of $q + \overline{q} \to \gamma^*/Z \to l + \overline{l}$ process where $q\overline{q}$ direction is assumed parallel to the beam.[21]

For measurements of the asymmetry, however, we need to define the emitted lepton pair angle produced with finite Q_{T}. To minimize the effect of the transverse momentum in the extraction of the asymmetry, the standard choice is to adopt the Collins–Soper frame (CS frame) [96], where the z-axis is defined in the rest frame of the lepton pair as the bisector of the incoming beam momentum $\boldsymbol{P}'_A$ and negative of the target momentum $-\boldsymbol{P}'_B$ (Figure 2.24a). The sine/cosine of the polar and azimuthal angle in the Collins–Soper frame can be defined by using variables in the CM frame ($\boldsymbol{P}_A = -\boldsymbol{P}_B$) as

$$\cos\theta = 2Q^{-1}\left(Q^2 + Q_{\mathrm{T}}^2\right)^{-1/2}\left(k_1^+ k_2^- - k_1^- k_2^+\right) \tag{2.85a}$$

$$\sin^2\theta = Q^{-2}\boldsymbol{\Delta}_{\mathrm{T}}^2 - Q^{-2}\left(Q^2 + Q_{\mathrm{T}}^2\right)^{-1}(\boldsymbol{\Delta}_{\mathrm{T}} \cdot \boldsymbol{Q}_{\mathrm{T}})^2 \tag{2.85b}$$

$$\tan\phi = \frac{\left(Q^2 + Q_{\mathrm{T}}^2\right)^{1/2}}{Q}\frac{\boldsymbol{\Delta}_{\mathrm{T}} \cdot \hat{\boldsymbol{R}}_{\mathrm{T}}}{\boldsymbol{\Delta}_{\mathrm{T}} \cdot \hat{\boldsymbol{Q}}_{\mathrm{T}}} \tag{2.85c}$$

where

$$k^{\pm} = (k^0 \pm k_z)/\sqrt{2} \tag{2.86a}$$

$$Q = k_1 + k_2\,, \quad \Delta = k_1 - k_2 \tag{2.86b}$$

$$\hat{\boldsymbol{R}} = \frac{\boldsymbol{P}_A \times \boldsymbol{Q}}{|\boldsymbol{P}_A \times \boldsymbol{Q}|}\,, \quad \hat{\boldsymbol{Q}}_{\mathrm{T}} = \frac{\boldsymbol{Q}_{\mathrm{T}}}{|\boldsymbol{Q}_{\mathrm{T}}|} \tag{2.86c}$$

and the azimuthal angle ϕ is defined on the $x'y'$ plane in Figure 2.24a. Notice that the right-hand side of Eq. (2.85a) are manifestly invariant under z boosts, and the equations can be applied equally well using laboratory frame variables. Proof of Eq. (2.85) is given in Appendix F.

21) The soft gluon emission is treated by the DGLAP equation and makes the distribution an evolving function of Q^2. On the other hand, extraction of quarks from jets with effective mass Q^2 produced in the $e^- e^+$ reaction used the fragmentation function. Treatment of the parton distribution and the fragmentation function is symmetric in the sense that they both obey the same DGLAP equation despite the fact that the relevant variables are spacelike in the former and timelike in the latter. See Chapter 8 for a more detailed discussion.

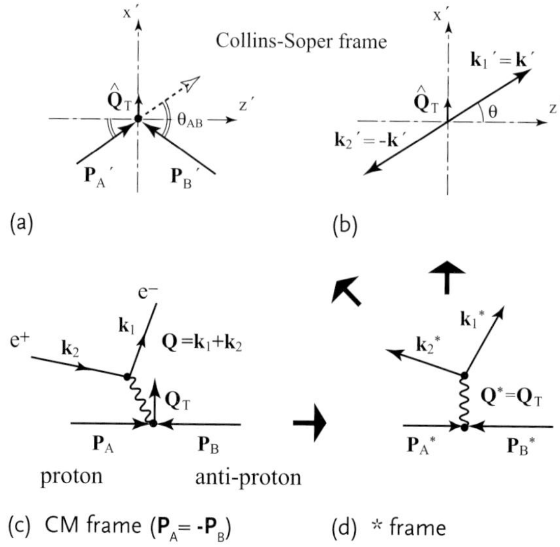

Figure 2.24 Kinematics defined in the (c) CM frame, (d) $*$ frame and (a, b) the Collins–Soper frame. The CS frame is the center of mass frame of the produced lepton pair, (b) but not of the incoming particles (a). Bold arrows: The Collins–Soper frame can be obtained by boosting the CM frame longitudinally to obtain the $*$ frame, and then Lorentz transforming the $*$ frame transverse to the beam axis.

The CS frame can be reached in two steps (Figure 2.24). First, starting from the proton–antiproton center of mass frame (CM frame), boost along the z-axis to an intermediate "$*$ frame" in which $Q_z^* = 0$ or $\boldsymbol{Q}^* = \boldsymbol{Q}_T$. The vectors $\boldsymbol{P}_A^*$ and $\boldsymbol{P}_B^*$ remain parallel to the z^* axis while $\boldsymbol{Q}_T^* = \boldsymbol{Q}_T$ remains unchanged. Second, boost in the $-\boldsymbol{Q}_T$ direction through boost parameters $\beta\gamma = |\boldsymbol{Q}_T|/Q, \gamma = \sqrt{Q^2 + \boldsymbol{Q}_T^2}/Q$. In the CS frame where variables are denoted with a dash, $\boldsymbol{Q}' = 0$ and $Q_0' = Q$. The vector $\boldsymbol{P}_A'$ and $\boldsymbol{P}_B'$ make equal angle $\theta_{AB}/2 = \tan^{-1}(|\boldsymbol{Q}_T|/Q)$ with the z' axis.

The forward–backward asymmetry in the hadronic reaction is defined similarly to the $e^-e^+ \to f\bar{f}$ reaction:

$$A \equiv \frac{\sigma_F - \sigma_B}{\sigma_F + \sigma_B} \tag{2.87}$$

in which $\sigma_{F/B}$ is the integrated cross section for the variable range $\cos\theta > 0$ and $\cos\theta < 0$, respectively. The asymmetry measured experimentally is given by

$$A_{FB} = \frac{N_F - N_B}{N_F + N_B} \tag{2.88}$$

in which $N_{F/B}$ is the acceptance, efficiency and background corrected lepton yields in the forward and backward directions respectively. Measuring the asymmetry rather than differential cross sections allows cancellation of many systematic uncertainties, particularly those affecting the overall normalization.

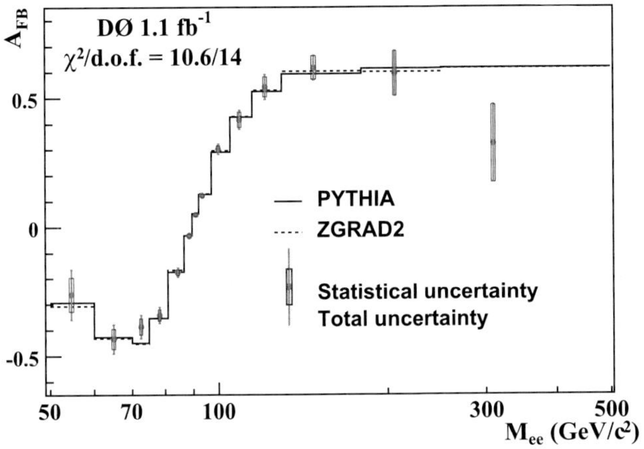

Figure 2.25 Dielectron forward–backward asymmetry as a function of dielectron mass. Comparisons are made between the unfolded AFB (points) and the PYTHIA (solid curve) and ZGRAD2 (dashed line) predictions. The inner (outer) vertical lines show the statistical (total) uncertainty. Compare this with Figure 2.19. Figure taken from [98, 99].

Experiments are carried out in the Tevatron [97–100]. The D0's result is shown in Figure 2.25.

The M_{ee} dependence of the asymmetry can be directly compared with the $\sqrt{s}$ dependence measured in the electron collider which is given in Figure 2.19. One notices the entire range of the asymmetry measured in the electron collider between $\sqrt{s} = 50$–200 GeV and beyond is covered by a single experiment in the hadron collider.

Using the hadronic lepton pair production asymmetry, the value of the Weinberg angle (including the higher order calculations) can be obtained:

$$\sin^2\theta_W^{\text{eff}} = \begin{cases} 0.2238 \pm 0.0040 & \text{CDF} \quad [97] \\ 0.2326 \pm 0.0018_{\text{stat}} \pm 0.0006_{\text{sys}} & \text{D0} \quad [98] \end{cases} \tag{2.89a}$$

which is to be compared with the world average:

$$\sin^2\theta_W^{\text{eff}} = 0.23149 \pm 0.00013 \quad [97] \tag{2.90}$$

Notice, the precision is coming near to that of other means and with more statistics which is coming soon, the hadronic data is expected to give the most precise values of the Weinberg angle for light quarks.

3
W

3.1
Discovery of W, Z

The Glashow–Weinberg–Salam (GWS) model seemed to have obtained a firm ground experimentally by discovery of the neutral current interaction. However, there were other models that predicted existence of the neutral current. The definitive evidence was the discovery of the weak force carriers, W and Z, with the right mass. Applying the GWS model to the neutral current reactions, one can obtain the value of the Weinberg angle $\sin^2\theta_W$ which, in turn, fixes the mass value of the gauge bosons to be around 80 GeV. The proton synchrotron accelerators (PS) with the highest energies ($\sim$ 400 GeV) in operation at that time were Fermilab PS and CERN SPS (super proton synchrotron), but they were not colliders. If one uses fixed targets, the required accelerator energy is $\sqrt{s} \sim \sqrt{2m_p E} \gtrsim m_W \rightarrow E > 3200\,\text{GeV}$. Therefore, a collider accelerator was the only choice. The only proton collider at that time was the ISR (Intersecting Storage Ring) whose energy $\sqrt{s} = 32.5 + 32.5\,\text{GeV}$ was not high enough. There were two accelerators in the planning stage: ISABELLE (pp collider with $\sqrt{s} = 400\,\text{GeV}$[1]) and LEP (large Electron Positron Collider (e^-e^+ collider with $\sqrt{s} = 100\,\text{GeV}$). They had enough energy, but its planned operation was scheduled for the late 1980s, too far into the future. Rubbia and company thought that the fixed target SPS machine could be converted to a collider [101]. Two beams are needed for a collider, but there was only one pipe for the beam in the fixed target accelerator. If one wants to circulate two beams in the opposite direction in one accelerator, the other beam has to have opposite charge, that is, the particle has to be the antiproton. The collider has far less luminosity[2] by its construction compared to fixed target accelerators. Unlike protons, the antiprotons have to be produced as secondary particles by bombarding the proton beam on a target. Unlike the primary proton beam, they are dispersed and have a continuous energy spectrum. In other words, the secondary antiproton flux is

1) It was canceled in 1983, but revived in 1990 as RHIC (Relativistic Heavy Ion Collider).
2) The luminosity L is defined by the equation $N = L\sigma$ where N is the event occurrence rate and σ is the cross section of the reaction. See Vol. 1, Sect. 12.2 for comparison with the fixed target accelerator.

Elementary Particle Physics, Volume 2: Foundations of the Standard Model, First Edition. Yorikiyo Nagashima.

not an ideal beam suitable for acceleration.[3] The difficulty was solved by Van de Meer's proposal of stochastic beam cooling and storing[4] [102, 103]. CERN adopted it, transformed SPS to $p\overline{p}$ collider ($Sp\overline{p}S$ with $\sqrt{s} = 540\,\text{GeV}$)[5] which led to the discovery of *W* and *Z*. Because of their accomplishment, Rubbia and Van de Meer received the Nobel Prize in 1984.

UA1 detector The $W^{\pm}$ and *Z* can be produced by reactions

$$\begin{aligned} &u+\overline{d} \to W^+\,, \quad u+\overline{s} \to W^+\,, \quad d+\overline{u} \to W^-\,, \quad s+\overline{u} \to W^+ \\ &u+\overline{u} \to Z\,, \quad d+\overline{d} \to Z\,, \quad s+\overline{s} \to Z \end{aligned} \tag{3.1}$$

Their decay modes are reverse actions of Eq. (3.1) plus leptonic branches:

$$\begin{aligned} W^- &\to e^-\overline{\nu}_e, \mu^-\overline{\nu}_\mu, \tau^-\overline{\nu}_\tau\,, \quad \overline{u}d', \overline{c}s' \\ W^+ &\to e^+\nu_e, \mu^+\nu_\mu, \tau^+\nu_\tau, \quad u\overline{d}', c\overline{s}' \\ Z &\to e^+e^-, \mu^+\mu^-, \tau^+\tau^-, \nu_i\overline{\nu}_i \ (i = e, \mu, \tau)\,, \quad u\overline{u}, d\overline{d}, c\overline{c}, s\overline{s}, b\overline{b} \end{aligned} \tag{3.2}$$

where d', s' are Cabibbo rotated d, s. Production of a (tb') pair is prohibited because of high top mass (it was not discovered at that time). Production of the *W*, *Z* can be verified by identifying their decay modes. Branching rate to each pair is equal in the massless limit, except that the quarks get an enhancement factor three because of the color degrees of freedom. The lepton's share is 3 while that of the quarks is 2×3. As the quarks appear as jets, it is hard to distinguish their flavor and accurate reconstruction of the kinematic variables is more difficult. The lepton's branching ratio is smaller, but identification as well as reconstruction of the kinematics is rather straight forward, as is obtaining clean samples with a small background. Because of this, it is customary in the high energy experiments to put more emphasis on the lepton identification in the detector design.

Reconstruction of the *W*, *Z* variables can be made by identifying the two decay particles and measuring their energy-momenta $(E_1, \boldsymbol{p}_1)$, $(E_2, \boldsymbol{p}_2)$. Their parent's energy momenta $(E_V, \boldsymbol{p}_V)$ and mass can be reconstructed using the formulas

$$\begin{aligned} &E_V = E_1 + E_2\,, \quad \boldsymbol{p}_V = \boldsymbol{p}_1 + \boldsymbol{p}_2 \\ &m_V^2 = (E_1 + E_2)^2 - (\boldsymbol{p}_1 + \boldsymbol{p}_2)^2 \end{aligned} \tag{3.3}$$

Two general purpose detectors UA1 and UA2 were constructed in order to discover the *W* and *Z*. They were the first major mainstream collider detectors, that

3) A good beam contains enough particles which have fixed velocity and directions, or more professionally stated, has small emittance. Denoting deviations of the position x and direction $x' = dx/dz$ of the particles from that of the central orbit as $\Delta x(\Delta y)$, $\Delta x'(\Delta y')$, it is defined as the phase space volume in which they are distributed. Here, z is defined as the beam direction and x, y are perpendicular to it.

4) The cooling means reduction of the beam phase space volume, that is, emittance. A random and large transverse motion of particles in the beam is, figuratively speaking, supposed to be at a high temperature. In the stochastic cooling, information on the x, x' and y, y' is read at one point of the beam orbit and fed to kickers located at another point to correct it.

5) Later upgraded to 670 GeV.

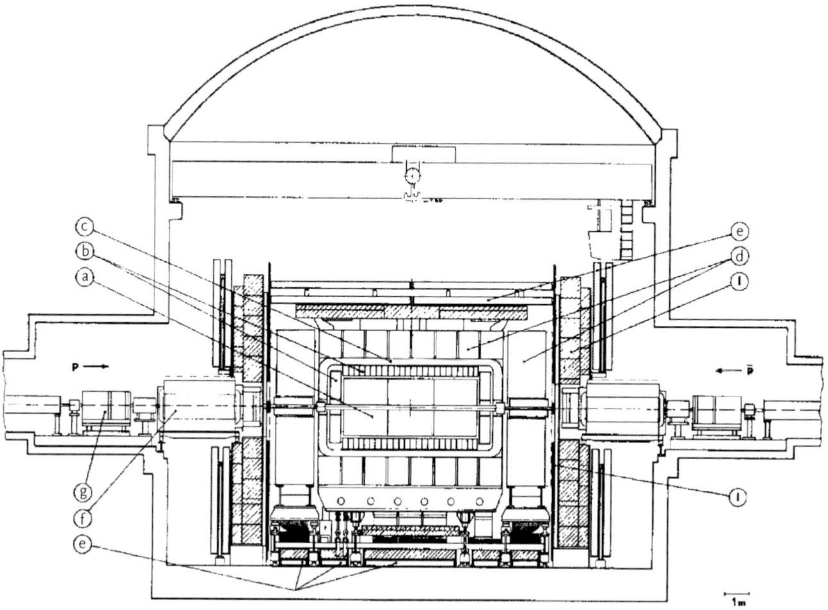

Figure 3.1 Side view of the UA1 detector in 1983: (a) central tracking detector, (b) electromagnetic calorimeter, (c) coil, (d) hadronic calorimeter, (e) muon chambers, (f) forward calorimeters, (g) very forward calorimeters. Note, the magnet is of dipole type and the field is horizontal [104, 105].

is, high energy measuring devices whose construction was made possible by organized efforts of large groups. Figure 3.1 illustrates the UA1 detector. Arrangement of the detector components is typical of the conventional collider detectors, but unlike most of its successors which have cylindrical symmetry around the beam, UA1 used a dipole magnet to generate a horizontal magnetic field of 7 kG (0.7 T) [104, 106]. In the following, we describe basic principles of particle identification of the collider detector (see Vol. 1, Sects. 12.5 and 12.6 for more details).

a. electrons and photons: Electrons are charged and their momentum can be determined by measuring the curvature of tracks they make in the central tracking detector. They also make showers in the electromagnetic calorimeter (ECAL) and the total energy is measured. The ratio E/p should be one for electrons, but pions which are the overwhelming source of backgrounds give values far less than one because they deposit very little energy in the ECAL. It is made of lead/scintillator plates with a total thickness of $\sim$ 30 radiation lengths, but only a few interaction lengths. The photons (mostly from π^0 decays) behave similarly to the electrons, except they do not leave tracks in the tracking chamber.
b. hadrons: Hadrons make showers not in the electromagnetic calorimeter, but in the hadronic calorimeter (HCAL). The HCAL is a combination of iron and thin plate scintillators in sandwich formation. In the UA1 detector, the

iron layers play a double role of magnetic flux return yoke. Almost all of the hadrons are pions. Their energy and direction can be measured.
c. muons: The muons only interact electromagnetically with detector materials and unlike electrons, they do not shower because they are much heavier. They lose energy only by ionizing atoms in their path. Therefore, they penetrate all the detector materials and are recorded in the muon chamber located at the periphery of the whole apparatus.
d. neutrinos: The neutrinos are recorded as invisible energy (denoted as $\not{E}$). In the center of mass frame, the sum of the momenta should vanish. Therefore, the minus of the sum of all the visible momentum vectors, including charged as well as neutral particles, gives the missing momentum. When it is known that only one neutrino is emitted (in this case, the missing mass should center around zero), it gives the neutrino energy. As identification of all the particles is impossible, all the hadrons are regarded as pions or massless particles and their energy vectors instead of momentum are used in the reconstruction.

In order to measure the missing energy accurately, hermiticity of the detector is essential. Namely, the detector should be designed to have no dead angle. However, it is unavoidable to miss particles that are emitted in the beam direction. Consequently, a standard procedure is to measure momentum or energy balance only in the transverse direction to the beam.

Data Figure 3.2 shows an event display of $W \to e\nu$ and Figure 3.3 illustrates its energy deposit in the three-dimensional plot on the η–ϕ plane where $\eta = \ln\cot(\theta/2)$ is the pseudorapidity[6] and ϕ is the azimuthal angle.

One sees many tracks are recorded in the tracking chamber, but most of them are low energy hadron backgrounds generated by soft gluons in the QCD reaction. They are invisible in the Lego plot (three-dimensional display of the transverse momentum on the η–ϕ plane) and the prompt electron from the $W \to e\nu$ decay stands out. Only when the energy scale is magnified (Figure 3.3b), does one see the hadronic background.

Figure 3.4 shows an event display of $Z \to ee$ observed in the UA2 detector. One sees that two electrons of equal energy are emitted back to back in the ϕ plane. For the Z, the formula Eq. (3.3) can be used to reconstruct the mass of the Z boson. It is shown in Figure 3.4b and the Z mass was determined to be ~ 90 GeV.

Compared to the Z boson, it is harder to determine the W mass as one of the decay products is invisible. To show that the missing energy measured by the apparatus matches that of the neutrino from W decay, we show distributions of direction of the missing energy vector and its magnitude in Figure 3.5. One sees that the di-

6) It is equivalent to the polar angle Lorentz boosted along the beam direction. The rapidity defined by $\eta = (1/2)\ln[(E + p_z)/(E - p_z)] \xrightarrow{E \gg p_z} (1/2)\ln[(1 + \cos\theta)/(1 - \cos\theta)]$ becomes the pseudorapidity in the massless limit. Then, the relativistic phase space volume for a emitted particle can be rewritten as $d^3p/E \sim (1/2)dp_T^2 d\eta d\phi$, where $p_T = p\sin\theta$. Namely, the phase space is flat on the η–ϕ plane and the hadron background is expected to distribute uniformly (see Section 11.1).

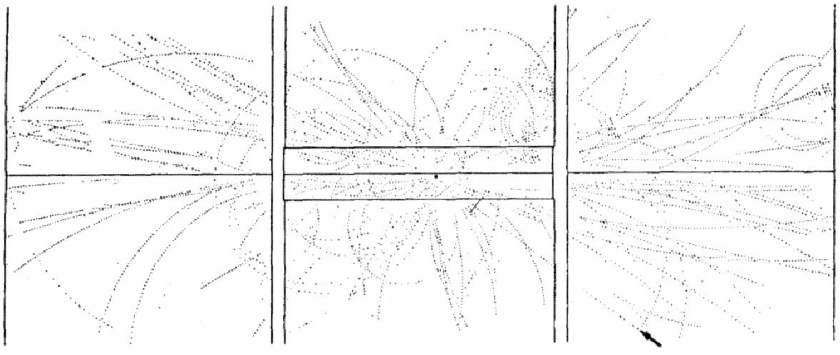

Figure 3.2 A $W \to e\nu$ event as seen in the UA1 tracker. The digitization from the central detector for the tracks in an event which has an identified, isolated, well-measured high-p_T electron, and the electron track is indicated by an arrow [107].

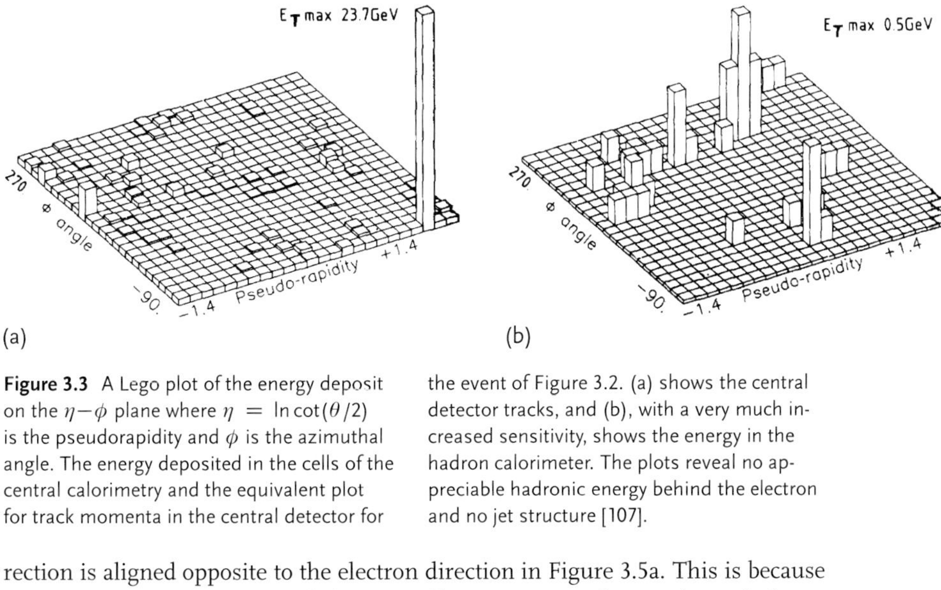

Figure 3.3 A Lego plot of the energy deposit on the $\eta-\phi$ plane where $\eta = \ln\cot(\theta/2)$ is the pseudorapidity and ϕ is the azimuthal angle. The energy deposited in the cells of the central calorimetry and the equivalent plot for track momenta in the central detector for the event of Figure 3.2. (a) shows the central detector tracks, and (b), with a very much increased sensitivity, shows the energy in the hadron calorimeter. The plots reveal no appreciable hadronic energy behind the electron and no jet structure [107].

rection is aligned opposite to the electron direction in Figure 3.5a. This is because the total energy of the $\overline{p}p$ collider was $\sqrt{s} = 540\,\text{GeV}$. The quarks and gluons in the hadron share their momenta equally between them and carry about half of the total momentum which is divided into three valence quarks giving $\langle x \rangle \sim 1/6$. Therefore, the average total energy of the quarks is $\sim 540/6 = 90\,\text{GeV}$. Namely, the W, Z bosons are produced nearly at rest. Then, the electron and the neutrino are in back to back position. In Figure 3.5b, the magnitude of the transverse missing energy ($\not{E}_T$) and the electron energy E_{eT} is shown to be correlated and about equal. The maximum value of the sum of $\not{E}_T$ and E_{eT} is about equal to $\sim 80\,\text{GeV}$, which is the predicted value of the W. All things considered, we can conclude that $W \to e\nu$ was observed.

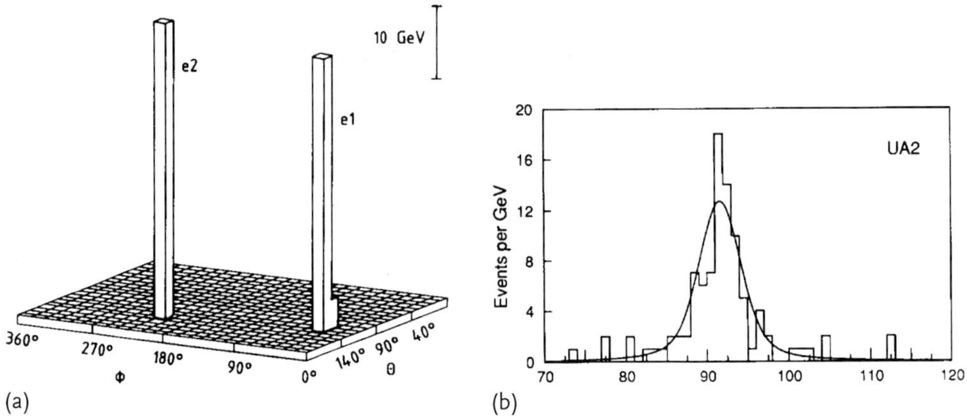

Figure 3.4 An event display of UA2 data. (a) Lego plot of $Z \to ee$ decay event. Two electrons are emitted back to back. (b) An invariant mass distribution of the two electrons [108, 109].

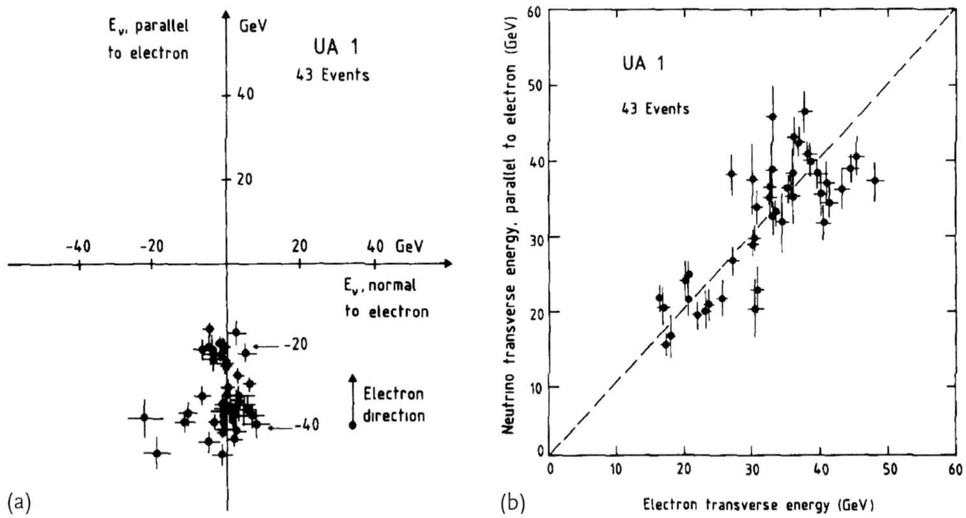

Figure 3.5 Two-dimensional plot of the transverse components of the missing energy (neutrino momentum). The neutrino direction is opposite to the electron. (b) Electron versus neutrino transverse momentum distributions for the $W \to e\nu$ samples collected by UA1 [110].

Thus, the most important particles predicted by the GWS model were discovered and their value agreed with the prediction ($m_W = 37.5/\sin\theta_W \sim 80\,\text{GeV}$, $m_Z = m_W/\cos\theta_W \sim 90\,\text{GeV}$ where $\sin\theta_W \sim 0.24$ was obtained from the neutral current reactions. One may say that the GWS model was firmly established. From then on, the GWS model was referred to as the Standard Model of the electroweak interaction.

3.2
Basic Formulas

3.2.1
Decay Width

According to the GWS model, the interaction Lagrangian of the gauge bosons and fermions are given by (see Eq. (1.50))

$$-\mathcal{L}_I = QeA_\mu J^\mu_{\rm EM} + g_Z Z_\mu J^\mu_{\rm NC} + \frac{g_W}{\sqrt{2}}\left(W^+_\mu J^{+\mu}_{\rm CC} + W^-_\mu J^{-\mu}_{\rm CC}\right) \tag{3.4a}$$

$$g_Z = \frac{g_W}{\cos\theta_W} = \frac{e}{\sin\theta_W\cos\theta_W} \tag{3.4b}$$

$$J^\mu_{\rm EM} = \overline{\psi}\gamma^\mu\psi \tag{3.4c}$$

$$J^\mu_{\rm NC} = \overline{\psi}_L\gamma^\mu I_3\psi_L - \sin^2\theta_W J^\mu_{\rm EM} = \frac{1}{2}\overline{\psi}\gamma^\mu(v_f - a_f\gamma^5)\psi \tag{3.4d}$$

$$v_f = I_{3L} - 2Q\sin^2\theta_W\,, \quad a_f = I_{3L} \tag{3.4e}$$

$$J^{+\mu}_{\rm CC} = \overline{\psi}_L\gamma^\mu I_+\psi_L = \frac{1}{2}\sum V_{ab}\overline{\psi}_a\gamma^\mu(1-\gamma^5)I_+\psi_b\,, \quad J^{-\mu}_{\rm CC} = \left(J^{+\mu}_{\rm CC}\right)^\dagger \tag{3.4f}$$

$$m^2_W = \frac{\pi\alpha}{\sqrt{2}G_F\sin^2\theta_W} \simeq (80.2\,\text{GeV})^2\,, \quad m^2_Z = \frac{m^2_W}{\cos^2\theta_W} \simeq (91.2\,\text{GeV})^2 \tag{3.4g}$$

I_{3L} means that it operates only on the left-handed component of the fermion field ψ. For instance, $I_3 e_L = (-1/2)e_L$, $I_3 e_R = 0$. Q is the electric charge in units of the proton charge and V_{ab} ($= \delta_{ab}$ if a, b are the leptons) are the Cabibbo–Kobayashi–Maskawa matrix elements. As the top quark ($m_{\rm top} = 172$ GeV) is heavier than W, the W can decay only to (ν_e, e), (ν_μ, μ), (ν_τ, τ), (u, d'), and (c, s') where $d' = \sum_b V_{1b}d_b$, $s' = \sum_b V_{2b}d_b$.

Let us first consider the leptonic decay width of the W. Denoting variables as $W^+(q) \to l^+(p_1)\nu(p_2)$ and the polarization vector of the W as ϵ^μ, the decay transition amplitude can be calculated from the Lagrangian. Following the standard procedure given in Vol. 1, Chapt. 7, we obtain

$$\langle l^+(p_1), \nu(p_2)|S|W^+(q, \epsilon(\lambda))\rangle - \delta_{fi} = -i(2\pi)^4\delta^4(q - p_1 - p_2)\mathcal{M}(W \to l\nu) \tag{3.5a}$$

$$\mathcal{M}(W \to l\nu) = \frac{g_W}{\sqrt{2}}\overline{u}(p_2)\gamma^\mu\left(\frac{1-\gamma^5}{2}\right)v(p_1)\epsilon_\mu(q, \lambda) \tag{3.5b}$$

The decay width is given by

$$\Gamma(W^+ \to l^+\nu) = \frac{1}{2m_W}\overline{\sum}_{\rm spin,pol}|\mathcal{M}|^2 dLIPS \tag{3.6a}$$

$$dLIPS = (2\pi)^4\delta^4(q - p_1 - p_2)\frac{d^3p_1}{(2\pi)^3 2E_1}\frac{d^3p_2}{(2\pi)^3 2E_2} \tag{3.6b}$$

where $\overline{\sum}$ denotes the average of the initial polarization and sum of spin orientation in the final state. Following the standard γ matrix calculation, the sum over fermion spin is obtained as

$$U \equiv \sum_{\text{spin}} |\overline{u}(p_2)\gamma^\mu(v - a\gamma^5)v(p_1)\epsilon_\mu|^2 = K_{\mathrm{L}}^{\mu\nu}\epsilon_\mu\epsilon_\nu^*$$
$$= 4\Big[(|v|^2 + |a|^2)\{(p_2\cdot\epsilon)(p_1\cdot\epsilon^*) + (p_2\cdot\epsilon^*)(p_1\cdot\epsilon) - (\epsilon\cdot\epsilon^*)(p_1\cdot p_2)\}$$
$$+ 2i\mathrm{Re}(va^*)\epsilon^{\mu\nu\rho\sigma}\epsilon_\mu\epsilon_\nu^* p_{1\rho}p_{2\sigma} - (|v|^2 - |a|^2)(\epsilon\cdot\epsilon^*)m_1m_2\Big] \tag{3.7}$$

where we used $v - a\gamma^5$ instead of $1 - \gamma^5$ for later reference. For the $W \to e\nu$, we set $v = a = 1/\sqrt{2}$.

Problem 3.1

Show that for the case $W^- \to l^-\nu$, the only difference is the sign of the second term in $[\cdots]$.

Problem 3.2

Show that if one neglects the lepton mass, Eq. (3.7) for $v = a = 1/\sqrt{2}$ becomes, in the W^+ rest frame,

$$U = 2m_W^2[1 - (\hat{\boldsymbol{p}}_1\cdot\boldsymbol{\epsilon})(\hat{\boldsymbol{p}}_1\cdot\boldsymbol{\epsilon}^*) - i(\boldsymbol{\epsilon}\times\boldsymbol{\epsilon}^*\cdot\hat{\boldsymbol{p}}_1)] \tag{3.8a}$$

$$\hat{\boldsymbol{p}}_1 \equiv \frac{\boldsymbol{p}_1}{|\boldsymbol{p}_1|}, \quad \epsilon^\mu = (0, \boldsymbol{\epsilon}) \tag{3.8b}$$

Show that corresponding to the W in polarization state $\lambda = \pm 1, 0$, the lepton angular distribution behaves like $(1\mp\cos\theta)^2$, $\sin^2\theta$ with respect to the polarization axis.

When the W is not polarized, we need to take the polarization average:

$$\frac{1}{3}\sum_{\lambda=\pm,0}\varepsilon^\mu(\lambda)\varepsilon^\nu(\lambda) = \frac{1}{3}\left(-g^{\mu\nu} + \frac{q^\mu q^\nu}{m_W^2}\right) \tag{3.9}$$

This is symmetric in suffixes μ, ν, and the second term in Eq. (3.7) does not contribute. Furthermore, using $q_\mu K_{\mathrm{L}}^{\mu\nu} = 0$ in the massless approximation, one can show that $U = (4/3)(v^2 + a^2)q^2$. In this case, the cross section does not contain the angular dependence, $dLIPS$ can be integrated by itself giving $dLIPS = 1/8\pi$. Putting all things together, one gets

$$\Gamma(W^\pm \to l^\pm\nu) = \frac{g_W^2 m_W}{48\pi} = \frac{G_F m_W^3}{6\sqrt{2}\pi} \equiv \Gamma_l \tag{3.10}$$

Table 3.1 Partial decay widths and branching ratios of the W decay.

Decay mode	Partial width (GeV)	Branching ratio (%)
$e\overline{\nu}$	0.223	10.8
$\mu\overline{\nu}$	0.233	10.8
$\tau\overline{\nu}$	0.233	10.8
$\overline{u}d(g)$	0.622	32.1
$\overline{c}s(g)$	0.661	32.1
$\overline{u}s(g)$	0.034	1.6
$\overline{c}d(g)$	0.034	1.6
$\overline{c}b(g)$	0.001	0.05
$\overline{u}b(g)$	0.000	0.00

Total width $\Gamma_W = 2.09\,\text{GeV}$.

Similar calculations for the quark pairs give in the massless approximation

$$\Gamma(W \to q_a\overline{q}_b) = N_c|V_{ab}|^2\Gamma_l = \frac{G_F m_W^3}{2\sqrt{2}\pi}|V_{ab}|^2 \tag{3.11}$$

Problem 3.3

Show that if the mass of the final fermions is not neglected, the decay width becomes

$$\begin{aligned}
\Gamma(W \to f_1\overline{f}_2) &= \Gamma(W \to f_1\overline{f}_2 : m_1 = m_2 = 0) \\
&\quad \times \lambda(1, x, y)\left[1 - \frac{x+y}{2} - \frac{(x-y)^2}{2}\right] \\
\lambda(1, x, y) &= (1 + x^2 + y^2 - 2xy - 2x - 2y)^{1/2} \\
x &= m_1^2/m_W^2\,, \quad y = m_2^2/m_W^2
\end{aligned} \tag{3.12}$$

For comparisons with data, one needs to take into account QCD corrections for the decay to the quark pairs (and higher order electroweak corrections which we neglect here).

$$\Gamma(W \to q_a q_b) \to \Gamma(W \to q_a q_b)\left(1 + \alpha_s\frac{m_W^2}{\pi} + \cdots\right) \tag{3.13}$$

The QCD correction is $\alpha_s(m_W^2)/\pi \simeq 0.036 \pm 0.005$. Table 3.1 shows the results of the above calculations with QCD correction.

3.2.2 Hadronic Production of W, Z

Production of the Z by e^-e^+ reaction will be described in detail in the next chapter. Here, we describe the hadronic production of the W, Z which played a historically important role and still is the main production mechanism for the W. Let us begin with the production of the Z first. The transition amplitude and the cross section for the process

$$q(p_1) + \overline{q}(p_2) \to Z(q, \epsilon(\lambda)) \tag{3.14}$$

is expressed as

$$\langle Z(q,\epsilon)|S-1|q(p_1)\overline{q}(p_2)\rangle = -i(2\pi)^4\delta(p_1+p_2-q) \\ \times \frac{g_Z}{2}\left[\overline{v}(p_2)\gamma^\mu\left(v_f - a_f\gamma^5\right)u(p_1)\epsilon_\mu^*\right] \tag{3.15a}$$

$$d\hat{\sigma}(q\overline{q}\to Z) = \frac{1}{2\hat{s}\lambda}\frac{g_Z^2}{4}\frac{1}{3}\frac{1}{4}\sum_{\text{spin, pol}}|\overline{v}(p_2)\gamma^\mu\left(v_f - a_f\gamma^5\right)u(p_1)\epsilon_\mu^*|^2 dLIPS \tag{3.15b}$$

$$\hat{s} = (p_1+p_2)^2 = q^2 = m_Z^2 \tag{3.15c}$$

$$dLIPS = (2\pi)^4\delta^4(p_1+p_2-q)\frac{d^3q}{(2\pi)^3 2\omega} \tag{3.15d}$$

Here, $2\hat{s}\lambda(1, m_1^2/\hat{s}, m_2^2/\hat{s})$ is the flux factor and $\lambda = 1$ if the quark mass is neglected. The reason to attach the ˆ is to remind one that the variable is not that of hadrons, but of partons. The factors 1/3 [$= 3\times(1/3)\times(1/3)$] and 1/4 in front of the spin sum is due to the color and spin average. The spin/polarization sum can be carried out in a similar manner as the decay width calculation and we obtain

$$\hat{\sigma}(q_a\overline{q}_a \to Z) = \frac{\sqrt{2}\pi}{3}G_N m_Z^2\left(v_f^2 + a_f^2\right)\delta\left(\hat{s} - m_Z^2\right) \tag{3.16}$$

The cross section of the W production [$q_a(p_1)\overline{q}_b(p_2) \to W(q)$] can be obtained by substituting

$$g_Z \to g_W V_{ab} \quad \left(\text{or} \quad G_N m_Z^2 \to G_F m_W^2 |V_{ab}|^2\right) \\ v_f \to 1/\sqrt{2}, \quad a_f \to 1/\sqrt{2} \tag{3.17}$$

into Eq. (3.16).

$$\hat{\sigma}(q_a\overline{q}_b \to W) = \frac{\sqrt{2}\pi}{3}G_F m_W^2 |V_{ab}|^2\delta\left(\hat{s} - m_W^2\right) \tag{3.18}$$

In reality, where the $p\overline{p}$ reaction is used, the initial $q_a, \overline{q}_b$ carry energy-momenta $x_a p_A, x_b p_B$. We denote the total energy of the hadron system as $s = (p_A + p_B)^2$ and that of the parton system as $\hat{s} = (x_a p_A + x_b p_B)^2 \simeq 2x_a x_b(p_A \cdot p_B) \simeq x_a x_b s$ and the parton cross section as $\hat{\sigma}$ in which $\delta(\hat{s} - m_Z^2)$ is replaced by $\delta(x_a x_b s - m_Z^2)$[7]. Expressing the parton's probability to carry fractional momentum x_a in the hadron A as $f_A^a(x_a)dx_a$, the production cross section of the vector particle V is expressed as (refer to Vol. 1, Sect. 17.9)

$$\begin{aligned}
&\sigma[P(p_A) + \overline{P}(p_B) \to VX] \\
&\quad = \sum_{a,b} \int dx_a dx_b \left[\left\{ f_A^a(x_a) f_B^b(x_b) + (a \leftrightarrow b) \right\} \hat{\sigma}(ab \to V) \right]
\end{aligned} \tag{3.20a}$$

$$\hat{\sigma}(ab \to Z) = \frac{\pi}{3} \sqrt{2} G_N m_Z^2 \left(v_f^2 + a_f^2 \right) \delta\left(s x_a x_b - m_Z^2 \right) \tag{3.20b}$$

$$\hat{\sigma}(ab \to W) = \frac{\pi}{3} \sqrt{2} G_F m_W^2 |V_{ab}|^2 \delta\left(s x_a x_b - m_W^2 \right) \tag{3.20c}$$

Sometimes, rapidity (η) distribution of the vector boson is required.

$$\eta \equiv \frac{1}{2} \ln \frac{E_V + P_V}{E_V - P_V} = \frac{1}{2} \ln \frac{E_{ab} + P_{ab}}{E_{ab} - P_{ab}} = \frac{1}{2} \ln \frac{x_a}{x_b} \tag{3.21}$$

where we used the relations

$$\begin{aligned}
E_{ab} &= x_a E_A + x_b E_B \xrightarrow{E_A = E_B} (x_a + x_b) E \\
P_{ab} &= x_a p_A + x_b p_B \xrightarrow{p_A = -p_B \simeq E} (x_a - x_b) E
\end{aligned} \tag{3.22}$$

in the $p\overline{p}$ center of mass system. Defining $\tau \equiv x_a x_b$, we have

$$\begin{aligned}
x_a &= \sqrt{\tau} e^{\eta}, \quad x_b = \sqrt{\tau} e^{-\eta} \\
dx_a dx_b &= d\tau d\eta
\end{aligned} \tag{3.23a}$$

Then, the differential cross section is expressed as

$$\frac{d^2\sigma}{d\tau d\eta}[P(p_A) + \overline{P}(p_B) \to VX] = \hat{\sigma}(ab \to V) \left[f_A^a(x_a) f_B^b(x_b) + (a \leftrightarrow b) \right] \tag{3.24}$$

Equations (3.20a)–(3.24) are the essential equations we need in order to understand the basic electroweak process in the hadron collision.

Measurements of the W, Z production processes were carried out at S$\overline{\text{p}}$pS/CERN and Tevatron/Fermilab by observing sum of the decay modes $W^\pm \to e^\pm + \nu$ and $W^\pm \to \mu^\pm + \nu$ (branching ratio = 10.8% × 2) and the decay mode $Z \to l\bar{l}(l = e, \mu)$ (BR = 3.34 × 2%). Results of the calculation and experimental values are shown in Figure 3.6 and Table 3.2. They agree well with each other.

7) The actual W, Z have Breit–Wigner width and one has to replace the δ function by

$$\delta(\hat{s} - m^2) \to \frac{m\Gamma}{\pi} \frac{1}{(\hat{s} - m)^2 + m^2\Gamma^2} \tag{3.19}$$

However, for slowly varying x distributions, this replacement produces the same result as the δ function.

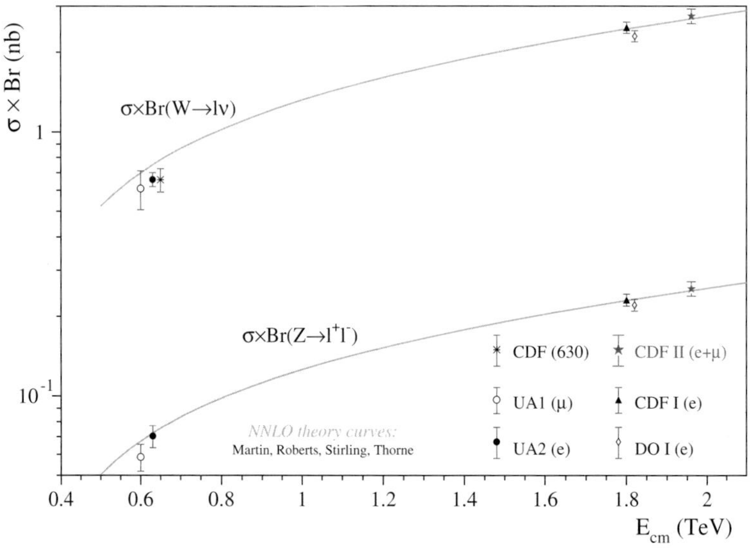

Figure 3.6 $W \to l\nu$ and $Z \to l\bar{l}$ cross section measurements as a function of the $p\bar{p}$ center of mass energy, $E_{CM} = \sqrt{s}$ [111, 112]. The solid lines correspond to the theoretical NNLO QCD calculations [113–117].

Table 3.2 $\sigma(p\bar{p} \to WX)\text{BR}(W \to l\nu)$, $\sigma(p\bar{p} \to ZX)\text{BR}(Z \to l\bar{l})$ Data and theoretical predictions. UA2: [109, 146], CDF,D0 at 1800 GeV: [111, 147–149], CDF at 1960 GeV: [112]. Theory at 630,1800: [16, 18, 150, 151] at 1960 GeV [113–117].

Mode	$\sqrt{s}$ GeV	Group	$\sigma \cdot B(pb)$
			(stat) (syst) (lum.)
$W \to l\nu$	630	UA2	682±12±40
	1800	CDF	2190±40±210
	1960	CDF	2749±10±53±165
$Z \to l\bar{l}$	630	UA2	56±4±3.8
	1800	CDF	209±13±17
	1960	CDF	254.9±3.3±4.6±15.2

Notice that for hadronic reactions, higher order QCD corrections are indispensable for quantitative treatments. For the W, Z production where the leading process is electroweak, the QCD corrections appear as the initial state radiation like those shown in Figure 3.7. Equation (3.20a) already contains such effects, that is, provided suitable (i.e., scale dependent) parton distribution functions are chosen and the normalization adjusted accordingly. The theoretical curves used in Figure 3.6 and Table 3.2 include higher order QCD corrections. However, the hard gluon emission has to be treated differently. It gives a transverse momentum kick to the *W*, *Z* which is one of the main QCD themes and will be treated separately in detail in Section 11.8.2.

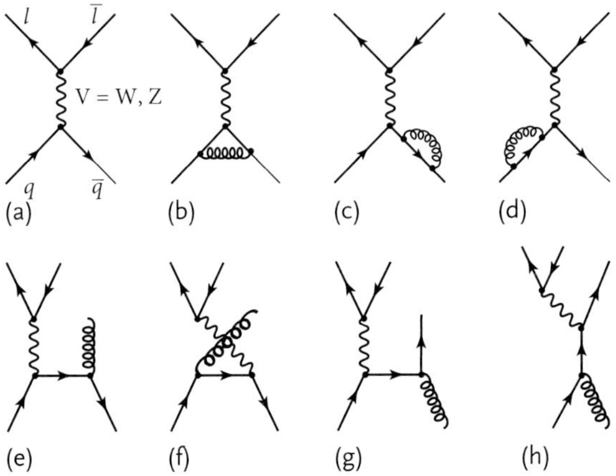

Figure 3.7 Diagrams for production and leptonic decay of a vector boson $V = W$, Z at leading (a) and higher order (b–d). Diagrams (e–h) emit extra gluons, but if they are soft cannot be distiguished from (a–d). Interference of (a) with (b–d) give contribitions at the same order as (e–h) and cancles the infrared divergence.

3.3 Properties of *W*

3.3.1 Asymmetry of Decay Leptons from *W*, *Z*

In order to determine the spin of the *W*, it is necessary to look at angular distributions of the leptons from the *W* decay. If we calculate the two-body cross sections for

$$d(p_1) + \overline{u}(p_2) \to e^-(p_3) + \overline{\nu}(p_4) \tag{3.25a}$$

$$u(p_1) + \overline{d}(p_2) \to e^+(p_3) + \nu(p_4) \tag{3.25b}$$

with a constraint to go through *W* intermediate state, the transition amplitude becomes

$$\begin{aligned}
&\langle e(p_3)\overline{\nu}(p_4)|S|d(p_1)\overline{u}(p_2)\rangle \\
&= (2\pi)^4\delta^4(p_1 + p_2 - p_3 - p_4)\left(-i\frac{g_W}{\sqrt{2}}\right)^2 V_{ud}^* \\
&\times \overline{u}(p_3)\gamma^\mu\frac{1-\gamma^5}{2}v(p_4)\frac{i\left(-g_{\mu\nu} + \frac{q_\mu q_\nu}{m_W^2}\right)}{\hat{s} - m_W^2 + i m_W\Gamma_W}\overline{v}(p_2)\gamma^\nu\frac{1-\gamma^5}{2}u(p_1)
\end{aligned} \tag{3.26}$$

Neglecting the mass of the quarks and leptons, one obtains

$$d\sigma_{W\to e} = \frac{1}{2\hat{s}}\left(\frac{g_W^2}{8}\right)^2\left(\frac{1}{3\cdot 4}\right)\frac{K_L^{\mu\nu}(p_4,p_3)K_{L\mu\nu}(p_1,p_2)}{(\hat{s}-m_W^2)^2+(m_W\Gamma_W)^2}dLIPS \tag{3.27a}$$

$$K_L^{\mu\nu}(p_1,p_2) = \sum_{\text{spin}}\left[\overline{v}(p_2)\gamma^\mu(1-\gamma^5)u(p_1)\right]\left[\overline{v}(p_2)\gamma^\nu(1-\gamma^5)u(p_1)\right]^* \tag{3.27b}$$

Carrying out the standard γ traces, one obtains

$$K_L^{\mu\nu}(p_4,p_3)K_{L\mu\nu}(p_1,p_2) = 256(p_1\cdot p_4)(p_2\cdot p_3) = 64\hat{u}^2 \tag{3.28a}$$

$$\hat{u} = (p_1-p_4)^2 = -\frac{\hat{s}}{2}(1+\cos\theta) \tag{3.28b}$$

Substituting Eq. (3.28) in Eq. (3.27), one gets the cross section:

$$\frac{d\sigma}{d\Omega}(q\overline{q}'\to W\to e\nu) = \frac{1}{3}\left(\frac{G_F m_W^2}{\sqrt{2}}\right)^2|V_{ud}|^2\frac{1}{16\pi^2}\frac{\hat{s}(1+\cos\theta)^2}{\left(\hat{s}-m_W^2\right)^2+(m_W\Gamma_W)^2} \tag{3.29}$$

Here, θ is the angle that the electron makes with respect to the incoming d (i.e., proton). The production cross section for the positron can be obtained by replacing $\hat{u}\to\hat{t}$ or $1+\cos\theta\to 1-\cos\theta$. The total cross section can be obtained by integrating Eq. (3.29):

$$\sigma_{TOT}(q\overline{q}'\to W\to e\nu) = \frac{|V_{ud}|^2}{9\pi}\left(\frac{G_F m_W^2}{\sqrt{2}}\right)^2\frac{\hat{s}}{(\hat{s}-m_W^2)^2+(m_W\Gamma_W)^2} \tag{3.30}$$

Problem 3.4

Show that the total cross section given by Eq. (3.30) agrees with $\sigma(ab\to W)\times(\Gamma(W\to e\nu)/\Gamma_W)$ given by Eqs. (3.20c) and (3.10) Hint: use Eq. (3.19).

Notice that the angular distribution $(1\pm\cos\theta)^2/4$ in Eq. (3.29) agrees with square of the rotational function $d^1_{\mp1,-1}(\theta)$. This can easily be understood from the $V-A$ structure of the weak interaction. The electron is produced by the process $d\overline{u}\to W^-\to e^-\overline{\nu}_e$ and the positron by $u\overline{d}\to W^+\to e^+\nu_e$. The $V-A$ structure forces the helicity of the participating quarks to be negative for u, d and positive for $\overline{u}$, $\overline{d}$. Therefore, $W^\mp$ is made with $S_z=-1$ where the z axis is set along the incoming proton (see Figure 3.8).

The electron (positron) is also forced to have negative (positive) helicity, hence the $e^-(e^+)$ emitted in the forward ($\theta=0$) (backward ($\theta=\pi$)) direction conserves the third component of the spin, and when emitted in the backward (forward) direction, it does not conserve which manifests as $(1\pm\cos\theta)^2$ distribution.

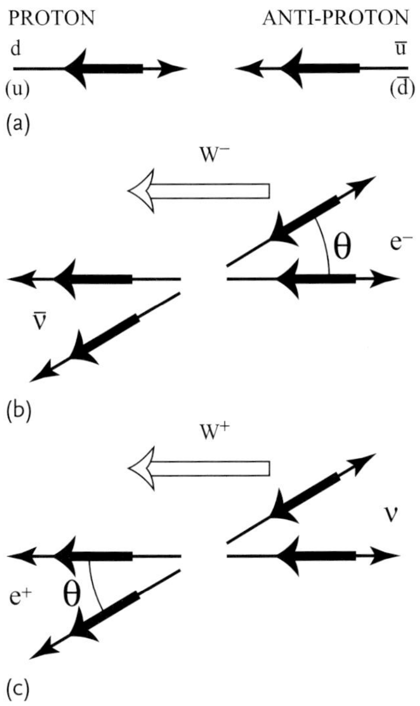

Figure 3.8 Qualitative explanation for the decay electron (positron) from $p\overline{p} \rightarrow W^{\mp}$ to make angular distribution $(1 \pm \cos\theta)^2$. Thin arrows denote momenta and thick arrows, spin. (a) $V-A$ structure of the weak interaction forces u, d to have negative helicity and the opposite is true for $\overline{u}, \overline{d}$. (b,c) The intermediate $W^{\mp}$ has spin $S_Z = -1$ where z is in the proton direction. The $V-A$ interaction is also at work for the $W^{\mp} \rightarrow e^{\mp} + \nu$ decay, so that decay electron (positron) has negative (positive) helicity. Thus, the electron is preferentially emitted in the forward direction and is forbidden to be emitted backward. The opposite is true for the positron.

The cross section for $q\overline{q} \rightarrow Z \rightarrow e\overline{e}$ can be calculated similarly, giving

$$\frac{d\sigma}{d\Omega}(q\overline{q} \rightarrow Z \rightarrow e\overline{e}) = \frac{1}{3}\left(\sqrt{2}G_{\text{N}} m_Z^2\right)^2 \frac{\hat{s}}{16\pi^2} \frac{A(1+\cos\theta)^2 + B\cos\theta}{\left(\hat{s} - m_Z^2\right)^2 + (m_Z \gamma_Z)^2} \tag{3.31a}$$

$$A = \frac{1}{4}\left(v_e^2 + a_e^2\right)\left(v_f^2 + a_f^2\right), \quad B = 2 v_e a_e v_f a_f \tag{3.31b}$$

3.3.2
Spin of W, Z

The basic production cross section of $q\overline{q}' \rightarrow W^{\mp} \rightarrow e^{\mp} + \nu$ was given in Eq. (3.29). The angular distribution of the leptons from W decay is shown in Figure 3.9.

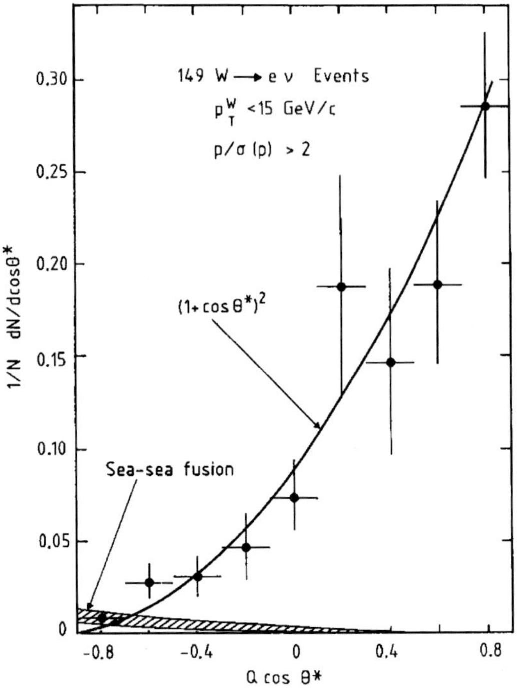

Figure 3.9 Decay angular distributions of *W*s from the electron decay mode. The shaded band shows the expected contribution from annihilation processes involving wrong polarity sea quarks only [118, 119].

The early data [119] agree with the theoretical predictions and confirmed the spin of the W, Z gauge bosons to be one. They also show the $V-A$ structure for the W and the mixture of V and A interaction for the Z. However, for value $\sin^2\theta_W$ close to 0.25, the asymmetry of $Z \to e^+e^-$ decay distributions is very small. We defer showing Z data to the next chapter.

3.3.3
Mass of W

Transverse momentum distribution From Eqs. (3.29) and (3.30), one can express the differential cross section for the electron as

$$\frac{d\sigma}{d\Omega}(q\overline{q}' \to W^{\mp} \to e^{\mp} + \nu) = \sigma_{\text{TOT}}\frac{3}{16\pi}(1 \pm \cos\theta)^2 \tag{3.32}$$

By using the expression for the transverse momentum $\hat{p}_{\text{T}}$ of the electron

$$\hat{p}_{\text{T}}^2 = (\hat{p}\sin\theta)^2 = \frac{\hat{s}}{4}(1 - \cos^2\theta) \tag{3.33}$$

the cross section can be rewritten as

$$\frac{d\sigma}{d\hat{p}_T^2} = \frac{2}{\hat{s}\cos\theta}\frac{d\sigma}{d\cos\theta} = \frac{3}{4}\sigma_{TOT}\frac{1+\cos^2\theta}{\hat{s}\cos\theta}$$
$$= \frac{3}{2}\frac{\sigma_{TOT}}{\hat{s}}\frac{1-2\hat{p}_T^2/\hat{s}}{\left(1-4\hat{p}_T^2/\hat{s}\right)^{1/2}} \tag{3.34}$$

The cos θ term does not contribute because $\hat{p}_T$ is symmetric for the exchange $\theta \leftrightarrow \pi-\theta$. The above formula becomes infinite as $\hat{p}_T \to \sqrt{\hat{s}}/2$, which is referred to as the Jacobian peak. It is characteristic of two-body reactions (Figure 3.10a). As the distribution is cut sharply at $\hat{p}_T = m_W/2$, the position of the Jacobian peak can be used to determine the W mass. If the initial partons do not have transverse momentum, p_T of the electron measured in the laboratory equals $\hat{p}_T$ given above. In reality, the cross section has to be integrated over the parton distribution inside the hadron as given by Eq. (3.20), the singularity disappears and the peak height becomes finite. Note that the transverse motion of the partons is very small, that is, $\langle p_T \rangle_{\text{parton}} \sim 300$ MeV, and in many cases can be neglected. However, if the parent W has its own transverse momentum q_T, it is added to that of the electron and further smearing occurs.

Figure 3.10b shows a measured distribution obtained by the D0 group at Fermilab [120, 121]. The solid line indicates smearing due to the parton distribution, the dotted line takes into account the transverse momentum q_T of the W, and finally the shaded area is the one folded with the detector resolution. One can also use the transverse mass which is less affected by the transverse momentum of the W.

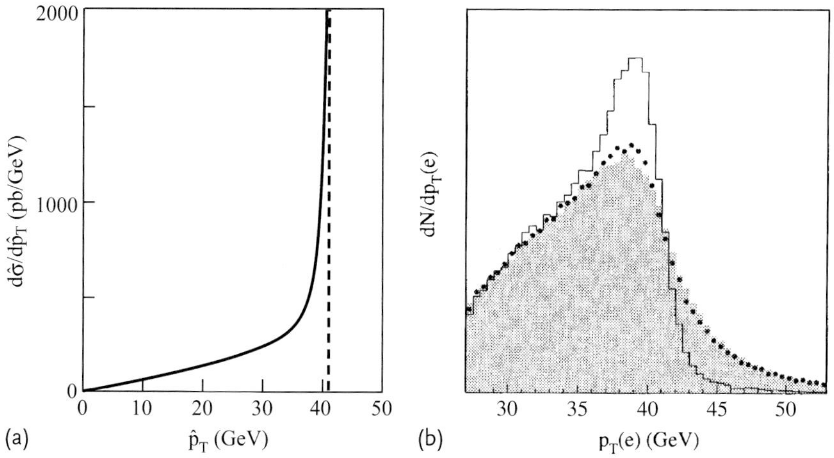

Figure 3.10 (a) Jacobian peak $d\bar{u} \to W^- \to e^-\bar{\nu}$. (b) Actual p_T spectrum folded with the parton distribution for W bosons with the transverse momentum of the W $q_T = 0$ (solid line), with the correct q_T distribution (•), and with a detector resolutions shaded area [120, 121].

Transverse Mass of the *W* In collider experiments, because of the beam pipe that penetrates through the detector, it is hard to catch jets emitted in the forward (backward) direction. The longitudinal momentum balance is hard to obtain. On the other hand, the transverse momentum can be measured with much better accuracy, the missing momentum $\not{p}_T = -$ (vector sum of all visible particles) is a valuable information source. Specially for the $p + \overline{p} \to W + X \to e\nu + X$ mode, it can be considered as the momentum of the neutrino. Namely, $\not{p}_T = \boldsymbol{p}_{\nu T}$. Therefore, we define the transverse mass by

$$m_T^2(e,\nu) \equiv (|\boldsymbol{p}_{\nu T}| + |\boldsymbol{p}_{eT}|)^2 - (\boldsymbol{p}_{\nu T} + \boldsymbol{p}_{eT})^2 = 2|\boldsymbol{p}_{\nu T}||\boldsymbol{p}_{eT}|(1 - \cos\Delta\phi) \tag{3.35}$$

where $\Delta\phi$ is the angle between $\boldsymbol{p}_{\nu T}$ and $\boldsymbol{p}_{eT}$ in the plane perpendicular to the beam. Compared with the exact invariant mass

$$m_W^2 = m^2(e,\nu) = (|\boldsymbol{p}_\nu| + E_e)^2 - (\boldsymbol{p}_\nu + \boldsymbol{p}_e)^2 \tag{3.36}$$

one can show that m_T lies between 0 and m_W. For the special case $q_{WT} = 0$, $m_T = 2|\not{p}_{eT}|$. Therefore, the m_T distribution is identical to Eq. (3.34) with $\hat{p}_T$ replaced by $m_T/2$ and exhibits a similar Jacobian peak. It has the form

$$\frac{d\sigma}{dm_T^2} \sim \frac{1}{(\hat{s} - m_W^2)^2 + (m_W\Gamma_W)^2}\left[\frac{2 - m_T^2/\hat{s}}{(1 - m_T^2/\hat{s})^{1/2}}\right] \tag{3.37}$$

If the W carries the transverse momentum, the equation $\max(m_T) = m_W/2$ does not change. Therefore, the value of m_W obtained from the near edge points of m_T distribution is more immune to the smearing due to finite q_T. The smearing due to the finite decay width of the Breit–Wigner formula still needs to be considered. Because of its advantage, the m_T distribution is usually used to obtain the most accurate value of the W mass. Note, however, that the p_T resolution is intrinsically better than that of m_T because it can be obtained from direct measurements, and should be considered as a complementary method. Values determined from p_{eT} and $p_{\nu T}$ distribution are very competitive with that determined from m_T method, and can be combined to obtain better resolution. Figure 3.11 illustrates D0 data for both distributions [10]. The present world average is

$$m_W = 80.420 \pm 0.031\ \text{GeV}/c^2 \tag{3.38}$$

$e^-e^+ \to W^-W^+$ When the LEP II began operation in 1995 at the energy above 161 GeV, which is the threshold of W pair production, it provided another possibility to determine the W mass. Precise knowledge of the e^-e^+ center of mass energy (see Section 4.3.3) enables one to reconstruct the W mass, even if one of them decays leptonically. Two methods were applied. One is to fit the theoretical cross section of the $e^-e^+ \to W^-W^+$ with the data near threshold, and the other

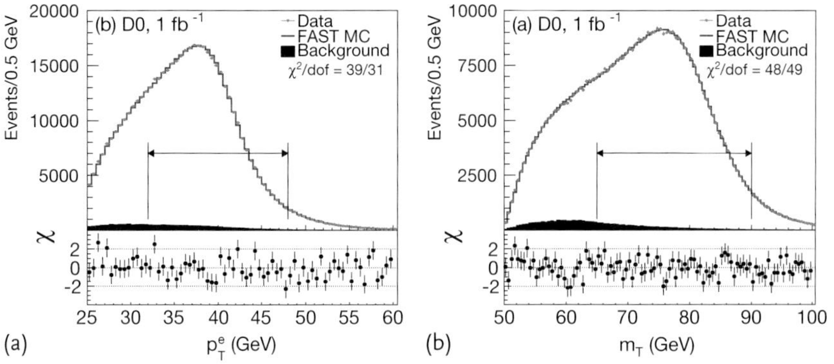

Figure 3.11 Distributions of (a) p_{eT} and (b) m_T, and FASTMC simulation with backgrounds. The χ values are shown below each distribution where $\chi_i = [N_i - (\text{FASTMC}_i)]/\sigma_i$ for each point in the distribution, N_i is the data yield in bin i, and only the statistical uncertainty is used. The fit ranges are indicated by the double-ended horizontal arrows [10]. For a color version of this figure, please see the color plates at the beginning of the book.

is to reconstruct the W mass directly from the decay products ($q\bar{q}q\bar{q}$, $q\bar{q}l\nu_l$ ($l = e, \mu, \tau$)). Figure 3.12 shows variation of theoretical cross sections for three different values of m_W as a function of the $\sqrt{s}$.[8)] A combined data of the LEP four groups is shown as the horizontal band. The determined value is [122]

$$m_W = 80.4 \pm 0.220_{\text{exp}} \pm 0.025_{E_{\text{beam}}} \text{ GeV} \tag{3.39}$$

Figure 3.13 shows two direct reconstructions of m_W using fully hadronic ($W \to q\bar{q}q\bar{q}$) and semileptonic decay modes ($W \to q\bar{q}e\nu$). The direct reconstruction gives far better resolution than that determined from the threshold behavior. The LEP four groups combined their data to produce [3]

$$80.376 \pm 0.033 \text{ GeV} \tag{3.40}$$

which is comparable with the Tevatron data. The combined world average by the particle data group gives [3]

$$80.399 \pm 0.023 \text{ GeV} \tag{3.41}$$

3.3.4
Decay Width of W

Direct Method In principle, the decay width can be determined by fitting the m_T spectrum, treating m_W and Γ_W as parameters. The method, referred to as the direct method, was not accurate in the early days due to the difficulty of obtaining

8) The theoretical expression for the differential cross section will be given later when we discuss the triple gauge coupling constant.

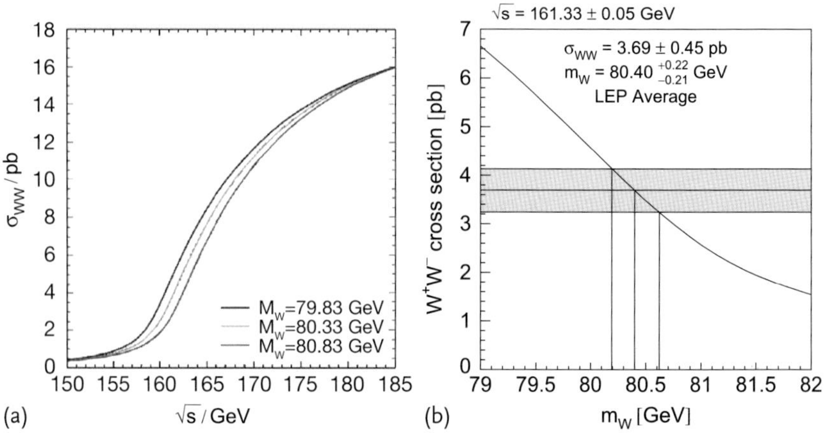

Figure 3.12 (a) The $e^-e^+ \to W^-W^+$ cross section as a function of $\sqrt{s}$ for various m_W values. (b) The combined measurement of σ_{WW} (shaded band) near threshold at LEP is compared to a semianalytic calculation of $\sigma_{WW}(MW, \sqrt{s})$ using the LEP average center of mass energy to extract a value m_W [122].

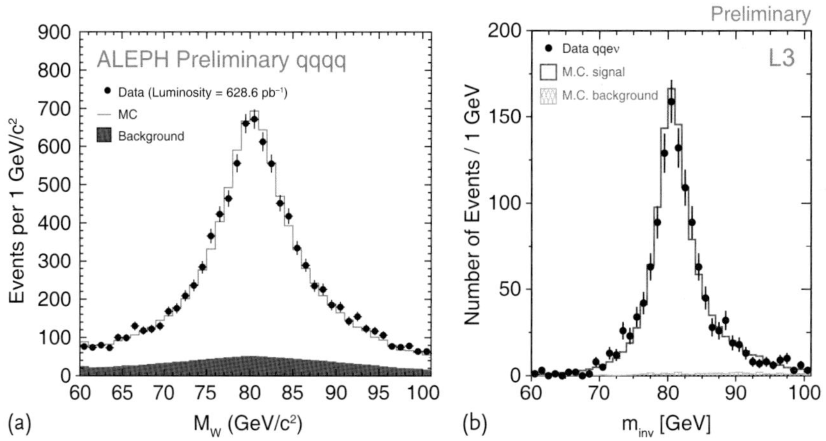

Figure 3.13 (a) W-boson mass spectra reconstructed for fully hadronic events (ALEPH) and (b) semileptonic events with electrons (L3) [123].

precise missing energy and poor statistics. Only recently, with the advent of high luminosity techniques on the accelerator side, it started to provide high quality data. We argued that m_T distribution is sharply cut off at the Jacobian peak. Tails beyond the Jacobian peak are completely determined by the Breit–Wigner broadening and the detector resolution. As the experimental resolution is small compared to the width, an accurate measurement of the high energy tail offers the most unbiased value of the width.

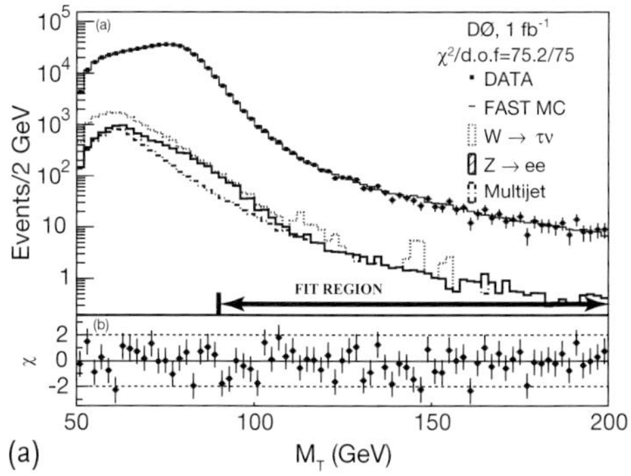

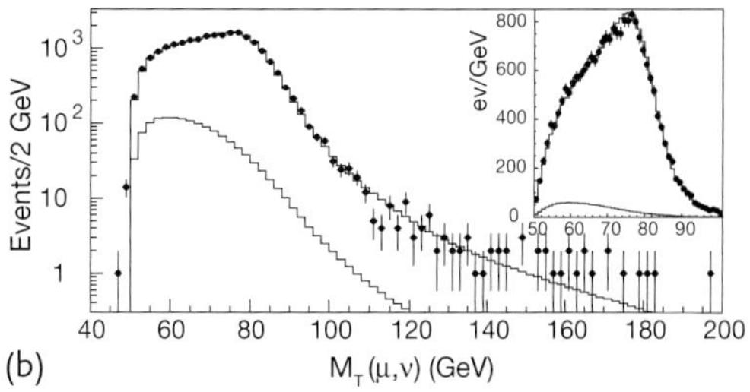

Figure 3.14 (a) $W \to e\nu$ data by D0. The distribution of the fast MC simulation, including the cumulative contributions of the different backgrounds, is normalized to the data in the region $50 < m_T < 100$ GeV [124]. (b) Transverse mass spectra (filled circles) for $W \to \mu\nu$ data by CDF, with best fits superimposed as a solid curve. The lower curve in each graph shows the sum of the estimated backgrounds. The inset shows the 50–100 GeV region on a linear scale [125, 126].

Figure 3.14 shows data by D0 [124] and CDF [125, 126]. The CDF data shows a nice contrast of the log scale data to emphasize the fit region with the linear data, similar to those in Figure 3.11, to determine the mass.

The combined data of CDF and D0 obtained this way gives the value [3]

$$\Gamma_W = 2.046 \pm 0.049\,\text{GeV} \tag{3.42}$$

Indirect Method The indirect method needs extra information to determine the decay width, but has offered more accurate values historically. Denoting R as the ratio of the electron numbers obtained from the W and Z decays,

$$R \equiv \frac{(\#\text{ of } W \to e\nu)}{(\#\text{ of } Z \to e\overline{e})} = \frac{\sigma^W \cdot \text{BR}(W \to e\nu)}{\sigma^Z \cdot \text{BR}(Z \to ee)} = \frac{\sigma^W}{\sigma^Z}\frac{\Gamma(W \to e\nu)}{\Gamma(Z \to e\overline{e})} \cdot \frac{\Gamma_Z}{\Gamma_W} \tag{3.43}$$

We may trust theoretical calculations for the ratio of the leptonic width. The ratio of the cross sections σ^W/σ^Z can be obtained experimentally. It contains theoretical errors which originate from the estimation of the accurate parton distributions. We trust it because the error estimate is more reliable for the ratio. Since a very accurate value of Γ_Z can be obtained from LEP data, one may use R, σ^W/σ^Z, and Γ_Z to determine Γ_W. Another method is to derive BR$(W \to e\nu)$ experimentally using LEP data for BR$(Z \to ee)$ and calculate Γ_W using a theoretical prediction of $\Gamma(W \to e\nu)$. The D0 group used this method. Using their data [127] to obtain

$$\sigma(p\overline{p} \to WX)\text{BR}(W \to e\nu) = 2310 \pm 10 \pm 50 \pm 100\,\text{pb} \tag{3.44a}$$

$$\sigma(p\overline{p} \to ZX)\text{BR}(Z \to ee) = 221 \pm 3 \pm 4 \pm 10\,\text{pb} \tag{3.44b}$$

$$R = 10.43 \pm 0.15 \pm 0.20 \pm 0.10 \tag{3.44c}$$

and quoting the result from

$$\mathrm{BR}(Z \to ee) = 0.033\,67 \pm 0.000\,06 \quad [128] \tag{3.44d}$$

$$\sigma^W / \sigma^Z = 3.362 \pm 0.053 \quad [113, 116] \tag{3.44e}$$

they obtained

$$\mathrm{BR}(W \to e\nu) = 0.1044 \pm 0.0015_{\mathrm{stat}} \pm 0.0020_{\mathrm{syst}} \pm 0.0017_{\mathrm{other}} \tag{3.45a}$$

to be compared with the Standard Model prediction

$$\mathrm{BR}(W \to e\nu)_{\mathrm{SM}} = 0.1084 \pm 0.0002 \tag{3.45b}$$

Using the Standard Model prediction of $\Gamma(W \to e\nu) = 0.2270 \pm 0.0011$ GeV [129], the value of Γ_W was determined to be

$$\Gamma_W = \frac{\Gamma(W \to e\nu)}{\mathrm{BR}(W \to e\nu)} = 2.169 \pm 0.031_{\mathrm{stat}} \pm 0.042_{\mathrm{syst}} \pm 0.041_{\mathrm{theory}} \pm 0.022_{\mathrm{NLO}}\ \mathrm{GeV} \tag{3.46}$$

to be compared with the Standard Model prediction of $\Gamma_W = 2.085 \pm 0.042$ GeV (Table 3.1, [3]).

$e^-e^+ \to W^-W^+$ At LEP2, the direct reconstruction of the m_W and its distribution were obtained which was given in Figure 3.13. The same data that determined the m_W was also used to determine the width to give

$$\Gamma_W = 2.196 \pm 0.083 \tag{3.47}$$

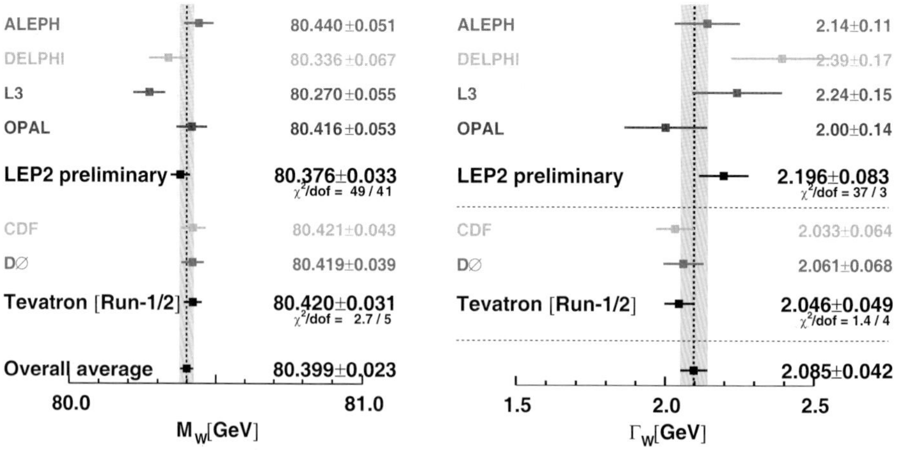

Figure 3.15 Measurements of the W-boson mass and decay width by the LEP and Tevatron experiments [3].

The combined world average of Γ_W is given by [3]

$$\Gamma_W = 2.085 \pm 0.042 \tag{3.48}$$

Figure 3.15 shows a summary of the W-boson mass and width measurements.

3.3.5 Triple Gauge Coupling

Confirmation through $e^-e^+ \to W^-W^+$ As the Standard Model is based on the symmetry $SU(2) \times U(1)$, the W, Z bosons have self-interactions which are characteristics of the non-Abelian gauge theory. W pair production by the reaction $e^-e^+ \to W^-W^+$ goes through four diagrams in the tree approximation described in Figure 3.16. The traditional contribution (a) alone gives a cross section which grows like $\sim s$ and violates unitarity at large s. In the gauge theory, new processes are necessary (Figure 3.16b–d). The process with nonlinear gauge coupling of the W with γ and Z (Figure 3.16b, c) works in such a way as to cancel it at large s and ameliorates the catastrophe (see discussions in Section 1.7). Therefore, comparison of the measured cross section of the process $e^-e^+ \to W^-W^+$ with the theory is an excellent test for the existence of the triple gauge coupling (TGC for short) and its magnitude. Experimentally, we need not consider the diagram Figure 3.16d because the strength of the Higgs coupling to the fermion is proportional to the fermion mass and at the W pair production energy, the factor m_e^2/m_W^2 reduces its contribution to a negligible level.

The Feynman amplitude of the gauge boson–fermion coupling and that of the gauge self-coupling are given in Eqs. (1.51a) and (1.56). The amplitude for the WWV ($V = \gamma, Z$) vertex is given by

$$\begin{aligned} \langle f|i\mathcal{L}_{WWV}|i\rangle &= ig_V[g_{\mu\nu}(k_1-k_2)_\lambda + g_{\nu\lambda}(k_2-k_3)_\mu + g_{\lambda\mu}(k_3-k_1)_\nu] \\ g_V &= e \qquad \text{for } V=\gamma \\ g_V &= e\cot\theta_W \text{ for } V=Z \end{aligned} \tag{3.49}$$

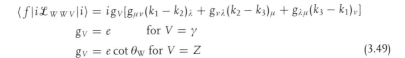

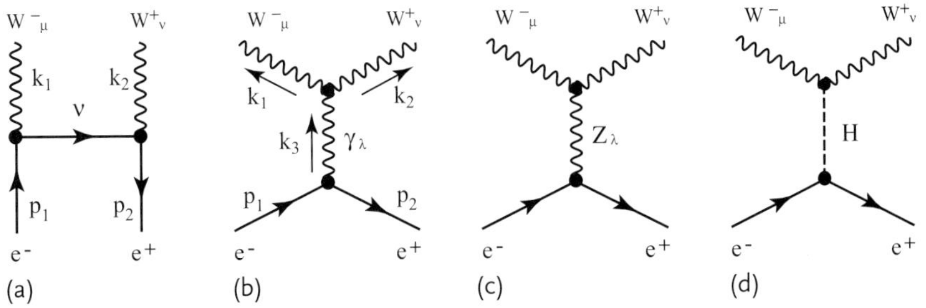

Figure 3.16 Feynman diagrams for $e^-e^+ \to W^-W^+$. (a) Charged current (b) Electromagnetic (c) Neutral current (d) Higgs exchange processes. Suffixes μ, ν, λ denote polarization states of the vector bosons. The diagram (d) is negligible because $m_e^2/m_W^2 \approx 0$.

where all the momenta are defined as incoming vectors. For the diagram in Figure 3.16b, replacements $k_1 \to -k_1, k_2 \to -k_2$ are to be understood. Using Eq. (3.49), the cross section $\sigma(e^-e^+ \to W^-W^+)$ can be calculated. Using variables

$$\beta \equiv \sqrt{1-\lambda}\,, \quad \lambda \equiv m_W^2/s\,, \quad x_W = \sin^2\theta_W \tag{3.50}$$

defined in the WW center of mass frame, the integrated total cross section is expressed as [130–132]

$$\begin{aligned}\sigma(e^-e^+ \to W^-W^+) &= \frac{\pi\alpha^2\beta}{2sx_W^2}\left[\left\{(1+2\lambda+2\lambda^2)\frac{1}{\beta}\ln\left(\frac{1+\beta}{1-\beta}\right)-\frac{5}{4}\right\}\right.\\ &+\frac{m_Z^2(1-2x_W)}{(s-m_Z^2)}\left\{2(2\lambda+\lambda^2)\frac{1}{\beta}\ln\left(\frac{1+\beta}{1-\beta}\right)-\frac{1}{12\lambda}-\frac{5}{3}-\lambda\right\}\\ &\left.+\frac{m_Z^4\left(1-4x_W+8x_W^2\right)}{\left(s-m_Z^2\right)^2}\frac{\beta^2}{48}\left(\frac{1}{\lambda^2}+\frac{20}{\lambda}+12\right)\right]\end{aligned} \tag{3.51}$$

The first line is the contribution of the t-channel ν exchange, the third line, that of the s-channel Z exchange, and the second line, that of interference. In the vicinity of the threshold [$s = (2m_W)^2$] where $\beta \ll 1$, the t-channel contribution dominates and the cross section is approximated by

$$\sigma_{\text{threshold}} \simeq \left(\frac{\pi\alpha^2}{4m_W^2 x_W^2}\right)\beta \simeq 46\beta\ (\text{pb}) \tag{3.52}$$

As the maximum value of Eq. (3.51) is 17 pb, the cross section rises sharply. For instance, $\sigma \sim 7$ pb at $s = (2m_W)^2 + 1$ GeV ($\beta \sim 0.16$). In such a case, the effect of the decay width Γ_W cannot be neglected, but its effect is to lower the threshold and to smooth the slope and the overall effect is small. Of course, this small change has to be taken into account in determining the mass of W as discussed in Section 3.3.3. For $s \to \infty$, again the first term dominates:

$$\sigma(s \gg m_W^2) \to \left(\frac{\pi\alpha^2}{2x_W^2}\frac{1}{s}\ln\frac{s}{m_W^2}\right) \tag{3.53}$$

If the TGC (triple gauge coupling) does not exist, the cross section increases like $\sigma \sim G_F^2 s$ and violates unitarity at the high energy. Figure 3.17 shows a comparison of the cross section with and without the triple gauge coupling. The cross section rises sharply at the threshold and after taking a maximum value of 17 pb, decreases slowly (logarithmically). The data [134] agrees well with the theoretical prediction [135, 136] and confirmed the existence of the TGC.

TGC beyond the Standard Model The existence of the triple gauge coupling was confirmed experimentally. Its coupling form in the Standard Model given by Eq. (3.49) is very restrictive and defines the shape of the TGC and its strength uniquely. They could be different if the charged boson has an anomalous magnetic

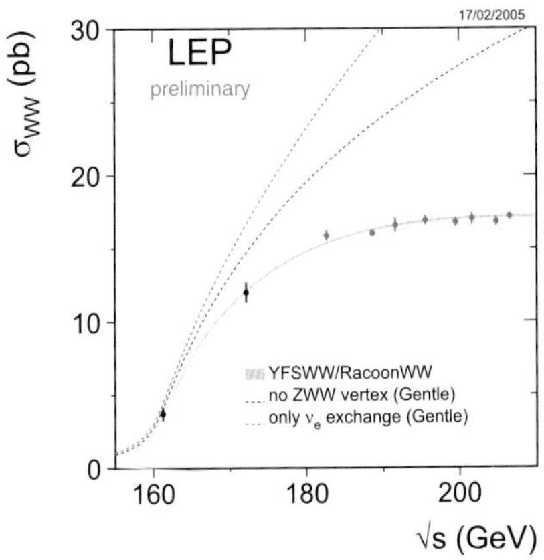

Figure 3.17 Cross section for $e^-e^+ \to W^-W^+$ compared with the SM expectation. Also shown is the expectation from the t-channel ν_e exchange only (top), and inclusion of additional Z exchange (middle) [133, 134].

moment and/or the quadrupole moment. A new physics may also appear as a difference from the SM prediction. It is worthwhile to determine the shape of the TGC phenomenologically and look for a possible clue beyond the SM. The most general Lagrangian for the triple gauge boson coupling consistent with the Lorentz invariance is expressed as [123, 137].

$$\begin{aligned}
i\mathcal{L}_{\text{TGC}} = g_{VWW}\Bigg[& g_1^V V^\mu (W^-_{\mu\nu} W^{+\nu} - W^+_{\mu\nu} W^{-\nu}) \\
& + \kappa_V W^+_\mu W^-_\nu V^{\mu\nu} + \frac{\lambda_V}{m_W^2} V^{\mu\nu} W^{+\rho}_\nu W^-_{\rho\mu} \\
& + i g_5^V \epsilon_{\mu\nu\rho\sigma} \left\{ (\partial^\rho W^{-\mu}) W^{+\nu} - W^{-\mu} (\partial^\rho W^{+\nu}) \right\} V^\sigma \\
& + i g_4^V W^+_\mu W^-_\nu (\partial^\mu V^\nu + \partial^\nu V^\mu) \\
& - \frac{\tilde{\kappa}_V}{2} W^-_\mu W^+_\nu \epsilon^{\mu\nu\rho\sigma} V_{\rho\sigma} - \frac{\tilde{\lambda}_V}{2m_W^2} W^-_{\rho\mu} W^{+\mu}_\nu \epsilon^{\nu\rho\alpha\beta} V_{\alpha\beta} \Bigg]
\end{aligned} \tag{3.54}$$

where $V_{\mu\nu} = \partial_\mu V_\nu - \partial_\nu V_\mu$. The overall coupling constant is fixed by $g_{VWW} = e$ for $V = \gamma$ and $g_{VWW} = e \cot\theta_W$ for $V = Z$ where θ_W is the weak mixing angle. Once the overall couplings are defined, there are seven complex parameters for two choices of $V = Z, \gamma$, adding to fourteen in total. They are too many to be measured simultaneously and some hypotheses have to be introduced. Let us see first how they transform under P, C and T transformations. The vector fields transform according to

$$\begin{aligned} &W_\mu^- \xrightarrow{C} -W_\mu^\dagger\,, \quad A_\mu \xrightarrow{C} -A_\mu \\ &V_\mu(t,\boldsymbol{x}) \xrightarrow{P} V^\mu(t,-\boldsymbol{x})\,, \quad \partial_\mu \xrightarrow{P} \partial^\mu \\ &V_\mu(t,\boldsymbol{x}) \xrightarrow{T} V^\mu(-t,\boldsymbol{x})\,, \quad \partial_\mu \xrightarrow{T} -\partial^\mu \end{aligned} \tag{3.55}$$

where A_μ is the electromagnetic potential (real vector) and $V_\mu = W_\mu^{\pm,0}, A_\mu$. One sees that the first three terms in Eq. (3.54) respect all the three P, C and T symmetries. The fourth term respects CP symmetry, but the last three terms do not. Thus, the first logical constraint to start with is to require the CP symmetry. The electromagnetic gauge invariance fixes $g_1^\gamma = 1$ and $g_5^\gamma = g_5^Z = 0$. The remaining $g_1^Z, \kappa_\gamma, \kappa_Z, \lambda_\gamma, \lambda_Z$ are assumed to be real (i.e., T invariant). Custodial $SU(2)$ symmetry implies [138, 139]

$$\kappa_Z = g_1^Z - \tan^2\theta_W(\kappa_\gamma - 1)\,, \quad \lambda_Z = \lambda_\gamma \tag{3.56}$$

and reduces the number of the TGC parameters to three which are g_1^Z, κ_γ and λ_γ. The three sets of parameters g_1^Z, κ_γ and λ_γ are the starting point of conventional experimental analysis. The Standard Model values are $g_1^Z = \kappa_\gamma = \kappa_Z = 1$, and $\lambda_\gamma = \lambda_Z = 0$ at the tree level.

The custodial $SU(2)$ is a global $SU(2)$ symmetry and is approximately valid in the electroweak phenomenology. The reasoning is as follows: start from a global chiral symmetry $SU(2)_L \times SU(2)_R$ and gauge the $SU(2)_L$ group. The Standard Model is obtained if the $U(1)$ is also gauged to produce the hypercharge force. The custodial symmetry is obtained by spontaneous breaking of the chiral symmetry $SU(2)_L \times SU(2)_R$ down to $SU(2)_V$. It is valid in QCD, but it is not the symmetry of the electroweak force because the $SU(2)_R$ mixes particles with different Y violating the Standard Model assumption. However, as long as g_B (coupling of $U(1)$ gauge boson) is small, the global $SU(2)_R$ is approximately valid up to the mass generating Higgs sector. Namely, it is also broken by the mass splitting ($m_t \neq m_b$) of the doublets. The custodial symmetry protects ρ, the relative strength of the neutral current to the charged current from acquiring a large radiative correction, hence its name. Phenomenologically, $\Delta\rho \equiv \rho - 1$ is indeed small ($\sim$ a few percent; see Section 4.5).

Let us take the expectation value of the Lagrangian, between the physical $W^\pm$ states $|p, \epsilon_h\rangle$ and $|p', \epsilon_{h'}\rangle$ where $\epsilon_{h,h'}$ are polarization vectors. Using the fact that $(p \cdot \epsilon_h) = (p' \cdot \epsilon_{h'}) = 0$, $p^2 = p'^2 = m_W^2$, the vertex function for $V = \gamma, Z$ in momentum space can be expressed as [140]

$$g_{VWW}\Gamma^\mu_{h',h} \equiv g_{VWW}\langle p's'h'|J_\mu(0)|psh\rangle \tag{3.57a}$$

$$\Gamma^{\mu}_{h',h} = i(\epsilon^*_{h'})_\beta V^{\mu\alpha\beta}(\epsilon_h)_\alpha \overset{(p^2=m^2,p'^2=m'^2)}{=\!=\!=} -\Big[G_1^V (p+p')^\mu (\epsilon^*_{h'}\cdot\epsilon_h)$$
$$+ G_2^V \{\epsilon^{*\mu}_{h'}(q\cdot\epsilon_h) - \epsilon_h{}^\mu (q\cdot\epsilon^*_{h'})\} + G_3^V (p+p')^\mu (q\cdot\epsilon^*_{h'})(q\cdot\epsilon_h)/m^2\Big] \tag{3.57b}$$

$$G_1^V(q^2) = 1 + \frac{1}{2}\lambda_V \frac{q^2}{m^2}, \quad G_2^V(q^2) = 1 + \kappa_V + \lambda_V, \quad G_3^V(q^2) = -\lambda_V \tag{3.57c}$$

$$q^\mu = (p-p')^\mu \tag{3.57d}$$

These parameters are related to variables familiar in the nonrelativistic electromagnetic phenomena. κ_γ and λ_γ can be expressed by the magnetic dipole μ_W and the electric quadrupole Q_W moments of the charged W bosons [140].[9)]

$$\mu_W = \frac{e}{2m_W}(1+\kappa_\gamma+\lambda_\gamma) \tag{3.59a}$$

$$Q_W = -\frac{e}{m_W^2}(\kappa_\gamma-\lambda_\gamma) \tag{3.59b}$$

Proof of the Eqs. (3.59) is given in Appendix G.

The differential cross section of the W-boson pair production exhibits a strong dependence on g_1^Z, κ_γ and λ_γ. For the $e^-e^+ \to W^-W^+$, $W^\mp \to l\nu$ process with unpolarized beam, it is possible to determine five angles completely. They are $\theta_{W\mp}$ which are the production angle of the $W^\pm$ with respect to the incoming electron, polar and azimuthal angles of fermions and antifemions. For semileptonic and fully hadronic events, some variables have to be folded. The largest sensitivity to the TGC comes from $\cos\theta_{W\pm}$ distribution of the cross section [100, 123].

$e^-e^+ \to W^-W^+$ The existence of the TGC coupling was confirmed by measurements of the $e^-e^-+ \to W^-W^+$ cross section as stated in the previous paragraph. Experimental searches for the anomalous behavior of the TGC is still in its infancy. Here, we discuss some constraints obtained by the same experiments. We show one example in Figure 3.18 obtained at the LEP2 [141, 142, 144] and the resultant allowed region for the g_1^Z and κ_γ in Figure 3.19 [145]. The combined LEP

9) Note that the electric dipole d_W and magnetic quadrupole moments $\tilde{Q}_W$ are parametrized as

$$d_W = \frac{e}{2m_W}(\tilde{\kappa}_\gamma + \tilde{\lambda}_\gamma),$$
$$\tilde{Q}_W = -\frac{e}{m_W^2}(\tilde{\kappa}_\gamma - \tilde{\lambda}_\gamma) \tag{3.57}$$

However, both $\tilde{\kappa}$ and $\tilde{\lambda}$ violate P and T.

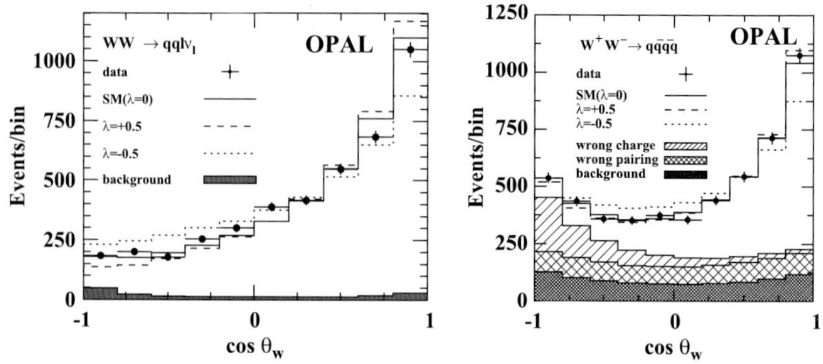

Figure 3.18 Differential distribution for the cosine of the W^- polar angle for semileptonic and fully hadronic events. Predictions of the Standard Model and those in the presence of an anomalous value of the coupling λ_γ are also given [123, 141].

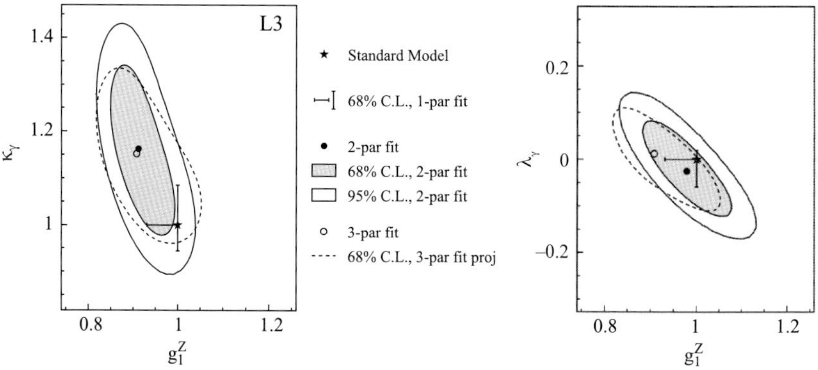

Figure 3.19 Results of determinations of the couplings g_1^Z and κ_γ, λ [145].

results are given by [134]

$$g_1^Z = 0.991^{+0.022}_{-0.021}\,, \quad \kappa_\gamma = 0.984^{+0.042}_{-0.047}\,, \quad \lambda_\gamma = -0.016^{+0.021}_{-0.023} \tag{3.60}$$

in agreement with predictions of the Standard Model.

4
Physics at Z Resonance

In Chapter 2, we saw that the neutral current data confirmed the Standard Model of the electroweak theory and established its validity beyond doubt. However, comparisons with experiments were at tree levels. The true test is if the theory including higher order corrections can reproduce experimental data accurately. The e^-e^+ reaction, unlike the hadronic reaction, is free of the major background (QCD process) and offers ideal work benches for testing the theoretical predictions. The process $e^-e^+ \to \gamma, Z \to f\overline{f}, (f = \nu, e, \mu, \tau$ and $q(\overline{q}) \to$ hadrons) among them, play a crucial role in testing the mathematical consistency of the Standard Model, i.e., higher order corrections of the renormalizable theory by comparing with precision data. The tests play a similar role as the Lamb shift or $g-2$ in QED. That was the expectation for the LEP[1] experiments at the Z resonance. To state conclusions of this chapter in advance, experiments at the Z pole firmly established credibility of the Standard Model and the focus of research after the LEP shifted to the discovery of the Higgs and new physics beyond the Standard Model.

Figure 4.1 shows the total cross section of e^-e^+ collision as a function of the total center of mass energy $\sqrt{s}$ between 20 and 200 GeV. Two prominent features are immediately apparent. A huge Z resonance peak and otherwise monotonic $1/s$ behavior[2] up to threshold for W^+W^- production as exactly predicted by the Standard Model. For comparison, cross sections of a pure QED process $\sigma(e\overline{e} \to \gamma\gamma)$ are also shown.

In this chapter, we first derive tree level Born formulas for various processes, including the total cross section, branching ratios to $f\overline{f}$, and forward–backward asymmetry. We then investigate how they are modified by the higher order corrections and compare them with experimental data. Formalism and detailed calculations of the radiative corrections will be discussed in the next chapter.

1) The large electron–positron collider at CERN with $\sqrt{s} = 91$ GeV started operation in 1989 and later upgraded to LEP II with $\sqrt{s} = 200$ GeV.
2) That is, if we ignore resonances appearing in the neighborhood of $q\overline{q}(q = s, c, b)$ production threshold.

Elementary Particle Physics, Volume 2: Foundations of the Standard Model, First Edition. Yorikiyo Nagashima.

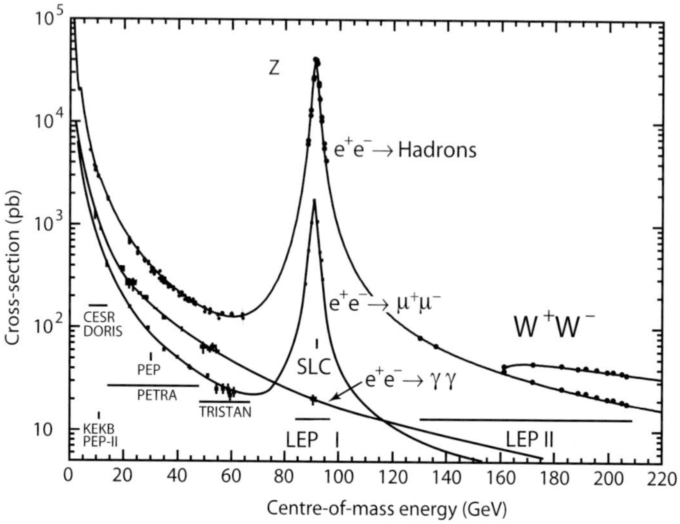

Figure 4.1 Cross sections for $e^-e^+ \to$ hadrons and $e^-e^+ \to \mu^-\mu^+$ as a function of center of mass energy. Also shown is a pure QED process ($e^-e^+ \to \gamma\gamma$) for comparison. Solid lines show predictions of the SM, and points are experimental measurements. Also indicated are the energy ranges of various e^-e^+ accelerators. The cross sections have been corrected for the effects of photon radiation. Figure adapted from [7, 152].

4.1 Born Approximation

Decay Width The Lagrangian to describe the neutral current processes was given in Eq. (1.50) and its coupling form in Eq. (1.51a). Setting kinematic variables as

$$Z(q,\epsilon) \to f(p_1) + \overline{f}(p_2) \tag{4.1}$$

where ϵ [3] (taken to be real) is the polarization vector of the Z boson. The decay width can be derived by calculating an invariant matrix element $\mathcal{M}$ and Lorentz invariant phase space volume (LIPS)

$$\Gamma(Z \to f\overline{f}) = \frac{1}{2m_Z}\overline{\sum_{\text{spin}}}|\mathcal{M}|^2 dLIPS \tag{4.2a}$$

$$\mathcal{M} = g_Z \overline{u}(p_1)\gamma^\mu \left(\varepsilon_L(f)\frac{1-\gamma^5}{2} + \varepsilon_R(f)\frac{1+\gamma^5}{2}\right) v(p_2)\epsilon_\mu \tag{4.2b}$$

$$g_Z = \frac{e}{\sin\theta_W \cos\theta_W} \tag{4.2c}$$

$$\varepsilon_L(f) = I_{3L} - Q_f \sin^2\theta_W\,, \quad \varepsilon_R(f) = -Q_f \sin^2\theta_W \tag{4.2d}$$

3) The polarization vector $\epsilon_\mu(\lambda)$ should not be confused with the strength of the neutral current coupling to fermions $\varepsilon_L(f), \varepsilon_R(f)$.

Alternative expressions in terms of vector (v_f) and axial vector (a_f) coupling coefficients are

$$\mathcal{M} = \frac{g_Z}{2}\overline{u}(p_1)\gamma^\mu\left(v_f - a_f\gamma^5\right)v(p_2)\epsilon_\mu \tag{4.3a}$$

$$v_f = \varepsilon_{\mathrm{L}}(f) + \varepsilon_{\mathrm{R}}(f) = I_{3f} - 2Q_f\sin^2\theta_{\mathrm{W}} \tag{4.3b}$$

$$a_f = \varepsilon_{\mathrm{L}}(f) - \varepsilon_{\mathrm{R}}(f) = I_{3f} \tag{4.3c}$$

Calculating the average of the polarization and the sum of the spin, we immediately obtain

$$\begin{aligned}\Gamma(Z\to f\overline{f}) &= \frac{N_{\mathrm{c}}^f g_Z^2 m_Z}{48\pi}(1-4x)^{1/2}\left[v_f^2(1+2x)+a_f^2(1-4x)\right]\\ &= \frac{N_{\mathrm{c}}^f G_{\mathrm{N}} m_Z^3}{6\sqrt{2}\pi}(1-4x)^{1/2}\left[v_f^2(1+2x)+a_f^2(1-4x)\right]\end{aligned} \tag{4.4a}$$

$$N_{\mathrm{c}}^f = \begin{cases}(1+\delta_{\mathrm{QED}}) & \text{(leptons)}\\ 3(1+\delta_{\mathrm{QED}})(1+\delta_{\mathrm{QCD}}) & \text{(quarks)}\end{cases} \tag{4.4b}$$

where N_{c}^f is the color factor and $x = m_f^2/m_Z^2$. Here, $G_{\mathrm{N}} \equiv \rho G_{\mathrm{F}}$ ($\rho = 1$ for the Born approximation) is the Fermi coupling constant. Note that, as far as the electroweak interaction is concerned, this is the Born (tree) approximation, except for the final state corrections.

$$\delta_{\mathrm{QED}} = \frac{3\alpha}{4\pi}Q_f^2 + \cdots \tag{4.5a}$$

$$\delta_{\mathrm{QCD}} = \frac{\alpha_s(m_Z)}{\pi} + 1.409\left(\frac{\alpha_s(m_Z)}{\pi}\right)^2 + \cdots \tag{4.5b}$$

If one neglects the fermion mass, the expressions are simplified:

$$\Gamma(Z\to f\overline{f}) = 2N_{\mathrm{c}}^f(v_f^2 + a_f^2)\Gamma_\nu \tag{4.6a}$$

$$\Gamma_\nu = \frac{G_{\mathrm{N}} m_Z^3}{12\sqrt{2}\pi} \tag{4.6b}$$

In Table 4.1, we compare the decay branching ratios obtained by the Born formula Eq. (4.4a) including the final state interaction. From this table, one sees that the branching ratio to $e\overline{e}$ is 3.3% and that to all $\nu\overline{\nu}$s (the so-called invisible branch), it is 20.5%. As will be clarified, the tree level calculation already gives quite a good approximation. The task of this chapter, however, is to prove that the small deviations of the Born terms from the experimental data can be well reproduced by the radiative corrections of the Standard Model.

Table 4.1 Decay width and Branching ratio for $Z \to f\overline{f}$. $m_Z = 91.19$ GeV at tree level.

f	v_f	a_f	Γ_f/Γ_ν	Width (GeV)	Branching Ratio (%)
ν	$\frac{1}{2}$	$\frac{1}{2}$	1	0.167×3	6.84×3
e, μ, τ	$-\frac{1}{2} + 2x_W$	$-\frac{1}{2}$	$\frac{1}{2}[1 + (1 - 4x_W)^2]$	0.084×3	3.34×3
u, c	$\frac{1}{2} - \frac{4}{3}x_W$	$\frac{1}{2}$	$\frac{N_c^f}{2}\left[1 + \left(1 - \frac{8}{3}x_W\right)^2\right]$	0.300×2	12.0×2
d, s	$-\frac{1}{2} + \frac{2}{3}x_W$	$-\frac{1}{2}$	$\frac{N_c^f}{2}\left[1 + \left(1 - \frac{4}{3}x_W\right)^2\right]$	0.383×2	15.4×2
b	$-\frac{1}{2} + \frac{2}{3}x_W$	$-\frac{1}{2}$	$\frac{N_c^f}{2}\left[1 + \left(1 - \frac{4}{3}x_W\right)^2\right]$	0.376	15.2
All hadrons				1.742	70
	$x_W = \sin^2\theta_W$		Total decay width $\Gamma_Z = 2.495$ GeV		

Cross Section We learned in Section 2.6 that the cross section for $e^-e^+ \to f\overline{f}$ (f = any fermion) in its general form is expressed as sum of the cross sections for a polarized beam by a polarized target. The expression for unpolarized target is given by Eq. (2.69).

$$\frac{d\sigma}{d\Omega} = \frac{1}{2}(d\sigma_L + d\sigma_R) = \frac{\alpha^2}{4s} N_c^f \left[(1 + \cos^2\theta) F_1 + 2\cos\theta\, F_2\right] \tag{4.7}$$

where

$$d\sigma_L = \frac{1}{2}(d\sigma_{LL} + d\sigma_{LR}), \quad d\sigma_R = \frac{1}{2}(d\sigma_{RL} + d\sigma_{RR}) \tag{4.8}$$

are cross sections with polarized beams. $d\sigma_{ab}$, $(a, b$ = LR with b denoting the target polarization) are given by Eq. (2.67a). The cross section for a polarized beam with polarization P is expressed as

$$\begin{aligned}\frac{d\sigma}{d\Omega}^{\text{pol}}(P) &= \frac{1-P}{2} d\sigma_L + \frac{1+P}{2} d\sigma_R \\ &= \frac{\alpha^2}{4s} N_c^f \left[\{(1+\cos^2\theta)F_1 + 2\cos\theta\, F_2\} - P\{(1+\cos^2\theta)F_3 + 2\cos\theta\, F_4\}\right]\end{aligned} \tag{4.9}$$

where

$$\begin{aligned}F_1 &= (1/4)(|F_{LL}|^2 + |F_{LR}|^2 + |F_{RL}|^2 + |F_{RR}|^2) \\ &= \left[(Q_f Q_e)^2 - 2Q_f Q_e|\chi(s)|\cos\delta_R v_e v_f + |\chi(s)|^2(v_e^2 + a_e^2)(v_f^2 + a_f^2)\right] \\ F_2 &= (1/4)(|F_{LL}|^2 - |F_{LR}|^2 - |F_{RL}|^2 + |F_{RR}|^2) \\ &= \left[-2Q_f Q_e|\chi(s)|\cos\delta_R a_e a_f + 4|\chi(s)|^2 v_e a_e v_f a_f\right] \\ F_3 &= (1/4)(|F_{LL}|^2 + |F_{LR}|^2 - |F_{RL}|^2 - |F_{RR}|^2) \\ &= \left[-2Q_f Q_e|\chi(s)|\cos\delta_R a_e v_f + 2|\chi(s)|^2 v_e a_e(v_f^2 + a_f^2)\right]\end{aligned}$$

$$F_4 = (1/4)(|F_{LL}|^2 - |F_{LR}|^2 + |F_{RL}|^2 - |F_{RR}|^2)$$
$$= \left[-2Q_f Q_e|\chi(s)|\cos\delta_R v_e a_f + 2|\chi(s)|^2(v_e^2 + a_e^2)(v_f a_f)\right] \quad (4.10a)$$

with

$$\chi(s) = \frac{G_N m_Z^2}{2\sqrt{2}\pi\alpha} \frac{s}{s - m_Z^2 + i m_Z \Gamma_Z} \quad (4.10b)$$

$$\tan\delta_R = \frac{m_Z \Gamma_Z}{m_Z^2 - s} \quad (4.10c)$$

F_{ab} are given in Eq. (2.66b). Here, θ is the scattering angle of the outgoing fermion with respect to the direction of e^-. $\chi(s)$ is the Z propagator where we included the decay width for phenomenological treatment.

The integrated $e\bar{e} \to f\bar{f}$ cross section around the Z resonance with the unpolarized beam is obtained from Eq. (4.9) by setting $P = 0$ and integrating over the solid angle. It consists of three parts, namely, the pure QED part which is the photon exchange process, Z production and their interference. At the Z pole, the interference term vanishes and the Z exchange dominates the process. The cross section can be expressed as

$$\sigma(s) = \frac{12\pi\Gamma_e\Gamma_f}{\left|s - m_Z^2 + i m_Z\Gamma_Z\right|^2}\left(\frac{s}{m_Z^2} + R_f\frac{s - m_Z^2}{m_Z^2}\right) + N_c^f(Q_f Q_e)^2\frac{4\pi\alpha^2}{3s} \quad (4.11a)$$

$$R_f = \frac{2\sqrt{2}\pi\alpha}{G_N m_Z^2}\frac{2Q_e Q_f v_e v_f}{(v_e^2 + a_e^2)\left(v_f^2 + a_f^2\right)} \quad (4.11b)$$

$$\Gamma_f = \frac{N_c^f G_N m_Z^3}{6\sqrt{2}\pi}\left(v_f^2 + a_f^2\right) \quad (4.11c)$$

We have used Eq. (4.6a) and (4.6b) to express the formula in terms of the decay widths. In the vicinity of the Z resonance,

$$\sigma_{res}(s) = \sigma^0\frac{s\Gamma_Z^2}{(s - m_Z^2)^2 + m_Z^2\Gamma_Z^2} \quad (4.12a)$$

$$\sigma^0 = \sigma_{res}\left(s = m_Z^2\right) = \frac{12\pi}{m_Z^2}\left(\frac{\Gamma_e}{\Gamma_Z}\right)\left(\frac{\Gamma_f}{\Gamma_Z}\right) \quad (4.12b)$$

At $s = m_Z^2$, the ratio of the total cross section to that of the QED μ pair production $\sigma_{QED} = 4\pi\alpha^2/3s$ becomes

$$\frac{\sigma(e\bar{e} \to all)}{\sigma_{QED}(e\bar{e} \to \mu\bar{\mu})} = \frac{9}{\alpha^2}\mathrm{BR}(Z \to e\bar{e}) \sim 6000 \quad (4.13)$$

This is a very large number as is demonstrated by the huge peak in Figure 4.1. The experimental data behaved exactly as expected and confirmed the giant Z resonance. Notice that the cross section is $O(1)$ at the resonance peak, to be contrasted with the $O(\alpha^2)$ expression.

4.2
Improved Born Approximation

Since precise data with large statistics are available at the Z peak, we consider effects of higher order radiative corrections and see how they modify the simple Born formula. We treat three kinds of radiative corrections:

1. Pure QED radiative corrections (mainly photon radiation from initial states)
2. QCD radiative corrections (mainly gluon radiation from final states)
3. The electroweak correction, which is our main interest.

The QCD radiative correction is one of the main subjects in QCD and will be discussed in detail later. Here, we merely treat it as an ad hoc small correction and include it in N_c^f whenever necessary as given in Eqs. (4.4b) and (4.5). The QED correction is well known and offers no new information. However, as its existence changes the line shape of the Z resonance production cross section as well as other observables, we first investigate its effect.

Pure QED Corrections The pure QED correction can be separated from other effects and is independently calculable [153–155]. The formula is by no means simple, but its treatment is well understood. Therefore, the experimental data are usually provided after unfolding the QED deformation. Among them, the most important is the photon radiation from the initial state. Its effect can be expressed in the following form.[4)]

$$\sigma_{\text{EW}}(s) \to \sigma_{\text{observed}} = \int_0^{\omega_{\max}} \sigma_{\text{EW}}(E-\omega)H(\omega)d\omega \tag{4.14}$$

$\omega_{\max}$ is the maximum energy of the emitted photon which is constrained by the experimental condition. The QED effects appear as

1. Reduction of the resonance height
2. Shift of the resonance peak
3. Long tail on the high energy side of the peak.

Their effects are demonstrated in Figure 4.2. Figure 4.2 shows the observed QED corrected cross section and the asymmetry as obtained by and averaged over four experimental groups [7]. The photon emission in the initial state introduces a large tail on the high energy side of the resonance peak. This is because electrons (or positrons) with higher energy lose their energy by radiation just to form the Z resonance. However, the change in the width is small.

4) The final state radiation is usually provided in the form given in Eq. (4.4b). Therefore, in using Eq. (4.14), one must be careful of double counting.

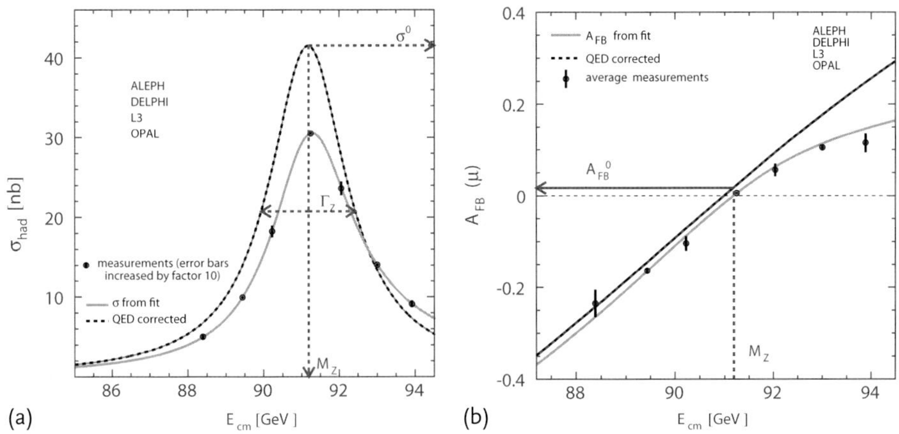

Figure 4.2 Average over measurements of the hadronic cross sections (a) and of the muon forward–backward asymmetry (b) by the four experiments at the LEP, as a function of center of mass energy. The full line represents the results of model-independent fits to the measurements. Correcting for QED photonic effects yields the dashed curves, which define the *Z* parameters described in the text. For a color version of this figure, please see the color plates at the beginning of the book.

If only the dominant $O(\alpha)$ contribution term is considered in the QED calculation, the peak value of the cross section is shown to be lower by

$$1-\beta\ln\frac{m_Z}{\Gamma_Z}\,, \quad \beta=\frac{2\alpha}{\pi}\left[\ln\frac{m_Z^2}{m_e^2}-1\right] \tag{4.15}$$

As the value of m_e^2 is small compared to m_Z^2, the reduction is large (~ 0.6). Compared to the initial state radiation, the final state radiation only gives a 0.17% correction. However, it is known that the $O(\alpha)$ effect alone is an overcorrection and it is argued that the proper treatment is to exponentiate Eq. (4.15) [153, 155–159], which gives the reduction factor

$$\sigma=\left(\frac{\Gamma_Z}{m_Z}\right)^{\beta}\sigma_0 \tag{4.16}$$

To understand qualitative features of the radiative corrections, let us consider a simple analytical expression. The cross section of one photon emission when the electron receives some kick can be factorized as a product of the cross section of the kick and the photon emission probability. For the case that the kick is given by the electromagnetic field of nuclei (i.e., Bremsstrahlung), it is given by Vol. 1,

Eq. (7.79) which is reproduced here.

$$\frac{d\sigma}{d\Omega_f} = \left(\frac{d\sigma}{d\Omega}\right)_{\text{elastic}} \frac{2\alpha}{\pi} A_{\text{IR}}(q^2) \ln \frac{\omega_{\max}}{\mu}$$

$$A_{\text{IR}}(q^2) = \int_0^1 dx \frac{m^2 - q^2/2}{m^2 - q^2 x(1-x)} - 1$$

$$\simeq \ln\left(\frac{-q^2}{m^2}\right) - 1 + O\left(\frac{m^2}{-q^2}\right), \quad (q^2 \gg m^2)$$

$$-q^2 = -(p_f - p_i)^2 \simeq |\boldsymbol{q}|^2 = 4|\boldsymbol{p}|^2 \sin^2(\theta/2) = s \sin^2(\theta/2) \tag{4.17}$$

where μ is the infrared cut off and p_i, p_f are the four-momenta of the electron before and after photon emission. It is a universal formula for soft photon radiation valid if one replaces the kick cross section by other processes. As the emission is in the extreme forward direction, the cross section after the photon emission can be approximated by

$$\sigma = (1-\delta)\sigma_0, \quad \delta = \frac{2\alpha}{\pi}\left(\ln \frac{s}{m_e^2} - 1\right) \int_{\Delta\omega}^{\sqrt{s}} \frac{d\omega}{\omega} = \beta \int_{\Delta\omega}^{E} \frac{d\omega}{\omega} \tag{4.18}$$

where δ is the angle integrated expression for one photon emission probability in Eq. (4.17) and s is the crossed momentum transfer squared to e^-e^+, i.e., the total energy squared. $\Delta\omega$ is the minimum photon energy that is not detected by the detector. Exponentiating Eq. (4.18), $\sigma = (1-\delta)\sigma_0$ is replaced by $e^{-\delta}\sigma_0$, giving

$$\sigma = \left(\frac{\Delta\omega}{E}\right)^{\beta} \sigma_0 \tag{4.19}$$

The formula is valid for the cross section, not appreciably varying with energy, which is not the case for the resonance production. The generalization was obtained by [160, 161]:

$$\sigma = \beta \int_0^{\Delta\omega} \frac{d\omega}{\omega} \left(\frac{\Delta\omega}{E}\right)^{\beta} \sigma_0(E-\omega) \tag{4.20}$$

The integral can be carried out for the Breit–Wigner shape. For a narrow resonance, it becomes

$$\sigma(E = \sqrt{s}) = \left[\frac{(E-m_Z)^2 + (\Gamma_Z/2)^2}{m_Z^2/4}\right]^{\beta/2} \times \left[1 + 2\beta \frac{E-m_Z}{\Gamma_Z}\left\{\frac{\pi}{2} + \tan^{-1}\left(\frac{E-m_Z}{\Gamma_Z/2}\right)\right\}\right] \sigma_0(E) \tag{4.21}$$

One notices that the $\Delta\omega$ has disappeared from the result. The width Γ_Z acted as the effective cut off. At $E = m_Z$, the formula reduces to Eq. (4.16). The second bracket in Eq. (4.21) induces an asymmetry in the shape of the resonance.

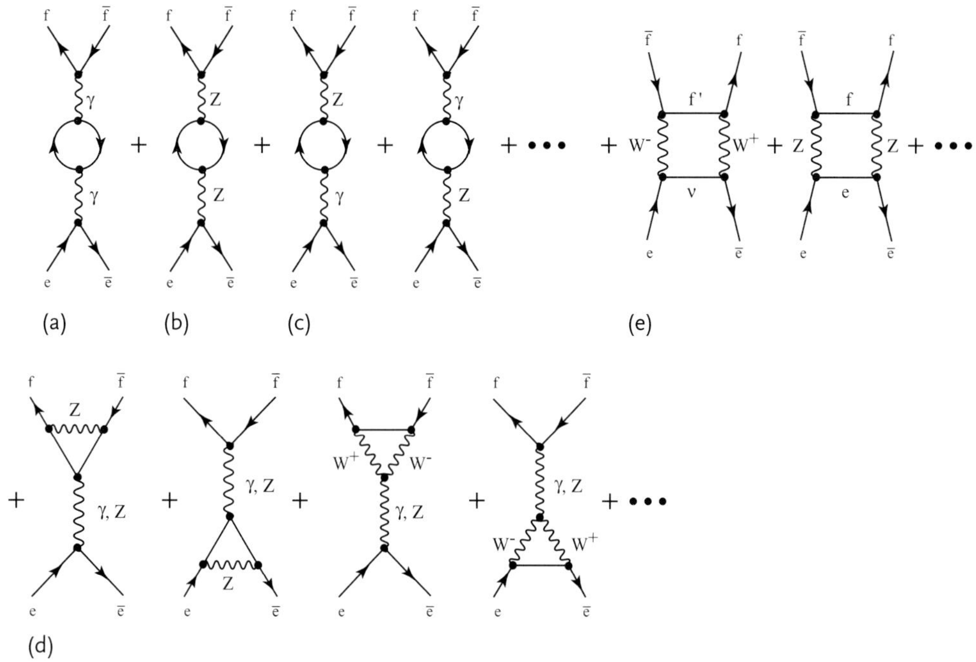

Figure 4.3 Examples of one-loop radiative corrections to $e\overline{e} \rightarrow f\overline{f}$ reactions. (a), (b) Self-energy of the gauge bosons. (c) $\gamma - Z$ mixing. (d) Vertex corrections and (e) box diagrams.

The shift of the peak position is given approximately by

$$\Delta s_{\max} \simeq \frac{\beta\pi}{8}\Gamma_Z \sim 106\ \text{MeV} \tag{4.22}$$

Electroweak Corrections Corrections due to the photon emission are large, but offer no interesting physics information. The electroweak correction,[5] on the other hand, is our main interest.

The electroweak radiative corrections can be classified in three categories:

a) Corrections to the gauge boson propagator (universal correction)
b) Vertex corrections (depends on particle species)
c) Box diagrams (specific to each process).

Examples of corresponding Feynman diagrams are shown in Figure 4.3a–d.

The bulk of the electroweak corrections in category (a) and (b) at the Z-pole is absorbed into complex form factors, $\mathcal{R}_f$ for the overall scale and $\mathcal{K}_f$ for the on-shell electroweak mixing angle, resulting in complex effective couplings [162, 163]

$$v_f = I_3 - 2Q_f \sin^2\theta_W \rightarrow \mathcal{V}_f = \sqrt{\mathcal{R}_f}\left(I_3 - 2Q_f\mathcal{K}_f \sin^2\theta_W\right) \tag{4.23a}$$

5) Radiative corrections, including both the photon and Z, W, are treated as the electroweak corrections.

$$a_f = I_3 \to \mathcal{A}_f = \sqrt{\mathcal{R}_f}\, I_3 \tag{4.23b}$$

where $\mathcal{R}_f$ and $\mathcal{K}_f$ are flavor dependent radiative corrections and are generally complex numbers. However, the imaginary part is insignificant and in our treatment, they are neglected. Box diagram corrections are process dependent, but they can be neglected generally or added as an ad hoc explicit correction.

Self-energy corrections: Category (a) is the dominant contribution which amounts to $\sim 6\%$. As discussed in Vol. 1, Chapt. 8 (radiative corrections to QED), the effect of self-energy corrections of the gauge boson due to fermion loops (Figure 4.3a,b) is to make the coupling constant run, i.e., to convert the constant $\alpha = e^2/4\pi$ to μ dependent $\alpha(\mu)$. It is also true for the electroweak correction and it changes $G_N \to G_N(\mu) = \rho(\mu^2) G_F$.[6] The scale μ is taken as the total energy $\sqrt{s}$ in $e\bar{e}$ annihilation cross section and is set $\mu = \sqrt{s} = m_Z$ for discussing data at the Z resonance.

The fermion loops which appear as the self-energy contribution to the gauge boson can couple to both γ and Z. Therefore, the radiative correction induces $\gamma - Z$ mixing (Figure 4.3c). It is possible to fold this effect into the electromagnetic part of the neutral current J_N

$$J_N^\mu = J_{3L}^\mu - e Q_f \sin^2\theta_W J_\gamma^\mu \tag{4.24}$$

The combined effect due to the self-energy correction to the gauge boson (Figure 4.3a–c) is to change the Weinberg angle as well as the coupling constant in the neutral current:

$$v_f = I_3 - 2Q_f \sin^2\theta_W \to \overline{v}_f = \sqrt{\rho_{se}}\left(I_3 - 2Q_f \kappa_{se} \sin^2\theta_{W,\text{tree}}\right) \tag{4.25a}$$

$$a_f = I_3 \to \overline{a}_f = \sqrt{\rho_{se}}\, I_3 \tag{4.25b}$$

$$\sin^2\theta_W \to \sin^2\overline{\theta}_W \equiv \kappa_{se} \sin^2\theta_{W,\text{tree}} \tag{4.25c}$$

where $\sin^2\theta_{W,\text{tree}}$ denotes the Weinberg angle at the tree level.

The parameters ρ, κ are written as

$$\rho_{se} = \rho_0(1 + \Delta\rho_{se}) \tag{4.26a}$$

$$\kappa_{se} = 1 + \Delta\kappa_{se} \tag{4.26b}$$

We have attached suffix "se" to denote the origin of the corrections and to remind the reader later that they are universal and process independent.

ρ_0 is the ρ in the Born approximation. In the Standard Model,

$$\begin{aligned}\rho_0 &= \left.\frac{G_N}{G_F}\right|_{\text{Born}} = \frac{g_Z^2}{8m_Z^2}\left(\frac{g_W^2}{8m_W^2}\right)^{-1} \\ &= \frac{e^2}{\sin^2\theta_W \cos^2\theta_W}\left(\frac{e^2}{\sin^2\theta_W}\right)^{-1}\left(\frac{m_W^2}{m_Z^2}\right) = 1\end{aligned} \tag{4.27}$$

6) Figure 4.3a is a pure QED correction. Due to equations $G_F m_W^2/\sqrt{2} = e^2/8\sin^2\theta_W$ at the tree level, the weak coupling constants are closely related to α which was, by definition, evaluated at $\mu^2 = 0$. As it will become clear later, G_F is, in a sense, a running constant evaluated at $\mu = m_W$ and the effect of Figure 4.3b appears as a correction to the neutral coupling constant G_N relative to G_F.

where $\sin^2\theta_W$ is defined by

$$\sin^2\theta_W = 1 - \frac{m_W^2}{m_Z^2} \tag{4.28}$$

This is the relation that holds at the tree level. If we extend the relation in Eq. (4.28) to define the Weinberg angle by physical masses of the W and Z, then all the radiative corrections to the Weinberg angle are already built in it by definition and receive no further corrections. This is referred as the on-shell Weinberg angle. Later, we will introduce other types of Weinberg angles, but nomenclature $\sin^2\theta_W$ is exclusively saved to denote the on-shell Weinberg angle, unless otherwise noted throughout this book.

The radiative corrections due to the self-energy of the gauge boson, which will be described in more detail in the next chapter, depend on the top and Higgs masses which enter in intermediate states:

$$\Delta\rho_{se} = \frac{3G_F m_W^2}{8\sqrt{2}\pi^2}\left[\frac{m_t^2}{m_W^2} - \frac{\sin^2\theta_W}{\cos^2\theta_W}\left(\ln\frac{m_H^2}{m_W^2} - \frac{5}{6}\right)\right] + \cdots \tag{4.29a}$$

$$\Delta\kappa_{se} = \frac{\cos^2\theta_W}{\sin^2\theta_W}\Delta\rho_{se} + \cdots \tag{4.29b}$$

These expressions can be used to test the validity of the Standard Model. As the correction due to the Higgs is relatively weak, one can infer the top mass by tuning the formula with precision data. Therefore, to produce the top quark directly and compare its mass with that inferred from the precision Z data provides one of the most stringent tests of the Standard Model. Knowing the top mass, the formula can be used to predict the Higgs mass.

Energy dependent width: The production cross section of the Z boson is $O(\alpha^2)$, but becomes $O(1)$ at the resonance peak. Therefore, the correction to the propagator has to include $O(\alpha^2)$ contributions. The effect of the higher order corrections to the Z width is shown to be represented by

$$\frac{1}{s - m_Z^2 + i m_Z \Gamma_Z} \rightarrow \frac{1}{s - m_Z^2 + i s\Gamma_Z/m_Z} \tag{4.30}$$

with very good approximation [155, 164]. In order to see the effect of the s dependent width, we set

$$s - m_Z^2 + is\frac{\Gamma_Z}{m_Z} = (1 + i\gamma)\left(s - \hat{m}_Z^2 + i\hat{m}_Z\hat{\Gamma}_Z\right) \tag{4.31a}$$

$$\hat{m}_Z = m_Z/(1+\gamma^2)^{-1/2} \tag{4.31b}$$

$$\hat{\Gamma}_Z = \Gamma_Z(1+\gamma^2)^{-1/2} \tag{4.31c}$$

$$\gamma = \frac{\Gamma_Z}{m_z} \tag{4.31d}$$

then the cross section in Eq. (4.12a) is modified to

$$\sigma_{\text{observed}} = \sigma^0 \frac{s\hat{\Gamma}_Z^2}{(s - \hat{m}_Z^2)^2 + (\hat{m}_Z \hat{\Gamma}_Z)^2} \tag{4.32}$$

and reproduces the same Breit–Wigner formula with modified mass and width. In other words, the effect of s dependent width is to change the apparent mass and width to [155]

$$\begin{aligned} m_Z &\to \hat{m}_Z \simeq m_Z - \frac{1}{2}\frac{\Gamma_Z^2}{m_Z} \simeq m_Z - 34\,\text{MeV} \\ \Gamma_Z &\to \hat{\Gamma}_Z = \Gamma_Z - 1\text{MeV} \end{aligned} \tag{4.33}$$

but keep the original peak height. The nominal peak position (determined by $d\sigma/ds = 0$) expressed by Eq. (4.30) is not at $\sqrt{s} = m_Z$, but at $\sqrt{s}_{\text{max}} = m_Z[1 + (\Gamma_Z/m_Z)^2]^{1/4} \simeq m_Z + (1/4)(\Gamma_Z^2/m_Z) \simeq m_Z + 17\,\text{MeV}$. Therefore, the apparent peak position after introducing the s dependent width becomes $\sqrt{s}_{\text{max}} = m_Z + 17\,\text{MeV} - 34\,\text{MeV}$. What is meant by the good approximation mentioned before is that one can get the accurate value of m_Z if one uses Eq. (4.30) and fit with the observed experimental data (QED deconvoluted) with m_Z as a parameter.

Equation (4.30) is the defining formula to obtain m_Z and Γ_Z. Theoretically, the mass of unstable particles is defined as the pole position $\overline{s}$ of the *S*-matrix in the complex *s*-plane [165–168].[7)]

$$\overline{s} = \overline{M}^2 - i\overline{M\Gamma} \tag{4.36}$$

As the pole is that of Eq. (4.30), the theoretical mass and width agree with those given in Eq. (4.31), namely [155, 164],

$$\overline{M} = \hat{m}_Z \simeq m_Z - 34.1\,\text{MeV}\,, \quad \overline{\Gamma} = \hat{\Gamma}_Z \simeq \Gamma_Z - 0.9\,\text{MeV} \tag{4.37}$$

In summary, the observed peak position including the QED as well as the electroweak effect is given by

$$\sqrt{s}_{\text{max, observed}} = m_Z + 106 + 17 - 34\,\text{MeV} = m_Z + 89\,\text{MeV} \tag{4.38}$$

7) Some authors [169] choose to define the *Z* mass and width by

$$\overline{s} = \left(M_Z - \frac{i}{2}\Gamma_Z\right)^2 \tag{4.34}$$

in which case

$$\begin{aligned} M &= m_Z \left\{1 - \frac{3}{8}\left(\frac{\Gamma_Z}{m_Z}\right)^2 + O\left(\frac{\Gamma_Z}{m_Z}\right)^4\right\} \simeq m_Z - 26\,\text{MeV} \\ \Gamma &= \Gamma_Z\left(1 - \frac{5}{8}\left(\frac{\Gamma_Z}{m_Z}\right)^2 + \cdots\right) \simeq \Gamma_Z - 1.2\,\text{MeV} \end{aligned} \tag{4.35}$$

Notice that the radiative corrections are different, depending on the mode one wants to observe because $\Gamma(Z \to f\overline{f})$ receives flavor dependent corrections. Therefore, Γ_Z is related to the observed total width $(\sqrt{s_+} - \sqrt{s_-})$ of each channel by equalities:

$$\Gamma_Z = (\sqrt{s_+} - \sqrt{s_-})/1.16 \quad (e\overline{e} \to \mu\overline{\mu}) \tag{4.39a}$$

$$\Gamma_Z = (\sqrt{s_+} - \sqrt{s_-})/1.15 \quad (e\overline{e} \to \text{hadrons}) \tag{4.39b}$$

$$(\sqrt{s_+} - \sqrt{s_-})_{\text{had}} = (\sqrt{s_+} - \sqrt{s_-})_{\mu\overline{\mu}} - 45\ \text{MeV}^{8)} \tag{4.39c}$$

where $\sqrt{s_\pm}$ is the position at half maximum on the upper (lower) side of the peak,

Vertex and box corrections: The vertex correction is dependent on the fermion flavor and the box correction is process dependent. Therefore, in addition to the self-energy correction of the gauge boson, the effect of the vertex correction can be written as

$$v_f \to \overline{v}_f = \sqrt{\rho_f}\left(I_{3f} - 2Q_f\kappa_f \sin^2\theta_{\text{W, tree}}\right) \tag{4.40a}$$

$$a_f \to \overline{a}_f = \sqrt{\rho_f}\, I_{3f} \tag{4.40b}$$

where

$$\rho_f = \rho_0(1 + \Delta\rho_{se} + \Delta\rho_f) \tag{4.41a}$$

$$\kappa_f = 1 + \Delta\kappa_{se} + \Delta\kappa_f \tag{4.41b}$$

The flavor dependent effective Weinberg angle is defined by

$$\sin^2\theta^f_{\text{eff}} \equiv \kappa_f \sin^2\theta_{\text{W, tree}} = (1 + \kappa_f)\sin^2\overline{\theta}_{\text{W}} \equiv (1 + \kappa_f)\overline{s}^2_{\text{W}} \tag{4.42}$$

Since this effective Weinberg angle is a part of the neutral current that couples to the Z boson and hence appears in the transition amplitude, it is directly related to the experimental data and is flavor dependent. The reader is reminded that this is different from the on-shell Weinberg angle given in Eq. (4.28) which is process independent.

The box corrections are process dependent and generally added explicitly as an ad hoc correction. As they give only small corrections, they are generally neglected in the following discussions. The vertex corrections are also small except for the b quark production where the large mass of the intermediate top quark gives a sizable correction. With the assumption of lepton universality and small vertex corrections, $\sin^2\theta^f_{\text{eff}}$ is actually an almost flavor independent universal variable as long as one-loop corrections are considered. Thus, $\sin^2\theta_{\text{W}}$ (on-shell) and $\sin^2\theta^f_{\text{eff}}$ are the two

8) The numbers are not quite accurate because $m_t = 120$ GeV, $m_H = 100$ GeV was assumed for this calculation. Nonetheless, they are shown to give an idea about the magnitude of the radiative corrections.

Weinberg angles directly related to the measured values of the mass and cross sections. In addition, the Weinberg angle denoted as $\sin^2\theta_{\overline{MS}}$ (or $\sin^2\hat{\theta}_W$) also appears in the literature. It is introduced for theoretical convenience and will be treated in the next chapter.

Asymmetry: If we rewrite the Born approximation formula in terms of the effective parameters $\overline{v}_f$ and $\overline{a}_f$ (or equivalently $\sin^2\theta_W \to \sin^2\theta^f_{\text{eff}}$, $G_N \to \rho G_F$ as will be shown soon), and bring in energy dependent width ($m_Z\Gamma_Z \to s\Gamma_Z/m_Z$), then all the important radiative corrections are included in this modified Born approximation formula. It is shown that it agrees with the more accurate formula with all the $O(\alpha)$ corrections included within 0.5% [170, 171]. The more formal treatment of the radiative corrections are given in the next chapter. Here, we proceed with the improved Born approximation and relate fundamental variables with experimentally observable quantities.

The cross section expressed in terms of the improved Born approximation is given by

$$\sigma(e\overline{e} \to f\overline{f}) = \frac{12\pi\Gamma_e\Gamma_f}{(s-m_Z)^2 + (s\Gamma_Z/m_Z)^2}\left(\frac{s}{m_Z^2} + R_f\frac{s-m_Z^2}{m_Z^2}\right) + N_{cf}Q_f^2Q_e^2\frac{4\pi\overline{\alpha}^2}{3s} \tag{4.43a}$$

$$R_f = \frac{2\sqrt{2}\pi\overline{\alpha}}{G_F m_Z^2}\frac{2Q_eQ_f\overline{v}_e\overline{v}_f}{(\overline{v}_e^2+\overline{a}_e^2)(\overline{v}_f^2+\overline{a}_f^2)} \tag{4.43b}$$

$$\Gamma_e = \frac{N_{ce}G_F m_Z^3}{6\sqrt{2}\pi}\left(\overline{v}_e^2+\overline{a}_e^2\right), \quad \Gamma_f = \frac{N_{cf}G_F m_Z^3}{6\sqrt{2}\pi}\left(\overline{v}_f^2+\overline{a}_f^2\right) \tag{4.43c}$$

where

$$\overline{\alpha} = \alpha(m_Z)^{9)} \tag{4.43d}$$

$$\overline{v}_e = \sqrt{\rho_e}\left[-1/2 + 2\sin^2\theta^e_{\text{eff}}\right], \quad \overline{a}_e = -(1/2)\sqrt{\rho_e} \tag{4.43e}$$

$$\overline{v}_f = \sqrt{\rho_f}\left[I_{3f} - 2Q_f\sin^2\theta^f_{\text{eff}}\right], \quad \overline{a}_f = \sqrt{\rho_f}I_{3f} \tag{4.43f}$$

The formulas are different from those of Eq. (4.11a) in the propagator, but otherwise keep the same form with parameters replaced by effective ones. Inspection of Eqs. (4.43b) and (4.43c) enables us to separate $\sqrt{\rho}$ from the effective couplings and attach it to G_F. Therefore, the effect of the radiative correction is equivalent to changing the neutral current coupling strength G_N from $\rho_0 G_F = G_F \to \rho G_F$ and $\sin^2\theta_{W,\text{tree}} \to \sin^2\theta^f_{\text{eff}}$. Using this formula and comparing it with experimental data, values of ρ and $\sin^2\theta^f_{\text{eff}}$ can be determined. In the following discussions, we adopt this formalism and take the bar out of the vector and the axial vector effective coupling constants with the understanding that $\sin^2\theta^f_{\text{eff}}$ is to be used. Note, however, that actual data analyses are carried out using programs like TOPAZ0 [172, 173] or ZFITTER [174–176], which include all the corrections.

9) QED corrected running α evaluated at $\mu = m_Z$.

An expression for 'the asymmetry parameter'[10] can be obtained from Eq. (4.9). Defining

$$\mathcal{A}_f \equiv \frac{2v_f a_f}{v_f^2 + a_f^2} = \frac{2(v_f/a_f)}{1 + (v_f/a_f)^2} \tag{4.44}$$

The forward–backward asymmetry of the $f\overline{f}$ final state using the unpolarized beam is given by

$$A_{\mathrm{FB}}^f = \frac{\sigma_\mathrm{F} - \sigma_\mathrm{B}}{\sigma_\mathrm{F} + \sigma_\mathrm{B}} = \frac{\left(\int_0^1 - \int_{-1}^0\right) d\Omega \frac{d\sigma}{d\Omega}}{\int_{-1}^1 d\sigma} = \frac{3}{4}\frac{F_2}{F_1} \tag{4.45a}$$

where the upper and lower limit of the integral denote the value of $\cos\theta$. Use of the effective parameters is to be understood for the expressions of F_i. In the vicinity of the Z peak, it can be approximated by

$$A_{\mathrm{FB}}^f \simeq A_{\mathrm{FB}}^f(s = m_Z^2) + \frac{s - m_Z^2}{s} \frac{3\pi\alpha}{\sqrt{2} G_\mathrm{F} m_Z^2} \frac{2 Q_f Q_e a_e a_f}{(v_e^2 + a_e^2)\left(v_f^2 + a_f^2\right)} \tag{4.45b}$$

$$A_{\mathrm{FB}}^f(s = m_Z^2) = \frac{3}{4}\mathcal{A}_e \mathcal{A}_f = \frac{3}{4}\frac{2v_e a_e}{v_e^2 + a_e^2}\frac{2v_f a_f}{v_f^2 + a_f^2} \tag{4.45c}$$

At $s = m_Z^2$, the interference term vanishes and the result is the simple form in Eq. (4.45c).

Measurement of the slope is important for determining the vector and axial vector coupling constants unambiguously. This is because there is exchange symmetry $v_e \leftrightarrow a_e, v_f \leftrightarrow a_f$ in the expressions of the decay width and the asymmetry parameter (see Eqs. (4.6a) and (4.44)) which can be resolved by looking at the slope in Eq. (4.45b).

Note that the asymmetry measured by the unpolarized beam is a product of two asymmetry parameters. On the other hand, if one uses a polarized beam or measure polarizations in the final state, one can obtain their values separately.

For the polarized beam, the L–R asymmetry A_{LR} is expressed as

$$A_{\mathrm{LR}} = \frac{1}{\langle P\rangle}\frac{\sigma_\mathrm{L} - \sigma_\mathrm{R}}{\sigma_\mathrm{L} + \sigma_\mathrm{R}} = \frac{1}{\langle P\rangle}\frac{\int_{-1}^1 [d\sigma(-P) - d\sigma(P)]}{\int_{-1}^1 [d\sigma(-P) + d\sigma(P)]} = \frac{F_3}{F_1} \overset{s=m_Z^2}{=\!=\!=} \mathcal{A}_e \tag{4.46}$$

where $\langle P\rangle$ is the polarization of the incident beam and $d\sigma(P)$ is given in Eq. (4.9). One nice feature of A_{LR} measurement is a fact that it only depends on knowing the beam polarization and is insensitive to final states or acceptance of the detector. It is considered to give the most reliable value of the asymmetry parameter $\mathcal{A}_e$.

10) The author apologizes for profuse use of the word "asymmetry" in what follows. Some authors use A_f for the asymmetry parameter as well as measured asymmetries defined here as A_{FB}^f, for example. Since it can be confusing, the reader is asked to proceed with care.

Furthermore, the forward–backward asymmetry obtained from the polarized beam gives

$$A_{\text{FB}}^{f,\text{pol}} = \frac{1}{\langle P\rangle}\frac{(\sigma_\text{F}-\sigma_\text{B})(-P)-(\sigma_\text{F}-\sigma_\text{B})(P)}{(\sigma_\text{F}+\sigma_\text{B})(-P)+(\sigma_\text{F}+\sigma_\text{B})(P)}$$
$$= \frac{1}{\langle P\rangle}\frac{\left(\int_0^1-\int_{-1}^0\right)d\sigma(-P)-\left(\int_0^1-\int_{-1}^0\right)d\sigma(P)}{\left(\int_0^1+\int_{-1}^0\right)d\sigma(-P)+\left(\int_0^1+\int_{-1}^0\right)d\sigma(P)}$$
$$= \frac{3}{4}\frac{F_4}{F_1} \overset{s=m_Z^2}{=\!=\!=} \frac{3}{4}\mathcal{A}_f \tag{4.47}$$

If one can measure the final state polarization as is the case for the τ production, one can get the same asymmetry as Eq. (4.46) without using the polarized beam, but with P and $\mathcal{A}_e$ replaced with (-1) and $\mathcal{A}_f$. This can be seen as follows. Defining

$$d\sigma[\text{L}] = \frac{1}{4}(d\sigma_\text{LL}+d\sigma_\text{RL}), \quad d\sigma[\text{R}] = \frac{1}{4}(d\sigma_\text{LR}+d\sigma_\text{RR})^{11)} \tag{4.48}$$

and using Eqs. (2.67a)–(2.67d), the cross section can be obtained as

$$d\sigma[\text{L(R)}] = \frac{\overline{\alpha}^2}{4s}\left[\{(1+\cos^2\theta)F_1+2\cos\theta\, F_2\}\pm\{(1+\cos^2\theta)F_4+2\cos\theta\, F_3\}\right] \tag{4.49}$$

Then, the τ asymmetry is given by

$$A_{\text{FB}}^{\tau\text{pol}} = \frac{\int_{-1}^1(d\sigma[\text{L}]-d\sigma[\text{R}])}{\int_{-1}^1(d\sigma[\text{L}]+d\sigma[\text{R}])} = \frac{F_4}{F_1} \overset{s=m_Z^2}{=\!=\!=} \mathcal{A}_\tau \tag{4.50}$$

which reproduces the desired formula.

As v_f/a_f contains $\sin^2\theta_{\text{eff}}^f$ and no other variables, one can obtain the radiatively corrected $\sin^2\theta_{\text{eff}}^f$ directly from asymmetry measurements. One sees that the asymmetry obtained using polarization measurements can determine $\mathcal{A}_e$ and $\mathcal{A}_f$ directly and separately, and as they are related to $\sin^2\theta_{\text{eff}}^f$ without any other theoretical complications, one can make a clear test of the Standard Model. The relation between $\sin^2\theta_{\text{eff}}^f$ and $\sin^2\theta_\text{W}$ (on-shell) is discussed in the next chapter, but both of them can be measured independently, the former from the asymmetry, the latter from the mass of W and Z. Experiments at Z resonance are precise enough to distinguish the small difference between them. By comparing with theoretical predictions, one can make precision tests of the Standard Model at the loop level. As mentioned before, the radiative corrections contain the largest contributions from the intermediate top quark and the Higgs. Therefore, the precision data with theoretical radiative corrections can constrain their values. The fact that the top mass measured by direct production at TEVATRON agreed with the prediction based on the radiatively corrected measurements at LEP is a major triumph of the Standard Model and gives credibility to the prediction of the yet to be discovered Higgs.[12)]

11) $d\sigma[\text{L, R}]$ are not to be confused with $d\sigma_\text{L} = d\sigma_\text{LL}+d\sigma_\text{LR}$, $d\sigma_\text{R} = d\sigma_\text{RL}+d\sigma_\text{RR}$.

12) The Higgs was discovered at LHC in 2012 with mass $m_H = 126$ GeV when this textbook was just about to be published.

4.3
Experimental Arrangements

4.3.1
Detectors

Five detectors, four at LEP (ALEPH, DELPHI, L3, OPAL) and one at SLC (SLD), are used to extract precise values of the asymmetry parameters, for example, at the Z resonance. Though they differ in details, they share common features characteristic to collider detectors. Their general structure is shown in Figure 4.4 with illustrations of basic elements and particle's behavior. Since basics of the particle measurements in the collider experiments are given in explaining the UA1 detector in Section 3.1 [for more details see Vol. 1, Chapt. 12], we restrict our discussions in the following to a few remarks common to all the general purpose collider detectors and some specific features of the Z decay measurements.

1. Capability to identify particle species. Specially important is to pick up leptons (e, μ, τ) from dominant QCD backgrounds (i.e., hadrons). So many hadrons are produced in one interaction that it is impossible to identify all the hadron species. Under favorable conditions, some of them like π^0, K^0, $\overline{K}^0$, Λ^0 can be identified by their specific decay modes as, for instance, those used for c-quark identification. In most cases, however, hadrons are treated as pions or massless particles.
2. Uniformity of sensitivity and accuracy over entire solid angle coverage.
3. Hermeticity: ideally, the detector should cover a 4π solid angle. Unfortunately, the beam pipe creates dead angles in the extreme forward–backward region. However, at least momentum and energy components of all the particles trans-

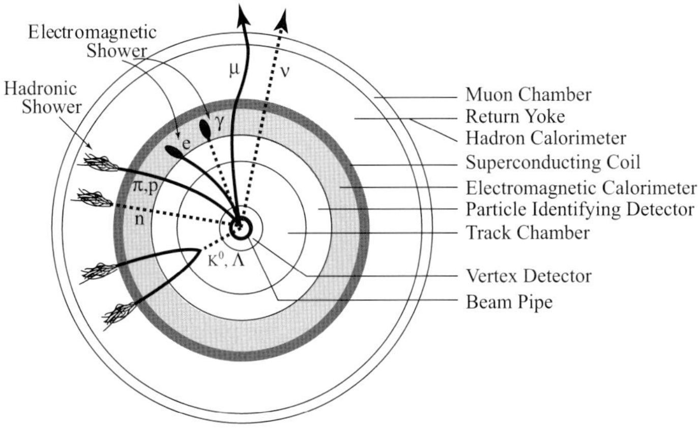

Figure 4.4 A schematic diagram of a typical collider detector is described in a cross section perpendicular to the beam axis. Dotted lines indicate neutral and hence invisible particles. In the examples shown below, the ALEPH group uses the TPC (time projection chamber (see Vol. 1, Figure 12.37) which has a combined function of a track detector as well as a particle identifier.

verse to the beam direction have to be measured. This is critical in defining the missing energy which contains important information about the production of the neutrino or some suggested new particles.

We skip details of identifying basic particles like e, μ, π, but show some examples of event displays to illustrate how their interaction patterns are used for particle identification. We will discuss later how to select c- and b-quark jets as they are important ingredients of physics at the Z resonance.

In Figures 4.5 and 4.6, we show typical lepton pairs and jet production event displays obtained in ALEPH and SLD detectors. Figure 4.6 shows a SLD event display for the reaction $e^- e^+ \to b\bar{b}$ with both b-quarks identified by their secondary vertices which will be discussed in more details later.

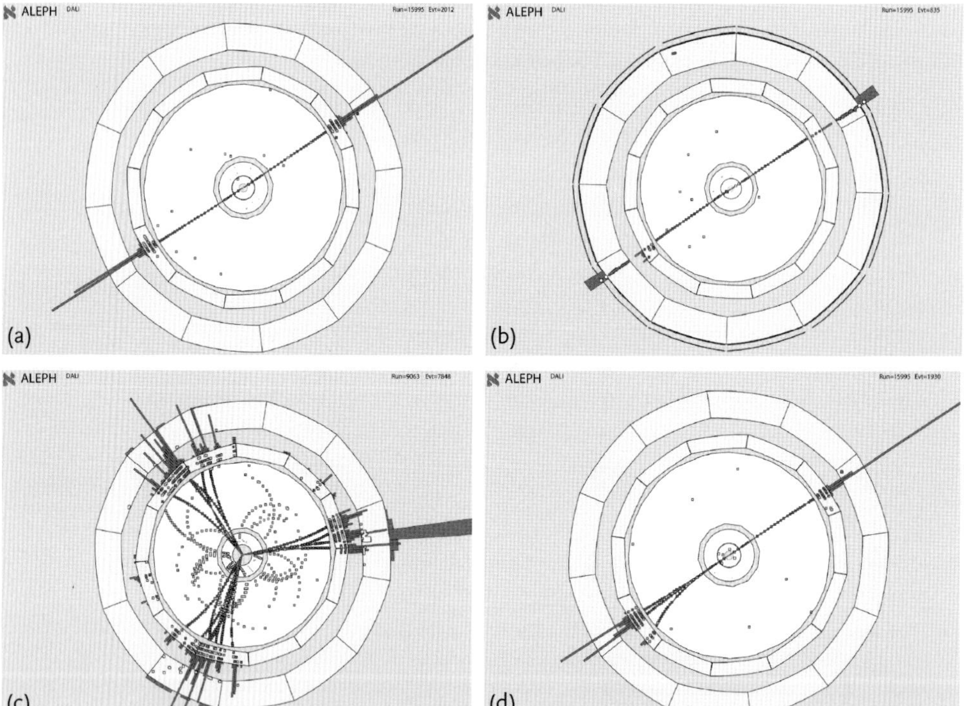

Figure 4.5 Examples of event display in $r-\phi$ view by the ALEPH detector. (a) $e^- e^+ \to ee$, (b) $\mu\mu$, (c) $3jets$ and (d) $\tau\tau$. The electrons are seen to stop in the electromagnetic calorimeter (inner dodecagon) while the muons penetrate the hadron calorimeter (outer dodecagon) and are recorded in the muon detector. In the τ event, one τ decays to an electron and the other to three pions. Energy deposits of the particles are displayed as histograms sticking out from the electromagnetic calorimeter and/or the hadron calorimeter. Detectors in the display are deformed for fish-eye viewing. The inner most circle is the beam pipe. Three white areas inside the electromagnetic calorimeter are active parts of the vertex detector, inner track chamber and TPC. A superconducting coil (thin dark color) to generate the magnetic field is placed between the electromagnetic and hadron calorimeters [177, 178].

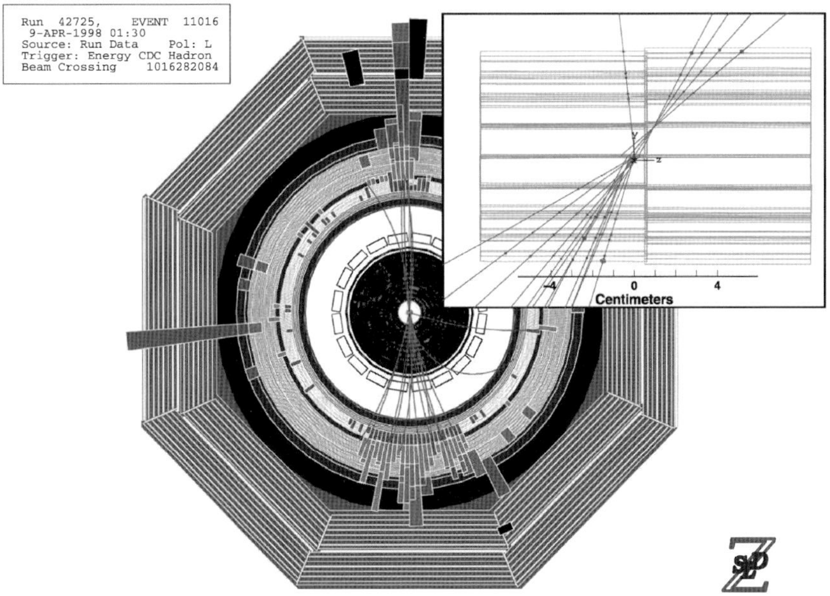

Figure 4.6 $e^- e^+ \to b\bar{b}$ event display by SLD. The inset is an enlarged r–z view of the vertex detector showing the secondary vertices [7]. For a color version of this figure, please see the color plates at the beginning of the book.

4.3.2 Luminosity Monitor

The luminosity measurement is essential in determining absolute values of the cross section (and here the decay width) which is one of the most difficult observables to determine accurately. We recollect that historically there were many examples of contradicting results obtained by different groups whose difference far exceeded each of their quoted error. Fortunately, for the case of $e^- e^+$ colliders, one can use typical QED events like small angle (typically 25–60 mrad at LEP) Bhabha scattering ($e^+ e^- \to e^+ e^-$) events [see Figure 4.7 for a typical event distribution], whose cross section can be calculated with good precision. The luminosity detector is made of electromagnetic calorimeters whose structure is similar to that in the main detector. A major challenge in the luminosity measurements is caused by strong dependence of the Bhabha cross section on the small scattering angle

$$\frac{d\sigma}{d\theta} \sim \frac{32\pi\alpha^2}{s} \frac{1}{\theta^3} \tag{4.51}$$

A careful arrangement and accurate determination of the detector edge were instrumental in improving the accuracy. With experience, final accuracy of the monitor better than $< 0.1\%$ was obtained by the four groups (ALEPH [179], DELPHI [180], L3 [181], OPAL [183]) as compared to that of $\sim 0.5\%$ in the early stage. This was better than theoretical uncertainty at that time. The largest error comes from in-

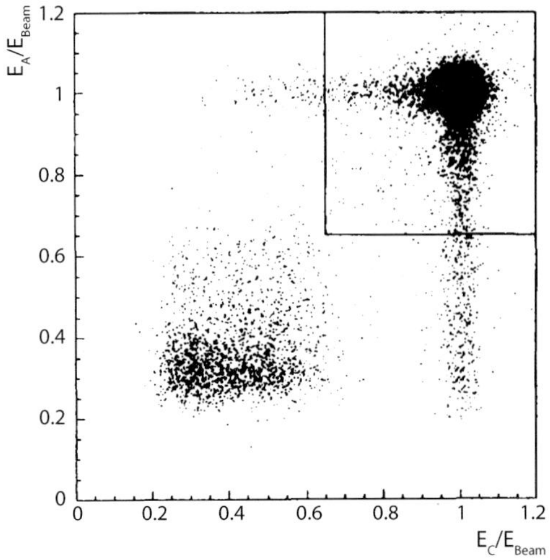

Figure 4.7 A fraction of the beam energy observed in the left and right luminosity monitors (STIC: Small Angle TIle Calorimeter) of the DELPHI experiment. The line indicates the acceptance region for the signal events. The initial state photon radiation leads to tails toward lower energies. Background events from accidental coincidences populate the low-energy regions in both calorimeters [182].

complete treatment of the radiative correction and its interplay with experimental cuts. Subsequent efforts to reduce theoretical uncertainties improved the accuracy to $\pm 1.1 \times 10^{-3}$ [184].

4.3.3 Energy Determination

The value of the Z mass obtained at LEP is one of the most precisely measured fundamental constants [185]. Its determination is based on a scan of the center of mass energy around the Z resonance pole. Accuracy of the Z mass is directly related to that of the beam energy. The present accuracy ($\Delta E = 2\,\text{MeV}$, $\Delta E/E = 2\times 10^{-5}$) is a result of a series of improvements dotted with unexpected incidents. The whole history is so interesting that they deserve a special section of its own.

Average momentum of particles circulating in a storage ring is proportional to the magnetic bending field $B(s)$ integrated over a circulating path of the particles [185].

$$E_{\text{beam}} = \frac{ec}{2\pi} \oint B(s) ds \tag{4.52}$$

The basic formula is simple, but techniques to determine the beam energy are different depending on the energy. One uses more than one method and combine them to obtain a final result. The energy measurement at LEP was carried out as follows [185–187, 189].

Depolarization Method At 20 GeV, the energy of the injection accelerator, the proton is not ultrarelativistic and its momentum can be measured by determining RF (radio frequency) acceleration voltage and its frequency. Then, one measures revolution frequency of positrons and protons in the same 20 GeV injection accelerator. From the difference of the revolution frequency, one knows the absolute value of the positron energy in the same orbit. Relative precision of this method is about 10^{-4}. Then, by integrating the current in the main accelerator from $20 \rightarrow 45$ GeV, one obtains the absolute value of the energy at 45 GeV with somewhat reduced accuracy. The error obtained is 2×10^{-4} [186].

Further improvement on the accuracy was obtained by measuring the spin precession frequency of the beam which has intrinsic accuracy of 2×10^{-5} [190]. The electron beam builds up the transverse polarization relative to the plane of rotation by synchrotron radiation (Sokolov–Ternov effect [191]). By applying a weak oscillating field radially, the polarization disappears at a certain resonance frequency determined by the beam energy. The polarization of the beam can be measured by using the spin dependence of the Compton scattering of circularly polarized laser beam off the electron beam. The angular distribution of the backward scattered photons is a sensitive measure of the beam polarization (Compton polarimeter).

The horizontal component of the spin rotates around the vertical axis, i.e., precesses due to the same magnetic field which keeps the beam in orbit. The spin precession angular frequency relative to the beam orbit is $\omega_a = \omega_s - \omega_c = (g_e - 2)/2 \times qB/m_e = aqB/m_e$ (see Vol. 1, Eq. (8.94) in the discussion of muon $g-2$) where ω_s is the spin precession frequency of the circulating electron beam, $\omega_c = qB/m_e\gamma$ is the cyclotron frequency and a is the anomalous magnetic moment of the electron. The precession frequency per turn $a\gamma$ is obtained by dividing ω_a by ω_c. Its average value over all electrons (denoted as ν) is referred to as the spin tune. It is directly proportional to the average beam energy. The energy is accurately determined if the spin tune is known because both $a = (g_e - 2)/2$ and the electron mass are accurately known and they give [186, 193]

$$\nu = a\gamma = a\frac{E}{m_e c^2} = \frac{E\,(\text{MeV})}{440.6486[1]} \tag{4.53}$$

The weak radial magnetic field of the RF perturbs the spin to rotate in the vertical plane. In standard conditions at LEP, the rotation angle is $\sim$ 140 μrad per turn about the radial direction due to $\int B_x dz = 2.4 \times 10^{-4}$ T m[13] [193]. If the perturbation from the RF is in phase with the spin precession, i.e., if the RF frequency coincides with a certain value referred to as the depolarization frequency

$$f_{\text{RF}} = f_{\text{depolarization}} = (\nu - \text{Int}[\nu])\, f_{\text{revolution}} \tag{4.54}$$

where Int[ν] (= 105 at 46.5 GeV) is the integer part of ν. The spin rotations on the radial direction adds up coherently from turn to turn. About 10^4 turns or roughly one second is necessary to integrate to 90°, i.e., to bring the spin orientation from

13) Variables used in the accelerator are: z is along the beam, y is vertical and x is radial.

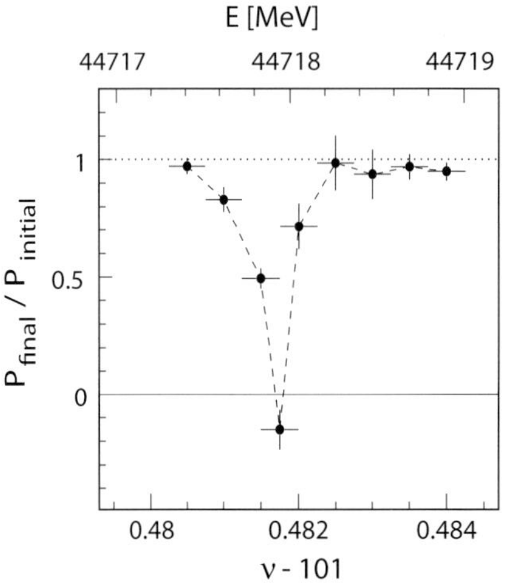

Figure 4.8 A measurement of the width of artificially excited spin resonance which is used for the energy calibration of LEP. The drop in the observed polarization level is shown as a function of the "fractional spin tune" (see Eq. (4.54)), i.e., the spin tune minus its integer part of 101 [192, 193].

the vertical position into the horizontal plane or twice its time to reverse the polarization because the revolution frequency at LEP is 11.25 kHz. By varying the RF frequency artificially, one can find the depolarization frequency. The depolarization can be measured by the Compton polarimeter. It disturbs the beam, and therefore the measurements cannot be done during the standard physics data taking period. Instead, it was carried out routinely at the end of LEP fills. Figure 4.8 shows the depolarization measured at LEP. The intrinsic accuracy of the method is within 0.2 MeV. However, the beam spread due to various effects which will be discussed in the following is typically 50 MeV. As the calibration was done outside the normal physics data taking, one needs to extrapolate from the depolarization measurement to the average beam energy under physics condition. The final overall accuracy was ~ 2 MeV.

Some important effects are described below.

Beam Energy Variation due to Detector Positions The center of mass energy at intersection points (IPs) is not necessarily twice the average beam energy. Depending on position and configuration of the accelerator cavities, IP specific shifts of up to 20 MeV can occur (see Figure 4.9).

Daily Beam Variation due to Earth Tides The Earth tides due to the sun and moon attraction changes the radius of Earth twice a day. Earth's surface expands and shrinks accordingly. The radius ($R = 4.24$ km) of the LEP ring changes 0.15 mm which causes a shift of the beam acceleration frequency f [193].

$$\frac{\Delta f}{f} \approx \frac{\Delta R}{R} = 3.5 \times 10^{-8}\,, \qquad \frac{\Delta E}{E} = \frac{1}{\alpha_c}\frac{\Delta f}{f} \simeq 10^{-4} \tag{4.55}$$

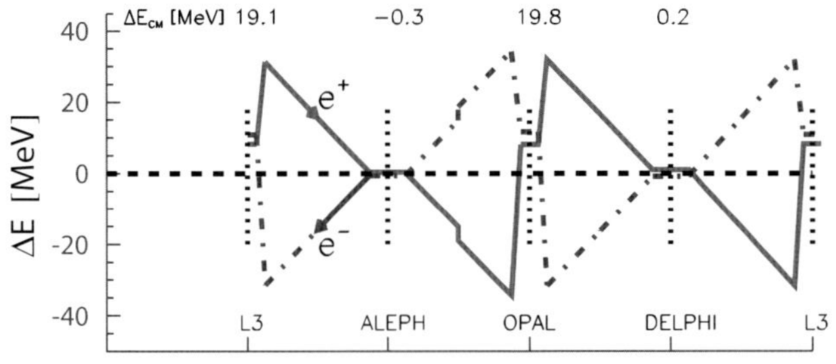

Figure 4.9 Typical variations of the beam energy around the LEP ring during the 1993 run. Energy losses from synchrotron radiation in the arcs, and in wiggler magnets between the ALEPH and OPAL IPs, are compensated by acceleration in the radio frequency (RF) cavities mounted in the straight sections on both sides of L3 and OPAL. Energy gains provided by the RF are clearly visible. A detailed modeling results in significant corrections in the center of mass energy at the IPs between acceleration sections, as indicated by the numbers on the top [7, 186].

$\alpha_c = 1.86 \times 10^{-4}$ is the momentum compaction factor. $\Delta E \simeq 8\,\text{MeV}$ is observed between high and low tides (Figure 4.10). Once the effect is clarified, it is a systematic shift that can be compensated.

Long Term Beam Energy Variation due to Climates Besides the periodic tidal movements, LEP is also subject to much slower long term environmental changes. During a typical run lasting from May to November, the LEP ring experiences circumference changes of up to 2 mm, as shown in Figure 4.11. The circumference usually increases during the summer months, some changes being clearly correlated

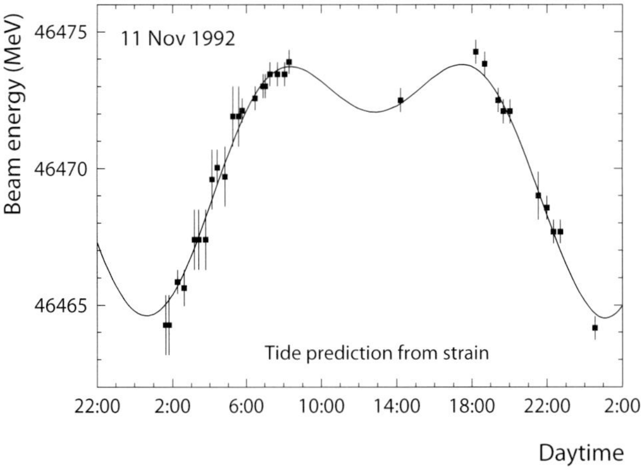

Figure 4.10 Energy variation of the LEP beams during a full moon day. The curve is the energy change from horizontal strain induced by Earth's tides [190].

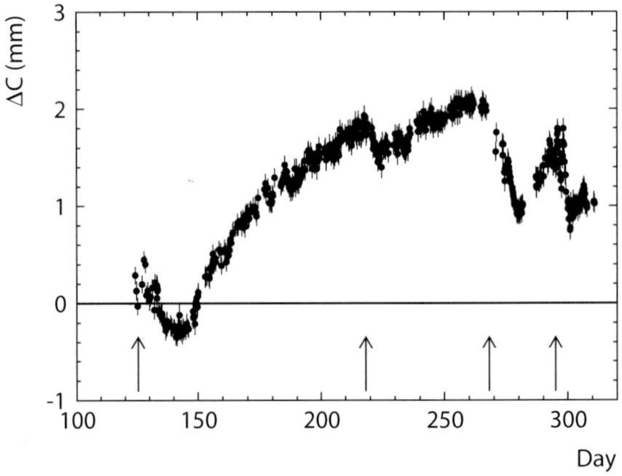

Figure 4.11 Evolution of the LEP circumference (corrected for tidal changes) as a function of the day in 1999. A drift of over 2 mm is observed during the LEP run. During summer, the circumference increases gradually. Following periods of heavy rainfall, indicated by arrows, the circumference shrinks for some time before expanding again [185].

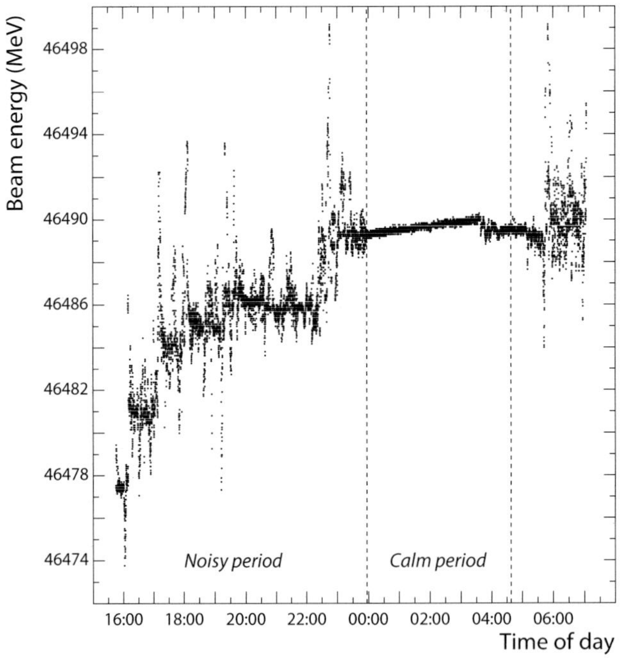

Figure 4.12 Magnetic field measured in a LEP dipole by an NMR probe over 10 h. Large short term fluctuations and a slow rise of the field are clearly visible. Between midnight and 4:30 a.m., the fluctuations disappear and the field is stable [185].

to rainfall and to fluctuations in the underground water table height. Monitoring of those seasonal variations of the circumference turned out to be very important to understand the evolution of the LEP beam energy since the associated energy variations reach $\Delta E_{\text{beam}}/E_{\text{beam}} \approx 5 \times 10^{-4}$.

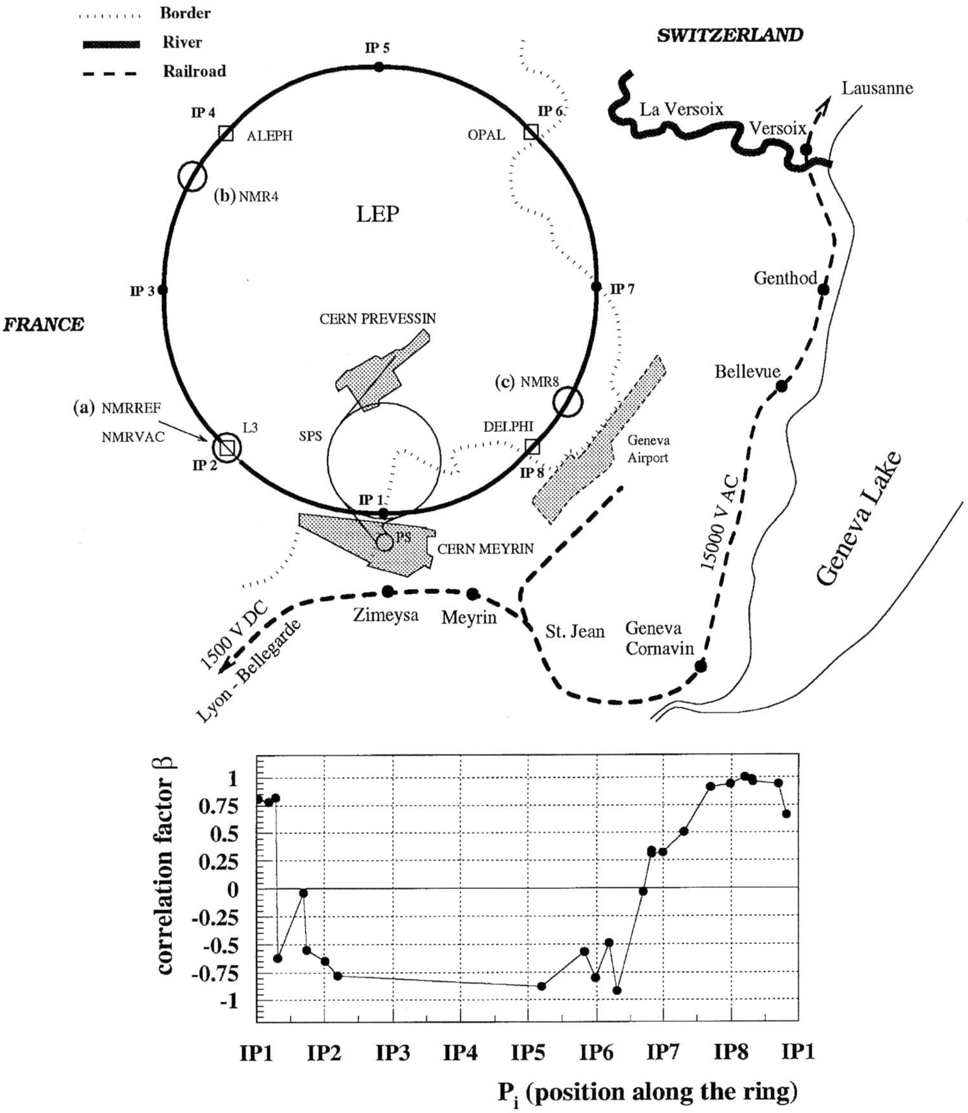

Figure 4.13 The LEP ring surrounded by the French and Swiss railroads with the location of the NMR probes and the four experiments. The measured correlation pattern of the parasitic currents along the ring is shown, proving that they enter and leave LEP near IP6 and IP1 [188, 189].

Daily Beam Variation due to Train Operation The beam energy also changes due to variation of the magnetic field strength. A perturbation of the dipole magnetic field was observed when nuclear magnetic resonance (NMR) probes were installed in two tunnel dipoles to monitor their fields. Figure 4.12 shows a systematic rise of the field of typically 8 MeV during fills.

In contrast, probes installed in a reference magnet located in a surface building did not show this rise. One source of the rise is parasitic currents through the beam pipe which were flowing from the southern part of the ring in both directions to

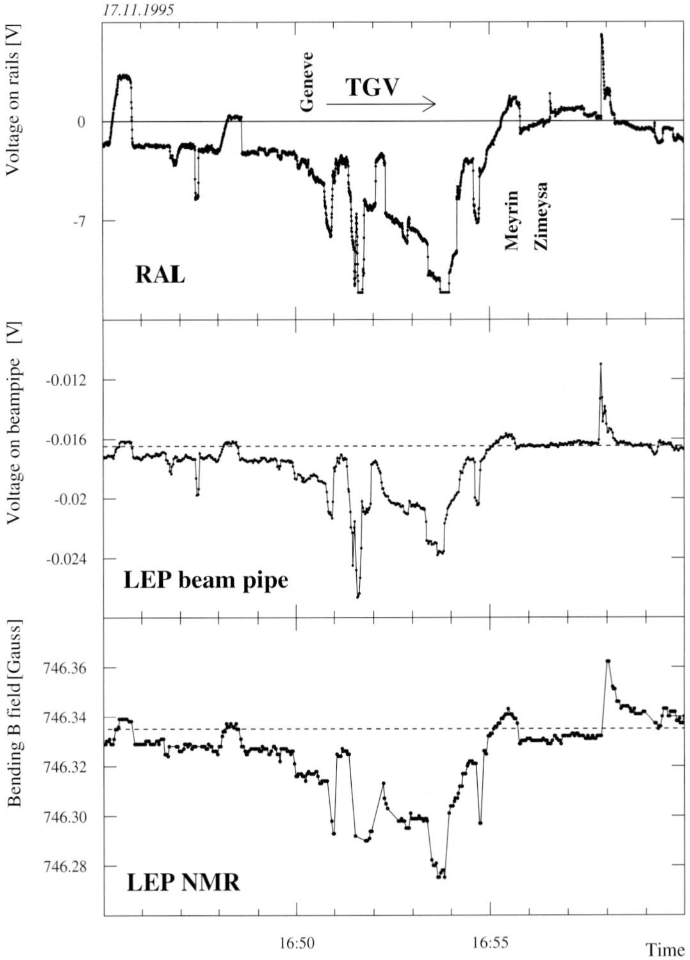

Figure 4.14 The synchronous measurement of the voltage difference between the ground and the train rails (a), the voltage difference between the LEP beam pipe and ground (b) and the NMR readings (c). The correlation is obvious. The label "Geneva" marks the time of the departure of the TGV (Train à Grande Vitesse') from Geneva central station. The label "Zimeysa" indicates the time when the TGV went past the measuring device [188, 189].

the north eastern part (Figure 4.13) and it had a very particular time structure, it only occurred from 5:00 a.m. to 12:00 p.m., as indicated in Figure 4.12.

These patterns finally helped to identify the source: electric trains arriving at and leaving Geneva south of the LEP ring cause a leakage current to the ground which finds its way to the LEP ring and returns via a small river northeast of LEP to another power station. A dedicated experiment clearly demonstrated the correlation between a departing TGV express, the beam pipe current and the dipole field as measured by NMR (Figure 4.14). The quiet period between midnight and 5:00 a.m. is a consequence of there being no trains running in the area at this time.

The leakage current themselves do not induce energy changes of the beam because the net current averages to zero over the circumference. The dipole field, however, does not drop back to its initial value after a current spike, but remains at a slightly higher value due to the magnet hysteresis curve. A succession of current spikes induces slow increase of the field and the beam energy over time. The field in the magnets finally saturates due to a finite amplitude of the spikes.

The above examples are only important disturbances. There are other complications including temperature dependence $\Delta E/E \simeq 10^{-4}\Delta T\,(°\mathrm{C})$, misalignments of the RF cavities etc. For more details, we refer to [185, 189].

4.3.4
Heavy Quark Tagging

The heavy quark (c, b) reaction contains valuable information on the top and Higgs as well as yet to be discovered new physics. However, unlike leptons, obtaining their clean sample is hindered by confinement. Some of the c- and b-quark tagging were already mentioned in discussing the neutral current interaction at lower energy (Section 2.6). Here, we will give more details of the heavy quark tagging relevant at the LEP energy.

Reconstruction of Charmed Hadrons Identification of the c-quark by full reconstruction of decay products and D^* tagging were described in Section 2.6. The method is useful at LEP as well and works better. Therefore, here, we only show how good the method works, demonstrating the data at LEP. Figure 4.15 shows reconstructed charmed particles by ALEPH [194]. Other groups obtained similar results.

Figure 4.16 shows $\Delta m = m(\pi^{\pm} D^0) - m(D^0)$ from DELPHI [201]. D^0 is fully reconstructed from $D^0 \to K^-\pi^+$ and $D^0 \to K^-\pi^+\pi^+\pi^-$. One can see a clear peak in the distribution of $\Delta M = m(\pi^+ D) - m(D^0)$. Pions in this peak have low p_{T} relative to D^* and are a good indicator of accompanying D^0, and hence their parent D^*. Many of the D^* can be tagged this way which constitutes a statistically independent sample from fully reconstructed charms as was described in Section 2.6. However, statistics are limited. To gain statistics, one needs to rely on the following methods which are very effective with the b-quark, but work as well if not better for the c-quark.

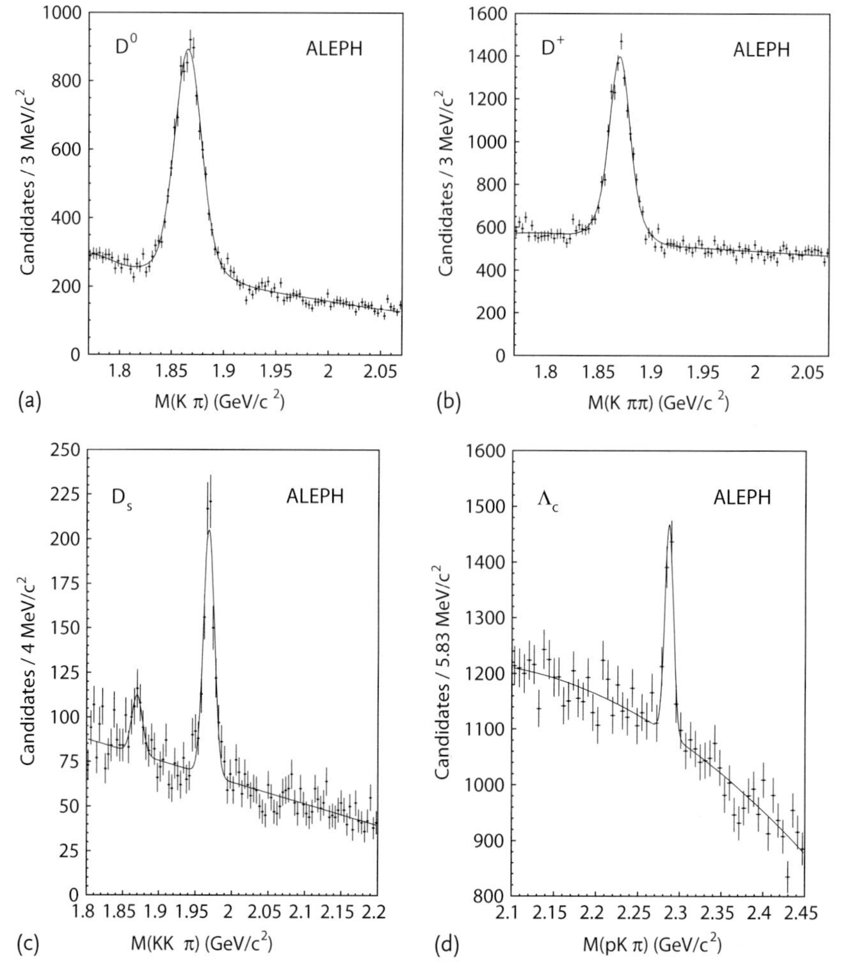

Figure 4.15 Mass spectra for (a) $D^0 \to K^-\pi^+$, (b) $D^+ \to K^-\pi^+\pi^+$, (c) $D_s \to K^+K^-\pi^+$ and (d) $\Lambda_c^+ \to pK^-\pi^+$ obtained by ALEPH [194, 201].

Lepton Tagging Leptonic decays of the c, b quarks provide both identification and clean separation between the two. The main leptonic decay modes are

- semileptonic b decays: $b \to l^-$ (BR = 10.6%)
- semileptonic c decay: $c \to l^+$ (BR = 9.8%)
- cascade b decays: $b \to c \to l^+$ (BR = 8.0%)

(BR is to be doubled, if both $l = e^-, \mu^-$ are tagged.)
Diagrams of each decay pattern are shown in Figure 4.17. Charge signs of the leptons are distinct depending on their parent. Therefore, lepton tagging provides unique identification for c or b. The cascade decay $b \to c \to l^+$ gives the wrong sign. Since its branching ratio is comparable to that of $b \to c + l^-$, it has to

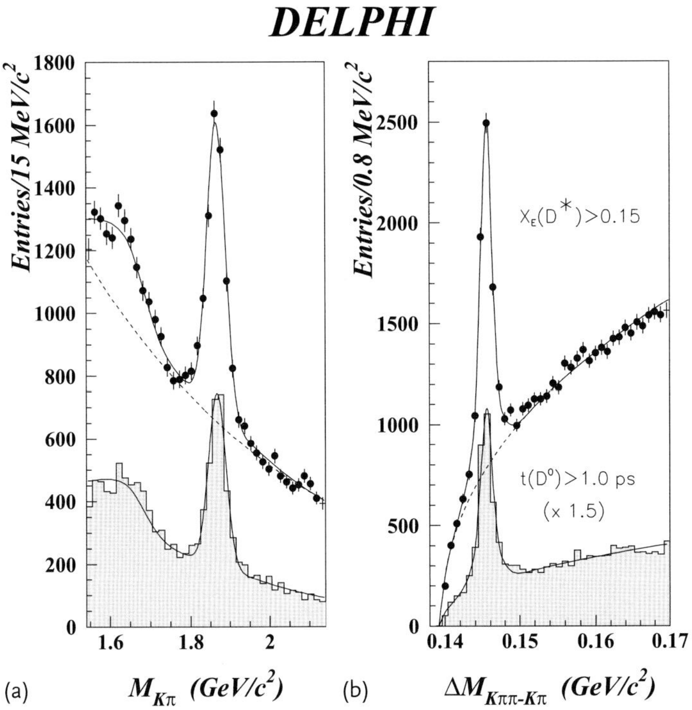

Figure 4.16 Mass difference spectrum $m(\pi^+ D^0) - m(D^0)$ from DELPHI. The D^0 is fully reconstructed in the decay modes $D^0 \to K^-\pi^+$ and $D^0 \to K^-\pi^+\pi^+\pi^-$ [201].

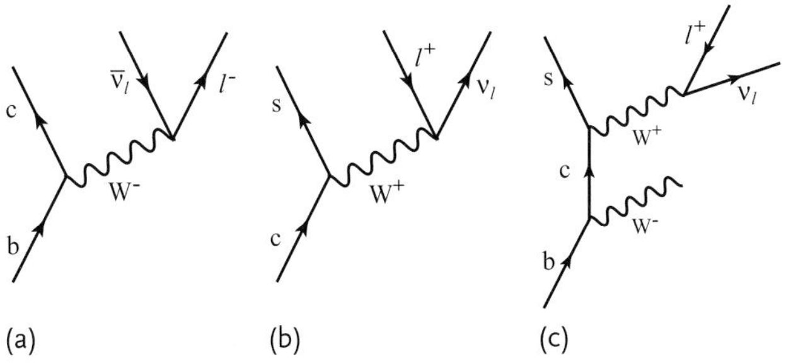

Figure 4.17 Diagram of $b \to l^-$ decay (a), $c \to l^+$ decay (b) and cascade decay $b \to c \to l^+$ (c).

be removed. This can be achieved by looking at the transverse momentum $\boldsymbol{p}_\mathrm{T}$ of the decay lepton relative to the jet axis. The jet axis is defined by the thrust axis[14] of the accompanying hadrons in the decay. Figure 4.18 shows three-dimensional plots on the $p-p_\mathrm{T}$ plane where p is the lepton momentum with p_T its transverse

14) The thrust axis $\hat{\boldsymbol{e}}_\mathrm{T}$ is defined by maximizing the thrust defined by $T \equiv \sum_i |\boldsymbol{p}_i \cdot \hat{\boldsymbol{e}}_\mathrm{T}| / \sum_i |\boldsymbol{p}_i|$ (Eq. (9.11a)) where $\boldsymbol{p}_i$ is the momentum vector of particle i in an event. Details of the thrust are given in Section 9.1.2.

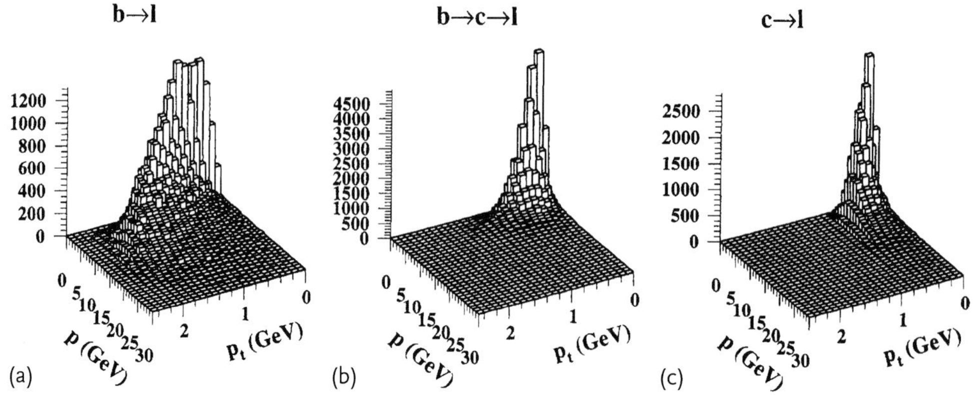

Figure 4.18 Distributions of p vs. p_T predicted by the JETSET Monte Carlo [196, 197] for leptons. (a) Semileptonic b decays; (b) Cascade decays; and (c) Semileptonic c decays in $Z \to c\bar{c}, b\bar{b}$. The momenta are defined in the experimental frame, and p_T is measured relative to the decaying b hadron direction (a,b), or relative to the decaying c hadron direction (c). The vertical scale is arbitrary [198].

component. One can see their characteristics, which are (a) high p, high p_T, (b) not so high p, low p_T, (c) low p, low p_T. The difference was produced by different masses of their parent, $m_b \simeq 5\,\text{GeV}$, $m_c \simeq 2\,\text{GeV}$. In the rest frame of the parent quark, the decay lepton typically carry one-third of the mass which is the average transverse momentum when boosted along the jet flight direction.

Lifetime Tagging Addition of the SSD (solid state detector) microvertex device to the collider detector at high energy provides, by far, a powerful tagger of the heavy quarks.

Decay length method: The most effective method of tagging the heavy quarks relies on the finite lifetime of the b and c quarks. The average lifetime of the b quarks is ~ 1.5 ps ($= 10^{-12}$ s) (see Table 4.2).

When b quarks are produced in the $Z \to b\bar{b}$ decay, each b has an average energy $E = 30\,\text{GeV}$ which has the Lorentz boost factor $\beta\gamma \sim 6$, giving a decay length of $c\beta\gamma\tau = 2.7$ mm on the average. This is more than one order of magnitude larger

Table 4.2 Lifetime tagging: At $\sqrt{s} = m_Z$, typical boost factor $\beta\gamma = 7$ for b and 12 for c hadrons.

	Lifetime τ (in ps $= 10^{-12}$ s)	Impact parameter $c\tau$ (mm)	Decay length $c\beta\gamma\tau$ (mm)
$\tau(D^+)$	1.057 ± 0.015	0.32	3.8
$\tau(D^0)$	0.415 ± 0.004	0.12	1.5
$\tau(D_s^+)$	0.467 ± 0.017	0.14	1.7
$\tau(B^+)$	1.638 ± 0.011	0.46	3.2
$\tau(B^0)$	1.530 ± 0.009	0.45	3.1

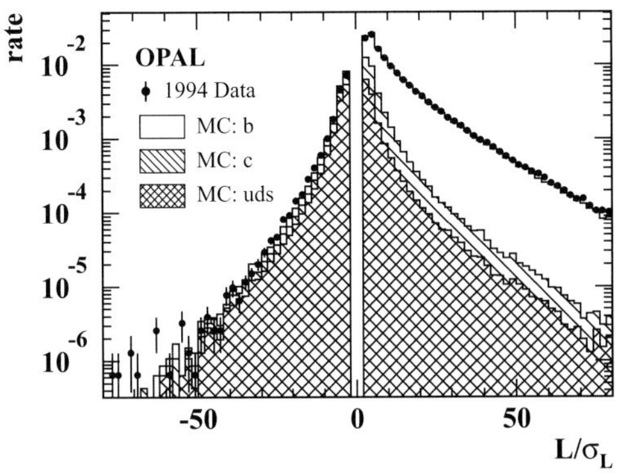

Figure 4.19 Decay length significance $S = L/\sigma_L$ for the data and b, c, uds Monte Carlo events [199].

than the vertex detector resolution. An example of secondary vertex identification was illustrated in Figure 4.6. The decay length L is the distance between the primary vertex (interaction point) and the secondary vertex (decay point of the heavy particle).

Figure 4.19 shows decay length distribution of the OPAL data [199]. It is plotted as a function of the decay length significance defined as $S = L/\sigma_L$ where σ_L is the resolution of the decay length determination. It takes the primary vertex position resolution and the track reconstruction uncertainty into account. Sign of the decay length is defined as positive if the extrapolated position of the secondary vertex lies nearer than the primary vertex and negative otherwise. The negative decay length arise from finite vertex detector resolution. If there is no secondary vertices, the distribution is that of the primary vertices which should distribute symmetrically around the interaction point. Therefore, the vertex length distribution on the $L < 0$ side can be used as a control sample to determine the resolution of the vertex position.

Charmed particles also have a significant lifetime, typically around 0.5 ps. The same lifetime tagging can be applied to the c quark as well. However, because of its shorter lifetime together with the smaller mass which makes lepton tagging less efficient, the identification power is not so strong as the b-quark.

Backgrounds for the lepton tagging include:

1. B^0–$\overline{B}^0$ mixing
2. Cascade $b \to c \to l^+$ causes a sign flop for the observed lepton
3. Misidentified leptons
4. For the b-quark, $c \to l$ decays.

Mass Tagging by SLD Although the SLD/SLC suffered from poor statistics due to the accelerator luminosity compared to LEP experiments, there were observables that they excelled in producing precise values due to their unique features of the accelerator/detector. One is the use of a polarized beam being able to produce the value of $\mathcal{A}_e$ directly which will be described in the next section.

Another is the fine primary vertex resolution, which was made possible by the small size of the beam. At the SLC, the transverse beam size is $(1.5 \times 0.8)\ \mu m^2$, smaller by an order of magnitude than that of the LEP. This is also considerably smaller than the resolution achievable by the vertex detector. Therefore, there was no need for measurement of the transverse position. The small beam size also made it possible to use a smaller beam pipe. The SLD vertex detector had the smallest radius, that is, 3 cm, as compared to 6 cm in the LEP, which resulted in better vertex resolution. Overall performance of the vertex detector can be demonstrated in the impact parameter (δ in Figure 4.20) resolution. It is defined as the minimum distance of a track from the interaction point. It should populate around zero for tracks originating from the primary vertex, but finite for those from the secondary vertex. The average decay length is given by $L = \beta\gamma c\tau$ and the average impact parameter by $\delta = L \sin\phi$ where ϕ is the decay angle relative to the flight direction of the parent quark. As $\langle \sin\phi \rangle = p_T/p \sim (m/2)/m\beta\gamma = 1/(2\beta\gamma)$, $\delta \sim L/2$. The impact parameter resolution at the LEP was typically $16 \sim 30\ \mu m$ in $r\phi$ view, between 30 and 100 μm in z while those of SLD are less than 10 μm in both views and b-tagging efficiency $\sim 60\%$ as compared to $25 \sim 30\%$ at LEP [7].

Excellent c-tag efficiency of 15% with 69% purity based on the same lifetime-mass tag as used for b-tagging was obtained as well. The excellent vertex resolution has provided a powerful mean to separate the b–c quarks. Thus, the mass tagging unique to the SLD shows an excellent separation between b, c and uds quarks as shown in Figure 4.21, with an additional check provided by the mass-momentum correlation of the jets originating from the secondary vertex (shown in Figure 4.22).

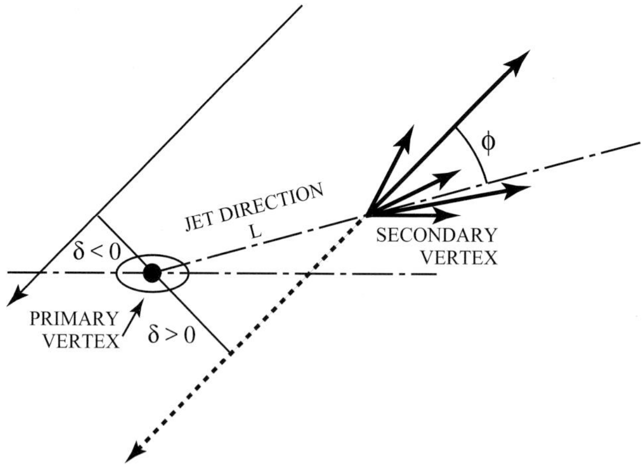

Figure 4.20 Definition of impact parameter δ, of its sign and decay length L.

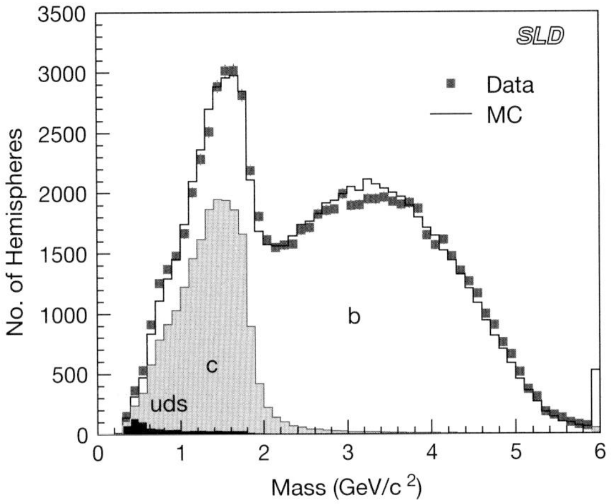

Figure 4.21 Reconstructed vertex mass from SLD for data and simulation [7, 200].

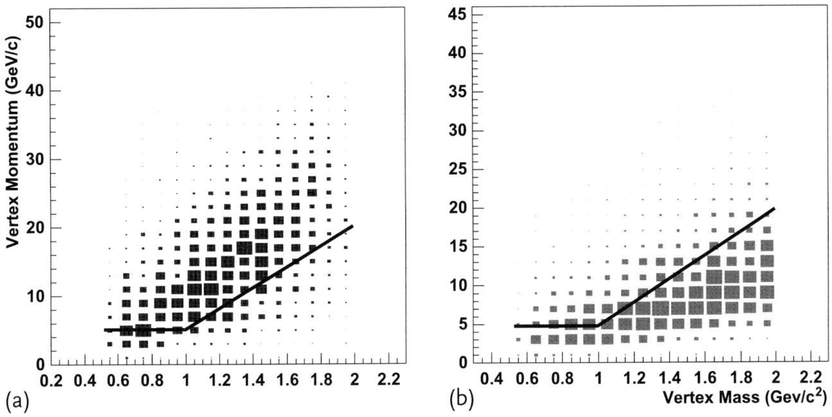

Figure 4.22 Reconstructed momentum versus invariant mass for vertices reconstructed by SLD. (a) is for hemispheres containing a charm quark and (b) is for those containing a b-quark. The line indicates the cut used in the R_c-analysis [201].

Because of the excellent separation of c from b, the SLD obtained a very competitive value of $R_c = \Gamma(Z \to c\bar{c})/\Gamma(Z \to \text{all})$

$$R_c = 0.169 \pm 0.005_{\text{stat}} \pm 0.004_{\text{sys}} \tag{4.56}$$

The statistical error is comparable to that of LEP groups, but the systematic error is half of them.

4.4
Observations

The contents of this section are largely based on the LEP/SLC combined analysis published in [7].

4.4.1
Event Characteristics

Measurements at the Z peak represent an exceptionally clean experiment, free from backgrounds which is a nightmare for experimentalists working on hadronic reactions.This is because the Z production cross section is two orders of magnitude larger than any other processes and essentially all products from Z decay constitute signals that contain important physics in them. The only major background is the two-photon process which are interactions of two virtual γ s emitted from incoming beams. They are forward going events with low visible energy, low multiplicity and easy to separate from the Z decays. If there is a small contamination of them, they can be removed easily by varying the center of mass energy or by applying simple cuts because they are nonresonance. Separation of leptonic decays from hadronic decays is also very clear.

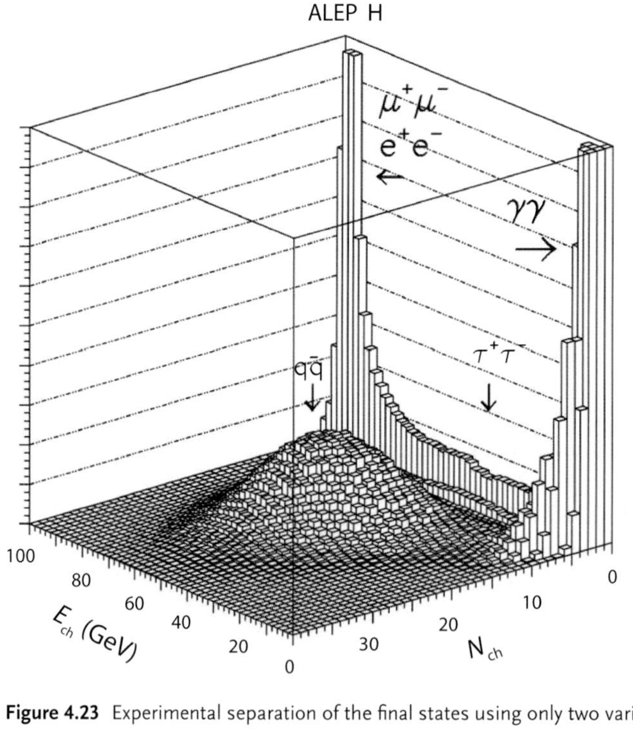

Figure 4.23 Experimental separation of the final states using only two variables, the sum of the track momenta E_{ch} and the track multiplicity N_{ch} in the central detector of the ALEPH experiment [7, 202].

Figure 4.23 shows a three-dimensional plot of events on the $N_{ch} - E_{ch}$ plane where N_{ch}, E_{ch} are multiplicity and the total energy of charged particles. A peak from $e^- e^+$ and $\mu^- \mu^+$ events at high momenta and low multiplicities is clearly separated from the background of two-photon reactions at relatively low multiplicities and momenta. The intermediate momentum region at low multiplicities is populated by $\tau^- \tau^+$ events. Separation of electrons and muons is achieved using information from the electromagnetic and hadronic calorimeters as well as the muon chambers.

Hadronic events populate the high multiplicity region at energies below the center of mass energy. They account for about 70% of the total cross section and their detection efficiency after cuts exceeded 97%. Each group collected about four million hadronic Z decays which amount to a statistical accuracy of less than 0.1%.

The charged leptonic branching ratio is about 10%, but they have distinct signals of low multiplicity and high energy as shown in Figure 4.5. Detection efficiency for the electron and muon is larger than 95%. The τ event, despite its hadronic branching ratio of about 60%, contains at least two neutrinos in the final product, which are mostly emitted back-to-back, resulting in a large missing mass which can be used efficiently to pick up the τ signals. The identification efficiency was about 85%. Each group collected about a half million charged lepton decays.

4.4.2 Mass, Widths and Branching Ratios

Parameters chosen to determine experimentally are often not exactly the same as those used for theoretical analysis. Experimental parameters are usually chosen to avoid correlations among them so that different measurements constitute statistically independent samples.

Among various final states, the hadron production is statistically dominant. The total width is the sum of all partial decay widths:

$$\Gamma_Z = \Gamma_e + \Gamma_\mu + \Gamma_\tau + \Gamma_{\text{had}} + \Gamma_{\text{inv}} \tag{4.57}$$

where $\Gamma_{\text{inv}} = \sum_f \Gamma_{\nu_f \overline{\nu}_f}$ is the partial decay width to all neutrinos, denoted as "invisible" as they are only observed as missing energy in the detector. Since the measured cross sections depend on products of the partial widths and also the total width, the widths constitute a highly correlated parameter set. In order to reduce correlations among the fit parameters, an experimentally-motivated set of six parameters are used to describe the total hadronic and leptonic cross sections around the Z peak. They are

$$m_Z\,, \quad \Gamma_Z\,, \quad \sigma^0_{\text{had}}\,, \quad R^0_e\,, \quad R^0_\mu, \quad R^0_\tau \tag{4.58}$$

where σ^0 is given by Eq. (4.12b) with f = hadron.

$$\sigma^0_{\text{had}} = \frac{12\pi}{m_Z^2}\frac{\Gamma_e \Gamma_{\text{had}}}{\Gamma_Z^2}$$

$$R_e^0 = \frac{\Gamma_{\text{had}}}{\Gamma_e}, \quad R_\mu^0 = \frac{\Gamma_{\text{had}}}{\Gamma_\mu}, \quad R_\tau^0 = \frac{\Gamma_{\text{had}}}{\Gamma_\tau} \tag{4.59}$$

If lepton universality is assumed, the last three ratios reduce to a single parameter

$$R_l^0 \equiv \frac{\Gamma_{\text{had}}}{\Gamma_l} \tag{4.60}$$

where Γ_l is the partial width to decay into one massless charged lepton flavor. The mass of τ is not negligible, but the correction to Γ_l is $\delta_\tau = -0.23\%$ (refer to Eq. (4.5) [7]) so it can be treated as an ad hoc correction. For heavy quark states where the primary quarks can be identified experimentally by detecting secondary decay vertices, the following ratios are defined:

$$R_c^0 = \frac{\Gamma_c}{\Gamma_{\text{had}}}, \quad R_b^0 = \frac{\Gamma_b}{\Gamma_{\text{had}}} \tag{4.61}$$

Values of these parameters obtained by fitting experimental data with theoretical formulas are shown in Figure 4.24 and Table 4.3. One sees that they all agree with predictions of the radiatively corrected Standard Model which include, among many, the top and Higgs loops. Notice also that the Born approximations (Table 4.1) already reproduce the experimental data with pretty good accuracy.

Table 4.3 Parameters determined by LEP and SLC observations. Data are averages of four groups ALEPH, DELPHI, L3, OPAL and SLD. R_x's ($x = i, b, c$) are denoted as R_x^0s in the text. Adopted from [3].

Quantity	Value	Standard Model
M_Z (GeV)	91.1876 ± 0.0021	91.1874 ± 0.0021
Γ_Z (GeV)	2.4952 ± 0.0023	2.4954 ± 0.0009
Γ (had) (GeV)	1.7444 ± 0.0020	1.7418 ± 0.0009
Γ (in ν) (MeV)	499.0 ± 1.5	501.69 ± 0.07
$\Gamma(\ell^+\ell^-)$ (MeV)	83.984 ± 0.086	84.005 ± 0.015
σ_{had} (nb)	41.541 ± 0.037	41.484 ± 0.008
R_e	20.804 ± 0.050	20.735 ± 0.010
R_μ	20.785 ± 0.033	20.735 ± 0.010
R_τ	20.764 ± 0.045	20.780 ± 0.010
R_b	0.21629 ± 0.00066	0.21578 ± 0.00005
R_c	0.1721 ± 0.0030	0.17224 ± 0.00003

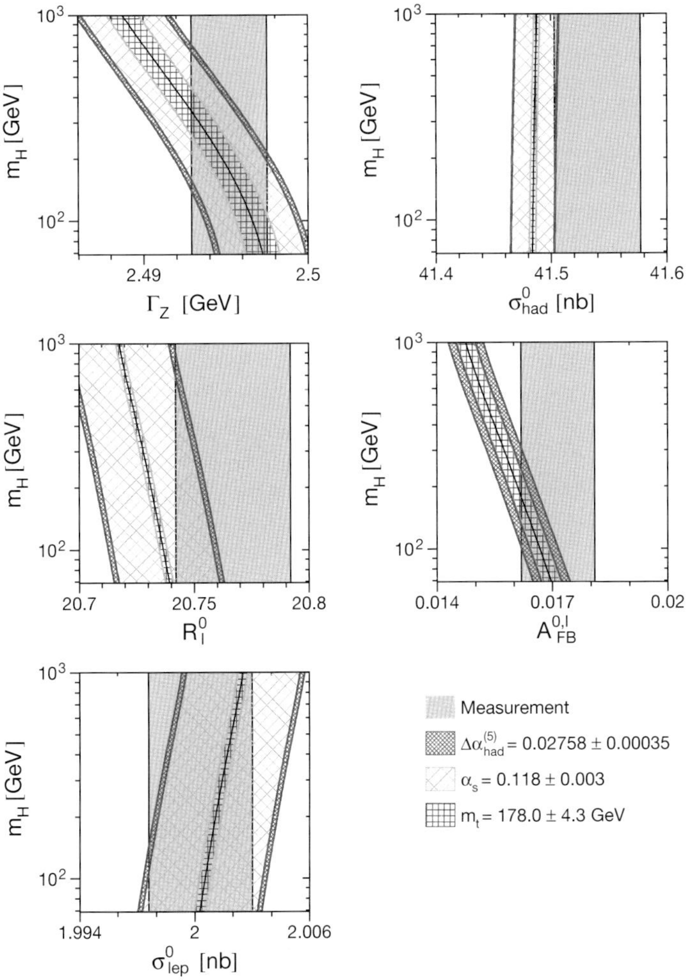

Figure 4.24 Comparisons of the LEP combined measurements of Γ_Z, σ^0_{had}, R^0_ℓ, $A^{0,\ell}_{\text{FB}}$, and σ^0_{lep} with the Standard Model prediction as a function of the Higgs boson mass. $\sigma^0_{\text{lep}} = (12\pi/m_Z^2)(\Gamma_l/\Gamma_Z)^2$ is the lepton analog of the hadronic cross section σ^0_{had}. The measured values with their uncertainty are shown as vertical bands. The width of the SM band arises due to uncertainties in $\Delta\alpha^{(5)}_{\text{had}}(m_Z^2)$, $\alpha_s(m_Z^2)$ and m_t in the ranges indicated. $\Delta\alpha^{(5)}_{\text{had}}(m_Z^2)$ is a radiative correction to the fine structure constant which contains contributions from five flavors of fermions [7]. For a color version of this figure, please see the color plates at the beginning of the book.

4.4.3
Invisible Width

Combining Eqs. (4.58) and (4.59) and assuming the lepton universality, the invisible width can be obtained as

$$R^0_{\text{inv}} \equiv \frac{\Gamma_{\text{inv}}}{\Gamma_l} = \left(\frac{12\pi R^0_l}{\sigma^0_{\text{had}} m_Z^2}\right)^{1/2} - R^0_l - (3 + \delta_\tau) \tag{4.62}$$

If all the decay channels are visible, all the partial widths can be determined without knowledge of the absolute scale of the cross sections, but Eq. (4.62) shows that it is necessary to determine the invisible width. Assuming that invisible Z decays are entirely due to neutrinos, the number of light neutrino generations N_ν can be determined by comparing the measured R^0_{inv} with the Standard Model prediction. Figure 4.25 shows the hadronic cross sections as a function of N_ν.

$$N_\nu = 2.940 \pm 0.0082 \tag{4.63}$$

The limit obtained in these measurements can also be used to constrain contributions on any invisible Z decays originating from sources other than the three known neutrino species. An example is the majoron, a supersymmetric partner of the neutrino whose existence in its simple form was denied by the measurement of the invisible width.

Until then, the number of neutrino species was inferred as $N_\nu = 3 \sim 4$ from cosmological argument and W, Z production data by hadrons [203]. The LEP experiment has proven the number to be three unambiguously. This strongly suggests that the number of generations among leptons and quarks is limited to three as well. If $m_\nu > m_Z/2 \sim 45$ GeV, it does not contribute to the decay width of the Z. Therefore, a possibility of the generation number being greater than three is not absolutely excluded. However, considering the fact that the hitherto known neutrinos all have mass below around 1 MeV, it is unlikely that the fourth generation exists with the neutrino having a mass greater than 45 GeV.

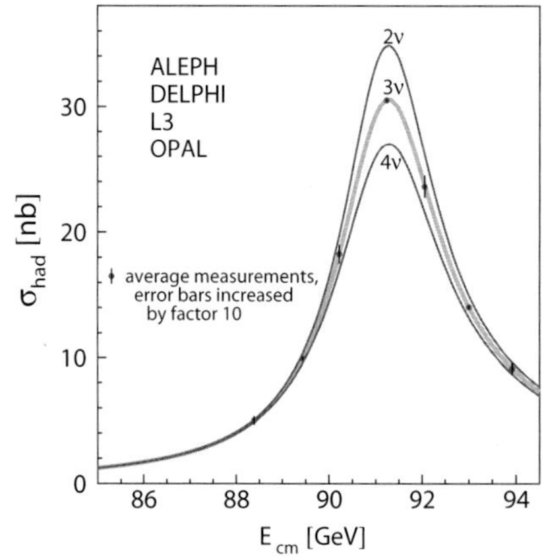

Figure 4.25 Measurements of the hadron production cross section around the Z resonance. The curves indicate the predicted cross section for two, three and four neutrino species with SM couplings and negligible mass.

4.4.4 Asymmetries

Lepton Forward-Backward Asymmetry For the charged lepton final state, the asymmetry obtained by the unpolarized beam almost vanishes at the Z pole. This is because the asymmetry is expressed as

$$A_{\mathrm{FB}}^{f} = \frac{3}{4}\mathcal{A}_e \mathcal{A}_f$$
$$\mathcal{A}_f = \frac{2v_f a_f}{v_f^2 + a_f^2} = \frac{1 - 4|Q_f|^2 \sin^2\theta_{\mathrm{eff}}^{f}}{1 - 4|Q_f|\sin^2\theta_{\mathrm{eff}}^{f} + 8|Q_f|^2 \sin^2\theta_{\mathrm{eff}}^{f}} \tag{4.64}$$

The value of $\sin^2\theta_{\mathrm{eff}}^{l}$ is very close to one-fourth. Figure 4.26 shows the asymmetry parameter as a function of the Weinberg angle. One can see that $\mathcal{A}_l(l = e, \mu, \tau) \simeq 0$ at the experimentally measured value of the Weinberg angle. Since the value of the Weinberg angle is not constrained theoretically, this is an accidental coincidence. The asymmetry parameter $\mathcal{A}_c, \mathcal{A}_b$ are sizable, but the observed asymmetry $A_{\mathrm{FB}}^{c,b}$ is a product of $\mathcal{A}_e$ and $\mathcal{A}_{c,b}$, and accurate values of both $\mathcal{A}_e$ and $\mathcal{A}_{c,b}$ have to be determined separately.

The top graph in Figure 4.27 shows DELPHI's data on $e\overline{e} \to \mu\overline{\mu}$ at $\sqrt{s} = m_Z$. One hardly sees asymmetry in the angular distribution. It can be partially recovered by going off resonance as shown by the two lower distributions at $\sqrt{s} = m_Z \pm 2\,\mathrm{GeV}$ in Figure 4.27. Asymmetries obtained by them are

$$\begin{aligned} A_{\mathrm{FB}}^{\mu} &= 0.0187 \pm 0.0019 \quad \text{at } \sqrt{s} = m_Z \\ &= -0.141 \pm 0.0114 \quad \text{at } \sqrt{s} = m_Z - 2\,\mathrm{GeV} \\ &= 0.095 \pm 0.0086 \quad \text{at } \sqrt{s} = m_Z + 2\,\mathrm{GeV} \end{aligned} \tag{4.65}$$

The combined results of the four LEP experiments give

$$A_{\mathrm{FB}}^{e} = 0.0145 \pm 0.0025 \tag{4.66a}$$

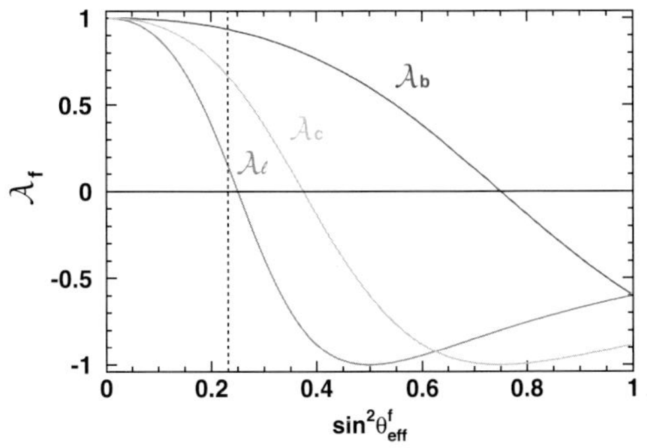

Figure 4.26 The asymmetry parameter $\mathcal{A}_f$ as a function of the Weinberg angle. At the Z pole, $A_{\mathrm{FB}}(f) = \mathcal{A}_e\mathcal{A}_f$ [7].

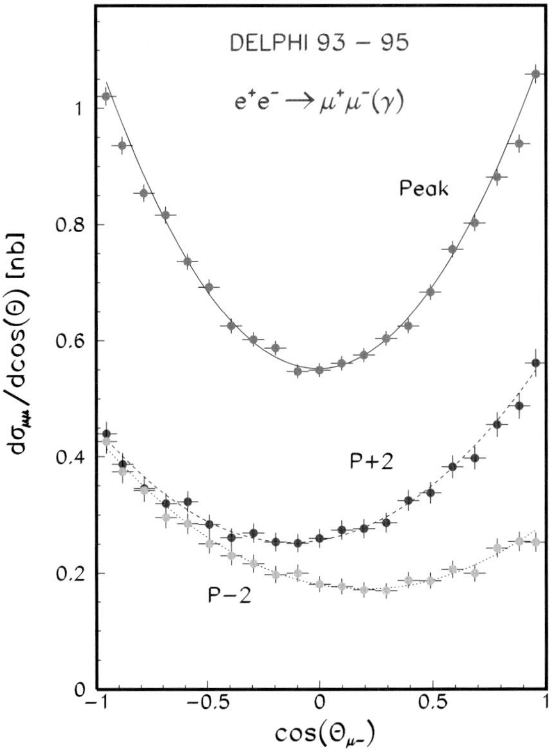

Figure 4.27 Distribution of the production polar angle $\cos\theta$, for $e^-e^+ \rightarrow \mu^-\mu^+$ and events at the three energies $\sqrt{s} = 90, 90 \pm 2\,\text{GeV}$ measured by DELPHI detectors [209]. The curves show the SM prediction from ALIBABA [168] for e^-e^+ and a fit to the data for $\mu^-\mu^+$, assuming the parabolic form of the differential cross section given in the text.

$$A_{\text{FB}}^{\mu} = 0.0169 \pm 0.0013 \tag{4.66b}$$

$$A_{\text{FB}}^{\tau} = 0.0188 \pm 0.0017 \tag{4.66c}$$

Assuming lepton universality, these values can be combined to give

$$A_{\text{FB}}^{l} = 0.0171 \pm 0.0010 \tag{4.66d}$$

which can be converted into

$$\sin^2\theta_{\text{eff}}^{l} = 0.23099 \pm 0.00053 \tag{4.67}$$

Left-Right Asymmetry by Polarized Beam The asymmetry parameter can be determined directly and independently if one uses a polarized beam or measures the polarization state of the final lepton pairs. The polar-angular distributions of a fermion pair produced by a polarized beam can be obtained from Eq. (4.9). At the

Z resonance peak, it is proportional to

$$\frac{d\sigma}{d\Omega}^{\text{pol}}(P_e) \propto (1 - P_e\mathcal{A}_e)(1+\cos^2\theta) + (\mathcal{A}_e - P_e)\mathcal{A}_f 2\cos\theta \tag{4.68}$$

where P_e is the polarization of the electron beam and $\mathcal{A}_e, \mathcal{A}_f$ are the asymmetry parameters defined by Eq. (4.44). Figure 4.28 shows data obtained by SLD/SLC group [205].

The left-right asymmetry of the polar angular distributions is quite apparent which is to be contrasted with the forward-backward asymmetry by unpolarized beams. From the measured distributions, one can determine the asymmetry parameters of all flavors independently. Here, we discuss specifically that of the electron. By integrating over the angle and taking difference of the left- and right-handed polarized beam, one can obtain the value of $\mathcal{A}_e$ directly [see also Eq. (4.46)].

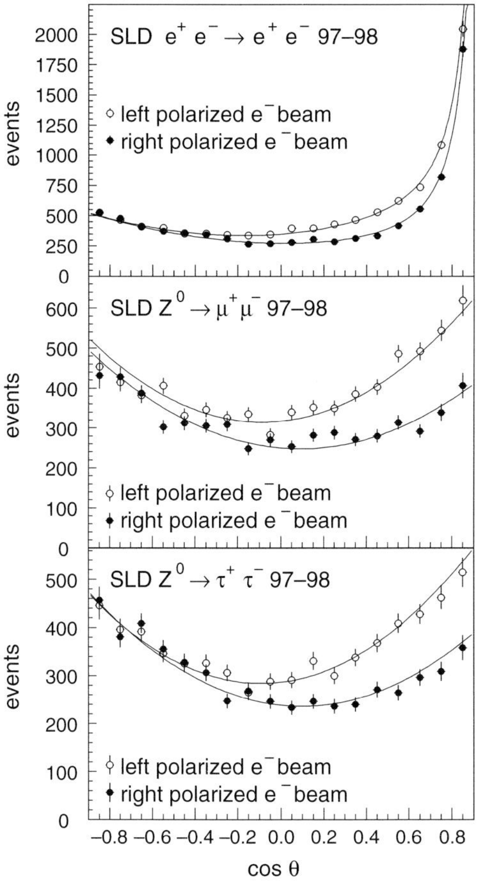

Figure 4.28 Polar angle distributions for Z-decays to e, μ and τ pairs produced by a polarized beam for the 1997-1998 SLD run. The solid line represents the fit- The distribution of the $e^-e^+$is different from others because of the t-channel exchange contribution [205].

Its value depends only on knowing the beam polarization and is insensitive to acceptance of the detector or final states of the Z decay.

Because of this advantage, the SLD/SLC detector at SLAC, despite its far smaller statistics compared to the LEP (total number of collected Zs is $\sim$ 60 000 as compared to total of $\sim 17 \times 10^6$ collected by the four groups at LEP), it still could offer the most accurate value of $\mathcal{A}_e$ and $\sin^2\theta^e_{\text{eff}}$ [204].

$$\mathcal{A}_e = 0.15138 \pm 0.00216 \tag{4.69a}$$

$$\sin^2\theta^e_{\text{eff}} = 0.23097 \pm 0.00027 \tag{4.69b}$$

One may compare this value with that obtained by LEP four groups (0.23099 $\pm$ 0.00053 $\rightarrow$ Eq. (4.67)) using the forward–backward symmetry.

A longitudinally polarized electron beam was obtained by illuminating GaAs photocathode with circularly polarized light produced by a Ti-sapphire laser. Using the high polarization strained lattice GaAs photocathodes [7, 206, 207], where mechanical strain induced in a 0.1 μm GaAs layer lifts an angular momentum degeneracy in the valence band of the material, the average electron polarization at the e^-e^+ interaction point (IP) was in the range of 73–77%. The longitudinal polarization was measured by a Compton scattering polarimeter [208] located 33 m downstream of the interaction point (IP). After the electron beam passes through the IP and before it is deflected by dipole magnets, it collides with a circularly polarized photon beam produced by a laser of wavelength 532 nm. After passing through a pair of dipole magnets, the scattered electrons are dispersed and its spectrum is measured by Cherenkov counters and proportional tubes.

τ Polarization Measurement One can extract an independent information on the tau asymmetry parameter $\mathcal{A}_\tau$ directly by measuring the polarizations of τ leptons produced by an unpolarizd bean. The polarization of the τ at any angle is defined as

$$P_\tau(\theta_\tau) = \frac{d\sigma[\text{R}] - d\sigma[\text{L}]}{d\sigma[\text{R}] + d\sigma[\text{L}]} \tag{4.70}$$

where $d\sigma[\text{R(L)}]$ are cross sections to produce right(left)-handed τ^- defined by Eq. (4.48). The polarization of the τ which is a function of $\cos\theta_\tau$ was measured on the entire $\cos\theta_\tau$ range. If one uses Eqs. (4.45), (4.46) and (4.50) to rewrite F_1, F_4 in terms of the asymmetry parameters, then Eq. (4.49) allows one to express the production cross section at the Z peak of the left- and right-handed τ as

$$\frac{d\sigma[\text{L}] - d\sigma[\text{R}]}{d\cos\theta_\tau} = \frac{3}{8}\sigma^{\tau\bar{\tau}}_{\text{TOT}}\left[\mathcal{A}_\tau(1+\cos^2\theta_\tau) + 2\mathcal{A}_e\cos\theta_\tau\right] \tag{4.71a}$$

$$\frac{d\sigma[\text{L}] + d\sigma[\text{R}]}{d\cos\theta_\tau} = \frac{3}{8}\sigma^{\tau\bar{\tau}}_{\text{TOT}}\left[(1+\cos^2\theta_\tau) + 2\mathcal{A}_e\mathcal{A}_\tau\cos\theta_\tau\right] \tag{4.71b}$$

The ratio of the above two expressions gives

$$P_\tau = \frac{\mathcal{A}_\tau(1+\cos^2\theta_\tau) + 2\mathcal{A}_e\cos\theta_\tau}{(1+\cos^2\theta_\tau) + 2\mathcal{A}_\tau\mathcal{A}_e\cos\theta_\tau} \tag{4.72}$$

The τ polarization P_τ can be determined using characteristic asymmetry in its weak decay. Figure 4.29 shows the measured polarization distribution. The data can be used to determine $\mathcal{A}_e$ and $\mathcal{A}_\tau$ simultaneously or assuming the lepton universality to determine $\sin^2 \theta^l_{\text{eff}}$.

The principle of the τ polarization analysis can be most easily understood by taking the simplest channel $\tau \to \pi \nu_\tau$. The τ decay goes via V–A interaction which constrains helicity of the ν 100% left-handed and that of $\overline{\nu}$ right-handed. This means the right-handed τ boosts the pion in the direction of τ direction (see Figure 4.30). The opposite is true for the left-handed τ. The differential decay distribution of the τ, expressed in terms of the pion energy $x_\pi = E_\pi / E_{\text{beam}}$ in the laboratory frame, is given by

$$\frac{1}{\Gamma_\tau}\frac{d\Gamma_\tau}{dx} = 1 + P_\tau(2x_\pi - 1) \tag{4.73}$$

The τ polarization can be extracted by performing a binned maximum likelihood fit of the measured distribution to the sum of the corresponding simulated distributions normalized by the coefficients $N(1 + P_\tau)$ and $N(1 - P_\tau)$.

Heavy Quark Asymmetry Asymmetry of the quarks in the final state is much larger than leptons. However, because of confinement, it is much harder to identify an

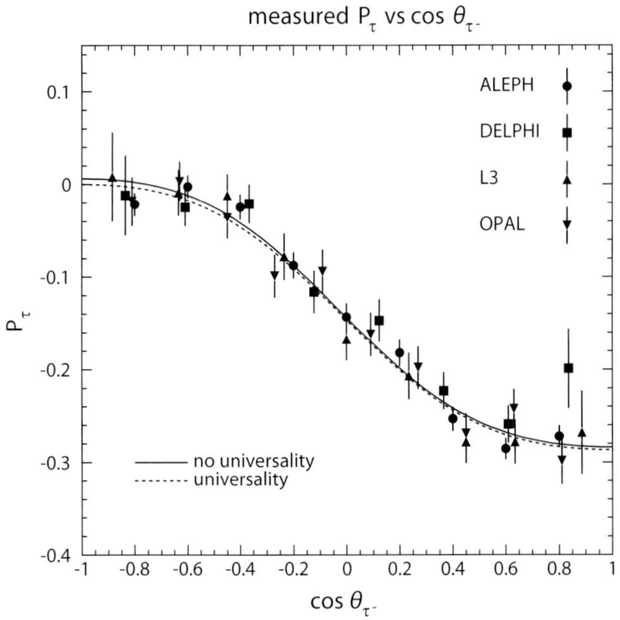

Figure 4.29 Values of P_τ as a function of $\cos\theta_\tau$ as measured by each of the LEP experiments. Only the statistical errors are shown. The values are not corrected for radiation, interference or pure photon exchange. The solid curve overlays Eq. (4.72) for the LEP values of A^τ_{FB} and $\mathcal{A}_e$. The dashed curve overlays Eq. (4.72) under the assumption of lepton universality for the LEP value of $\mathcal{A}_l$ [2, 7].

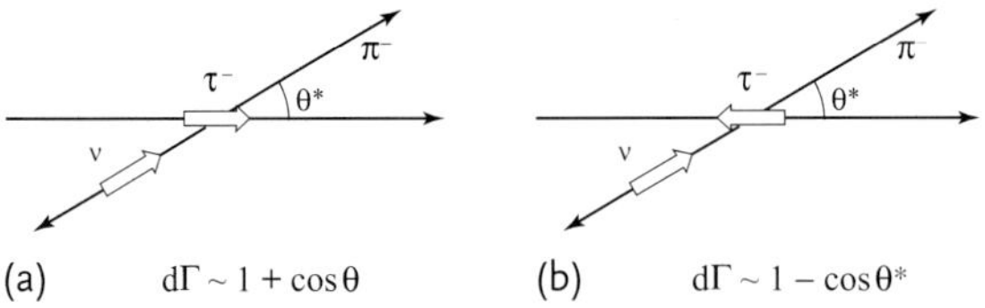

Figure 4.30 Angular distribution of the decay π in the rest frame of τ is proportional to (a) $|d^{1/2}_{1/2,\pm 1/2}|^2 = 1 \pm \cos\theta^*$, i.e., (b) $d\Gamma \propto 1 + P_\tau \cos\theta^*$. The expression can be converted to that in the laboratory frame, $\cos\theta^* \simeq 2(E_\pi/E_\tau) - 1$. Therefore, in the laboratory frame, one gets $d\Gamma \propto 1 + P_\tau\{2(E_\pi/E_\tau) - 1\}$.

individual quark. A sample tagged with high p_T leptons yields visible forward–backward asymmetry (see Figure 4.18). Hadronic decays of the b quark can be used as well. However, as they appear as QCD jets, the sign of the parent quark is hard to determine which is the crucial factor for measuring the asymmetry.

Jet charge tagging: At high energies, the jets are well collimated and their separation becomes easier. The average charge of all particles in a jet, that is, the jet charge, retains some information on the original quark charge. The jet charge is defined as

$$Q_{\text{jet}} = \frac{\sum_i Q_i p^{\kappa}_{\parallel i}}{\sum_i p^{\kappa}_{\parallel i}} \tag{4.74}$$

where Q_i is the electric charge of particle "i" in the hemisphere[15] to which the jet belongs and $p_\parallel$ is the longitudinal momentum along the jet thrust axis. κ is a tunable parameter with typical values between 0.3 and 1. The idea is that the charge of the primary quark manifests itself mostly in the leading hadrons, which have high momentum along the thrust axis. The exponent is chosen to optimize the charge resolution: $\kappa = \infty$ is equivalent to using only the highest-momentum particle and $\kappa = 0$ corresponds to unweighted charge summing. A principal problem of the jet charge algorithm is its strong dependence on the fragmentation model (which will be described in detail in Section 9.3). Only when single quark flavor is isolated by additional tags, it is possible to control this dependence on the data; otherwise, this method introduces large systematic uncertainties. With $Q_{\text{F/B}}$, $N_{\text{F/B}}$ being the jet charge and the number of particles in the forward/backward hemisphere,

$$\begin{aligned}\langle Q_{\text{F}} - Q_{\text{B}}\rangle &= \frac{(N_{\text{F}}\langle Q_b\rangle + N_{\text{B}}\langle Q_{\bar{b}}\rangle) - (N_{\text{B}}\langle Q_b\rangle + N_{\text{F}}\langle Q_{\bar{b}}\rangle)}{N_{\text{F}} + N_{\text{B}}} \\ &= \frac{N_{\text{F}} - N_{\text{B}}}{N_{\text{F}} + N_{\text{B}}}\langle Q_b - Q_{\bar{b}}\rangle = A^b_{\text{FB}}\delta \end{aligned} \tag{4.75}$$

Figure 4.31 shows measurements by ALEPH [212, 213]. The mean charge separation δ is defined as the mean difference between the jet charges of a negative b and a positive $\bar{b}(\delta \equiv \langle Q_b - Q_{\bar{b}}\rangle)$. Assuming no correlation between the forward and

15) Defined by a plane orthogonal to the thrust axis which defines the jet direction.

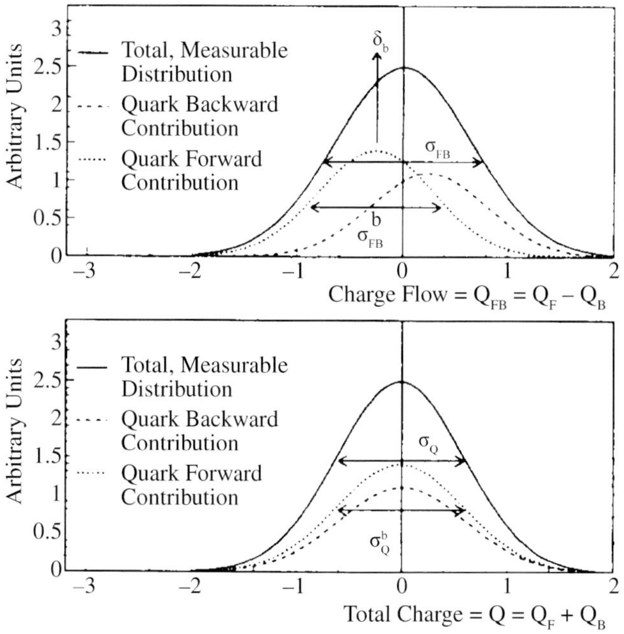

Figure 4.31 Charge separation of the ALEPH neural net tag using jet charge, vertex charge and charged kaons [212, 213]. The asymmetry reflects A^b_{FB} diluted by the nonperfect charge tagging.

backward jet charges and no bias for the average charge, δ is given by

$$\frac{\delta^2}{4} = \langle Q_b \rangle \langle -Q_{\bar{b}} \rangle = -\langle Q_b Q_{\bar{b}} \rangle = -\langle Q_{\mathrm{F}} Q_{\mathrm{B}} \rangle \tag{4.76}$$

In practice, a slight charge correlation between the two hemispheres arises and the background from other flavors must be taken into account, which is evaluated using simulations. With this technique, the observables A^b_{FB} and the main correction factor δ_b are determined simultaneously from the data by both measuring $\langle Q_{\mathrm{F}} - Q_{\mathrm{B}} \rangle$ and the product $\langle Q_{\mathrm{F}} Q_{\mathrm{B}} \rangle$. Note, in this method, the B^0–$\overline{B}^0$ mixing effect is already included in the estimate of δ_b from $\langle Q_{\mathrm{F}} Q_{\mathrm{B}} \rangle$. Precision of the measurement is comparable to that of using semileptonic decays. A large gain in the number of events due to good efficiency of the lifetime tag is compensated by much worse resolution of the jet charge measurement compared with the leptonic charge identification.

Summary of c, b asymmetry measurements: The present world average for the b and c forward–backward asymmetry at the Z pole as given in [7] are

$$A^b_{\mathrm{FB}} = \frac{3}{4}\mathcal{A}_e \mathcal{A}_b = 0.0992 \pm 0.0016 \tag{4.77a}$$

$$A^c_{\mathrm{FB}} = \frac{3}{4}\mathcal{A}_e \mathcal{A}_c = 0.0707 \pm 0.0035 \tag{4.77b}$$

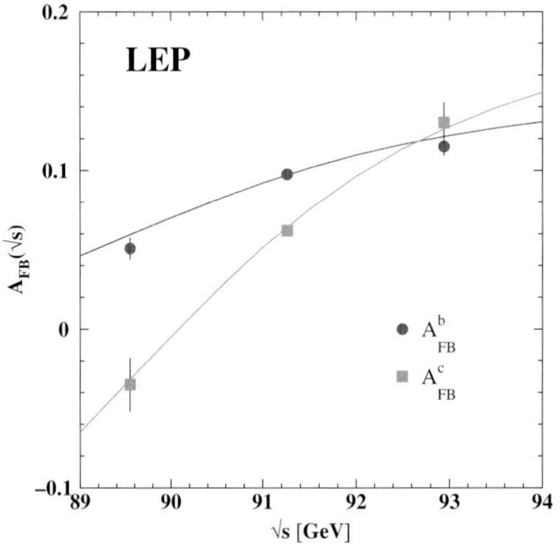

Figure 4.32 Energy dependence of A^b_{FB} and A^c_{FB}. The solid line represents the Standard Model prediction for $m_t = 178\,\text{GeV}$, $m_H = 300\,\text{GeV}$ [134, 201].

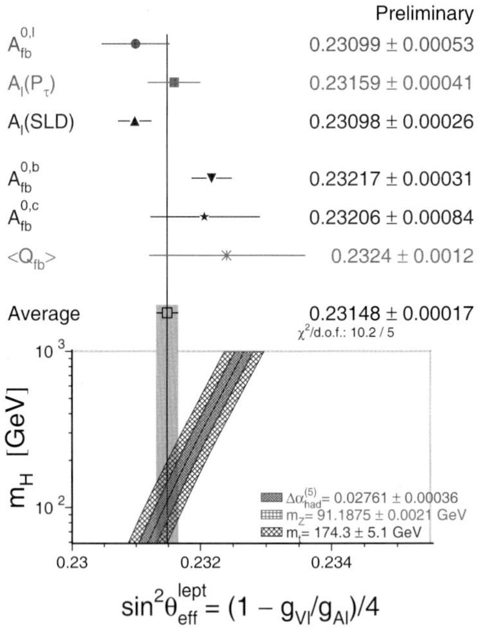

Figure 4.33 Comparison of the effective electroweak mixing angle $\sin^2\theta_{\text{eff}}$ derived from measurements depending on lepton couplings only (top) and also quark couplings (bottom). Also shown is the SM prediction for $\sin^2\theta_{\text{eff}}$ as a function of m_H. The additional uncertainty of the Standard Model prediction is parametric and dominated by the uncertainties in $\Delta\alpha^{(5)}_{\text{had}}(m_Z^2)$ and m_t, shown as the bands. The total width of the band is the linear sum of these effects [2, 7]. For a color version of this figure, please see the color plates at the beginning of the book.

To extract values of $\mathcal{A}_b, \mathcal{A}_c$, one has to divide those in Eq. (4.77) by $\mathcal{A}_e$ in Eq. (4.69a). The result with the SLD measurement combined yields

$$\begin{aligned}\mathcal{A}_b &= 0.923 \pm 0.020\\ \mathcal{A}_c &= 0.670 \pm 0.027\end{aligned} \tag{4.78}$$

Figure 4.32 shows the energy dependence of the b, c asymmetry. As Eq. (4.45b) shows, the slope of the energy dependence is $\sim Q_e Q_f a_e a_f = Q_f I_{3f}/2$, positive for both b and c, but twice as large for c because of its charge. This behavior is confirmed by the figure.

Finally, we present a world summary of the asymmetry values converted to the Weinberg angle in Figure 4.33.

4.5 Weinberg Angle and ρ Parameter

The Weinberg angle can be determined solely from asymmetry measurements using

$$\mathcal{A}_f = \frac{2v_f a_f}{v_f^2 + a_f^2} = \frac{1 - 4|Q_f|^2 \sin^2\theta_{\text{eff}}^f}{1 - 4|Q_f|\sin^2\theta_{\text{eff}}^f + 8|Q_f|^2\sin^4\theta_{\text{eff}}^f} \tag{4.79}$$

where we quote the average of all asymmetry data from Figure 4.33.

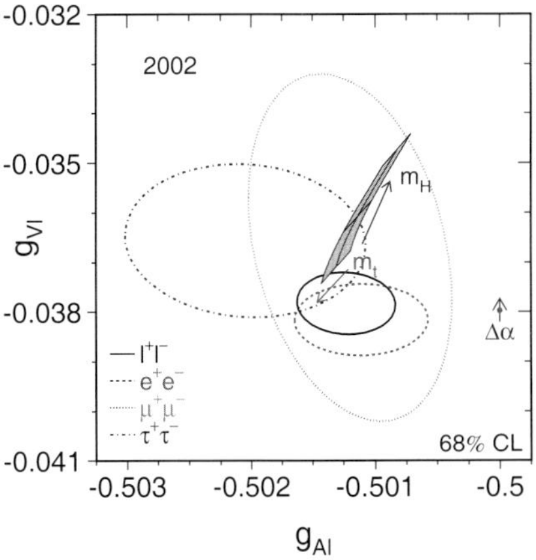

Figure 4.34 Comparison of the effective vector and axial-vector coupling constants for leptons. Here, a_l, v_l is denoted as g_{Al}, g_{Vl}. The shaded region in the lepton plot shows the predictions within the Standard Model for $m_t = 178.0 \pm 4.3$ GeV and $m_H = 300^{+700}_{-186}$ GeV. Varying the hadronic vacuum polarization by $\Delta\alpha^{(5)}_{\text{had}} = 0.02758 \pm 0.00035$ yields an additional uncertainty on the SMs prediction shown by the arrow labeled $\Delta\alpha$. This is the one shown in Figure 2.7 in Section 2.2 [2, 7]. For a color version of this figure, please see the color plates at the beginning of the book.

$$\sin^2\theta^l_{\text{eff}} = 0.23113 \pm 0.00021 \quad \text{pure leptonic} \tag{4.80a}$$

$$\sin^2\theta_{\text{eff}} = 0.23148 \pm 0.00017^{16)} \quad \text{all asymmetries combined} \tag{4.80b}$$

$$\sin^2\theta^b_{\text{eff}} = 0.281 \pm 0.016 \tag{4.80c}$$

To determine ρ, one has to use Γ_l, Γ_b.

$$\rho_l = 1.005 \pm 0.0010 \quad \text{pure leptonic} \tag{4.81a}$$

$$\rho_b = 1.059 \pm 0.021 \tag{4.81b}$$

Finally, we show a two-dimensional plot of v_f–a_f (denoted as g_{Al}–g_{vl}), for $f = e, \mu, \tau$ in Figure 4.34.

In conclusion, the precision data at the Z resonance agreed well with predictions of the Standard Model, including higher order radiative corrections and established its validity beyond doubt. Our next task, then, is to detect deviations from the Standard Model and find clues to new physics. For instance, the b-asymmetry data (Eq. (4.80c)) and the Weinberg angle measured in the deep inelastic scattering (Eq. (2.41)) are more than 3σ (standard deviation) off from the world average. Some people look for their origin in new physics like the supersymmetric extension of the Standard Model.

16) LEP experimental groups [7] use nomenclature $\sin^2\theta^{\text{lept}}_{\text{eff}}$ despite extension of the average to include hadronic data. This is made possible by using the known relation between $\mathcal{A}^q_{\text{FB}}$ and $\mathcal{A}^l_{\text{FB}}$ in the Standard Model, $\sin^2\theta^q_{\text{eff}}$ can be converted to $\sin^2\theta^l_{\text{eff}}$. We chose not to use the suffix "lept" to avoid confusion.

5
Precision Tests of the Electroweak Theory

The GWS (Glashow–Weinberg–Salam) model is a unified electroweak theory by its construction, is mathematically self-consistent and reproduces various phenomena quite well. The fact that the Weinberg angle takes the same value in a variety of processes is one strong piece of evidence illustrating that the electromagnetic and weak force are unified. As technology advances, more precision data are provided from the experimental side. Credibility of the unified theory will be firmly established if it could reproduce those precision data theoretically. In that sense, a test of radiative corrections plays a crucial role in establishing the validity of the theory, similar to the Lamb shift or $g-2$ experiments in QED. The electroweak theory has stood this severe test as was amply demonstrated in the previous chapter.

In this chapter, we go one step further. We already saw that the SM prediction of the top mass agreed well with the observed mass. We clarify its mechanism. Using those precision data, we can make constraints on the mass value of the unknown Higgs particle which will soon be confirmed by the Large Hadron Collider (LHC)[1]. It can also place severe constraints on the possible new physics beyond the Standard Model. We emphasize the importance of improving precision of both experiments and theories to find a small discrepancy of the data from the SM prediction because it could lead to the discovery of new physics. This is reason enough to make a detailed discussion of the radiative corrections.

5.1
Input Parameters

The Weinberg Angle and Δr Radiative corrections appear in every process, though here we consider relations among fundamental constants of the SM, the mass of W and Z boson, the coupling constant of the neutral as well as charged current and the Weinberg angle $\sin^2\theta_W$. Given the SM Lagrangian in Chapter 1, we had

1) The Higgs was discovered in 2012, just prior to the publication of this textbook, with a mass value of $m_H = 126\,\text{GeV}$.

Elementary Particle Physics, Volume 2: Foundations of the Standard Model, First Edition. Yorikiyo Nagashima.

the following relations in the Born approximation (see Eqs. (1.39) and (1.44)).

$$m_W^2 = \frac{g_W^2 v^2}{4} \tag{5.1a}$$

$$m_Z^2 = \frac{g_Z^2 v^2}{4} = \frac{(g_B^2 + g_W^2) v^2}{4} \tag{5.1b}$$

$$4\pi\alpha = e^2 = \left(\frac{g_B g_W}{\sqrt{g_B^2 + g_W^2}} \right)^2 = g_W^2 \sin^2\theta_W \tag{5.1c}$$

$$g_Z = \sqrt{g_W^2 + g_B^2} = \frac{e}{\sin\theta_W \cos\theta_W} \tag{5.1d}$$

$$\frac{G_F}{\sqrt{2}} = \frac{1}{2v^2} = \frac{g_W^2}{8m_W^2} = \frac{\pi\alpha}{2m_W^2 \sin^2\theta_W} \tag{5.1e}$$

$$\frac{G_N}{\sqrt{2}} = \frac{g_Z^2}{8m_Z^2} = \frac{G_F}{\sqrt{2}} \tag{5.1f}$$

where e, g_B, g_W, g_Z are coupling constants of the gauge bosons γ, B, W, Z, and v is the vacuum expectation value of the Higgs field. Independent variables are (g_B, g_W, v). Others are expressed by their combination. When radiative corrections are taken into account, the original parameters, that is, the bare mass m_0 and the bare coupling constant which appeared in the Lagrangian are modified. Since those with all the corrections included should correspond to the observables, we rewrite them in terms of real (i.e., observed) variables m, e, g_W:

$$m_0^2 = m^2 + \delta m^2, \quad e_0 = e + \delta e, \quad g_{W0} = g_W + \delta g_W \tag{5.2}$$

As the degree of freedom is not changed by the radiative corrections, the number of independent variables is still three. If we choose them as (α, m_W, m_Z), $\sin^2\theta_W$ and G_F are dependent variables that can be derived from them, namely,

$$G_F = G_F(\alpha, m_W, m_Z, m_t, m_H) \tag{5.3a}$$

$$\sin^2\theta_W = \sin^2\theta_W(\alpha, m_W, m_Z, m_t, m_H) \tag{5.3b}$$

The reason we added m_t and m_H is that they enter as virtual particles in the radiative corrections. Other particles also appear, but in the theory where the symmetry is spontaneously broken, the coupling constants are proportional to their mass and numerically they are not important.

Experimentally, α, G_F, m_Z were determined precisely [3].

$$\alpha^{-1} = 137.035\,999\,084(51) \quad \frac{\delta\alpha}{\alpha} = 0.4 \times 10^{-9} \tag{5.4a}$$

$$G_F = 1.166\,364(5) \times 10^5\ \text{GeV}^{-2} \quad \frac{\delta G_F}{G_F} = 0.4 \times 10^{-5} \tag{5.4b}$$

$$m_Z = 91.1876 \pm 0.0021\ \text{GeV} \quad \frac{\delta m_Z}{m_Z} = 2.3 \times 10^{-5} \tag{5.4c}$$

Therefore, it is convenient to use them as inputs and determine others.

Using Eq. (5.1a)–(5.1c), we have the relation [215, 216]:

$$\text{Definition 1:} \quad \sin^2\theta_W = 1 - \frac{m_W^2}{m_Z^2} \tag{5.5a}$$

This is valid in the Born approximation, but if we use experimentally observed values in the right-hand side, it is a constant and does not depend on the renormalization scheme or its scale μ. All the radiative corrections are already contained in the observed values. Therefore, the $\sin^2\theta_W$ defined by Eq. (5.5a) using observed values does not receive further corrections. Conceptually, the definition is simple and clear. It is referred to as the on-shell $\sin^2\theta_W$. However, by definition, it is the reparametrization of the gauge boson masses and does not offer new information. Our second definition of the Weinberg angle is to use $\sin^2\theta_W$ which appears in the neutral current.

$$\text{Definition 2:} \quad J_N^\mu = J_3^\mu - Q_f \sin^2\theta_{\text{eff}}^f J_\gamma^\mu \tag{5.5b}$$

As the current appears in the transition amplitude, the $\sin^2\theta_{\text{eff}}^f$ can be determined directly from measurements of the cross section or the asymmetry. This Weinberg angle changes as the radiative correction is added and is a function of the renormalization scale μ (typically the momentum transfer Q of the reaction). This definition is very convenient in discussing the experimental observables obtained at Z peak or from deep inelastic neutrino scattering data. The $\sin^2\theta_W(\nu e)$ determined in Chapter 2 from the $\nu_e e$ scattering is the low energy version of this Weinberg angle[2]. We can use another Weinberg angle which is defined as the ratio of the two coupling constants.

$$\text{Definition 3:} \quad \sin^2\theta_W(\mu) = \frac{e^2(\mu)}{g_W^2(\mu)} = \frac{g_B^2(\mu)}{g_B^2(\mu) + g_W^2(\mu)} \tag{5.5c}$$

The $\sin^2\theta_W(\mu)$, in which the ultraviolet divergence is treated using the $\overline{MS}$ renormalization scheme, is written as $\sin^2\theta_{\overline{MS}} = \sin^2\hat{\theta}_W = \hat{s}^2$. This one is not directly connected with observables, but it uses the same renormalization scheme as used in QCD and is convenient in theoretical discussions including those of grand unified theories (GUTS).

At the Born approximation level, the three definitions are identical, giving the same value. However, when the radiative corrections are included, different processes contribute and are numerically different. In the following discussions, we use the one defined by Eq. (5.5a) as the base. If we start form (α, m_W, m_Z), G_F and $\sin^2\theta_W$ are derived quantities and we introduce the radiative correction Δr to be inserted in Eq. (5.1e) by [216]

$$m_W^2 \sin^2\theta_W = m_Z^2 \cos^2\theta_W \sin^2\theta_W = \frac{\pi\alpha}{\sqrt{2}G_F}\frac{1}{1-\Delta r} = \frac{A}{1-\Delta r}$$
$$A = (37.280\,61(8)\ \text{GeV})^2 \tag{5.6}$$

2) See footnote 5 in Chapter 2.

G_F can be determined precisely from the muon decay lifetime (Vol. 1, Sect. 15.5.1), and α from the Thomson scattering. As a function of the renormalization scale, we can consider that $G_F = G_F(\mu = m_\mu)$, $\alpha = \alpha(\mu = m_e \simeq 0)$. On the other hand, $\sin^2\theta_W$ is defined by the mass of the gauge bosons, and its natural scale should be considered as $\mu = m_Z$(or m_W^2). The Δr can be considered as a bridge to connect the two energy scales. As will be shown soon, its value is ~ 0.04 which is big enough to be tested by experiments.

Experimentally, m_Z can be determined very precisely compared with m_W or $\sin^2\theta_W$. Thus, we can regard Eq. (5.6) as equations to determine precise values of m_W and $\sin^2\theta_W$. In this case,

$$m_W^2 = \frac{m_Z^2}{2}\left(1 + \sqrt{1 - \frac{4A}{m_Z^2(1-\Delta r)}}\right) \tag{5.7a}$$

$$\sin^2\theta_W = \frac{1}{2}\left(1 - \sqrt{1 - \frac{4A}{m_Z^2(1-\Delta r)}}\right) \tag{5.7b}$$

In fact, the most precise value of m_W and $\sin^2\theta_W$ at present is obtained from experimentally determined m_Z by using theoretical calculations to evaluate Δr.

The sensitivity of m_W^2 to the radiation correction is

$$\frac{\Delta m_W^2}{m_W^2} = -\frac{\sin^2\theta_W}{\cos^2\theta_W - \sin^2\theta_W}\Delta r \tag{5.8}$$

Assuming $\Delta r = 0.04$,

$$\Delta m_W = -\frac{m_W \sin^2\theta_W/2}{\cos^2\theta_W - \sin^2\theta_W}\Delta r \sim -16.4\Delta r\,\text{GeV} \sim -623\,\text{MeV} \tag{5.9}$$

The present experimental accuracy of m_W being about 23 MeV, we see that the radiative correction is ~ 30 times, big enough to test the higher order correction of the GWS theory.

ρ Parameter At this point, we define another important radiative correction parameter ρ which appeared in the neutral current reactions. It is defined as the ratio of the coupling strength of the neutral and charged current.

$$\rho \equiv \rho_{\text{Born}}(1 + \Delta\rho) \equiv \frac{G_N}{G_F} \tag{5.10}$$

An alternative definition is given by

$$\rho = \rho_{\text{mass}} = \frac{m_W^2}{m_Z^2 \cos^2\theta_W} \tag{5.11}$$

It becomes one if an on-shell Weinberg angle is used, but it differs from unity in other definitions. If there are more than one Higgs particles unlike the minimum Standard Model, values of the gauge bosons are modified to

$$m_W^2 = \frac{g_W^2}{2}\sum_i v_i^2\left[I_i(I_i+1) - I_{3i}^2\right] \tag{5.12a}$$

$$m_Z^2 = \frac{g_W^2}{2\cos^2\theta_W} \sum_i (2I_{3i}^2) \tag{5.12b}$$

and hence at the Born level,

$$\rho_{\text{mass}} = \frac{\sum_i \left[I_i(I_i+1) - I_{3i}^2\right]}{\sum_i (2I_{3i}^2)} \tag{5.12c}$$

If the extra Higgs field contains an isospin triplet, $1/2 \leq \rho_{\text{mass}} \leq 1$, but as long as only isospin doublets are included, the value of ρ_{mass} at the Born level is

$$\rho = \rho_{\text{mass}} = 1 \tag{5.13}$$

This fact can be used to test if new physics exist beyond the Standard Model.

5.2 Renormalization

5.2.1 Prescription

The radiative corrections we want to calculate include loop integrals which are generally divergent. Renormalization is a prescription to extract a physically meaningful result out of the divergent integral. Since it is a mathematical procedure to subtract infinity from infinity to extract finite parts, there is an ambiguity as to how to choose the infinity which is the origin of the variety of renormalization schemes. Probably, conceptually, the most simple one is the method of minimum subtraction. We start with it in order to have some idea about the renormalization.

Minimum Subtraction ($\overline{MS}$) Scheme The radiative correction contains divergent integrals which typically appear in the form $\int d^4k/k^4$. It can be made finite formally by introducing cut offs or by going from four-dimensional momentum space to D-dimensional space. Nowadays, it is conventional to adopt the latter method because the gauge invariance as well as the Lorentz invariance are retained in the formalism. If $D < 4$, the integral $\mu^{4-D}\int d^Dk/k^4$ is convergent. The scale μ is introduced to keep the original physical dimension. The divergence reappears when one goes back to the original four-dimensional space. If we write $D = 4 - \epsilon$, the divergence takes the form of a pole in the complex D-plane, and behaves as $\sim 1/\epsilon$ as ε goes to zero. The divergent term in the dimensional regularization always appears in the combination

$$\Delta = \frac{2}{\epsilon} - \gamma_E + \ln 4\pi + \ln\mu \tag{5.14}$$

where γ_E denotes the Euler constant. The minimum subtraction method which was adopted originally by Veltman and 't Hooft simply removes $2/\epsilon$ from the radiative corrections. In the modified minimum subtraction scheme which is the standard scheme in QCD adopted by most authors nowadays, the whole Δ is replaced with $\ln \mu$.

This method is mathematically the simplest. The price to pay, however, is the appearance of the scale parameter μ which has the dimension of the energy. Once the value of μ is specified, observable values can easily be evaluated. In QCD, where the masses of the quarks or gluons cannot be specified because of confinement, this is the standard practice. A fundamental difficulty in predicting precise values of observables in QCD originates from ambiguity of choosing the right scale value.

On-Shell Renormalization The situation is different in the electroweak theory where the charge and the mass of the gauge bosons are easily identifiable and their values can be measured experimentally. These fundamental constants can serve as the reference to fix the otherwise indeterminate scale variable μ. The original parameters (mass m_0 and coupling constant g_0) in the Lagrangian change after adding the radiative corrections $(\delta m, \delta g)$. They contain divergent integrals which are dependent on the cut off (or $\varepsilon = 4 - D$) and on how to subtract the divergent terms. Hence, they are not true observables, but should be considered as mere parameters in the theory. They are referred to as bare charge or bare coupling constants. The true physical observables

$$m = m_0 + \delta m \,, \quad g = g_0 + \delta g \tag{5.15}$$

should include all those radiative corrections. In the on-shell scheme, one replaces the mass m_0 and the coupling constant g_0 in the original Lagrangian by their observed values. As a consequence, the divergences disappear formally. The whole expression is reformulated in terms of observables only and experimentally testable predictions can be made.

In Vol. 1 of this textbook, we saw that all the divergences in the QED can be absorbed by redefinition (renormalization) of the mass, the coupling constant, and the wave function normalization (see Vol. 1, Chapt. 8). In the electroweak theory, it is more complicated because of the non-Abelian and chiral nature of the theory. The left-handed and right-handed fermions have to be treated separately. In a relativistic treatment, one needs to add a Gauge fixing term to the Lagrangian and take into account ghosts in order to remove unphysical contributions of the gauge bosons.

A full mathematical formulation of the renormalization is beyond the scope of this book. We limit our discussions to a minimum treatment to understand the necessary physical background of the renormalization prescription. We limit our case studies to the four-fermion processes $[\mu^- \to (W^- + \nu_\mu) \to e^- + \overline{\nu}_e + \nu_\mu, e^- e^+ \to (\gamma Z) \to f \overline{f}]$ which are our major topics of interest. In this case, the important radiative corrections mainly come from the self-energy part of the gauge bosons, especially those of fermion loops which greatly simplify the matter. For the full treatment, we refer to the original papers by [170, 171, 216–218].

The Counter Term Before jumping into the details of the mathematics, we first illustrate a general idea of renormalization prescriptions using a toy model, which only includes one field φ_0 with mass m_0 and one coupling constant g_0. By writing

$$\begin{aligned}\mathcal{L}(m_0, \varphi_0, g_0) &= \mathcal{L}(Z_1^{1/2} Z_1^{-1/2} m_0, Z_2^{1/2} Z_2^{-1/2} \varphi_0, Z_3^{1/2} Z_3^{-1/2} g_0) \\ &= \mathcal{L}(Z_1^{1/2} m, Z_2^{1/2} \varphi, Z_3^{1/2} g)\end{aligned} \tag{5.16}$$

where we have introduced the renormalized field φ, the renormalized mass m and renormalized coupling constant g, that is, without changing the physical content of the Lagrangian. Writing the renormalization constant as

$$Z_1 = 1 + \delta Z_1\,, \quad Z_2 = 1 + \delta Z_2\,, \quad Z_3 = 1 + \delta Z_3 \tag{5.17}$$

and inserting the above expression in $\mathcal{L}$, which is a polynomial in φ and g, we obtain

$$\begin{aligned}\mathcal{L}(m_0, \varphi_0, g_0) &= \mathcal{L}(m, \varphi, g) + \delta \mathcal{L}(m, \varphi, g, \delta m, \delta Z, \delta g) \\ \delta m &= \frac{1}{2} \delta Z_1 m, \quad \delta Z = \delta Z_2, \quad \delta g = \frac{1}{2} \delta Z_3 g\end{aligned} \tag{5.18}$$

In this way, $\mathcal{L}$ is decomposed into a part which has the same form as the original Lagrangian, but involves only renormalized variables and fields and a counter term Lagrangian which generates amplitudes containing the renormalization constants which are adjusted to absorb divergent parts. In the actual calculation of the transition amplitude, one includes Feynman diagrams which result from the counter term. Then, the subtraction terms $(\delta m, \delta g, \delta Z)$ appear in the right combination with the corresponding higher order corrections and all the divergent terms disappear automatically from final expressions.

The decomposition of $\mathcal{L}(\varphi_0, g_0)$ into $\mathcal{L}(\varphi, g)$ and $\delta\mathcal{L}$ is to a large extent arbitrary dependent on the adopted scheme and this makes the end results scheme dependent. This is because one can always add a finite term to δZ_i and still realize a finite expression for the amplitude. If the calculation is carried out to all orders in the power expansion, the different renormalization scheme should give the same result. However, at finite order, the result can be scheme dependent. As the residual terms contain higher powers of the expansion parameter, the difference is (hopefully) small.

The Minimum Scheme Some comments for the field renormalization are worthwhile. As we saw in Vol. 1, Sect. 8.1.8, the renormalization constant of the field appears only in the external field, that is, as renormalization of the external wave function. This shows that we have a freedom to renormalize the internal field (which appear as propagators) in any way we want.[3] As normalization of the wave functions is a technical matter which does not contain any physical content, we can get away from the wave field renormalization and apply the scheme only to masses and coupling constants. Only a formal rewriting of the external wave function at the very end is necessary to keep mathematical consistency. The simplest on-shell

3) This fact can be most easily seen in the path integral expression of the transition matrix. There, the internal field is integrated over all paths and the change of the field $\phi \to s\phi$ can simply be absorbed by redefinition of integration variables.

renormalization uses this scheme [215, 216]. However, in this scheme, the propagator and the vertex function are left with divergent terms which is inconvenient in many ways. In the other extremes, it is also possible to use this freedom to adjust the propagator so that not only is it finite, but also its residue remains to be unity as is the case for the free propagator.

Here, we use the field renormalization, but keep its number to a minimum to avoid mathematical complications. We take the independent parameters as e, m_W, m_Z and apply the wave field renormalization only to B and W, which appear in the original Lagrangian and make the individual propagator finite. Renormalization of the Z, γ and $\gamma - Z$ mixing are derived from that of B and W.

Renormalization of the fermion field is necessary, but as long as we limit our treatment to the four-fermion interactions, the extended Ward identity (namely, the Slavnov–Taylor relation) guarantees that the external field renormalization compensates the divergent part of the vertex function and no manifest divergent terms appear. Therefore, we need not worry about it. The Higgs field contributes as a loop only to the self-energy of the gauge field. So, in the following, we apply the field renormalization only to the gauge bosons.

5.2.2
Self-Energy of Gauge Bosons

Mass Renormalization and the Running Coupling Constant We first consider how the renormalization constant for the field appears in the propagator. The propagator of the gauge boson $V(V = W^{\pm}, Z, \gamma)$ is defined in the first approximation as

$$D_{V0}^{\mu\nu}(q^2) \equiv \int d^4 x e^{iq\cdot x} T\langle V_0^{\mu}(x) V_0^{\nu}(0)\rangle = -i\frac{\left(g^{\mu\nu} - \frac{q^{\mu}q^{\nu}}{m_V^2}\right) + B(q^2)q^{\mu}q^{\nu}}{q^2 - m_V^2 + i\epsilon} \tag{5.19}$$

where B is dependent on the gauge. However, the propagators in the transition amplitude always appear sandwiched by source currents. Therefore, the $q^{\mu}q^{\nu}$ term vanishes for the conserved vector current and are proportional to the mass of the particle for the axial current. For the electron and/or neutrino of our interest, it is negligible compared to the W or Z mass and can also be assumed to vanish. Therefore, in the following, we only consider

$$D_{V0} = \frac{-i}{q^2 - m_V^2} \tag{5.20}$$

We express this pictorially as Figure 5.1g.

We consider how the first approximation Eq. (5.19) is modified. Let us consider including order $O(g_V^2)$ corrections, for example, the fermion loop term like Figure 5.1a, which is explicitly expressed as

$$\frac{-i\left(g_{\rho\mu} - \frac{p_{\rho}p_{\mu}}{m_{V1}^2}\right)}{p^2 - m_{V1}^2}\left[-i\Sigma_{V_1V_2}^{\mu\nu}(p^2)\right]\frac{-i\left(g_{\nu\sigma} - \frac{p_{\nu}p_{\sigma}}{m_{V2}^2}\right)}{p^2 - m_{V2}^2} \tag{5.21a}$$

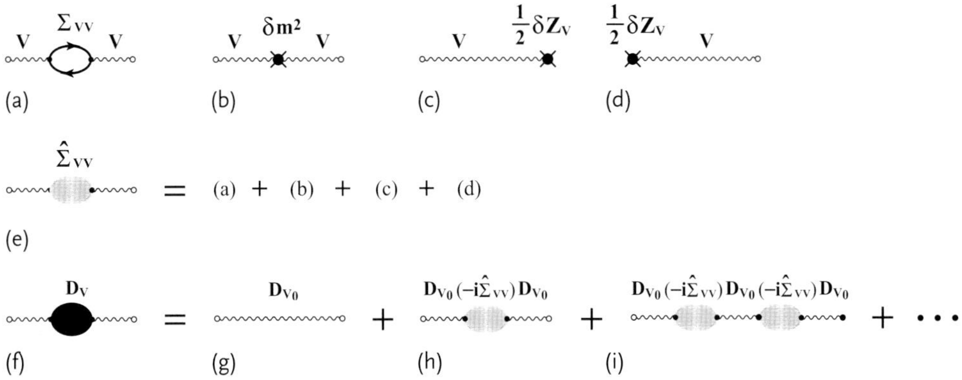

Figure 5.1 Pictorial representation of the Feynman diagrams for the gauge field propagator including the higher order self-energy and the subtraction term. (a) The self-energy due to a fermion loop which together with the mass counter term (b) and the wave field renormalization term (c), (d) is made finite. (e) is the sum of (a–d). The renormalization group equation resums the dominant ladder type contributions and the net result (f) is the sum of (g–i).

$$-i\Sigma^{\mu\nu}_{V_1 V_2}(p^2) = -g_{V_1} g_{V_2} \int \frac{d^4 k}{(2\pi)^4} \mathrm{Tr}\left[\frac{i(\not{k} + m_1)}{k^2 - m_1^2} i\gamma^{\nu}(v_2 - a_2\gamma^5) \right.$$
$$\left. \times \frac{i(\not{p} + \not{k} + m_2)}{(k+p)^2 - m_2^2} i\gamma^{\mu}(v_1 - a_1\gamma^5)\right] \tag{5.21b}$$

$$\Sigma^{\mu\nu}_{V_1 V_2} \equiv \left(g^{\mu\nu} - p^{\mu}p^{\nu}/m_V^2\right)\Sigma_{V_1 V_2}(p^2) \tag{5.21c}$$

where g_i, v_i, a_i $(i = W, \gamma, Z)$ are the coupling constant, its vector and axial vector components. For the propagator with no mixing, $V_1 = V_2 = V$. Here, $\Sigma_{VV}(p^2)$ are referred to as the self-energy of the particle. Explicit expressions for $\Sigma_{VV}(p^2)$ are given in Appendix C. Equations (5.21) give corrections to the Born term in Eq. (5.20). If one neglects the gauge dependent terms, the Born propagator corrected with the loop contribution will be modified to

$$D_{V0} \to \frac{-i}{q^2 - m_V^2} + \frac{-i}{q^2 - m_V^2}\left(-i\Sigma_{VV}(q^2)\right)\frac{-i}{q^2 - m_V^2} \tag{5.22}$$

The loop integral Σ_{VV} contains divergent terms which should be canceled by the counter terms. Namely, one adds the Feynman diagram that can be obtained from $\delta\mathcal{L}$. The mass correction term in $\delta\mathcal{L}$ has the form $\delta m_V^2 V_\mu V^\mu$ and connects two propagators directly. Its contribution has the same form as the second term in Eq. (5.22) with Σ_{VV} replaced by $-\delta m_V^2$ which is pictorially represented by Figure 5.1b.

The wave field renormalization constant enters in the following way.

$$D_V^{\mu\nu} = \langle T[V^\mu V^\nu]\rangle = \left\langle T\left[Z_V^{-1} V_0^\mu V_0^\nu\right]\right\rangle = (1 - \delta Z_V) D_{V0}^{\mu\nu} \tag{5.23}$$

Then, the propagator after adding all the $O(g_V^2)$ correction terms becomes

$$D_{V0}(q^2) \to D_V(q^2) = \frac{-i}{q^2 - m_V^2}\left[1 - \delta Z_V + \left(\delta m_V^2 - \Sigma_{VV}(q^2)\right)\frac{1}{q^2 - m_V^2}\right] \tag{5.24}$$

The contribution of δZ_V is pictorially represented by Figure 5.1c,d. Each $(1/2)\delta Z_V$ at both ends of the propagator contributes to vertices in the transition amplitude. $\Sigma_{VV}(q^2)$ in Eq. (5.21b) can be expanded in powers of q^2. As one can see easily by counting the powers of q^2 in the integrand, the divergent terms are only contained in the first and second terms.[4] In other words, it has the form like

$$\Sigma_{VV}(q^2) = A + B(q^2 - m_V^2) + \text{finite term} \tag{5.25}$$

By inspecting Eq. (5.24), one can see that by choosing δm_V^2 and δZ_V carefully, the divergence in the Σ_{VV} can be canceled and it becomes finite. Therefore, by introducing subtraction terms, we can define a finite renormalized loop function

$$\hat{\Sigma}_{VV}(q^2) = \Sigma_{VV}(q^2) - \delta m_V^2 + \delta Z_V\left(q^2 - m_V^2\right) \tag{5.26}$$

which is represented pictorially as in Figure 5.1e. Substitution of Eq. (5.26) into Eq. (5.24) changes the propagator to the same form back as Eq. (5.22), though this time the propagator only contains finite terms.

As the self-energy loop can be inserted anywhere in the propagator, summation of ladder type self-energy contributions (Figure 5.1h,i, and their repetition) gives[5]

$$\begin{aligned} D_{V0} &= \frac{-i}{q^2 - m_V^2} + \frac{-i}{q^2 - m_V^2}\left(-i\hat{\Sigma}_{VV}(q^2)\right)\frac{-i}{q^2 - m_V^2} \\ &\quad + \frac{-i}{q^2 - m_V^2}\left(-i\hat{\Sigma}_{VV}(q^2)\right)\frac{-i}{q^2 - m_V^2}\left(-i\hat{\Sigma}_{VV}(q^2)\right)\frac{-i}{q^2 - m_V^2} + \cdots \\ &= \frac{-i}{(q^2 - m_V^2)\left[1 + \hat{\Sigma}_{VV}(q^2)\frac{1}{q^2 - m_V^2}\right]} = \frac{-i}{q^2 - m_V^2 + \hat{\Sigma}_{VV}} \end{aligned} \tag{5.27}$$

In summary, the final result of the fermion loop contribution renormalization is given by

$$i D_{V0}(q^2) = \frac{1}{q^2 - m_V^2} \to i D_V(q^2) = \frac{1}{q^2 - m_V^2 + \hat{\Sigma}_{VV}(q^2)} \tag{5.28}$$

As the observed physical mass is defined as the real part of the pole, the subtraction term δm_V^2 has to satisfy the condition

$$\delta m_V^2 = \text{Re}\Sigma_{VV}\left(m_V^2\right) \tag{5.29}$$

4) If one counts the power of q^2 in Eq. (5.21b) formally, they give first and second order divergences. However, in the gauge theory, it can be proved that the divergence is at most logarithmic. Refer to discussions in Chapter 1.

5) The procedure is referred to as resummation. The renormalization group equation discussed in Section 7.2 automatically collects the dominant contribution, enhances convergence of the power expansion and justifies the procedure.

Determination of δZ_V will be described later. At this point, we define another finite self-energy function $\hat{\Pi}_V$ by

$$\mathrm{Re}\hat{\Sigma}_{VV}(q^2) = (q^2 - m_V^2)\hat{\Pi}_V(q^2) \tag{5.30a}$$

$$\hat{\Pi}_V(q^2) = \mathrm{Re}\frac{\Sigma_{VV}(q^2) - \Sigma_{VV}(m_V^2)}{q^2 - m_V^2} + \delta Z_V \equiv \Pi_V(q^2) + \delta Z_V \tag{5.30b}$$

and rewrite Eq. (5.28) to obtain

$$i D_V(q^2) = \frac{1}{1 + \hat{\Pi}(q^2)} \frac{1}{q^2 - m_V^2 + i m_V \Gamma_V} \tag{5.31a}$$

$$m_V \Gamma_V = \mathrm{Im}\frac{\hat{\Sigma}_{VV}(q^2)}{1 + \hat{\Pi}(q^2)} \tag{5.31b}$$

In the actual process, for example, in the W propagator of the muon decay, the above propagator appears sandwiched by the weak current and is multiplied by the coupling constant g_V^2. This means that the effect of the higher order correction to the gauge boson propagator is equivalent to changing the coupling constant g_V^2 to q^2 dependent running coupling constant

$$g_V^2 \to g_V^2(q^2) = \frac{g_V^2}{1 + \hat{\Pi}(q^2)} \tag{5.32}$$

and at the same time, introducing the decay width $m_V \Gamma_V$ in the denominator.

$\gamma - Z$ Mixing In the electroweak theory, both γ and Z couple to the same fermions, and hence they mix via a process depicted in Figure 5.2a. Denoting $\gamma - Z$ mixing energy as $\Sigma_{\gamma Z}$, there are four functions $\Sigma_{WW}, \Sigma_{ZZ}, \Sigma_{\gamma\gamma}, \Sigma_{\gamma Z}$ that need subtraction. However, considering γ and Z are mixtures of W and B, they are related. In order to see the relation, we use

$$Z_\mu = \cos\theta_W W_{3\mu} - \sin\theta_W B_\mu \tag{5.33a}$$

$$A_\mu = \sin\theta_W W_{3\mu} + \cos\theta_W B_\mu \tag{5.33b}$$

and the renormalized field relations

$$W_{0\mu} = Z_W^{1/2} W_\mu = \left(1 + \frac{1}{2}\delta Z_W\right) W_\mu \tag{5.34a}$$

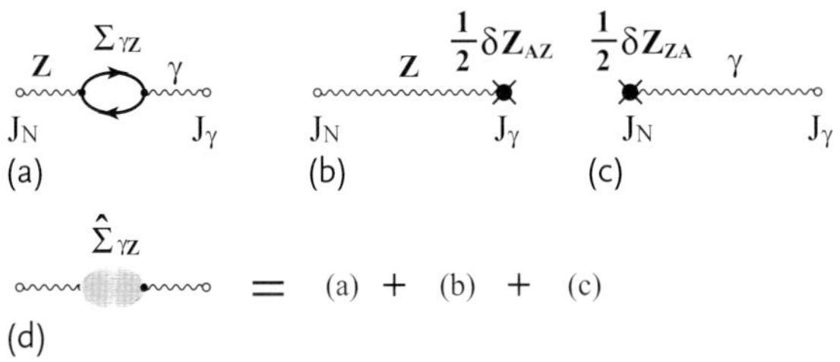

Figure 5.2 $\gamma - Z$ mixing.

$$B_{0\mu} = Z_B^{1/2} B_\mu = \left(1 + \frac{1}{2}\delta Z_B\right) B_\mu \tag{5.34b}$$

which lead to the following equalities.

$$\begin{bmatrix} Z_{0\mu} \\ A_{0\mu} \end{bmatrix} = \begin{bmatrix} 1 + \frac{1}{2}\delta Z_Z, & \frac{1}{2}\delta Z_{ZA} \\ \frac{1}{2}\delta Z_{AZ}, & 1 + \frac{1}{2}\delta Z_A \end{bmatrix} \begin{bmatrix} Z_\mu \\ A_\mu \end{bmatrix} \tag{5.35a}$$

$$\begin{bmatrix} \delta Z_Z \\ \delta Z_A \end{bmatrix} = \begin{bmatrix} c_W^2, & s_W^2 \\ s_W^2, & c_W^2 \end{bmatrix} \begin{bmatrix} \delta Z_W \\ \delta Z_B \end{bmatrix} \tag{5.35b}$$

$$\delta Z_{ZA} = s_W c_W (\delta Z_W - \delta Z_B) - \frac{\delta s_W^2}{c_W s_W} \tag{5.35c}$$

$$\delta Z_{AZ} = s_W c_W (\delta Z_W - \delta Z_B) + \frac{\delta s_W^2}{c_W s_W} \tag{5.35d}$$

$$\frac{\delta s_W^2}{c_W s_W} = \frac{c_W}{s_W}\left(\frac{\delta m_Z^2}{m_Z^2} - \frac{\delta m_W^2}{m_W^2}\right) \tag{5.35e}$$

where we denoted $s_W = \sin\theta_W, c_W = \cos\theta_W$ and we used Eq. (5.5a) in deriving the last equality. δZ_{AZ} means that the photon mixes with Z and gives direct coupling of the Z to the electromagnetic current J_γ^μ (Figure 5.2b). Similarly, δZ_{ZA} means the direct coupling of the photon with the neutral current J_N^μ (Figure 5.2c). Both of them appear as subtraction terms for the $\Sigma_{\gamma Z}$. Writing $\Sigma_{\gamma Z}$ and the subtraction terms in the form of Eq. (5.24), we can summarize the relation as

$$i D_{ab} = \frac{1}{2}\left[\frac{\delta_{ab} - \delta Z_{ba}}{q^2 - m_a^2} + \frac{\delta_{ab} - \delta Z_{ab}}{q^2 - m_b^2}\right] + \frac{\delta_{ab}\delta m_a^2 - \Sigma_{ab}(q^2)}{(q^2 - m_a^2)(q^2 - m_b^2)} \tag{5.36}$$

Here, a, b represent Z, A and $\delta Z_{aa} = \delta Z_a$. For $a = b$, Eq. (5.36) reduces to Eq. (5.24). For the photon, the gauge invariance guarantees vanishing of the mass, that is, $\Sigma_{\gamma\gamma}(0) = 0$ and no mass subtraction is necessary. We list the regularized self-energy terms in the following.

$$\hat{\Sigma}_{\gamma\gamma}(q^2) = \Sigma_{\gamma\gamma}(q^2) + \delta Z_A q^2 \equiv q^2\left[\Pi_\gamma(q^2) + \delta Z_A\right] \tag{5.37a}$$

$$\hat{\Sigma}_{\gamma Z}(q^2) = \Sigma_{\gamma Z}(q^2) + \frac{1}{2}\left[\delta Z_{AZ} q^2 + \delta Z_{ZA}(q^2 - m_Z^2)\right] \tag{5.37b}$$

$$\hat{\Sigma}_{ZZ}(q^2) = \Sigma_{ZZ}(q^2) - \delta m_Z^2 + \delta Z_Z (q^2 - m_Z^2) \tag{5.37c}$$

$$\hat{\Sigma}_{WW}(q^2) = \Sigma_{ZWW}(q^2) - \delta m_W^2 + \delta Z_W (q^2 - m_W^2) \tag{5.37d}$$

The subtraction terms can be determined by the following four conditions.

1. The propagators of W have a pole at $q^2 = m_W^2$.
2. The propagators of Z have a pole at $q^2 = m_Z^2$.
   The wave field renormalization is determined similarly to QED, that is,
3. The photon propagator should become $-g^{\mu\nu}/q^2$ in the limit $q^2 \to 0$. Namely, the residue of the pole should agree with that of the free propagator.
4. The real γ should have no Z component.

Conditions (1) and (2) give

$$\delta m_W^2 = \mathrm{Re}\Sigma_{WW}\left(m_W^2\right) \tag{5.38a}$$

$$\delta m_Z^2 = \mathrm{Re}\Sigma_{ZZ}\left(m_Z^2\right) \tag{5.38b}$$

Condition (3) gives

$$\hat{\Pi}_\gamma(0) = 0 \rightarrow \delta Z_A = -\Pi_\gamma(0) \tag{5.39}$$

Condition (4) gives

$$\hat{\Sigma}_{\gamma Z}(0) = 0 \rightarrow \delta Z_{ZA} = \frac{2\Sigma_{\gamma Z}(0)}{m_Z^2} \tag{5.40}$$

Other subtraction terms can be derived from relations Eq. (5.35a)–(5.35e). We summarize all the self-energy expressions as

$$\hat{\Sigma}_{\gamma\gamma}(q^2) = q^2\hat{\Pi}_\gamma(q^2) = \Sigma_{\gamma\gamma}(q^2) - q^2\Pi_\gamma(0) = q^2\left(\Pi_\gamma(q^2) - \Pi_\gamma(0)\right) \tag{5.41a}$$

$$\hat{\Sigma}_{\gamma Z}(q^2) = q^2\hat{\Pi}_{\gamma Z}(q^2) = \Sigma_{\gamma Z}(q^2) - \Sigma_{\gamma Z}(0) + q^2\frac{\delta' s_W^2}{c_W s_W} \tag{5.41b}$$

$$\hat{\Sigma}_{ZZ}(q^2) = \left(q^2 - m_Z^2\right)\hat{\Pi}_Z(q^2) = \Sigma_{ZZ}(q^2) - \mathrm{Re}\Sigma_{ZZ}\left(m_Z^2\right) + \delta Z_Z(q^2 - m_Z^2) \tag{5.41c}$$

$$\begin{aligned}\hat{\Sigma}_{WW}(q^2) &= \left(q^2 - m_W^2\right)\hat{\Pi}_W(q^2)\\ &= \Sigma_{WW}(q^2) - \mathrm{Re}\Sigma_{WW}\left(m_W^2\right) + \delta Z_W(q^2 - m_W^2)\end{aligned} \tag{5.41d}$$

$$\delta Z_Z = -\Pi_\gamma(0) + \frac{c_W^2 - s_W^2}{c_W^2 s_W^2}\delta' s_W^2 \tag{5.41e}$$

$$\delta Z_W = -\Pi_\gamma(0) + \frac{\delta' s_W^2}{s_W^2} \tag{5.41f}$$

$$\delta' s_W^2 = c_W^2\,\mathrm{Re}\left[\frac{\Sigma_{ZZ}\left(m_Z^2\right)}{m_Z^2} - \frac{\Sigma_{WW}\left(m_W^2\right)}{m_W^2} + 2\frac{s_W}{c_W}\frac{\Sigma_{\gamma Z}(0)}{m_Z^2}\right] \tag{5.41g}$$

Charge Renormalization As the result of radiative corrections in the propagator, the bare coupling constant is also modified. The observed coupling constant e is primarily defined as the value obtained from the Thomson scattering, namely, the vertex function at $q^2 = 0$. To obtain the charge renormalization condition, we need the vertex function and the fermion field renormalization as well. However, the extended Ward identity (Slavnov–Taylor relation) relates both and they disappear from the final expression. The end result is a universal relation independent of the fermion flavor:

$$\frac{\delta e}{e} = \frac{1}{2}\Pi_\gamma(0) + \frac{s_W}{c_W}\frac{\Sigma_{\gamma Z}(0)}{m_Z^2} \tag{5.42}$$

If the self-energy consists entirely of fermion loops, $\Sigma_{\gamma Z}(0) = 0$ (see Eq. (C.31a)) and the above condition is identical to that of the QED (see Vol. 1, Eq. (8.58b)). Corrections to other coupling constants ($g_W = e/\sin\theta_W$ and $g_Z = e/\sin\theta_W\cos\theta_W$) follow from the above expression.

5.3
μ Decay

5.3.1
$\Delta r, \rho, \Delta\alpha$

To evaluate the correction of the muon decay Δr, let us recollect the definition of the coupling constant G_F. It is defined by the formula that the Born approximation for the expression of the muon decay rate with the QED correction gives the correct experimental value (see Vol. 1, Eq. (15.86) and [219]).[6]

$$\frac{1}{\tau_\mu} = \frac{G_F^2 m_\mu^5}{192\pi^3} f\left(\frac{m_e^2}{m_\mu^2}\right)\left(1 + \frac{3}{5}\frac{m_\mu^2}{m_W^2}\right)\left[1 + \frac{\alpha(m_\mu)}{2\pi}\left(\frac{25}{4} - \pi^2\right)\right] \tag{5.43a}$$

$$\alpha_\mu^{-1}(m_\mu) = \alpha^{-1}(0) - \frac{2}{3\pi}\ln\left(\frac{m_\mu}{m_e}\right) + \frac{1}{6\pi} \approx 135.9 \tag{5.43b}$$

$$f(x) = 1 - 8x + 8x^3 - x^4 - 12x^2\ln x\,, \quad f\left(\frac{m_e^2}{m_\mu^2}\right) = 0.999\,813 \tag{5.43c}$$

In this definition, pure electromagnetic correction is included, but not that of the electroweak interaction.[7] By the electroweak interaction, we mean processes including both γ and W/Z. Among the electroweak corrections to the muon decay, there are three types of contributions:

1. The self-energy correction to the W propagator which is sometimes referred to as the oblique correction → Figure 5.3b–e.
2. The vertex correction → Figure 5.3g–i.
3. The box type correction → Figure 5.3k–n.

(2) and (3) are sometimes collectively referred to as the direct correction. Since the contribution of (2) and (3) turned out to be small, we do not make any detailed discussion of them except to add to the final result when necessary.

When one writes down the transition amplitude for the muon decay using the Standard Model, and includes the self-energy effect of the W boson, it is expressed

6) As more detailed theoretical calculations have become available, the defining equation for G_F gets more complicated with small change numerically. For the most recent expression, see [3].

7) The third factor $[1 + (3/5)(m_\mu^2/m_W^2)]$ should be included in the definition of Δr from the spirit of the Fermi theory, but the Particle Data Group chose to leave it for historical consistency.

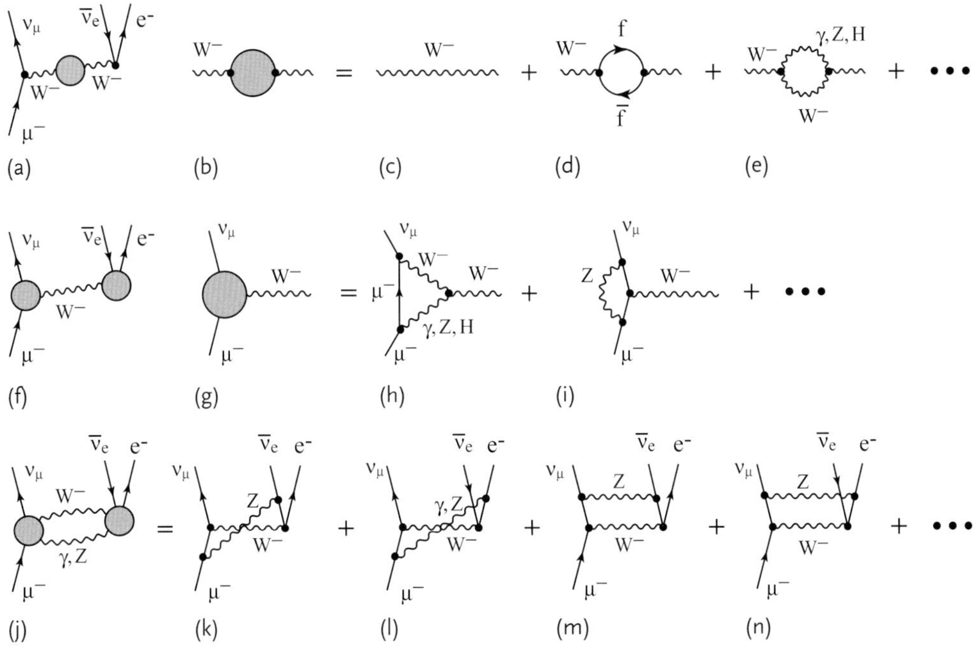

Figure 5.3 Various corrections to the muon decay process (a). (b)–(e) Self-energy of the $W^{\pm}$ boson, (f)–(i) Vertex corrections and (j)–(n) box diagrams. The dominant contribution came from the gauge boson self-energy.

as follows:

$$\mathcal{M} \sim -\frac{g_W^2}{8\left[1+\hat{\Pi}_W(q^2)\right]} \frac{J_W^- \otimes J_W^+}{q^2 - m_W^2 + i m_W \Gamma_W}\Bigg|_{q^2=0} + \cdots = \frac{g_W^2}{8 m_W^2} \frac{J_W^- \otimes J_W^+}{1+\hat{\Pi}_W(0)} + \cdots$$
$$g_W^2 = e^2/s_W^2 \tag{5.44}$$

Here, we used Eq. (5.31a) to express the renormalized loop correction to the self-energy of W. We also used a formal expression $J_W^- \otimes J_W^+$ to denote the product of currents made by plane wave solutions, namely,

$$J_W^- \otimes J_W^+ = \left[\overline{u}(p_e)\gamma_\mu(1-\gamma^5)v(p_{\nu_e})\right]\left[\overline{u}(p_{\nu_\mu})\gamma^\mu(1-\gamma^5)u(p_\mu)\right] \tag{5.45}$$

On the other hand, the Born approximation to the transition amplitude is given by

$$\mathcal{M} = \frac{G_F}{\sqrt{2}} J_W^- \otimes J_W^+ \tag{5.46}$$

Comparing Eqs. (5.44) and (5.46) with Eq. (5.6), one can obtain the expression for Δr:

$$\Delta r = -\hat{\Pi}_W(0) + \cdots \tag{5.47a}$$

$$= \mathrm{Re}\frac{\Sigma_{WW}(0) - \Sigma_{WW}(m_W^2)}{m_W^2} + \Pi_\gamma(0) - \frac{\delta' s_W^2}{s_W^2} + \cdots \tag{5.47b}$$

Here, $\cdots$ represents contributions other than the W self-energy, and in deriving the second equation, we used Eqs. (5.41d) and (5.41f). We can further rewrite the expression using Eqs. (5.41c)–(5.41g) to obtain

$$\begin{aligned}\Delta r = \Pi_\gamma(0) &- \frac{c_W^2}{s_W^2}\left(\frac{\Sigma_{ZZ}(m_Z^2) - \Sigma_{ZZ}(0)}{m_Z^2}\right)\\ &+ \frac{c_W^2 - s_W^2}{s_W^2}\left(\frac{\Sigma_{WW}(m_W^2) - \Sigma_{WW}(0)}{m_W^2}\right)\\ &- \frac{c_W^2}{s_W^2}\left(\frac{\Sigma_{ZZ}(0)}{m_Z^2} - \frac{\Sigma_{WW}(0)}{m_W^2} + 2\frac{s_W}{c_W}\frac{\Sigma_{\gamma Z}(0)}{m_Z^2}\right)\end{aligned} \tag{5.47c}$$

Here, we understand that Σ means $\mathrm{Re}\Sigma$. Inspection of Eq. (5.47c) tells us that Δr includes, in addition to the W self-energy, the photon and Z self-energy. They are all related to each other. These functions Σ_{ab} were calculated for the fermion loop (see Appendix C). The last term can be identified with $\Delta\rho$ we defined in Eq. (5.10). To see it, we notice that the charged current coupling constant G_F receives the W propagator correction and the neutral current coupling constant G_N receives correction from the Z propagator. From the definition of ρ in Eq. (5.10), we obtain

$$\rho = \frac{G_N}{G_F} = \frac{1 + \hat{\Pi}_W(q^2)}{1 + \hat{\Pi}_Z(q^2)} \simeq 1 + \hat{\Pi}_W(q^2) - \hat{\Pi}_Z(q^2) \tag{5.48}$$

Conventionally, the value of ρ is evaluated at $q^2 = 0$. Then, using Eqs. (5.30b) and (5.41c)–(5.41g), Eq. (5.48) evaluated at $q^2 = 0$ gives

$$\Delta\rho = \frac{\Sigma_{ZZ}(0)}{m_Z^2} - \frac{\Sigma_{WW}(0)}{m_W^2} + 2\frac{s_W}{c_W}\frac{\Sigma_{\gamma Z}(0)}{m_Z^2} \tag{5.49}$$

It can be shown that as long as $q^2 \ll m_W^2$, the difference between $\Delta\rho(q^2)$ and $\Delta\rho(0)$ is negligible. Below the Z resonance, $\Delta\rho$ plays the major role in the radiative corrections. The self-energy of the Z boson contains the contribution from the isospin current which belongs to the same multiplet, and that from the electromagnetic current and their interference term. Therefore, its structure is more apparent if one separates their contributions into isospin components. Noticing that $J_N = J_3 - Qs_W^2 J_\gamma$ and separating the coupling constants, we rewrite the self-energy functions as follows:

$$\Sigma_{\gamma\gamma} = e^2 Q^2 \Sigma_{QQ} \tag{5.50a}$$

$$\Sigma_{\gamma Z} = e g_Z (Q\Sigma_{3Q} - Q^2 s_W^2 \Sigma_{QQ}) \tag{5.50b}$$

$$\Sigma_{ZZ} = g_Z^2 (\Sigma_{33} - 2Qs_W^2 \Sigma_{3Q} + Q^2 s_W^4 \Sigma_{QQ}) \tag{5.50c}$$

$$\Sigma_{WW} = g_W^2 \Sigma_{11} \tag{5.50d}$$

Σ_{11}, Σ_{33} represent the isospin component of the fermion current that couples to the gauge bosons. Substituting Eqs. (5.50a)–(5.50d) into Eq. (5.49), we obtain

$$\Delta\rho = \frac{e^2}{m_W^2 s_W^2}(\Sigma_{33}(0) - \Sigma_{11}(0)) \tag{5.51}$$

which demonstrates that the $\Delta\rho$ is a measure of the isospin breaking in the electroweak interaction. As the coupling in the isospin part of the interaction is chiral, the contribution from $\Sigma_{11}(0)$ and $\Sigma_{33}(0)$ are proportional to the mass of fermions in the isospin multiplet, that is, $m_{f\pm}^2/m_W^2$. Therefore, contributions other than the top and the bottom quarks are negligible. If the isospin invariance holds ($m_t = m_b$), $\Sigma_{11} = \Sigma_{33}$ and $\Delta\rho = 0$.[8] The actual calculation (Eq. (C.39c)) gives

$$\begin{aligned}\Delta\rho_t &\approx \frac{3e^2}{64\pi^2 m_W^2 s_W^2}\left(m_t^2 + m_b^2 - \frac{2m_t^2 m_b^2}{m_t^2 - m_b^2}\ln\frac{m_t^2}{m_b^2}\right)\\ &\approx \frac{3G_F m_t^2}{8\sqrt{2}\pi^2} \approx 0.009\,39\left(\frac{m_t}{173.1\ \text{GeV}}\right)^2 {}^{9)}\end{aligned} \tag{5.52}$$

Defining Σ'_{ab} function by

$$\Sigma'_{ab}(q^2) = \frac{\Sigma_{ab}(q^2) - \Sigma_{ab}(0)}{q^2} \tag{5.53}$$

and noting $\Pi_\gamma(q^2) = \Sigma_{\gamma\gamma}/q^2 = e^2 Q^2 \Sigma'_{QQ}(q^2)$, the expression for Δr in Eq. (5.47c) can be rewritten as

$$\begin{aligned}\Delta r = {}& e^2 Q^2 \Sigma'_{QQ}(0) - \frac{e^2}{s_W^4}\left[\Sigma'_{33}\left(m_Z^2\right) - 2Q s_W^2 \Sigma'_{3Q}\left(m_Z^2\right) + Q^2 s_W^4 \Sigma'_{QQ}\left(m_Z^2\right)\right]\\ &+ \frac{c_W^2 - s_W^2}{s_W^4} e^2 \Sigma'_{11}\left(m_Z^2\right) - \frac{c_W^2}{s_W^2}\Delta\rho + \cdots\end{aligned} \tag{5.54}$$

As the running coupling constant of the electromagnetic interaction is given by

$$e^2 + \Delta e^2 = \frac{e^2}{1 + \hat{\Pi}_\gamma} \approx e^2\left(1 - \hat{\Pi}_\gamma(q^2)\right) = e^2\left(1 - \Pi_\gamma(q^2) + \Pi_\gamma(0)\right) \tag{5.55}$$

we rewrite it using $\Sigma'_{QQ}(m_Z^2)$ and obtain

$$\Delta\alpha \equiv \frac{\alpha\left(m_Z^2\right) - \alpha(0)}{\alpha(0)} = -\hat{\Pi}_\gamma\left(m_Z^2\right) = -Q^2 e^2\left(\Sigma'_{QQ}\left(m_Z^2\right) - \Sigma'_{QQ}(0)\right) \tag{5.56}$$

8) Phenomenologically, $\Delta\rho$ is small which means that the isospin symmetry is at work. This is referred to as the "custodial $SU(2)$ symmetry." See arguments in the box in Section 3.3.5.

9) Quantities expressed in terms of G_F instead of α include certain correction terms and is sometimes written as $\Delta\bar{\rho}$ to differentiate it (see Section 5.4.2).

Substituting Eq. (5.56) in Eq. (5.54), we have

$$\begin{aligned}\Delta r &= \Delta\alpha - \frac{c_W^2}{s_W^2}\Delta\rho - \frac{e^2}{s_W^4}\left[\Sigma'_{33}\left(m_Z^2\right) - 2Qs_W^2\Sigma'_{3Q}\left(m_Z^2\right)\right] \\ &\quad + \frac{c_W^2 - s_W^2}{s_W^4}e^2\Sigma'_{11}\left(m_W^2\right) + \cdots \\ &= \Delta\alpha - \frac{c_W^2}{s_W^2}\Delta\rho + \frac{c_W^2 - s_W^2}{s_W^2}\epsilon_2 + 2\epsilon_3 + \cdots \end{aligned} \tag{5.57}$$

where ϵ_2 and ϵ_3 are defined by [220]

$$\epsilon_2 \equiv \frac{e^2}{s_W^2}\left[\Sigma'_{11}\left(m_W^2\right) - \Sigma'_{33}\left(m_Z^2\right)\right] \tag{5.58a}$$

$$\epsilon_3 \equiv \frac{e^2}{s_W^2}\left[\Sigma'_{3Q}\left(m_W^2\right) - \Sigma'_{33}\left(m_Z^2\right)\right] \tag{5.58b}$$

Here, we have used the fact that the coefficient of Σ'_{3Q} is really QI_3 and that sum over isospin doublets gives the same value as that of Σ'_{33}. It can be shown that all the terms in Eq. (5.58) are finite (see Appendix C, Paragraph ϵ_2 and ϵ_3). As the ϵ_2 and $\epsilon_3 \cdots$ are small terms in the Standard Model, we combine them and denote them as r_1. Note, they may be significant if a certain new physics is in operation which will be discussed later.

$$\Delta r = \Delta\alpha - \frac{c_W^2}{s_W^2}\Delta\rho + r_1 \tag{5.59}$$

In summary, the main contribution to the Δr is two-fold; one from pure QED correction, the change of the fine structure constant $\Delta\alpha$ evolved from $q^2 = 0$ to $q^2 = m_Z^2$ and the other $\Delta\rho$ which is the vacuum polarization effect of the gauge boson generated by the top-bottom fermion loop.

5.3.2 Running Electromagnetic Constant

Contributions of the fermion loops to the fine structure constant were calculated and are given by Eq. (C.32). The correction is given by

$$\Delta\alpha = -\hat{\Pi}_\gamma\left(m_Z^2\right) = \frac{\alpha\left(m_Z^2\right) - \alpha(0)}{\alpha(0)} \overset{m_f \ll m_Z}{=\!=\!=} \frac{\alpha}{3\pi}\sum_f Q_f^2\left(\ln\frac{m_Z^2}{m_f^2} - \frac{5}{3}\right) \tag{5.60}$$

This formula can give accurate values for the lepton contribution, but not for the quarks because of the confinement. Fortunately, one can use the dispersion relation

$$\hat{\Pi}_\gamma(s) = \frac{1}{\pi}\int ds'\frac{\mathrm{Im}\hat{\Pi}_\gamma(s')}{s' - s} \tag{5.61}$$

to express the hadronic part in terms of experimental data. Using a unitarity relation, one can express the imaginary part of the self-energy function in terms of the e^-e^+ total cross section (see Eq. (I.39)[10]):

$$\mathrm{Im}\hat{\Pi}_\gamma(s) = -\frac{s}{e^2}\sigma_{\mathrm{TOT}}(e^-e^+ \to \text{hadrons}) = -\frac{\alpha}{3}R_\gamma(s) \tag{5.62}$$

where R_γ is the ratio of the hadronic to $\mu^-\mu^+$ production cross sections by e^-e^+ annihilation. Since the $R_\gamma \sim$ constant as $s \to \infty$, one uses the once subtracted dispersion relation to give

$$\mathrm{Re}\left[\hat{\Pi}_\gamma\left(m_Z^2\right)\right] = -\frac{\alpha}{3\pi}m_Z^2\int_{4m_\pi^2}^{\infty} ds\frac{R_\gamma}{s\left(s - m_Z^2\right)} \tag{5.63}$$

Using experimental values of the cross section, one can calculate the hadronic part of $\Delta\alpha$ [11, 221, 222].[11]

$$\Delta\alpha(\text{hadron}) = 0.027\,93 \pm 0.000\,11 \tag{5.64}$$

The top quark is heavy and its contribution is proportional to $(m_Z^2/m_t^2)\alpha$ and can be neglected. Combining the above value with the leptonic part gives [3, 222, 223]

$$\begin{aligned}\Delta\alpha\left(m_Z^2\right) &= \Delta\alpha(\text{lepton}) + \Delta\alpha(\text{hadron})\\ &= 0.031\,497\,69 + 0.027\,93 \pm 0.000\,11\\ &= 0.059\,43 \pm 0.000\,11\end{aligned} \tag{5.65}$$

Using the value of α which is determined accurately from $g-2$ gives

$$\begin{aligned}\alpha(0)^{-1} &= 137.035\,999\,084(51)\\ \alpha\left(m_Z^2\right)^{-1} &= 128.91 \pm 0.02\end{aligned} \tag{5.66}$$

5.3.3
Residual Corrections and Numerical Evaluation

Here, we give a brief summary of the residual corrections and show the nonnegligible contribution to $\Delta\rho$ due to the top and Higgs loop effect [170, 171].

$$\Delta\rho \approx \Delta\rho_t + \Delta\rho_H \tag{5.67a}$$

$$\Delta\rho_t = \frac{G_F m_W^2}{8\sqrt{2}\pi^2}\left[\frac{3m_t^2}{m_W^2} + 2\left(1 - \frac{s_W^2}{3c_W^2}\right)\ln\frac{m_t^2}{m_W^2} + \cdots\right] \tag{5.67b}$$

$$\Delta\rho_H = -\frac{3G_F m_Z^2 s_W^2}{8\sqrt{2}\pi^2}\left(\ln\frac{m_H^2}{m_t^2} - \frac{5}{6}\right) + \cdots \tag{5.67c}$$

10) R_γ in Eq. (I.39) is defined as $-R_\gamma/e^2$ given here.
11) Since the measurements do not reach the top pair production, the integral only includes contributions of five flavors. In this sense, $\Delta\alpha$(hadron) is sometimes denoted as $\alpha_{\mathrm{had}}^{(5)}$ [3].

Their contribution to Δr is given by

$$\Delta\rho_t = 0.009\,39\left(\frac{m_t}{173.1}\right)^2 \rightarrow \Delta r_t = -\frac{c_W^2}{s_W^2}\Delta\rho = -0.0327 \tag{5.68}$$

In addition to the gauge boson and Higgs self-energy, there are contributions to Δr from the vertex correction and the box diagrams (Figure 5.3f–n). Collectively, they contribute [170, 171, 223]

$$\Delta r_{\text{rem}} = \Delta r_{\text{vertex+box}} = \frac{\alpha}{4\pi s_W^2}\left(6 + \frac{7-4s_W^2}{2s_W^2}\ln c_W^2\right) \approx 0.0064 \tag{5.69}$$

The contribution of the Higgs is proportional to $\ln m_H$ and the sensitivity of the radiative correction to the Higgs mass is not high.

As one can see, the two largest contributions come from evolution of the fine structure constant $\alpha(\mu)$ from $\mu = 0$ to $\mu = m_Z$ and the loop correction due to the top-bottom pair. All contributions collected, a recent value is quoted as [3]

$$\Delta r \approx 0.0362 \pm 0.0005 \pm 0.000\,11 \tag{5.70}$$

where the second error is due to uncertainty in $\Delta\alpha$.

5.4
Improved Born Approximation

Next we consider the process for $e^-e^+ \rightarrow Z \rightarrow f\overline{f}$ (f = fermion) relevant at LEP experiments. In the lowest order, one photon or one Z boson is exchanged (Figure 5.4a,b). Among the higher order corrections, we have already described the self-energy part of the contributions (Figure 5.4c,d). Here, we consider additional contributions due to $\gamma-Z$ mixing. The mixing occurs because both the photon and the Z couple to the same fermions (Figure 5.4e,f) and hence one cannot argue γ and Z separately. One has to include fermion's couplings to $J_{N\mu}J_\gamma^\mu$ and $J_\gamma^\mu J_{N\mu}$.

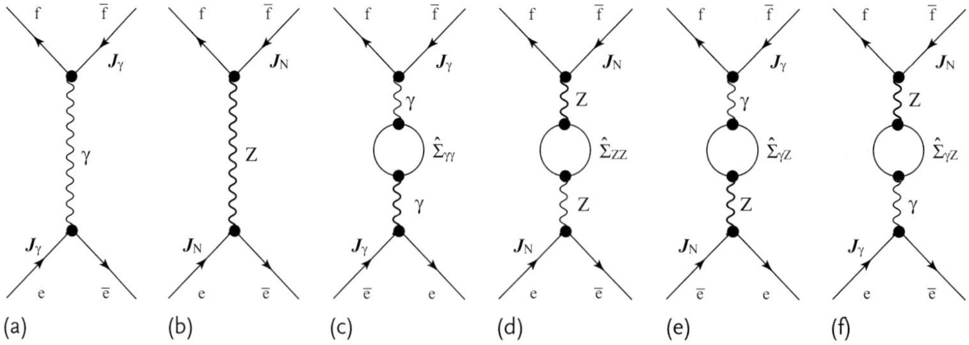

Figure 5.4 $e^-e^+ \rightarrow f\overline{f}$ process. (a) and (c) are the γ exchange, (b) and (d) are the Z exchange and (e) and (f) are the $\gamma-Z$ mixing term.

5.4.1 $\gamma - Z$ Mixing

Let us write down the scattering amplitude including all the Feynman amplitudes in Figure 5.4. Denoting the current product of the fermion fields like $J_\gamma \otimes J_\gamma$ (the left current is out going and the right one is in coming), and using Eqs. (5.28) and (5.41a)–(5.41g), we can write down the amplitude as follows:

$$\begin{aligned}
\mathcal{M}_{e\bar{e}\to f\bar{f}} &= e^2 \frac{Q_f J_\gamma \otimes Q_i J_\gamma}{s + \hat{\Sigma}_{\gamma\gamma}} + g_Z^2 \frac{J_N \otimes J_N}{s - m_Z^2 + \hat{\Sigma}_{ZZ}} \\
&\quad - e g_Z \frac{Q_f J_\gamma \otimes J_N + J_N \otimes Q_i J_\gamma}{\left[s - m_Z^2 + \hat{\Sigma}_{ZZ}\right]} \frac{\hat{\Sigma}_{\gamma Z}}{\left[s + \hat{\Sigma}_{\gamma\gamma}\right]} \\
&= e^2(s) \frac{Q_f J_\gamma \otimes Q_i J_\gamma}{s} + g_Z^2(s) \frac{\overline{J}_N \otimes \overline{J}_N}{s - m_Z^2 + i m_Z \Gamma_Z} + O(\alpha^3) \qquad (5.71)
\end{aligned}$$

In deriving the last equality, we defined

$$\overline{J}_N = J_N - \frac{Qe}{g_Z} \frac{\hat{\Sigma}_{\gamma Z}(s)}{s + \hat{\Sigma}_{\gamma\gamma}(s)} J_\gamma = J_N - \frac{Qe}{g_Z} \frac{\hat{\Pi}_{\gamma Z}(s)}{1 + \hat{\Pi}_\gamma(s)} J_\gamma \qquad (5.72)$$

and used Eqs. (5.31a) and (5.32). Equation (5.71) shows that the effect of $\gamma - Z$ mixing is to change the neutral current $J_N \to \overline{J}_N$ at the given order. Considering $J_N = J_3 - Q s_W^2 J_\gamma$, it translates to modification of the Weinberg angle

$$\sin^2\theta_W \to \overline{s}_W^2 \equiv \kappa_{se} \sin^2\theta_W = \sin^2\theta_W + c_W s_W \frac{\hat{\Pi}_{\gamma Z}(s)}{1 + \hat{\Pi}_\gamma(s)} \qquad (5.73)$$

At the Z resonance, the radiative correction due to the mixing is expressed as

$$\begin{aligned}
\hat{\Pi}_{\gamma Z}\left(m_Z^2\right) &= \frac{\hat{\Sigma}_{\gamma Z}\left(m_Z^2\right)}{m_Z^2} = \frac{\Sigma_{\gamma Z}\left(m_Z^2\right) - \Sigma_{\gamma Z}(0)}{m_Z^2} + \frac{\delta' s_W^2}{c_W s_W} \\
&= \Sigma'_{\gamma Z}\left(m_Z^2\right) + \frac{\delta' s_W^2}{c_W s_W} \\
&= \frac{c_W}{s_W} \Delta\rho + r_2 \qquad (5.74a)
\end{aligned}$$

$$r_2 = \Sigma'_{\gamma Z}\left(m_Z^2\right) + \frac{c_W}{s_W}\left[\Sigma'_{ZZ}\left(m_Z^2\right) - \Sigma'_{WW}\left(m_W^2\right)\right] = -\frac{c_W}{s_W}\epsilon_2 - \frac{s_W}{c_W}\epsilon_3 \qquad (5.74b)$$

$$\hat{\Pi}_\gamma\left(m_Z^2\right) \overset{(5.60)}{=\!=} -\Delta\alpha \qquad (5.74c)$$

As the contribution of ϵ_2, ϵ_3 is small in the SM, we have

$$\begin{aligned}
\overline{s}_W^2 &\approx s_W^2 + c_W s_W \hat{\Pi}_{\gamma Z}\left(m_Z^2\right) = s_W^2 + c_W^2 \Delta\rho - c_W^2 \epsilon_2 - s_W^2 \epsilon_3 \\
&\approx s_W^2 + c_W^2 \Delta\rho + \cdots \qquad (5.75a)
\end{aligned}$$

$$\kappa_{se} \approx 1 + \frac{c_W^2}{s_W^2} \Delta\rho + \cdots \qquad (5.75b)$$

From the above discussions, one sees that the $\gamma - Z$ mixing effect amounts to the redefinition of the Weinberg angle ($\sin^2\theta_W \to \bar{s}_W^2$) which appears in the neutral current.

5.4.2
α or G_F?

To gain more insight into the effective Weinberg angle which has been brought in by the mixing, let us compare it with the change that the Δr has induced.

$$m_W^2 s_W^2 \equiv \frac{\pi\alpha}{\sqrt{2}G_F} \frac{1}{1 - \Delta\alpha + \frac{c_W^2}{s_W^2}\Delta\rho + \cdots} \tag{5.76}$$

Substituting Eq. (5.75a) into Eq. (5.76) and using $\alpha\left(m_Z^2\right) = \alpha(1 + \Delta\alpha)$, we obtain

$$m_W^2 \bar{s}_W^2 \approx \pi\frac{\alpha\left(m_Z^2\right)}{\sqrt{2}G_F} \equiv \frac{\pi\bar{\alpha}}{\sqrt{2}G_F} \tag{5.77}$$

Considering

$$\bar{c}_W^2 = 1 - \bar{s}_W^2 \approx c_W^2(1 - \Delta\rho) = \frac{c_W^2}{\rho} \tag{5.78}$$

we have an alternative relation to Eq. (5.6)

$$m_Z^2 \bar{s}_W^2 \bar{c}_W^2 \approx \frac{\pi\bar{\alpha}}{\sqrt{2}G_F\rho} = \frac{\pi\bar{\alpha}}{\sqrt{2}G_N} \tag{5.79}$$

We next evaluate the running coupling constant $g_Z^2(s) = g_Z^2/(1 + \hat{\Pi}_Z(s))$. From Eq. (5.48) evaluated at $q^2 = 0$ and Eq. (5.47a), we have

$$\Delta\rho = -\Delta r - \hat{\Pi}_Z\left(m_Z^2\right) + \left[\hat{\Pi}_Z\left(m_Z^2\right) - \hat{\Pi}_Z(0)\right] \simeq -\Delta r - \hat{\Pi}_Z\left(m_Z^2\right) \tag{5.80}$$

where we have used the fact that for $s \le m_Z^2$, the main contribution to $\hat{\Pi}_Z(s) \approx \hat{\Pi}_Z(m_Z^2)$ lies in the subtraction term. The above relation is rewritten as

$$\hat{\Pi}_Z\left(m_Z^2\right) = -\Delta\alpha + \frac{c_W^2 - s_W^2}{s_W^2}\Delta\rho + \cdots \tag{5.81}$$

Therefore, the running coupling constant can be expressed as

$$\begin{aligned} g_Z^2(s) &= \frac{g_Z^2}{1 + \hat{\Pi}_Z\left(m_Z^2\right)} = \frac{4\pi\alpha}{s_W^2 c_W^2} \frac{1}{1 - \Delta\alpha + \frac{c_W^2 - s_W^2}{s_W^2} + \cdots} \\ &\approx \frac{4\pi\bar{\alpha}}{\bar{s}_W^2 \bar{c}_W^2} = 8\frac{\rho G_F}{\sqrt{2}} m_Z^2 \end{aligned} \tag{5.82}$$

Using all the above relations, we finally have

$$\mathcal{M}(e\overline{e} \to f\overline{f}) = 4\pi\overline{\alpha}\frac{J_\gamma \otimes J_\gamma}{s} + 4\sqrt{2}\rho G_F m_Z^2 \frac{\overline{J}_N \otimes \overline{J}_N}{s - m_Z^2 + i m_Z \Gamma_Z} \tag{5.83}$$

This has the same form as the Born approximation, except the running variables ($\alpha \to \overline{\alpha}$, $\sin^2\theta_W \to \overline{s}_W$) are used and the width of the gauge boson is correctly taken into account. Here, we remark that if one chooses $g_Z = e/s_W c_W$ as the starting point, the radiative correction is given by

$$\frac{\overline{\alpha}}{\overline{s}_W^2 \overline{c}_W^2} = \frac{\alpha}{s_W^2 c_W^2}\left(1 + \Delta\alpha - \frac{c_W^2 - s_W^2}{s_W^2}\Delta\rho + \cdots\right) \tag{5.84}$$

which is large ($\sim 4\%$) because it contains $\Delta\alpha$. On the other hand, if one chooses $g_Z^2 = 8(G_F/\sqrt{2})m_Z^2$ as the starting coupling constant, the correction is to add $\Delta\rho$ only,

$$G_F m_Z^2 \rho = G_F m_Z^2 (1 + \Delta\rho) \tag{5.85}$$

and the amount of correction is $\Delta\rho \sim 1\%$. In other words, when the renormalization scale is changed from $\mu = 0$ to $\mu = m_Z$, α receives a large correction, but the correction to G_F is small. This means that if one used G_F as the starting point, the pure QED correction is already included.

We have learned that if we use $\overline{s}_W^2$, $\overline{\alpha}$ or equivalently $G_N m_Z^2 = \rho G_F m_Z^2$, namely, fundamental constants at the scale $\mu = m_Z$, the Born term expression already contains all the important radiative corrections except the mass width term which one conventionally adopts from phenomenological consideration. Thus, we have justified the use of the improved Born approximation in the previous section in analyzing the LEP/CERN data with considerable success.

s_{Mz}^2: Let us reconsider our old convention. We started with fundamental constants determined at the low energy scale simply because they are the familiar ones. When we discuss high energy phenomena, those determined at the proper energy scale should also be eligible as a standard. Thinking in this way, we could define a new $\sin^2\theta_W$ with the same Born expression, but using the fundamental constants evaluated at the scale $\mu = m_Z$ as the new reference which we refer to as s_{MZ}^2.

$$s_{MZ}^2 c_{MZ}^2 \equiv \frac{\pi\alpha\left(m_Z^2\right)}{\sqrt{2}G_F m_Z^2}, \quad c_{MZ}^2 = 1 - s_{MZ}^2 \tag{5.86}$$

Since the value of $\alpha(m_Z^2) = (128.91 \pm 0.02)^{-1}$, G_F, m_Z can be determined very accurately, its numerical value is determined to be

$$s_{MZ}^2 = 0.231\,08 \mp 0.000\,05 \quad [3] \tag{5.87}$$

The small uncertainty in s_{MZ}^2 compared to the other scheme is because the m_t dependence has been removed by definition. Corrections to s_{MZ}^2 is sensitive to the top quark, Higgs or other unknown new physics, but corrections due to light fermions which do not offer new information are already built into its structure. As the value of s^2_{MZ} is that of s^2_W with Δr replaced with $\Delta\alpha$, its relation with the conventional $\sin^2\theta_W$ can be obtained from

$$
\begin{aligned}
s^2_W - s^2_{MZ} &\simeq \frac{\partial s^2_W}{\partial \Delta r}(\Delta r - \Delta\alpha) \overset{\text{Eq. (5.9)}}{=\!=} -\frac{s^2_{MZ} c^2_{MZ}}{c^2_{MZ} - s^2_{MZ}}(\Delta r - \Delta\alpha) \\
&\overset{\text{Eq. (5.57)}}{=\!=} -\frac{c^4_{MZ}}{c^2_{MZ} - s^2_{MZ}}\Delta\rho + c^2_{MZ}\epsilon_2 + \frac{2 s^2_{MZ} c^2_{MZ}}{c^2_{MZ} - s^2_{MZ}}\epsilon_3 \\
s^2_W &= (0.9661 \pm 0.0005) s^2_{MZ}
\end{aligned}
\tag{5.88}
$$

where we used $\delta s^2_W \simeq -c^2_W(\delta m^2_W / m^2_W)$. ϵ_2 and ϵ_3 are small in the Standard Model and were neglected.

5.4.3 Vertex Correction

The vertex correction depends on the flavor. Therefore, after the vertex corrections, the Weinberg angle as well as the ρ parameter are modified to

$$
\begin{aligned}
\sin^2\theta_W &\to \sin^2\theta^f_{\text{eff}} = \kappa_{f,\text{vtx}}\bar{s}^2_W = \kappa_f s^2_W = (1 + \Delta\kappa_{f,\text{vtx}} + \Delta\kappa_{se}) s^2_W \\
\rho_0 &\to \rho_0(1 + \Delta\rho_{f,\text{vtx}} + \Delta\rho_{se}) \\
\Delta\kappa_{se} &= \frac{c^2_W}{s^2_W}\Delta\rho, \quad \Delta\rho_{se} = \Delta\rho
\end{aligned}
\tag{5.89}
$$

This is a version of the Weinberg angle that is almost independent of m_t. Precision measurements at the Z-pole including not only $A_{LR} = \mathcal{A}_e, A^l_{FB}$ and P_τ, but also $A^q_{FB}(q = c, b, s)$ and hadronic asymmetry are mainly sensitive to the $\sin^2\theta^l_{\text{eff}}$.

Notice that although the vertex correction is small under normal circumstances, there is an exception. When $f = b$, the top quark enters in the intermediate state (Figure 5.5) and the correction is sizable [223–225].

$$
\Delta\kappa_{b,\text{vtx}} = \frac{G_F m^2_W}{8\sqrt{2}\pi^2}\left[2\frac{m^2_t}{m^2_W} + \frac{1}{3}\left(16 + \frac{1}{c^2_W}\ln\frac{m^2_t}{m^2_W}\right)\right] \tag{5.90a}
$$

$$
\Delta\rho_{b,\text{vtx}} = -2\Delta\kappa_{b,\text{vtx}} \tag{5.90b}
$$

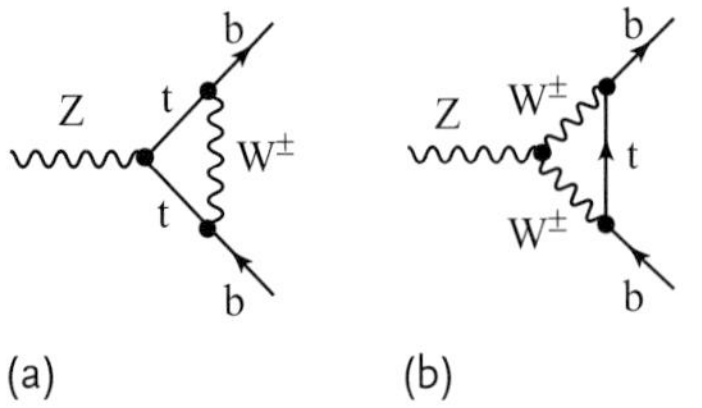

Figure 5.5 $Zb\bar{b}$ vertex correction. The top quark enters in the intermediate state and the correction is large.

Their effect can be expressed in terms of known corrections we experienced for other fermions.

$$\rho \to \rho_b = \rho \left(1 - \frac{4}{3}\Delta\rho\right) \tag{5.91a}$$

$$\overline{s}_W^2 = \kappa_{se} s_W^2 \to \kappa_b s_W^2 = \left(1 + \frac{2}{3}\Delta\rho\right) \overline{s}_W^2 \tag{5.91b}$$

When the b-vertex exists only on one side of the Z propagator as is the case for $\nu - e$ scattering, one may use

$$\rho \to \sqrt{\rho\rho_b} \tag{5.92}$$

5.5 Effective Weinberg Angle

At this stage, we consider the Weinberg angle $\sin^2\theta_{\text{eff}}^f$ by the second definition (Eq. (5.5b)) because due to large uncertainty of m_W^2, the value of on-shell s_W^2 is hard to determine. The leptonic $\sin^2\theta_{\text{eff}}^l$ is particularly useful because it can be determined precisely from the leptonic decay of the Z and asymmetry. Neglecting small vertex corrections, expression of $\sin^2\theta_{\text{eff}}^l$ is given by Eqs. (5.73) and (5.75a). Substituting Eq. (5.88) into Eq. (5.75a), we obtain

$$\sin^2\theta_{\text{eff}}^l \approx s_{M_Z}^2 - \frac{s_{M_Z}^2 c_{M_Z}^2}{c_{M_Z}^2 - s_{M_Z}^2}\Delta\rho + \frac{s_{M_Z}^2}{c_{M_Z}^2 - s_{M_Z}^2}\epsilon_3 \tag{5.93}$$

The third term is small in the Standard Model and will be neglected here. The top and Higgs contribution to $\Delta\rho$ is given in Eqs. (5.67b) and (5.67c). The correction due to the top is about 1%. By carrying out precision experiments more accurate than that and using the known top mass value, it is possible to infer the Higgs mass. If it is produced directly by the LHC, it serves as the test of the Standard Model or possibly give a clue to new physics.

$\nu_e - e$ Scattering and Z Decay In Chapter 2, when we discussed the leptonic neutral current processes, we derived the Weinberg angle $\sin^2\theta_W = \sin^2\theta_W(\nu_e e)$. It is actually a low energy version (at different renormalization scale μ) of $\sin^2\theta_{\text{eff}}^l$ ($l = e$) which can be obtained at the Z resonance. There are subtle differences though. For the neutrino vertex, the charge radius of the neutrino comes in while it is $\gamma - Z$ mixing that affects to $\sin^2\theta_e$ of the $e\overline{e}$ production. However, both effects almost cancel each other and the difference turned out to be small. The end result of theoretical consideration is [170, 171]

$$\sin^2\theta(\nu e) = 1.002 \sin^2\theta_{\text{eff}}^e \tag{5.94}$$

The result of the CHARM experiment [55]

$$\sin^2\theta(\nu e) = 0.2324 \pm 0.083 \tag{5.95}$$

agrees well with $\sin^2\theta_{\text{eff}}^e = 0.231\,53 \pm 0.000\,16$ [7] obtained from forward–backward asymmetry A_{FB}^e.

5.6 Weinberg Angle in the $\overline{MS}$ Scheme

The Weinberg angles $\sin^2\theta_W$, $s^2_{M_Z}$, $\sin^2\theta^f_{\text{eff}}$ are convenient as long as we confine our discussions within the framework of the electroweak theory and experimentally easy to interpret these variables. However, when we try to extend our discussion to the grand unified theory or when we deal with new physics beyond the Standard Model, it is not necessarily a convenient tool. There is an obvious advantage to use the same renormalization scheme as QCD in discussing the grand unified theories. In QCD, there is no clear directly observable reference point due to confinement, and the method of the (modified) minimum subtraction scheme ($\overline{MS}$) is the standard. In the $\overline{MS}$ prescription, each divergence is separately handled, making the calculation much easier. The price to pay is that each physics observable has its own scale μ dependence and relations to observables are not so apparent.

In the following, we attach hat mark "^" to variables that are treated in the $\overline{MS}$ scheme and attach index $\overline{MS}$ to radiative corrections that are renormalized according to the prescription. Their relation to those in the on-shell scheme can be obtained by requiring that the scale dependent mass, coupling constants which are obtained after applying the radiative corrections should give the same observables at the appropriate scale [226–231]. Namely,

$$e = \frac{\hat{e}}{1 - \frac{1}{2}\Pi_\gamma(0)|_{\overline{MS}} - \frac{\hat{s}}{\hat{c}}\frac{\Sigma_{\gamma Z}(0)|_{\overline{MS}}}{\hat{m}_Z^2}} \tag{5.96a}$$

$$m_Z^2 = \hat{m}_Z^2 - \text{Re}\Sigma_{ZZ}\left(m_Z^2\right)|_{\overline{MS}} \tag{5.96b}$$

$$m_W^2 = \hat{m}_W^2 - \text{Re}\Sigma_{WW}\left(m_W^2\right)|_{\overline{MS}} \tag{5.96c}$$

These relations correspond to those of on-shell prescription in Eqs. (5.42) and (5.38). From the above expressions, we see that $\hat{e}$, $\hat{m}$ are also functions of μ (i.e., $\hat{e} = \hat{e}(\mu)$, $\hat{m} = \hat{m}(\mu)$). Adopting $\mu = m_Z$ as a logical scale to handle data at Z resonance, $\hat{\alpha}$ obtained from $\hat{e}$ almost agrees with $\overline{\alpha} = \alpha\left(m_Z^2\right)$, but is not identical ($\overline{\alpha}^{-1} = 128.91 \pm 0.02$, $\hat{\alpha}^{-1} = 127.916 \pm 0.015$).

Parameters in the $\overline{MS}$ scheme in the Born approximation satisfy the same relations as those of on-shell parameters. Therefore, defining $\sin^2\theta_W$ by the ratio of the coupling constants (definition 3 in Eq. (5.5c)) and denoting it as $\sin^2\theta_{\overline{MS}}$, it satisfies the relation

$$\sin^2\theta_{\overline{MS}} = \frac{\hat{e}^2(\mu)}{\hat{g}_W^2(\mu)} = 1 - \frac{\hat{m}_W^2(\mu)}{\hat{m}_Z^2(\mu)} \tag{5.97}$$

$\sin^2\theta_{\overline{MS}}$ is also denoted as $\sin^2\hat{\theta}_W$, $\hat{s}_W^2$ and specially $\hat{s}_Z^2 \equiv \hat{s}_W^2(m_Z)$ is also used. Its relation to the on-shell s_W^2 can be obtained by substituting Eqs. (5.96a)–(5.96c) in Eq. (5.97). To $O(\alpha)$, they are given by [228, 229]

$$\hat{s}_Z^2 = s_W^2 + c_W^2 \Delta\hat{\rho} + \cdots \tag{5.98a}$$

$$\hat{\rho} = \frac{c^2}{\hat{c}^2} = \frac{m_W^2}{m_Z^2 \hat{c}^2} \tag{5.98b}$$

$$\Delta\hat{\rho} = \left(\frac{\mathrm{Re}\Sigma_{ZZ}(m_Z^2)}{m_Z^2} - \frac{\mathrm{Re}\Sigma_{WW}(m_W^2)}{m_W^2} \right)\Bigg|_{\overline{MS}} \left(1 + \frac{\mathrm{Re}\Sigma_{ZZ}(m_Z^2)\,|_{\overline{MS}}}{m_Z^2} \right)^{-1} \tag{5.98c}$$

$$\simeq \frac{3\hat{\alpha} m_t^2}{16\pi \hat{s}_Z^2 m_W^2} + \cdots \quad \simeq \frac{3 G_F m_t^2}{8\sqrt{2}\pi^2} + \cdots \simeq \Delta\rho \tag{5.98d}$$

The difference between $\Delta\hat{\rho}$ and $\Delta\rho$ is $O(\alpha^2)$. Comparing with Eq. (5.75), we have

$$\hat{s}_Z^2 \simeq \overline{s}_W^2 \tag{5.99}$$

To give an idea of the relative size of $\hat{s}_Z^2$, we list its relation to on-shell s_W^2 and s_{mZ}^2 [11].

$$\begin{aligned}
\hat{s}_Z^2 &\equiv c(m_t, m_H) s_W^2 \equiv \overline{c}(m_t, m_H) s_{mZ}^2 \sim \sin^2\theta_{\mathrm{eff}}^f - 0.00029(f \neq b) \\
c(m_t, m_H) &= 1.0361 \pm 0.0005\,, \quad \overline{c}(m_t, m_H) = 1.0010 \mp 0.0002 \\
c(m_t, m_H) &\simeq 1 + \frac{c_W^2}{s_W^2}\Delta\rho\,, \quad \overline{c}(m_t, m_H) \simeq 1 - \frac{c_W^2}{c_W^2 - s_W^2}\Delta\rho
\end{aligned} \tag{5.100}$$

The relation of $\hat{s}_Z^2$ with on-shell s_W^2 can also be made by using the Δr defining relation Eq. (5.6). Namely,

$$\hat{s}_Z^2 \hat{c}_Z^2 (1 - \Delta\hat{r}) = s_W^2 c_W^2 (1 - \Delta r) = (37.2806)^2 / m_Z^2 \tag{5.101a}$$

$$\hat{s}_Z^2 (1 - \Delta\hat{r}_W) = s_W^2 (1 - \Delta r) = (37.2806)^2 / m_W^2 \tag{5.101b}$$

$$(1 - \Delta\hat{r}) = \hat{\rho}(1 - \Delta\hat{r}_W) \tag{5.101c}$$

$\Delta\hat{r}$ is the same Δr in the on-shell scheme, but treated by $\overline{MS}$ prescription. The difference between $\Delta\hat{r}$ and $\Delta\hat{r}_W$ is whether one adopts (α, G_F, m_Z) or (α, G_F, m_W) as the starting point. If we only write down the dominant terms

$$\Delta\hat{r} \simeq \Delta\hat{\alpha} - \Delta\hat{\rho} + \cdots \tag{5.102a}$$

$$\Delta\hat{r}_W \simeq \Delta\hat{\alpha} + O(\ln m_t / m_W) + \cdots \tag{5.102b}$$

and we see that m_t dependence of $\Delta\hat{r}$, $\Delta\hat{r}_W$ is different from that of Δr. Figure 5.6 shows $\Delta\hat{r}$ and $\Delta\hat{r}_W$ as a function of the top mass compared with Δr. One sees that they have practically no m_t dependence.

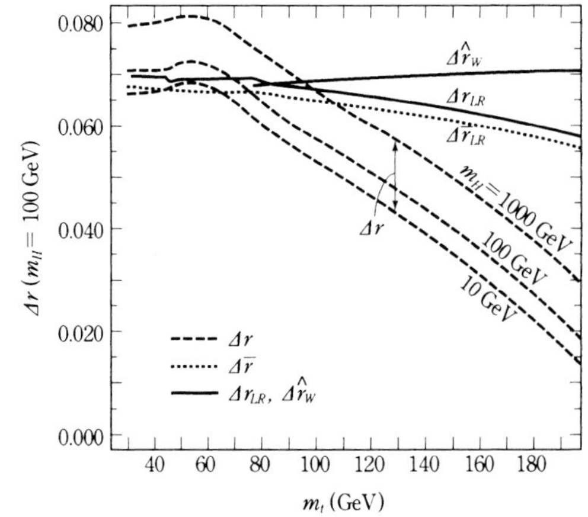

Figure 5.6 The radiative correction Δr as a function of m_t [171, 223, 229]. Δr is defined by $m_W^2 \sin^2\theta_W^2(1-\Delta r) = \pi\alpha/\sqrt{2}G_F$. The different definition of $\sin^2\theta_W$ gives a different Δr. Δr : $\sin^2\theta_W = 1-(m_W^2/m_Z^2)$, $\Delta\hat{r}_W$: $\sin^2\theta_W = \sin^2\hat{\theta}_W$, $\Delta\bar{r}$: $\sin^2\theta_W = \sin^2\overline{\theta}$, Δr_{LR} : $\sin^2\theta_l$ derived from A_{LR}. Note that a description of $\Delta\hat{r}$, $\Delta\hat{r}_W$ is given in Section 5.6.

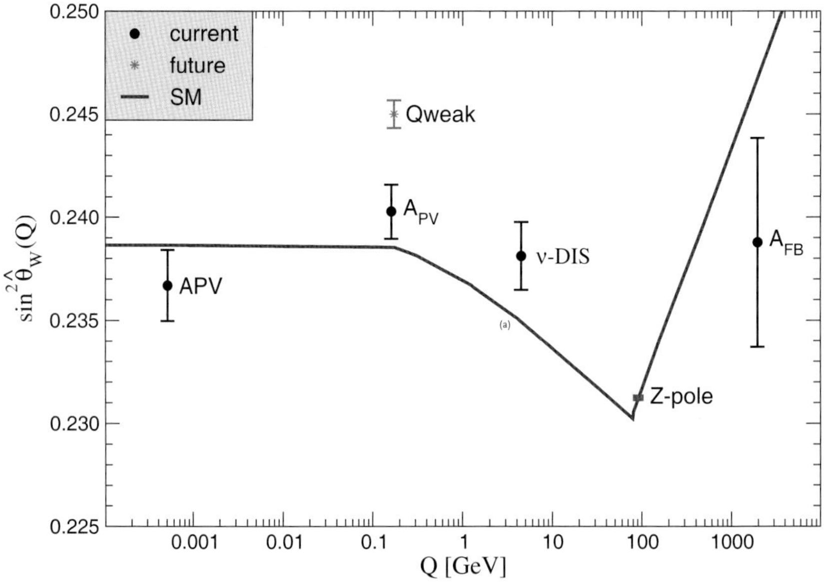

Figure 5.7 Scale dependence of the weak mixing angle in the MS renormalization scheme. The minimum of the curve corresponds to $\mu = m_W$, below which it is switched to an effective theory with the $W^{\pm}$ bosons integrated out, and where the β-function for the weak mixing angle changes sign [11, 232].

Considering the universal applicability of $\sin^2\hat{\theta}_W$, we show s dependence of $\sin^2\hat{\theta}_W$ evaluated at all energies thus far available in Figure 5.7.

Table 5.1 Values of $\hat{s}_Z^2 = \sin^2\theta_{\overline{MS}}(m_Z^2)$, s_W^2 (on-shell) for various (combinations of) observables. Unless otherwise indicated, the top quark mass, $m_t = 170.9 \pm 1.9\,\text{GeV}$, is used as an additional constraint in the fits [11].

Data	$\hat{s}_Z^2$	s_W^2
All data	0.231 19(14)	0.223 08(30)
All indirect (no m_t)	0.231 23(16)	0.222 97(36)
Z pole (no m_t)	0.231 21(17)	0.223 12(59)
LEP 1 (no m_t)	0.231 52(21)	0.223 77(67)
SLD + M_Z	0.230 67(30)	0.222 16(54)
$A_{FB}^{(b,c)} + M_Z$	0.231 93(28)	0.224 89(75)
$M_W + M_Z$	0.230 95(28)	0.222 65(55)
M_Z	0.231 33(7)	0.223 37(21)
polarized Møller	0.2331(14)	0.2252(14)
DIS (isoscalar)	0.2345(17)	0.2267(17)
Q_W (APV)	0.2291(19)	0.2212(19)
elastic $\nu\mu(\overline{\nu}\mu)e$	0.2310(77)	0.2232(77)
SLAC eD	0.222(18)	0.213(19)
elastic $\nu\mu(\overline{\nu}\mu)p$	0.211(33)	0.203(33)

$\sin^2\hat{\theta}_W$ vs. $\sin^2\theta_W$ In Table 5.1, we compare values of $s_W^2 = \sin^2\theta_W$ (on-shell) and $\hat{s}_Z^2 = \sin^2\theta_{\overline{MS}}(m_Z^2)$ derived from various reactions.

We also show m_t dependence of $\sin^2\hat{\theta}_W$ in Figure 5.8. Figure 5.8 shows that $\sin^2\hat{\theta}_W$ has much smaller dependence on m_t compared to on-shell $\sin^2\theta_W$. Optimizing the value of $\hat{s}_Z^2$ to be consistent with various reactions, one gets $m_t = 175 \pm 5\,\text{GeV}$ which agrees well with the CDF/D0 data $173.1 \pm 1.3\,\text{GeV}$ obtained by direct productions. The optimized QCD coupling constant was also obtained

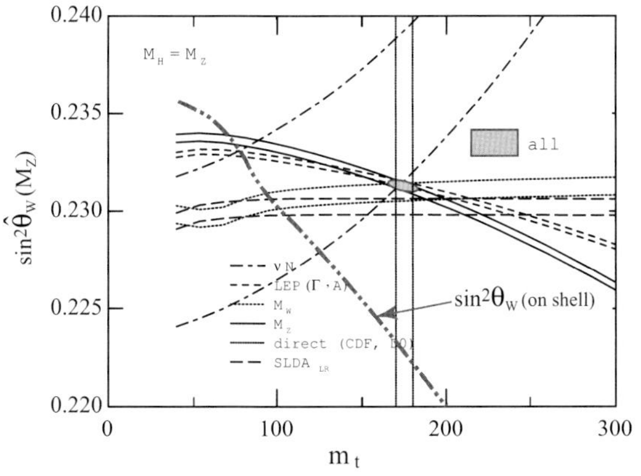

Figure 5.8 Standard deviation uncertainties in $\sin^2\hat{\theta}_W$ as a function of m_t, the direct CDF and D0 result range $175 \pm 5\,\text{GeV}$, and the 90% CL region in $\sin^2\hat{\theta}_W - m_t$ allowed by all data, assuming $m_H = m_Z$. The overall fit gives $\hat{s}_Z^2 = 0.23141 \pm 0.00031$. The m_t variation of the on-shell $\sin^2\theta_W$ is shown for comparison [233, 234].

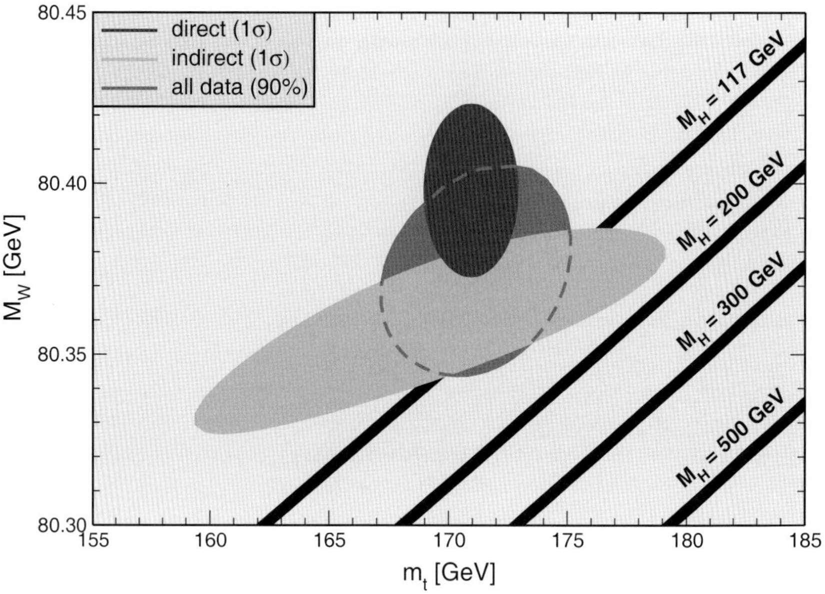

Figure 5.9 One standard deviation region in m_W as a function of m_t for the direct production and indirect (inferred from radiative correction) data and 90% region allowed by all data. The Standard Model prediction as a function of m_H is also indicated. Widths of the m_H lines reflect the theoretical uncertainty from $\alpha(m_Z)$ [11]. For a color version of this figure, please see the color plates at the beginning of the book.

to be $\alpha_s(m_Z^2) = 0.1198 \pm 0.0028$, which agrees well with a value $\alpha_s\left(m_Z^2\right) = 0.1184 \pm 0.0007$ [3] determined by jet phenomena (see Section 9.4).

Finally, we mention the data-vs-SM prediction consistency of the mass of the W boson which cannot be determined as precisely as that of m_Z, yet offers an equally useful test bench of the SM. Figure 5.9 shows the effect of radiative corrections on the mass of W. As Eq. (5.7a) and (5.7b) show, the correction mechanism in m_W is quite similar to $\sin^2\theta_W$. Figure 5.9 indicates a relatively small value of the Higgs mass (see also Figure 4.33).[12)]

In summary, all of the Weinberg angles determined from various reactions agree with the predictions of the Standard Model within experimental errors. One can say that the validity of the GWS electroweak theory is firmly established and its credibility matches that of QED.

Relations among the various Weinberg Angles We have seen in Eq. (5.100) that $\sin^2\hat{\theta}_W$ is very close to $\sin^2\theta^f_{\text{eff}}$, independent of flavors except for $f = b$. $\sin^2\theta^f_{\text{eff}}$ is also very close to $s^2_{M_Z}$, whose value can be accurately determined. The uncertainty of $s^2_{M_Z}$ mainly comes from uncertainty of $\alpha(m_Z)^{-1} = 128.91 \pm 0.02$, yielding $s^2_{M_Z} = 0.231\,08 \pm 0.000\,05$. Since this uncertainty is common to all versions

12) Prior to the publication of this textbook, the discovery of the Higgs with mass 126 GeV was announced by the ATLAS and CMS group at LHC.

of the Weinberg angle, we may use $\hat{s}_Z^2 \simeq \sin^2\theta_{\text{eff}}^f(f \neq b)$ for all practical applications. We have seen that many Weinberg angles are defined and are applied to fit the data. If one determines the value of one, then that of others can be calculated and therefore mathematically equivalent. People use different definitions depending on the process in consideration and it is very confusing. Among the radiative corrections, the large contribution comes from the pure QED correction of $\Delta\alpha = (\alpha(m_Z) - \alpha(0))/\alpha(0)$ and $\Delta\rho$ whose origin is the self-energy of the gauge bosons. All other corrections are small. If we ignore all the small corrections other than $\Delta\alpha$ and $\Delta\rho_t$,

$$\hat{s}_Z^2 \simeq \bar{s}_f^2(f \neq b) \simeq s_{M_Z}^2 \simeq s_W^2 + c_W^2\Delta\rho + \cdots \tag{5.103}$$

Numerically speaking, only the on-shell $\sin^2\theta_W$ is significantly different and all other Weinberg angles are said to be degenerate for all practical discussions.

In short, the Weinberg angle should be considered as a useful parameter rather than a fundamental constant. The on-shell Weinberg angle is conceptually clearest, but it does not give any extra information once m_W and m_Z are known. $\bar{s}_W^2, \sin^2\theta_{\text{eff}}^f, s_{M_Z}^2$ can be considered as convenient experimental parameters useful for limited discussions of phenomena at the Z resonance. In that sense, the $\sin^2\hat{\theta}_W$ $(= \sin^2\theta_{\overline{MS}})$, which is scale dependent, is not easily connected to observables, though theoretically, a useful tool, especially in discussing grand unified theories.

5.7 Beyond the Standard Model

From the various discussions we have made thus far, we may safely say that the validity of the Standard Model has been proved beyond doubt experimentally. Our next task, therefore, is to pursue a possible deviation from the Standard Model and explore new physics or a theory which cannot be explained by the Standard Model. New phenomena are considered to be effective in the higher energy scale and typically appear as new particles with large mass. The large mass, if not directly produced, appears indirectly in the loop of the radiative corrections, in particular, in the vacuum polarization effect. Several convenient parameters sensitive to the existence of such particles are proposed [220, 235, 236].

$$S = \frac{4s_W^2}{\alpha}\epsilon_3 = 16\pi\left[\Sigma'_{3Q}\left(m_Z^2\right) - \Sigma'_{33}\left(m_Z^2\right)\right]^{13)} \tag{5.104a}$$

13) In the $\overline{MS}$ scheme, they are defined by [3, 237]

$$\frac{\hat{\alpha}(m_Z)}{4\hat{s}_Z^2\hat{c}_Z^2}S_{\text{new}} = \left(-\frac{\Sigma_{ZZ}^{\text{new}}\left(m_Z^2\right) - \Sigma_{ZZ}^{\text{new}}(0)}{m_Z^2} + \frac{\hat{c}_Z^2 - \hat{s}_Z^2}{\hat{c}_Z\hat{s}_Z}\frac{\Sigma_{\gamma Z}^{\text{new}}\left(m_Z^2\right)}{m_Z^2} + \frac{\Sigma_{\gamma\gamma}^{\text{new}}\left(m_Z^2\right)}{m_Z^2}\right)_{\overline{MS}} \tag{5.105a}$$

$$\hat{\alpha}(m_Z)T_{\text{new}} = \left(\frac{\Sigma_{ZZ}^{\text{new}}(0)}{m_Z^2} - \frac{\Sigma_{WW}^{\text{new}}(0)}{m_W^2}\right)_{\overline{MS}} \tag{5.105b}$$

$$\frac{\hat{\alpha}(m_Z)}{4\hat{s}_Z^2}(S+U)_{\text{new}} = \left(-\frac{\Sigma_{WW}^{\text{new}}\left(m_W^2\right) - \Sigma_{WW}^{\text{new}}(0)}{m_W^2} + \frac{\hat{c}_Z}{\hat{s}_Z}\frac{\Sigma_{\gamma Z}^{\text{new}}\left(m_Z^2\right)}{m_Z^2} + \frac{\Sigma_{\gamma\gamma}^{\text{new}}\left(m_Z^2\right)}{m_Z^2}\right)_{\overline{MS}} \tag{5.105c}$$

$$T = \frac{1}{\alpha}\epsilon_1 \equiv \frac{1}{\alpha}\Delta\rho = \frac{4\pi}{s_W^2 m_W^2}[\Sigma_{33}(0) - \Sigma_{11}(0)] \tag{5.104b}$$

$$U = -\frac{4s_W^2}{\alpha}\epsilon_2 = 16\pi\left[\Sigma'_{33}\left(m_W^2\right) - \Sigma'_{11}\left(m_Z^2\right)\right] \tag{5.104c}$$

$\Delta\rho$ and ϵ_2, ϵ_3 have already been introduced (Eq. (5.58)). A contribution of the Standard Model is given in Appendix C, Eq. (C.46a).

Why just three? There are four vacuum-polarization functions: $\Sigma_{\gamma\gamma}(q^2)$, $\Sigma_{\gamma Z}(q^2)$, $\Sigma_{WW}(q^2)$, and $\Sigma_{ZZ}(q^2)$. Measurements have been made at two energy scales: $q^2 = 0, m_Z^2$. Thus, there are eight correlators. Of these eight, $\Sigma'_{\gamma\gamma}(0) = \Sigma'_{\gamma Z}(0) = 0$ due to the Ward identity. Of the remaining six, three linear combinations are absorbed in the redefinition of the experimental inputs: α, G_F and m_Z. The remaining three independent combinations are S, T, U.

Newly introduced S, T, U (or $\epsilon_1, \epsilon_2, \epsilon_3$) are defined as the vacuum polarization effects, excluding those of the Standard Model (i.e., $S_{\text{new}} = S - S_{\text{SM}}$). Each one is sensitive to different aspects of the new physics.

S, T, and U can be experimentally determined if three observables, for instance, m_W^2, Γ_Z, A_{FB} are measured. If one gives Γ_l and A_{FB}, v_l and a_l are determined to become (see Eq. (4.25))

$$v_l = -\left(1 + \frac{\Delta\rho}{2}\right)\left(\frac{1}{2} - 2\sin^2\theta_l\right) \tag{5.106a}$$

$$a_l = \frac{1}{2}\left(1 + \frac{\Delta\rho}{2}\right) \tag{5.106b}$$

which determine $\Delta\rho(= \alpha T)$ and $\sin^2\theta_l$. S and U can be determined from the relations (refer to Eqs. (5.8) and (5.88))

$$\frac{m_W^2 - m_Z^2 c_{MZ}^2}{m_Z^2 c_{MZ}^2} = -\frac{s_{MZ}^2}{c_{MZ}^2 - s_{MZ}^2}(\Delta r - \Delta\alpha) \tag{5.107a}$$

$$= \frac{\alpha}{c_{MZ}^2 - s_{MZ}^2}\left(c_{MZ}^2 T - \frac{S}{2}\right) + \frac{\alpha}{4s_{MZ}^2}U \tag{5.107b}$$

$$\sin^2\theta_{\text{eff}}^f - s_{MZ}^2 = -\frac{s_{MZ}^2}{c_{MZ}^2 - s_{MZ}^2}(c_{MZ}^2\Delta\rho - \epsilon_3) \tag{5.107c}$$

$$= \frac{\alpha}{c_{MZ}^2 - s_{MZ}^2}\left(-s_{MZ}^2 c_{MZ}^2 T + \frac{S}{4}\right) \tag{5.107d}$$

There are other observables to derive S, T, and U. Under the assumption of the self-energy dominance in the radiative corrections, they are universal parameters that appear in observables at the Z resonance. Under normal circumstances, U is small compared to S and T.

The present constraints on T_{new} and S_{new} are given in Figure 5.10. The thick oval third from the top around $T_{\text{new}} \simeq S_{\text{new}} \simeq 0$ is the overall constraint with $m_H = 117$ GeV. The obtained limits on S_{new}, T_{new}, U_{new} are [3, 238]

$$\begin{aligned} S_{\text{new}} &= 0.01 \pm 0.10(-0.08) \\ T_{\text{new}} &= 0.03 \pm 0.11(+0.09) \\ U_{\text{new}} &= 0.06 \pm 0.10(+0.01) \end{aligned} \tag{5.108}$$

The central value corresponds to $m_H = 117$ GeV and numbers in parentheses denote their shift if $m_H = 300$ GeV is assumed.

How do the new physics appear? T represents the magnitude of the isospin breaking and is a sensitive parameter if a doublet exists with a large mass difference. S characterizes chiral symmetry breaking by the mass term. It gives a large value if the doublet masses are large even if they are degenerate. For instance, if

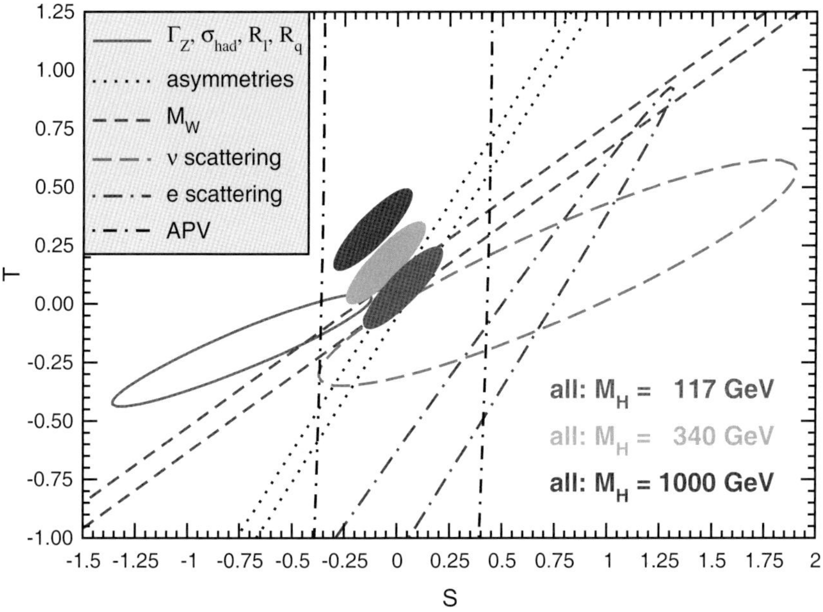

Figure 5.10 1σ constraints (39.35%) on S_{new} and T_{new} from various inputs combined with m_Z. S and T only represent contributions of new physics (The contour assumes $U = 0$). (Uncertainties from m_t are included in the errors.) The contours assume $m_h = 117$ GeV [3]. For a color version of this figure, please see the color plates at the beginning of the book.

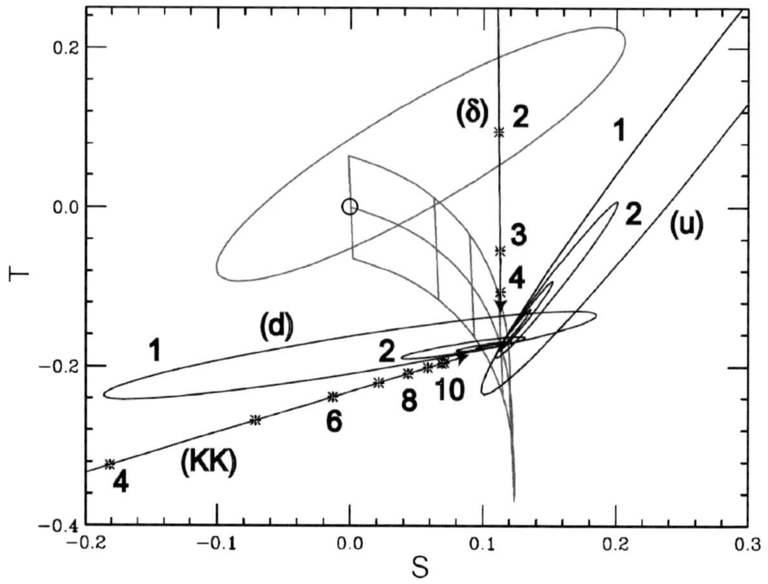

Figure 5.11 Constraints on (S, T) of various models. The upper ellipse shows the 68% contour for the minimum Standard Model (MSM). The small circle in the ellipse corresponds to (m_H = 100 GeV, m_t = 174.3 GeV). The vertical bar going through the circle corresponds to the variation of m_t = 174.3 GeV ± 5.15 GeV. The banana shaped figure shows the central value of a fit to the MSM for m_t = 174.3 GeV ± 5.15 GeV and m_h varying from 100 to 1000 GeV, with m_h = 200, 300, and 500 GeV marked with vertical bands. Compatibility of an extra Z_0' boson in various models assuming m_H = 500 GeV for MSM is examined. δ (vertical line) MSM+extra Z_0' boson. (u) and (d) (contours) MSM+E6 superstring-inspired Z_0' model with mixing H_u or H_d. (KK) (line) Z_0' in the extra-dimensional model of [643]. The numbers indicate the values of the Z_0' mass M. All of the $M_{Z_0'}$ predictions tend to MSM with m_H = 500 GeV as its mass goes to infinity [240].

there exists a new doublet (N, E) with large mass, they give [239]

$$S_{\text{new}} \sim \frac{N_c^f}{6\pi}, \quad T_{\text{new}} \sim \frac{N_c^f \Delta M^2/m_Z^2}{12\pi s_W^2 c_W^2}, \quad U_{\text{new}} \sim \frac{2N_c^f}{15\pi}\frac{\Delta M^2}{M_E^2} \sim O\left(\frac{m_Z^2}{M_E^2}\right) \tag{5.109}$$

where $\Delta M = M_E - M_N \ll M_E$ is assumed. If the mass is degenerate, $T_{\text{new}}, U_{\text{new}} \sim 0$, but S becomes significant. For instance, the technicolor models[14] of various symmetry can be rejected by examining the S value [137]. Figure 5.11 shows an example of testing the new model in which T ($\sim \epsilon_1$) and S ($\sim \epsilon_2$)

14) The technicolor model considers the Higgs particle as a bound state of yet to be discovered fundamental constituents (U, D) referred to as the techniquarks. The source of the force is referred to as the technicolor which has the degree of freedom N_{TC} and works similarly as the color in QCD. The technipion which is the bound state of $U\overline{D}$ is identified as the Higgs particle. If it exists, it predicts a variety of bound states in the TeV region which can be tested at the LHC. The model was popular at one time, but no viable model has been proposed yet.

values are plotted on a two-dimensional plane. It examined a possibility if a heavier Higgs is compatible with observations by extending the minimum Standard Model (MSM) before its discovery. Since the precision electroweak data already constrained the value of the Higgs severely to be less than 150 GeV, compability of data with heavier mass with extended SM containing an additional neutral boson Z' [240] was examined.[15)]

15) As the Higgs was found to have the mass value 126 GeV, the hypothesis of heavy Higgs has lost its relevance. However the graph serves to illustrate how the STU parameters could be used to probe new physics.

6
Cabibbo–Kobayashi–Maskawa Matrix

6.1
Origin of the CKM Matrix

We discussed the CP violation in the K-meson sector in Vol. 1 from a phenomenological point of view. In this chapter, we discuss it in the framework of the Standard Model and mainly address phenomena in the B-meson sector. Fields that appear in the Standard Model are quarks, leptons, gauge bosons (γ, $W^{\pm}$, Z, gluons) and the Higgs bosons. All the fields have definite transformation properties as to the P, C or T transformations. The Lagrangian for the free fields is CP invariant by its construction. The gauge interaction in its original form, (i.e., before spontaneous symmetry breaking) is also made to be CP invariant. However, something happens as we modify them to describe realities. Let us take a look at the mass generating interaction of the Higgs sector. Thus, we write the quark and Higgs fields belonging to $SU(2)$ doublets as

$$\Psi_{kL} = \begin{bmatrix} u'_{kL} \\ d'_{kL} \end{bmatrix} , \quad \Phi = \begin{bmatrix} \phi^{+} \\ \phi^{0} \end{bmatrix} , \quad \Phi^{c} = \begin{bmatrix} \phi^{0\dagger} \\ -\phi^{-} \end{bmatrix} , \tag{6.1}$$

where $u'_{kL}(k = 1 \sim 3) = u'_L, c'_L, t'_L$ and $d'_{kL}(k = 1 \sim 3) = d'_L, s'_L, b'_L$. The right-handed components u'_{kR}, d'_{kR} are singlets. The fermions are thought to be massless. The reason to attach dashes is because the quark fields that interact with the Higgs before the symmetry breaking are not necessarily the same as the physical quarks with definite mass. We refer the dashed fields as the weak eigenstates and the physical fields as the mass eigenstates.

A representative Yukawa interaction which generates masses was given in Eq. (1.45). Extending the formula to include all the quarks in the SM, the $SU(2)$ invariant Yukawa interaction can be written as

$$-\mathcal{L}_I(\Psi, \Phi) = \sum_{j,k} \left[Y_{jk} \overline{\Psi}'_{jL} \Phi d'_{kR} + Y'_{jk} \overline{\Psi}'_{jL} \Phi^{c} u'_{kR} \right] \tag{6.2}$$

Terms with $j \neq k$ appear because there is no reason to assume that the Higgs interaction is flavor diagonal. The above Lagrangian is arranged in such a way as to give masses to the quarks u'_j and d'_k when $\phi^0, \phi^{0\dagger}$ acquire the vacuum expectation

Elementary Particle Physics, Volume 2: Foundations of the Standard Model, First Edition. Yorikiyo Nagashima.

value. Y_{jk}, Y'_{jk} are the coupling constants of the Yukawa interaction between the quark and Higgs fields.

When the symmetry is spontaneously broken, the Higgs fields acquire the vacuum expectation values ($\langle\phi^{\pm}\rangle = 0$, $\langle\phi^0\rangle = v/\sqrt{2}$) and Eq. (6.2) becomes

$$\begin{aligned} -\mathcal{L}_I(\Psi, \Phi) &= \sum_{j,k}\left[\overline{u}'_{jL} M'_{jk} u'_{kR} + \overline{d}'_{jL} M_{jk} d'_{kR} + \text{h.c.}\right] + \cdots \\ &\equiv \left[\overline{U}'_L M^U U'_R + \overline{D}'_L M^D D'_R + \text{h.c.}\right] + \cdots , \end{aligned} \tag{6.3}$$

where h.c. denotes the hermitian conjugate.

$$M_{jk} = Y_{jk}\frac{v}{\sqrt{2}}, \quad M'_{jk} = Y'_{jk}\frac{v}{\sqrt{2}} \tag{6.4a}$$

$$M^U = [M'_{jk}], \quad M^D = [M_{jk}] \tag{6.4b}$$

$$U' = \begin{bmatrix} u' \\ c' \\ t' \end{bmatrix}, \quad D' = \begin{bmatrix} d' \\ s' \\ b' \end{bmatrix} \tag{6.4c}$$

The content of the brackets is the quark mass term in its general form.[1] The ellipses after the bracket are the Higgs interaction terms which do not concern us in the following discussion. As the mass matrices M_{jk} and M'_{jk} are not diagonal in general, the weak eigenstates do not necessarily coincide with the mass eigenstates. To describe physical processes, we have to diagonalize Eq. (6.3) and express the Lagrangian in terms of the mass eigenstates.

According to the matrix theory, any $N \times N$ matrix (need not be hermitian) can be diagonalized using two unitary matrices. Then, denoting A_L, $A_R^\dagger$, B_L, $B_R^\dagger$ as four unitary matrices, we can write

$$M^U_{\text{dia}} = A_L M^U A_R^\dagger, \quad M^D_{\text{dia}} = B_L M^D B_R^\dagger \tag{6.5a}$$

$$M^U_{\text{dia}} = \begin{bmatrix} m_u & & \\ & m_c & \\ & & m_t \end{bmatrix}, \quad M^D_{\text{dia}} = \begin{bmatrix} m_d & & \\ & m_s & \\ & & m_b \end{bmatrix} \tag{6.5b}$$

It is possible to choose unitary matrices A_L, A_R, B_L, B_R to make all the masses m_u, m_c, m_t and m_d, m_s, m_b real.[2] Since the product of the diagonal matrices

$$M^U_{\text{dia}} M^{U\dagger}_{\text{dia}} = \begin{bmatrix} m_u^2 & 0 & 0 \\ 0 & m_c^2 & 0 \\ 0 & 0 & m_t^2 \end{bmatrix}, \quad M^D_{\text{dia}} M^{D\dagger}_{\text{dia}} = \begin{bmatrix} m_d^2 & 0 & 0 \\ 0 & m_s^2 & 0 \\ 0 & 0 & m_b^2 \end{bmatrix} \tag{6.6}$$

1) In principle, the same argument can be applied to the lepton sector. However, within the conventional Standard Model framework in which all the neutrinos are assumed massless, the flavor eigenstates and the mass eigenstates of the leptons are the same. We will defer the CP violation problem in the lepton sector until later when we discuss the neutrino oscillation.

2) In general, m_js are complex numbers. However, one can always make them real by applying the chiral transformation.

can be expressed as

$$M_{\rm dia}^U M_{\rm dia}^{U\dagger} = A_{\rm L} M^U A_{\rm R}^\dagger A_{\rm R} M^{U\dagger} A_{\rm L}^\dagger = A_{\rm L} M^U M^{U\dagger} A_{\rm L}^\dagger$$
$$M_{\rm dia}^{U\dagger} M_{\rm dia}^U = A_{\rm R} M^{U\dagger} A_{\rm L}^\dagger A_{\rm L} M^U A_{\rm R}^\dagger = A_{\rm R} M^{U\dagger} M^U A_{\rm R}^\dagger \tag{6.7}$$

$$M_{\rm dia}^D M_{\rm dia}^{D\dagger} = B_{\rm L} M^D B_{\rm R}^\dagger B_{\rm R} M^{D\dagger} B_{\rm L}^\dagger = B_{\rm L} M^D M^{D\dagger} B_{\rm L}^\dagger ,$$
$$M_{\rm dia}^{D\dagger} M_{\rm dia}^D = B_{\rm R} M^{U\dagger} B_{\rm L}^\dagger B_{\rm L} M^U B_{\rm R}^\dagger = B_{\rm R} M^{U\dagger} M^U B_{\rm R}^\dagger \tag{6.8}$$

one sees that $A_{\rm L}, A_{\rm R} \cdots$ are unitary matrices that diagonalize the hermitian matrices $M^U M^{U\dagger}$, $M^{U\dagger} M^U \cdots$. Using Eq. (6.5), Eq. (6.3) can be expressed as

$$\begin{aligned}
-\mathcal{L}_M &= \overline{U}'_{\rm L} M^U U'_{\rm R} + \overline{D}_{\rm L} M^D D'_{\rm R} + \text{h.c.} \\
&= \overline{U}'_{\rm L} A_{\rm L}^\dagger A_{\rm L} M^U A_{\rm R}^\dagger A_{\rm R} U'_{\rm R} + \overline{D}'_{\rm L} B_{\rm L}^\dagger B_{\rm L} M^D B_{\rm R}^\dagger B_{\rm R} D'_{\rm R} + \text{h.c.} \\
&= (\overline{A_{\rm L} U'_{\rm L}}) M_{\rm dia}^U (A_{\rm R} U'_{\rm R}) + (\overline{B_{\rm L} D'_{\rm L}}) M_{\rm dia}^D (B_{\rm R} D'_{\rm R}) + \text{h.c.} \\
&= \sum \left[m_j (\overline{u}_{j\rm L} u_{j\rm R} + \overline{u}_{j\rm R} u_{j\rm L}) + m'_j (\overline{d}_{j\rm L} d_{j\rm R} + \overline{d}_{j\rm R} d_{j\rm L}) \right]
\end{aligned} \tag{6.9}$$

Equation (6.9) has exactly the desired form of the fermion mass Lagrangian. Therefore, we see that the quarks in the mass eigenstates (U, D) are obtained from the weak eigenstates (U', D') by the unitary transformations

$$U_{\rm L} = A_{\rm L} U'_{\rm L}, \quad U_{\rm R} = A_{\rm R} U'_{\rm R}, \quad D_{\rm L} = B_{\rm L} D'_{\rm L}, \quad D_{\rm R} = B_{\rm R} D'_{\rm R} \tag{6.10}$$

The free Lagrangian $\mathcal{L}_{\rm free}$ and the neutral current interaction Lagrangian $\mathcal{L}_{\rm nc}$ which are flavor diagonal do not change their form under the above unitary transformations, but merely change their components from $q'_j \to q_j$. Therefore, they can be expressed equally as well using the mass eigenstates from the beginning. This is a generalized statement of the GIM mechanism, that is, no flavor changing neutral current exists (see Vol. 1, Sect. 15.7). On the other hand, the charged current interaction Lagrangian $\mathcal{L}_{\rm cc}$ connects quarks of different flavors and changes its form depending whether it is written in terms of the weak eigenstates or the mass eigenstates. Rewriting the original $\mathcal{L}_{\rm cc}$ given in terms of the weak eigenstates by those of the mass eigenstates,

$$\begin{aligned}
-\mathcal{L}_{\rm cc} &= (g_W/\sqrt{2}) \sum \left[\overline{U}'_{\rm L} \gamma^\mu D'_{\rm L} W_\mu^+ + \text{h.c.} \right] \\
&= (g_W/\sqrt{2}) \sum \left[\overline{U}_{\rm L} \gamma^\mu \left(A_{\rm L}^\dagger B_{\rm L} \right) D_{\rm L} W_\mu^+ + \text{h.c.} \right] \\
&= (g_W/\sqrt{2}) \sum \left[\overline{U}_{\rm L} \gamma^\mu V D_{\rm L} W_\mu^+ + \text{h.c.} \right] \\
&= (g_W/\sqrt{2}) \sum \left[\overline{u}_{j\rm L} \gamma^\mu V_{jk} d_{k\rm L} W_\mu^+ + \overline{d}_{k\rm L} \gamma^\mu V_{jk}^* u_{j\rm L} W_\mu^- \right]
\end{aligned} \tag{6.11}$$

we see that the interaction does not close within each doublet, but induces mixing between different generations. The matrix $V = A_{\rm L}^\dagger B_{\rm L}$ is called the Cabibbo–Kobayashi–Maskawa (CKM) matrix.

By applying the CP transformation, the vector fields transform according to (See Vol. 1, Appendix F)

$$\overline{\psi}_1 \gamma^\mu \psi_2 \to -\overline{\psi}_2 \gamma_\mu \psi_1 \,, \quad W_\mu^+ \to -W^{-\mu} \tag{6.12}$$

Accordingly, the Lagrangian changes to

$$-\mathcal{L}_{cc} \xrightarrow{\text{CP}} (g_W/\sqrt{2}) \sum \left[\overline{d}_{kL} \gamma_\mu V_{jk} u_{jL} W^{-\mu} + \overline{u}_{jL} \gamma_\mu V_{jk}^* d_{kL} W^{+\mu} \right] \tag{6.13}$$

Therefore, CP invariance requires that the CKM matrix be real.

Note, we defined the CKM matrix as the mixing matrix of the down quark states. However, it is clear from Eq. (6.13) that we can treat it equally as that of mixing the up quarks with down quarks unchanged. The convention is historical.

Why Three Generations? The reason why we need at least three generations to have a CP violating Lagrangian was discussed in Vol. 1, Sect. 16.6. Briefly, a $N \times N$ unitary matrix has N^2 independent real variables. In this real variable space of N^2 dimensions, one can define ${}_N C_2 = N(N-1)/2$ rotations and $N(N+1)/2$ phases. On the other hand, the $2N$ quarks can have $(2N-1)$ degrees of freedom to change their phase, keeping the mass Lagrangian invariant which can be used to absorb the phase of the matrix V[3]. The minus 1 corresponds to the change of overall phase (if all the quarks are changed to $q_j \to q'_j = e^{i\phi} q_j$, V does not change). Therefore, the number of phases in the $N \times N$ unitary matrix reduces to

$$N^2 - \frac{N(N-1)}{2} - (2N-1) = \frac{(N-1)(N-2)}{2} \tag{6.14}$$

This means that for the case $N \le 2$, the matrix can be written using only real variables and no CP violating term appears. In the real world, the CP symmetry is violated, therefore we must have $N \ge 3$. We also know that the number of light neutrinos ($m_\nu < m_Z/2$) is three from measurements of the invisible width of the Z decay (see Section 4.4.3). Therefore, it is natural to assume that the number of generation is three.

Notice that if any two of the quarks are mass degenerate, there is no longer CP violation for $N = 3$. The reason is as follows. Suppose d and s quarks are degenerate, then any combination of d and s (denote d' and s') is again the mass eigenstate and physically represents the same particle. Therefore, the quarks (D) has larger freedom than simple phase transformation, namely, a freedom to transform as

$$D \to F_D D = \begin{bmatrix} U & \\ & e^{i\phi} \end{bmatrix} D \tag{6.15}$$

and is able to reduce one more phase to make the CKM matrix V real.

3) This is true only for Dirac particles. The Majorana particles can absorb only N phases and for this reason CMNS matrix which is the equivalent of CKM matrix in the lepton sector, has extra two Majorana phases.

Parametrization of the CKM Matrix There are many ways to express the CKM matrix, except that it is parametrized by three rotation angles and one phase angle. Here, we follow the convention adopted by the particle data group [3] and express it as

$$
\begin{aligned}
V_{\mathrm{CKM}} &\equiv \begin{bmatrix} V_{ud} & V_{us} & V_{ub} \\ V_{cd} & V_{cs} & V_{cb} \\ V_{td} & V_{ts} & V_{tb} \end{bmatrix} = R_{23}(I_{\delta_D} R_{13} I^{\dagger}_{\delta_D}) R_{12} \\
&= \begin{bmatrix} 1 & 0 & 0 \\ 0 & c_{23} & s_{23} \\ 0 & -s_{23} & c_{23} \end{bmatrix} \begin{bmatrix} c_{13} & 0 & s_{13}e^{-i\delta} \\ 0 & 1 & 0 \\ -s_{13}e^{i\delta} & 0 & c_{13} \end{bmatrix} \begin{bmatrix} c_{12} & s_{12} & 0 \\ -s_{12} & c_{12} & 0 \\ 0 & 0 & 1 \end{bmatrix} \\
&= \begin{pmatrix} c_{12}c_{13} & s_{12}c_{13} & s_{13}e^{-i\delta} \\ -s_{12}c_{23} - c_{12}s_{23}s_{13}e^{i\delta} & c_{12}c_{23} - s_{12}s_{23}s_{13}e^{i\delta} & s_{23}c_{13} \\ s_{12}s_{23} - c_{12}c_{23}s_{13}e^{i\delta} & -c_{12}s_{23} - s_{12}c_{23}s_{13}e^{i\delta} & c_{23}c_{13} \end{pmatrix}
\end{aligned}
\tag{6.16}
$$

where $I_{\delta_D} = \mathrm{diag}(1, 1, e^{i\delta})$ is the diagonal phase matrix and c_{ij} and s_{ij} denote $\cos\theta_{ij}$ and $\sin\theta_{ij}$ respectively. It has the property that in the limit $\theta_{23} = \theta_{13} \to 0$, the third generation is separated and reduces to the Cabibbo rotation matrix with $\theta_C = \theta_{12}$. By suitable redefinition of the quark fields, all the elements c_{ij}, s_{ij} can be made positive, namely,

$$
0 \le \theta_{ij} \le \pi/2\,, \quad 0 \le \delta \le 2\pi \tag{6.17}
$$

Experimentally, $c_{13} - 1 < 10^{-5}$ is observed (see next section), and therefore we approximately have

$$
\begin{aligned}
|V_{us}| &= s_{12}c_{13} \approx s_{12} \\
|V_{ub}| &= s_{13} \\
|V_{cb}| &= s_{23}c_{13} \approx s_{23}
\end{aligned}
\tag{6.18}
$$

Therefore, by parametrizing

$$
s_{12} = \lambda\,, \quad s_{23} = A\lambda^2\,, \quad s_{13}e^{-i\delta} = A\lambda^3(\rho - i\eta)^{4)} \tag{6.19}
$$

We can express the CKM matrix as

$$
V_{\mathrm{CKM}} = \begin{bmatrix} 1-\lambda^2/2 & \lambda & A\lambda^3(\rho - i\eta) \\ -\lambda & 1-\lambda^2/2 & A\lambda^2 \\ A\lambda^3(1-\rho-i\eta) & -A\lambda^2 & 1 \end{bmatrix} + O(\lambda^4) \tag{6.21}
$$

4) Historically, these parameters were introduced as convenient approximations. It is now customary to adopt Eq. (6.19) as the definition. Then, if we define

$$
\rho = \frac{s_{13}}{s_{12}s_{23}}\cos\delta\,, \quad \eta = \frac{s_{13}}{s_{12}s_{23}}\sin\delta \tag{6.20}
$$

Eq. (6.19) becomes exact to all orders in λ, which, theoretically, is a very convenient feature [243].

which is referred to as the Wolfenstein's parametrization [244]. In this expression, the near diagonality of the CKM matrix is manifest and very convenient for the data analysis. Inserting experimentally observed values, their values are given by [3]

$$\lambda = 0.2253 \pm 0.0007\,, \quad A = 0.808^{+0.022}_{-0.015}\,,$$
$$\overline{\rho} = 0.132^{+0.022}_{-0.014}\,, \quad \overline{\eta} = 0.341 \pm 0.013 \tag{6.22}$$

where

$$\overline{\rho} + i\overline{\eta} \equiv -\frac{V_{ub}^* V_{ud}}{V_{cb}^* V_{cd}} = (\rho + i\eta)\left(1 - \frac{\lambda^2}{2}\right) + O(\lambda^4)^{5)} \tag{6.23}$$

6.2 CKM Matrix Elements

In the Standard Model, the CP violation stems from the phase of the CKM matrix. Notice that the SM has prepared a framework to investigate the CP violation process, but it does not give any explanation for the dynamical origin of the complex phase. Therefore, investigation of its structure is a necessary step to find clues to its origin. It could go beyond the Standard Model. As the experimental reproducibility of the Standard Model is very good, the CP conserving part of the CKM matrix (i.e., the absolute value of the matrix elements) is expected to change very little if the origin of the CP violation is beyond the Standard Model. Indeed, in many models proposed thus far, this is the case. In the following, we determine the absolute values of the CKM matrix elements assuming $N = 3$ [245, 246].

$|V_{ud}|$ It can be determined from the nuclear beta decay . The ft value is given by (see Vol. 1, Eq. (15.22))

$$f t_{1/2} = \frac{2\pi^3 \ln 2}{G_F^2 |V_{ud}|^2 |\langle \mathbf{1} \rangle|^2} \tag{6.25}$$

where f is the phase space factor for the beta decay, $t_{1/2}$ the half-life and $\langle \mathbf{1} \rangle$ is the nuclear matrix element. Combining data of several superallowed $0^+ \to 0^+$ transitions and with suitable refinement of the theoretical formula with radiative corrections, one obtains

$$|V_{ud}| = 0.974\,25 \pm 0.000\,22 \quad [247] \tag{6.26}$$

5) The inverse relation is given by

$$\rho + i\eta = \sqrt{\frac{1 - A^2\lambda^4}{1 - \lambda^2}} \frac{\overline{\rho} + i\overline{\eta}}{1 - A^2\lambda^4(\overline{\rho} + i\overline{\eta})} \tag{6.24}$$

$|V_{us}|$ It can be determined from the leptonic decay of the strange particles. The decay formula is given by (see Vol. 1, Eqs. (15.140) and (15.141))

$$\frac{d\Gamma}{dq^2}(\overline{K}^0 \to \pi^+ l \nu_l) = \frac{G_F^2 |V_{us}|^2}{24\pi^3} |\boldsymbol{p}_\pi|^3 |f_+(q^2)|^2 \tag{6.27a}$$

$$q^2 = (p_K - p_\pi)^2 = m_K^2 + m_\pi^2 - 2m_K\sqrt{|\boldsymbol{p}_\pi|^2 + m_\pi^2} \tag{6.27b}$$

where $|\boldsymbol{p}_\pi|$ is the pion momentum in the kaon rest frame. Equation (6.27a) determines $|V_{us}|^2 f_+^2(q^2)$. The form factor $f_+(q^2)$ is defined by [3]

$$\begin{aligned}\langle \pi^+(p_\pi)|\overline{u}_L \gamma^\mu s_L|\overline{K}^0(p_K)\rangle &= f_+(q^2)\left[p_K^\mu + p_\pi^\mu - \frac{m_K^2 - m_\pi^2}{q^2} q^\mu\right] \\ &\quad + f_0(q^2)\frac{m_K^2 - m_\pi^2}{q^2} q^\mu\end{aligned} \tag{6.28}$$

and dominance of f_+ term over f_0 is assumed because q^μ multiplied with the leptonic current produces a multiplicative factor m_l^2. All the nonperturbative contribution is included in the form factor. One has to rely on theory to evaluate it and separate it to obtain $|V_{us}|$. It has become increasingly conventional to use the lattice QCD calculated values and one obtains $f_+(0) = 0.9644 \pm 0.0049$ [248]. For the case of the semileptonic K decay, $f_+(q^2) \simeq f_+(0)$ gives a good approximation as the value of q^2 is small. Using a compilation of world data on the experimental data and a theoretically improved formula together, one obtains $|V_{us}| = 0.2246 \pm 0.0012$ [249].

Another method is to use the pure leptonic decay process. The formula is given as (see Vol. 1, Eq. (15.47))

$$\Gamma(K^\pm \to \mu^\pm \nu) = G_F^2 |V_{us}|^2 f_K^2 \frac{m_\mu^2}{8\pi} \frac{\left(m_K^2 - m_\mu^2\right)^2}{m_K^3} \tag{6.29}$$

where f_K is the decay constant of the kaon defined by

$$\langle 0|\overline{s}_L \gamma^\mu \gamma^5 u_L|K^+\rangle = i f_K q^\mu \tag{6.30}$$

Again, only the combination of $|V_{ud}|^2 f_K^2$ is directly determined from experiments and one has to rely on theory to obtain the value of f_K. The situation is quite similar to the semileptonic decay. However, in the case of $K \to l\nu$, by taking the ratio

$$\frac{\Gamma(K^+ \to l\nu_l(\gamma))}{\Gamma(\pi^+ \to l\nu_l(\gamma))} = \frac{|V_{us}|^2 f_K^2 m_K \left(1 - m_l^2/m_K^2\right)^2}{|V_{ud}|^2 f_\pi^2 m_\pi \left(1 - m_l^2/m_\pi^2\right)^2} \tag{6.31}$$

both the experimental and theoretical accuracy can be improved. Here, $K^+ \to l\nu_l(\gamma)$ means that an experimentally extra photon emission process is included. Using the lattice QCD calculated $f_K/f_\pi = 1.189 \pm 0.007$ [248] and the KLOE result [250], one can determine the ratio $|V_{us}|^2/|V_{ud}|^2$ which in turn leads to $|V_{us}| = 0.2259 \pm 0.0014$. The average of the two determinations gives

$$|V_{us}| = 0.2252 \pm 0.0009 \quad [3] \tag{6.32}$$

$|V_{cd}|$ It can be determined from $D \to \pi l \nu_l$ using the same method stated above. However, as of now, another method to use charm production by the neutrino is still the best. The neutrino production cross section on an isoscalar target (see Vol. 1, Eq. (17.116)) is generalized to include flavors other than u and d. They can be written as

$$\begin{aligned}
\sigma(\nu d) &\equiv \frac{d\sigma}{dx}\left[\nu_\mu + d(\overline{u}) \to \mu^- + u(\overline{d})\right] = K|V_{ud}|^2\left[q(x) + \frac{\overline{q}(x)}{3}\right] \\
\sigma(\overline{\nu} u) &\equiv \frac{d\sigma}{dx}\left[\overline{\nu}_\mu + u(\overline{d}) \to \mu^+ + d(\overline{u})\right] = K|V_{ud}|^2\left[\overline{q}(x) + \frac{q(x)}{3}\right] \\
\sigma(\nu c) &\equiv \frac{d\sigma}{dx}\left[\nu_\mu + d(s) \to \mu^- + c\right] = K\left[|V_{cd}|^2 q(x) + 2|V_{cs}|^2 s(x)\right] \\
\sigma(\overline{\nu c}) &\equiv \frac{d\sigma}{dx}\left[\overline{\nu}_\mu + \overline{d}(\overline{s}) \to \mu^+ + \overline{c}\right] = K\left[|V_{cd}|^2 \overline{q}(x) + 2|V_{cs}|^2 \overline{s}(x)\right] \qquad (6.33)
\end{aligned}$$

where x is the fractional momentum of the quarks relative to their parent hadron and K is the common factor. The functions $q(x) \equiv u(x) + d(x)$, $\overline{q}(x) \equiv \overline{u}(x) + \overline{d}(x)$, $s(x) = \overline{s}(x)$ are the momentum distribution functions of u, $d(\overline{u}, \overline{d})$ and $s(\overline{s})$ quarks in the proton ($u \leftrightarrow d$ for the neutron). The charm (anticharm) particles decay to leptons of definite sign $c(\overline{c}) \to l^+(l^-)$, and therefore the signal of the charm production is the existence of two opposite sign leptons in the final state. The two opposite sign lepton production cross sections contain the branching fraction $\text{BR}(\mu) = \Gamma(c \to \mu)/\Gamma(c \to all)$ which can be determined experimentally by comparing semileptonic and total charm hadronic decays of the produced charms in the emulsion [251] giving $\text{BR}(\mu) = 0.087 \pm 0.005$. Using Eq. (6.33), $|V_{cd}|$ can be obtained from the formula

$$\frac{|V_{cd}|^2}{|V_{ud}|^2} = \frac{2}{3}\frac{\sigma(\nu c)/\sigma(\nu d) - R\sigma(\overline{\nu c})/\sigma(\overline{\nu} u)}{1 - R} \qquad (6.34a)$$

$$R \equiv \frac{\sigma^{\overline{\nu}}}{\sigma^{\nu}} = \frac{\int dx[\overline{q}(x) + q(x)/3]}{\int dx[q(x) + \overline{q}(x)/3]} = 0.48 \pm 0.02 \qquad (6.34b)$$

From combined experimental data [252–255], the Particle Data Group [256] quotes $\text{BR}(l)|V_{cd}|^2 = (0.463 \pm 0.034) \times 10^{-2}$ which, in turn, using the above $\text{BR}(l)$ value gives

$$|V_{cd}| = 0.230 \pm 0.011 \quad [3] \qquad (6.35)$$

$|V_{cs}|$ It can be determined from both semileptonic and pure leptonic decays of charmed meson D just like $|V_{us}|$ was obtained from the kaon decay. The rate is given by the same formula as Eq. (6.27) with replacement of $K \to D, \pi \to K$, $V_{us} \to V_{cs}$, $f_+(0) \to f_{+D}^K(0)$. Using a theoretical value $f_{+D}^K(0) = 0.73 \pm 0.03 \pm 0.07$ [257] and combination of experimental results ([258–260]), one obtains the value $|V_{cs}| = 0.98 \pm 0.01_{\text{stat}} \pm 0.10_{\text{syst}}$. Again, by combining the result with that obtained from the pure leptonic decay rate $\Gamma(D_s^+ \to \mu^+\nu_\mu)$, one can improve the accuracy.

$$|V_{cs}| = 1.023 \pm 0.036 \quad [3] \qquad (6.36)$$

$|V_{cb}|$ In principle, it can be determined similarly using semileptonic decay rate Eq. (6.27a) for the K-meson with suitable replacement of variables.

$$\frac{d\Gamma}{dq^2}(B \to D l \nu_l) = \frac{G_F^2 |V_{cb}|^2}{24\pi^3} |\boldsymbol{p}_D|^3 |f_{+B}^D(q^2)|^2 \tag{6.37}$$

where $f_{+B}^D(q^2)$ is defined analogously to Eq. (6.28). However, due to the heavy mass of the B meson, the q^2 range is sizable and one encounters the problem of determining the shape of the form factor as a function of q^2 which is a nonperturbative quantity.

Fortunately, in this case, one can rely on the HQET (heavy quark effective theory [261, 262]). Both the heavy quarks b (in B) and c (in D) are treated as nonrelativistic particles. The transition from b to c occurs in the weak time scale $\Delta t \sim 1/m_W$ which is much faster compared to hadron formation time $\Delta t \sim 1/\Lambda_{\text{QCD}}$. Therefore, the heavy quark acts as a static pointlike color source with fixed four-velocity which cannot be altered by the soft gluons responsible for hadronizing the b, c quarks to B, D mesons. Hence, the light quark degrees of the freedom become independent of the spin and flavor (i.e., mass) of the heavy quarks. The form factor which describes overlap of the wave functions between the initial and final state becomes independent of the Dirac structure of the weak current. It only depends on a scalar quantity $w = v_b \cdot v_c$, where v_b, v_c are four velocities of the b, c quarks. Thus, $B \to D^* l \overline{\nu}_l$ and $B \to D l \overline{\nu}_l$ can be treated on common arguments as far as the nonperturbative part is concerned and their form factors are related by a single Isgur–Wise function [263] which contains all the low-energy nonperturbative hadronic physics relevant for the decays. In short, the relevant variable for the process $\overline{B} \to D l \nu_l$ is the particle's four velocity:

$$p_B^\mu = m_B v_B^\mu + O(\Lambda_{\text{QCD}}), \quad p_D^\mu = m_D v_D^\mu + O(\Lambda_{\text{QCD}}) \tag{6.38a}$$

$$q^2 \simeq m_B^2 + m_D^2 - 2 m_B m_D (v_B \cdot v_D) + \cdots \tag{6.38b}$$

$$|\boldsymbol{p}_D| = m_D \left[(v_B \cdot v_D)^2 - 1\right]^{1/2} + \cdots \tag{6.38c}$$

The form factors are reexpressed as a function of $w = (v_B \cdot v_D) = (m_B^2 + m_D^2 - q^2)/(2 m_B m_D)$ as [243, 261, 264]

$$\langle D^+(v_D)| V^\mu |\overline{B}^0(v_B)\rangle = \left[h_+(w)(v_B + v_D)^\mu + h_-(w)(v_B - v_D)^\mu\right] \tag{6.39a}$$

$$h_+(w) = 1 + O\left(\frac{m_B - m_D}{m_B m_D}\right), \quad h_-(w) = O\left(\frac{1}{m_b}, \frac{1}{m_c}\right) \tag{6.39b}$$

where normalization of the meson state $|B(v_B)\rangle$ in the HQET is related to the conventional relativistic state $|B(p_B)\rangle$ by

$$|B(v_B)\rangle = \lim_{m_B \to \infty} \frac{1}{\sqrt{m_B}} |B(p_B)\rangle \tag{6.40}$$

Then, Eq. (6.37) is altered to

$$\frac{d\Gamma}{dw}(\overline{B} \to D l \overline{\nu}_l) = \frac{G_F^2 |V_{cb}|^2}{48\pi^3}(m_B + m_D)^2 m_D^3 (w^2 - 1)^{3/2} |G(w)|^2$$
$$G(w) = R f_+(q^2) = h_+(w) - \frac{m_B - m_D}{m_B + m_D} h_-(w)$$
$$= 1 + O\left(\frac{m_B - m_D}{m_B + m_D}\frac{1}{m_c}\right)$$
$$R = \frac{2\sqrt{m_B m_D}}{m_B + m_D} \approx 0.88 \tag{6.41}$$

The lattice QCD gives $G(1) = 1.074 \pm 0.018 \pm 0.016$ [265]. The combined experimental values [266–268] give $|V_{cb}|G(1) = (42.3 \pm 0.7 \pm 1.3) \times 10^{-3}$ and $|V_{cs}| = (38.7 \pm 1.1) \times 10^{-3}$. The value of $|V_{cb}|$ obtained from exclusive decays is given by

$$|V_{cb}| = (38.7 \pm 1.1) \times 10^{-3} \quad [3] \tag{6.42}$$

$|V_{ub}|$ It can be determined by using the same data which was used to extract $|V_{cb}|$. The b-quark has three distinct semileptonic decay modes: $b \to c l^- \overline{\nu}$, $b \to u l^- \overline{\nu}$ and $b \to c \to l^+ \nu$. Their spectra are shown in Figure 6.1a.

One can extract the branch process $b \to u l \overline{\nu}_l$ from the upper edge of the lepton spectrum to determine the ratio $|V_{ub}/V_{cb}|$. Experimentally, the signal $\overline{B}^0 \to \pi^+ l \overline{\nu}_l$ suffers from large backgrounds due to $b \to c l \overline{\nu}_l$ and $\overline{B}^0 \to X_u l \overline{\nu}$ (X_u = any meson including the u-quark) as well. At the B-factory, where $B^0\overline{B}^0$ pair is produced

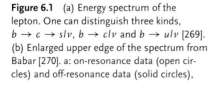

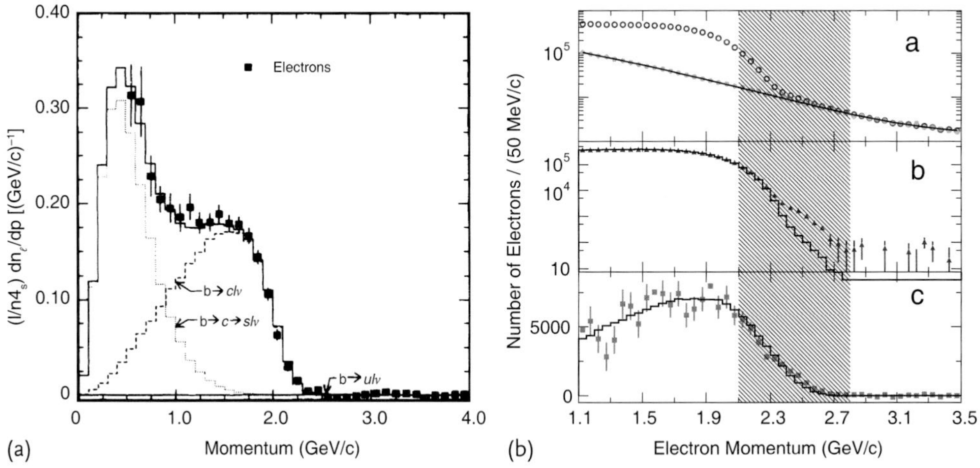

Figure 6.1 (a) Energy spectrum of the lepton. One can distinguish three kinds, $b \to c \to sl\nu$, $b \to cl\nu$ and $b \to ul\nu$ [269]. (b) Enlarged upper edge of the spectrum from Babar [270]. a: on-resonance data (open circles) and off-resonance data (solid circles), b: on resonance data subtracted for non-$B\overline{B}$ backgrounds (triangles) and the simulated contribution from B decays other than $b \to ul\nu$ (histogram) and c: background-subtracted data (points) compared with a model of the $b \to ul\nu$ spectrum (histogram).

through the process $e^- e^+ \to \Upsilon(4S) \to B^0\overline{B}^0$, clean signals can be obtained by tagging one of the B^0–$\overline{B}^0$. Figure 6.1b shows the upper end of the momentum spectrum for $b \to u l \overline{\nu}_l$ obtained at the B-factory [268, 271]. The formula to derive the $|V_{ub}|$ is obtained from Eq. (6.27) by replacing $K \to B$, $V_{us} \to V_{ub}$ and $f_+ \to f^u_{+B}$

$$\frac{d\Gamma}{dq^2}(B^0 \to \pi^+ l \nu_l) = \frac{G_F^2 |V_{ub}|^2}{24\pi^3} |\boldsymbol{p}_\pi|^3 |f^u_{+B}(q^2)|^2 \tag{6.43a}$$

$$q^2 = (p_B - p_\pi)^2 = m_B^2 + m_\pi^2 - 2m_B\sqrt{|\boldsymbol{p}_\pi|^2 + m_\pi^2} \tag{6.43b}$$

The difficulty in this case is that q^2 goes up to $(m_B - m_\pi)^2 \sim 26.4\,\text{GeV}^2$. The shape of the form factor is needed to fit the decay rate as a function of q^2 or to integrate to obtain the total rate. To determine the form factor, one has to rely on the nonperturbative treatment. It is model dependent and hard to make an accurate estimate. The region that can be determined accurately by the lattice QCD is limited to high q^2 end ($q^2 > 16$–$18\,\text{GeV}^2/c^2$) where the data suffers from poor statistics. Figure 6.2 compares data with various models. The obtained value is

$$|V_{ub}| = (3.89 \pm 0.44) \times 10^{-3} \quad [3] \tag{6.44}$$

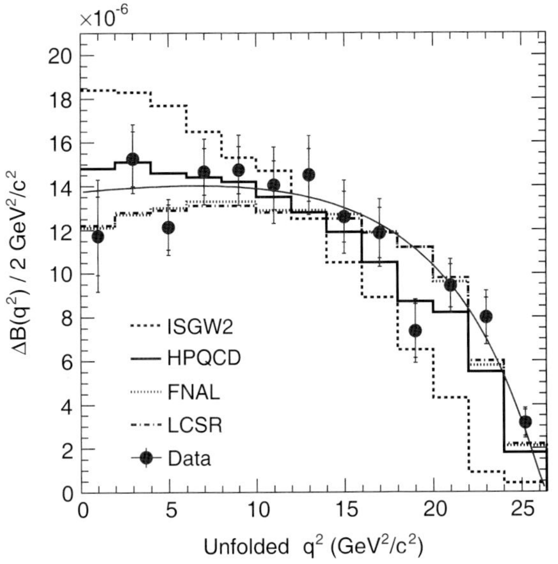

Figure 6.2 Measured q^2 spectrum of $B \to \pi l \nu$ compared with a fit of the BK parametrization [273] and with theory predictions from LQCD [265, 274], LCSR [275] and the ISGW2 [276] quark model. Figure taken from [277, 278].

$|V_{tb}|$ It can be determined from the branching ratio

$$\mathrm{BR}(t \to bW) = \frac{\Gamma(t \to b + W)}{\Gamma(t \to \text{all})} = \frac{|V_{tb}|^2}{|V_{tb}|^2 + |V_{ts}|^2 + |V_{td}|^2} \tag{6.45}$$

which can be measured at the hadron collider. However, the accuracy is not yet satisfactory. If one assumes $N = 3$ unitarity, it can be derived indirectly from the formula

$$|V_{tb}| = (1 - |V_{ub}|^2 - |V_{cb}|^2)^{1/2} = 0.9991 \pm 0.0001 \tag{6.46}$$

$|V_{ts}|$ and $|V_{td}|$ Values of $|V_{tb}V_{td}^*|$ and $|V_{tb}V_{ts}^*|$ can be estimated from the $B^0_{d(s)} - \overline{B}^0_{d(s)}$ oscillation. Its oscillation period is determined by the mass difference which is expressed as (see Eq. (6.165))

$$|\Delta m_{d(s)}| \approx 2|M_{12}| = \frac{G_F^2 m_W^2}{6\pi^2} m_B B_B f_B^2 \eta_B S\left(\frac{m_t^2}{m_W^2}\right) |V_{tb} V_{td(s)}^*|^2 \tag{6.47}$$

where η_B is a QCD correction factor and $B_{B_{d(s)}} f_{B_{d(s)}}$ are the bag parameter (see Eq. (6.150)) and the decay constant. Using measured values

$$\Delta m_d = 0.507 \pm 0.005\ \mathrm{ps}^{-1} \tag{6.48a}$$

$$\Delta m_s = 17.77 \pm 0.10_{\text{stat}} \pm 0.07_{\text{syst}}\ \mathrm{ps}^{-1} \tag{6.48b}$$

together with theoretical values for $\eta_B = 0.55$, $f_{B_d}\sqrt{B_{B_d}} = (216 \pm 9 \pm 13)$ MeV, $f_{B_s}\sqrt{B_s} = (275 \pm 7 \pm 15$ MeV [279] and assuming $|V_{tb}| = 1$, one obtains [11]

$$|V_{td}| = (8.4 \pm 0.6) \times 10^{-3}\,, \quad |V_{ts}| = (38.7 \pm 2.1) \times 10^{-3} \tag{6.49}$$

The uncertainties are dominated by the lattice QCD calculation. Some of them are reduced by taking ratios $\xi = (f_{B_s}\sqrt{B_{B_s}})/(f_{B_d}\sqrt{B_{B_d}}) = 1.243 \pm 0.021 \pm 0.021$, and therefore the constraint on $|V_{td}/V_{ts}|$ from $\Delta m_d/\Delta m_s$ is more reliable theoretically. These provide an improved constraint

$$\frac{|V_{td}|}{|V_{ts}|} = 0.211 \pm 0.001(\text{exp}) \pm 0.005(\text{lattice}) \tag{6.50}$$

In summary, numerical values of the matrix elements are given by

$$\begin{bmatrix} 0.97425 \pm 0.00022 & 0.2252 \pm 0.0009 & 0.0039 \pm 0.00044 \\ -0.230 \pm 0.011 & 1.023 \pm 0.036 & 0.0415 \pm 0.0007 \\ 0.0084 \pm 0.0006 & -0.0387 \pm 0.0023 & 0.9991 \pm 0.0001 \end{bmatrix} \tag{6.51}$$

$$s_{12} \approx |V_{us}| = 0.220 \pm 0.0018\,, \tag{6.52}$$

$$s_{23} \approx |V_{cb}| = 0.0415 \pm 0.0007\,, \tag{6.53}$$

$$s_{13} = |V_{ub}| = 0.0039 \pm 0.00044 \tag{6.54}$$

and no restriction for δ.

6.3 The Unitarity Triangle

By the unitarity condition of the CKM matrix, we have

$$V_{ub}^* V_{ud} + V_{cb}^* V_{cd} + V_{tb}^* V_{td} = 0 \tag{6.55a}$$

which makes a triangle in the complex plane (Figure 6.3a) and is referred to as the unitarity triangle. Dividing by $V_{cb}^* V_{cd}$,

$$1 + \frac{V_{ub}^* V_{ud}}{V_{cb}^* V_{cd}} + \frac{V_{tb}^* V_{td}}{V_{cb}^* V_{cd}} = 0 \tag{6.55b}$$

and using Eq. (6.16), we obtain the unitarity triangle expressed as in Figure 6.3b. This is a triangle which has the base length 1 and the side lengths $|\overline{\rho} + i\overline{\eta}|$ and $|1 - \overline{\rho} - i\overline{\eta}|$. The angles are given by

$$\begin{aligned}
\phi_1 &= \beta^{6)} = \arg\left(-\frac{V_{cd} V_{cb}^*}{V_{td} V_{tb}^*}\right) \\
\phi_2 &= \alpha = \arg\left(-\frac{V_{td} V_{tb}^*}{V_{ud} V_{ub}^*}\right) \\
\phi_3 &= \gamma = \arg\left(-\frac{V_{ud} V_{ub}^*}{V_{cd} V_{cb}^*}\right)
\end{aligned} \tag{6.56}$$

Note that when parametrization or phases of the CKM matrices are changed, the unitarity triangles rotate in the complex plane without changing the shape and size of the triangle. Experimentally measurable quantities, therefore, are related to angles and sides of the triangles. The CP problem in the Standard Model reduces to determining the size and shape of the unitarity triangle. In addition to the one we just described, there are a total of six triangles corresponding to three combinations

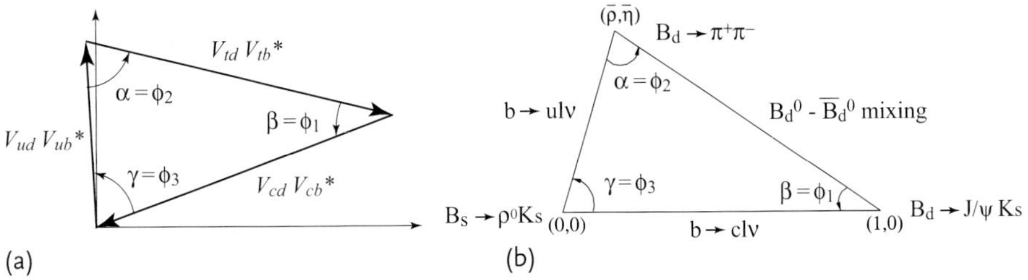

Figure 6.3 The unitarity triangle. (a) Definition of angles. The triangle rotates in the complex plane depending on the phase convention. (b) The phases as well as the lengths are fixed by taking $V_{cd} V_{cb}^*$ as a reference and dividing others by it. The figure also lists a few representative decay modes that can determine the angles and sides (see text).

6) It is confusing to define two sets of variables, but both nomenclatures ϕ_is and α, β, γ appear in the literature. We use ϕ_1, ϕ_2, ϕ_3 unless otherwise noted.

of the two rows and two columns, but they almost collapse as one side is much shorter than the other two. Only two triangles, one given above and the other made from

$$V_{td}^* V_{ud} + V_{ts}^* V_{us} + V_{tb}^* V_{ub} = 0 \tag{6.57}$$

have three sides comparable in length so that the angles (relative phases) are large, giving a large asymmetry in CP odd observables. The triangle given by Eq. (6.57) is useful under some special circumstance. All six triangles, however, have the same area.

Problem 6.1

Show that the area of the triangle in Figure 6.3a is $(1/2)\text{Im}[V_{ud} V_{ub}^* V_{cb} V_{cd}^*]$.

The fact that observables should be independent of the way the CKM matrix is expressed or the choice of the phases can be summarized in the following statements.

1. Absolute values of the CKM matrix elements are physical and do not depend on the parametrization.
2. Magnitude of the CP violation effect is proportional to the Jarlskog parameter [280, 281] J defined by

$$\text{Im}\left[V_{im} V_{jn} V_{jm}^* V_{in}^*\right] = J \sum_{s,t=1}^{3} \epsilon_{ijs}\epsilon_{mnt} \tag{6.58}$$

We already determined the modulus of V_{ij} experimentally. The Jarlskog parameter J has magnitude twice the area of the unitarity triangle. Adopting the parametrization in Eq. (6.16), it is expressed as

$$|J| = c_{12} c_{13}^2 c_{23} s_{12} s_{13} s_{23} \sin\delta \tag{6.59}$$

Therefore, for the CP violation to exist, any angle or any phase should not vanish. We also argued that the quarks should not be mass degenerate to produce CP violation by the CKM matrix. Denoting $[A, B] = AB - BA$, the following commutator made of the mass matrices introduced by Eq. (6.4b) turned out to be a convenient CP violation indicator [280, 281]

$$\begin{aligned}\left|\text{Im}\{\det[M^D M^{D\dagger}, M^U M^{U\dagger}]\}\right| = \Big| J \cdot \left(m_t^2 - m_c^2\right)\left(m_c^2 - m_u^2\right)\left(m_u^2 - m_t^2\right) \\ \times \left(m_b^2 - m_s^2\right)\left(m_s^2 - m_d^2\right)\left(m_d^2 - m_b^2\right)\Big|\end{aligned} \tag{6.60}$$

All the necessary conditions for having the CP violation can be combined in a single equation

$$\text{Im}\left\{\det\left[M^D M^{D\dagger}, M^U M^{U\dagger}\right]\right\} \neq 0 \tag{6.61}$$

The size of the CP violation can be estimated as

$$|J| = c_{12}c_{13}^2c_{23}s_{12}s_{13}s_{23}\sin\delta \sim A^2\lambda^6\eta \sim 10^{-5} \tag{6.62}$$

which is much smaller than mathematical maximum $1/6\sqrt{3} \simeq 0.1$. The experimental value is given by

$$J = 2.91^{+0.19}_{-0.11} \times 10^{-5} \quad [3] \tag{6.63}$$

Smallness of the CP violation effect can be ascribed to that of J. Let us estimate the size of asymmetry of CP odd variables. Its size is of the order

$$\text{degree of CP violation} \sim \text{asymmetry} \times \text{branching ratio} \sim |J| \lesssim 10^{-4} \tag{6.64}$$

which means that processes with large branching ratio have small asymmetry and those with large asymmetry have small branching ratios.

6.4 Formalism of the Two Neutral Meson System

6.4.1 Mass Matrix, Mixing and CP Parameters

The mechanism for the CP violation in the B meson decays has many common features to that of the K meson, and it can be treated in a similar formalism as the one we studied in Vol. 1, Chapt. 16. Their eigenstates in the weak decays are mixtures of the neutral B-meson and its antiparticle, and are characterized by their mass and width which are different from those of the strong interaction. A major difference to distinguish the B from the K mesons is that the mass of the B meson ($\sim$ 5 GeV) is much larger than that of the K meson ($\sim$ 0.5 GeV). This means that the phase space volume of the B meson decay is much larger and so is the decay width. Namely, the B meson lifetime is much shorter. The mass of the two neutral B mesons is nearly equal open to many channels which makes their decay width nearly equal. Namely,

$$\begin{aligned} &\Gamma_B \gg \Gamma_K \quad \text{or} \quad \tau_B \approx 10^{-12}\,\text{s} \ll \tau_K = 10^{-10} \sim 10^{-8}\,\text{s} \\ &\Gamma_{\text{long}} \approx \Gamma_{\text{short}} \gg \Delta\Gamma = |\Gamma_{\text{short}} - \Gamma_{\text{long}}| \end{aligned} \tag{6.65}$$

where suffixes "long", "short" stand for a long or short living meson of the two. Because of this, it is difficult to separate the two B mesons (B_{long}, B_{short}) and measure corresponding CP violation parameters ϵ, ϵ' directly. Therefore, phenomenological treatment of the B meson system is very different from that of the K meson system. Let us define the B eigenstates by their mass and denote the heavy one as B_H and the light one as B_{L}. Assuming CPT invariance, the two states can be expressed as

$$|B_{\mathrm{L}}\rangle = p|B^0\rangle + q|\overline{B}^0\rangle = \frac{1}{\sqrt{1+|\tilde{\epsilon}|^2}}\left[(1+\tilde{\epsilon})|B^0\rangle + (1-\tilde{\epsilon})|\overline{B}^0\rangle\right]$$
$$|B_H\rangle = p|B^0\rangle - q|\overline{B}^0\rangle = \frac{1}{\sqrt{1+|\tilde{\epsilon}|^2}}\left[(1+\tilde{\epsilon})|B^0\rangle - (1-\tilde{\epsilon})|\overline{B}^0\rangle\right]$$
$$|p|^2 + |q|^2 = 1 \tag{6.66}$$

where we have defined $B_q = \overline{b}q, \overline{B}_q = b\overline{q}(q = d, s)$. The definition was made to be consistent with $K^0 = d\overline{s}, \overline{K}^0 = s\overline{d}$. The variables $\tilde{\epsilon}, p, q$ should really be denoted as $\tilde{\epsilon}_B, p_B, q_B$, but we omit the suffix whenever there is no confusion. Just to remind the reader that $\tilde{\epsilon}$ is the CP violating parameter in mixing, because if one uses the $CP = \pm$ eigenstate $|B_\pm\rangle = (|B_0\rangle \pm |\overline{B}^0\rangle)/\sqrt{2}$, B_{L}, B_H can be expressed as

$$|B_{\mathrm{L}}\rangle = \frac{1}{\sqrt{1+|\tilde{\epsilon}|^2}}[|B_+\rangle + \tilde{\epsilon}|B_-\rangle], \quad |B_H\rangle = \frac{1}{\sqrt{1+|\tilde{\epsilon}|^2}}[|B_-\rangle + \tilde{\epsilon}|B_+\rangle] \tag{6.67}$$

We also define

$$\mathrm{CP}|B^0\rangle = e^{i\xi_B}|\overline{B}^0\rangle, \quad \mathrm{CP}|\overline{B}^0\rangle = e^{-i\xi_B}|B^0\rangle \tag{6.68}$$

and under normal circumstances we set $\xi_B = 0$.[7] Writing an arbitrary B state as

$$|\Psi(t)\rangle = \alpha(t)|B_0\rangle + \beta(t)|\overline{B}^0\rangle \tag{6.70}$$

and the Schrödinger equation as

$$i\frac{\partial}{\partial t}\begin{bmatrix}\alpha\\ \beta\end{bmatrix} = \begin{bmatrix}\Lambda_{11} & \Lambda_{12}\\ \Lambda_{21} & \Lambda_{22}\end{bmatrix}\begin{bmatrix}\alpha\\ \beta\end{bmatrix} \tag{6.71}$$

where the mass matrix Λ is defined by[8] (see Vol. 1, Eq. (16.11) and Appendix H)

$$\Lambda_{ij} = M_{ij} - i\Gamma_{ij}/2 \tag{6.72a}$$

$$M_{ij} = M_i\delta_{ij} + \langle i|H_{\mathrm{W}}|j\rangle + P\sum_{n\neq j}\frac{\langle i|H_{\mathrm{W}}|n\rangle\langle n|H_{\mathrm{W}}|j\rangle}{M_j - E_n} \tag{6.72b}$$

$$\Gamma_{ij} = \sum_n \rho_n\langle i|H_{\mathrm{W}}|n\rangle\langle n|H_{\mathrm{W}}|j\rangle|_{E_n=m_j} \tag{6.72c}$$

where i, j $(i, j = 1, 2)$ stands for $B^0, \overline{B}^0$ or B_{L}, B_H depending on the situation and ρ_n is a phase space factor.

7) The K and B mesons have a quantum number strangeness S and B (B-ness) which conserves in the strong interaction. Hence, the equation of motion does not change if the states are phase transformed

$$|B\rangle \to e^{-i\alpha}|B\rangle, \quad |\overline{B}\rangle \to e^{+i\alpha}|\overline{B}\rangle \tag{6.69}$$

8) This mass matrix Λ should not be confused with the mass matrices M^U, M^D.

Decay amplitudes of particle i to final state f and their antiparticle (CP conjugate) states $\overline{i} \to \overline{f}$ can be written as (see Vol. 1, Sect. 16.2.1)

$$\langle f|H_W|i\rangle = A(i \to f)e^{i\delta}, \quad \langle \overline{f}|H_W|\overline{i}\rangle = A(\overline{i} \to \overline{f})e^{i\delta} \tag{6.73}$$

where δ is the strong scattering phase in the final state. The symmetry requirements constrain the amplitudes:

$$\begin{aligned}
\text{CPT}: &\quad A(\overline{i} \to \overline{f}) \equiv \overline{A}_{\overline{f}} = A^*(i \to f) \\
\text{T}: &\quad \begin{cases} A(i \to f) \equiv A_f = A^*(i \to f) \\ A(\overline{i} \to \overline{f}) \equiv \overline{A}_{\overline{f}} = A^*(\overline{i} \to \overline{f}) \end{cases} \\
\text{CP}: &\quad A(\overline{i} \to \overline{f}) = A(i \to f)
\end{aligned} \tag{6.74}$$

Applying the constraints Eq. (6.74) to Eq. (6.72), we have

$$\begin{aligned}
\text{CPT}: &\quad \Lambda_{11} = \Lambda_{22} \\
\text{T}: &\quad |\Lambda_{12}| = |\Lambda_{21}| \quad \text{or} \quad \text{Im}(M_{12}^* \Gamma_{12}) = 0 \\
\text{CP}: &\quad |\Lambda_{12}| = |\Lambda_{21}| \quad \text{and} \quad \Lambda_{11} = \Lambda_{22}
\end{aligned} \tag{6.75}$$

We assume the CPT invariance in the following discussions. Unlike the K mesons, we do not know which of the $B_{H,\text{L}}$ corresponds to the long lived B_{long} or short lived B_{short}.[9] The mass eigenvalues $\lambda_{H,\text{L}}$ can be obtained by solving the secular equation and can be expressed as

$$\lambda_{\text{L},H} = M_{\text{L},H} - i\frac{\Gamma_{\text{L},H}}{2} = \frac{\Lambda_{11} + \Lambda_{22}}{2} \pm \sqrt{\Lambda_{12}\Lambda_{21}} \tag{6.76}$$

The mixing parameter ε is expressed as

$$\tilde{\epsilon} = \frac{p - q}{p + q} \tag{6.77}$$

$$\frac{q}{p} = \frac{1 - \tilde{\epsilon}}{1 + \tilde{\epsilon}} = \sqrt{\frac{\Lambda_{21}}{\Lambda_{12}}} = \sqrt{\frac{M_{12}^* - i\left(\Gamma_{12}^*/2\right)}{M_{12} - i(\Gamma_{12}/2)}} \tag{6.78}$$

Denoting the average of the mass and the width as M, Γ and difference as $\Delta m, \Delta\Gamma$, we have the following relations (see Vol. 1, Eq. (16.81)):

$$M = (M_H + M_{\text{L}})/2, \quad \Gamma = (\Gamma_H + \Gamma_{\text{L}})/2 \tag{6.79a}$$

$$\Delta \equiv \lambda_H - \lambda_{\text{L}} = \Delta m - i\frac{\Delta\Gamma}{2} = \sqrt{\Lambda_{12}\Lambda_{21}} \tag{6.79b}$$

$$\Delta m = M_H - M_{\text{L}} = 2\text{Re}\left[\left(M_{12} - i\frac{\Gamma_{12}}{2}\right)\left(M_{12}^* - i\frac{\Gamma_{12}^*}{2}\right)\right]^{1/2} \tag{6.79c}$$

9) For the K meson, $K_H = K_{\text{long}}$, $K_{\text{L}} = K_{\text{short}}$.

$$\Delta\Gamma^{10)} = \Gamma_H - \Gamma_L = -4\mathrm{Im}\left[\left(M_{12} - i\frac{\Gamma_{12}}{2}\right)\left(M_{12}^* - i\frac{\Gamma_{12}^*}{2}\right)\right]^{1/2} \tag{6.79d}$$

$$\Lambda_{12} = -\frac{p}{2q}\Delta\,, \quad \Lambda_{21} = -\frac{q}{2p}\Delta \tag{6.79e}$$

$$\frac{\Lambda_{21} - \Lambda_{12}}{2\Delta} = \frac{1}{4}\left(\frac{p}{q} - \frac{q}{p}\right) = \frac{\tilde{\epsilon}}{1 - \tilde{\epsilon}^2} \tag{6.79f}$$

6.4.2 CP Parameters of the K Meson

We already discussed details of the CP violation phenomena of the K meson in Vol. 1, Chapt. 16. Here, we restrict ourselves to defining parameters that become necessary in discussing the CP phenomena in the B meson system. The three important parameters of the CP violation in the K meson system that were directly observed in experiments are

$$\eta_{+-} \equiv \frac{A(K_L \to \pi^+\pi^-)}{A(K_S \to \pi^+\pi^-)} = |\eta_{+-}|\exp(i\phi_{+-}) = \frac{\epsilon_K + \epsilon'_K}{1 + \omega} \simeq \epsilon_K + \epsilon'_K \tag{6.80a}$$

$$\eta_{00} \equiv \frac{A(K_L \to \pi^0\pi^0)}{A(K_S \to \pi^0\pi^0)} = |\eta_{00}|\exp(i\phi_{00}) = \frac{\epsilon_K - 2\epsilon'_K}{1 - 2\omega} \simeq \epsilon_K - 2\epsilon'_K \tag{6.80b}$$

$$A_L \equiv \frac{\Gamma(K_L \to \pi^-\bar{l}\nu_l) - \Gamma(K_L \to \pi^+ l\bar{\nu}_l)}{\Gamma(K_L \to \pi^-\bar{l}\nu_l) + \Gamma(K_L \to \pi^+ l\bar{\nu}_l)} = \frac{1 - |q/p|^2}{1 + |q/p|^2} \simeq \frac{2\mathrm{Re}[\tilde{\epsilon}_K]}{1 + |\tilde{\epsilon}_K|^2} \tag{6.80c}$$

Their world values are [3]

$$\begin{aligned} |\eta_{+-}| &= (2.232 \pm 0.011) \times 10^{-3} \\ |\eta_{00}| &= (2.221 \pm 0.011) \times 10^{-3} \\ \phi_{+-} &= (43.4 \pm 0.7)^\circ \\ \Delta\phi &= \phi_{00} - \phi_{+-} = (0.2 \pm 0.4)^\circ \\ A_L &= (3.32 \pm 0.06) \times 10^{-3} \end{aligned} \tag{6.81}$$

Theoretically, the parameters ϵ_K, ϵ'_K are defined by

$$\epsilon_K \equiv \frac{A[K_L \to \pi\pi(I = 0)]}{A[K_S \to \pi\pi(I = 0)]} \simeq \tilde{\epsilon}_K + i\frac{\mathrm{Im}A_0}{\mathrm{Re}A_0} \tag{6.82a}$$

$$\epsilon_{K2} \equiv \frac{1}{\sqrt{2}}\frac{A[K_L \to \pi\pi(I = 2)]}{A[K_S \to \pi\pi(I = 0)]} \equiv \epsilon'_K + \epsilon_K\omega \tag{6.82b}$$

10) Δm is, by definition, positive, but the sign of $\Delta\Gamma$ is to be determined experimentally. For the K meson, $\Delta\Gamma < 0$ and D meson $\Delta\Gamma > 0$ are established. The Standard Model predicts $\Delta\Gamma < 0$ for B_d and B_s. Therefore, we follow the same indexing as the K meson.

$$\epsilon'_K \simeq \frac{1}{\sqrt{2}}\frac{A[K_2 \to \pi\pi(I=2)]}{A[K_1 \to \pi\pi(I=0)]} - i\frac{\mathrm{Im}A_0}{\mathrm{Re}A_0}\omega = i\omega\left[\frac{\mathrm{Im}A_2}{\mathrm{Re}A_2} - \frac{\mathrm{Im}A_0}{\mathrm{Re}A_0}\right] \quad (6.82c)$$

$$\omega \equiv \frac{1}{\sqrt{2}}\frac{A[K_S \to \pi\pi(I=2)]}{A[K_S \to \pi\pi(I=0)]} \simeq \frac{1}{\sqrt{2}}\frac{\mathrm{Re}A_2}{\mathrm{Re}A_0}e^{i(\delta_2-\delta_0)} \quad (6.82d)$$

Note that $\tilde{\epsilon_K}$ is a phase dependent mixing parameter, but ϵ_K, ϵ'_K are phase independent observables. A_0, A_2 are decay amplitudes defined by

$$A_{0,2} = A[K^0 \to \pi\pi(I=0,2)] \quad (6.83)$$

ω is a measure of $\Delta I = 1/2$ rule violation and δ_I is the $\pi\pi$ scattering phase shifts with isospin I. An important parameter not directly related to the CP violation is the superweak phase ϕ_{SW} which is defined by

$$\phi_{SW} \equiv \tan^{-1}\frac{2\Delta m_K}{\Delta\Gamma_K} = \frac{2(m_{K_L} - m_{K_S})}{\Gamma_{K_S} - \Gamma_{K_L}} \simeq 43.51 \pm 0.05^\circ \quad (6.84)$$

The experimental values show

$$|\eta_{+-}| \simeq |\eta_{00}|, \quad \phi_{+-} \simeq \phi_{00} \simeq \phi_{SW} \quad (6.85)$$

which means

$$\epsilon_K \simeq \eta_{+-} \simeq \eta_{00}, \quad |\epsilon'_K| \ll |\epsilon_K| \quad (6.86)$$

The numerical value of the ϵ_K is

$$|\epsilon_K| = (2.228 \pm 0.011) \times 10^{-3} \quad (6.87)$$

The CPT invariance and unitarity restricts the phase of ϵ_K approximately to

$$\phi_{\epsilon_K} = \arg\epsilon_K \simeq \phi_{SW} \quad (6.88)$$

and from Eq. (6.82c)

$$\phi_{\epsilon'_K} = \delta_2 - \delta_0 + \frac{\pi}{2} \simeq 42.3 \pm 1.5^\circ \quad (6.89)$$

The real part of the ϵ'_K relative to ϵ_K can be determined directly by measuring the double ratio

$$R = \frac{\Gamma(K_L \to \pi^+\pi^-)/\Gamma(K_S \to \pi^+\pi^-)}{\Gamma(K_L \to \pi^0\pi^0)/\Gamma(K_S \to \pi^0\pi^0)} = \left|\frac{\eta_{+-}}{\eta_{00}}\right|^2 \simeq 1 + 6\mathrm{Re}\left(\frac{\epsilon'_K}{\epsilon_K}\right)$$

$$\mathrm{Re}\left(\frac{\epsilon'_K}{\epsilon_K}\right) \simeq \frac{\epsilon'_K}{\epsilon_K} \simeq (1.65 \pm 0.26) \times 10^{-3} \quad (6.90)$$

6.4.3
Mixing in the B^0–$\overline{B}^0$ System

Mixing of a two neutral meson system is a phenomenon independent of the CP violation, but is closely related. For this reason, we first discuss a mixing formalism and introduce (in addition to ϵ, ϵ') the third CP violation parameter λ_f which plays an important role in the B meson system.[11] Note that we discuss the B meson in the following, but the formalism is general and can be applied equally to the K and D meson system. To describe time evolution of the $B^0, \overline{B}^0$ system, we first expand them in terms of B_H, B_L and insert the time evolution factor and transform back to $B^0, \overline{B}^0$. Using Eq. (6.66), we have

$$\begin{aligned} |B^0(t)\rangle &= \frac{1}{2p}\left[|B_L\rangle e^{-i\lambda_L t} + |B_H\rangle e^{-i\lambda_H t}\right] \\ &= g_+(t)|B^0\rangle + (q/p)g_-(t)|\overline{B}^0\rangle \end{aligned} \tag{6.93a}$$

$$\begin{aligned} |\overline{B}^0(t)\rangle &= \frac{1}{2q}\left[|B_L\rangle e^{-i\lambda_L t} - |B_H\rangle e^{-i\lambda_H t}\right] \\ &= (p/q)g_-(t)|B^0\rangle + g_+(t)|\overline{B}^0\rangle \end{aligned} \tag{6.93b}$$

where

$$\lambda_{L,H} = M_{L,H} - i\frac{\Gamma_{L,H}}{2} = M - i\frac{\Gamma}{2} \mp \frac{1}{2}\left(\Delta m + i\frac{\Delta\Gamma}{2}\right) \tag{6.94a}$$

$$g_\pm = \frac{1}{2}\left[e^{-i\lambda_L t} \pm e^{-i\lambda_H t}\right] \tag{6.94b}$$

The following formulas will be convenient to remember for later use.

$$|g_\pm(t)|^2 = \frac{e^{-\Gamma t}}{2}\left[\cosh\left(\frac{\Delta\Gamma}{2}t\right) \pm \cos\Delta m t\right] \tag{6.95a}$$

$$g_+^*(t)g_-(t) = \frac{e^{-\Gamma t}}{2}\left[-\sinh\left(\frac{\Delta\Gamma t}{2}\right) + i\sin\Delta m t\right] \tag{6.95b}$$

We define decay amplitudes:

$$A_f \equiv \langle f|H|B^0\rangle, \quad \overline{A}_f \equiv \langle f|H|\overline{B}^0\rangle \tag{6.96a}$$

$$A_{\overline{f}} \equiv \langle \overline{f}|H|B^0\rangle, \quad \overline{A}_{\overline{f}} \equiv \langle \overline{f}|H|\overline{B}^0\rangle \tag{6.96b}$$

Then, time variation of decay rates to a final state f are obtained by multiplying Eq. (6.93) by $\langle f|, \langle\overline{f}|$ from the left. For instance, the time dependence of the decay

11) Actually, using λ_f defined by Eq. (6.97b), the CP violating amplitude in the K decay can be expressed as

$$\eta_{+-} = \frac{1-\lambda_{\pi^+\pi^-}}{1+\lambda_{\pi^+\pi^-}} \simeq \epsilon_K + \epsilon'_K, \quad \eta_{00} = \frac{1-\lambda_{\pi^0\pi^0}}{1+\lambda_{\pi^0\pi^0}} \simeq \epsilon_K - 2\epsilon'_K \tag{6.91}$$

The smallness of $|\eta_{+-}|, |\eta_{00}|$ allows us to express ϵ_K, ϵ'_K as

$$\epsilon_K \simeq \frac{1}{2}\left(1-\lambda_{\pi\pi(I=0)}\right), \quad \epsilon'_K \simeq \frac{1}{6}\left(\lambda_{\pi^0\pi^0} - \lambda_{\pi^+\pi^-}\right) \tag{6.92}$$

Namely, the λ_f would have surfaced if we had discussed the imaginary part of ϵ_K and ϵ'_K.

rate $\Gamma(B^0 \to f)$ is written as

$$\Gamma(B^0 \to f; t) = \left| g_+(t)\langle f|H|B^0\rangle + \frac{q}{p} g_-(t)\langle f|H|\overline{B}^0\rangle \right|^2$$
$$= |A_f|^2 \left| g_+(t) + \lambda_f g_-(t) \right|^2 \tag{6.97a}$$

$$\lambda_f \equiv \frac{q}{p}\frac{\overline{A}_f}{A_f} = \frac{q}{p}\frac{\langle f|H|\overline{B}^0\rangle}{\langle f|H|B^0\rangle} \tag{6.97b}$$

Note that the parameter λ_f is an important CP violation parameter to remember. It plays a central role in the B meson sector with final states $|\overline{f}\rangle = |f\rangle$. Even if the CP violation does not appear in mixing ($|q/p| = 1$) or in decays ($|\overline{A}_f/A_f| = 1$), it can manifest itself through $\text{Im}\lambda_f \neq 0$ (see Section 6.6.3). Similarly,

$$\Gamma(\overline{B}^0 \to f; t) = \left| g_+(t)\langle f|H|\overline{B}^0\rangle + \frac{p}{q} g_-(t)\langle f|H|B^0\rangle \right|^2$$
$$= |\overline{A}_f|^2 \left| g_+(t) + \lambda_f^{-1} g_-(t) \right|^2 \tag{6.98}$$

$$\Gamma(B^0 \to \overline{f}; t) = \left| g_+(t)\langle \overline{f}|H|B^0\rangle + \frac{q}{p} g_-(t)\langle \overline{f}|H|\overline{B}^0\rangle \right|^2$$
$$= |A_{\overline{f}}|^2 \left| g_+(t) + \lambda_{\overline{f}} g_-(t) \right|^2 \tag{6.99a}$$

$$\Gamma(\overline{B}^0 \to \overline{f}; t) = \left| g_+(t)\langle \overline{f}|H|\overline{B}^0\rangle + \frac{p}{q} g_-(t)\langle \overline{f}|H|B^0\rangle \right|^2$$
$$= |\overline{A}_{\overline{f}}|^2 \left| g_+(t) + \lambda_{\overline{f}}^{-1} g_-(t) \right|^2 \tag{6.99b}$$

$$\lambda_{\overline{f}} \equiv \frac{q}{p}\frac{\overline{A}_{\overline{f}}}{A_{\overline{f}}} = \frac{q}{p}\frac{\langle \overline{f}|H|\overline{B}^0\rangle}{\langle \overline{f}|H|B^0\rangle} \tag{6.99c}$$

Multiplying $\langle B^0|$ or $\langle \overline{B}^0|$ with Eq. (6.93) from the left, the probability for the state B^0 at $t = 0$ to remain as B^0 or to convert to $\overline{B}^0$ becomes

$$P(B^0 \to B^0; t) = |\langle B^0|B^0(t)\rangle|^2 = |g_+(t)|^2 \tag{6.100a}$$

$$= \frac{e^{-\Gamma t}}{2}\left[\cosh(\Delta\Gamma t/2) + \cos \Delta m t\right] \tag{6.100b}$$

$$\xrightarrow{\Delta\Gamma/\Gamma \to 0} \frac{e^{-\Gamma t}}{2}\left[1 + \cos \Delta m t\right] \tag{6.100c}$$

$$P(B^0 \to \overline{B}^0; t) = |\langle \overline{B}^0|B^0(t)\rangle|^2 = |q/p|^2 |g_-(t)|^2 \tag{6.100d}$$

$$= \left|\frac{q}{p}\right|^2 \frac{e^{-\Gamma t}}{2}\left[\cosh(\Delta\Gamma t/2) - \cos \Delta m t\right] \tag{6.100e}$$

$$\xrightarrow{\Delta\Gamma/\Gamma\to 0} \left|\frac{q}{p}\right|^2 \frac{e^{-\Gamma t}}{2}[1-\cos\Delta m t] \tag{6.100f}$$

Similarly,

$$P(\overline{B}^0 \to \overline{B}^0; t) \simeq \frac{e^{-\Gamma t}}{2}[1+\cos\Delta m t] \tag{6.101a}$$

$$P(\overline{B}^0 \to B^0; t) \simeq \left|\frac{p}{q}\right|^2 \frac{e^{-\Gamma t}}{2}[1-\cos\Delta m t] \tag{6.101b}$$

In leptonic decays, $b, \overline{b}$ emit leptons of opposite sign:

$$\begin{aligned} &b \to l^-\overline{\nu}_l q\,, \quad \overline{b} \to l^+\nu_l\overline{q} \quad (q = u, c)\\ &b \nrightarrow l^+\nu_l q\,, \quad \overline{b} \nrightarrow l^-\overline{\nu}_l\overline{q} \end{aligned} \tag{6.102}$$

and in hadronic decays

$$\begin{aligned} &b \to c + W^-, \quad \overline{b} \to \overline{c} + W^+\\ &\qquad\quad \hookrightarrow s\overline{u}\,(K^-) \qquad \hookrightarrow \overline{s}u\,(K^+) \end{aligned} \tag{6.103}$$

Therefore, by looking at the sign of the decay leptons or which of $K^\pm$ is emitted, one can identify $B^0(\overline{b}q)$ and $\overline{B}^0(b\overline{q})$. Putting $x \equiv \Delta m/\Gamma$, $y \equiv \Delta\Gamma/(2\Gamma)$ and rewriting

$$\begin{aligned} \Delta m t &= (\Delta m/\Gamma)\Gamma t \equiv x s\\ \Delta\Gamma t/2 &= y\Gamma t = y s \end{aligned} \tag{6.104}$$

where $s = \Gamma t = t/\tau$ is the time measured in units of the mean lifetime. One sees that the B^0–$\overline{B}^0$ mixing oscillation has period τ_B/x. As the lifetime of the b mesons is short ($\sim 10^{-12}$ s), measurement of the decay time was difficult in the early days.

When the decay points are not measured, one has to integrate over time. The total probabilities are

$$P(B^0 \to B^0) = \int dt e^{-\Gamma t}\frac{\cosh ys + \cos xs}{2} = \frac{1}{2\Gamma}\left(\frac{2+x^2-y^2}{(1+x^2)(1-y^2)}\right) \tag{6.105a}$$

$$P(\overline{B}^0 \to \overline{B}^0) = P(B^0 \to B^0) \tag{6.105b}$$

$$P(B^0 \to \overline{B}^0) = \left|\frac{q}{p}\right|^2 \int dt e^{-\Gamma t}\frac{\cosh ys - \cos xs}{2} = \left|\frac{q}{p}\right|^2 \frac{1}{2\Gamma}\left(\frac{x^2+y^2}{(1+x^2)(1-y^2)}\right) \tag{6.105c}$$

$$P(\overline{B}^0 \to B^0) = \left|\frac{p}{q}\right|^4 P(B^0 \to \overline{B}^0) = \left|\frac{p}{q}\right|^2 \frac{1}{2\Gamma}\left(\frac{x^2+y^2}{(1+x^2)(1-y^2)}\right) \tag{6.105d}$$

At this point, we define the ratio of mixing to nonmixing by

$$r \equiv \frac{P(B^0 \to \overline{B}^0)}{P(B^0 \to B^0)} = \frac{\Gamma(B^0 \to \overline{B}^0 \to l^-)}{\Gamma(B^0 \to l^+)} = \left|\frac{q}{p}\right|^2 \left(\frac{x^2+y^2}{2+x^2-y^2}\right) \tag{6.106a}$$

$$\overline{r} \equiv \frac{P(\overline{B}^0 \to B^0)}{P(\overline{B}^0 \to \overline{B}^0)} = \frac{\Gamma(\overline{B}^0 \to B^0 \to l^+)}{\Gamma(\overline{B}^0 \to l^-)} = \left|\frac{p}{q}\right|^2 \left(\frac{x^2+y^2}{2+x^2-y^2}\right) \tag{6.106b}$$

The fractional mixing ratio to all is also used frequently:

$$\chi \equiv \frac{P(B^0 \to \overline{B}^0)}{P(B^0 \to all)} = \frac{r}{1+r} = \left|\frac{q}{p}\right|^2 \frac{x^2+y^2}{2(1+x^2)}$$
$$\overline{\chi} \equiv \frac{P(\overline{B}^0 \to B^0)}{P(\overline{B}^0 \to all)} = \frac{\overline{r}}{1+\overline{r}} = \left|\frac{p}{q}\right|^2 \frac{x^2+y^2}{2(1+x^2)} \tag{6.107}$$

Mass difference $\Delta m_d, \Delta m_s$: ARGUS was the first group to discover a large mixing and later it was confirmed by CLEO [282, 283]. The mixing was detected by observing

$$e^+ + e^- \to \Upsilon(4S) \to B^0\overline{B}^0 \to l_1 l_2 + X \tag{6.108}$$

$\Upsilon(4S)$ (see Figure 6.4) is the fourth level of the bottomonium and has $J^{PC} = 1^{--}$, $M = 10.5794 \pm 0.0012$ GeV, $\Gamma = 20.5 \pm 2.5$ MeV. Its main decay modes are $B_d^0\overline{B}_d^0$ (BR = 48.4%) and B^+B^- (BR = 51.6%). Note that there is no decay mode to $B_s^0\overline{B}_s^0$, and hence it is an ideal place to examine properties of $B_d^0\overline{B}_d^0$. When there is no mixing, the two leptons of the decay product have opposite sign, but they have the same sign when there is mixing. Assuming CP invariance,

$$r = \overline{r} = \frac{1}{2}(r+\overline{r}) = \frac{N(B^0B^0) + N(\overline{B}^0\overline{B}^0)}{N(B^0\overline{B}^0)} \tag{6.109}$$

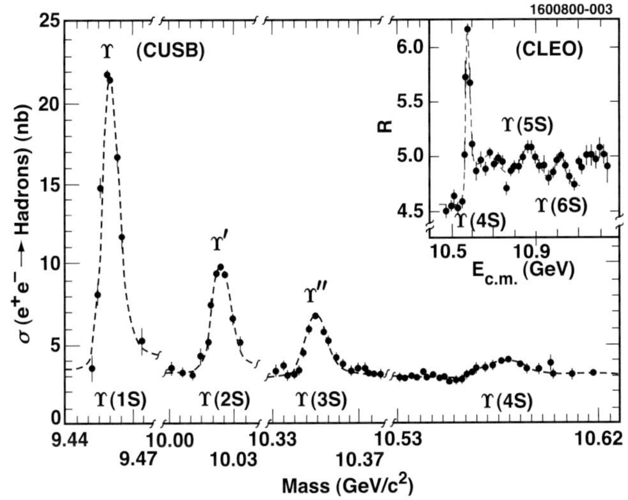

Figure 6.4 Bottomonium spectrum. Three narrow resonances and the broad 4S ($\Gamma \sim 20$ MeV) are seen [284].

The value to be obtained at $\Upsilon(4S)$ is $r = r_d$.[12]

In order to establish that the mixing has occurred, one has to verify that $B^0 B^0$ or $\overline{B}^0\overline{B}^0$ are produced. The most unambiguous sample can be obtained by reconstructing all the particles that are produced in the decay. The B decay can be identified, for example, by observing the following cascade decays.

$$B^0 \to D^{*-} + \pi^+ , \quad D^{*-} + \rho^+ , \quad D^{*-} + \pi^+\pi^+\pi^- , \quad D^{*-}l^+\nu_l$$

$$\begin{array}{l} D^{*-} \\ \quad \hookrightarrow \overline{D}^0 \pi^- \\ \qquad\quad \hookrightarrow K^+\pi^-,\ K^+\pi^-\pi^0,\ K^+\pi^+\pi^-\pi^-,\ K_S^0\pi^+\pi^- \end{array} \tag{6.110}$$

We show an event display of ARGUS in Figure 6.5. This is not the first event, though it was chosen because it identified the hitherto unconfirmed decay mode $b \to u\mu^-\overline{\nu}_\mu$, thus establishing the nonzero V_{ub}.

$$\begin{array}{l} \Upsilon(4S) \to \overline{B}_d^0 + B_d^0 \\ \qquad\qquad\quad \hookrightarrow \overline{B}_d^0 \\ \qquad\qquad\qquad\quad \hookrightarrow D^{*+}\rho^- \\ \qquad\qquad\qquad\qquad\quad \hookrightarrow \pi_\rho^-\pi^0 \\ \qquad\qquad\qquad\quad \hookrightarrow \pi^+ D^0 \\ \qquad\qquad\qquad\qquad\quad \hookrightarrow \pi^0 K^0 \\ \qquad\qquad\qquad\qquad\quad \hookrightarrow \pi_K^+\pi_K^- \\ \qquad\quad \hookrightarrow \pi^+\mu^-\overline{\nu}_\mu \end{array} \tag{6.111}$$

The lower indices attached to the decay products indicate their parent.

To determine the mixing parameter r_d quantitatively, we need to count the number of $l^\pm l^\pm$ from $B^0\overline{B}^0$. However, the $\Upsilon(4S)$ decays to B^+B^- as well as to $B^0\overline{B}^0$. So, we write decay probabilities to $B_d\overline{B}_d$, B^+B^- as p_0, p_+ and leptonic branching ratios of $B^0(B^+)$ as $b_0(b_+)$. Then, numbers of the leptons can be written as

$$\begin{aligned} N(l^+l^+) &= N(B^0B^0)b_0^2 , \quad N(l^-l^-) = N(\overline{B}^0\overline{B}^0)b_0^2 \\ N(l^+l^-) &= N(B^0\overline{B}^0)b_0^2 + N(B^+B^-)b_+^2 \end{aligned} \tag{6.112}$$

By defining a parameter to indicate the relative contribution of charged to neutral mesons

$$\lambda \equiv \frac{p_+}{p_0}\frac{b_+^2}{b_0^2} \tag{6.113}$$

the mixing parameter r_d can be determined from measured number N using the relation

$$r_d = \frac{\left[N(l^+l^+) + N(l^-l^-)\right](1+\lambda)}{N(l^+l^-) - \lambda\{N(l^+l^+) + N(l^-l^-)\}} \tag{6.114}$$

12) To be rigorous, one has to consider the correlation between B^0 and $\overline{B}^0$ because they are in a coherent state if produced via $\Upsilon(4S)$. However, use of Eq. (6.106) can be justified for the reason to be described later (see Eq. (6.216)).

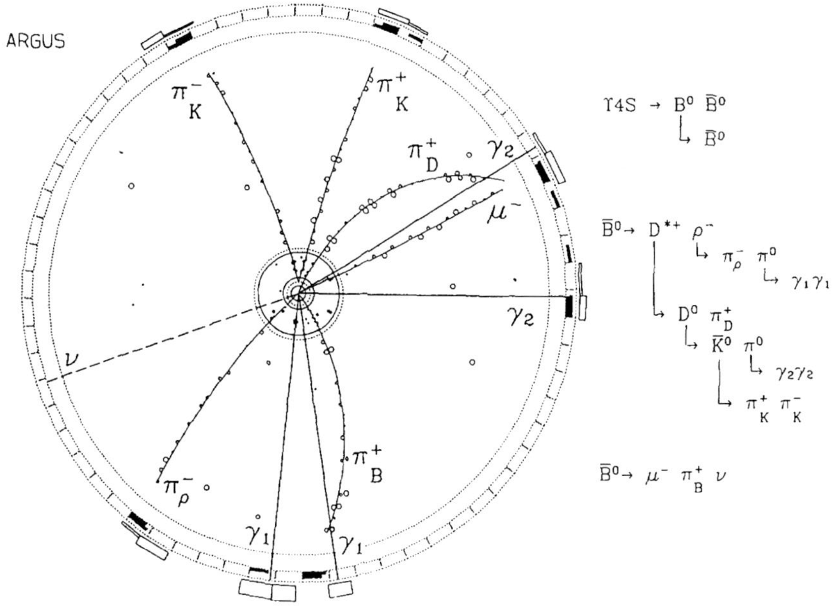

Figure 6.5 An event display to exhibit the B^0–$\overline{B}^0$ mixing. It is a fully reconstructed process which contains $b \to u\mu^-\overline{\nu}_\mu$. See Eq. (6.111) for the decay chain [286].

Problem 6.2

Derive Eq. (6.114).

The λ was an unknown parameter then,[13] but theoretically estimated to be $1 \sim 1.3$ (ARGUS chose $\lambda = 1.2$). Furthermore, the value of x_d is not very sensitive to the choice of λ. ARGUS obtained the value [282]

$$r_d = 0.22 \pm 0.09 \pm 0.04 \tag{6.116}$$

Nowadays, direct determination of Δm is available. By observing time variation of the oscillation in decay, much better values of x, r can be obtained using various tagging methods. We show an example of oscillation curves of $B_d^0 - \overline{B}_d^0$ obtained by Belle group at B-factory/KEK and $B_s^0 - \overline{B}_s^0$ by CDF at TEVATRON in Figure 6.6. The time variation of the oscillation was obtained by plotting the asymmetry of the same flavor (SF) events against the opposite flavor (OS) events as a function of time, namely,

$$A_{\text{ch}}(t) \equiv \frac{N_{\text{OF}}(t) - N_{\text{SF}}(t)}{N_{\text{OF}}(t) + N_{\text{SF}}(t)} \tag{6.117}$$

13) Present values are

$$\begin{aligned} p_0 &= 48.4 \pm 0.6\%, \quad p_+ = 51.6 \pm 0.6\% \\ b_+ &= 10.99 \pm 0.26\%, \quad b_0 = 10.33 \pm 0.28\% \end{aligned} \tag{6.115}$$

giving $\lambda = 1.21$.

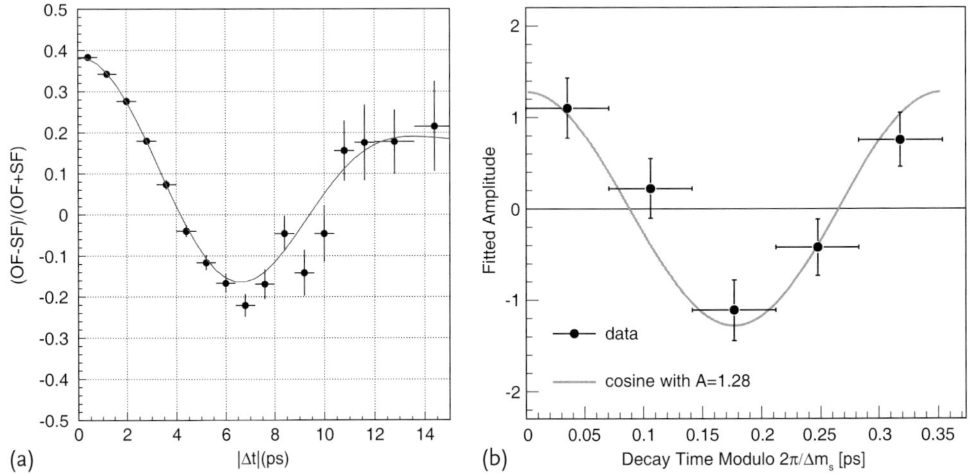

Figure 6.6 (a) Oscillation curves of $B_d^0 - \overline{B}_d^0$ [14]. Time dependence of the asymmetry between OF (opposite flavor) and SF (same flavor) events is plotted. The flavors are tagged by $B_d^0(\overline{B}_d^0) \to D\pi, D \cdot \rho$ for different charge mode decays. The curve shows the result of the m_d fit. (b) $B_s^0 - \overline{B}_s^0$ [287] oscillation signals measured in five bins of proper decay-time modulo the measured oscillation period $2\pi/\Delta m_s$ are plotted.

Note that at Tevatron, the energy of the production reactions $q\overline{q} \to B^0\overline{B}^0$ is sufficiently high so that production channels for $B^0\overline{B}^0$ pairs are widely open. In such cases, B^0 and $\overline{B}^0$ are incoherent and decay independently, and hence the following r_{BB} instead of Eq. (6.106) has to be used to derive the ratio $N(l^\pm l^\pm)/N(l^\pm l^\mp)$.

$$r_{BB} = \frac{\chi(1-\overline{\chi}) + \overline{\chi}(1-\chi)}{\chi\overline{\chi} + (1-\chi)(1-\overline{\chi})} = \frac{r + \overline{r}}{1 + r\overline{r}} \sim \frac{2r}{1+r^2} \tag{6.118}$$

The measured quantity is a relative ratio of the same sign lepton pairs to all the lepton pairs, namely,

$$\frac{N(l^+l^+) + N(l^-l^-)}{N(l^-l^+) + N(l^+l^+) + N(l^-l^-)} = \chi(1-\overline{\chi}) + \overline{\chi}(1-\chi) \tag{6.119}$$

The quarks are observed as jets at high energies. In order to observe $B\overline{B}$ pairs, one looks for two-jet events which contain a lepton with large p, p_T relative to the jet axis in both jets (see Section 4.3.4 for details of heavy quark tagging). $B\overline{B}$ pairs contain $B_u\overline{B}_u(= B^+B^-)$ as well as $B_d^0\overline{B}_d^0$ and $B_s^0\overline{B}_s^0$. The former does not mix and reduces the value of χ. The observed value of χ can be expressed as

$$\chi = f_d\chi_d + f_s\chi_s + (f_u \cdot 0 + f_{\text{other}} \cdot 0) \tag{6.120}$$

where f_i is the probability of producing $\overline{b}q_i b\overline{q}_i$ ($q_i = u, d, s$ and others) which are determined from experiments. The numerical values are

$$f_u = 0.397\,, \quad f_d = 0.397\,, \quad f_s = 0.10\,, \quad f_{\text{other}} = 0.1 \tag{6.121}$$

As a summary, we list recent values of the mixing and mass parameters [3]

$$\begin{aligned}
r_d &= 0.2305 \pm 0.0036 \\
\chi_d &= 0.1873 \pm 0.0024 \\
x_d &= \frac{\Delta m_d}{\Gamma_d} = 0.774 \pm 0.008 \\
\Delta m_d &= 0.507 \pm 0.005\ \text{ps}^{-1} \\
x_s &= \frac{\Delta m_s}{\Gamma_s} = 26.2 \pm 0.5 \\
\Delta m_s &= 17.77 \pm 0.10_{\text{stat}} \pm 0.07_{\text{syst}}\ \text{ps}^{-1}
\end{aligned} \tag{6.122}$$

Width Difference $\Delta\Gamma_d$, $\Delta\Gamma_s$ Knowing precise values of the mixing and the mass difference parameters χ, x, the width difference $\Delta\Gamma_B = 2y\Gamma_B$ could, in principle, be determined from Eq. (6.107). However, for the B-mesons, direct time dependent studies provide stronger constraints and give

$$\frac{\Delta\Gamma_{B_d}}{\Gamma_{B_d}} = 0.010 \pm 0.037 \quad [3, 288, 289] \tag{6.123}$$

which does not contradict with theoretical predictions of

$$\frac{\Delta\Gamma_{B_d}}{\Gamma_{B_d}} = (4.1 \pm 0.9 \pm 1.2) \times 10^{-3} \quad \text{theory [243, 307]} \tag{6.124}$$

under the assumption of no CP violation in the mixing (see Eq. (6.175)).

Qualitative features of the width difference can be understood as follows. From Eqs. (6.79d) and (6.72c), one sees that only states to which both B^0 and $\overline{B}^0$ can decay contribute to $\Delta\Gamma$. Their constituents are $u\overline{q}q\overline{u}$, $u\overline{q}q\overline{c}$, $c\overline{q}q\overline{c}$ and are doubly suppressed states for $q = d$ because of the CKM matrix elements ($\sim (V_{ib}V_{id}^* V_{jb}V_{jd}^*) i, j = u, c$, see Figure 6.11). Therefore, their fractional contributions to the decay width are small. This is a restatement of Eq. (6.65).

However, for $q = s$, they are Cabibbo-allowed and the branch to the common state is considerably larger than that for $q = d$. This theoretical prediction was confirmed by experiment.

$$\begin{aligned}
\frac{\Delta\Gamma_{B_s}}{\Gamma_{B_s}} &= (13 \pm 2 \pm 4) \times 10^{-2} && \text{theory [243]} \\
&= +0.092^{+0.051}_{-0.054} && \text{exp [3]} \\
\frac{1}{\Gamma_{B_s}} &= 1.472^{+0.024}_{-0.026}\ \text{ps}
\end{aligned} \tag{6.125}$$

or equivalently

$$\frac{1}{\Gamma_{L,B_s}} = 1.408^{+0.033}_{-0.030}\ \text{ps} \quad \text{and} \quad \frac{1}{\Gamma_{S,B_s}} = 1.543^{+0.058}_{-0.060}\ \text{ps} \tag{6.126}$$

At this stage, y_d is still consistent with zero. Therefore, for the $B_d^0 - \overline{B}_d^0$ mixing, one may set $y = 0$ with reasonable approximation.

6.4.4 D^0–$\overline{D}^0$ Mixing

Before discussing the CP violation in the B sector, we briefly describe the D^0–$\overline{D}^0$ mixing [290, 291]. In the Standard Model D^0–$\overline{D}$ mixing occurs through processes like the one depicted in Figure 6.7. The mixing is suppressed by the small mass of the intermediate states (refer to Eq. (6.148)). The one that has the largest mass (i, $j = b$) contains CKM matrix elements $|V_{cb}^* V_{ub}|^2 \sim \lambda^{10}$, which is another suppression factor. Therefore, the Standard Model predicts very small mixing and indeed it was not observed until very recently. The signal is still not good enough to determine the value of the parameters accurately, but at least the existences of the mixing has been established. We describe here how to identify the mixing. For that, we start with Eq. (6.93) where $B^0\overline{B}^0$ is replaced by $D^0, \overline{D}^0$.[14)]

In the following, we consider three distinct cases.

Semileptonic Decays Identification of c or $\overline{c}$ decays relies on the SM constraints. In the Standard Model, the semileptonic decays occur through

$$\begin{aligned} &c \to l^+ \nu_l q\,, \quad \overline{c} \to l^- \overline{\nu}_l \overline{q} \quad \text{right sign; (=rs)} \\ &c \nrightarrow l^- \overline{\nu}_l q\,, \quad \overline{c} \nrightarrow l^+ \nu_l \overline{q} \quad \text{wrong sign; (=ws)} \end{aligned} \tag{6.127}$$

where q denotes either d or s. For the wrong sign decays, $A_f = \overline{A}_{\overline{f}} = 0$. They can only be realized through mixing. Using

$$\begin{aligned} |g_\pm(t)|^2 &= \frac{e^{-\Gamma t}}{2}\left[\cosh\frac{\Delta\Gamma t}{2} \pm \cos\Delta m t\right] = \frac{e^{-s}}{2}[\cosh y s \pm \cos x s] \\ y &\equiv \frac{\Delta\Gamma}{2\Gamma}\,, \quad x \equiv \frac{\Delta m}{\Gamma} \end{aligned} \tag{6.128}$$

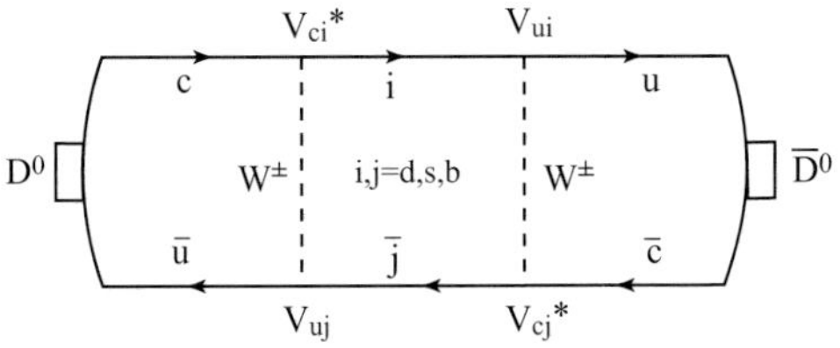

Figure 6.7 A process that induces D^0–$\overline{D}^0$ mixing in the Standard Model. The intermediate states are d, s, b quarks.

14) By definition, D_L, D_H denote the light and heavy mesons in this text book. However, nomenclature for the D mesons is somewhat confusing. D_1, D_2 are used instead of D_L, D_H by the particle data group [3] and $\Delta m \equiv m_1 - m_2$ and $\Delta\Gamma \equiv \Gamma_1 - \Gamma_2$. Also note that $D^0 \equiv c\overline{u}, \overline{D}^0 \equiv \overline{c}u$ are already widely adopted as compared to $K^0 = \overline{s}d, \overline{K}^0 = s\overline{d}$, $B^0 = \overline{b}q, \overline{B}^0 = b\overline{q}(q = d, s)$.

where $s = \Gamma t = t/\tau$ is the time in units of average lifetime and assuming $y, x \ll 1$, we expand only up to the second order and obtain

$$|g_+(s)|^2 = e^{-s}\left(1 + \frac{y^2 - x^2}{4}s^2\right)$$
$$|g_-(s)|^2 = e^{-s}\left(\frac{x^2 + y^2}{4}s^2\right) \tag{6.129}$$

Denoting the relative decay rate of the wrong sign to the right sign states as $r(t)$, and integrating to obtain the total rate ratio, we have

$$r(t) \equiv \left|\frac{q}{p}\right|^2 \frac{|g_-(t)|^2}{|g_+(t)|^2}$$
$$R_M \equiv \left|\frac{q}{p}\right|^2 \frac{\int_0^\infty dt|g_-(t)|^2}{\int_0^\infty dt|g_+(t)|^2} = \left|\frac{q}{p}\right|^2 \frac{(x^2 + y^2)/2}{1 + (y^2 - x^2)/2} \simeq \frac{1}{2}(x^2 + y^2) \tag{6.130}$$

Experimentally, the wrong sign semileptonic decay has not been observed yet and [3]

$$R_M = (0.13 \pm 0.27) \times 10^{-3} \tag{6.131}$$

Doubly Cabibbo-suppressed decays We saw that the mixing rate is proportional to the square of the mixing parameters ($\sim x^2 + y^2$) in the semileptonic decays. We wonder, then, if we can find a process which has linear dependence on x or y. For this purpose, we consider a wrong sign process which is not completely forbidden, but highly suppressed so that its transition amplitude is comparable to that of mixing. We expect that the two competing processes will interfere and provide the desired linear dependence on the mixing parameter. Let us consider a hadronic decay process $D^0 \to K^+\pi^-$ [292, 293]. This is doubly Cabibbo-suppressed (DCS) $\sim V_{cd}^* V_{us} \sim -\sin^2\theta_c$ (see Figure 6.8a). The relative decay rate normalized to Cabibbo favored (CF) process $\sim V_{cs}V_{ud}^* \sim \cos^2\theta_c$ can be calculated from Eq. (6.97) by setting $f = K^+\pi^-$. Normalizing the decay rate by the Cabibbo favored amplitude squared, we obtain

$$r(t) \equiv \frac{\Gamma(D^0 \to K^+\pi^-; t)}{|\langle K^+\pi|H|\overline{D}^0\rangle|^2} = \frac{1}{|\overline{A}_f|^2}\left|g_+(t)\langle f|H|D^0\rangle + \frac{q}{p}g_-(t)\langle f|H|\overline{D}^0\rangle\right|^2$$
$$= \left|\frac{q}{p}\right|^2 \left|g_+(t)\lambda_f^{-1} + g_-(t)\right|^2, \quad \lambda_f = \frac{q}{p}\frac{\overline{A}_f}{A_f} \tag{6.132}$$

If we ignore the CP violation effect, we may set $q/p = 1$ and write

$$\frac{A_f}{\overline{A}_f} = \frac{\langle K^+\pi^-|H|D^0\rangle}{\langle K^+\pi^-|H|\overline{D}^0\rangle} = -\sqrt{R_D}e^{-i\delta_f}, \quad \left|\frac{A_f}{\overline{A}_f}\right| = \sqrt{R_D} \sim O(\tan^2\theta_c) \tag{6.133}$$

where R_D is the doubly Cabibbo-suppressed decay rate relative to the Cabibbo fa-

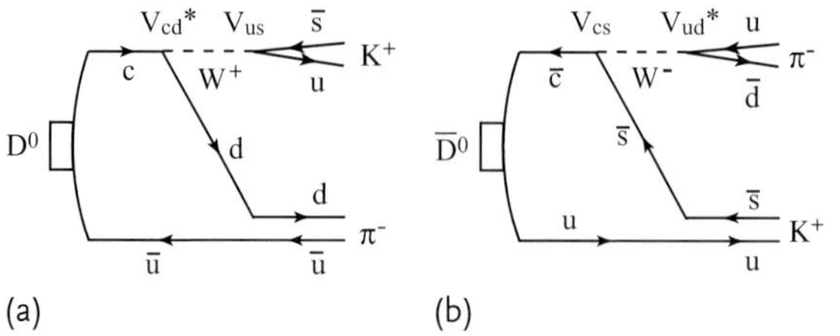

Figure 6.8 Decay processes that are used to establish the D^0–$\overline{D}^0$ mixing. (a) Doubly Cabibbo-suppressed decay mode. (b) Cabibbo favored decay process.

vored rate, and δ_f is the strong phase difference between DCS and CF processes. The minus sign originates from the sign of V_{us} relative to V_{cd}. Using

$$|g_\pm(t)|^2 = \frac{e^{-\Gamma t}}{2}\left[\cosh(y\Gamma t) \pm \cos(x\Gamma t)\right] \simeq e^{-\Gamma t}\begin{cases}1+\left(\frac{y^2-x^2}{4}\right)(\Gamma t)^2\\ \left(\frac{x^2+y^2}{4}\right)(\Gamma t)^2\end{cases}$$

$$2g_+(t)^* g_-(t) = e^{-\Gamma t}\left[-\sinh(y\Gamma t) + i\sin(x\Gamma t)\right] \simeq e^{-\Gamma t}(-y+ix)(\Gamma t) \quad (6.134)$$

we obtain to leading order

$$\begin{aligned} r(t) &= e^{-\Gamma t}\left[R_D + \sqrt{R_D}y'\Gamma t + \frac{1}{2}R_M(\Gamma t)^2\right]\\ y' &= y\cos\delta_f - x\sin\delta_f\\ x' &= x\cos\delta_f + y\sin\delta_f \end{aligned} \quad (6.135)$$

where $R_M = (x'^2 + y'^2)/2$ is defined in Eq. (6.130) and $x^2 + y^2 = x'^2 + y'^2$ has been used. By integrating over time and normalizing by the total CF rate, one obtains

$$R = \Gamma\int_0^\infty dt r(t) = R_D + \sqrt{R_D}y' + R_M \quad (6.136)$$

Experimental values obtained from different groups vary, but a finite value of x and y has been obtained, and hence the evidence for D^0–$\overline{D}^0$ mixing was marginally established [294–296]. For extraction of the mixing parameters x, y from measured R value, knowledge of the relative strong phase δ_f is required. The phase shift $\delta_{K\pi}$ can be obtained separately from isospin analysis of $K\pi$ data. Using the relation

$$\sqrt{2}A\left(D_\pm \to K^-\pi^+\right) = A(D^0 \to K^-\pi^+) \pm A(\overline{D}^0 \to K^-\pi^+) \quad (6.137)$$

where $D_\pm$ denote the CP-even and CP-odd eigenstate, one obtains for $\sqrt{R_D} \ll 1$

$$\cos\delta_{K\pi} = \frac{|A(D_+ \to K^-\pi^+)|^2 - |A(D_- \to K^-\pi^+)|^2}{2\sqrt{R_D}|A(D^0 \to K^-\pi^+)|^2} \quad (6.138)$$

Combining other measurements with different final states, the following gives the world average for x and y with no CP violation assumption. Those analyzed with the CP violation gives a similar result [3].

$$\begin{aligned} R &= 3.80 \pm 0.05 \times 10^{-3} \\ R_D &= 3.35 \pm 0.09 \times 10^{-3} \\ x &= 0.80 \pm 0.29 \\ y &= 0.33 \pm 0.24 \end{aligned} \tag{6.139}$$

Decays to CP eigenstates The previous method was useful in discovering the D^0–$\overline{D}^0$ mixing. However, it involves an extra variable, the strong phase shift. For direct measurement of the mixing parameter, one has to find a process that measures x or y directly. For this purpose, we consider the case where the final state is the CP eigenstate accessible from both D^0 and $\overline{D}^0$. Then $|A_f| = |A_{\overline{f}}|$ and $\lambda_f = \pm 1$ without CP violation. For $f = K^+K^-$ [15] the decay rate becomes

$$\begin{aligned} \Gamma(D^0 \to K^+K^-) &= |\langle K^+K^-|H|\overline{D}^0\rangle|g_+(t) \pm g_-(t)|^2 \\ &= \frac{|\overline{A}_f|^2}{2} e^{-\Gamma t}[\cosh(y\Gamma t) \mp \sinh(y\Gamma t)] \\ &\sim e^{-\Gamma t}(1 \mp y\Gamma t) \end{aligned} \tag{6.140}$$

where $\pm$ corresponds to decays regarding CP $= \pm$eigenstates. This is equivalent to changing the lifetime of D^0:

$$\frac{1}{\tau} \to \frac{1}{\tau_\pm} = \frac{1}{\tau(1 \mp y)} \tag{6.141}$$

This y is customarily denoted as y_{CP} because CP invariance is assumed. If it is not assumed, by writing $\lambda_f = \pm|q/p|e^{i\phi}$ and expressing the lifetime in units of reference decay mode, Eq. (6.141) is modified to

$$\begin{aligned} \frac{1}{\tau_\pm} &= 1 \pm \left|\frac{q}{p}\right|(y\cos\phi - x\sin\phi) \\ \frac{1}{\overline{\tau}_\pm} &= 1 \pm \left|\frac{p}{q}\right|(y\cos\phi + x\sin\phi) \end{aligned} \tag{6.142}$$

Experimentally, the Cabibbo favored process $\Gamma(D^0 \to K^-\pi^+)$ was used as the reference. Therefore, y_{CP} can also be defined as

$$\begin{aligned} y_{CP} &= \frac{\langle\tau_{K^-\pi^+}\rangle}{\langle\tau_{h\overline{h}}\rangle} - 1 \\ \langle\tau_{h\overline{h}}\rangle &= \left(\tau_{h\overline{h}}^{D^0} + \tau_{h\overline{h}}^{\overline{D}^0}\right)/2, \quad h\overline{h} = K^+K^-, \pi^+\pi^- \end{aligned} \tag{6.143}$$

If decays to K^+K^- have a shorter lifetime than those to $K^-\pi^+$, y is positive. The lifetime in Eq. (6.143) should be considered as an effective one as it is the

15) Other modes ($f = \pi^+\pi^-$, $K_s K^+K^-$) are also measured.

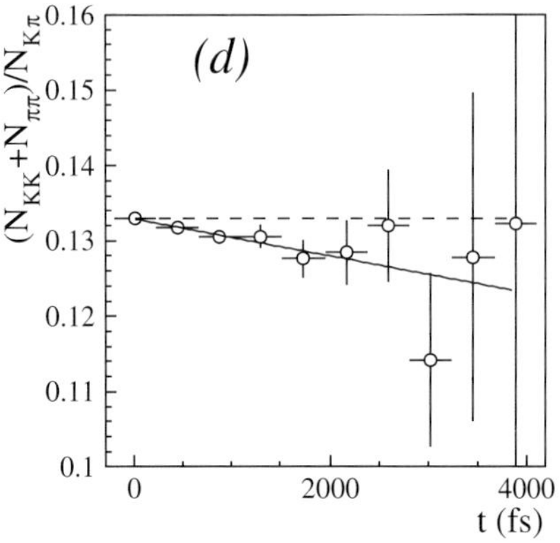

Figure 6.9 Ratio of decay-time distributions between CP-even eigenstates (K^+K^-, $\pi^+\pi^-$) and $D^0 \to K^-\pi^+$ decays. The solid line is a fit to the data points. The value of y_{CP} is determined to be $1.31 \pm 0.32 \pm 0.25\%$ [297].

average of lifetimes and also the mixture of Cabibbo-suppressed as well as Cabibbo favored processes. As the mixing parameter is small, the lifetime distribution for $h\bar{h}$ and $K\pi$ can be considered as exponential to a good approximation. Figure 6.9 shows one such relative lifetime measurement for $h\bar{h} = K^+K^-, \pi^+\pi^-$ [297, 298]. Combining the world data, its average on y_{CP} gives [3]

$$y_{CP} = 1.11 \pm 0.22\% \tag{6.144}$$

Summary of D^0–$\overline{D}^0$ Mixing The current world values after averaging all the experimental measurements using all the methods can be summarized as [3]

$$\begin{aligned} x &= 0.98^{+0.24}_{-0.26}\,\% \\ y &= 0.83 \pm 0.16\,\% \\ R_D &= 0.337 \pm 0.009\,\% \end{aligned} \tag{6.145}$$

6.5 Theoretical Evaluation of the Mass Matrix

In this section, we evaluate the CP violation effects in the framework of the Standard Model and show that it can reproduce experiments qualitatively. There is a qualitative difference in the way the CP violation manifests itself in the K meson sector and in the B meson sector. In the former, we saw $\epsilon \gg \epsilon'$, namely, the effect has mainly appeared in the mixing (indirect violation) and the effect in the decay (direct violation) was small. By investigating properties of the CKM matrix, it has become clear that the reverse is realized in the B sector and that there is a third possibility of the violation through combination of mixing and decay (i.e., $\lambda_f = (q/p)\overline{A}_f/A_f$). The reason for the large direct CP violation is that the measure of the CP violation is given by the Jarlskog parameter $|J| \sim \lambda^6$ which is universal. Experimentally, however, the effect appears in the form of asymmetry between CP odd decays. It is a number that is divided by the partial decay rate. In the K meson sector, it is proportional to $|V_{ud}^* V_{us}|^2 \sim \lambda^2$, but it is given by $|V_{td}^* V_{tb}|^2 \sim \lambda^6$ in the B sector which is of the same order as $|J|$. Therefore, the asymmetry can be as large as $\sim O(1)$. Note, however, that the large asymmetry does not necessarily mean that it is an easy experiment. The B meson has a large mass and large Q value, and hence many decay channels are open, making the branching ratio very small.

6.5.1 Mass Matrix of the K Meson

Here, we evaluate the mass matrix

$$\Lambda_{12} = \langle K^0|\Lambda|\overline{K}^0\rangle = M_{12} - i\frac{\Gamma_{12}}{2} \tag{6.146}$$

which has been treated in general terms so far. The CP violation effect appears in the imaginary part of M_{12} which contributes to the mixing, that is, ε and of Γ_{12}, which contributes to both ε and ϵ'. In the Standard Model the process $K^0 \leftrightarrow \overline{K}^0$ is given by box diagrams depicted in Figure 6.10a,b. Therefore, an expression for the mass matrix is obtained by calculating the diagram using the Feynman rules. The kaon decay amplitudes can be obtained by cutting the box diagram along the dot-dashed line into halves as shown in Figure 6.10c,d. Therefore, the decay rate and

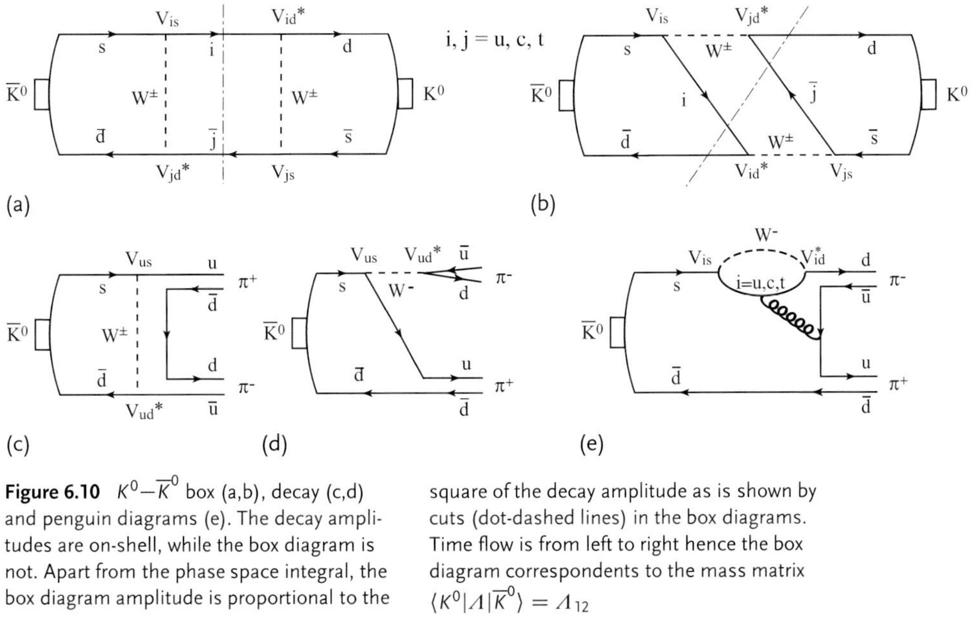

Figure 6.10 K^0–$\overline{K}^0$ box (a,b), decay (c,d) and penguin diagrams (e). The decay amplitudes are on-shell, while the box diagram is not. Apart from the phase space integral, the box diagram amplitude is proportional to the square of the decay amplitude as is shown by cuts (dot-dashed lines) in the box diagrams. Time flow is from left to right hence the box diagram correspondents to the mass matrix $\langle K^0|\Lambda|\overline{K}^0\rangle = \Lambda_{12}$

the box amplitude is proportional if other processes do not contribute. However, the decay is only realized when the energy-momentum is conserved between the initial and final states, that is, the decay amplitude has to be on-shell. Therefore, among the possible intermediate states $i, j = u, c, t$ in the box diagram, the top quark does not contribute to the decay amplitude. As the CP violation is contained only in V_{td} up to $O(\lambda^3)$ (V_{ub} does not appear in the K meson decay), this means that we need to introduce other loop diagrams of the top to have CP violation in the decay amplitude. An example which is commonly referred to as the penguin diagram is shown in Figure 6.10e.

Δm_K The box diagrams are evaluated to give [299]

$$M_{12} = -\frac{G_F^2 m_W^2}{12\pi^2} m_K B_K f_K^2 \left[\eta_{cc}\lambda_c^2 S(x_c) + \eta_{tt}\lambda_t^2 S(x_t) + 2\eta_{ct}\lambda_c\lambda_t S(x_c, x_t)\right]$$
$$\lambda_i = V_{id}^* V_{is}\,, \quad x_i = m_i^2/m_W^2 \quad (i = u, c, t) \tag{6.147}$$

$S(x)$, $S(x, y)$ appear as a result of the loop integral and are given by

$$S(x) = xF(x) = x\left[1 - \frac{3}{4}\frac{(x + x^2)}{(1-x)^2} - \frac{3}{2}\frac{x^2 \ln x}{(1-x)^3}\right]$$
$$\sim \begin{cases} x & x \ll 1 \\ 2.46\left(\frac{m_t}{170\,\text{GeV}}\right)^{1.52} & x \gtrsim x_t \end{cases}$$
$$S(x, y) = xy\left[\left\{1 - \frac{3}{4}\frac{x^2}{(1-x)^2}\right\}\frac{\ln x}{x - y} + (x \leftrightarrow y) - \frac{3}{4}\frac{1}{(1-x)(1-y)}\right]$$
$$\sim x\left[\ln\left(\frac{y}{x}\right) - \frac{3}{4}\frac{y}{(1-y)} - \frac{3}{4}\frac{y^2 \ln y}{(1-y)^2}\right] \quad \text{for } x \ll 1, x \ll y \tag{6.148}$$

Numerically, their approximate values are:

$$S(x_c) \sim 3 \times 10^{-4}\,, \quad S(x_t) \sim 2.5\,, \quad S(x_c, x_t) \sim 2.7 \times 10^{-3} \tag{6.149}$$

and $\eta_{cc} = 1.38 \pm 0.2$, $\eta_{tt} = 0.57 \pm 0.01$, $\eta_{ct} = 0.47 \pm 0.05$ are QCD corrections [300–302]. $f_K = 157 \pm 2$ MeV [303] is the decay constant of the $K \to \mu\nu$. $B_K = 0.725 \pm 0.026$ [279] is commonly referred to as the bag parameter defined by

$$\frac{8}{3} B_K f_K^2 m_K^2 = \langle K^0 | \bar{d}\gamma^\mu(1-\gamma^5)s\bar{d}\gamma_\mu(1-\gamma^5)s|\overline{K}^0\rangle \tag{6.150}$$

$B_K = 1$ by definition when the intermediate states are replaced with vacuum. The coefficient 8/3 is the color factor (quark color average) × 8(gluon multiplicity).

As one can see from Eq. (6.147), each intermediate state (u,c,t) contributes in proportion to the product of its mass squared m_i^2 and the CKM matrix element squared λ_i^2. Therefore, contribution of the u quark is negligible.[16] If not suppressed

16) This is true only when we consider short distance contributions. Actually, the long distance contribution which is due to the u-quark loop effect and phenomenologically can be interpreted as due to intermediate states $K^0 \to \pi, \eta, \eta', \pi\pi, \cdots \to \overline{K}^0$ is quite sizable [304–306]. Writing $\Delta m_K(1 - D) = 2\text{Re}[-M_{12}]$, it is estimated that D can be as large as 0.5. In the following discussion, the long range effect is ignored.

by the CKM matrix elements, the top quark's contribution is dominant. This is certainly the case for the B meson, but for the K meson, it is suppressed by

$$\lambda_c^2 \sim \lambda^2 , \quad \lambda_t^2 \sim A^4\lambda^{10}(1-\rho-i\eta)^2 , \quad \lambda_c\lambda_t \sim A^2\lambda^6(1-\rho-i\eta) \tag{6.151}$$

which makes the charm contribution dominant and the top contribution is a small correction to it. As the imaginary part is provided by the CKM matrix elements,

$$\mathrm{Re}M_{12} \gg \mathrm{Im}M_{12} , \quad \mathrm{Re}\Gamma_{12} \gg \mathrm{Im}\Gamma_{12} \tag{6.152}$$

Referring to Eq. (6.78), it means

$$q/p \simeq 1 - 2\tilde{\epsilon} , \quad |\tilde{\epsilon}| \ll 1 \tag{6.153}$$

Therefore, in the approximation $m_c^2 \ll m_t^2, m_W^2$, one obtains

$$\Delta m_K \simeq -2\mathrm{Re}M_{12} \simeq \frac{G_F^2 m_c^2}{6\pi^2} m_K B_K f_K^2 \eta_{cc}(V_{cs}V_{cd}^*)^2 \tag{6.154}$$

When the Δm_K was calculated for the first time, the charm quark was not yet discovered, but knowing the value of $|\Delta m_K|$, its mass was predicted to be approximately $1 \sim 2$ GeV [307].

ϵ_K and ϵ'_K The expression for ϵ_K is obtained by combining Eqs. (6.82a), (6.79f), (6.72a) and (6.84) as

$$\begin{aligned}\epsilon_K &= \frac{\sin\phi_{SW} e^{i\phi_{SW}}}{\Delta m_K}\left(-\mathrm{Im}M_{12} + i\frac{\mathrm{Im}\Gamma_{12}}{2}\right) + i\frac{\mathrm{Im}A_0}{\mathrm{Re}A_0} \\ &\simeq \frac{e^{i\pi/4}}{\sqrt{2}}\left(-\frac{\mathrm{Im}M_{12}}{\Delta m_K} - \frac{\mathrm{Im}\Gamma_{12}}{\Delta\Gamma}\right) + i\left(\frac{\mathrm{Im}\Gamma_{12}}{\Delta\Gamma} + \frac{\mathrm{Im}A_0}{\mathrm{Re}A_0}\right)\end{aligned} \tag{6.155}$$

First, we can show that the second term is negligible. This can be proved as follows. It is a good approximation to only use the $n = 2\pi$ state because experimentally $\mathrm{BR}(K_s \to 2\pi) = 99.89\%$. Then, using Eq. (6.72c),

$$\begin{aligned}\mathrm{Im}\Gamma_{12} &= \mathrm{Im}\left[\langle K^0|H|\pi\pi(I=0)\rangle\langle\pi\pi(I=0)|H|\overline{K}^0\rangle + (I=2)\right]\rho_{2\pi} \\ &= \mathrm{Im}\left[A_0^{*2} + A_2^{*2}\right]\rho_{2\pi} \simeq -2[\mathrm{Im}A_0\mathrm{Re}A_0]\rho_{2\pi}\end{aligned} \tag{6.156a}$$

$$\Delta\Gamma = \Gamma_s - \Gamma_L \simeq \Gamma_s = \sum_n |\langle n|H|K_s\rangle|^2\rho_n \approx 2(\mathrm{Re}A_0)^2\rho_{2\pi} \tag{6.156b}$$

$$\therefore \quad \frac{\mathrm{Im}\Gamma_{12}}{\Gamma_s} \simeq -\frac{\mathrm{Im}A_0}{\mathrm{Re}A_0} \tag{6.156c}$$

Smallness of $|\mathrm{Im}\Gamma_{12}/\Delta\Gamma| \simeq |\mathrm{Im}A_0/\mathrm{Re}A_0|$ will be justified by Eq. (6.163) in the discussion of ϵ'_K. Thus, the dominant contribution to ϵ_K is given by $\mathrm{Im}M_{12}$.

Substituting Eq. (6.147) into Eq. (6.155) and approximating $\phi_{\rm SW} \sim \pi/4$, one obtains

$$\epsilon_K = \frac{G_F^2 m_W^2 m_K B_K f_K^2}{12\sqrt{2}\pi^2 \Delta m_K} e^{i\phi_{\rm SW}} \Big[\eta_{cc} S(x_c) \mathrm{Im}\left[\left(V_{cs} V_{cd}^*\right)^2\right] + \eta_{tt} S(x_t) \mathrm{Im}\left[\left(V_{ts} V_{td}^*\right)^2\right] + 2\eta_{ct} S(x_c, x_t) \mathrm{Im}\left[V_{cs} V_{cd}^* V_{ts} V_{td}^*\right] \Big] \quad (6.157)$$

For evaluation of the imaginary part, we need to go beyond $O(\lambda^3)$ to express the CKM matrix which is given by [243].

$$V_{\rm CKM} = \begin{bmatrix} 1-\frac{1}{2}\lambda^2-\frac{1}{8}\lambda^4 & \lambda+O(\lambda^7) & A\lambda^3(\rho-i\eta) \\ -\lambda+\frac{1}{2}A^2\lambda^5[1-2(\rho+i\eta)] & 1-\frac{1}{2}\lambda^2-\frac{1}{8}\lambda^4(1+4A^2) & A\lambda^2+O(\lambda^8) \\ A\lambda^3(1-\overline{\rho}-i\overline{\eta}) & -A\lambda^2+\frac{1}{2}A\lambda^4[1-2(\rho+i\eta)] & 1-\frac{1}{2}A^2\lambda^4 \end{bmatrix} \quad (6.158)$$

One obtains

$$\begin{aligned} \mathrm{Im}\left(V_{cs} V_{cd}^*\right)^2 &\approx -2A^2\lambda^6\eta \\ \mathrm{Im}\left(V_{ts} V_{td}^*\right)^2 &\approx \ 2A^4\lambda^{10}(1-\overline{\rho})\overline{\eta} \\ 2\mathrm{Im}\left(V_{cs} V_{cd}^* V_{ts} V_{td}^*\right) &\approx \ 2A^2\lambda^6\overline{\eta} \end{aligned} \quad (6.159)$$

Then, $|\epsilon_K|$ can be expressed as

$$|\epsilon_K| \simeq C_\epsilon A^2 B_K \overline{\eta} \left[D(x_c, x_t) + E(x_t) A^2 (1-\overline{\rho})\right] \quad (6.160a)$$

The constants C_ϵ, $D(x_c, x_t)$, $E(x_c, x_t)$ are given by

$$C_\epsilon = \frac{G_F^2 m_W^2 f_K^2 B_K m_K}{6\sqrt{2}\pi^2 \Delta m_K} = 3.837 \times 10^4$$

$$D(x_c, x_t) = \lambda^6[\eta_{ct} S(x_c, x_t) - \eta_{cc} S(x_c)], \quad E(x_t) = \lambda^{10}\eta_{tt} S(x_t) \quad (6.160b)$$

Equation (6.160a) gives a hyperbola on the $\overline{\rho}, \overline{\eta}$ plane.

The direct CP violation parameter ϵ_K' is given by Eq. (6.82c) and (6.82d)

$$\epsilon_K' = i\omega \left(\frac{\mathrm{Im}A_2}{\mathrm{Re}A_2} - \frac{\mathrm{Im}A_0}{\mathrm{Re}A_0}\right), \quad \omega = \frac{e^{i(\delta_2-\delta_0)}}{\sqrt{2}} \frac{\mathrm{Re}A_2}{\mathrm{Re}A_0} \quad (6.161)$$

where A_I's are decay amplitudes for $K^0 \to \pi\pi$ $(I = 0, 2)$ and $|\omega| \simeq 0.03$ is a measure of $\Delta I = 1/2$ rule violation. δ_I's are $\pi\pi$ scattering phase shifts. Accidentally, the phase of the ϵ_K' is very close to that of ϵ_K and can be set as $\pi/4$ [see Eq. (6.89)]. As discussed previously, the ϵ_K' gets a contribution from the penguin diagrams Figure 6.10e. It is a process with $\Delta I = 1/2$ and gives no contribution to A_2. The main contribution to the real part of A_0 comes from the tree diagram Figure 6.10c,d. Writing the amplitudes of the penguin and tree diagrams after separating the CKM matrix elements as A_p and A_t, their relative magnitude is estimated as

$$\frac{\mathrm{Im}A_0}{\mathrm{Re}A_0} \sim \frac{\mathrm{Im}(V_{ts}V_{td}^*)}{\mathrm{Re}(V_{us}V_{ud}^*)}\frac{A_p}{A_t} \sim \frac{\lambda^5\overline{\eta}}{\lambda}\frac{A_p}{A_t} \sim \text{a few} \times 10^{-4}\frac{A_p}{A_t} \tag{6.162}$$

The penguin diagram contains the QCD process which is hard to estimate, but can be considered to be $|A_p/A_t| \lesssim O(1)$. Experimentally,

$$\frac{|\mathrm{Im}A_0/\mathrm{Re}A_0|}{\epsilon_K} \simeq \frac{|\epsilon'_K/\omega|}{\epsilon_K} = \frac{1}{\omega}\left|\frac{\epsilon'_K}{\epsilon_K}\right| \simeq \frac{1.6\times 10^{-3}}{0.03} < 0.1 \tag{6.163}$$

Detailed calculations give a numerical value consistent with the experimental value:

$$\mathrm{Re}\left[\frac{\epsilon'_K}{\epsilon_K}\right] \approx \frac{\epsilon'_K}{\epsilon_K} = (1.65 \pm 0.23)\times 10^{-3} \quad [3] \tag{6.164}$$

Historically, the ϵ'_K played an important role in establishing the existence of the direct CP violation. However, in the Standard Model the QCD and electroweak penguin (gluon replaced with Z) compete with opposite sign. Therefore, its quantitative evaluation is very hard and it is not used to constrain CKM matrix elements on the $\overline{\rho}$–$\overline{\eta}$ plane.

6.5.2 Mass Matrix of the *B* Meson

Δm_d, Δm_s The mass difference of the neutral B mesons can be calculated similarly to that of the K meson by estimating $M_{12} = \langle B^0|\Lambda|\overline{B}^0\rangle$ (see Figure 6.11). It has exactly the same functional form, but this time the contribution of the top quark intermediate state is overwhelmingly large. Therefore, to a good approximation,

$$M_{12} = -\frac{G_F^2 m_W^2}{12\pi^2} m_B B_B f_B^2 \eta_{tt} S(x_t)(V_{tb}V_{ti}^*)^2 \quad i = d, s \tag{6.165}$$

where [3, 278]

$$\begin{aligned} m_B &= 5279.50 \pm 0.30\ \text{MeV} \\ f_{B_d} &= 192.8 \pm 9.9\ \text{MeV} \\ f_{B_d}\sqrt{B_{B_d}} &= 216 \pm 15\ \text{MeV} \\ f_{B_s} &= 238.8 \pm 9.5\ \text{MeV} \\ f_{B_s}\sqrt{B_{B_d}} &= 275 \pm 13\ \text{MeV} \end{aligned} \tag{6.166}$$

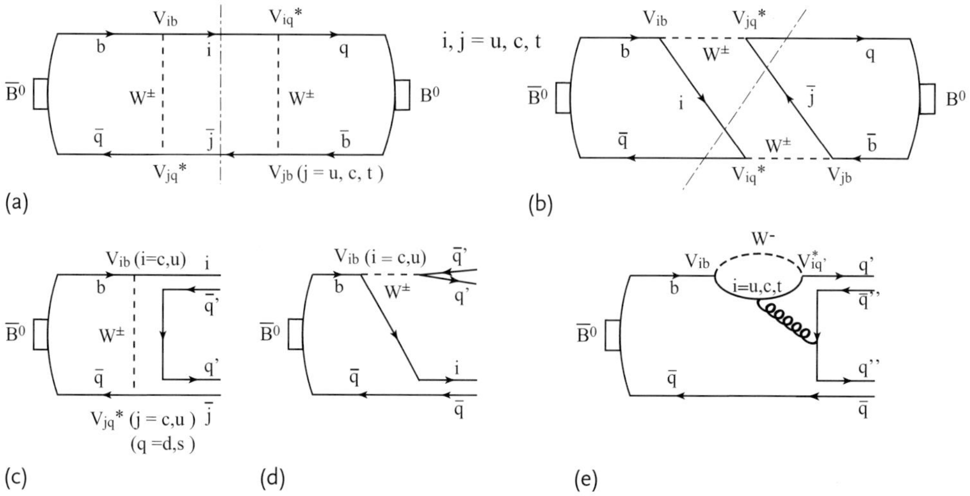

Figure 6.11 B^0–$\overline{B}^0$ box (a,b), decay (c,d) and penguin diagrams (e). Time flow is from left to right hence the box diagram corresponds to the mass matrix $\langle B^0|\varLambda|\overline{B}^0\rangle = \varLambda_{12}$.

$\Delta\Gamma$ As the decay amplitude has to be on-shell, the energy at the cut line on the box diagrams in Figure 6.11a,b have to be m_B, which means the top quark does not contribute. Only u and c contribute and according to calculations [308, 309],

$$\begin{aligned}
\Gamma_{12} &\approx \frac{G_F^2}{8\pi} m_B B_B f_B^2 m_b^2 \left[\lambda_c^2 p_{cc} + \lambda_u^2 p_{uu} + 2\lambda_u\lambda_c p_{uc}\right] \\
p_{ij} &= \left[\left(1 - \frac{(m_i + m_j)^2}{m_b^2}\right)\left(1 - \frac{(m_i - m_j)^2}{m_b^2}\right)\right]^{1/2} \\
&\quad \times \left[1 - \frac{1}{3}\frac{m_i^2 + m_j^2}{m_b^2} - \frac{2}{3}\frac{(m_i^2 - m_j^2)^2}{m_b^4}\right]
\end{aligned} \tag{6.167}$$

where $\lambda_i = V_{iq}^* V_{ib}$. Using the approximation $m_u^2 \ll m_c^2 \ll m_b^2$

$$p_{uu} \sim 1, \quad p_{cc} \sim 1 - \frac{8}{3}\frac{m_c^2}{m_b^2}, \quad p_{uc} \sim 1 - \frac{4}{3}\frac{m_c^2}{m_b^2} \tag{6.168}$$

one obtains

$$\Gamma_{12} = \frac{G_F^2}{8\pi} m_B B_B f_B^2 m_b^2 \left[(\lambda_c + \lambda_u)^2 - (\lambda_u + \lambda_c)\lambda_c \frac{8}{3}\frac{m_c^2}{m_b^2}\right] \tag{6.169}$$

Using the unitarity constraint

$$\lambda_u + \lambda_c + \lambda_t = 0 \tag{6.170}$$

one gets

$$\Gamma_{12} = \frac{G_F^2}{8\pi} m_B B_B f_B^2 m_b^2 \left[\lambda_t^2 + \frac{8}{3}\lambda_c\lambda_t \frac{m_c^2}{m_b^2}\right] \tag{6.171}$$

Comparing with Eq. (6.165), one obtains

$$\frac{\Gamma_{12}}{M_{12}} \approx -\frac{3\pi}{2}\frac{m_b^2}{m_t^2}\frac{x_t}{S(x_t)}\left(1+\frac{8}{3}\frac{\lambda_c}{\lambda_t}\frac{m_c^2}{m_b^2}\right) \tag{6.172}$$

The first term gives $\mathrm{Re}(\Gamma_{12}/M_{12})$ and the second $\mathrm{Im}(\Gamma_{12}/M_{12})$. As the former is much larger than the latter, M_{12} and Γ_{12} almost has the opposite phase, implying that the mass eigenstates have mass and width differences of opposite sign. This means that, like in the K^0–$\overline{K}^0$ system, the heavy state is expected to have a smaller decay width than that of the light state, that is, $\Gamma_H < \Gamma_L$.

Furthermore, if the phases of M_{12} and Γ_{12} are exactly opposite, q/p is a pure phase as can be seen from Eq. (6.78), and hence ϵ_B becomes pure imaginary. As the phase of q/p can be changed, the modulus can be made to coincide with one by choosing the phase of $|\overline{B}^0\rangle$ properly. In this case, there is no CP violation in the mixing.

In summary, there are two ways to make ε small. Both $\mathrm{Im}M_{12}$ and $\mathrm{Im}\Gamma_{12}$ are small, which makes $|\epsilon|$ small or M_{12}, and Γ_{12} have nearly equal phases which makes $\mathrm{Re}\epsilon$ small. In our convention of the CKM parametrization, the former is realized in the K sector and the latter in the B sector. Therefore, for the B mesons, Eq. (6.79c) and Eq. (6.79d) tell

$$\Delta m \approx 2|M_{12}|\,, \quad \Delta\Gamma \approx 2|\Gamma_{12}| \tag{6.173}$$

If one uses the experimental facts $\Delta\Gamma_{B_d} \ll \Gamma_{B_d}$ and $x_d = \Delta m_{B_d}/\Gamma_{B_d} = 0.774 \pm 0.008$,

$$\left|\frac{\Delta\Gamma_{B_d}}{\Delta m_{B_d}}\right| \approx \left|\frac{\Gamma_{12}}{M_{12}}\right| \approx \frac{m_b^2}{m_t^2} \lesssim 10^{-3} \tag{6.174}$$

is predicted. Historically, when the large value of $x_d \sim 1$ is demonstrated, it was realized that the top quark should have a much larger value ($m_t \gg m_b$) than most people expected.

Today, precise QCD corrected theoretical estimates for the width difference exist and give [243, 310]

$$\frac{\Delta\Gamma_{B_d}}{\Gamma_{B_d}} = (4.1 \pm 0.9 \pm 1.2) \times 10^{-3}\,, \quad \frac{\Delta\Gamma_{B_s}}{\Gamma_{B_s}} = (13 \pm 2 \pm 4) \times 10^{-2} \tag{6.175}$$

Evaluation of x_d and x_s As x is the ratio of Δm and Γ, it can be expressed by using the expressions Eqs. (6.173) and (6.165) as

$$x_q = \frac{\Delta m_q}{\Gamma_{B_q}} = \tau_{B_q}\frac{G_F^2 m_W^2}{6\pi^2} m_B f_B^2 B_B \eta_{tt} |V_{tb} V_{tq}^*|^2 S(x_t) \tag{6.176}$$

The formula is common to B_d and B_s with replacement of $q = d \leftrightarrow s$. Several uncertainties cancel in the mass difference ratio

$$\frac{\Delta m_s}{\Delta m_d} = \frac{m_{B_s}}{m_{B_d}}\xi^2\left|\frac{V_{ts}}{V_{td}}\right|^2 = \frac{m_{B_s}}{m_{B_d}}\xi^2\frac{1}{\lambda^2(|1-\overline{\rho}|^2+|\overline{\eta}|^2)} \tag{6.177}$$

where $\xi \equiv (f_{B_s}\sqrt{B_{B_s}})/(f_{B_d}\sqrt{B_{B_d}}) = 1.243 \pm 0.028$ is a $SU(3)$ flavor-symmetry breaking factor obtained from lattice QCD calculations [279].

As is clear from the above arguments, $|V_{ts}/V_{td}|^2$ determines the apex position of the unitarity triangle on a circle which is centered at $(\overline{\rho}, \overline{\eta}) = (1, 0)$ and has the radius determined by the mass difference ratio. Similarly, the value of $|V_{ub}|^2 \sim \sqrt{|\overline{\rho}|^2 + |\overline{\eta}|^2}$ determines the apex position on a circle centered at the origin. As $|V_{ub}|^2$ can be obtained by measurement of the decay rate $\Gamma(b \to u l^- \overline{\nu}_l)$ (Eq. (6.43a)), knowledge of x_s/x_d and $\Gamma(b \to u l^- \overline{\nu}_l)$ alone can determine the unitarity triangle. Therefore, even if the CP violation effect is not detected directly, its existence can be determined indirectly using the CP conserving observables only.

6.6 CP Violation in the *B* Sector

6.6.1 Indirect CP Violation

The CP violation in the mixing can be searched for in samples where the initial flavor state is tagged. In the case of semileptonic decays, the final-state tagging is also available. In this case,

$$A_f = \frac{N(l^-l^-) - N(l^+l^+)}{N(l^-l^-) + N(l^+l^+)} = \frac{\overline{r} - r}{\overline{r} + r} = \frac{\left|\frac{p}{q}\right|^2 - \left|\frac{q}{p}\right|^2}{\left|\frac{p}{q}\right|^2 + \left|\frac{q}{p}\right|^2}$$
$$\approx 4\mathrm{Re}\epsilon_B \tag{6.178}$$

where relations Eq. (6.106) and $q/p = (1 - \epsilon_B)/(1 + \epsilon_B)$ have been used. Furthermore, for $|\Gamma_{12}/M_{12}| \ll 1$, the power expansion of Eq. (6.78) gives

$$\left|\frac{q}{p}\right|^2 \approx 1 - 4\mathrm{Re}\epsilon_B \approx 1 - \mathrm{Im}\left(\frac{\Gamma_{12}}{M_{12}}\right) \tag{6.179}$$

Substituting Eq. (6.172) in the above expression, we obtain

$$\mathrm{Re}\,\epsilon_B \approx \frac{1}{4}\mathrm{Im}\left(\frac{\Gamma_{12}}{M_{12}}\right) \approx -\frac{\pi x_c}{S(x_t)}\mathrm{Im}\left(\frac{V_{cb}V_{cq}^*}{V_{tb}V_{tq}^*}\right)$$
$$\approx \begin{cases} O(10^{-4}) & B_d \\ O(10^{-5}) & B_s{}^{17)} \end{cases} \tag{6.180}$$

They are at least one order of magnitude smaller than ϵ_K. Since the lifetime is much shorter and it is almost impossible to separate B_H from B_L, it is hard to

17) Today's QCD corrected calculations give [243]

$$|(q/p)_{B_d}| - 1 = (2.96 \pm 0.67) \times 10^{-4}\,, \quad |(q/p)_{B_s}| - 1 = (1.28 \pm 0.28) \times 10^{-5} \tag{6.181}$$

measure ϵ_B directly. So far, no group has observed the effect and the world value for B_d and B_s is given by

$$
\begin{aligned}
A_f^{B_d} &= -0.0005 \pm 0.0056\,, \quad \text{or} \quad |q/p|_{B_d} = 1.0002 \pm 0.0028 \\
A_f^{B_s} &= -0.0036 \pm 0.0094\,, \quad \text{or} \quad |q/p|_{B_s} = 1.0018 \pm 0.0047
\end{aligned}
\tag{6.182}
$$

6.6.2 Direct CP Violation

After evaluating the CP violation of the first kind (indirect violation) which exists in the mixing, we discuss the second kind (direct violation) which is the CP violation in the decay. It can be observed when the decay rate is different from that of CP conjugate mode. Under the assumption of CPT invariance, the decay amplitude can be written as (see Vol. 1, Section 16.2)

$$
\begin{aligned}
A(B \to f) &= \sum_i D_i e^{i\delta_{sc}} = \sum_i |D_i| e^{i\phi_W} e^{i\delta_{sc}} \\
A(\overline{B} \to \overline{f}) &= \sum_i D_i^* e^{i\delta_{sc}} = \sum_i |D_i| e^{-i\phi_W} e^{i\delta_{sc}}
\end{aligned}
\tag{6.183}
$$

where suffix i denotes different intermediate states. ϕ_W is referred to as the weak phase and δ_{sc} denotes the scattering phase shift in the final states. The direct CP violation means $D_i \neq D_i^*$. A necessary condition for experimental verification is

$$
\delta\Gamma = \Gamma(B \to f) - \Gamma(\overline{B} \to \overline{f}) \neq 0 \tag{6.184}
$$

For this inequality to be satisfied, existence of the interference is necessary and it requires at least two intermediate states. Writing

$$
\begin{aligned}
A(B \to f) &= |D_1| e^{i\phi_{W1}} e^{i\delta_{s1}} + |D_2| e^{i\phi_{W2}} e^{i\delta_{s2}} \\
A(\overline{B} \to \overline{f}) &= |D_1| e^{-i\phi_{W1}} e^{i\delta_{s1}} + |D_2| e^{-i\phi_{W2}} e^{i\delta_{s2}}
\end{aligned}
\tag{6.185}
$$

the difference of the decay rate becomes

$$
\Gamma(B \to f) - \Gamma(\overline{B} \to \overline{f}) = |D_1 D_2| \sin(\phi_{W1} - \phi_{W2}) \sin(\delta_{s1} - \delta_{s2}) \tag{6.186}
$$

This demonstrates that for the direct CP violation to be observed, existence of at least two intermediate states having different strong phase as well as the weak phases are required. For the case of neutral K mesons, they were provided by two different isospin states ($I = 0, 2$) of the 2π final states. On the other hand, $K^\pm \to \pi^\pm \pi^0$ has only $I = 2$ and no CP violation effect can be observed.

$B^+ \to K^+\pi^0$ Turning to the B meson sector, let us consider, for instance, $B^+ \to K^+\pi^0$. It has a tree process by emission of $W^\pm$ and the penguin process as well (Figure 6.12a,b). Therefore, the amplitude $A(B^+ \to K^+\pi^0)$ has at least two intermediate states, and hence there is a possibility to observe the direct

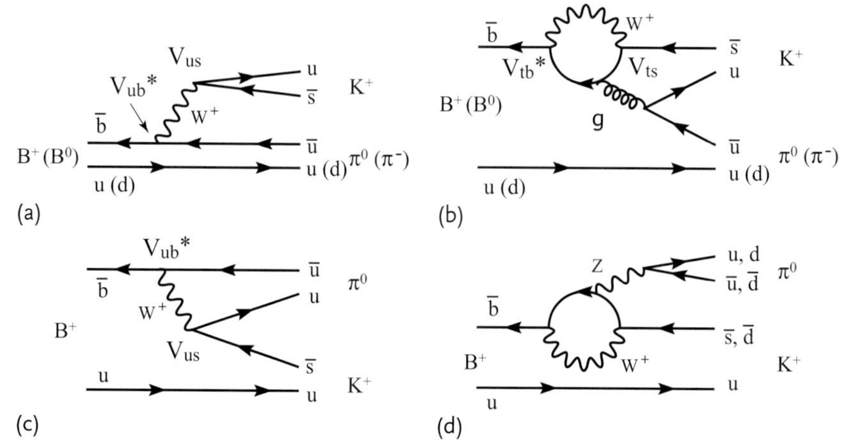

Figure 6.12 $B^+ \to K^+\pi^0(\pi^-)$ process has both tree (a) and penguin diagrams (b). Other processes such as (c) (tree process, but color suppressed)[18] and (d) (electroweak penguin) contribute, but are expected to be subdominant.

CP violation. The former is suppressed compared to the dominant decay mode $\overline{b} \to \overline{c} + W^+$ by the CKM matrix elements $|V_{ub}^*/V_{cb}|$ and the latter by presence of loop processes. Then, both are of the same order. Therefore, the branching ratio may be small (experimentally $\sim 10^{-5}$), but large interference is expected. The effect was observed at the B-factory [12, 311].

Figures 6.13c,d show experimental data of $B^- \to K^-\pi^0$ and $B^+ \to K^+\pi^0$ obtained by the Belle group. More events from $B^- \to K^-\pi^0$ are observed than those from $B^+ \to K^+\pi^0$, and the asymmetry has been determined to be [12]

$$\begin{aligned} A_{K^\pm\pi^0} &= \frac{\Gamma(B^- \to K^-\pi^0) - \Gamma(B^+ \to K^+\pi^0)}{\Gamma(B^- \to K^-\pi^0) + \Gamma(B^+ \to K^+\pi^0)} \\ &= +0.07 \pm 0.03_{\text{stat.}} \pm 0.01_{\text{syst.}} \end{aligned} \tag{6.187}$$

Thus, existence of the direct CP violation was confirmed in the $B^\pm \to K^\pm\pi^0$ mode. Magnitude of the asymmetry is consistent with calculations assuming dominance of the two diagrams in Figure 6.12a,b.

One can consider a similar process with the spectator u quark replaced with the d quark, and indeed the asymmetry in the yield between the CP conjugate modes $B^0 \to K^+\pi^-$ and $\overline{B}^0 \to K^-\pi^+$ were observed [311].

$$\begin{aligned} A_{K^\pm\pi^\mp} &= \frac{\Gamma(\overline{B}^0 \to K^-\pi^+) - \Gamma(B^0 \to K^+\pi^-)}{\Gamma(\overline{B}^0 \to K^-\pi^+) + \Gamma(B^0 \to K^+\pi^-)} \\ &= -0.094 \pm 0.018_{\text{stat.}} \pm 0.008_{\text{syst.}} \end{aligned} \tag{6.188}$$

18) The color conservation in the process Figure 6.12a is automatic, but in Figure 6.12c, the color of $\overline{s}$ from W^+ decay and that of spectator u have to match, thus giving a factor $1/N_c$ reduction.

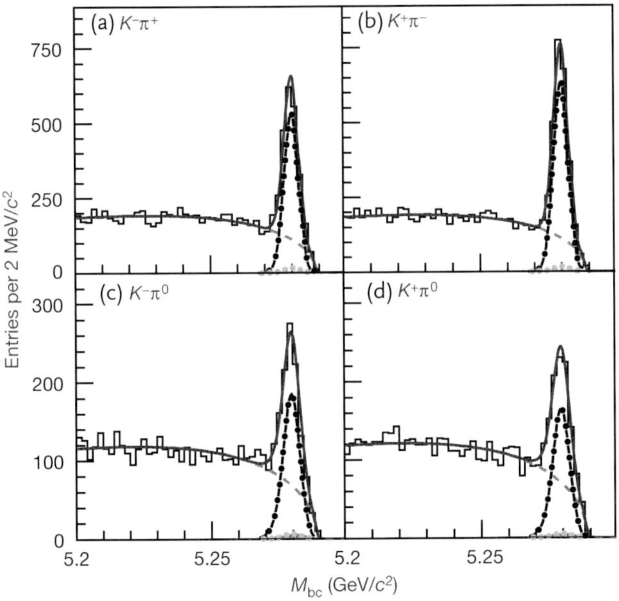

Figure 6.13 M_{bc} projections for $K^-\pi^+$ (a), $K^+\pi^-$ (b), $K^-\pi^0$ (c) and $K^+\pi^0$ (d). Histograms are data, solid blue lines are the fit projections, point-dashed lines are the signal components, dashed lines are the continuum background, and gray dotted lines are the $\pi^\pm\pi$ signals that are misidentified as $K^\pm\pi$. M_{bc} is a variable that should agree with the B meson mass (= 5.279 GeV) if the observed $K\pi$ events are decay products of the B meson. It is defined by $\sqrt{E_{\text{beam}}^2 - (\boldsymbol{p}_K + \boldsymbol{p}_\pi)^2}$, $E_{\text{beam}} = m\{\Upsilon(4S)\}/2$ [12]. For a color version of this figure, please see the color plates at the beginning of the book.

The data are shown in Figure 6.13a,b. Note the asymmetry is taken between the particle b in $B^-(b\overline{u})$ or $\overline{B}^0(b\overline{d})$ and the antiparticle $\overline{b}$ in $B^+(\overline{b}u)$ or $B^0(\overline{b}d)$.

One may notice some anomaly between the two modes. If the asymmetry of both the charged and neutral modes are due to interference of the diagrams in Figure 6.12a,b with the spectator as either u or d, they should be similar. Experimentally, they are not only different in magnitude, but also in sign. The difference is given by

$$\Delta A = A_{K^\pm\pi^0} - A_{K^\pm\pi^\mp} = 0.164 \pm 0.037 \tag{6.189}$$

Some argue that processes like Figure 6.12c,d may be important, contrary to common thinking, and some even argue that elements of new physics may exist. Quantitatively, the experimental evidence is not yet firm and the subject is under debate.

Other decay modes where the direct CP violation is observed are $B \to \pi^+\pi^-$, $B^0 \to \eta K^{*0}$ and $B^+ \to \rho^0 K^+$ [3].

Extraction of ϕ_3 The previous example exhibits existence of the direct CP violation. However, comparisons with the theory are hampered by the penguin contribution which is a QCD process and hard to estimate quantitatively. Processes free

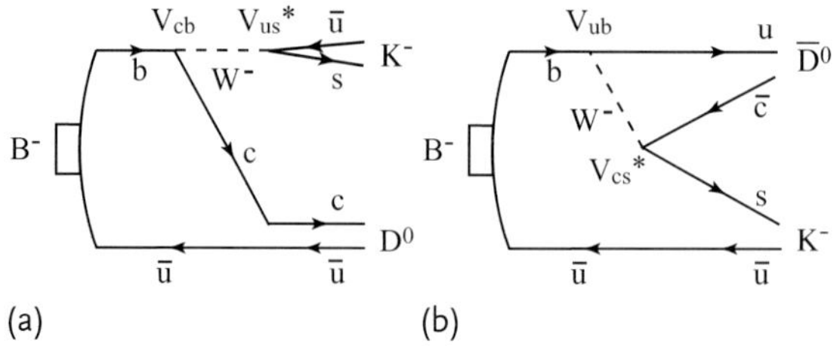

Figure 6.14 $B^- \to K^- D(\overline{D}^0)$ process which can be used to extract CKM phase ϕ_3 using a common final state $|fK^-\rangle$ $(D^0(\overline{D}^0) \to f)$.

of penguin pollution are desirable. An example is the interference of $B^- \to D^0 K^-$ and $B^- \to \overline{D}^0 K^-$ transitions (Figure 6.14) which further decay to common final states. It can be used to extract the unitarity angle $\phi_3(=\gamma)$ [312, 313].[19] As Figure 6.14a,b contains $V_{cb}V_{us}^* \sim A\lambda^2 \times \lambda$ and $V_{ub}V_{cs}^* \sim A\lambda^3(\overline{\rho} - i\overline{\eta})$ respectively, the interference term contains $\sim V_{ub} \sim e^{-i\phi_3}$.

A variety of practical choices for the common final state $D^0(\overline{D}^0) \to f$: $f = \pi^+\pi^-$ [314, 315], $K^-\pi^0$ [316, 317] and $K_s\pi^+\pi^-$ [318, 319] were discussed and measured at the B factories (KEKB and PEPII/SLAC). Here, we consider $f = K_s\pi^+\pi^-$ because it provides the best statistics. Note that the K_s in the final state is produced either as K^0 or $\overline{K}^0$, but the observed state is K_s, which is the short living neutral K. Diagrams of the process $D^0(\overline{D}^0) \to K_s\pi^+\pi^-$ are shown in Figure 6.15. The strong phase is large because of resonances, but it can be eliminated by observing the CP conjugate process $B^+ \to \overline{D}^0(D^0) \to K_s\pi^+\pi^-$. Disadvantage of the $K_s\pi\pi$ mode compared to others is that the state contains three particles, and hence one has to rely on the Dalitz plot for the analysis. Denoting invariant mass of the $K_s\pi^\pm$ as $m_\pm^2$, the decay amplitude for $B^- \to K_s\pi^+\pi^-$ can be written as [318, 319]

$$M_- \sim f(m_-^2, m_+^2) + r_B e^{i(\delta+\phi_3)} f\left(m_+^2, m_-^2\right) \tag{6.190}$$

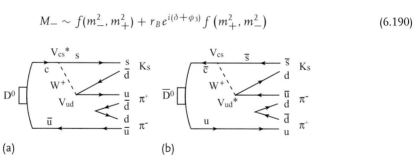

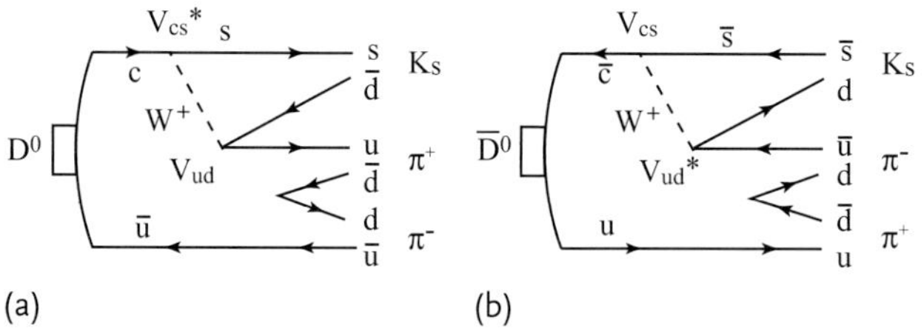

Figure 6.15 $K_s\pi^+\pi^-$ is a common final state to which both D^0 and $\overline{D}^0$ can decay. Another example is $K^+\pi^-$, see Figure 6.8.

19) A general method of extracting all three angles $\phi_1 \sim \phi_3$ will be discussed in Section 6.7. This is a special process that currently gives a better value for ϕ_3.

where $f(m_+^2, m_-^2)$ is the complex decay amplitude of $\overline{D}^0 \to K_s\pi^+\pi^-$. The corresponding charge conjugate amplitude for $B^+ \to K_s\pi^+\pi^-$ is

$$M_+ \sim f(m_+^2, m_-^2) + r_B e^{i(\delta-\phi_3)} f\left(m_-^2, m_+^2\right) \qquad (6.191)$$

The function $f(m_+^2, m_-^2)$ can be determined from a large sample of flavor-tagged $D^0(\overline{D}^0) \to K_S^0\pi^+\pi^-$ decays produced in continuum e^-e^+ annihilation using a resonance model. The resonance dominant model is suggested by strong resonance enhancement in the Dalitz plot as seen in Figure 6.16a–c [320, 322]. Then, the Dalitz distributions of both decay modes are fitted to determine r_B, δ and ϕ_3.

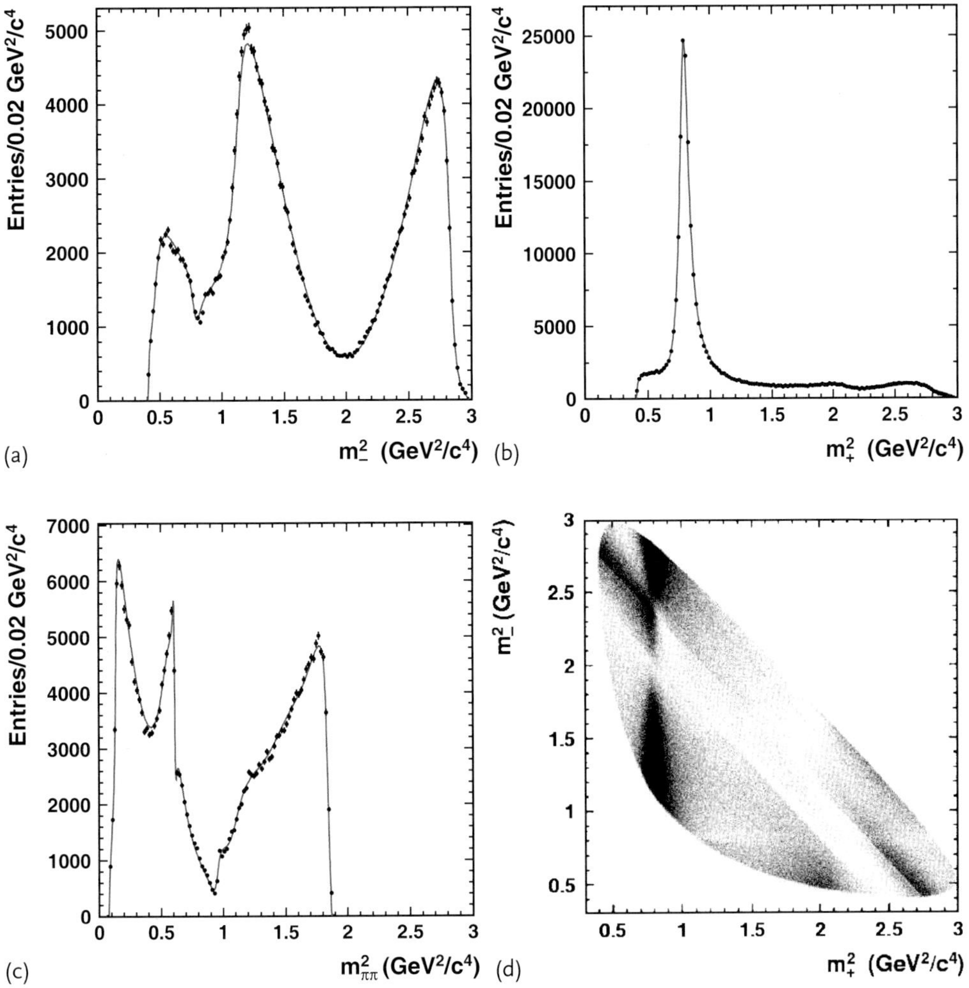

Figure 6.16 (a) m_+^2, (b) m_-^2, (c) $m_{\pi^+\pi^-}^2$ distributions and (d) Dalitz plot for the $\overline{D}^0 \to K_s\pi^+\pi^-$ decay from the $D^{*\pm} \to D\pi_s^\pm$ process. The points with error bars show the data; the smooth curve is the fit result [318, 320, 321].

In the actual analysis, decay modes which contain D^* and K^* as well as D and K discussed here are included to increase statistics. The obtained value of the ϕ_3 is [3]

$$\phi_3(=\gamma) = (73^{+23}_{-22})^\circ \tag{6.192}$$

6.6.3
CP Violation in Interference

In the previous section, we saw that at least two intermediate states are required to test the direct CP violation effect. At this point, we notice that for the case of neutral B meson decays $B^0, \overline{B}^0 \to f$ with common final states, the mixing effect provides the CP conjugate amplitude $B \to \overline{B} \to f$ that can interfere. We will show that by choosing CP eigenstates for the final state $(\mathrm{CP}|f\rangle = |\overline{f}\rangle = \pm|f\rangle)$, it is possible to extract the weak phases without requiring two interfering intermediates states [214, 323].

Our starting point is Eqs. (6.97a) and (6.99b) which are reproduced here.

$$\Gamma(B^0 \to f; t) = |A_f|^2 |g_+(t) + \lambda_f g_-(t)|^2 \tag{6.193a}$$

$$\Gamma(\overline{B}^0 \to \overline{f}; t) = |\overline{A}_{\overline{f}}|^2 |g_+(t) + \lambda_{\overline{f}}^{-1} g_-(t)|^2 \tag{6.193b}$$

$$\lambda_f \equiv \frac{q}{p}\frac{\overline{A}_f}{A_f} = \frac{q}{p}\frac{\langle f|H|\overline{B}^0\rangle}{\langle f|H|B^0\rangle}, \quad \lambda_{\overline{f}} \equiv \frac{q}{p}\frac{\overline{A}_{\overline{f}}}{A_{\overline{f}}} = \frac{q}{p}\frac{\langle \overline{f}|H|\overline{B}^0\rangle}{\langle \overline{f}|H|B^0\rangle} \tag{6.193c}$$

Let us assume that the final state f is a CP eigenstate, that is, $|\overline{f}\rangle = \eta_{\mathrm{CP}}|f\rangle$ and that only one intermediate state contributes. Then,

$$\overline{A}_{\overline{f}} = A_f^* = |A_f| e^{-i\phi_{\mathrm{W}}}, \quad \overline{A}_f = \eta_{\mathrm{CP}}\langle \overline{f}|H|\overline{B}^0\rangle = \eta_{\mathrm{CP}} A_f^*$$

$$\lambda_f = \lambda_{\overline{f}} = \eta_{\mathrm{CP}}\frac{q}{p} e^{-2i\phi_{\mathrm{W}}} \tag{6.194}$$

where ϕ_{W} is the weak phase defined in Eq. (6.183). The strong phase does not come into the λ_f. We have already seen that $|q/p|$ is very close to one. Therefore, it can be written as

$$\frac{q}{p} \sim e^{-2i\phi_M} \approx \left(\frac{M_{12}^*}{M_{12}}\right)^{1/2} \approx \begin{cases} (V_{td}V_{tb}^*)/(V_{td}^*V_{tb}) & B_d \\ (V_{ts}V_{tb}^*)/(V_{ts}^*V_{tb}) \approx 1 & B_s \end{cases} \tag{6.195}$$

In this case, λ_f is a pure phase and can be expressed as

$$\lambda_f \equiv \eta_{CP}|\lambda_f| e^{-2i\phi_\lambda} \simeq \eta_{\mathrm{CP}} e^{-2i(\phi_M + \phi_{\mathrm{W}})} \tag{6.196}$$

Let us further assume that $y = \Delta\Gamma/2\Gamma = 0$, which is a good approximation for B_d^0–$\overline{B}_d^0$ system [see Eq. (6.175)]. In this case,

$$|g_\pm(t)|^2 = \frac{e^{-\Gamma t}}{2}[1 \pm \cos\Delta m t]$$

$$g_+(t)^* g_-(t) = i\frac{e^{-\Gamma t}}{2}\sin\Delta m t \tag{6.197}$$

Substituting Eq. ((6.197)) into Eq. (6.193), we have

$$\begin{aligned}\Gamma(B^0 \to f;t) &= |A_f|^2 e^{-\Gamma t}[1 - \mathrm{Im}\lambda_f \sin \Delta m t] \\ &= |A_f|^2 e^{-\Gamma t}[1 + \eta_{\mathrm{CP}} \sin 2\phi_\lambda \sin \Delta m t]\end{aligned} \tag{6.198a}$$

$$\Gamma(\overline{B}^0 \to f;t) = |\overline{A}_f|^2 e^{-\Gamma t}[1 - \eta_{\mathrm{CP}} \sin 2\phi_\lambda \sin \Delta m t] \tag{6.198b}$$

The asymmetry becomes

$$A_{fCP}(t) \equiv \frac{\Gamma(\overline{B} \to f;t) - \Gamma(B \to f;t)}{\Gamma(\overline{B} \to f;t) + \Gamma(B \to f;t)} = -\eta_{\mathrm{CP}} \sin 2\phi_\lambda \sin \Delta m t \tag{6.199}$$

If the time variation is not observed, we integrate the numerator and denominator, giving

$$A_{fCP}(\mathrm{int}) = -\left[\frac{x}{1+x^2}\right] \eta_{\mathrm{CP}} \sin 2\phi_\lambda \tag{6.200}$$

The above formula shows that even if both $|q/p| = 1$ and $|\overline{A}_f/A_f| = 1$ hold, which means absence of the indirect as well as the direct CP violation, one can still detect the CP violation in the phase of λ_f. It is referred to as the CP violation in interference between a decay without mixing ($M^0 \to f$) and a decay with mixing ($M \to \overline{M}^0 \to f$) [or CP violation in the mixing for short]. By choosing appropriate final states, we can extract the desired weak phases, including those that determine the unitarity triangle.

If $A_f = \overline{A}_f$, $y = 0$ are not assumed, the decay rate is modified to

$$\Gamma(B^0 \to f;t) = |A_f|^2 \frac{(1+|\lambda_f|^2)}{2} e^{-\Gamma t} \left[\cosh y\Gamma t + \frac{2\mathrm{Re}\lambda_f}{1+|\lambda_f|^2} \sinh y\Gamma t + C_f \cos \Delta m t - S_f \sin \Delta m t\right] \tag{6.201a}$$

$$\Gamma(\overline{B}^0 \to f;t) = |\overline{A}_f|^2 \frac{(1+|\lambda_f|^2)}{2|\lambda_f|^2} e^{-\Gamma t} \left[\cosh y\Gamma t + \frac{2\mathrm{Re}\lambda_f}{1+|\lambda_f|^2} \sinh y\Gamma t - C_f \cos \Delta m t + S_f \sin \Delta m t\right] \tag{6.201b}$$

$$S_f \equiv \frac{2\mathrm{Im}\lambda_f}{1+|\lambda_f|^2}, \quad C_f \equiv \frac{1-|\lambda_f|^2}{1+|\lambda_f|^2} \tag{6.201c}$$

If further $y = \Delta\Gamma/2\Gamma = 0$ is assumed, the asymmetry is given by

$$A_{fCP}(t) = S_f \sin \Delta m t - C_f \cos \Delta m t \tag{6.202}$$

The second term represents a direct CP violation. Notice that if $\Delta\Gamma = 0$ and $|q/p| = 1$ are assumed as expected to be a good approximation for B mesons but not for K mesons and if, in addition, $|\overline{A}_f| = |A_f|$ is assumed, the interference between decays with and without mixing is the only source of the asymmetry.

As a final remark, we remind the reader that the value of q/p in Eq. (6.195) is valid only when we adopt the convention for CKM matrix in the form of Eq. (6.16). The phase of q/p can be changed by redefinition of $|\overline{B}^0\rangle$, and it is not an observable by itself. However, λ_f is phase convention independent because by rephasing $|B^0\rangle \to e^{-i\xi}|B^0\rangle$, $|\overline{B}\rangle \to e^{i\xi}|\overline{B}^0\rangle$, the mass matrix changes accordingly to $\Lambda_{12} = \langle B^0|\Lambda|\overline{B}^0\rangle \to e^{2i\xi}\Lambda_{12}$. Therefore, $q/p = \sqrt{\Lambda_{21}/\Lambda_{12}} \to e^{-2i\xi}(q/p)$ and $\overline{A}_f/A_f = \langle f|H|\overline{B}^0\rangle/\langle f|H|B^0\rangle \to e^{2i\xi}\overline{A}_f/A_f$ which keeps the phase of λ_f unchanged.

6.6.4
Coherent $B^0\overline{B}^0$ Production

Thus far, in discussing decay processes, we always treated the $|B^0\rangle$ or $|\overline{B}^0\rangle$ states as given. Experimentally, however, the particle identification is made by observing its decay modes or secondary scatterings. For instance, in the semileptonic decay mode of the neutral B mesons, sign of the lepton could tell its parent. However, in measuring common final states to which both B^0 and $\overline{B}$ can decay, it is imperative to identify their parent state.

In the B-factory, coherent $B^0\overline{B}^0$ pairs are produced through the process $e^- + e^+ \to \Upsilon(4S) \to B_d^0 + \overline{B}_d^0$ in $J^{PC} = 1^{--}$ state. They are produced with definite orbital angular momentum ($L = 1$) and extraction of theoretically pure samples is possible. In observing decay particles, one can identify their parent state by tagging the other member of the pair which makes a clean experiment possible. Taking into account the correlation between the tagged state and observed CP eigenstate, the time evolution equations are modified, but have a similar form to Eq. (6.201).

Depending on the orbital angular momentum (L) state, the coherent two-particles state can be expressed as

$$|\psi\rangle = |B^0\overline{B}^0\rangle + CP|B^0\overline{B}^0\rangle = |B^0\overline{B}^0\rangle \pm |\overline{B}^0 B^0\rangle \tag{6.203}$$

where the sign $\pm$ corresponds to $L =$ even/odd states. The $\Upsilon(4S)$ is a $L =$ odd state, but we retain both signs in the following discussions. We write time dependence of $|\psi\rangle$ as

$$|\psi(t_1, t_2)\rangle = |B^0; k_1, t_1\rangle|\overline{B}^0; k_2, t_2\rangle \pm |\overline{B}^0; k_1, t_1\rangle|B^0; k_2, t_2\rangle \tag{6.204}$$

where $|B^0; t\rangle, |\overline{B}^0; t\rangle$ are given by Eq. (6.93).

For tagged measurements, we identify one particle as either B^0 or $\overline{B}^0$ and observe the other to decay into final state $|f\rangle$. The final state g which can be used for tagging

$B^0(\overline{B}^0)$ has to satisfy

$$\begin{aligned}&\langle g|H|B^0\rangle \neq 0\,, \quad \langle g|H|\overline{B}^0\rangle = 0 \quad \text{tag } B^0\\ &\langle \overline{g}|H|B^0\rangle = 0\,, \quad \langle \overline{g}|H|\overline{B}^0\rangle \neq 0 \quad \text{tag } \overline{B}^0\end{aligned} \tag{6.205}$$

Examples of such states were given in Eqs. (6.102) and (6.103).

We consider two cases in the following: a semileptonic decay where $f, g = l^{\pm}\nu X$, namely, the second particle can also be identified as B^0 or $\overline{B}^0$ and the other where f is a CP eigenstate to which both $B^0, \overline{B}^0$ can decay.

Semileptonic Decays We can use $g = l^+$ to tag the B^0. Suppose the particle 1 decays to B^0 at time t_1 in a quantum state k_1, then

$$\langle g(t_1)|\psi(t_1,t_2)\rangle = \langle g|H|B^0\rangle \left[g_+(t_1)|\overline{B}^0; k_2, t_2\rangle \pm (p/q) g_-(t_1)|B^0; k_2, t_2\rangle\right] \tag{6.206}$$

Furthermore, let us suppose that the second particle is observed as $\overline{g}$ at t_2 in a quantum state k_2, then the transition amplitude can be written as

$$\langle g(t_1)\overline{g}(t_2)|\psi(t_1,t_2)\rangle = \langle g|H|B^0\rangle\langle \overline{g}|H|\overline{B}\rangle F_{\pm}(t_1,t_2) \tag{6.207}$$

where

$$F_{\pm}(t_1,t_2) = g_+(t_1)g_+(t_2) \pm g_-(t_1)g_-(t_2) \tag{6.208}$$

If the second particle is also identified as g at t_2, the amplitude becomes

$$\langle g(t_1)g(t_2)|\psi(t_1,t_2)\rangle = (\langle g|H|B^0\rangle)^2 (p/q) G_{\pm}(t_1,t_2) \tag{6.209}$$

$$G_{\pm}(t_1,t_2) = g_+(t_1)g_-(t_2) \pm g_-(t_1)g_+(t_2) \tag{6.210}$$

Similarly,

$$\langle \overline{g}(t_1)g(t_2)|\psi(t_1,t_2)\rangle = \langle g(t_1)\overline{g}(t_2)|\psi(t_1,t_2)\rangle \tag{6.211a}$$

$$\langle \overline{g}(t_1)\overline{g}(t_2)|\psi(t_1,t_2)\rangle = \pm(\langle \overline{g}|H|\overline{B}^0\rangle)^2 (q/p) G_{\pm}(t_1,t_2) \tag{6.211b}$$

The functions $F_{\pm}, G_{\pm}$ are expressed as

$$\begin{aligned}F_{\pm}(t_1,t_2) &= \quad e^{-iM(t_1+t_2)}e^{-(\Gamma/2)(t_1+t_2)}\cos\left[i\frac{\Delta\Gamma}{4} + \frac{\Delta m}{2}\right](t_1 \pm t_2)\\ G_{\pm}(t_1,t_2) &= \pm i e^{-iM(t_1+t_2)}e^{-(\Gamma/2)(t_1+t_2)}\sin\left[i\frac{\Delta\Gamma}{4} + \frac{\Delta m}{2}\right](t_1 \pm t_2)\end{aligned} \tag{6.212}$$

If we approximate $y = \Delta\Gamma/\Gamma = 0$, then the functions $F_{\pm}, G_{\pm}$ are expressed as

$$\begin{aligned}F_{\pm}(t_1,t_2) &\approx \quad e^{-iM(t_1+t_2)}e^{-(\Gamma/2)(t_1+t_2)}\cos\frac{\Delta m}{2}(t_1 \pm t_2)\\ G_{\pm}(t_1,t_2) &\approx \pm i e^{-iM(t_1+t_2)}e^{-(\Gamma/2)(t_1+t_2)}\sin\frac{\Delta m}{2}(t_1 \pm t_2)\end{aligned} \tag{6.213}$$

For semileptonic decays, we set $g = l^+\nu h$, $\overline{g} = l^-\overline{\nu}\overline{h}$, and we obtain

$$dN(l^+l^-) = dN(l^-l^+) \sim e^{-\Gamma(t_1+t_2)}[1 + \cos\Delta m(t_1 \pm t_2)] \tag{6.214a}$$

$$dN(l^+l^+) \sim \left|\frac{p}{q}\right|^2 e^{-\Gamma(t_1+t_2)}[1 - \cos\Delta m(t_1 \pm t_2)] \tag{6.214b}$$

$$dN(l^-l^-) \sim \left|\frac{q}{p}\right|^2 e^{-\Gamma(t_1+t_2)}[1 - \cos\Delta m(t_1 \pm t_2)] \tag{6.214c}$$

If the time evolution is not measured, the decay rates are to be integrated. Considering

$$\int dt_1 dt_2 e^{-\Gamma(t_1+t_2)} \cos\Delta m(t_1 \pm t_2) = \begin{cases} \frac{1}{\Gamma}\frac{1-x^2}{(1+x^2)^2} & L = \text{even} \\ \frac{1}{\Gamma}\frac{1}{(1+x^2)} & L = \text{odd} \end{cases} \tag{6.215}$$

we can deduce values of the mixing ratio r

$$r = \frac{N(l^+l^+) + N(l^-l^-)}{N(l^+l^-) + N(l^-l^+)} = \begin{cases} \frac{1}{2}\left(\left|\frac{p}{q}\right|^2 + \left|\frac{q}{p}\right|^2\right)\frac{3x^2+x^4}{2+x^2+x^4} & L = \text{even} \\ \frac{1}{2}\left(\left|\frac{p}{q}\right|^2 + \left|\frac{q}{p}\right|^2\right)\frac{x^2}{2+x^2} & L = \text{odd} \end{cases} \tag{6.216}$$

The expression for r, the ratio for the same sign to opposite sign lepton pairs for $L =$ odd agrees with Eq. (6.106b) derived without considering the correlation if we set $y = 0$. Since $\Upsilon(4S)$ has $L = 1$, we have justified the statement in footnote 12 made just after Eq. (6.109). The coincidence is not accidental, but can be derived from the Bose–Einstein statistics of the B mesons. If $L =$ odd, the two-body wave function is antisymmetric and B^0B^0, $\overline{B}^0\overline{B}^0$ states cannot exist. Therefore, once the particle 1 is identified as B^0, the particle 2 can only be in $\overline{B}^0$ state. Therefore, the probability of decaying to $\overline{B}^0$ after the partner is identified as B^0 is the same as that of $\Gamma(B^0 \to \overline{B}^0)$ of uncorrelated one particle parent.

EPR paradox: Incidentally, this is one counter example to the famous EPR (Einstein–Podolsky–Rosen) paradox [324–327] which claims that if two particles were identified independently within time interval Δt at a distance far enough ($\Delta x > c\Delta t$), then no correlation should exist between the two states because it violates the causality.

In the measurements of the $B^0\overline{B}^0$ pair productions, only decay products of $B\overline{B}$ can be observed and no particles to help identify the interaction point are emitted, namely, one does not know the collision point. Therefore, t_1 and t_2 cannot be determined independently. Only the time difference $t = t_1 - t_2$ (actually, the interval of two decay points) can be measured. In other words, even when time evolution of the decaying B mesons are measured, one has to integrate Eq. (6.214) over the total time $T = t_1 + t_2$. For fixed Δt, the result is

$$dN(l^{\pm}l^{\pm}) = \begin{cases} \frac{e^{-\Gamma|\Delta t|}}{1+x^2}\left[x^2 + 2\sin^2\frac{x}{2}\frac{|\Delta t|}{\tau} + x\sin x\frac{|\Delta t|}{\tau}\right] & L = \text{even} \\ e^{-\Gamma|\Delta t|}\cdot 2\sin^2\frac{x}{2}\frac{|\Delta t|}{\tau} & L = \text{odd} \end{cases} \tag{6.217}$$

At the B-factory, the $B\overline{B}$ system is produced almost at rest. Considering the B lifetime $\sim 1.5\times 10^{-12}$ ps, in the symmetric $e^- e^+ \to \Upsilon(4S)$ collider ($E_{e^-} = E_{e^+}$), the mean decay length ℓ is

$$\langle\ell\rangle = c\beta\gamma\tau \approx 30\ \mu\text{m} \tag{6.218}$$

It is small and the produced B mesons decay before they show any sign of flight. To determine the decay vertex points, one has to extrapolate tracks of the decay particles measured outside the beam pipe. It is hard to obtain the required accuracy. Making the colliding energy asymmetric ($E_{e^+} \gg E_{e^-}$ or $E_{e^+} \ll E_{e^-}$), the decay length can be prolonged due to the Lorentz boost. Chosen values were $E_{e^+} =$ 3.5 GeV, $E_{e^-} = 8$ GeV at KEK/B and $E_{e^+} = 3.1$ GeV, $E_{e^-} = 9$ GeV at PEPII/SLAC. The decay length becomes $\langle\ell\rangle \approx 380\ \mu$m at KEKB. Assuming the vertex position resolution $\sigma \simeq 40\ \mu$m, one expects to measure the time evolution curve up to $x = \sim 8$–10. As $x_d \simeq 0.8$, one can expect the experiments at the B-factory good enough to measure B_d oscillation.

Measurements of the B_s oscillation is harder compared to B_d because

1. the production rate of B_s is smaller than B_d by 1/2–1/3
2. the mixing effect is at its maximum
3. $x_s \sim 26$ as compared to $x_d \sim 0.77$, hence the oscillation period is ~ 40 times shorter.

Here, the hadron collider has the advantage because of their higher energy beams. The produced $B_s^0\overline{B}_s^0$ are more boosted with longer decay length which makes observation of the time variation easier. The fact that the $B_s^0 - \overline{B}_s^0$ oscillation was observed in the Tevatron reflects this fact (see Figure 6.6).

Decays to CP Eigenstates Let us consider the second case where we want to measure the CP violation in the mixed $B^0\overline{B}^0$ decays. Assuming $\overline{B}^0$ is tagged at $t = t_1$ and the other particle ($= B^0$) decayed to a CP eigenstate f at time $t = t_2$, the amplitude is given by

$$\begin{aligned}
\langle g(t_1) f(t_2)|\psi\rangle &= A(B^0 \to f; t_2) \\
&= \langle \overline{g}|H|\overline{B}^0\rangle \Bigg[\frac{q}{p} g_-(t_1) \left\{ \frac{p}{q} g_-(t_2)\langle f|H|B^0\rangle + g_+(t_2)\langle f|H|\overline{B}^0\rangle \right\} \\
&\quad \pm g_+(t_1) \left\{ g_+(t_2)\langle f|H|B^0\rangle + \frac{q}{p} g_-(t_2)\langle f|H|\overline{B}^0\rangle \right\} \Bigg] \\
&= \pm \overline{A}_{\overline{g}} A_f [F_\pm(t_1, t_2) + \lambda_f G_\pm(t_1, t_2)]
\end{aligned} \tag{6.219}$$

If the particle 1 is tagged as B^0 at $t = t_1$ and the particle 2 ($= \overline{B}^0$) decays to f at $t = t_2$, the decay amplitude is obtained from Eq. (6.206) by multiplying $\langle f(t_2)|$ from the left.

$$\begin{aligned}
\langle g(t_1) f(t_2)|\psi\rangle &= A(\overline{B}^0 \to f; t_2) \\
&= \langle g|H|B^0\rangle \Bigg[g_+(t_1) \left\{ \frac{p}{q} g_-(t_2)\langle f|H|B^0\rangle + g_+(t_2)\langle f|H|\overline{B}^0\rangle \right\} \\
&\quad \pm \frac{p}{q} g_-(t_1) \left\{ g_+(t_2)\langle f|H|B^0\rangle + \frac{q}{p} g_-(t_2)\langle f|H|\overline{B}^0\rangle \right\} \Bigg] \\
&= A_g A_f \frac{p}{q} \big[G_\pm(t_1, t_2) + \lambda_f F_\pm(t_1, t_2) \big]
\end{aligned} \tag{6.220}$$

By making square of Eqs. (6.220) and (6.219), and substituting Eq. ((6.213)), we obtain

$$\Gamma[B^0(t_1), f(t_2)] \sim e^{-\Gamma(t_1+t_2)} \big[1 \pm \mathrm{Im}\lambda_f \sin \Delta m(t_2 \pm t_1)\big] \tag{6.221a}$$

$$\Gamma[\overline{B}^0(t_1), f(t_2)] \sim e^{-\Gamma(t_1+t_2)} \big[1 \mp \mathrm{Im}\lambda_f \sin \Delta m(t_2 \pm t_1)\big] \tag{6.221b}$$

where

$$\lambda_f = \frac{q}{p} \frac{\langle f|H|\overline{B}\rangle}{\langle f|H|B\rangle} \tag{6.222}$$

and we have set $|q/p| = 1$. The sign $\pm$ corresponds to $L =$ even (+), odd (−). The asymmetry is given by

$$A_{fCP}(t) \equiv \frac{\Gamma(\overline{B}^0, f) - \Gamma(B^0, f)}{\Gamma(\overline{B}^0, f) + \Gamma(B^0, f)} = \mp \mathrm{Im}\lambda_f \sin \Delta m(t_2 \pm t_1) \tag{6.223}$$

One recognizes that the asymmetry is an odd function of $\Delta t = t_2 - t_1$ for the case L is odd. Therefore, if the time t_1 and t_2 are treated symmetrically, or integrated, the asymmetry compensates by itself to vanish. In order to observe the asymmetry, one has to set the collider energy at the resonance, make it asymmetric like the B-factory and detect the time evolution of the decaying B mesons. One can observe the time difference $\Delta t = t_2 - t_1$ as the difference of the decay flight paths, but not the sum $(t_1 + t_2)$. Therefore, by integrating the L=odd part of Eq. (6.221) over total time $t_1 + t_2$, one obtains

$$\begin{aligned} \Gamma(B^0, f : |\Delta t|) &\sim e^{-\Gamma|\Delta t|}\left[1 - \mathrm{Im}\lambda_f \sin \Delta m|\Delta t|\right] \\ \Gamma(\overline{B}^0, f : |\Delta t|) &\sim e^{-\Gamma|\Delta t|}\left[1 + \mathrm{Im}\lambda_f \sin \Delta m|\Delta t|\right] \end{aligned} \tag{6.224}$$

and the asymmetry is given by

$$A_{fCP}(t) = \frac{\Gamma(\overline{B}^0, f) - \Gamma(B^0, f)}{\Gamma(\overline{B}^0, f) + \Gamma(B^0, f)} = \mathrm{Im}\lambda_f \sin \Delta m \Delta t \tag{6.225}$$

Integrating over the asymmetry, one gets

$$\begin{aligned} A_{f\mathrm{CP}} &= \frac{\int d\Delta t\left[\Gamma(\overline{B}^0, f) - \Gamma(B^0, f)\right]}{\int d\Delta t\left[\Gamma(\overline{B}^0, f) + \Gamma(B^0, f)\right]} \\ &\sim \mathrm{Im}\lambda_f \frac{x}{1 + x^2} \end{aligned} \tag{6.226}$$

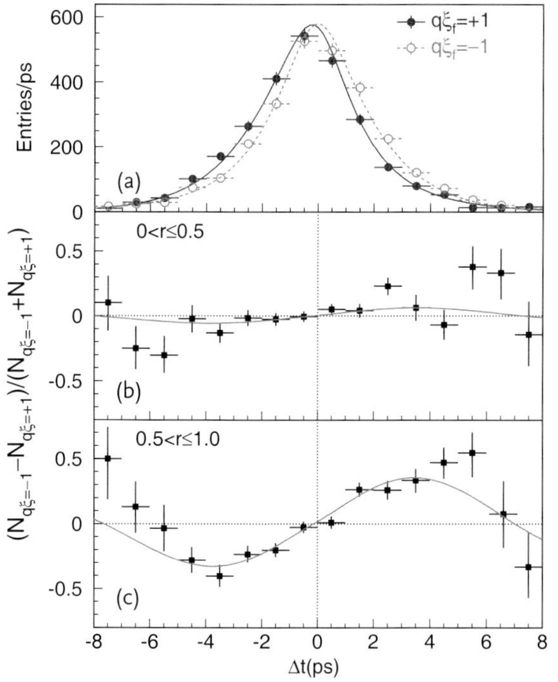

Figure 6.17 (a) Δt distributions for events with $q\xi_f = -1$ (open points) and $q\xi_f = +1$ (solid points) for various decay modes combined. $\xi_f = \eta_{\mathrm{CP}}$ in the text, and $q = \pm 1$. $q = 1$ means that B^0 is tagged and $\overline{B}^0$ decayed to a CP eigenstate while $q = -1$ means $\overline{B}^0$ is tagged. (b) Asymmetry of events for which tagging constraint is loose ($0 < r < 0.5$). (c) Asymmetry of clearly tagged ($0.5 < r < 1$) events. The results of the global unbinned maximum likelihood fit ($\sin 2\phi_1 = 0.728$) are also shown [14].

The asymmetry is reduced by the factor $[x/(1+x^2)]$ which is referred to as the dilution factor.

Figure 6.17 shows a time variation of the asymmetry for the decay process $B^0(\overline{B}^0) \to J/\psi + K_s$.

6.7
Extraction of the CKM Phases

6.7.1
Using B^0–$\overline{B}$ Mixing

As can be seen from Eq. (6.199) or (6.225), one can determine the phase of $\lambda_f = (q/p)\overline{A}_f/A_f$ by looking at the asymmetry in the B^0–$\overline{B}^0$ oscillation. It is a sum of the phase of the mass matrix and the phase difference of the amplitudes $\langle f|H|\overline{B}^0\rangle$ and $\langle f|H|B^0\rangle$. By choosing appropriate decay modes, we can determine angles of the unitarity triangle.

$\phi_1 = \beta$ As an example, let us consider the process $\overline{B}_d^0 \to J/\psi + K_s$ where K_s denotes the short life neutral kaon. A dominant diagram that contributes to the process is given in Figure 6.18a. As is clear from Figure 6.18, the process is $b \to c$ transition and the amplitude ratio $\overline{A}_f/A_f$ contains $(V_{cb}V_{cs}^*)/(V_{cb}^*V_{cs})$. To order λ^3 in the Wolfenstein parametrization, it does not contain any phase. However, $(q/p)_{B_d} = \sqrt{\Lambda_{21}/\Lambda_{12}}$ contains $(V_{tb}^*V_{td})/(V_{tb}V_{td}^*) \sim e^{i2\phi_1}$, where $\phi_1, \phi_2, \phi_3 (= \beta, \alpha, \gamma)$ are apex angles of the unitarity triangle defined by Eq. (6.56) (see Figure 6.11 for $i = t$). As we defined $\arg[q/p] \equiv -2i\phi_M$, the phase that can be measured by $\overline{B}_d \to J/\psi + K_s$ is $\arg[\lambda_f] = -2\phi_M = 2\phi_1$.

$$\overline{B}_d \to J/\psi + K_s : 2\phi_\lambda = -2\phi_M = 2\phi_1 \tag{6.227}$$

The current world average is given by [3]

$$\sin 2\phi_1 (= \sin 2\beta) = 0.673 \pm 0.023 \tag{6.228}$$

Other modes that are used to determine ϕ_1 are $\psi(2S)K_s, \eta_c K_s (\eta_{CP} = -)$ and $J/\psi K_L, J/\psi K^{*0}$ $(\eta_{CP} = +)$, and so on.

$\phi_2 = \alpha$ Next, let us take a look at Figure 6.18b which describes the process $\overline{B}_d \to \pi^+\pi^-(\rho^+\rho^-)$. This is a transition $b \to u$ and contains $V_{td}V_{ub}V_{ud}^* \sim V_{ub}V_{td}$ or ϕ_2. The phase that can be extracted is;

$$\overline{B}_d \to \pi^+\pi^- : \phi_\lambda = \arg\left[\frac{V_{td}}{V_{td}^*}\right] + \arg\left[\frac{V_{ub}}{V_{ub}^*}\right] = 2(\phi_1+\phi_3) \simeq -2\phi_2 \tag{6.229}$$

Experimentally, accuracy of ϕ_2 is not good yet. Including the $\rho^+\rho^-$ mode, the current world average is given by [3]

$$\phi_2(=\alpha) = (89.0^{+4.4}_{-4.2})^\circ \tag{6.230}$$

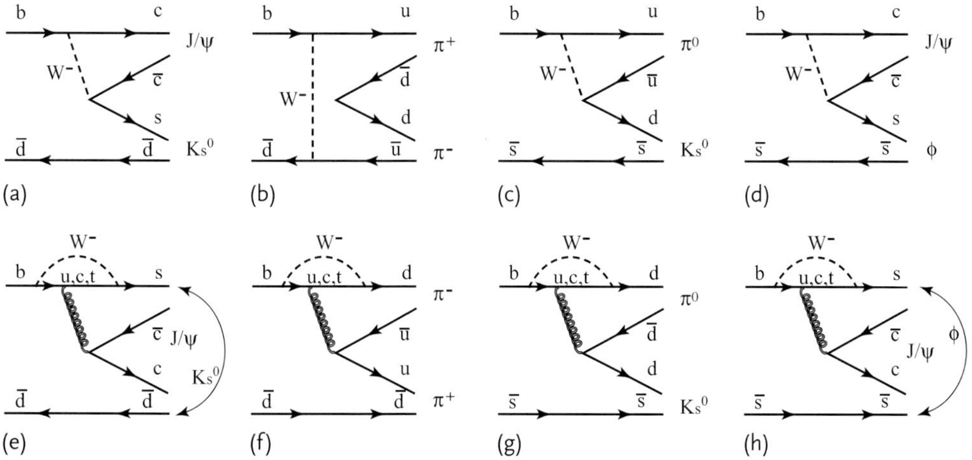

Figure 6.18 Processes which are useful for detecting the CP violation in the interference and extracting phases of the CKM matrix. (a) $\overline{B}_d \to J/\psi + K_s$ measures arg$[V_{td}]$ or ϕ_1. (b) $\overline{B}_d \to \pi^+\pi^-$ measures arg$[(V_{td}V_{ub})]$ or ϕ_2. (c) $\overline{B}_s \to \pi^0 + K_s$ measures arg$[V_{ub}V_{td}]$ or ϕ_3. (d) $\overline{B}_s \to J/\psi + \phi$ measures arg$[(V_{tb}V_{cb})/(V_{ts}^* V_{cb}^*)]$ which should vanish to $O(\lambda^2)$. (e)–(h) are corresponding penguin polution diagrams.

$\phi_3 = \gamma$ Similarly, by measuring $\overline{B}_s \to \rho^0(\pi^0) + K_s$ and $\overline{B}_s \to J/\psi + \phi$ (Figure 6.18c,d), one can extract the other angle.

$$\overline{B}_s \to \rho^0(\pi^0) + K_s : \quad \phi_\lambda = \arg\left[\left(\frac{q}{p}\right)_{B_s}\right] + \arg\left[\frac{V_{ub}V_{ud}^*}{V_{ub}^*V_{ud}}\right] \simeq 2\arg[V_{ub}] = 2\phi_3 \tag{6.231a}$$

$$\begin{aligned}\overline{B}_s \to J/\psi \cdot \phi : \quad \phi_\lambda &= \arg\left[\left(\frac{q}{p}\right)_{B_s}\right] + \arg\left[\frac{V_{cb}V_{cs}^*}{V_{cb}^*V_{cs}}\right] \simeq 2\arg\left[\frac{V_{tb}^*V_{ts}}{V_{cb}^*V_{cs}}\right] \\ &\equiv -2\phi_s = O(\lambda^2)\end{aligned} \tag{6.231b}$$

Regarding ϕ_3, the current best result comes from comparisons of $B^\mp \to D^0(\overline{D}^0)$ $K^\mp$ which was already discussed (Eq. (6.192)). The process Eq. (6.231b) does not produce any asymmetry at order (λ^2), but its absence is equally essential to validify the CKM formalism for the CP violation. The value of the angle is estimated from global fits to give $\phi_s = 0.018 \pm 0.001$ [328], but the measurements are far from decisive [329, 330].

We summarize the four processes and some others in Table 6.1. If one can measure the asymmetry in all the processes together with others to determine the three sides, the unitarity triangle can be over constrained. If they are all consistent, it validates the CKM formalism unambiguously. If one or more of the measurements do not agree with others, it means something is missing and is potentially a clue to new physics.

Table 6.1 Processes to extract CKM phases.

Parent	transition	final state	$-\arg[\lambda_f]$	CP	angle
$\overline{B}_d$	$b \to cs\overline{c}$	$J/\psi K_s$	$\arg[(V_{tb}^* V_{td})(V_{cb} V_{cs}^*)] \sim \arg[V_{td}]$	$-$	ϕ_1
$\overline{B}_d$	$b \to ud\overline{u}$	$\pi^+\pi^-$	$\arg[(V_{tb}^* V_{td})(V_{ub} V_{ud}^*)] \sim \arg[V_{td} V_{ub}]$	$+$	ϕ_2
$\overline{B}_d$	$b \to us\overline{u}$	$\pi^0 K_s$	$\arg[(V_{tb}^* V_{td})(V_{ub} V_{us}^*)] \sim \arg[V_{td} V_{ub}]$	$-$	ϕ_2
$\overline{B}_d$	$b \to cd\overline{c}$	$D^+ D^-$	$\arg[(V_{tb}^* V_{td})(V_{cb} V_{cd}^*)] \sim \arg[V_{td}]$	$+$	ϕ_1
$\overline{B}_s$	$b \to cs\overline{c}$	$J/\psi\phi$	$\arg[(V_{tb}^* V_{ts})(V_{cb} V_{cs}^*)] \sim 0$	$\sim +$	$O(\lambda^2)$
$\overline{B}_s$	$b \to ud\overline{u}$	$\pi^0 K_s$	$\arg[(V_{tb}^* V_{ts})(V_{ub} V_{ud}^*)] \sim \arg[V_{ub}]$	$-$	ϕ_3
$\overline{B}_s$	$b \to us\overline{u}$	$K^+ K^-$	$\arg[(V_{tb}^* V_{ts})(V_{ub} V_{us}^*)] \sim \arg[V_{ub}]$	$+$	ϕ_3
$\overline{B}_s$	$b \to cd\overline{c}$	$J/\psi K_s$	$\arg[(V_{tb}^* V_{ts})(V_{cb} V_{cs}^*)] \sim 0$	$-$	$O(\lambda^2)$

6.7.2
Penguin Pollution

In the discussion to determine the phases, we deliberately assumed that Figure 6.18 $(a) \sim (d)$ are the dominant diagrams to the above mentioned processes. Unfortunately, the world is not that simple. In most cases, there always exist competing background processes. Among them, the simplest ones are the penguin diagrams which are listed in Figure 6.18e–h right below the corresponding main tree processes Figure 6.18a–d. Writing the tree process amplitude as A_T and the corresponding penguin amplitude as A_P, the asymmetry is modified by the amount

$$\sim \left|\frac{A_\mathrm{P}}{A_\mathrm{T}}\right| \sin(\alpha_\mathrm{T} - \alpha_P) \tag{6.232}$$

where $\alpha_\mathrm{T}, \alpha_P$ are phases of the amplitudes. The penguin diagrams contain loop contributions due to u, c, t intermediate states. The momentum transfer of the u and c loops is $\sim O(m_b)$ and its contribution is proportional to and of the same order as [331]

$$\sim \frac{\alpha_s(m_b)}{12\pi} \ln\left(\frac{m_W^2}{m_b^2}\right) \tag{6.233}$$

Because of the unitarity condition

$$V_{ub} V_{uq}^* + V_{cb} V_{cq}^* = -V_{tb} V_{tq}^* \tag{6.234}$$

the amplitudes of the u, c loops have roughly the same phase as that of the t loop. Namely, the phase of the penguin diagrams are determined essentially by the top intermediate state. For the case $b \to cs\overline{c}$ decays ($\overline{B}_d \to J/\psi K_s, \overline{B}_s \to J/\psi\phi$),

$$\left.\begin{aligned} A_\mathrm{T} &\sim V_{cb} V_{cs}^* \sim A\lambda^2 \\ A_\mathrm{P} &\sim V_{tb} V_{ts}^* \sim A\lambda^2 \end{aligned}\right\} \to \alpha_\mathrm{T} - \alpha_P \approx 0 \tag{6.235}$$

Therefore, the tree and penguin processes have the same phase and we can expect that modification due to penguin background processes is small. However, for the case $b \to du\bar{u}$ decays ($\overline{B}_d \to \pi^+\pi^-$, $\overline{B}_s \to \rho^0 K_s$), one sees that

$$\left.\begin{array}{l} A_\mathrm{T} \sim V_{ub}V_{ud}^* \sim \lambda^3 e^{-i\phi_3} \\ A_\mathrm{P} \sim V_{tb}V_{td}^* \sim \lambda^3 e^{-i\phi_1} \end{array}\right\} \to \alpha_\mathrm{T} - \alpha_P \neq 0 \tag{6.236}$$

and the two amplitudes have different phases. For this case, depending on the relative size of the penguin processes, there is a possibility that the asymmetry changes significantly.

The gluon coupling in the penguin diagrams which decreases asymptotically, could suppress their contributions. However, relation between the bare quark process and the experimentally observed hadronic process is nonperturbative and hard to estimate accurately. Depending on models, the relative size $|A_\mathrm{P}/A_\mathrm{T}|$ could be that of order of one.

For the case of $\overline{B}_d \to J/\psi + K_s$, the penguin contribution is estimated to be small, below a few %. It is a gold-plated process because of small and in-phase background. However, size change of the $\overline{B}_d \to \pi^+\pi^-$ asymmetry could be as large as 20% [332, 333]. Fortunately, in this case, by doing isospin analysis, namely, by comparing six decay rates of $B^\pm \to \pi^+\pi^0$, $B^0(\overline{B}^0) \to \pi^+\pi^-, \pi^0\pi^0$, the penguin contribution can be separated [334]. The amplitude relation is

$$A_{\pi^+\pi^+}/\sqrt{2} + A_{\pi^0\pi^0} - A_{\pi^+\pi^0} = 0 \tag{6.237}$$

Besides, the charged $B^\pm$ has spin zero and because of Bose statistics, its decay product $\pi^\pm\pi^0$ is in pure $I = 2$ state. The penguin contributes to only $I = 0$ state.

6.7.3 Experiments at the *B*-factory

Two specialized accelerators, KEKB at KEK and PEPII at SLAC, were constructed to clarify the CKM matrix structure. As the mass of the B mesons is considerably larger than that of the K mesons and many decay channels are open, the branching ratio to a specific decay mode is very small. One needs a high luminosity machine to do viable experiments. In order to measure the asymmetries with significance, let us estimate the required luminosity (L) of the accelerator. As a case study, we consider the gold-plated process, $B \to J/\psi + K_s$. Let us aim for at least statistical significance $n = 3$, namely, the statistical error should be at least smaller than one-third of the measured asymmetry. As the measured asymmetry (A_{fCP}) has Poisson distribution, the statistical error of the asymmetry is given by $\sigma_A^2 = (1 - A_{fCP}^2)/N$ (see Vol. 1, Eq. (12.146)), the number N of required events is given by

$$N > n^2(1 - A_{fCP}{}^2)/A_{fCP}{}^2 \tag{6.238}$$

The necessary integrated luminosity $\int dt L$ is given by

$$\begin{aligned} &\int dt L > \frac{(n/A_\mathrm{eff})^2}{\sigma f(2B)\epsilon_r\epsilon_t} \\ &A_\mathrm{eff} = (1-2w)d\frac{\sqrt{1-A_{fCP}{}^2}}{A_{fCP}} \end{aligned} \tag{6.239}$$

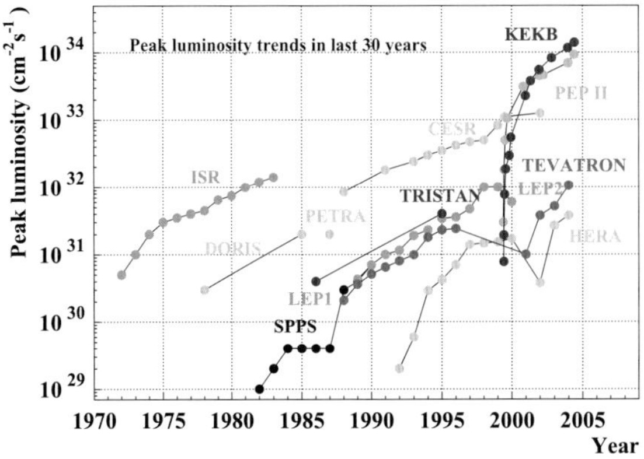

Figure 6.19 Peak luminosity trends achieved by the world collider accelerators which aimed for luminosity frontiers in the last 30 years. TRISTAN (e^-e^+), LEP (e^-e^+) aimed the energy frontier at the construction. No simple comparisons can be made with colliders of different particle species, but ISR (p–p), SppS (p–p), TEVATRON ($\overline{p}$–p), and HERA (e–p) are also listed for reference. One sees a quantum jump in the luminosity achieved by the *B*-factories (KEKB and PEPII). For a color version of this figure, please see the color plates at the beginning of the book [6].

where the variables are given by $n = 3$: statistical significance, A_{eff}: effective asymmetry, $w \sim 0.07$: misidentification rate (B^0 as $\overline{B}^0$ or vice versa), $d = x_d/(1+x_d^2) = 0.47$: dilution factor[20], $A_{fCP} = |\text{Im}\lambda| \sim 0.5$: physical asymmetry to be measured, $\sigma = 1.15\,\text{nb} = 1.15 \times 10^{-33}\,\text{cm}^2$: cross section for $e^-e^+ \to \Upsilon(4S)$, $f = 0.5$: Branching ratio for $\Upsilon(4S) \to B^0\overline{B}^0$, B: Product of all the branching ratios other than f. $\sim 4 \times 10^{-4}$ (BR($B^0 \to J/\psi K_s$)); ~ 0.14 (BR($J/\psi \to l^+l^-$)); ~ 0.69 (BR($K_s \to \pi^+\pi^-$)), $\epsilon_t \sim 0.65$: Tagging efficiency ($l^\pm$ or $K^\pm$), $\epsilon_r \sim 0.48 = \epsilon(J/\psi)\epsilon(K_s)$: Reconstruction efficiency of B decay.

The listed numbers were estimates before the experiment. Actual numbers are much improved. The reduction factor $1 - 2w$ of the asymmetry due to misidentification is because the number change in the numerator of the asymmetry is $\Delta N = 1 - (-1) = 2$. Solving the above equation, we obtain

$$\int dt L = (1\text{–}10) \times 10^{40}/\text{cm}^2 \tag{6.240}$$

Assuming one can operate $10^7\,s$ (= 115 full days) a year, this requires $L = 10^{34}/\text{cm}^2/\text{s}$. This is about $1 \sim 2$ orders of magnitude higher luminosity hitherto realized with the electron collider (DORIS, PETRA, CESR, TRISTAN, LEP1, 2) (see Figure 6.19). One sees that both KEKB/KEK and PEPII/SLAC achieved the desired luminosity shortly after the start of operation in 2000.

Two detectors, BaBar/SLAC [335] and Belle/KEK, [336] are very similar in concept and design. The only major difference is the Cherenkov detector used for

20) The dilution factor results from integrating the asymmetry at least over the time resolution of the experiment.

particle identification. Belle used aerogel [337] and Babar used the DIRC [338] which only differ in the method of light collection. They both produced most of the physics results discussed in this chapter.

Figure 6.20 shows a side view of the upper half of the Belle detector.

Figure 6.21 shows a typical event display. Many events are fully reconstructed. The desired decay modes are usually identified by reconstructing kinematics of

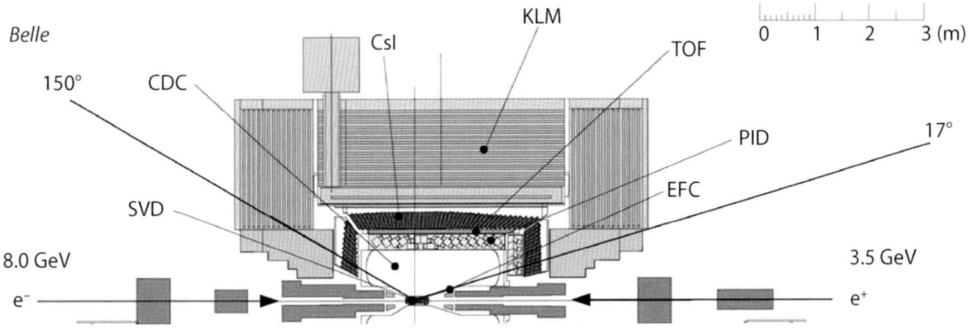

Figure 6.20 Side view of the Belle/KEK detector. Only upper half is shown. SVD:silicon vertex detector, CDC: central drift chamber for tracking, CsI: electromagnetic calorimeter made of CsI crystal modules ($17^\circ < \theta < 150^\circ$), KLM: long life kaon and muon detector made of iron plates and charge particle detectors (glass-resistive plate counters), TOF; time of flight counter, PID: particle identification detector made of aerogel Cherenkov counter modules, EFC: extreme forward calorimeter to cover angle region $6.4^\circ < \theta < 11.5^\circ$, $163.3^\circ < \theta < 171.2^\circ$ [336].

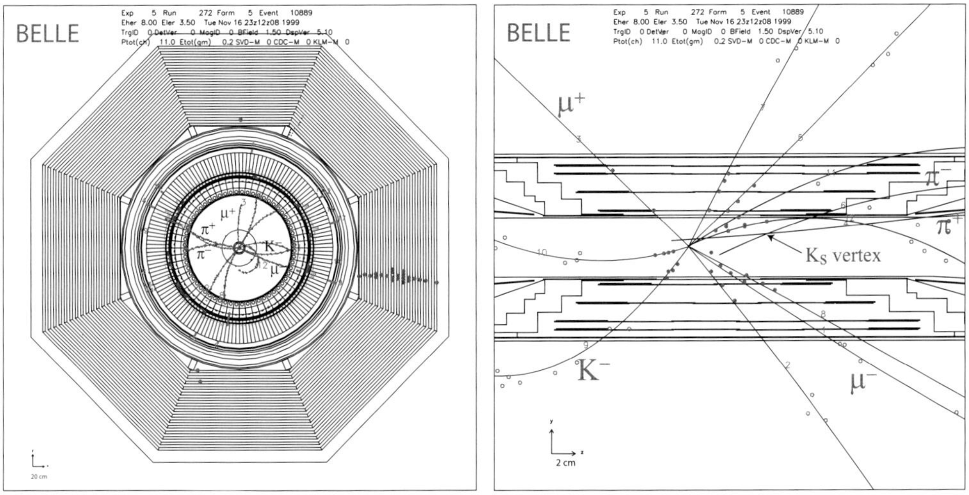

(a) (b)

Figure 6.21 A typical Belle event $B^0(\overline{B}^0) \to J/\psi + K_s$, $J/\psi \to \mu^-\mu^+$, $K_s \to \pi^+\pi^-$. (a) shows $x-y$ view, (b) expanded $y-z$ view of the vertex detector [13]. For a color version of this figure, please see the color plates at the beginning of the book.

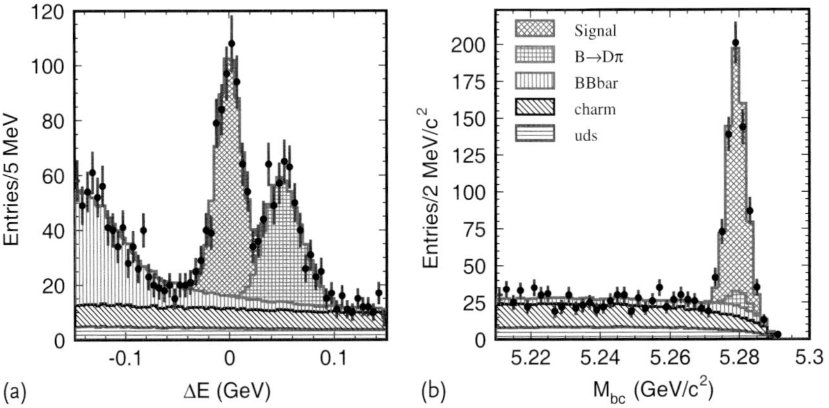

Figure 6.22 (a) ΔE and (b) M_{bc} distributions for the $B^+ \to DK^+$. One sees the desired signal $B \to DK$ is at the right position while $B \to D\pi$ is slightly shifted in ΔE plot because of wrong mass assignment for the K meson [321].

their parent particles and applying cuts in the two-dimensional plane $M_{bc} - \Delta E$. Figure 6.22 shows an example that was used to identify the $B^\pm \to DK^\pm(D \to K_s\pi^+\pi^-)$ mode [321]. M_{bc} is a variable that should agree with the B meson mass ($= 5.279$ GeV) if observed $K\pi$ events are decay products of the B meson and ΔE is the difference of the reconstructed and observed beam energy which should vanish. They are defined by $M_{bc} = \sqrt{E_{\text{beam}}^2 - \boldsymbol{p}_{\text{B}}^2}$ and $\Delta E = E_{\text{B}} - E_{\text{beam}}$ where $E_{\text{beam}} =$

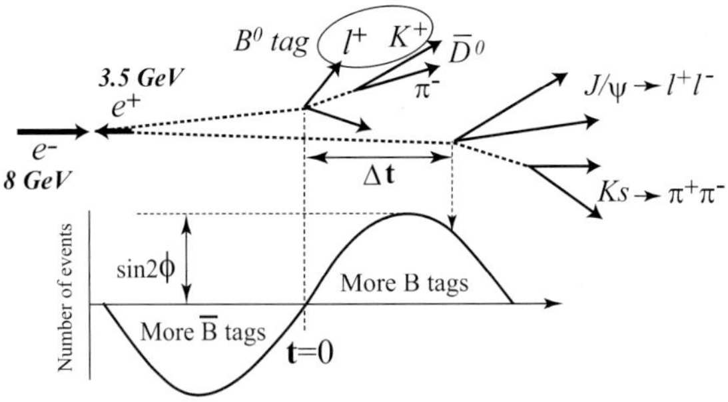

Figure 6.23 An event topology for the process produced in the asymmetric collider. $B^0\overline{B}^0$ pair is produced through $e^-e^+ \to \Upsilon(4S) \to B^0\overline{B}^0$. $B^0(\overline{b}d)$ is tagged by its emission of positive lepton ($\overline{b} \to \overline{c}l^+\nu$) and associated $\overline{D}^0(\overline{c}d)(\to K^-\pi^+$ etc) decay. The other member $\overline{B}(b\overline{d})$ decays to $J/\psi + K_s$, followed by $J/\psi \to l^+l^-$ and $K_S \to \pi^+\pi^-$. $\Delta t = 0$ is defined by the decay vertex point of the tagged B^0 and Δt ($\Delta s = c\beta\gamma\Delta t$ to be exact) is measured. For the CP $= +$ final state, more B tagged events are observed at $\Delta t > 0$, and for $\overline{B}^0$ tagged events, more events are observed at $\Delta t < 0$. For the CP $= -$ final state, the opposite is true. The height of the sine distribution peak determines the unitarity angle $\sin 2\phi_1$ (see text). The asymmetry appears as deviation from the normal exponential decay curve.

$m\{\Upsilon(4S)\}/2$ and $E_{\rm B}, \boldsymbol{p}_{\rm B}$ are the reconstructed B meson energy momentum, that is, $E_{\rm B} = E_D + E_{K^+}, \boldsymbol{p}_{\rm B} = \boldsymbol{p}_D + \boldsymbol{p}_{K^+}$.

Figure 6.23 shows how an event of $B^0\overline{B}$ decay used in extracting the unitarity angle ϕ_1 looks like in the asymmetric collider.

An example of the asymmetry measurement is already given in Figure 6.17.

6.8
Test of Unitarity

There are processes which cannot be accessed by the B-factory, but have important implications to understand the CP violation effects. We discuss only those that can be expected to be realized in the near future.

6.8.1
Rare Decays of the K Meson

$K^+ \to \pi^+\nu\overline{\nu}$ Here, $\nu\overline{\nu}$ means $\sum_{i=e,\mu,\tau}\nu_i\overline{\nu}_i$. The process is free from the so-called long range effect (see footnote 16). The main contribution comes from the box diagram depicted in Figure 6.24. Derivation of the theoretical expressions for the decay rate is similar to that of the neutral meson mixing. Conveniently, they can be calculated from the low energy effective Hamiltonian [302, 339, 340]

$$\begin{aligned}\mathcal{H}_{\rm eff} &= \frac{G_{\rm F}}{\sqrt{2}}\frac{\alpha}{2\pi\sin^2\theta_{\rm W}}\sum_{l=e,\mu,\tau}\left[V_{cs}^* V_{cd} X_{\rm NL}^l + V_{ts}^* V_{td} X(x_t)\right]\\ &\quad\cdot(\overline{s}_{\rm L}\gamma^\mu(1-\gamma^5)d_{\rm L})(\overline{\nu}_{\rm L}\gamma_\mu(1-\gamma^5)\nu_{\rm L})\\ x_c &= m_c^2/m_{\rm W}^2\,, \quad x_t = m_t^2/m_{\rm W}^2\end{aligned} \tag{6.241}$$

which includes NLLA QCD corrections. $X(x)$ is a function similar to $S(x)$ in Eq. (6.148) which originates from the loop calculation of the Feynman diagrams depicted in Figure 6.24 with QCD corrections. $X_{\rm NL}^l$ is the charm contribution similar to $X(x_t)$. The difference arises because while the box contribution involving the leptons is negligible for the top intermediate state, this is not so for the charm because the relative size of the lepton mass is not negligible. By writing the Hamiltonian in this form, the GIM mechanism is already incorporated and the up-quark contribution is understood to be subtracted from the top and charm loops. The $X(x)$ function has the property $X(0) = 0$ and monotonically increases as a function of x. Under normal circumstances, the top intermediate state dominates because $X(x_c) \ll X(x_t)$. For the K meson case, however, the strong CKM suppression makes the charm contribution comparable to that of the top. This suppression is absent for the B meson case. It is also absent for $K_{\rm L} \to \pi^0\nu\overline{\nu}$ to be discussed soon because it is CP violating and only the imaginary part contributes. Therefore, except for the $K^+ \to \pi^+\nu\overline{\nu}$, the top dominance assumption gives a good approximation in the following discussions. Inclusion of NLLA QCD corrections to the

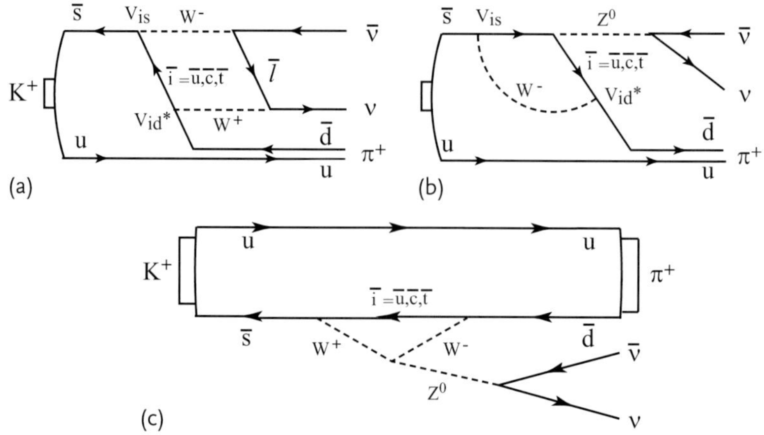

Figure 6.24 Box diagrams (a) and Z-penguin diagrams (b,c) that contribute to the process $K^+ \to \pi^+ \nu\overline{\nu}$. Diagrams for $K_L \to \pi^0 \nu\overline{\nu}$ can be obtained by replacing the spectator u quark with d quark.

electroweak diagrams + LLA in Figure 6.24 typically reduced ambiguities of the branching ratio from ~ 25% down to ~ 3%.

Calculations yield the decay branching ratio expressed as [302, 339–341]

$$\begin{aligned}
\mathrm{BR}(K^+ \to \pi^+ \nu\overline{\nu}) &= \frac{\alpha^2 \mathrm{BR}(K^+ \to \pi^0 e^+ \nu_e)}{|V_{us}|^2 2\pi^2 \sin^4\theta_W} \sum_{l=e,\mu,\tau} |V_{cs}^* V_{cd} X_{\mathrm{NL}}^l + V_{ts}^* V_{td} X(x_t)|^2 \\
&\simeq 4.11 \times 10^{-11} A^4 X^2(x_t) \frac{1}{\sigma} \left[(\sigma\overline{\eta})^2 + (\rho_0 - \overline{\rho})^2\right] \\
&\simeq (8.5 \pm 0.7) \times 10^{-11} \quad [342] \\
\sigma &= 1/(1 - \lambda^2/2)^2 , \quad \rho_0 \simeq 1.4
\end{aligned} \tag{6.242}$$

The theoretical estimate predicts the value ~ 10^{-10} and confines the apex position of the unitarity triangle in a circle band on the $\overline{\rho} - \overline{\eta}$ plane centered at $\overline{\rho}_c \simeq 1.4$ (see Figure 6.25). Thus far, three events were observed and the obtained result is [343, 344]

$$0.173^{+0.115}_{-0.105} \times 10^{-11} \tag{6.243}$$

which does not contradict the Standard Model but also does not constrain the value of CKM matrix.

$K_L^0 \to \pi^0 \nu\overline{\nu}$ This is a CP violating process in the leading order because

$$\begin{aligned}
A(K_L^0 \to \pi^0 \nu\overline{\nu}) &\sim \langle \pi^0 | J_\mu | K_L \rangle \langle \nu\overline{\nu} | \overline{\nu}\gamma^\mu (1 - \gamma^5) \nu | 0 \rangle \\
&\propto A(K^0 \to \pi^0) - A(\overline{K}^0 \to \pi^0) \\
&\sim A(K^0 \to \pi^0) - A^*(K^0 \to \pi^0) \\
&\sim \mathrm{Im} A(K^0 \to \pi^0)
\end{aligned} \tag{6.244}$$

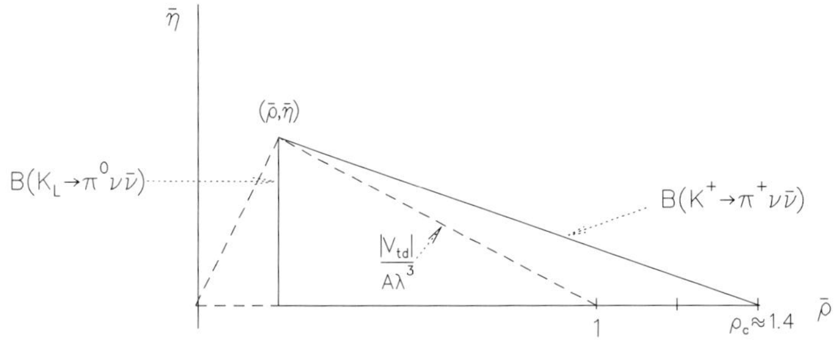

Figure 6.25 $K^+ \to \pi^+\nu\nu$ and $K_L^0 \to \pi^0\nu\nu$ can determine the apex of the unitarity triangle and provide independent information from that of the B^0–$\overline{B}^0$ system.

which is nonzero only when the CP is violated. Therefore, discovery of a single unambiguous event can establish the existence of the direct CP violation in the $\Delta S = 1$ process. Although the honor of proving the direct CP violating process went to ϵ'_K, it is theoretically a clean process and can determine the value of $\overline{\eta}^2$ definitively, namely, the height of the unitarity triangle. The contamination due to the CP violation in the mixing (i.e., effect of ϵ_K) is estimated smaller by three orders of magnitude [341, 345].

$$\begin{aligned} B(K_L \to \pi^0 \nu\overline{\nu}) &= 3\frac{\tau(K_L)}{\tau(K^+)}\frac{\alpha^2 \mathrm{BR}(K^+ \to \pi^0 e^+ \nu)}{|V_{us}|^2 2\pi^2 \sin^4\theta_W}\left(\mathrm{Im}[V_{td}^* V_{ts}]X(x_t)\right)^2 \\ &\approx 7.6 \times 10^{-5} |V_{cb}|^4 \overline{\eta}^2 \\ &\approx (2.6 \pm 0.3) \times 10^{-11} \quad [340] \end{aligned} \tag{6.245}$$

Experimentally, the measurement of this decay mode is challenging. The signal is "two photons from π^0 decay plus nothing," which is plagued by vast amount of backgrounds. The current experimental upper limit is [346]

$$\mathrm{BR}(K \to \pi^0 \nu\overline{\nu}) < 0.67 \times 10^{-7} \tag{6.246}$$

A semiphenomenological upper limit can be obtained from the measured rate of $K^+ \to \pi^+\nu\overline{\nu}$. The decay amplitude $A(K_L \to \pi^0\nu\overline{\nu})$ is related to $A(K^+ \to \pi^+\nu\overline{\nu})$ by isospin rotation. The experimental value of the decay rate $\mathrm{BR}(K^+ \to \pi^+\nu\overline{\nu})$ set an upper limit on the $\mathrm{BR}(K_L \to \pi^0\nu\overline{\nu})$ and gives

$$\mathrm{BR}(K_L \to \pi^0 \nu\overline{\nu}) < 1.46 \times 10^{-9} \quad [3] \tag{6.247}$$

Note that the combination of $\mathrm{BR}(K^+ \to \pi^+\nu\overline{\nu})$ and $\mathrm{BR}(K^0 \to \pi^0\nu\overline{\nu})$ can give an independent and theoretically clean value of the $\sin\phi_1$ and the apex height of the unitarity triangle (Figure 6.25). It is also sensitive to new physics different from the precision B^0–$\overline{B}^0$ oscillation measurements.

6.8.2
Global Fit of the Unitarity Triangle

Thus far, by using measured CKM elements independently, the unitarity of the CKM matrix can be checked.

The first row gives: [3]

$$|V_{ud}|^2 + |V_{us}|^2 + |V_{ub}|^2$$
$$= |0.974\,25 \pm 0.000\,22|^2 + |0.2252 \pm 0.0009|^2 + |0.003\,89 \pm 0.000\,44|^2$$
$$= 0.9999 \pm 0.0001 \tag{6.248a}$$

The second row gives:

$$|V_{cd}|^2 + |V_{cs}|^2 + |V_{cb}|^2 = 1.002 \pm 0.027 \tag{6.248b}$$

The third row gives:

$$|V_{td}|^2 + |V_{ts}|^2 + |V_{tb}|^2 = 1.098 \pm 0.074 \tag{6.248c}$$

The sum of the three angles of the unitarity triangle gives:

$$\phi_1 + \phi_2 + \phi_3 = (183^{+22}_{-25})^\circ \tag{6.248d}$$

We conclude that within experimental errors, the unitarity is satisfied quite well. The Wolfenstein parameters of the CKM matrix are given by

$$\begin{aligned} \lambda &= 0.2253 \pm 0.0007\,, & A &= 0.808^{+0.022}_{-0.015} \\ \overline{\rho} &= 0.132^{+0.022}_{-0.014}\,, & \overline{\eta} &= 0.341 \pm 0.013 \end{aligned} \tag{6.249}$$

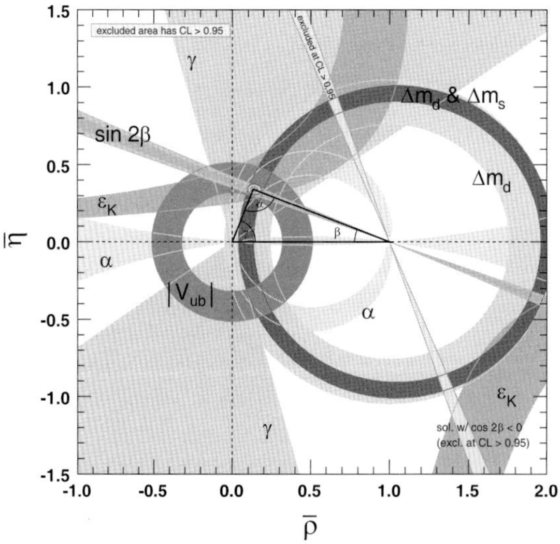

Figure 6.26 Constraints on the $\overline{\rho}$–$\overline{\eta}$ plane by various measurements. Note the angles are $\alpha = \phi_2, \beta = \phi_1, \gamma = \phi_3$. The shaded area is 95% CL bands. For a color version of this figure, please see the color plates at the beginning of the book [3].

Figure 6.26 illustrates an overall fit and the constraints on the $\overline{\rho}$–$\overline{\eta}$ plane. The shaded area are 95% CL allowed regions. All of them overlap and there are no inconsistencies within the Standard Model framework.

6.9 CP Violation beyond the Standard Model

So far, we have discussed phenomena of the CP violation within the framework of the Standard Model. There are some processes which are not solved within the Standard Model. Here, we will discuss a few models which go beyond the Standard Model.

More than Two Higgs Fields In the Standard Model, there is only one Higgs doublet and after spontaneous symmetry breaking, only one physical Higgs field remains. If one extends the Standard Model and tries to unify the strong and the electroweak interaction (GUT = grand unified theory), one faces the hierarchy problem.[21] A possible remedy to the problem is to assume the supersymmetry which is a symmetry to unify the fermion and the boson. In the supersymmetric extension of the GUT, two Higgs doublets are required. Another incentive to discuss the two Higgs schemes is the strong CP problem discussed below. A prominent model which can solve the problem requires existence of the axion. The model assumes the existence of two Higgs doublets. If there are more than two Higgs doublets, the relative phase of the two Higgs induces CP violation [347]. However, proposed models thus far face phenomenological problems, for instance, producing the neutron electric dipole moment too large to explain the observed value. Because of this, models based on the two Higgs doublet are not so popular for now, but remain a future possibility.

The Strong PC Problem In the conventional formalism, the QCD Lagrangian does not contain a self-interaction of the gluons expressed by the following Lagrangian

$$\mathcal{L} = \frac{g^2}{64\pi^2}\theta \sum_a \epsilon^{\mu\nu\rho\sigma} F^a_{\mu\nu} F^a_{\rho\sigma}$$
$$\theta = \theta_0 + \arg(\det M^U \det M^D) \tag{6.250}$$

The parameter θ_0 is related to the QCD vacuum and the matrices M^U, M^D are the mass matrices or the couplings of the Higgs to the weak eigenstate matter particles that appeared in Eq. (6.4). The Langrangian in Eq. (6.250) has the form of a total derivative and does not contribute to perturbative processes. However, it was shown that in certain circumstances, it can produce a finite effect, owing to the nonperturbative effect (θ vacuum) [348, 349]. The problem is, if it exists, the

21) Two energy scales, the GUT energy ($\sim 10^{16}$ GeV) and the electroweak energy ($\sim$ 1 TeV), have to be fine tuned at every order of the perturbation series to obtain physical results which are considered unnatural.

Lagrangian violates P, T and CP and as a result, produces a CP violation effect, for instance, the neutron electric dipole moment ($d_n \sim 10^{-16}\theta\, e\,\mathrm{cm}$). In order to not contradict the observed value, the value of θ is highly constrained ($\theta < 10^{-9}$). Therefore, it is logical to consider that it should not exist. However, the QCD vacuum and the mass matrices which are induced by the electroweak vacuum are of completely different origin and the reason why they can cancel each other is a mystery. This is the strong CP problem. A solution was proposed, but it requires the existence of a light boson referred to as the axion [350]. Despite extensive searches, it has not been observed.

Left–Right Symmetric Model The symmetry of the electroweak interaction in the Standard Model is $SU(2)_L \times U(1)$. There is a proposal to extend it to $SU(2)_L \times SU(2)_R \times U(1)$ which is referred to as the left–right symmetric model [351]. Its motivation is to recover the parity invariance at high energies. It considers the parity violation not as an intrinsic symmetry of the nature, but something that is produced by spontaneous symmetry breaking. It requires the existence of the gauge boson W_R which couples to right-handed currents of matter fields. In calculating cross sections, one has to include diagrams that contain W_R in addition to ordinary W, Z. In reality, reproducibility of the Standard Model is excellent. Therefore, to be consistent with observations, the mass of W_R must be large. Then, its coupling is reduced by the mass squared just like the Fermi coupling constant is reduced by the factor m_W^2 compared to the electromagnetic strength α. In the left–right symmetric model, the reduced right-handed current still participates in the charged current interaction, and thus the constraint to the CKM matrix is loosened and 2×2 matrix can generate the CP violation phase. For instance, the value of the box diagram calculated in Section 6.5 has to be modified. The mass of the W_R is estimated to be at least larger than a few TeV.

Cosmic Baryon Number The universe we live is matter dominant. The amount of matter and antimatter is very different. Sakharov showed that three conditions are necessary to produce the observed asymmetry of matter in the big-bang universe: (1) baryon number nonconservation, (2) CP violation and (3) departure of the universe from thermal equilibrium [352]. It is known that the CP violation generated by the CKM matrices are too small to account for the present matter-antimatter asymmetry. Namely, clarification of the CP problem is directly related to our very existence.

CP Violation in the Lepton Sector The discovery of the neutrino oscillation established a nonvanishing neutrino mass and mixing among different neutrino flavors. The mixing is characterized by the PMNS (Primakof–Maki–Nakagawa–Sakata) matrix which is quite similar to the CKM matrix in the quark sector, except that it contains two extra Majorana phases for Majorana neutrinos. All the three mixing angles $\theta_{12}.\theta_{23}, \theta_{13}$ are determined by the solar, atmospheric and terrestrial neutrino oscillation experiments. The CP violating phase δ, however, is still not determined. Current frontier neutrino experiments aim to determine it.

Part Two QCD Dynamics

7
QCD

Why QCD? Quantum chromodynamics (QCD) is the theory of the strong interaction based on the gauge symmetry of color $SU(3)$. The quark which comes in six flavors (u, d, s, c, b, t) is equipped with a kind of (electric) charge referred to as the color charge which generates the force field. Just as the electromagnetic field acts on electrically charged particles, the color field acts on particles that have color charges (i.e., quarks and gluons). It is a force to bind quarks to form hadrons, and acts or reacts to produce various kind of dynamic reactions as the quarks come near one another. The mechanism of generating the color field is very similar to that of electromagnetism, for instance, the static color charge generates a Coulomb like potential and the moving charge (i.e., the color current) generates the color magnetic field which acts on the spin (or more generally, the angular momentum) of colored particles. The difference is that there are three kinds of color charges (and their anticolor charges) commonly dubbed as red (R), green (G) and blue (B) as compared to only one kind of electric charge.

Although the quark has the color degrees of freedom, only color neutral (singlet) combinations have been observed. The color degrees of freedom are based on the symmetry $SU(3)$ which is sometimes referred to as the "color spin", just like the isospin degrees of freedom which obeys $SU(2)$. To understand the role of the color spin, it is illuminating to have a look at the role which the isospin played in clarifying basic features of the nuclear force. It is mediated by the pion which is an isospin carrier and because of this, the nuclear force becomes an isospin exchange force, meaning that the identity of the proton and the neutron is interchanged on force actions. For the two nucleon system, for instance, the isospin exchange operator changes sign as well as the strength of the force depending on whether they are in a singlet ($I = 0$) or in a triplet ($I = 1$) state. The reason why only the isosinglet combination of the nucleon (proton and neutron) is attractive and forms a bound state known as the deuteron is clarified by investigating structure of the $SU(2)$ symmetry. The sign of the exchange force acting on the triplet turned out to be positive, that is, repulsive and no bound states can be formed. It also clarified that the pion is not a scalar, but a pseudoscalar.

Elementary Particle Physics, Volume 2: Foundations of the Standard Model, First Edition. Yorikiyo Nagashima.

In exactly the same logic, the color force being mediated by the color carrier (i.e., gluon) is characterized by the color exchange force, its sign and strength dictated by its color spin state. Some examples of the color exchange force were given in Vol. 1, Section 14.6. It was shown that color singlets have a strong attractive force, but octets are repulsive, indicating (but not proving) the origin of confining force. The gluon cannot be a (pseudo)scalar. If it is, qq bound states as well as $q\overline{q}$ bound states (mesons) should be observed. So far, no qq bound states have been confirmed. If the gluon is a vector and massless, it is natural that it can be identified as the gauge particle of the strong force just like the photon in the electromagnetic force. This was the original motivation to consider the $SU(3)$ based gauge theory as a prime candidate for the strong interaction theory and discovery of the asymptotic freedom has solidified its foundation.

The gluon, the force mediator and the equivalent of the photon in QED, is massless, but comes in eight kinds (nine combinations of color-anticolor minus one which is color neutral). Although the basic mathematical framework of QCD is essentially identical to that of QED, the existence of the three charges makes equations of motion nonlinear and generates characteristic features very different from those of QED. The most conspicuous is a phenomenon referred to as the confinement which forbids the quarks and gluons to be isolated, but forces them to hadronize at and beyond distances of typical hadron size ($\sim 10^{-13}$ cm). Experimentally, therefore, direct observation of the quarks and gluons is impossible, which is the major cause of difficulties in comparing the QCD predictions with experiments. Nevertheless, making use of characteristic features of QCD, several techniques have been developed to make correspondences between them. In particular, the following concepts turned out to be very useful.

1. Asymptotic freedom
2. Renormalization group equation
3. Factorization

The asymptotic freedom and the confinement are two sides of the same coin. They state that the strength of the strong coupling constant becomes weak at a small distance (or equivalently at high Q^2 in the energy scale) and strong at a large distance. The asymptotic freedom enables us to use perturbation theories to calculate hard (i.e., large Q^2) reactions in QCD. The asymptotic freedom makes the strength of the force (coupling constant) vary as a function of Q^2 which makes it necessary to specify its value at a certain reference point. However, the confinement makes it difficult to define a clear-cut reference point at which one measures and determines the strength of the coupling constant. Therefore, we have to introduce a somewhat arbitrary scale "μ" as a reference point. Observables, however, should not depend on μ because it is an artificially introduced variable. The requirement leads to the renormalization group equation which enables us to relate expressions at scale μ to actual observables. It is a very powerful tool in engineering various techniques in QCD.

Partons (quarks and gluons) that have participated in hard processes are experimentally observed as jets, namely, as a bundle of hadrons flying more or less in the same direction. There exists no established theory to connect the parton at the theoretical level and jets and hadrons at observational level. However, after a long trial and error period, ways to fairly accurately reproduce and predict jet phenomena have been established.

The factorization enables us to separate hard processes from the soft (low energy or low impact to be exact) part of the QCD. $Q \sim \Lambda$ ($\sim$ 200 MeV) is the characteristic energy scale below which the confinement sets in. At the energy scale $Q \gtrsim m_{\text{hadron}} \sim 1\,\text{GeV}$, the perturbative treatment becomes viable. The low energy part cannot be calculated perturbatively, but can be separated and renormalized into scale dependent functions. They are generally treated as given inputs to perturbation calculations. Model constructions or lattice QCD (LQCD) are necessary to determine them. A typical example is the parton distribution inside the hadron which provides the initial conditions at some scale. Once it is given, then the parton reactions at higher scales can be calculated using perturbative QCD (pQCD). The partons become jets and are converted into hadrons referred to as the fragmentation process. Notice all the experimental observations are carried out at the hadron level. This is a nonperturbative process again. Therefore, nonperturbative treatments are required both in preparing initial conditions and in comparing data with QCD predictions.

Fortunately, it turned out that parton distribution in a hadron and the parton's hadronization process are universal in the sense that they do not depend on the reactions we want to investigate. It can be determined separately and independently from the hard process in question. The parton distribution function (pdf) and the jet formation function (referred as the fragmentation function) obey the same evolution equation known as the DGLAP equation and are closely related. Their treatment is a main theme of QCD, but its outcome is a universal function that everybody can use to investigate processes of one's own interest. Thus, the factorization has made it possible to connect QCD predictions with experiments within reasonable approximations.

The purpose of the present and next chapter is to introduce the basic tools in QCD calculations and to familiarize the reader with the use of the above three features of QCD. The formalism of evolution equations which connect the parton distributions in the hadron or hadrons in the outgoing jets with the perturbatively calculable hard process are developed. The latter was part of the QCD process one wants to investigate, but the factorization made it possible to separate the low energy process and evaluate independently from those specific to the hard process. This makes what is referred to as the QCD corrected parton model. One will see in later chapters that techniques developed here are instrumental in making solid predictions for inclusive reactions like the deep inelastic scattering or jet phenomena and testing QCD experimentally.

7.1 Fundamentals of QCD

7.1.1 Lagrangian

Active players of the QCD are the quarks (spin $s = 1/2$) and the gluons ($s = 1$). In a relativistic gauge, the ghost fields ($s = 0$) come in as well. They are multiplets in the color space[1)] denoted by

$$\text{quark field:} \quad \Psi(x) = (\psi_a(x), a = \text{R,G,B or } 1 \sim 3) \tag{7.1a}$$

$$\text{gluon field:} \quad \boldsymbol{A}_\mu(x) = \left(A_\mu^A(x), A = 1 \sim 8\right) \tag{7.1b}$$

$$\text{ghost field:} \quad \boldsymbol{\eta}(x) = (\eta^A(x), A = 1 \sim 8) \tag{7.1c}$$

where μ $(= 0 \sim 3)$ denotes the Lorentz (i.e., spacetime) index. The Lagrangian for a single flavor quark is expressed as

$$\begin{aligned}\mathcal{L}_{\text{QCD}} = &\sum_{a,b=1}^{3} \left[\overline{\psi}_a(x)\left(i\gamma^\mu[D_\mu]_{ab} - m\right)\psi_b(x)\right] - \frac{1}{4}\boldsymbol{F}_{\mu\nu}(x)\cdot\boldsymbol{F}^{\mu\nu}(x)\\ &- \frac{1}{2\lambda}\left(\partial_\mu \boldsymbol{A}^\mu(x)\cdot\partial_\nu \boldsymbol{A}^\nu(x)\right) + \sum_{A,B=1}^{8}(\partial^\mu \eta_A(x))[D_\mu]_{AB}\eta_B(x)\end{aligned} \tag{7.2}$$

where $\boldsymbol{A}\cdot\boldsymbol{B} = \sum_{A=1}^{8} A^A B^A$ denotes a scalar product of two vectors in the color space. The covariant derivative D_μ will be explained soon. We omit $\sum_{A,B}$ from now on, but it is understood that sum is to be taken when the double suffix appears.[2)] The first term in the second line is the gauge fixing term and the second term is the ghost field which is necessary if the covariant gauge

$$\langle f|\partial_\mu \boldsymbol{A}^\mu(x)|i\rangle = 0 \tag{7.3}$$

is adopted. The ghost is not a physical particle. Its role is to preserve the mathematical consistency of the theory (i.e., unitarity). It cancels unphysical contributions of longitudinal and scalar gluons which are artifacts created by adopting the covariant gauge. The conventional Feynman gauge is defined with $\lambda = 1$. The eight component field strength $\boldsymbol{F}_{\mu\nu}$ and covariant derivatives are defined by

$$F_{\mu\nu}^A(x) = [\partial_\mu A_\nu^A(x) - \partial_\nu A_\mu^A(x) - g_s f^{ABC} A_\mu^B(x) A_\nu^C(x)], \quad A = 1 \sim 8 \tag{7.4a}$$

$$[D_\mu]_{ab} = \delta_{ab}\partial_\mu + i g_s [\boldsymbol{t}]_{ab}\cdot\boldsymbol{A}_\mu(x) \tag{7.4b}$$

$$[D_\mu]_{AB} = \delta_{AB}\partial_\mu + i g_s [\boldsymbol{T}]_{AB}\cdot\boldsymbol{A}_\mu(x) \tag{7.4c}$$

1) The color space is the entire set of state vectors on which operators representing the $SU(3)$ symmetry group act. In the fundamental representation in which the state vectors represent the three color states of the quark, the degree of the color freedom is three. The states are often denoted as R (red), G (green) and B (blue) instead of insipid numbers. The gauge field belongs to the adjoint representation, hence the color degree of the freedom is eight. See Appendix H for more details.

2) As to suffixes of the color, no distinction is made between upper and lower suffix, it is placed conveniently for easy viewing.

where $\boldsymbol{t}$ and $\boldsymbol{T}$ are color matrices for the quark and gluon.

$$[\boldsymbol{t}]_{ab} = \left([t^A]_{ab},\, a, b = 1 \sim 3,\, A = 1 \sim 8\right) = [\lambda^A/2]_{ab} \tag{7.4d}$$

$$[\boldsymbol{T}]_{BC} = \left([T^A]_{BC},\, A, B, C = 1 \sim 8\right) = -i f^{ABC} \tag{7.4e}$$

Here, λ^A is the Gell–Mann matrix and f^{ABC} is the structure constant of $SU(3)$ (Eqs. (H.7)). The equation of motion for the quark field is given by

$$[i\gamma^\mu \partial_\mu - m]\psi_a(x) = g_s[\boldsymbol{t}]_{ab} \cdot \boldsymbol{A}_\mu \psi_b(x) \tag{7.5}$$

and the gluon field satisfies the Maxwell-type equation

$$\partial_\mu \boldsymbol{F}^{\mu\nu} - g_s \boldsymbol{A}_\mu \times \boldsymbol{F}^{\mu\nu} + \frac{1}{\lambda}\partial^\nu(\partial_\mu \boldsymbol{A}^\mu) = g_s \overline{\Psi} \boldsymbol{t} \gamma^\nu \Psi \,. \tag{7.6}$$

where $[\boldsymbol{A}_\mu \times \boldsymbol{F}^{\mu\nu}]_A = f_{ABC} A^B_\mu F^{C\mu\nu}$ is the vector product in the color space. If there is no nonlinear terms (one in $\boldsymbol{F}^{\mu\nu}$ (the third term in Eq. (7.4a)) and the other the second term in Eq. (7.6)), it reduces to the Maxwell equation.

The Axial Gauge The basic form of the QCD Lagrangian is given in Eq. (7.2). One can construct equations of motion from this Lagrangian using the Euler–Lagrange equation. The gauge condition is required to remove the unphysical (i.e., longitudinal and scalar) component of the gauge particle. Classically, it can be chosen and be fixed separately. In quantum theories, the gauge fixing processes is not a trivial endeavor, especially if one wants to construct the formalism in a Lorentz covariant way. In addition to the gauge fixing term (the first term on the second line of Eq. (7.2)), one also needs the ghost fields η_A which are included on the second line of Eq. (7.2).

In QED, there are ways to only quantize the physically meaningful transverse components using the physical gauge (Coulomb gauge). However, in this gauge, manifest Lorentz covariance is lost and calculations are generally more cumbersome. This is the main reason that the covariant gauge is usually adopted. In this gauge, the gauge field has four independent components, but the longitudinal and scalar parts are unphysical. The scalar component has a negative norm, but it is required in order to cancel the unphysical longitudinal component (see arguments in Vol. 1, Sections 5.3.4 or 11.3). In QED, they cancel each other by themselves, but in QCD, the cancellation is not complete because of the gluon self-coupling. The ghost field is required precisely to remove the unphysical part of the gluon coupling and maintain unitarity of the theory. It serves no other purpose. Since it is a mathematical artefact and appears only in internal lines of the Feynman diagrams, it has peculiar features not realized in the physical observables. The ghost is a scalar field but obeys Fermi statistics. It has the same color degrees of freedom as the gauge field, that is, $N^2 - 1$ components for $SU(N)$ or eight components in QCD. It only couples to the gauge field and only appears as internal lines. It never appears as physically meaningful quantities.

It is sometimes more convenient to use the axial gauge because there is no ghosts in this gauge (see Vol. 1, Sect. 11.7). The physics content of equations are more transparent and the gluon's behavior is very similar to that of the physical photon in the Coulomb gauge. The axial gauge condition is given by

$$A_0^A(x) = 0 \quad \text{or more generally} \quad \langle f|n^\mu A_\mu^A(x)|i\rangle = 0 \quad (\text{for } A = 1 \sim 8) \tag{7.7}$$

where n is a fixed spacelike or null vector ($n_\mu n^\mu = 0$) subject to $(k \cdot n) \neq 0$ where k is the gluon's four-momentum. If one adopts the axial gauge, one has to use the gauge fixing term

$$\mathcal{L}_{\text{gf}} = -\frac{1}{2\lambda}\left(n^\mu \boldsymbol{A}_\mu\right) \cdot \left(n^\nu \boldsymbol{A}_\nu\right) \tag{7.8}$$

for which the equation of motion for the gluon field is given by

$$\partial_\mu \boldsymbol{F}^{\mu\nu} - g_s \boldsymbol{A}_\mu \times \boldsymbol{F}^{\mu\nu} + \frac{1}{\lambda} n^\nu \left(n_\mu \boldsymbol{A}^\mu\right) = g_s \overline{\Psi} \boldsymbol{t} \gamma^\nu \Psi \,. \tag{7.9}$$

Treating the nonlinear part of the field operators as interactions, the coefficient of linear $\boldsymbol{A}$ terms in the equation gives the inverse propagator[3] which is expressed as

$$D_{F\mu\nu}^{-1} = i\delta_{AB}\left[-g_{\mu\nu}\partial_\rho\partial^\rho + \partial_\mu\partial_\nu + \frac{1}{\lambda}n_\mu n_\nu\right] \tag{7.11}$$

The propagator in momentum space is expressed as

$$\begin{aligned} D_{F,AB\mu\nu}(k) &= \delta_{AB}\frac{i}{k^2 + i\varepsilon} d_{\mu\nu}(k) \\ d_{\mu\nu}(k) &= \sum_{\lambda=\pm,3,0} \varepsilon_\mu(\lambda)\varepsilon_\nu^*(\lambda) \\ &= \left[-g_{\mu\nu} + \frac{n_\mu k_\nu + k_\mu n_\nu}{n \cdot k} - \frac{(n^2 + \lambda k^2)k_\mu k_\nu}{(n \cdot k)^2}\right] \end{aligned} \tag{7.12}$$

$d_{\mu\nu}$ is a polarization tensor summed over all four states.

Problem 7.1

Derive Eq. (7.12) from Eq. (7.11).

We can choose n and λ to satisfy ($n^2 = 0, \lambda = 0$) which is called the light cone gauge. Then,

$$d_{\mu\nu}(k) = -g_{\mu\nu} + \frac{n_\mu k_\nu + k_\mu n_\nu}{n \cdot k} \tag{7.13}$$

$d_{\mu\nu}(k)$ satisfies constraints

$$n^\mu d_{\mu\nu} = n^\nu d_{\mu\nu} = 0\,, \quad k^\mu d_{\mu\nu} = k^2 n_\nu/(n \cdot k) \tag{7.14}$$

3) Compare with the Klein–Gordon equation in which the inverse propagator Δ^{-1} is defined by

$$\Delta^{-1}\Delta \equiv -[\partial_\mu\partial^\mu + m^2]\Delta(x) = \delta^4(x - y)\,. \tag{7.10}$$

For physical gluons which satisfy $k^2 = 0$, $k^\mu d_{\mu\nu} = 0$, the $d_{\mu\nu}$ with $\lambda = 0$ agrees with the polarization sum of two transverse polarizations. Because of this, the gluons in the axial gauge can be treated as (almost) physical.

Problem 7.2

For the physical gluons, show that

$$\sum_{\lambda=\pm} \varepsilon_\mu(\lambda)\varepsilon_\nu(\lambda)^* = -g_{\mu\nu} + \frac{(k_\mu n_\nu + n_\mu k_\nu)}{(n\cdot k)} - n^2 \frac{k_\mu k_\nu}{(n\cdot k)^2} \tag{7.15}$$

where $n^\mu (n^2 \le 0, n\cdot k \ne 0)$ is a vector independent of k^μ and two transverse polarizations.
Hint: writing $\sum_{\lambda=\pm} \varepsilon_\mu(\lambda)\varepsilon_\nu(\lambda)^* \equiv T_{\mu\nu}$, it can be expanded by $g_{\mu\nu}$, $(k_\mu n_\nu + n_\mu k_\nu)$, $k_\mu k_\nu$, and $n_\mu n_\nu$ because of Lorentz invariance. Require also $k\cdot\varepsilon(\lambda) = n\cdot\varepsilon(\lambda) = 0$, $(\varepsilon(\lambda)\cdot\varepsilon(\lambda')) = -\delta_{\lambda\lambda'}$.

7.1.2 The Feynman Rules

Once the Lagrangian is fixed, establishing the Feynman rule is straightforward. Major rules are listed in Figure 7.1. Rules concerning the external lines are not listed because they are the same as QED. For each fermion and ghost loop, an extra "−" sign is to be multiplied.

Fermion Field

$$S_F(p) = \delta_{ab}\frac{i}{\not p - m + i\varepsilon} = \delta_{ab}\frac{i(\not p + m)}{p^2 - m^2 + i\varepsilon} \tag{7.16}$$

It is the same as QED, but usually light quarks are treated as massless.

Gluon propagator The gluon propagator is expressed as

$$\delta^{AB}\frac{i}{q^2+i\varepsilon} \times \begin{cases} -g^{\mu\nu} + (1-\lambda)\frac{q^\mu q^\nu}{q^2+i\varepsilon} & \text{Covariant gauges} \\ -g^{\mu\nu} + \frac{q^\mu n^\nu + n^\mu q^\nu}{n\cdot k} - \frac{n^2 q^\mu q^\nu}{(n\cdot k)^2} & \text{Axial gauges with } \lambda = 0 \end{cases} \tag{7.17}$$

The propagator changes for a different choice of λ. Physically meaningful quantities do not depend on λ, and hence it is usually chosen for calculational convenience. If the Feynman gauge ($\lambda = 1$) is adopted, the gluon propagator in the covariant gauge becomes identical to that of QED, except the color factor.

$$\delta^{AB}\frac{i}{q^2+i\varepsilon}\times\begin{cases}\text{Covariant gauges:}\\ \quad\left[-g^{\mu\nu}+(1-\lambda)\frac{q^\mu q^\nu}{q^2+i\varepsilon}\right]\\ \text{Axial gauges with } \lambda=0:\\ \quad\left[-g^{\mu\nu}+\frac{q^\mu n^\nu+n^\mu q^\nu}{n\cdot k}-\frac{n^2 q^\mu q^\nu}{(n\cdot k)^2}\right]\end{cases}$$

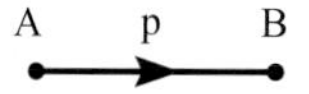

$$\delta^{ab}\frac{i(\not p+m)}{p^2-m^2+i\varepsilon}$$

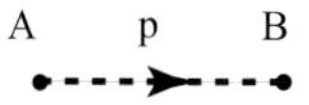

$$\delta^{AB}\frac{i}{p^2+i\varepsilon}\qquad\text{(Covariant gauge only)}$$

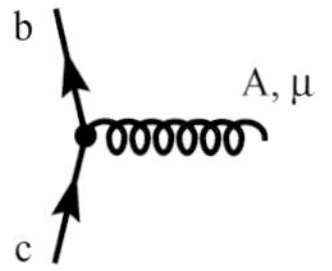

$$-ig_s[t^A]_{bc}\gamma^\mu$$

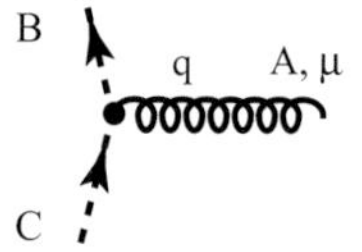

$$g_s f^{ABC} q^\mu\qquad\text{(Covariant gauge only)}$$

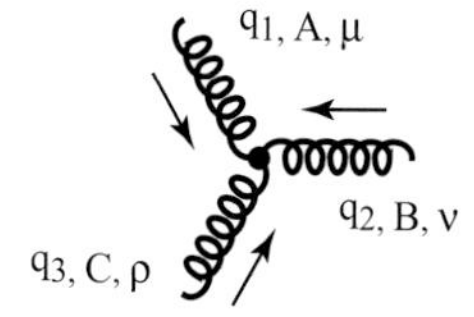

$$-g_s f^{ABC}\left[g^{\mu\nu}(q_1-q_2)^\rho + g^{\nu\rho}(q_2-q_3)^\mu + g^{\rho\mu}(q_3-q_1)^\nu\right]$$

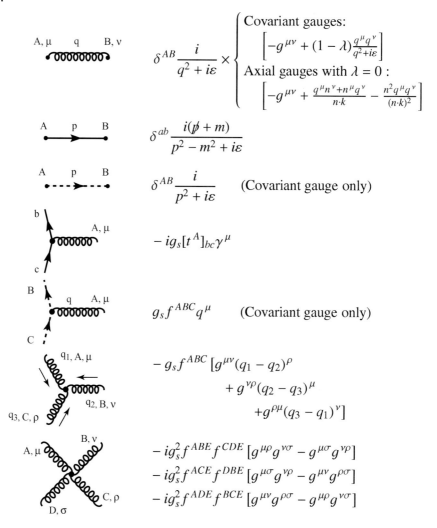

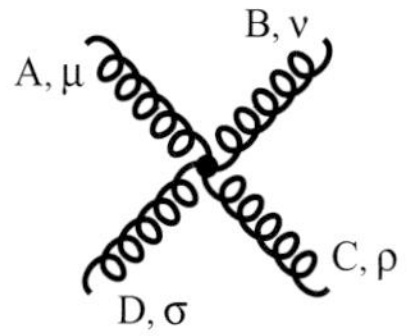

$$\begin{aligned}&-ig_s^2 f^{ABE} f^{CDE}\left[g^{\mu\rho}g^{\nu\sigma}-g^{\mu\sigma}g^{\nu\rho}\right]\\&-ig_s^2 f^{ACE} f^{DBE}\left[g^{\mu\sigma}g^{\nu\rho}-g^{\mu\nu}g^{\rho\sigma}\right]\\&-ig_s^2 f^{ADE} f^{BCE}\left[g^{\mu\nu}g^{\rho\sigma}-g^{\mu\rho}g^{\nu\sigma}\right]\end{aligned}$$

Figure 7.1 Feynman rules of QCD. Solid, dashed and coiled lines represent the quark, ghost and gluon respectively. Spacetime spinor indices are omitted. The ghosts only appear in the covariant gauges. All the momenta in the three- and four-gluon vertices are inward and satisfy $\sum q_i = 0$.

g–q–$\overline{q}$ Vertex

$$-ig_s\gamma^\mu[t^A]_{ab} = -ig_s\gamma^\mu\left(\frac{\lambda_A}{2}\right)_{ab} \tag{7.18}$$

Gluon Self-Coupling An important item which does not exist in QED but is essential in QCD is the gluon's self-coupling interaction. Partially expanding the term

$-(1/4)\boldsymbol{F}_{\mu\nu}\cdot\boldsymbol{F}^{\mu\nu}$, we have

$$\begin{aligned}
-\frac{1}{4}\boldsymbol{F}_{\mu\nu}\cdot\boldsymbol{F}^{\mu\nu} &= -\frac{1}{2}(\partial_\mu\boldsymbol{A}_\nu-\partial_\nu\boldsymbol{A}_\mu)\cdot\partial^\mu\boldsymbol{A}^\nu+\frac{1}{2}g_s\left(\boldsymbol{A}_\mu\times\boldsymbol{A}_\nu\cdot\partial^\mu\boldsymbol{A}^\nu-\partial^\nu\boldsymbol{A}^\mu\right)\\
&\quad-\frac{1}{4}g_S^2\left[\left(\boldsymbol{A}_\mu\cdot\boldsymbol{A}^\mu\right)^2-\left(\boldsymbol{A}_\mu\cdot\boldsymbol{A}_\nu\right)\left(\boldsymbol{A}^\mu\cdot\boldsymbol{A}^\nu\right)\right]\\
&= -\frac{1}{2}\left(\partial_\mu A_\nu^A-\partial_\nu A_\mu^A\right)\partial^\mu A^{A\nu}\\
&\quad+\frac{1}{2}g_s f^{ABC}A_\mu^A A_\nu^B\left(\partial^\mu A^{C\nu}-\partial^\nu A^{C\mu}\right)\\
&\quad-\frac{1}{4}g_s^2 f^{ABE}f^{CDE}A_\mu^A A_\nu^B A^{C\mu}A^{D\nu}
\end{aligned}\tag{7.19}$$

The Feynman rules corresponding to the above interactions can be obtained by taking matrix elements between initial and final states. Setting all the inward momenta as positive, we obtain

3-Gluon Vertex:

$$-g_s f^{ABC}\left[g^{\mu\nu}(q_1-q_2)^\rho+g^{\nu\rho}(q_2-q_3)^\mu+g^{\rho\mu}(q_3-q_1)^\nu\right]\tag{7.20}$$

4-Gluon Vertex:

$$\begin{aligned}
&-ig_s^2 f^{ABE}f^{CDE}\left[g^{\mu\rho}g^{\nu\sigma}-g^{\mu\sigma}g^{\nu\rho}\right]\\
&-ig_s^2 f^{ACE}f^{DBE}\left[g^{\mu\sigma}g^{\nu\rho}-g^{\mu\nu}g^{\rho\sigma}\right]\\
&-ig_s^2 f^{ADE}f^{BCE}\left[g^{\mu\nu}g^{\rho\sigma}-g^{\mu\rho}g^{\nu\sigma}\right]
\end{aligned}\tag{7.21}$$

7.1.3
Gauge Invariance and the Gluon Self-Coupling

Let us consider a gluon Compton scattering, that is, $g(q,A)+q(p,a)\to g(q',B)+q(p',b)$, to illustrate the role of nonlinear coupling in preserving the gauge invariance where variables inside the parentheses denote four-momenta and color indices. Denoting the color indices of the gluon and the quark as $A(B)$, and $a(b)$, the scattering amplitude for Figure 7.2a,b are expressed as

$$\begin{aligned}
-i\mathcal{M}(q+p\to q'+p') &= -i[\mathcal{M}_{(a)}+\mathcal{M}_{(b)}]\equiv\varepsilon^{*B}_{\mu}\left(-i\mathcal{M}^{\mu\nu}_{BA}\right)\varepsilon^A_\nu\\
&=(-ig_s)^2\left[\overline{u}_b(p')[t^B]_{bd}\,\not{\varepsilon}^{B*}\left(\frac{i\delta_{dc}(\not{q}+\not{p}+m)}{(q+p)^2-m^2}\right)\not{\varepsilon}^A[t^A]_{ca}u_a(p)\right.\\
&\quad\left.+\overline{u}_b(p')[t^A]_{bd}\,\not{\varepsilon}^A\left(\frac{i\delta_{dc}(\not{p}'-\not{q}+m)}{(p'-q)^2-m^2}\right)\not{\varepsilon}^{B*}[t^B]_{ca}u_a(p)\right]
\end{aligned}\tag{7.22}$$

where we retained the quark mass. The form of Eq. (7.22) is identical to that of the Compton scattering in QED except the color factor (see Vol. 1, Sect. 7.2). In QED, the gauge invariance means invariance under the transformation $A_\mu \to A'_\mu(x) = A_\mu(x) + \partial_\mu \chi(x)$ or in momentum space under $\varepsilon_\mu(k) \to \varepsilon_\mu(k) + \alpha k_\mu$. Namely, the amplitude has to satisfy conditions $k_\mu \mathcal{M}^{\mu\nu} = k_\nu \mathcal{M}^{\mu\nu} = 0$. Let us see if it can be applied to QCD processes too. Inserting $\varepsilon^A_\mu \to q_\mu$ and using the fact that $\not{p}\not{q} = 2(q\cdot p) - \not{q}\not{p}$, $(q+p)^2 - m^2 = 2(q\cdot p)$ and $\not{p}u(p) = mu(p), \overline{u}(p')\not{p}' = \overline{u}(p')m, \cdots$, we obtain

$$\varepsilon^{B*}_\mu \mathcal{M}^{\mu\nu}_{BA} q_\nu = -(-ig_s)^2 \overline{u}_b(p') \not{\varepsilon}_B u_a(p) \left([t^B t^A - t^A t^B]_{ba}\right) \tag{7.23}$$

In QED, this vanishes as the color matrices t^A, t^B are commutative. However, it does not in QCD. In QCD, the Lagrangian is invariant under the infinitesmal gauge transformation (see Eq. (1.10))

$$A^B_\mu(x) \to A^B{}_\mu(x)' = A^B{}_\mu(x) + \partial_\mu \alpha^B(x) + g_s f^{BCD} \alpha^C(x) A^D_\mu(x) \tag{7.24}$$

which contains an extra term that is absent in QED. If the gauge transformation only includes the second term $(\partial_\mu \alpha^B(x))$, the same rule would have to be applied to QCD as well. The third term induces nonlinear gluon interactions. If one takes into account such an interaction, one can conceive a Feynman diagram as depicted in Figure 7.2c. The quartic self-interaction does not contribute in the $O(g_s^2)$ approximation. The scattering amplitude for the triple gluon interaction (Figure 7.2c) is given by

$$\begin{aligned} -i\mathcal{M}_{(c)} &= \left(-g_s f^{ABD} T^{AB}_\mu\right) \delta_{DC} \frac{-ig^{\mu\nu}}{(q'-q)^2} \left(-ig_s \overline{u}_b(p')[t^C]_{ba} \gamma_\nu u_a(p)\right) \\ T^{AB}_\mu &= (\varepsilon^A \cdot \varepsilon^B)(q+q')_\mu + (\varepsilon^A \cdot q - 2q')\varepsilon^B_\mu + (\varepsilon^B \cdot q' - 2q)\varepsilon^A_\mu \end{aligned} \tag{7.25}$$

We have omitted the $q^\mu q^\nu$ part in the gluon propagator as it vanishes when multiplied with the conserved current. Content of the first parenthesis is the triple gluon interaction which can be obtained by taking contractions of the triple gluon vertex with the gluon polarizations. Substitution of $\varepsilon(q_1) \to \varepsilon^A, \varepsilon(q_2) \to \varepsilon^B$, $q_1 \to q, q_2 \to -q', q_3 \to q' - q$ in Eq. (7.20) gives T^{AB}_μ. Using $(\varepsilon^A \cdot q) = (\varepsilon^B \cdot q') = 0$ and $q' - q = p - p'$, replacement of $\varepsilon^A_\mu \to q_\mu$ in the triple gluon part reduces to

$$T^B_\mu = (q\cdot\varepsilon^B)(p-p')_\mu - 2(q\cdot q')\varepsilon^B_\mu \tag{7.26}$$

Substituting Eq. (7.26) in Eq. (7.25) and using $(q-q')^2 = -2(q\cdot q')$, $\overline{u}(p')(\not{p} - \not{p}')u(p) = 0$, we obtain

$$\mathcal{M}_{(c)} = (-ig_s)^2 \overline{u}_b(p') \not{\varepsilon}^B u(p)_a \left(-i f^{ABC} [t^C]_{ba}\right) \tag{7.27}$$

Therefore, if

$$[t^B, t^A] = i f^{BAC} t^C = -i f^{ABC} t^C \tag{7.28}$$

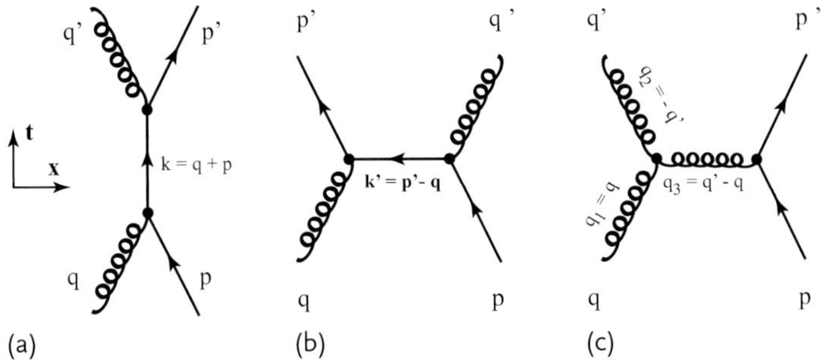

Figure 7.2 Gluon Compton scattering: (a) and (b) give the same contribution as QED except for the color factor. (c) includes the gluon self-coupling. All three amplitudes are necessary to keep the gauge invariance.

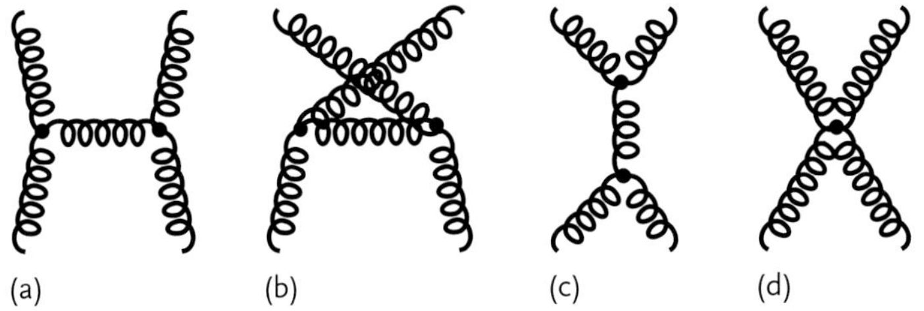

Figure 7.3 The leading Feynman diagrams contributing to gluon–gluon scattering.

the scattering amplitude vanishes when ε_μ^A is replaced with q_μ. However, the above equation is exactly what the $SU(3)$ gauge symmetry requires (see Eq. (H.12)). The Compton scattering is only an example. But it can be proven that the argument is valid in general.

In conclusion, the existence of the triple gluon vertices is vital in making the scattering amplitude gauge invariant which requires $k_\mu \mathcal{M}^{\mu\nu} = k_\nu \mathcal{M}^{\mu\nu} = 0$.

As a corollary to this conclusion, the rule that worked for the photons in QED

$$\sum_{\text{all } \lambda} \varepsilon_\mu(\lambda)\varepsilon_\nu(\lambda) = -g_{\mu\nu} \tag{7.29}$$

is no longer true in QCD, if there are more than two external gluons. One can, however, use Eq. (7.29) for one external gluon. When there are more than two gluons, one has to use Eq. (7.15) to sum physical polarizations.

Problem 7.3

Consider a gluon–gluon scattering in Figure 7.3. Figure 7.3a–c can be calculated using the triple interaction of Eq. (7.20). Show that they do not satisfy the gauge invariance condition alone, but they do if one adds the quartic interaction as shown in Figure 7.3d.

One may worry that more and more powers of self-interacting terms have to be added to preserve the gauge invariance order by order in the perturbation expansion. Fortunately, the triple and quartic self-interactions are all one needs. They are sufficient to guarantee the gauge invariance in all orders of the QCD processes.

7.1.4 Strength of the Color Charge

We discussed the strength of the color force generated by a quark acting on another quark in Vol. 1, Sect. 14.6. We review a few results derived there and extend discussions to gluons. We can show that the strength of the force is primarily determined by the coupling constant, but there is an additional factor because unlike the photon, the gluon carries eight colors. Firstly, we show that the strength of the force is universal among the colors, although it is not quite apparent from the structure of the Gell–Mann matrix.

Let us first consider the strength of the color force of the quark which carries the color charges (R,G,B). The interaction Lagrangian of the quark-gluon is obtained by substituting Eq. (7.4b) in $[D_\mu]_{ab}$ in Eq. (7.2), and extracting terms including A_μ.

$$\mathcal{L}_{\text{INT}} = -g_s \overline{\psi}_a [\boldsymbol{t}]_{ab} \cdot \boldsymbol{A}_\mu \psi_b \tag{7.30}$$

The quark charged with one kind of color "b" ($b = 1 \sim 3$) can emit a gluon with color A ($= 1 \sim 8$) and is converted to color "a." Its strength is determined by the $SU(3)$ generator $t^A = \lambda^A/2$. For instance, the quark with color "$b = 1$" (R) can emit a gluon with $A = 1, 2$ to become "$a = 2$" (G) or one with $A = 3, 8$ to stay in R. It is determined by nonzero elements of the Gell–Mann Matrix (Eq. (H.7a)) which we list here.

$$\lambda_{i,(i=1\sim3)} = \begin{bmatrix} \tau_i & & 0 \\ & & 0 \\ 0 & 0 & 0 \end{bmatrix}, \quad \lambda_4 = \begin{bmatrix} 0 & 0 & 1 \\ 0 & 0 & 0 \\ 1 & 0 & 0 \end{bmatrix}, \quad \lambda_5 = \begin{bmatrix} 0 & 0 & -i \\ 0 & 0 & 0 \\ i & 0 & 0 \end{bmatrix}$$
$$\lambda_6 = \begin{bmatrix} 0 & 0 & 0 \\ 0 & 0 & 1 \\ 0 & 1 & 0 \end{bmatrix}, \quad \lambda_7 = \begin{bmatrix} 0 & 0 & 0 \\ 0 & 0 & -i \\ 0 & i & 0 \end{bmatrix}, \quad \lambda_8 = \frac{1}{\sqrt{3}} \begin{bmatrix} 1 & 0 & 0 \\ 0 & 1 & 0 \\ 0 & 0 & -2 \end{bmatrix} \tag{7.31}$$

where τ_i's are Pauli matrices. In the following, we use 1, 2, 3 to represent R, G and B.

$$\begin{aligned}
&[t^1]_{21} = [t^1]_{12} = \frac{1}{2}, \quad [t^2]_{21} = -[t^2]_{12} = \frac{i}{2}, \quad [t^3]_{11} = -[t^3]_{22} = \frac{1}{2} \\
&[t^4]_{31} = [t^4]_{13} = \frac{1}{2}, \quad [t^5]_{31} = -[t^5]_{13} = \frac{i}{2}, \quad [t^6]_{32} = [t^6]_{23} = \frac{1}{2} \\
&[t^7]_{32} = -[t^7]_{23} = \frac{i}{2}, \quad [t^8]_{11} = [t^8]_{22} = \frac{1}{\sqrt{3}}, \quad [t^8]_{33} = -\frac{2}{\sqrt{3}}
\end{aligned} \tag{7.32}$$

The strength of the coupling that R-, G-, B-quark generate is given by

$$\begin{aligned} \mathrm{R} : &\left|[t^1]_{21}\right|^2 + \left|[t^2]_{21}\right|^2 + \left|[t^3]_{11}\right|^2 + \left|[t^4]_{31}\right|^2 + \left|[t^5]_{31}\right|^2 + \left|[t^8]_{11}\right|^2 \\ &= \frac{1}{4}\left[1+1+1+1+1+\frac{1}{3}\right] = \frac{4}{3} = C_F \end{aligned} \tag{7.33a}$$

$$C_F = \sum_{j,A} \left|[t^A]_{j1}\right|^2 = \sum_{j,A} t^A_{j1} t^{A*}_{j1} = \sum_{j,A} t^A_{1j} t^A_{j1} = \sum_{A} [t^A t^A]_{11} \tag{7.33b}$$

Similarly,

$$\mathrm{G} : \sum_{j,A} \left|[t^A]_{j2}\right|^2 = \sum_{A} [t^A t^A]_{22} = \frac{4}{3} \tag{7.33c}$$

$$\mathrm{B} : \sum_{j,A} \left|[t^A]_{j3}\right|^2 = \sum_{A} [t^A t^A]_{33} = \frac{4}{3} \tag{7.33d}$$

which shows that the strength is the same for all colors. The incoherent sum was taken since the final state is distinct by the color of the emitted gluon. The above algebra is summarized as a diagonal element of a matrix

$$\sum_{A} [t^A t^A]_{ij} \equiv C_F \delta_{ij} \tag{7.34}$$

which is proportional to the unit matrix, and hence all the elements are identical. This is the reason for the color universality. It is a consequence of a general theorem that holds in the $SU(N)$ group. Denoting the matrix representations of $SU(N)$ group generators in the n dimensional vector space as $F^A(n)$ $(A = 1 \sim N^2 - 1)$, operators defined by

$$[\boldsymbol{F} \cdot \boldsymbol{F}]_{st} = \sum_{A}^{N^2-1} [F^A(n) F^A(n)]_{st} \equiv C_2(n) \delta_{st} \quad (s, t = 1 \sim n) \tag{7.35a}$$

$$\mathrm{Tr}[F^A(n) F^B(n)] \equiv T_R(n) \delta_{AB} \tag{7.35b}$$

are Casimir operators which commute with any other matrices hence proportional to the unit matrix.

For the fundamental representation, we denote $n = N_\mathrm{F}$, $C_2(N_\mathrm{F}) \equiv C_F$, $T_R(N_\mathrm{F}) \equiv T_\mathrm{F} = 1/2$. The value of T_F determines the normalization of the matrices and is, by definition, set as 1/2.[4] For the adjoint representation, $n = N_\mathrm{A} = N_\mathrm{F}^2 - 1$. The group $SU(3)$ to which the color of the quark belongs has $N_\mathrm{F} = 3$ and $N_\mathrm{A} = 8$. For SU(N),

$$\begin{aligned} T_A &\equiv T_R(N_\mathrm{A}) = \mathrm{Tr}[F^A(N_\mathrm{A}) F^A(N_\mathrm{A})]_{A;\text{fixed}} \\ &= \sum_{D,E} (-i f^{ADE})(i f^{ADE}) = N_\mathrm{F} \end{aligned} \tag{7.36}$$

4) Sometimes, it is quoted that the QCD coupling strength $g_s^2/2$ corresponds to e^2 in QED.

Summing over all indices, Eqs. (7.35) yield the same result and one obtains

$$C_F N_F = T_F N_A \quad \text{for} \quad n = N_F \tag{7.37a}$$

$$C_A N_A = T_A N_A \quad \text{for} \quad n = N_A \tag{7.37b}$$

which yields

$$C_F = \frac{N_F^2 - 1}{2N_F} = \frac{4}{3} \tag{7.38a}$$

$$C_A = T_A = N_F = 3 \tag{7.38b}$$

The last number in each equation is the value for $SU(3)$. Applying the above discussion to the adjoint representation ($n = N_F^2 - 1$), we see that the strength of the gluon color charge S is given by

$$\begin{aligned} S\left(g_C \to \sum_{A,B} g_A g_B\right) &= \sum_{A,B} |f^{ABC}|^2 = \sum_{A,B} \left|[T^A]_{CB}\right|^2 \\ &= \sum_A [T^A T^A]_{CC}\Big|_{C:\text{fixed}} = C_A = 3 \end{aligned} \tag{7.39}$$

Problem 7.4

Derive Eq. (7.39) by explicit calculations using the gluon color matrix $[T^A]_{BC} = -i f^{ABC}$ where f^{ABC} is the totally antisymmetric structure constant of $SU(3)$ given in Table (H.1) and prove that all eight gluons have the same strength.

The strength of a gluon of fixed color charge to emit a quark–antiquark pair is given by $\sum_{b,c} t^A_{bc}(t^A_{bc})^* = \text{Tr}[t^A t^A]_{A\,\text{fixed}} = T_F = 1/2$.

In summary, the strength of the color charge is summarized as

$$\begin{array}{ccc} q \to qg & g \to gg & g \to q\overline{q} \\ C_F = \frac{4}{3} & C_A = 3 & T_F = \frac{1}{2} \end{array} \tag{7.40}$$

One can see that the strength of a gluon to emit a gluon is $(C_A/C_F)^2 = (9/4)^2 = 5.06$ times that of a quark to emit a gluon.

It is customary to leave C_F, C_A, T_R in the equation so that one can compare results with those of QED by setting $C_F = 1, C_A = 0, T_R = 1$ or to investigate possibilities of other gauge symmetries.

7.1.5 QCD Vacuum

We will show in this section that there are two kinds of vacuum in QCD. The conventional QCD vacuum[5] respects the color gauge symmetry and the so-called

5) It is also referred to as the Wigner mode when contrasted to the Nambu–Goldstone mode.

Nambu–Goldstone vacuum which arose as a consequence of the spontaneous breakdown of global chiral symmetry. In the high energy regime of QCD, in which this book deals mostly with, and where the asymptotic freedom is effective, we hardly need to worry about the latter. However, it is the basis of the chiral perturbation theory (ChPT) which deals with some nonperturbative aspects of QCD. In the discussion of low energy phenomena of the strong force including nuclear dynamics, ChPT is an indispensable tool and we should at least be aware of its existence and meaning.

Quark Condensate QCD is the theory of strong force which acts among the quarks[6]. The quarks exist in three colors and six flavors. We still do not understand why there are as many as six flavors of quarks. It looks like the whole world can live with only two kinds of quarks, namely, u and d. Therefore, let us restrict our discussion to u and d only for the moment. According to the Standard Model hypothesis, all the particles were massless in the beginning. Then, the gauge invariance holds independently for left-handed as well as right-handed fields which ensures the QCD to have the chiral gauge symmetry based on the color degrees of freedom. In addition, it also has a global $[SU(2)_L \times SU(2)_R]_{\text{flavor}}$ chiral symmetry based on the flavor degrees of freedom ($u \leftrightarrow d$). Generation of mass by the spontaneous breakdown of the electroweak symmetry breaks the axial part of the above symmetries, but not the flavor diagonal vector current.

$$\partial_\mu V_0^\mu(x) = \partial_\mu \overline{u}(x)\gamma^\mu u(x) = \partial_\mu \overline{d}(x)\gamma^\mu d(x) = 0 \quad \text{for any color} \tag{7.41a}$$

This preserves the $SU(3)_{\text{color}}$ symmetry and $U(1)$ flavor symmetry. However, the following vector and axial vectors do not conserve.

$$\partial_\mu V_+^\mu(x) = \partial_\mu(\overline{\psi}\gamma^\mu \tau_+ \psi) = \partial_\mu\left(\overline{u}(x)\gamma^\mu d(x)\right) = i(m_d - m_u)\overline{u}(x)d(x) \tag{7.41b}$$

$$\partial_\mu A_+^\mu(x) = \partial_\mu(\overline{\psi}\gamma^\mu\gamma^5\tau_+\psi) = \partial_\mu\left(\overline{u}(x)\gamma^\mu\gamma^5 d(x)\right) = i(m_d + m_u)\overline{u}(x)\gamma^5 d(x) \tag{7.41c}$$

Equation (7.41b) illustrates the breaking of global $SU(2)$ or isospin symmetry, and Eq. (7.41c) shows the chiral symmetry breaking.[7] The degree of isospin symmetry breaking is measured by $\Delta m = m_d - m_u$, while that of chiral symmetry breaking is measured by $m_{u+d} = m_d + m_u$. As $\Delta m \sim m_{u+d}$ (see Table 7.1 in Section 7.2.5), the degree of symmetry breaking is of the same order. The approximate isospin symmetry ($u \leftrightarrow d$) is apparent phenomenologically as is exemplified by the fact that $m_{\pi^+} \simeq m_{\pi^0}$, $m_p \simeq m_n$, mirror symmetry of the nuclear force and so on, but the chiral symmetry (L $\leftrightarrow$ R, that is, parity operation) is not so because there are no scalar 0^+ mesons with approximately equal mass as that of the pion. Experimentally, $m(0^+) \gg m(0^-)$, $m(1^+) \gg m(1^-)$. The difference lies in the vacuum, that

6) Here, we closely follow the arguments of [353].

7) The axial symmetry is also broken by the quantum anomaly, but its discussion is deferred to Volume 3.

is, the chiral symmetry breaking is spontaneous and is in the Nambu–Goldstone mode.[8] We can show it by the following argument.

As the right-hand side of Eq. (7.41c) has a quantum number of π^+, we write it as

$$\partial_\mu A^\mu_+(x) = f_\pi m_\pi^2 \phi_\pi(x) \tag{7.42}$$

which defines the constant f_π referred to as the pion decay constant. ϕ_π is the pion field which is considered as the asymptotic state of the composite $\overline{u}d$. The right-hand side of Eq. (7.42) vanishes in the limit $m_\pi^2 \to 0$ and is referred to as the partially conserved axial vector current hypothesis (PCAC) (see discussions in Vol. 1, Sect. 15.5.2). The value of $f_\pi = 132$ MeV is determined by the $\pi \to \mu\nu$ decay rate (see Vol. 1, Eq. (15.47)).

To demonstrate that the chiral symmetry is spontaneously broken, we consider a two-point function (we drop "+" index from A^μ, but it is understood to be there)

$$\Pi^{\mu\nu}(q) = i\int d^4x e^{iq\cdot x}\langle T(A^\mu(x)A^\nu(0)^\dagger)\rangle_0 \tag{7.43}$$

where $\langle\cdots\rangle_0$ means the vacuum expectation value. Contraction with q_μ, q_ν gives

$$\begin{aligned}
&q_\mu q_\nu \Pi^{\mu\nu}\\
&= -q_\nu \int d^4x e^{iq\cdot x}\partial_\mu\langle T(A^\mu(x)A^\nu(0)^\dagger)\rangle_0\\
&= -q_\nu \int d^4x \delta(x_0)\langle[A^0(x), A^\nu(0)^\dagger]\rangle_0\\
&\quad - q_\nu \int d^4x e^{iq\cdot x}\left\langle T\left(\partial_\mu A^\mu(x)A^\nu(0)^\dagger\right)\right\rangle_0\\
&= 2i\int d^4x \delta(x_0)\langle[A^0(x), \partial_\nu A^\nu(0)^\dagger]\rangle_0\\
&\quad + i\int d^4x e^{iq\cdot x}\left\langle T\left(\partial_\mu A^\mu(x)\partial_\nu A^\nu(0)^\dagger\right)\right\rangle_0
\end{aligned} \tag{7.44}$$

In going to the third equation, we have used the translational invariance

$$\langle A^\mu(x)B^\nu(0)\rangle_0 = \sum_n \langle e^{i\hat{p}\cdot x}A^\mu(0)e^{-i\hat{p}\cdot x}B^\nu(0)\rangle_0 = \langle A^\mu(0)B^\nu(-x)\rangle_0 \tag{7.45}$$

The first term in the fourth line of Eq. (7.44) is an equal time commutator and after substitution of $A^\mu(x) = \overline{u}(x)\gamma^\mu\gamma^5 d(x)$ and Eq. (7.41c) into Eq. (7.45), we obtain

$$\begin{aligned}
q_\mu q_\nu \Pi^{\mu\nu} &= 2(m_u + m_d)\langle[\overline{u}(0)u(0) + \overline{d}(0)d(0)]\rangle_0\\
&\quad + i f_\pi^2 m_\pi^4 \int d^4x e^{iq\cdot x}\langle T\left(\phi_\pi(x)\phi_\pi(0)^\dagger\right)\rangle_0
\end{aligned} \tag{7.46}$$

8) Actually, the first introduction of the spontaneous symmetry breaking in the particle field theory was Nambu's proposal of the chiral symmetry breaking and the pion as the Goldstone boson [45, 46].

or in the limit $q \to 0$,

$$
\begin{aligned}
&2(m_u + m_d)\langle \overline{u}(0)u(0) + \overline{d}(0)d(0)\rangle_0 \\
&= -i f_\pi^2 m_\pi^4 \int d^4x e^{iq\cdot x} \left\langle T\left(\phi_\pi(x)\phi_\pi^\dagger(0)\right)\right\rangle_0 |_{q\to 0}
\end{aligned}
\tag{7.47}
$$

One notices that the right-hand side of Eq. (7.47) is proportional to the Feynman propagator of the pion or more generally, an analytic function with a pole at $q^2 = m_\pi^2$ and a branch cut at $q^2 > (3m_\pi)^2$ (see Appendix I). Then, using dispersion relations, it can be expressed as

$$
\begin{aligned}
\Pi(q^2) &\equiv i \int d^4x e^{iq\cdot x} \langle T(\phi_\pi(x)\phi_\pi(0)^\dagger)\rangle_0 \\
&= \frac{1}{m_\pi^2 - q^2} + \frac{1}{\pi}\int dt \frac{\text{Im}\Pi(t)}{t - q^2} \text{ [9]}
\end{aligned}
\tag{7.48}
$$

In the above expression, the pole contribution dominates for $q^2 \simeq m_\pi^2$. Considering m_π^2 as a small parameter, Eq. (7.47) can be expressed as

$$
(m_u + m_d)\langle \overline{u}u + \overline{d}d\rangle_0 = -\frac{f_\pi^2 m_\pi^2}{2}\left\{1 + O\left(m_\pi^2\right) + \cdots\right\}
\tag{7.49}
$$

Experimentally, we know $m_u, m_d \neq 0$. Therefore, the above equation is a strong indication that $\langle \overline{u}u + \overline{d}d\rangle_0 \neq 0$. This means that the chiral symmetry is broken globally and spontaneously in QCD. Then, the Goldstone theorem dictates the appearance of the zero-mass Nambu–Goldstone boson which is identified as the pion. Actually, $\overline{s}s$ also condensates because the condensation is due to the color force and it is expected to be flavor blind. The order parameter of the condensation is defined by

$$
v = \langle \overline{u}u\rangle_0 \simeq \langle \overline{d}d\rangle_0 \simeq \langle \overline{s}s\rangle_0 < 0
\tag{7.50}
$$

The chiral symmetry is also broken explicitly by the finite mass of the quarks which adds mass to the Goldstone bosons. It acts as an external force to break the flavor symmetry. The measured kaon decay constant is very close to that of the pion ($f_K =$ 157 MeV $\sim f_\pi$ within 20%) which indicates that the same dynamics are acting on Ks. Therefore, the approximate flavor $SU(2)$ of (u, d) can be extended to $SU(3)$ of (u, d, s) and supports the notion that the whole 0^- octet (π, K, η) of $SU(3)_{\text{flavor}}$ [10] are the Goldstone bosons. It explains why there are no mirror 0^+ mesons and why 0^- octet members are light. Extension of the flavor symmetry beyond $SU(3)$ is not considered viable because of the large mass of the c, b, and so on.

9) Actually, one has to use the dispersion formula with subtraction to compensate the high energy growth of the function, but the conclusion does not change.

10) The suffix "flavor" is attached to distinguish it from $SU(3)_{\text{color}}$ which acts on colors possessed by every quark.

Chiral Perturbation Theory (ChPT) We introduce principles of the chiral perturbation theory which will be used for evaluating the light quark mass later. It is a QCD based phenomenological theory which handles low energy hadron phenomena where the asymptotic freedom cannot be applied. It starts with assumptions that the light 0^- scalar mesons are the massless Nambu–Goldstone bosons produced by spontaneous chiral symmetry breaking. Here, the symmetry is taken to be three flavor chiral symmetry $SU(3)_L \times SU(3)_R$ which can be applied to the three lightest quarks. The quark mass generated by the electroweak force works as an external symmetry breaking perturbation and produces the mass of the Goldstone bosons. The flavor $SU(3)$ violation is brought in by the different quark masses. Pions and kaons are treated as asymptotic fields induced by the quark pair condensation, that is, by Eq. (7.50). Namely,

$$\begin{aligned}\langle|\overline{q}^j_L q^i_R|\rangle_0 \rightarrow\ & \frac{v}{2}U^{ij}(\phi) \equiv \frac{v}{2}e^{i\sqrt{2}\Phi/F_\pi}\\ \Phi(x) \equiv \frac{1}{\sqrt{2}}\boldsymbol{\lambda}\cdot\phi(x) = & \begin{bmatrix} \frac{\pi^0}{\sqrt{2}}+\frac{\eta_8}{\sqrt{6}} & \pi^+ & K^+ \\ \pi^- & -\frac{\pi^0}{\sqrt{2}}+\frac{\eta_8}{\sqrt{6}} & K^0 \\ K^- & \overline{K}^0 & -\frac{2}{\sqrt{6}}\eta_8 \end{bmatrix}\end{aligned} \tag{7.51}$$

where $\boldsymbol{\lambda} = \lambda^i (i = 1 \sim 8)$ are the Gell–Mann matrices and $F_\pi = f_\pi/\sqrt{2}$. The form of Φ was worked out in Vol. 1, Eq. (14.15). The symmetry of $SU(3)_L \times SU(3)_R$ is expressed as the invariance under the transformation

$$\begin{aligned}&U \rightarrow U' = RUL^\dagger\\ &R = e^{-i\boldsymbol{\lambda}\cdot\boldsymbol{\alpha}_R/2}, \quad L = e^{-i\boldsymbol{\lambda}\cdot\boldsymbol{\alpha}_L/2}\end{aligned} \tag{7.52}$$

The adopted form of the field U is motivated by the nonlinear σ model with the heavy scalar meson integrated out from the Lagrangian (see Appendix J). In its exponential representation, the symmetry operation on U amounts to shift symmetry for $\boldsymbol{\phi}$

$$\boldsymbol{\phi} \rightarrow \boldsymbol{\phi}' = \boldsymbol{\phi} + F_\pi(\boldsymbol{\alpha}_L - \boldsymbol{\alpha}_R)/2 \tag{7.53}$$

which ensures that, in the absence of an explicit external force, $\boldsymbol{\Phi}$ can only have derivative interactions. Then, in the low energy regions, perturbative expansion in powers of momenta should give a good approximation. The constraint of the derivative interaction is essential for the viable effective Lagrangian at low energies. The symmetry breaking VEV (vacuum expectation value) is proportional to the unit matrix because it should conserve strangeness as well as electric charge.

$$\langle\boldsymbol{\phi}\rangle \equiv \boldsymbol{\phi}_0 = v\begin{bmatrix}1 & 0 & 0\\ 0 & 1 & 0\\ 0 & 0 & 1\end{bmatrix} \tag{7.54}$$

This VEV is left invariant under vector transformation for which $L = R = T$.

$$\boldsymbol{\phi}_0 \rightarrow T\boldsymbol{\phi}_0 T^\dagger = \boldsymbol{\phi}_0 \tag{7.55}$$

The explicit symmetry breaking term due to the quark mass is taken into account similarly:

$$\left\langle\left[m_i\left(q_L^i \overline{q}_R^j + \text{h.c.}\right)\right]\right\rangle \delta^{ij} \rightarrow \frac{v}{2}\mathcal{M}(U + U^\dagger)$$
$$\mathcal{M} = \text{dia}(m_u, m_d, m_s) \tag{7.56}$$

which does not respect the symmetry operation given in Eq. (7.52), but is invariant under the vector transformation, namely, it respects the symmetry $SU(3)_V$. Dynamics are dictated by the effective Lagrangian which contains all possible terms allowed by the assumed symmetries. Dynamical behavior of mesons and associated baryons below $\sim$ 1 GeV are treated in the power expansion of momenta (p/Λ or equivalently with an increasing number of derivatives) and m_q/Λ (m_q = quark mass) where Λ is the QCD energy scale below which the effective Lagrangian works. To the lowest order after expansion of U in powers of Φ, it is given by [354, 355]

$$\mathcal{L}_{\text{ChPT}} = \mathcal{L}_2 + \mathcal{L}_m$$
$$\mathcal{L}_2 = \frac{1}{2}\text{Tr}\left[\partial_\mu \Phi \partial^\mu \Phi\right] + \frac{1}{12F_\pi^2}\text{Tr}\left[\left(\Phi \overleftrightarrow{\partial_\mu} \Phi\right)\left(\Phi \overleftrightarrow{\partial^\mu} \Phi\right)\right] + O\left(\frac{\Phi^6}{F_\pi^4}\right)$$
$$\mathcal{L}_m = |v|\left\{-\frac{1}{F_\pi^2}\text{Tr}\left[\mathcal{M}\Phi^2\right] + \frac{1}{6F_\pi^4}\text{Tr}\left[\mathcal{M}\Phi^4\right] + O\left(\frac{\Phi^6}{F_\pi^4}\right)\right\} \tag{7.57a}$$

where

$$F_\pi = f_\pi/\sqrt{2} = 92.4\,\text{MeV}\,, \quad \hat{m}v = -F_\pi^2 m_\pi^2\,, \quad \hat{m} = \frac{1}{2}(m_u + m_d) \tag{7.57b}$$

Coefficients of the power expansion can only be determined phenomenologically. Once determined, they serve as universal constants to predict dynamics of various processes. The chiral perturbation theory is an infinite series of powers of Φ and $\partial_\mu \Phi (\sim q_\mu \Phi)$ and therefore unrenormalizable, but it can be made finite to the given order. It reproduces low energy meson dynamics quite well and all the current algebra consequences derived in the 1960s are built into it. It also works as a base for formulation of dynamical nuclear physics phenomena.

7.2 Renormalization Group Equation

7.2.1 Running Coupling Constant

The QCD prediction for an observable at energy scale Q^2 can be expressed as the sum of perturbative and nonperturbative terms. The perturbative terms are shown to have $\ln Q^2$ dependence while nonperturbative terms enter as power law corrections. There are no universal treatments to calculate the nonperturbative terms. Generally, they are negligible or subdominant in the energy region where perturbative treatments are valid. Therefore, we will ignore them and concentrate on the perturbative terms in the following. We will comment on the nonperturbative power corrections whenever necessary.

When a certain observable is expanded in a perturbation series, it is generically written as

$$P = c_1 \alpha_s + c_2 \alpha_s^2 + \cdots \tag{7.58}$$

The coefficients $c_1, c_2, \cdots$ can be obtained by calculating corresponding Feynman diagrams. In higher order corrections, divergent loop diagrams appear which are usually regulated by introducing some cut off parameter and absorbing them in the definition of the coupling constant, mass or normalization of the field, a process called renormalization (see discussions in Chapter 5 and Vol. 1, Chapt. 8). There are several schemes for renormalization depending on how the divergences are treated. Nowadays, it is customary to use dimensional regularization (See Appendix C) because it preserves the gauge invariance as well as the Lorentz invariance. There, integration is carried out in $D = 4 - \varepsilon$ dimension instead of four-dimension (i.e., $d^4 p \rightarrow \mu^{4-D} d^D p$) and the limit $\varepsilon \rightarrow 0$ is taken at the very end of calculation. μ is a scale with mass dimension to keep the physical dimension of the renormalized object invariant. Divergence appears as Δ in the following combination.

$$\Delta = \frac{2}{\varepsilon} - \gamma_{\mathrm{E}} + \ln(4\pi) + \ln \mu^2 \tag{7.59}$$

where γ_{E} is the Euler's constant. The way to subtract the infinity is not unique. The modern convention is to subtract the first three terms together and is called "modified minimum subtraction" [356, 357]. It is conventional to attach the suffix $\overline{\mathrm{MS}}$ to variables treated in this scheme. However, variations of the $\overline{\mathrm{MS}}$ scheme do appear for historical reasons as well as for convenience.

After subtraction, the last term $\ln \mu^2$ remains. This means the coefficients in Eq. (7.58) depend on $\ln \mu^2$ as well as choice of the subtraction constant (renormalization scheme). α_s itself becomes a function of μ. If the observable is specified at an energy scale Q, they are also a function of Q. Notice μ is an artifact that is introduced in the process of renormalization and physical observables should not depend on it.

Let us first consider a dimensionless observable P in the high energy limit, so that all the particles can be considered as massless. Then, only dimensionful variables that appear in the equations of motion are Q and μ. Since the observable is dimensionless, it is a function of Q^2/μ^2. As the observable should not depend on the scale μ, its change due to different choice of μ must be compensated for the corresponding change of α_s. It is a constant ($\alpha_s = g_s^2/4\pi$) in the first order, but acquires μ^2 dependence as higher order corrections are taken into account.

Therefore, the following equations must hold.

$$\mu^2 \frac{d}{d\mu^2} P\left(\mu^2, \alpha_s, Q^2\right) = \left[\mu^2 \frac{\partial}{\partial \mu^2} + \mu^2 \frac{\partial \alpha_s}{\partial \mu^2} \frac{\partial}{\partial \alpha_s}\right] P\left(\mu^2, \alpha_s, Q^2\right) = 0 \quad (7.60)$$

The equation is expressed in a form

$$\left[\mu^2 \frac{\partial}{\partial \mu^2} + \beta(\alpha_s) \frac{\partial}{\partial \alpha_s}\right] P\left(\mu^2, \alpha_s, Q^2\right) = 0 \quad (7.61a)$$

$$\beta(\alpha_s) = \mu^2 \frac{\partial \alpha_s}{\partial \mu^2} \quad (7.61b)$$

which is referred to as the renormalization group equation (or RGE for short). We try to solve the observable as a function of Q^2 rather than μ^2 treating β as given. In the above case, change of the variable is trivial since Q^2 only appears in a dimensionless variable Q^2/μ^2, namely, $dQ^2 = -d\mu^2$.

$$\left[-Q^2 \frac{\partial}{\partial Q^2} + \beta \frac{\partial}{\partial \alpha_s}\right] P\left(\mu^2, \alpha_s, Q^2\right) = 0 \quad (7.62)$$

The equation can be solved using a technique known as a method of characteristics. We introduce a new variable $\overline{\alpha}$ and a new parameter defined by

$$\tau = \ln(Q^2/\mu^2)\,, \quad d\tau = \frac{dQ^2}{Q^2} = \frac{d\overline{\alpha}_s(\tau)}{\beta(\tau)} \quad (7.63)$$

with boundary condition

$$\overline{\alpha}_s(0) = \alpha_s \quad (7.64)$$

A solution to Eq. (7.63) is easily obtained.

$$\tau = \int_{\alpha_s}^{\overline{\alpha}_s} \frac{d\alpha}{\beta(\alpha)} \quad (7.65)$$

Differentiating by τ and α_s, we have

$$\frac{d\overline{\alpha}_s(\tau)}{d\tau} = \beta(\overline{\alpha}_s(\tau))\,, \quad \frac{d\overline{\alpha}_s(\tau)}{d\alpha_s} = \frac{\beta(\overline{\alpha}_s(\tau))}{\beta(\alpha_s)} \quad (7.66)$$

Combining Eqs. (7.65) and (7.66), we obtain an equation for $\overline{\alpha}_s(\tau)$.

$$\left(-\frac{\partial}{\partial \tau} + \beta(\alpha_s) \frac{\partial}{\partial \alpha_s}\right) \overline{\alpha}_s(\tau) = 0 \quad (7.67)$$

which means, for an analytic function $F(x)$, we also have

$$\left(-\frac{\partial}{\partial \tau} + \beta(\alpha_s) \frac{\partial}{\partial \alpha_s}\right) F\left(\overline{\alpha}_s(\tau)\right) = 0 \quad (7.68)$$

Comparing the above equation with Eq. (7.62), we obtain a general solution

$$P(\tau, \alpha_s) = P(0, \overline{\alpha}_s(\tau)) \tag{7.69}$$

for Eq. (7.62). This means all the μ dependence of an observable only appears through the coupling constant $\overline{\alpha}_s(\mu)$.

This is a remarkable conclusion. It says that all the renormalization scale dependence comes in only through the scale dependent running coupling constant. Therefore, all we need to do to is to calculate the observable as a function of the coupling constant and substitute the running coupling constant in it to obtain the correct expression at the scale in consideration.

From now on, we omit the overline of the running coupling constant and reexpress it simply as $\alpha(\tau)$. Once one knows $P(0, \alpha_s)$, one knows its Q^2 dependence as well. For instance, the first order QCD correction to the "R" value is known to be (see Eq. (7.139))

$$R \equiv \frac{\sigma(e^-e^+ \to \text{hadrons})}{\sigma(e^-e^+ \to \mu^-\mu^+)} = R_0\left(1 + \frac{\alpha_s}{\pi} + \cdots\right) = 3\sum_q \lambda_q^2\left(1 + \frac{\alpha_s}{\pi} + \cdots\right) \tag{7.70}$$

where λ_q is the electric charge of the quarks. The Q^2 dependence of R is immediately obtained as

$$R(Q^2) = R_0\left(1 + \frac{\alpha_s(Q^2)}{\pi} + \cdots\right) \tag{7.71}$$

In the case of e^-e^+ annihilation reaction, $Q^2 = s$ is usually adopted.

7.2.2
Asymptotic Freedom

The scale dependent coupling constant $\alpha_s(\tau)$ can be calculated by using Eqs. (7.61b) or (7.65) once the β function is given. The β function, however, can only be obtained by the perturbation method and is expanded as

$$\beta(\alpha_s(\tau)) = \mu^2\frac{\partial\alpha_s(Q^2)}{\partial\mu^2} = -\alpha_s^2 b_0\left[1 + b_1\alpha_s + b_2\alpha_s^2 + \cdots\right] \tag{7.72a}$$

$$= -\frac{\beta_0}{4\pi}\alpha_s^2 - \frac{\beta_1}{8\pi^2}\alpha_s^3 - \frac{\beta_2}{128\pi^3}\alpha_s^4 + \cdots \text{[11)]} \tag{7.72b}$$

We learned in Vol. 1, Section 8.1.5 that a bare coupling constant in QED becomes Q^2 dependent when higher order (loop) corrections are included. It has a form (see

11) The nomenclature follows that of the Particle Data Group [11]. Some people use β_i for b_i and vice versa.

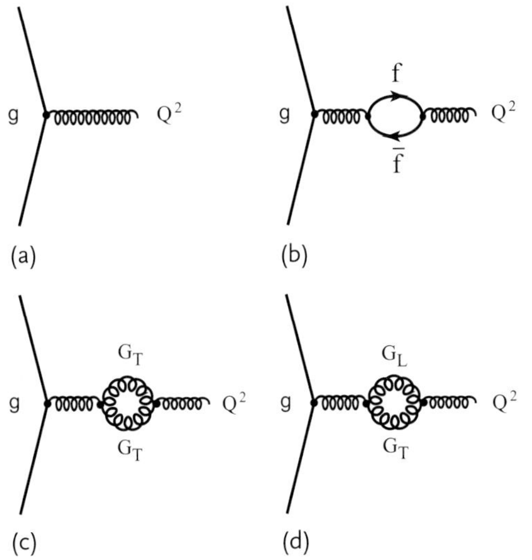

Figure 7.4 Feynman diagrams that contribute to the first order term in $\beta(\alpha_s)$.

Eqs. (C.31a) and (C.32))

$$\alpha_R = \alpha\left[1 - \Pi(0) - \hat{\Pi}(Q^2)\right], \quad Q^2 = -q^2 \tag{7.73a}$$

$$\Pi(0) = \frac{\alpha}{3\pi}\Delta = \frac{\alpha}{3\pi}\left[\frac{2}{\varepsilon} - \gamma_E + \ln 4\pi\right] \tag{7.73b}$$

$$\hat{\Pi}_\gamma(Q^2) = \begin{cases} -\frac{\alpha Q^2}{15\pi m^2} & Q^2 \to 0 \\ -\frac{\alpha}{3\pi}\ln\frac{Q^2}{m^2} & Q^2 \to \infty \end{cases} \tag{7.73c}$$

Comparing the above equations with Eq. (7.72), we obtain $b_0 = -1/(3\pi)$ for QED. This was obtained from the fermion loop diagram in Figure 7.4b. In QCD, additional gluon loops (Figure 7.4c, d) contribute as well. Let us consider this in the axial gauge, in which only the transverse components of the gluon are taken into account. In this gauge, the ghost does not appear and the physical content is the clearest, though the same result can be obtained in the covariant gauge.

$$\alpha_{\text{eff}}(Q^2) = \alpha_s\left[1 + b_0\alpha_s\{\Delta - \ln Q^2\} + \cdots\right], \quad Q^2 \to \infty \tag{7.74a}$$

$$b_0 = \frac{\beta_0}{4\pi} = \frac{11C_A - 2n_f}{12\pi} = \frac{11 - \frac{2}{3}n_f}{4\pi} \tag{7.74b}$$

where n_f is the number of active ($m_f < Q$) fermion species.[12] Unlike QED, we cannot use $Q^2 = 0$ as a reference point because α_s is divergent. This is not because $\hat{\Pi}(Q^2)$ is divergent, but due to the confinement which invalidates the perturbation treatment. Therefore, we choose a reference point at $Q = \mu$ and subtract it from

12) Justification is given later in Section 7.2.4.

$\alpha_s(Q^2)$. Namely,

$$\alpha_{\text{eff}}(Q^2) - \alpha_s(\mu^2) = \alpha_s^2 b_0 \ln \frac{Q^2}{\mu^2} + \cdots = \alpha_s^2 b_0 \tau + \cdots \tag{7.75}$$

b_0 of QED agrees with Eq. (7.74b) if we set the nonlinear gluon coupling $C_A = 0$, $n_f = 1$ and $\alpha_s/2 \to \alpha$ which originates in the different normalization of QED and QCD.[13] Substituting $\beta(\alpha_s) = -b_0\alpha_s^2$ in the renormalization group of Eq. (7.65), we obtain

$$\alpha_s(Q^2) = \frac{\alpha_s(\mu^2)}{1 + \alpha_s(\mu^2) b_0 \ln\left(\frac{Q^2}{\mu^2}\right)} \tag{7.76}$$

which is referred to as the leading log approximation (LLA) for α_s. The meaning of the word will be explained later. Unlike QED, where b_0 is negative, this is a decreasing function of Q^2 which is referred to as the asymptotic freedom. This is one of the outstanding characteristics of QCD.

The asymptotic freedom was the prime reason for the successful parton model in explaining the deep inelastic scattering phenomena which treated constituent particles inside hadrons as freely moving. The asymptotic freedom enables us to use the perturbative method for large Q^2 phenomena. Conversely, this also means that when Q^2 is small, the perturbative method breaks down. In QED, one could define a coupling constant at a long distance which is a measurable quantity. However, in QCD, there are no corresponding observables that can define a coupling constant unambiguously. The coupling constant $\alpha_s(Q^2)$ in QCD is dependent on the renormalization scale μ^2, though physical observables are not.

As mathematical handling of the obscure scale is cumbersome, one wonders if there is a way to get rid of it and use some universal constant to substitute for it. An alternative expression which does not depend on μ^2 can be obtained by rewriting Eq. (7.76) as

$$\frac{1}{\alpha_s(Q^2)} - b_0 \ln Q^2 = \frac{1}{\alpha_s(\mu^2)} - b_0 \ln \mu^2 \equiv -b_0 \ln \Lambda^2 \tag{7.77}$$

The left-hand side is a function of Q^2 and the central equation is a function of μ only, and hence it does not depend on Q^2 or μ^2 and is a constant. Thus, the right-hand side defines a universal constant Λ which does not depend on the renormalization point. In terms of Λ, the coupling constant is written as

$$\alpha_s(Q^2) = \frac{1}{b_0 \ln(Q^2/\Lambda^2)} \equiv \alpha_{s\,\text{LLA}} \tag{7.78}$$

When further higher order corrections are taken into account, Λ is defined by

$$\ln \frac{Q^2}{\Lambda^2} = -\int_{\alpha(\tau)}^{\infty} \frac{dx}{\beta(x)} \tag{7.79}$$

13) Sometimes, it is quoted that the QCD coupling strength $g_s^2/2$ corresponds to e^2 in QED.

The value of Λ is typically in the neighborhood of 200 MeV, although its precise value varies, depending on how it is treated. This will be discussed in the next section in more detail.

As can be seen from Eq. (7.78), Λ represents the scale at which the coupling will diverge or when α_s becomes strong. For perturbative treatments to be valid, $\alpha_s(Q^2)$ has to be small $\alpha_s(Q^2) \ll 1$, which happens if the energy scale in consideration is large $Q \gg \Lambda$. Phenomenologically, this happens at $Q \gtrsim 1\,\text{GeV}$ or at typical hadron mass scale. Low energy phenomena at scales small compared to Λ need nonperturbative treatments which are the realms of lattice QCD or nonrelativistic chiral perturbation theories.

The second (NLLA, that is, next leading log approximation) and higher order (NNLLA) contributions are calculated to be [11]

$$b_0 b_1 = \frac{17C_A^2 - (6C_F + 10C_A)T_F n f}{24\pi^2} = \frac{153 - 19n_f}{24\pi^2} \tag{7.80a}$$

$$b_0 b_2 = \frac{1}{128\pi^3}\left(2857 - \frac{5033}{9}n_f + \frac{325}{27}n_f^2\right) \tag{7.80b}$$

b_0 and b_1 do not depend on the renormalization scheme, but b_2 and higher order coefficients are scheme dependent.

Problem 7.5

Show that b_0 and b_1 are renormalization scheme independent.

Hint: Express the change of the coupling constant as

$$\alpha_s^A \to \alpha_s^B = \alpha_s^A(1 + c_1\alpha_s^A \cdots) \tag{7.81}$$

and use Eq. (7.72).

Substituting Eq. (7.80a) into Eq. (7.79), the second order Λ can be defined as

$$\frac{1}{\alpha_s} + b_1 \ln\left(\frac{b_1\alpha_s}{1 + b_1\alpha_s}\right) = b_0 \ln\frac{Q^2}{\Lambda^2} \tag{7.82}$$

The Λ defined here includes the second order correction and is different from that defined in Eq. (7.77). Equation (7.82) can be solved for α_s numerically, but it is inconvenient for routine use. Therefore, what is usually done is to expand $\alpha_s(Q^2)$ in terms of $\alpha_{s\,\text{LLA}}$ and drop terms higher than $\alpha_{s\,\text{LLA}}^2$. Replacing the right-hand side of Eq. (7.82) by $\alpha_{s\,\text{LLA}}^{-1}$, we obtain

$$\alpha_s(Q^2) = \alpha_{s\,\text{LLA}}\left[1 - b_1\alpha_{s\,\text{LLA}} \ln\left(\frac{b_1\alpha_s}{1 + b_1\alpha_s}\right)\right]^{-1} \tag{7.83}$$

and retain up to the second order expansion in inverse power of $\ln(Q^2/\Lambda^2)$. Then,

$$\begin{aligned}\alpha_{s\,\mathrm{NLLA}} &= \alpha_{s\,\mathrm{LLA}}\left[1 - b_1 \alpha_{s\,\mathrm{LLA}} \ln(\ln(Q^2/\Lambda^2))\right]^{-1} \\ &\simeq \frac{1}{b_0 \ln(Q^2/\Lambda^2)}\left[1 - \frac{b_1 \ln(\ln(Q^2/\Lambda^2))}{b_0 \ln(Q^2/\Lambda^2)}\right]\end{aligned} \tag{7.84a}$$

$$\alpha_{s\,\mathrm{LLA}} = \frac{1}{b_0 \ln(Q^2/\Lambda^2)} \tag{7.84b}$$

$\alpha_{s\,\mathrm{NLLA}}$ is conventionally referred to as the coupling strength of the next leading logarithmic approximation. The third order correction has also been obtained, but in comparing with experimental data presently available, the second order provides enough accuracy. Since $\alpha_s(Q^2)$ is a function of Q^2 and the value of Q^2 differs from one experiment to another, it is customary to translate it to one at $Q = m_Z$ using Eq. (7.84). The present world value is given by [11]

$$\alpha_s(m_Z) = 0.1170 \pm 0.0012 \tag{7.85}$$

7.2.3 Scale Dependence of Observables

We learned that according to the renormalization group equation (RGE), all the renormalization scale (μ) dependence is contained in the coupling constant and no μ^2 term will appear explicitly elsewhere. Well, that is true, provided we know the exact μ^2 dependence of $\alpha(\mu)$. In reality, we only know up to some powers of the perturbation series and scale dependence of one kind or another inevitably comes in. Consider an observable expanded in the power series of α_s.

$$P(\tau, \alpha) = r_0(\tau) + r_1(\tau)\alpha_s + r_2(\tau)\alpha_s^2 + r_3(\tau)\alpha_s^3 \cdots \tag{7.86}$$

where α_s is a constant evaluated at some scale $\mu = \mu_0$ and all the dependence on $\tau = \ln(Q^2/\mu^2)$ is in r_is. If RGE (Eqs. (7.62) and (7.72)) are satisfied at each order of α_s, Eq. (7.86) must have a form

$$\begin{aligned}P(\tau, \alpha_s) &= c_0 + c_1\alpha_s + \left[c_2 - c_1(b_0\tau)\right]\alpha_s^2 \\ &\quad + \left[c_3 - (c_1 b_1 + 2c_2)(b_0\tau) + c_1(b_0\tau)^2\right]\alpha_s^3 \\ &\quad + \text{(higher order terms)}\end{aligned} \tag{7.87}$$

where c_is are constants. From the above expression, one realizes that the nth term is expressed as

$$\alpha_s^n \sum_{i=1}^{n} d_i(n)\tau^{n-i} \tag{7.88}$$

where $d_i(n)$ is another constant. This means in order for the perturbation expansion to be valid, $\alpha_s\tau = \alpha_s \ln(\frac{Q^2}{\mu^2}) \ll 1$ is necessary as well as $\alpha_s \ll 1$. If $\alpha_s\tau \simeq 1$, the nth term in the perturbation expansion is still $\sim O(\alpha_s)$ and is meaningful only if all powers of $\alpha_s\tau$ are summed and included in $O(\alpha_s)$ term. Noticing the highest power of τ appears with the same coefficient in every order, we extract and combine them and rename $\alpha_{s\,\mathrm{LLA}}$. It coincides with $\alpha_{s\,\mathrm{LLA}}$ in Eq. (7.76). The remaining terms are $O[\alpha_s^2\{\alpha_s\tau\}^{n-2}]$, $n \geq 2$. This is the origin of the name LLA (leading log approximation). If one uses $\alpha_{s\,\mathrm{LLA}}(Q^2)$ instead of $\alpha_s(\mu^2)$, the power of the potentially dangerous logarithmic factor τ is decreased by one which also means μ dependence is at most $O(\alpha_s^2)$. Similarly, if one uses $\alpha_{s\,\mathrm{NLLA}}(Q^2)$, μ dependence will appear only in $O(\alpha_s^3)$ terms.

The solution to RGE Eq. (7.69) shows that as one includes higher and higher order terms, all the μ dependence will eventually be folded in $\alpha_s(Q^2)$. When the approximation is good only up to nth order, a term $\sim (\ln\mu^2)^{n+1}$ remains and an observable has residual dependence on μ.

Let us consider a case of $P = R(e^-e^+ \to \text{hadrons})/R_0$ again. From Eq. (7.71), we see $c_0 = 1, c_1 = 1/\pi$. The QCD correction to R has been calculated up to $O(\alpha_s^3)$ [358–361].

$$R = R_0\left[1 + \frac{\alpha_s}{\pi} + K_2\left(\frac{\alpha_s}{\pi}\right)^2 - 12.805\left(\frac{\alpha_s}{\pi}\right)^3 + \cdots\right] \tag{7.89}$$

$$\begin{aligned} K_2 &= \frac{1}{16}\left[-\frac{3}{2}C_F^2 + C_F C_A\left(\frac{123}{2} - 44\zeta_3\right) + C_F T_R N_F(-22 + 16\zeta_3)\right] \\ &\simeq 1.4092 \end{aligned} \tag{7.90}$$

where $\zeta_3 = 1.202\,056\,9\ldots$. In the above example, we set the $Q^2 = s =$ total center of mass energy squared. However, the choice of the scale variable is by no means clear. Depending on processes, Q may be set as the total energy, transverse momentum or associated heavy quark mass. It affects the magnitude of the remaining $\ln(\mu^2/Q^2)$ which could be sizable. Therefore, in a circumstance where the higher order calculation is not carried out and only an approximate estimate is possible, choice of the variable Q to substitute for μ is an important consideration to increase the credibility of the QCD calculation. Indeed, maximum ambiguity in comparing QCD predictions with experimental data is due to theoretical uncertainty in determining the scale μ. The RGE guarantees minimum μ dependence if higher and higher order corrections are included, but it is still there thus far as the calculation stops at finite order.

Let us show an example that when the corrections are terminated at finite order, the scale dependence does not vanish completely. Figure 7.5 shows scale dependences of the R-value in LLA (denoted as L), NLLA (L + NL) and NNLLA (L + NL + NNL) approximations [362, 363]. The assumption was $Q = \sqrt{s} = 34\,\text{GeV}$, $n_f = 5$, $\alpha_s(M_Z) = 0.117$. The scale was varied in the range $\mu = Q/8 \sim 4Q$. One can see that the scale dependence becomes smaller as one goes to higher order. Although accuracy of the calculation should be $\sim O(\alpha_s^3)$, the experimental value is not yet accurate enough to test the validity of $O(\alpha_s^3)$ calculation. A general statement one can deduce from the above consideration is that perturbative QCD calculations in LLA have a large dependence on μ and it is desirable to use at least

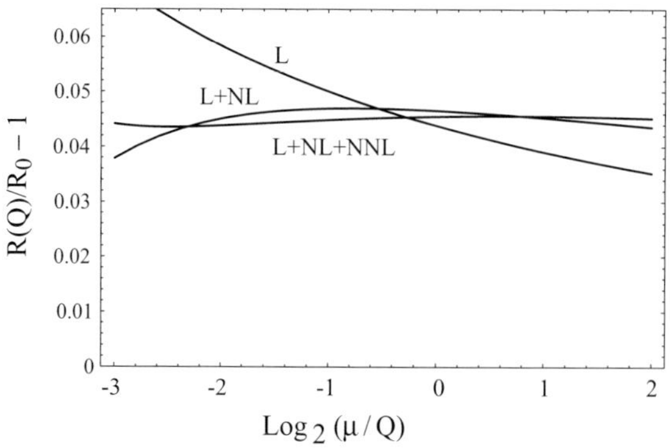

Figure 7.5 Scale dependence of R [362, 364].

the NLLA approximation in comparing theoretical calculations with experiments. Most of today's practical QCD treatments are all carried out at least to NLLA order.

Recipes to Minimize Scale Dependence Choice of which variables to use for Q depends on the process in consideration and has to be determined on a case by case basis. Even after the scale variable is chosen, the exact value at which to fix the μ remains to be determined. There are several recipes to try to minimize the residual scale dependence. We show a few examples below.

FAC The fastest apparent convergence approach [365] chooses μ such that if an observable is expressed as

$$P(\alpha, \tau) = \alpha_s^N \left[c_0 + c_1 \alpha_s + \cdots + c_n \alpha_s^n + \cdots\right] \tag{7.91}$$

Then, the first nontrivial term (i.e., $c_0 \alpha_s^N$) gives the same result if higher order terms (up to c_n) are included.

$$P^{(0)}(\mu_{\text{FAC}}) = P^{(n)}(\mu_{\text{FAC}}) \tag{7.92a}$$

Applied to Eq. (7.87) with $n = 2$, this yields

$$\mu_{\text{FAC}}^2 = s \exp\left[-\frac{c_2}{c_1 b_0}\right] \tag{7.92b}$$

PMS Principle of minimal sensitivity [366] requires that the truncated terms should be independent of μ. Applied to Eq. (7.87) with $n = 2$, this yields

$$\mu^2 \frac{d}{d\mu^2} P^{(n)}(\mu^2)\bigg|_{\mu_{\text{PMS}}} = 0 \Rightarrow \mu_{\text{PMS}}^2 = s \exp\left[-\frac{c_2}{c_1 b_0} - \frac{b_1}{2 b_0}\right] \tag{7.93}$$

BLM [367] requires that n_f dependence of the coefficients c_n should vanish.

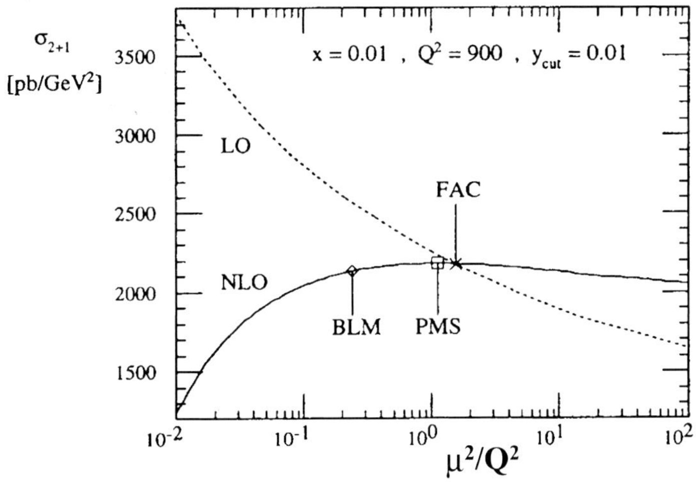

Figure 7.6 Scale dependence of the cross section for inclusive 2 + 1 jets in the deep inelastic scattering [368].

One should note that results of the above recipes vary, depending on processes. However, differences are generally not very large as is illustrated in Figure 7.6. The RGE at least guarantees that the residual error of calculations to $O(\alpha_s^n)$ is $O(\alpha_s^{n+1})$.

Problem 7.6

Derive Eq. (7.93).

7.2.4 Running Mass

In deriving the RGE Eq. (7.61a), we ignored contributions of particle masses. This is satisfactory for light quarks (u, d and s), but not for the top quark. The charm and bottom quarks are marginal, depending on what one is dealing with. Therefore, we need to estimate the mass effect. Let us assume, for simplicity, that only one flavor of quarks has a nonnegligible mass. Since the mass term appears in the Lagrangian as $m\overline{\psi}\psi = m(\overline{\psi}_L\psi_R + \overline{\psi}_R\psi_L)$, it can be considered as an interaction to convert the left-handed fermions to the right-handed fermions or vice versa. Namely, the mass can also be considered as the varying coupling constant just like α_s. It is apparent in the electroweak interaction because it is proportional to the strength of the coupling with the Higgs field. Besides, the mass is modified when higher order corrections are included and becomes a function of Q^2 [see Eq. (5.28) and arguments in Volume 1, Sect. 8.1.6]. The scale factor μ comes in just like it does in the running coupling constant. Therefore, the mass effect can be treated in a similar manner as α_s. Then, the RGE will take a form

$$\left[\mu^2 \frac{\partial}{\partial \mu^2} + \beta(\alpha_s, m/\mu) \frac{\partial}{\partial \alpha_s} + \gamma_m(\alpha_s, m/\mu) m \frac{\partial}{\partial m}\right] P\left(\mu^2, \alpha_s, m, Q^2\right) = 0 \tag{7.94a}$$

$$\beta(\alpha_s, m/\mu) = \mu^2 \frac{\partial \alpha_s}{\partial \mu^2}, \quad \gamma_m(\alpha_s, m/\mu) = \frac{\mu^2}{m} \frac{\partial m}{\partial \mu^2} \tag{7.94b}$$

It is possible to choose a renormalization scheme ($\overline{MS}$ scheme included) in which the higher order correction is independent of (m/μ) [369]. Then, both β and γ are functions of α_s (and μ^2) only.

The solution can be found similarly to the running coupling constant, namely, by treating $\ln m$ and γ_s just like α_s and β. To solve the equation, we again define new variables (running mass) $m(\tau)$ defined by

$$d\tau = \frac{1}{\gamma_m(\tau)} \frac{dm(\tau)}{m(\tau)} \tag{7.95a}$$

with boundary conditions

$$m(0) = m \tag{7.95b}$$

We also write $m(0) = m(\mu^2)$ because $\tau = 0$ means $Q^2 = \mu^2$. The solution is obtained as

$$m(\tau) = m(\mu^2) \exp\left[\int_0^\tau d\tau \gamma_m(\tau)\right] = m(\mu^2) \exp\left[\int_{\alpha_s}^{\alpha_s(\tau)} d\alpha \frac{\gamma_m(\alpha)}{\beta(\alpha)}\right] \tag{7.96}$$

The second equation follows from Eq. (7.65). In terms of the running coupling constant and running mass, the solution of the renormalization group equation takes a similar form as before. It is given by folding all the μ^2 dependence in $\alpha_s(\tau)$ and $m(\tau)$, namely, it is given by

$$P\left(\tau, \alpha_s, m, Q^2\right) \to P(0, \alpha_s(\tau), m(\tau), Q^2) \tag{7.97}$$

Problem 7.7

Show that Eq. (7.97) satisfies Eq. (7.94).

By using the same Feynman diagrams as was done for calculating the β function, we can calculate $\gamma(\alpha_s)$ and the running mass. It is given as [11]

$$\gamma_m(\alpha_s) = -\alpha_s\left(\gamma_0 + \gamma_1 \alpha_s + \gamma_3 \alpha_s^2 + \cdots\right) \tag{7.98a}$$

$$\gamma_0 = \frac{3}{4\pi} C_F = \frac{1}{\pi} \tag{7.98b}$$

$$\gamma_1 = \frac{C_F(97C_A + 9C_F - 20T_F n_f)}{96\pi^2} = \frac{303 - 10n_f}{72\pi^2} \tag{7.98c}$$

Here, γ_0 is independent of the renormalization scheme, but γ_1 is specific to the $\overline{MS}$ scheme. Substituting Eqs. (7.80) and (7.98) in Eq. (7.96), we have an expression for the running mass.

$$\begin{aligned}\ln\left[\frac{m(Q^2)}{m(Q_0^2)}\right] &= \int_{\alpha_s(Q_0^2)}^{\alpha_s(Q^2)} d\alpha \frac{\gamma_m(\alpha)}{\beta(\alpha)} = \int_{\alpha_s(Q_0^2)}^{\alpha_s(Q^2)} d\alpha \frac{\gamma_0 + \gamma_1\alpha + \cdots}{\alpha b_0(1 + b_1\alpha + \cdots)} \\ &\simeq \int_{\alpha_s(Q_0^2)}^{\alpha_s(Q^2)} d\alpha \left[\frac{1}{b_0}\left(\frac{\gamma_0}{\alpha} + \frac{\gamma_1 - b_1\gamma_0}{1 + b_1\alpha}\right)\right]\end{aligned} \tag{7.99}$$

$$\therefore \quad m(Q^2) = m_0 \left[\frac{\alpha_s(Q^2)}{\alpha_s(Q_0^2)}\right]^{\frac{\gamma_0}{b_0}} \left[1 + \frac{\gamma_1 - b_1\gamma_0}{b_0}\alpha_s(Q^2) + O(\alpha_s^2)\right] \tag{7.100}$$

where m_0 is some constant which is independent of the renormalization group. Specializing to the leading order result ($\gamma_1 = \beta_1 = 0$), we have

$$\frac{m(Q^2)}{m(Q_0^2)} = \left[\frac{\alpha_s(Q^2)}{\alpha_s(Q_0^2)}\right]^{\frac{\gamma_0}{b_0}} = \left[\frac{\ln(Q_0/\Lambda)}{\ln(Q/\Lambda)}\right]^{\frac{12}{33-2n_f}} \tag{7.101}$$

Thus, at high Q^2, we see that the mass effect is suppressed as the inverse power of logarithm as Q^2 increases. This originates from the fact that $\gamma_m < 0$. In addition, there is a further m/Q suppression which arise naturally as a result of the dimensional analysis.[14] Therefore, there is a theoretical justification to drop the mass at high Q^2. Consider the opposite case $Q^2 \ll m(Q^2)$. Then, the heavy particles only appear in the loops and it can be shown that the effect of the mass on cross sections is suppressed by powers of Q^2/m^2 [370, 371]. This decoupling theorem means that we can ignore the contribution of quarks if $m(Q^2) \gg Q$. On the other hand, if the quark mass is much smaller than Q ($Q \gg m_q$), then we have to include its effect. This is the reason we count the number (n_f) of active flavors in the renormalization equation.

14) If an observable $P(\mu, \alpha_s, m, Q)$ has proper dimension d like a fermion propagator which has $d = -1$, P should behave like

$$P(s\mu, \alpha_s, sm, sQ) = s^d P(\mu, \alpha_s, m, Q) \tag{7.102}$$

where s is a scale factor. This allows us to write

$$P(\mu, \alpha_s, m, sQ) = s^d P(\mu/s, \alpha_s, m/s, Q) \tag{7.103}$$

Namely, if $Q \to sQ$, then $\mu \to \mu/s, m \to m/s$, which means they appear in the observable as combinations $\mu/Q, m/Q$.

Values of the quark masses will be evaluated shortly in Section 7.2.5. The light quarks (u, d, s) have a much smaller mass ($m_q \ll \Lambda_{\text{QCD}}$, see Table 7.1 in Section 7.2.5) compared to Λ above in which the perturbative QCD is valid. In many cases, we observe that accuracy is improved by including the charm ($m_c = 1.2 \sim 1.5$ GeV). In the intermediate energies, where Q is more or less in the same range as the b-quark mass ($m_b \sim 5$ GeV), we should be more careful as to whether we should include it or not. It matters because the running couping constant Λ_{QCD} are also functions of the number of fermions that enter the equation (see Eq. (7.74b)). We may adopt $n_f = 4$ in one case (i.e., in the deep inelastic scattering processes where $Q \sim 1$–5 GeV), but we may have to use $n_f = 5$ in another (i.e., in R where $Q = \sqrt{s} \gg 5$ GeV). The effect of the top quark is usually ignored because in the presently available energy scale $Q \lesssim 100$ GeV (not applicable to LHC), the energy scale we deal with is much smaller than the top mass.

Continuity of the Coupling Constant Considerations in the previous section bring another ambiguity to the coupling constant which we now treat. If one uses α_+ defined at $Q \gg m$ and α_- at $Q \ll m$, α_+ and α_- have to be matched in the intermediate region. In the first order, they can be written as

$$\frac{1}{\alpha_+(Q^2)} - b_+ \ln \frac{Q^2}{m^2} = \frac{1}{\alpha_-(Q^2)} - b_- \ln \frac{Q^2}{m^2} \tag{7.104a}$$

where $b_- - b_+ = 1/6\pi$ from Eq. (7.74b). This gives

$$\alpha_+(Q^2) = \alpha_-(Q^2)\left[1 - \frac{x}{6\pi}\alpha_-(Q^2)\right]^{-1}, \quad x = \ln \frac{Q^2}{m^2} \tag{7.104b}$$

The next order calculation in $\overline{MS}$ scheme [372, 373] improves the above formula to

$$\alpha_+(Q^2) = \alpha_-(Q^2)\left[1 + \frac{x}{6\pi}\alpha_-(Q^2) + \frac{2x^2 + 33x - 11}{72\pi^2}\left(\alpha_-(Q^2)\right)^2\right] \tag{7.104c}$$

If we take into account the running mass and use $m_q = m(m_q^2)$, which means we set $x = \ln(Q^2/m^2)|_{Q^2=m_q^2} = 0$, Eq. (7.104c) reduces to almost requiring α_s to be continuous at the scale $Q = m_q$ [374].

$$\alpha_+\left(m_q^2\right) = \alpha_-\left(m_q^2\right) - \frac{11}{72\pi^2}\left(\alpha_-\left(m_q^2\right)\right)^3 \tag{7.105}$$

The difference is negligible for all practical purposes. One may argue that the matching should be done at the threshold ($Q = 2m_q$) rather than $Q = m_q$, but Eq. (7.104) is only valid at $Q^2 \gg m^2$ or $Q^2 \ll m^2$ and one is not sure how to treat them at $Q^2 \simeq m^2$. The above arguments can be justified from consistency of calculations in the full theory, including the effect of heavy quarks. Because, then, the calculations using the full theory can be extended smoothly to low energies.

Ambiguity of Λ Parameter The QCD constant Λ is conceptually a more important quantity than α_s because of its scale independence. It can be derived from α_s using Eq. (7.79) in principle, but more practically from Eq. (7.78) for LLA and (7.84a) for NLLA. Its precise value, however, is very difficult to determine. One should note that Λ is a scale independent parameter, but it depends on the renormalization scheme, n_f and others. We discuss some of the problems below.

1. The effective coupling constant $\alpha_s(Q^2)$ is not so small, and hence higher order terms which are ignored in comparing with observations are not necessarily small.
2. Experiments are not carried at $Q^2 \to \infty$, an ideal situation which the QCD usually assumes. Then, the so-called twist term, which is a correction term of the order $(O(m^2/Q^2))$ and is very difficult to calculate blurs the exact correspondence of experimental data to a theoretical prediction.
3. Even if a very precise experimental value is obtained, there is an inherent ambiguity in determining Λ if one uses the LLA approximation Eq. (7.78). If Λ is changed to $\Lambda + C$

$$\alpha_{s\,\text{LLA}}(Q^2) \to \frac{1}{b_0 \ln[Q^2/(\Lambda + C)^2]}$$
$$= \frac{1}{b_0 \ln(Q^2/\Lambda)^2} + \frac{\ln(2C/\Lambda)}{b_0\left[\ln(Q^2/\Lambda^2)\right]^2} + \cdots \tag{7.106}$$

   one can see that the second term in the right-hand side is $O[\alpha_s^2]$. This indefiniteness can be removed if one uses the next order expression in Eq. (7.84a).
4. If $\alpha_{s\,\text{NLLA}}$ is used, $\Lambda = \Lambda_1$ defined by Eq. (7.82) and $\Lambda = \Lambda_2$ defined by Eq. (7.84a) is not the same. For the same value of $\alpha_s(Q^2)$, the two Λs are related by

$$\Lambda_1 = \exp\left[\frac{b_1}{2b_0^2}\ln\left(\frac{b_0^2}{b_1}\right)\right]\Lambda_2 \simeq 1.15\Lambda_2 \quad (n_f = 5) \tag{7.107}$$

   However, within the context of this book, Eq. (7.84) has enough accuracy.[15)]
5. The fifth ambiguity originates from the renormalization scheme to define the expression. If calculated to infinite orders, the value of Λ should not depend on which scheme is used, but it does not apply to the finite order calculation. When the most conventional $\overline{MS}$ scheme is used, it is denoted as $\Lambda_{\overline{MS}}$.

15) In 2010, the Particle Data Group [3] adopted an approximate analytic solution to the third order which is expressed as

$$\alpha_s(Q^2) = \frac{1}{b_0 t}\left[1 - \frac{b_1}{b_0^2}\frac{\ln t}{t} + \frac{b_1^2(\ln^2 t - \ln t - 1)}{b_0^4 t^2}\right.$$
$$\left. - \frac{b_1^3\left(\ln^3 t - \frac{5}{2}\ln^2 t - 2\ln t + \frac{1}{2}\right) + 3b_0 b_1 b_2 \ln t - \frac{1}{2}b_0^2 b_3}{b_0^6 t^3}\right],$$
$$t \equiv \ln\frac{Q^2}{\Lambda^2} \tag{7.108}$$

6. Lastly, the value of Λ depends on n_f, the number of active fermions. One that is derived from the deep inelastic scattering data in the early days uses $n_f = 4$ and is denoted as $\Lambda = \Lambda(4)$. The value of Λ is in the range $200 \sim 300$ MeV. However, in the $e^- e^+$ collider experiments, usually $Q^2 = s$ is adopted. Experiments at PETRA, DESY, TRISTAN and LEP all exceed the b-quark production threshold and $n_f = 5$ is used ($\Lambda = \Lambda(5)$). When referring to Λ values, one should be alert as to which definition is used. $\Lambda(4)$ and $\Lambda(5)$ are determined from a condition that $\alpha_s(Q^2)$ be continuous (see Eq. (7.105)).

$$\alpha_s(m_b; n = 4) = \alpha_s(m_b; n = 5) \tag{7.109}$$

If one uses Eq. (7.84a), the relation is given by [374]

$$\Lambda_{\overline{MS}}(4) \approx \Lambda_{\overline{MS}}(5) \left[\frac{m_b}{\Lambda_{\overline{MS}}(5)}\right]^{2/25} \left[2 \ln\left(\frac{m_b}{\Lambda_{\overline{MS}}(5)}\right)\right]^{963/14\,375} \tag{7.110}$$

Numerically, the difference is rather large. When $\Lambda_{\overline{MS}}(5) = 200$ MeV, $\Lambda_{\overline{MS}}(4) = 289$ MeV. For these reasons, in determination of the coupling constant, it has become standard practice to quote the value of α_s at a given scale (typically m_Z) rather than to quote a value for the universal constant Λ.

Note that what is determined experimentally is $\alpha_s(Q^2)$ most of the time and Λ is calculated from it. Despite its many caveats, however, Λ is conceptually a fundamental constant and it is still widely used in phenomenological discussions. Therefore, one should be acquainted with methods of how to convert one to the other.

7.2.5 Quark Masses

As is clarified by Eq. (7.100), the quark mass runs. Since $\gamma(\alpha_s)/\beta(\alpha_s) \sim \gamma_0/b_0$ is positive, the quark mass is smaller at higher energies.

$$m_q\,(1\,\text{GeV}^2)/m_q(M_Z^2) = 2.30 \pm 0.05 \tag{7.111}$$

To evaluate running objects, we need a reference point to specify their values. It is conventional to express the heavy quark masses at the quark mass scale itself. However, for the light quarks, a reference scale $\mu_0 = 1$ GeV is chosen.

$$\overline{m}_q = m_q\left(m_q^2\right) \quad \text{for} \quad c, b, t \tag{7.112a}$$

$$\overline{m}_q = m_q\,(1\,\text{GeV}^2) \quad \text{for} \quad u, d, s \tag{7.112b}$$

The reason for choosing the reference point at 1 GeV is that this is the minimum energy that the asymptotic freedom is assumed to work so that the formula Eq. (7.100) can be used.

Relations among the quark mass of different flavors are independent of renormalization schemes. They can be fixed considering the chiral symmetry breaking mechanism. We notice that in the absence of the quark masses, QCD respect global chiral symmetry.

$$\begin{aligned} q_L &\to e^{-i\alpha_L} q_L\,, \quad q_L^\dagger \to e^{i\alpha_L} q_L^\dagger \\ q_R &\to e^{-i\alpha_R} q_R\,, \quad q_R^\dagger \to e^{i\alpha_R} q_R^\dagger \end{aligned} \tag{7.113}$$

In QCD phenomenology, the chiral symmetry is not respected. We argued in Section 7.1.5 that the symmetry is broken spontaneously because of the quark–antiquark pair condensation [45, 46] and that the 0^- octet mesons can be considered as the Goldstone bosons which should be massless. The Goldstone bosons acquire flavor dependent masses because of an external force provided by the Higgs mechanism. The quark mass relations discussed below are based on the assumption that the order parameter of the quark condensate is the dominant component.

To the first order, the mass of the 0^- mesons is given by Eq. (7.49). Its extension to other flavors can be obtained by explicitly calculating the first term of the Lagrangian $\mathcal{L}_m$ in Eq. (7.57a).

$$m_{\pi^\pm}^2 = B(m_u + m_d) + \delta_{EM}\,, \quad m_{\pi^0}^2 = B(m_u + m_d) - \varepsilon + O(\varepsilon^2) \tag{7.114a}$$

$$m_{K^\pm}^2 = B(m_u + m_s) + \delta_{EM}\,, \quad m_{K^0}^2 = B(m_d + m_s) \tag{7.114b}$$

$$m_{\eta_8}^2 = \frac{B}{3}(m_u + m_d + 4m_s) + \varepsilon + O(\varepsilon^2) \tag{7.114c}$$

$$\varepsilon = \frac{B}{2}\frac{(m_d - m_u)^2}{2m_s - m_u - m_d}\,, \quad B = -2\frac{v}{F_\pi^2} \tag{7.114d}$$

where v, F_π^2 are given by Eqs. (7.50) and (7.57b). η_8 is the eighth member of the 0^- octet before mixing with the singlet η_0 to make physical η and η' (see Vol. 1, Sect. 14.2.3). The $O(\varepsilon)$ correction originates from a small mixing between the π^0 and η_8 fields (Problem 7.8). δ_{EM} is an additional self-energy due to the electromagnetic interaction. Dashen's theorem [375] was used here which says that electrically neutral mesons do not acquire electromagnetic self-energy and that its contribution to $K^\pm$ and $\pi^\pm$ is the same. If we neglect δ_{EM}, ε and assume $m_u = m_d$, Eq. (7.114) reproduces the well known Gell–Mann–Okubo mass formula (Vol. 1, Sect. 14.2.1)

$$3m_{\eta_8}^2 = 4m_K^2 - m_\pi^2 \tag{7.115}$$

From Eq. (7.114), by neglecting the tiny ε, one can derive the light quark mass ratios [354, 376].

$$\frac{m_u}{m_d} = \frac{m_{K^\pm}^2 - m_{K^0}^2 + 2m_{\pi^0}^2 - m_{\pi^\pm}^2}{m_{K^0}^2 - m_{K^\pm}^2 + m_{\pi^\pm}^2} = 0.55 \tag{7.116a}$$

$$\frac{m_s}{m_d} = \frac{m_{K^\pm}^2 + m_{K^0}^2 - m_{\pi^0}^2}{m_{K^0}^2 - m_{K^\pm}^2 + m_{\pi^\pm}^2} = 20.3 \tag{7.116b}$$

The absolute value of the quark masses have been estimated with several different methods. Examples are from QCD sum rules, $\overline{m}_d + \overline{m}_u = 12.8 \pm 2.5$ MeV [377,

Table 7.1 Mass values of quarks and leptons [355].

	u	d	s
mass ($\mu = 1\,\text{GeV}$)	$4.6 \pm 0.9\,\text{MeV}$	$8.2 \pm 1.6\,\text{MeV}$	$164 \pm 33\,\text{MeV}$
	c	b	t
mass ($\mu = m_q$)	$\sim 1.2 \pm 0.1\,\text{GeV}$	$\sim 4.2 \pm 0.1\,\text{GeV}$	$165 \pm 5\,\text{GeV}$

378], from τ decay analysis,[16] $\overline{m}_s(m_\tau) = 119 \pm 24\,\text{MeV}$ [379, 380], and from lattice QCD, [$(\overline{m}_d + \overline{m}_u)\,(2\,\text{GeV}) = 8.46 \pm 0.58\,\text{MeV}$, $\overline{m}_s\,(2\,\text{GeV}) = 129 \pm 12\,\text{MeV}$ [381–383] and so on.] (see also [355, 376] and references therein). Although their precise numerical values are still somewhat controversial, they are converging and their representative values are given in Table 7.1.

The masses determined above are referred to as the current mass in contrast with the constituent mass which is determined from static properties of the quarks, like the mass and magnetic moment of hadrons (see Vol. 1, Eq. (14.57) and Table 14.7).

Problem 7.8

Derive ε in Eq. (7.114d).

7.3 Gluon Emission

Emission of gluons by a quark in any process appears as an important QCD correction. We derive the cross section for emitting a gluon in the process $e^- e^+ \to q\overline{q}g$. It is a basic expression that appears in many QCD processes and thus has many uses. For instance, it is the first order QCD correction to an important physical observable $R(e^- e^+ \to \text{hadrons})$ which actually is calculated to the third order in α_s and provides a test bench in determining the most accurate running coupling constant. Secondly, it has a practical use to determine 3-jet production cross section and provides an ideal environment for investigating (gluon-)jet properties. Thirdly, it provides a training ground for understanding the collinear as well as infrared divergences which are indispensable concepts in understanding the jet nature of partons (quarks and gluons) and subsequently the factorization. Fourthly, it provides a basic tool for the evolution equation of fragmentation functions and simulation programs for generating jet induced hadrons. Lastly, the process is closely connected by crossing to the deep inelastic scattering, the most thoroughly investigated QCD process and the theoretical consistency between the spacelike and timelike region can be directly tested.

16) See Eq. (9.89d) and a comment after them.

7.3.1 Emission Probability

Reduced Feynman diagrams for $e^-e^+ \to q\overline{q}g$ are depicted in Figure 7.7. By "reduced," we mean it is treated as a decay process of a timelike virtual photon (γ^*) created by e^-e^+ collision. It can also be treated as a crossed ($p_1 \to -p_1$) Compton scattering by an incident virtual photon and with the final photon replaced with a gluon. Hereafter, the gluon emission process by a photon scattering off a quark is referred to as QCD Compton scattering. The reasoning to justify this view is as follows. The transition matrix element of the process $e^-(k_1) + e^+(k_2) \to \overline{q}(p_1)q(p_2)g(p_3)$ can be expressed as

$$\begin{aligned}\mathcal{M}(e\overline{e} \to q\overline{q}g) &\equiv \mathcal{L}_\mu \frac{1}{Q^2}\mathcal{M}^\mu = \left[\overline{v}(k_2)(ie\gamma_\mu)u(k_1)\right]\frac{1}{Q^2} \\ &\times g_s\left[\overline{u}_a(p_2)\left(i\lambda_q e\gamma^\mu\right)\frac{-(\not{p}_1+\not{p}_3)+m}{(p_1+p_3)^2-m^2}\not{\varepsilon}^*[t_A]_{ab}v_b(p_1)\right. \\ &\left.+\,\overline{u}_a(p_2)\,[t_A]_{ab}\,\not{\varepsilon}^*\frac{\not{p}_2+\not{p}_3+m}{(p_2+p_3)^2-m^2}(i\lambda_q e\gamma^\mu)v_b(p_1)\right]\end{aligned} \tag{7.117}$$

where λ_q denotes the electric charge of the quark, g_s is the strong constant, $[t_A]_{ab}(A = 1-8, a, b = 1-3)$ are QCD color generator matrices with A, a, b the gluon and quark color index and $\varepsilon_\nu(p_3)$ polarization vector of the gluon having momentum p_3. Variables of the electrons are in the first square bracket. The minus sign in the anti-quark propagator arises because the momentum is flowing in the opposite direction to the fermion number. If one separates the electron part and carries out trace calculations after the matrix element is squared, one gets

$$\overline{\sum}_{\text{spin of } e\overline{e}} |\mathcal{M}(e\overline{e} \to q\overline{q}g)|^2 = e^2 L_{\mu\nu}(k_1, k_2)\frac{1}{Q^4}H^{\mu\nu}(p_1, p_2, p_3) \tag{7.118a}$$

$$\begin{aligned}L_{\mu\nu}(k_1, k_2) &= \frac{1}{4}\sum_{\text{spin}}[\overline{v}(k_2)\gamma_\mu u(k_1)][\overline{v}(k_2)\gamma_\nu u(k_1)]^* \\ &= \left[k_{1\mu}k_{2\nu} + k_{1\nu}k_{2\mu} - (k_1\cdot k_2)g_{\mu\nu}\right]\end{aligned} \tag{7.118b}$$

$$H^{\mu\nu}(p_1, p_2, p_3) = \mathcal{M}^\mu\mathcal{M}^{\nu*} \tag{7.118c}$$

$H^{\mu\nu}$ contains information on $q\overline{q}g$ while that of the electrons is contained in $L_{\mu\nu}$. It includes $Q^2 = q^2 = (k_1 + k_2)^2$ and the direction of $\boldsymbol{q}$. Since $\boldsymbol{q} = \boldsymbol{p}_1 + \boldsymbol{p}_2 + \boldsymbol{p}_3$ if one does not observe the orientation of the plane made by three final particles, it can be integrated. Then, by Lorentz invariance, the electron term $L_{\mu\nu}$ must be equal to constant $\times Q^2(-g_{\mu\nu} + Q^\mu Q^\nu/Q^2)$ because $Q^\mu L_{\mu\nu} = Q^\nu L_{\mu\nu} = 0$ as can be easily verified by calculating $Q^\mu L_{\mu\nu} = (k_1^\mu + k_2^\mu)L_{\mu\nu}$ and using $(\not{k}_2 + m_e)v(k_2) = \overline{u}(k_1)(\not{k}_1 - m_e) = 0$. The gauge invariance is at work here. The constant can be determined by making a contraction $-L_{\mu\nu}g^{\mu\nu} = e^2Q^2 = 3C$. Thus, $L_{\mu\nu}$ can be replaced with

$$\overline{L}_{\mu\nu} \equiv \frac{Q^2}{3}\left[-g_{\mu\nu} + \frac{Q_\mu Q_\nu}{Q^2}\right] \tag{7.119}$$

The gauge invariance is also at work on the hadronic part. Using $Q^\mu = p_1^\mu + p_2^\mu + p_3^\mu$ and omitting the color index, we obtain

$$Q_\mu \mathcal{M}^\mu \propto \overline{u}(p_2)\left[(\not p_1 + \not p_2 + \not p_3)\frac{1}{-(\not p_1 + \not p_3) - m}\not\varepsilon^* \right.$$
$$\left. + \not\varepsilon^* \frac{1}{(\not p_2 + \not p_3) - m}(\not p_1 + \not p_2 + \not p_3)\right] v(p_1) \tag{7.120}$$
$$= \overline{u}(p_2)\left[-(\not p_1 + \not p_3 + m)\frac{1}{\not p_1 + \not p_3 + m}\not\varepsilon^* \right.$$
$$\left. + \not\varepsilon^* \frac{1}{\not p_2 + \not p_3 - m}(\not p_2 + \not p_3 - m)\right] v(p_1) \tag{7.121}$$
$$= \overline{u}(p_2)\left[-\varepsilon^* + \varepsilon\right] v(p_1) = 0 \tag{7.122}$$

where we have used the identity $1/(\not p - m) = (\not p + m)/(p^2 - m^2)$.

If we use an equality $\sum_\lambda \varepsilon_\mu(\lambda)\varepsilon_\nu(\lambda) = -g_{\mu\nu}$ which is valid for the photon polarization in the Feynman gauge, the above procedure is equivalent to regarding the hadronic part as the process of a virtual photon decaying into $q\overline{q}g$ with sum over initial polarization states taken. Thus, replacement of $e\overline{e} \to q\overline{q}g$ with $\gamma^* \to q\overline{q}g$ can be justified.

Defining kinematic variables of the process as shown in Figure 7.7, they are related with the Mandelstam variables (s, t, u) in QED Compton scattering by crossing relations

$$\begin{aligned} s &= (k+p)^2 \to (q-p_1)^2 = (p_2+p_3)^2 \equiv \hat{t} \\ t &= (k-k')^2 \to (q-p_3)^2 = (p_1+p_2)^2 \equiv \hat{u} \\ u &= (k-p')^2 \to (q-p_2)^2 = (p_1+p_3)^2 \equiv \hat{s} \end{aligned} \tag{7.123}$$

where hats over t, s, u are attached to remind one that the variables are those of partons. The spin averaged transition matrix elements squared is given by (see

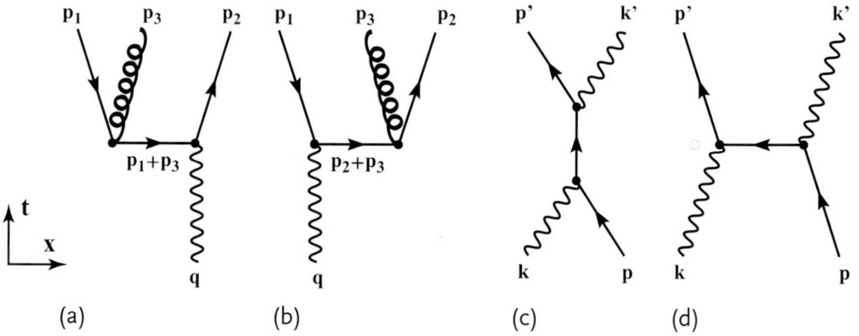

Figure 7.7 (a,b) Feynman diagrams for $\gamma^* \to q\overline{q}g$ and (c,d) for Compton scattering $(\gamma^* e \to \gamma e)$. $e^- e^+ \to q\overline{q}g$ can be considered as a timelike virtual photon decaying to $q\overline{q}g$. Diagrams (a) and (b) can be obtained from (c) and (d) by changing the electron to a quark, crossing the incoming quark and by changing the final $\gamma \to g$.

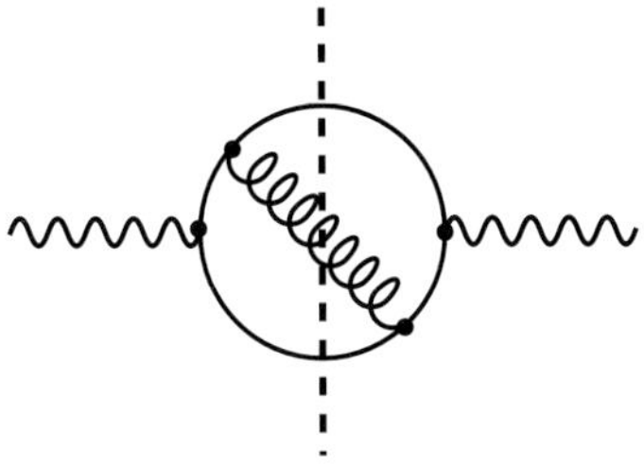

Figure 7.8 Crossing changes sign of the amplitude if it is a part of the fermion loop.

Vol. 1, Eq. (7.60))

$$|\overline{\mathcal{M}}|^2[\gamma^*(k) + e(p) \to \gamma(k') + e(p')] = 2e^4\left(-\frac{u}{s} - \frac{s}{u} - \frac{2k^2 t}{su}\right) \tag{7.124}$$

Substituting Eq. (7.123) in Eq. (7.124), replacing $k^2 \to q^2 = Q^2$, $e^4 \to \lambda_q^2 e^2 g_s^2$, and taking color factors into account, we obtain the transition matrix element squared of $\gamma^* \to q\overline{q}g$

$$\begin{aligned} &|\mathcal{M}|^2[\gamma^*(q) \to \overline{q}(p_1) + q(p_2) + g(p_3)] \\ &\quad = N_c C_F \times 4 \times 2\lambda_q^2 e^2 g_s^2 \left(\frac{\hat{s}}{\hat{t}} + \frac{\hat{t}}{\hat{s}} + \frac{2Q^2\hat{u}}{\hat{s}\hat{t}}\right)^{17)} \\ &N_c C_F = \sum_A \mathrm{Tr}[t^A t^A] = \underbrace{T_F}_{\substack{1/2 \\ \sim \\ e^2 \to g_s^2/2}} \underbrace{(N^2-1)}_{\substack{8 \\ \sim \\ \text{gluon color}}} \end{aligned} \tag{7.125}$$

where the factor 4 in the second line arises because no spin or polarization average were taken for the initial γ^* and crossed $\overline{q}$. The color factor is $\sum_{a,b}\sum_A [t^A t^A]_{ab} = N_c C_F = 4$, or in plain words, there are eight degrees of color freedom in the final state, and the QED coupling α corresponds to $\alpha_s/2$ in QCD.

Problem 7.9

Derive Eq. (7.125) by calculating the Feynman diagrams explicitly.

When one treats three body kinematics, it is convenient to introduce dimensionless energy variables in the center of mass frame defined by

$$x_i \equiv 2E_i/Q \tag{7.126}$$

17) The extra minus sign acquired by crossing can be understood from the optical theorem which states that the total cross section is proportional to the forward scattering amplitude. It is represented by a diagram in the Figure 7.8 which includes a fermion loop and hence an extra minus sign. The reason for the sign change is because crossing is a discrete jump in the s, t, u variables. Analytical continuation is the right way to get the correct expression.

which are expressed as (in the massless limit of the final particles)

$$\begin{aligned} \hat{s} &= (p_1 + p_3)^2 = 2p_1 \cdot p_3 = Q^2(1 - x_2) \\ \hat{t} &= (p_2 + p_3)^2 = 2p_2 \cdot p_3 = Q^2(1 - x_1) \\ \hat{u} &= (p_1 + p_2)^2 = 2p_1 \cdot p_2 = Q^2(1 - x_3) \end{aligned} \tag{7.127}$$

In terms of x_is, Eq. (7.125) takes the form of

$$\begin{aligned} \overline{|\mathcal{M}|^2} &= 24C_F\lambda_q^2 e^2 g_s^2 \left[\frac{1 - x_2}{1 - x_1} + \frac{1 - x_1}{1 - x_2} + \frac{2(1 - x_3)}{(1 - x_1)(1 - x_2)}\right] \\ &= 24C_F\lambda_q^2 e^2 g_s^2 \frac{x_1^2 + x_2^2}{(1 - x_1)(1 - x_2)} \end{aligned} \tag{7.128}$$

The decay rate of the virtual γ^* to $q\bar{q}g$ is given by

$$\begin{aligned} d\Gamma(\gamma^* \to q\bar{q}g) &= \frac{1}{2Q}\overline{|\mathcal{M}|^2} dLIPS \\ dLIPS &= (2\pi)^4\delta^4(q - p_1 - p_2 - p_3)\frac{d^3p_1}{(2\pi)^3 2E_1}\frac{d^3p_2}{(2\pi)^3 2E_2}\frac{d^3p_3}{(2\pi)^3 2E_3} \end{aligned} \tag{7.129}$$

Problem 7.10

Show that angular integration of the phase space factor gives

$$dLIPS = \frac{Q^2}{128\pi^3} dx_1 dx_2 \tag{7.130}$$

Problem 7.11

(a) Show that by energy conservation

$$x_1 + x_2 + x_3 = 2\,, \quad \hat{s} + \hat{t} + \hat{u} = Q^2 \tag{7.131}$$

(b) Show that the boundary of variables where x_is can vary is given by

$$\hat{s}\hat{t}\hat{u} = 0 \quad \text{or} \quad (1 - x_1)(1 - x_2)(1 - x_3) = 0 \tag{7.132}$$

Problem 7.12

Show that the total decay rate $\Gamma(\gamma^* \to q\bar{q})$ summed over all polarization states is given by $\Gamma = \lambda_q^2 \alpha Q N_c$.

Using results of the above problems, we can obtain one gluon emission probability which is given by

$$\frac{1}{\Gamma}\frac{d\Gamma}{dx_1 dx_2} = \frac{\alpha_s}{2\pi} C_F \frac{x_1^2 + x_2^2}{(1 - x_1)(1 - x_2)} \tag{7.133a}$$

$$0 \le x_1 \le 1\,, \quad 1 - x_1 \le x_2 \le 1 \tag{7.133b}$$

The total one gluon emission cross section is given by

$$\sigma_g = \sigma_0 \frac{\alpha_s}{2\pi} C_F \int_0^1 dx_1 \int_{1-x_1}^1 dx_2 \left[\frac{x_1^2 + x_2^2}{(1 - x_1)(1 - x_2)}\right] \tag{7.134a}$$

$$\sigma_0 = \sigma_{\text{tot}}(e^- e^+ \to \text{hadrons}) = \frac{4\pi\alpha^2}{3s} \sum_q N_c \lambda_q^2 \tag{7.134b}$$

Here, σ_0 is the lowest order total cross section of the hadron production. $N_c = 3$ is the color degrees of freedom of the quark.

7.3.2 Collinear and Infrared Divergence

The integral Eq. (7.134) diverges at $x_1 = 1, x_2 = 1$. This is the first encounter of the collinear singularity which occurs in QCD over and over again whenever one performs perturbative calculations.

If one does not neglect quark masses,

$$(p_2 + p_3)^2 - m_q^2 = \hat{t} - m_q^2 = 2E_3(E_2 - p_2 \cos\theta_{23}) \tag{7.135a}$$

$$(p_1 + p_3)^2 - m_q^2 = \hat{s} - m_q^2 = 2E_3(E_1 - p_1 \cos\theta_{13}) \tag{7.135b}$$

There are two circumstances that the denominator of propagators will vanish.

1. Emission of a soft gluon: $E_3 \to 0$
2. Emission of a collinear gluon: $m_q = 0$ and $\theta \to 0$

(1) is referred to as infrared divergence and (2) as collinear or mass singularity. Note that collinear divergence is further enhanced by growing $\alpha_s(Q^2)$ as the scale Q^2 is reduced. This is a characteristic of gauge theories which require that the mass of the gauge boson vanishes. The infrared divergence in QCD has exactly the same properties as that of QED (see arguments in Vol. 1, Chapt. 8). In QED, $m_e \neq 0$ and no collinear singularity exists. Strictly speaking, $m_q \neq 0$ in QCD, but as the energy scale is much larger, the (light) quarks are treated as massless. Moreover, because of confinement, the parton is converted to a bundle of hadrons (i.e., a jet). Two collinear partons produce overlapping jets which cannot be distinguished from a single jet. Therefore, the collinear configuration in QCD has to be treated with

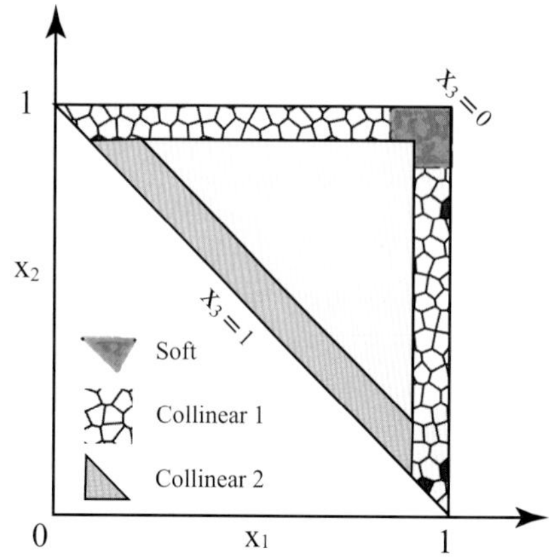

Figure 7.9 Allowed range of x_1, x_2. In the white area, 3-jets are distinguished from the 2-jet configuration. The white area is a divergence free region. The "Collinear 2" region is also divergence safe, but with two quarks being collinear, 3-jets are not resolved. The "soft" region is where the gluon energy becomes very small and the collinear 1 region is where one of x_1, x_2 approaches 1.

care. Figure 7.9 depicts soft and collinear regions. Those denoted as "Collinear 1" near $x_1 = 1$ and $x_2 = 1$ are regions where the gluon is emitted in the collinear configuration and diverge toward the edge. The one denoted "Collinear 2" at the $x_3 = 1$ edge has no singularity, but is a region where three partons cannot be resolved as three jets. In the white area, three partons are clearly resolved as three jets.

The infrared divergence (IR) occurs in the region denoted as "soft" where the gluon energy almost vanishes. It is not a real divergence, though. Just like QED, it disappears when the effect of virtual gluons is properly taken into account. When the IR divergence appears, it can be regularized in various ways. The most orthodox method is to use dimensional regularization. For strict theoretical treatments, it is most desirable, but here we adopt the mass regularization method for simplicity. It avoids the divergence by assigning a small mass m_g to the gluon. Then, by carrying out integration of Eq. (7.134a), one gets [356, 357]

$$\begin{aligned} \sigma_g(\text{real}) &= \frac{2\alpha}{3\pi}\sigma_0\left[\ln^2(Q_g) + 3\ln(Q_g) - \frac{\pi^2}{3} + 5\right] \\ Q_g &= m_g^2/Q^2 \end{aligned} \tag{7.136}$$

Although the real gluon emission process (Figure 7.10c,d) is physically different from that of no gluon emission Figure 7.10a,b), it cannot be separated in the soft gluon limit. The process to exchange a virtual gluon (Figure 7.10b) is of $O(\alpha_s^2)$, and the interference effect with the process without gluon emission is of $O(\alpha_s)$. Contribution of the virtual process is given by

$$\sigma_g(\text{virtual}) = \frac{2\alpha}{3\pi}\sigma_0\left[-\ln^2(Q_g) - 3\ln(Q_g) + \frac{\pi^2}{3} - \frac{7}{2}\right] \tag{7.137}$$

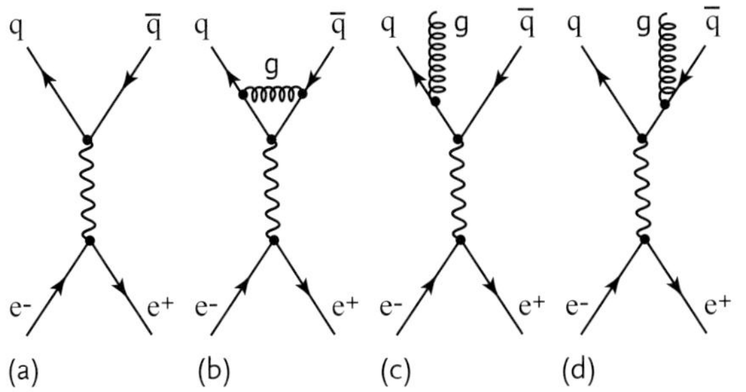

Figure 7.10 (a) $e^-e^+ \to q\bar{q}$. (b) Virtual one gluon exchange. (c,d) $e^-e^+ \to q\bar{q}g$.

Adding Eqs. (7.136) and (7.137) together, one obtains

$$\sigma_g(\text{real}) + \sigma_g(\text{virtual}) = \frac{\alpha_s}{\pi} \tag{7.138}$$

In other words, the total cross section for the process "$e^-e^+ \to$ hadrons + 1gluon" is given by

$$\sigma(e^-e^+ \to \text{hadrons}) = \sigma_0\left(1 + \frac{\alpha_s}{\pi}\right) \tag{7.139}$$

As one can see, Eq. (7.139) does not contain the infrared divergence. This is an example of the so-called infrared safe observables which are described in more detail in Section 9.1.2.

7.3.3 Leading Logarithmic Approximation

Equation (7.133a) can be interpreted as the probability for emitting a gluon as long as $(1 - x_1)$, $(1 - x_2)$ are not too small. If one is discussing the emission of only one hard gluon, it is an exact formula. However, in QCD, the parton (quarks and gluons) never appears as a single isolated particle, but as jets because of confinement. It is always accompanied by emission of many soft or collinear gluons. Equation (7.133a) is inconvenient to see through such effects. Therefore, we introduce the light cone variables defined by (see Fig. 7.11)

$$zQ \equiv E_2 + p_{2\parallel} = E_2(1 + \cos\theta_2) \tag{7.140a}$$

$$(1 - z)Q = E_3 + p_{3\parallel} \tag{7.140b}$$

$$\hat{t} = Q^2(1 - x_1) = \frac{p_T^2}{z(1 - z)} \tag{7.140c}$$

$$p_T = p_2 \sin\theta_2 = p_3 \sin\theta_3 \tag{7.140d}$$

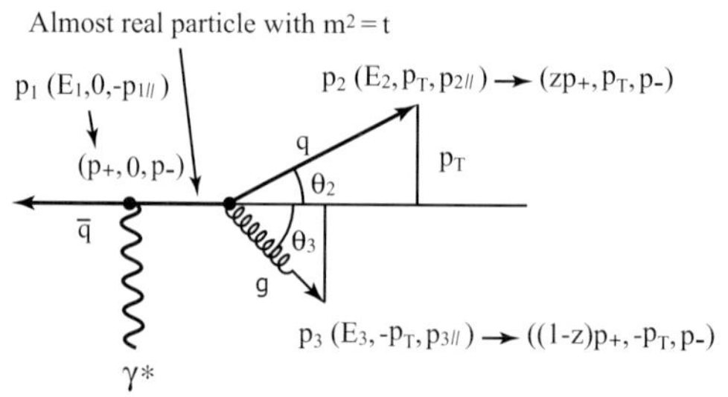

Figure 7.11 Kinematical variables in the center of mass frame of $e^- e^+ \to q\bar{q}g$ and introduction of the light cone variables $p_\pm = E \pm p_z$. In the infinite momentum frame, all particles are treated as massless real particles.

They are related with previously defined variables by

$$z = 1 - \frac{1-x_2}{x_1} = \frac{1-x_3}{x_1} \tag{7.141a}$$

$$x_T^2 \equiv \frac{4p_T^2}{Q^2} = 4z(1-z)\frac{\hat{t}}{Q^2} = \frac{4(1-x_1)(1-x_2)(1-x_3)}{x_1^2} \tag{7.141b}$$

$$\sin^2\left(\frac{\theta_2}{2}\right) = \frac{(1-x_1)(1-x_2)}{x_1 x_2} \tag{7.141c}$$

Using the light cone variables, Eq. (7.133a) can be rewritten as

$$\begin{aligned}\frac{1}{\sigma_0}\frac{d^2\sigma}{d\hat{t}dz} &= \frac{1}{\sigma_0}\frac{x_1}{Q^2}\frac{d^2\sigma}{dx_1 dx_2}\\ &= \frac{\alpha_s}{2\pi}C_F\frac{1}{\hat{t}}\left[\frac{1+z^2}{1-z} - \left(\frac{\hat{t}}{Q^2}\right)\frac{2\{1-z(1-z)\}}{1-z} + \left(\frac{\hat{t}}{Q^2}\right)^2\frac{\{1+(1-z)^2\}}{1-z}\right]\end{aligned} \tag{7.142}$$

When $\hat{t}$ is small or equivalently when $p_T^2 \ll Q^2$, the first term dominates and the probability of emitting a gluon is given by

$$\frac{1}{\sigma_0}\frac{d\sigma}{dz} = \frac{\alpha_s}{2\pi}P_{q\leftarrow q}(z)d\ln\hat{t} \tag{7.143a}$$

$$P_{q\leftarrow q}(z) = C_F\frac{1+z^2}{1-z} = \frac{4}{3}\frac{1+z^2}{1-z} \tag{7.143b}$$

This is also referred to as the LLA (leading logarithmic approximation) approximation of the gluon emission.[18] Note that we have separated the soft gluon and collinear part. The former happens when $z \to 1$ and the latter when $p_T^2 \to 0$. In the collinear region, $E_i + p_{i\parallel} \approx 2E_i$, Eq. (7.143a) can be considered as the probability for a virtual quark having mass $\hat{t}$ and momentum p, decays into another

18) This is equivalent to adopting the infinite momentum approximation and in the DGLAP evolution equation it amounts to summation of the leading $[\ln(Q^2/\mu^2)]^n$ terms, hence the name LLA. It is described in detail in Section 8.3.3–8.3.4.

quark and a gluon, each having a longitudinal momentum zp and $(1-z)p$, and transverse momentum $p_{\rm T}$.

One notices resemblance of the formula with virtual photon flux in the Weizsäcker–Williams formula (Vol. 1, Eq. (17.89)). In fact, if we make necessary transformation from QED to QCD, that is, $\alpha \to (4/3)\alpha_s$ and replace $y \to 1-z$ because y is the momentum of the virtual photon while z is that of the fermion, they are identical.

7.3.4 Transverse Kick

A main part of $e^- e^+ \to$ hadrons is 2-jet productions. Both jets are emitted back to back and has no relative transverse momentum. When a hard gluon is emitted, its kick produces the transverse momentum to the jets. Its behavior is obtained by rewriting Eq. (7.143a) in terms of the transverse momentum, but has to be handled with care because of the singularity at $p_{\rm T} \to 0$. Let $S(p_{\rm T}^2)$ be the integrated probability for producing a gluon with $0 \le k_{\rm T}^2 \le p_{\rm T}^2$. The probability to produce a gluon with $k_{\rm T}^2$ is given by $(dS/dk_{\rm T}^2)dk_{\rm T}^2$. To obtain $S(p_{\rm T}^2)$, one integrates Eq. (7.143) with constraints

$$k_{\rm T}^2 = z(1-z)\hat{t} \le p_{\rm T}^2\,, \quad 0 \le z, \frac{\hat{t}}{Q^2} \le 1 \tag{7.144}$$

Dominant contribution to the integral comes from neighborhood of $\hat{t} \approx 0, 1-z \approx 0$, but the integral is infrared divergent. One can avoid the divergence by considering $T(p_{\rm T}^2)$ instead of $S(p_{\rm T}^2)$, which is defined as the integrated probability for the gluon emission with arbitrary $k_{\rm T}^2 \ge p_{\rm T}^2$. In principle, $S(p_{\rm T}^2)$ should be divergent free if calculated correctly and satisfy the relation

$$S\left(p_{\rm T}^2\right) + T\left(p_{\rm T}^2\right) = 1 \tag{7.145}$$

$T(p_{\rm T}^2)$ can be easily calculated.

$$\begin{aligned} T(p_{\rm T}^2) &= \int \frac{1}{\sigma_0}\frac{d^2\sigma}{d\hat{t}dz}d\hat{t}dz = \frac{\alpha_s}{2\pi}C_F \int \frac{1+z^2}{1-z}\frac{dzd\hat{t}}{\hat{t}} \\ &\approx \frac{\alpha_s}{2\pi}C_F \int_{p_{\rm T}^2}^{Q^2}\frac{d\hat{t}}{\hat{t}}\int_{p_{\rm T}^2/\hat{t}}^{1}\frac{2d\bar{z}}{\bar{z}} = \frac{\alpha_s}{2\pi}C_F \ln^2\left(\frac{p_{\rm T}^2}{Q^2}\right) \\ \therefore \quad S\left(p_{\rm T}^2\right) &= 1 - \frac{\alpha_s}{2\pi}C_F \ln^2\left(\frac{p_{\rm T}^2}{Q^2}\right) \end{aligned} \tag{7.146}$$

where $\bar{z} = 1 - z$ and we approximated the numerator by 2 and the integration range by $\hat{t}\bar{z} \geq p_{\mathrm{T}}^2$. By differentiating $S(p_{\mathrm{T}}^2)$, we obtain

$$\frac{1}{\sigma_0}\frac{d\sigma}{dp_{\mathrm{T}}^2} = \frac{\alpha_s}{\pi} C_F \frac{\ln(Q^2/p_{\mathrm{T}}^2)}{p_{\mathrm{T}}^2} \tag{7.147}$$

As the transverse spread is produced by the gluon emission, this function appears universally wherever the gluon is emitted, that is, so long as the LLA approximation is taken.

8
Deep Inelastic Scattering

8.1
Introduction

The first and still the most successful application of the QCD is the deep inelastic scattering (to be referred to as DIS) where high energy leptons (e, μ, ν) are scattered by a nucleon at a large angle or equivalently large momentum transfer. The first successful explanation of the DIS was provided by the parton model proposed by Feynman and applied to DIS by Bjorken. It is widely recognized that the partons are quarks and gluons, and that the parton model can be considered as the lowest order QCD process. The essence of the parton model is that hadrons (nucleons, pions and others) are made of pointlike particles (hence its name) and that they are (almost) freely moving inside the hadrons. The DIS is a hard process where the incident lepton knocks off one of the partons without really affecting the rest of them. Such a picture is allowed if binding energy of the partons is small compared with the total hadron mass. However, the partons are confined, namely they have infinite binding energy if one tries to pull the parton slowly which invalidates the notion of freely moving partons. Nevertheless, the parton model worked and it was justified later by QCD's asymptotic freedom. It guarantees almost free action of the partons, provided the impact given to the parton is sufficiently large. The measure of the impact is given by the momentum transfer Q defined by

$$Q^2 \equiv -q^2 = -(k_1 - k_2)^2 \simeq 4\omega\omega' \sin^2\frac{\theta}{2} \tag{8.1}$$

where k_1, k_2 are the four-momenta of the incident and scattered leptons, ω, ω' their energy and θ scattering angle. It is also a measure of the depth ($d \sim 1/Q$) that the probes (e, μ, ν or their messengers, that is, photons or weak bosons) can penetrate into the hadron.

In inclusive processes at given incident energy where only scattered leptons are observed, the cross section depends on two variables, conventionally taken as the scattering angle θ, and the energy loss of the electron $\nu \equiv \omega - \omega'$. The Bjorken scaling, which is a mathematical expression to provide conditions for the parton model, states that at high energies, all the masses can be neglected and the cross section should be a function of a single dimensionless quantity $x = Q^2/2m_N\nu$

Elementary Particle Physics, Volume 2: Foundations of the Standard Model, First Edition. Yorikiyo Nagashima.

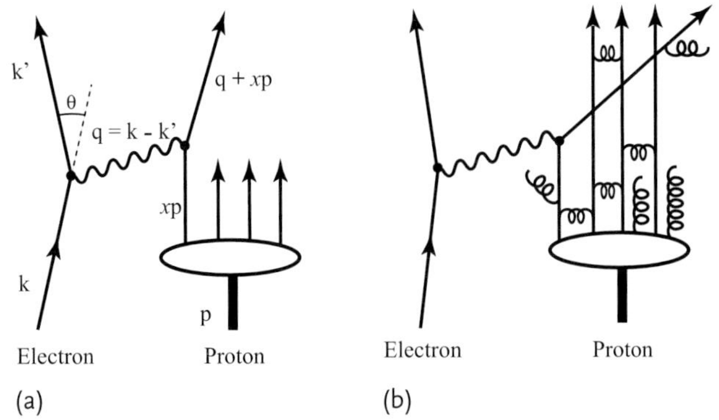

Figure 8.1 Deep inelastic scattering in the parton model (a) and in QCD (b). Partons are made of quarks and gluons. The quarks and gluons are constantly interacting with each other. Their effect has to be included for the complete description of the processes.

which is a ratio of two independent energy scale variables. It was originally introduced as a theoretical conjecture and was confirmed by the MIT-SLAC experiment in 1969 in deep inelastic scattering [384]. It then acquired a very physical interpretation in terms of Feynman's parton model. The variable x turned out to be the fractional momentum each parton carries inside the hadron. The kinematics of the parton model are depicted in Figure 8.1a. A basic assumption of the parton model was the impulse approximation which regards the scattering by the nucleon as an incoherent sum of those by its constituents. In QCD, the quark is not an independent object, that is, it accompanies gluon clouds just like the electron or the proton are surrounded by the photon or pion clouds. Then, the QCD description of the DIS process is obtained by dressing the partons by gluons as depicted in Figure 8.1b.

The parton model reproduces properties of the deep inelastic scattering data quite well, provided the incident energy is high ($\gtrsim$ 10 GeV) and the momentum transfer Q is large, though not too large, that is, $1 \lesssim Q^2 \lesssim 30\,\text{GeV}^2$. When the probe to look into the inner structure of the target is soft (meaning small Q), it does not have much resolving power. The virtual photon is simply absorbed by the target nucleon (Figure 8.2a). When Q is large, the resolving power increases and the photon begins to see the partons in the nucleon carrying fractional momentum x of its parent (Figure 8.2c) and interacts with one of them, expelling it outside the hadron. The interaction occurs instantaneously and the rest of the partons are hardly affected. Then, the cross section becomes a function of the fractional momentum x only and does not depend on Q^2 (Bjorken scaling).

The partons should have a momentum spectrum peaked at one-third of the parent nucleon because the proton is essentially made of three high momentum partons which are referred to as valence quarks. Pictorially, it is depicted as Figure 8.2b.

However, from the QCD point of view, the parton model description is just the zeroth-order approximation in which the gluon effect is neglected. The measured

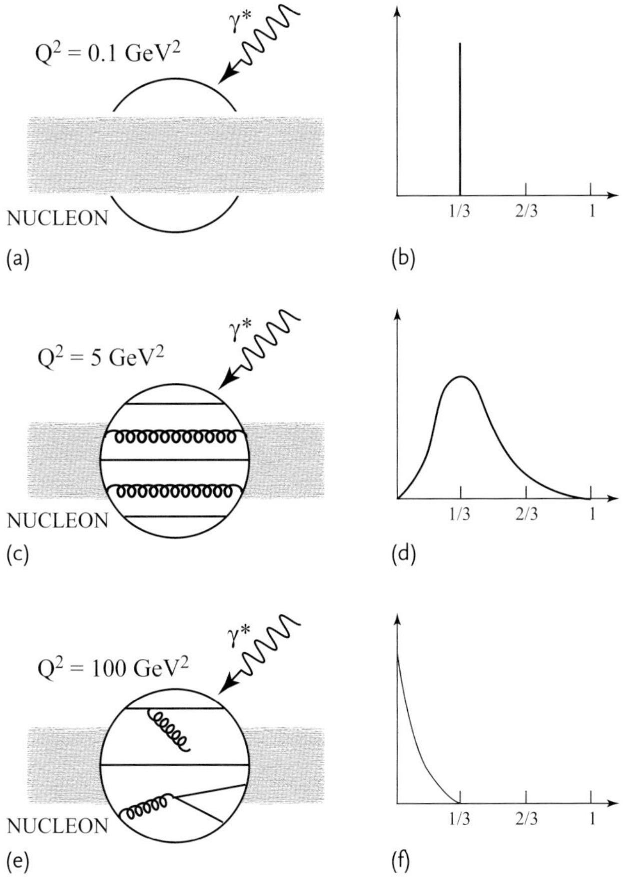

Figure 8.2 The momentum distributions of partons as viewed by a virtual photon microscope. (a) At $Q^2 = 0.1\,\text{GeV}^2$, the nucleon is not resolved to partons. It reacts as a whole, $x = 1$. (c) At $Q^2 = 5\,\text{GeV}^2$, the nucleon is seen as a collection of quarks and gluons. (b) If the partons are free, $x = 1/3$. (d) The quarks are constantly exchanging gluons which blurs the parton's momentum distribution. However, it is a slow process and they can be considered as quasifree in a time scale $t \sim 1/Q$. (e) At $Q^2 = 100\,\text{GeV}^2$, quarks are no longer isolated, but dressed with gluon clouds, that is, emission of gluons and the creation of quark pairs by gluons are inseparable. (f) Gluons tend to have small momenta ($x \simeq 0$).

momentum distribution of the quarks (Figure 8.2d) peaks at $x \sim 1/3$[1], but it is broad. It was interpreted as the evidence of interactions among the partons. Most momenta are carried by valence quarks (uud) in the proton or (udd) in the neutron. There are other components referred to as sea quarks which appear as virtual states of the gluon. However, the interaction among the quarks is a slow process, whose time scale is of the order of the inverse of nucleon binding energy which

1) Or 1/6 if one takes into account of the gluon. The gluon carries half of the total momentum of the parent hadron [see Eq. (8.110) and Vol. I Eq. (17.135)].

is at most $1/m_N$. It is long compared with the typical time scale of DIS which is $\sim 1/Q \ll 1/m_N$. This is the reason why the impulse approximation is valid.

The struck parton is instantaneously kicked off from the parent proton before surrounding partons can adjust themselves for the loss of the struck parton. Namely, the scattering takes place without influencing other partons. Therefore, the momentum distribution of the partons can be treated as a given external input which allows us to calculate the DIS cross section simply by convoluting that of free parton scattering (which is calculable perturbatively) with the parton momentum distribution. As broadening of the momentum distribution is a result of the exchange of many soft gluons by the quarks where the asymptotic freedom is inactive, its derivation needs the nonperturbative treatment.

The momentum transfer Q given to the nucleon is a measure of the strength of the impact, or size of the space domain that becomes an arena of the interaction. The larger the magnitude of Q^2 is, the smaller the size of the region that the interaction takes place. Thus, the momentum transfer Q^2 can be considered as a measure of the magnification of the probe.

As Q^2 is also a measure of time scale that the interaction takes place, the photon with large Q^2 can see events of short time duration which would have long disappeared if Q^2 is small. Namely, by making Q^2 large, the probe can penetrate deep inside the target and see[2)] gluons being exchanged or a process of the gluon making the sea of $q\bar{q}$ pairs on time as is shown in Figure 8.2e. As the gluon is massless, the emission pattern should be similar to that of soft bremsstrahlung in QED and be proportional to $\sim 1/x$. This means at higher Q^2, the distribution at small x becomes dominant like Figure 8.2f.

The above consideration prompts us to include higher order diagrams as Q^2 grows. The process of the photon probing the inner structure of the hadron is represented by Figure 8.3a. In the lowest order, the parton model of DIS is an electroweak process in which the parton absorbs the photon carrying the virtual mass Q^2 (Figure 8.3b). The effect of the gluon does not enter.

However, if the QCD is correct, its effect will reveal the existence of the gluons. Sometimes, the quark is seen after emission of gluons which is the exact meaning of the dressed quark accompanying the gluon cloud. If the emitted gluons are soft, they are hardly seen. Its effect appears as a shift of the original parton momentum distribution. In other words, as the probe becomes harder and penetrates deeper in the target, it begins to resolve the core parton from its cloud. In this case, the probability of hitting the quark inside the dressed quark is not 100%. We can consider that the core parton in the dressed parton carries fractional momentum "z" of the dressed parton before absorbing the virtual photon.

The whole process can be expressed pictorially as Figure 8.3a. The nucleon carrying momentum P supplies a parton with fractional momentum $p = yP$. The blob "B" represents the momentum distribution $F(y)$ of the partons that the nucleon supplies. The magnified parton carrying fractional momentum $p = yP$, in turn, becomes another supplier of gluon clouds (and if the parton is the gluon, sup-

2) Or should we say *feel* because the photon is color blind and cannot see the gluon.

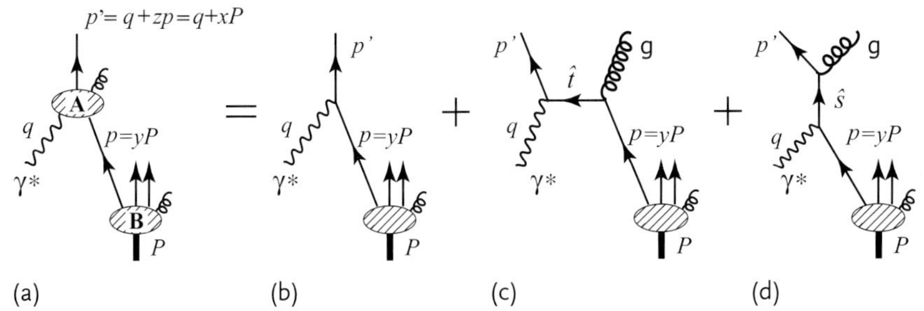

Figure 8.3 Deep inelastic scattering with QCD higher order diagrams. The whole process (a) can be conveniently separated into two parts. Blob "B" $= F(x)$ is the parton momentum distribution which is a nonperturbative part. Blob "A" is a hard process calculable using perturbation theories which is decomposed into (b–d). The diagram (b) which is a QCD Born process with $x = y\delta(y - x)dy$. (c,d) constitute the next order QCD correction. Factorization means that the low energy (low impact to be exact) part of QCD corrections (upper part of the Feynman diagrams in (c,d) $+\cdots$) can be separated and folded into $F(x)$ (blob B), making it the scale dependent but process independent parton distribution $F(x, \mu^2)$. The high energy part can be combined and folded into the blob A. The photon sees a parton with momentum $p = yP$, which, after interaction, obtains fractional momentum $zp = zyP = xP$.

plier of a $q\bar{q}$ pair, see Figure 8.9) which interacts with the photon carrying energy-momentum q. The interaction is represented by the blob "A" in Figure 8.3. The photon picks up the fraction z of the incoming parton, interacts with it and the parton comes out carrying $q + zp = q + xP(x = zy)$. Just like the parton model treated the proton as collection of partons, QCD sees a parton split into many partons as Q becomes large. The parton model is then the Born approximation (Figure 8.3b) of the whole reaction Figure 8.3a and one can think of the next order QCD process as depicted in Figure 8.3c,d.

8.2
The Parton Model Revisited

To prepare for the QCD treatment of the deep inelastic scattering, we review and list the basic variables and cross section formula for the process.[3]

To derive the parton model formula, we start with scattering cross sections of point particles having spin 1/2. As a representative, we consider scattering by polarized electrons by muons. The kinematical variables are defined as

$$e(k) + \mu(p) \to e(k') + \mu(p')$$

The usual relativistic variables are defined by

$$\begin{aligned} s &= (k+p)^2 \simeq 4\boldsymbol{p}^{*2} \\ t &= q^2 = (k-k')^2 \simeq -2\boldsymbol{p}^{*2}(1-\cos\theta^*) \\ u &= (k-p')^2 \simeq -2\boldsymbol{p}^{*2}(1+\cos\theta^*) \end{aligned} \tag{8.2}$$

3) The short summary of discussions is given in Vol. 1, Chapt. 17.

where $\boldsymbol{p}^*$ and θ^* denote the momentum and scattering angle in the CM (center of mass frame). The second approximate equalities of s, t, u hold at high energies. In the massless limit, the left-handed electron defined by the chirality operator $e_L \equiv [(1-\gamma^5)/2]e$ is in pure helicity $h = -$ state $[h \equiv (\boldsymbol{\sigma} \cdot \boldsymbol{p})/|\boldsymbol{p}|]$ while the antiparticle is in pure $h = +$ state. The opposite is true for chirality positive particles. We use left or right-handed particle as having the helicity minus or plus for discussing the parton model, but note that the nomenclature "left-handed" or "right-handed" may also mean the chirality state, in which case the opposite helicity for the antiparticle is to be understood.

Cross sections for polarized $e-\mu$ scattering were worked out in Vol. 1, Eqs. (7.31) and (7.32).

$$\begin{aligned}
d\sigma_{LL} &\equiv \left.\frac{d\sigma}{d\Omega}\right|_{CM}(e_L\mu_L \to e_L\mu_L) = \frac{\alpha^2}{t^2}s \\
d\sigma_{RR} &= d\sigma_{LL} = \frac{\alpha^2}{t^2}s \\
d\sigma_{LR} &= \left.\frac{d\sigma}{d\Omega}\right|_{CM}(e_L\mu_R \to e_L\mu_R) \\
&= d\sigma_{RL} = \frac{\alpha^2}{t^2}\frac{u^2}{s} = \frac{\alpha^2}{t^2}s\left(\frac{1+\cos\theta^*}{2}\right)^2
\end{aligned} \tag{8.3}$$

The difference between $d\sigma_{LL\,(RR)}$ and $d\sigma_{LR\,(RL)}$ can be understood as due to helicity conservation. In the CM frame, LL, RR are in total helicity zero state and the angular momentum is conserved at any angle. However, LR, RL are in the total helicity ± 1 state and the angular momentum does not conserve at $\theta = \pi$, and hence the cross section vanishes there. Notice that $(1+\cos\theta)/2 = d^1_{1,-1}(\theta)$ is the rotation matrix for spin 1 state (see boxed paragraph in Sect. 2.2).

The unpolarized cross section is obtained by summing all possible helicity states and dividing by four to account for average spin states.

$$\left.\frac{d\sigma}{d\Omega}\right|_{CM} = \frac{1}{4}\sum_{a,b} d\sigma_{ab} = \frac{\alpha^2}{t^2}\frac{s}{2}\left[1+\left(\frac{1+\cos\theta^*}{2}\right)^2\right] \tag{8.4}$$

It is convenient to rewrite the cross sections in terms of Lorentz invariant scaling variables defined by

$$\begin{aligned}
x &= -\frac{q^2}{2(p\cdot q)} = \frac{Q^2}{2M\nu} \simeq \frac{2\omega\omega'}{M(\omega-\omega')}\sin^2\frac{\theta}{2} = \frac{s}{M^2}\frac{1-y}{y}\sin^2\frac{\theta}{2} \\
y &= \frac{(p\cdot q)}{(p\cdot k)} = \frac{\omega-\omega'}{\omega} = \frac{1-\cos\theta^*}{2}
\end{aligned} \tag{8.5}$$

Here, variables without $*$ are those evaluated in the LAB (laboratory) frame in which the target is at rest. The second equality for y holds in the LAB frame and the third equality holds in the CM frame. Then, using $d\sigma/dy = 4\pi d\sigma/d\Omega_{CM}$, the cross sections can be reexpressed as

$$\frac{d\sigma_{\rm LL}}{dy} = \frac{d\sigma_{\rm RR}}{dy} = \frac{4\pi\alpha^2}{t^2} s$$
$$\frac{d\sigma_{\rm LR}}{dy} = \frac{d\sigma_{\rm RL}}{dy} = \frac{4\pi\alpha^2}{t^2} s(1-y)^2 \tag{8.6}$$

The scattering cross sections for the electron and parton can be obtained from Eq. (8.6) by simply multiplying appropriate quark charge (squared). Next, we extend the formulas to include the neutrino quark scattering. Since they are both spin 1/2 point particles, the scattering formulas are identical to the polarized cross sections.[4] In the region of our interest where the momentum transfer Q is much smaller than the mass of the W boson, all we have to do is to replace the electromagnetic coupling to the weak coupling $(e^2/t)^2 \to [g_W^2/(t-m_W^2)]^2 \sim 8(G_F/\sqrt{2})^2$ and not to take the spin average for the neutrino because it is naturally polarized. The net effect (at low energies $s \ll m_W^2$) is to replace

$$\frac{4\pi\alpha^2}{t^2} \to \frac{G_F^2}{\pi} \tag{8.7}$$

and we obtain

$$\frac{d\sigma}{dy}(\nu q, \overline{\nu q}) = \frac{G_F^2}{\pi}\hat{s}^{5)}$$
$$\frac{d\sigma}{dy}(\nu\overline{q}, \overline{\nu}q) = \frac{G_F^2}{\pi}\hat{s}(1-y)^2 \tag{8.8}$$

where we used $\hat{s}$ instead of s to remind that the total energy is that of the lepton and the parton system. Since the quarks live inside the hadron with fractional momentum x of the parent hadron, we define the quark distribution function $u(x)dx, d(x)dx$ as the probability for the u, d quark to have fractional momentum x in the proton p. Neglecting the s quark component or the charm production effect, the $\nu(\overline{\nu})p$ charged current (CC) reactions consist of the following processes.

$$\nu_l + d \to l^- + u\,, \quad \nu_l + \overline{u} \to l^- + \overline{d}$$
$$\overline{\nu}_l + u \to l^+ + d\,, \quad \overline{\nu}_l + \overline{d} \to l^+ + \overline{u} \tag{8.9}$$

The cross section for the proton target is then given by

$$\frac{d\sigma}{dy}(\nu p) = \frac{G_F^2}{\pi}\hat{s}\left[d(x) + \overline{u}(x)(1-y)^2\right]dx$$
$$\frac{d\sigma}{dy}(\overline{\nu}p) = \frac{G_F^2}{\pi}\hat{s}[u(x)(1-y)^2 + \overline{d}(x)]dx \tag{8.10}$$

4) For the neutrino, this applies only to charged current interactions. Scatterings by neutral current interactions will be treated separately.

5) Note that the neutrino is naturally polarized and hence no spin average in the initial state is taken. Furthermore, the weak interaction is chiral and thus is selective of the target polarization state. Consequently, νq and $\nu\overline{q}$ cross sections are obtained from $d\sigma_{\rm LL}$ and $d\sigma_{\rm LR}$ with suitable replacement of variables respectively.

The isospin symmetry dictates that the distributions in the neutron are obtained by exchanging $u \leftrightarrow d$. Then, for the isoscalar target, the cross section is given by the average of the proton and the neutron target. Using the fact that the target parton has fractional momentum x

$$\hat{s} = (k + xp)^2 \simeq 2x(k \cdot p) \simeq x(k + p)^2 = xs \tag{8.11}$$

where s is the total energy of the νN (N for nucleon) system, we have

$$\begin{aligned}
\frac{d\sigma^{CC}}{dxdy}(\nu N) &= \frac{G_F^2}{2\pi} s \left[x\{u(x) + d(x)\} + x\{\overline{u}(x) + \overline{d}(x)\}(1-y)^2 \right] \\
&= \frac{G_F^2}{2\pi} \hat{s} \sum_i \left[x q_i(x) + x\overline{q}_i(x)(1-y)^2 \right] \\
\frac{d\sigma^{CC}}{dxdy}(\overline{\nu} N) &= \frac{G_F^2}{2\pi} s \left[x\{u(x) + d(x)\}(1-y)^2 + x\{\overline{u}(x) + \overline{d}(x)\} \right] \\
&= \frac{G_F^2}{2\pi} \hat{s} \sum_i \left[x q_i(x)(1-y)^2 + x\overline{q}_i(x) \right]
\end{aligned} \tag{8.12}$$

where we attached superscript "CC" to denote that they are charged current cross sections due to $W^{\pm}$ exchange. The electron parton scattering cross section is given by

$$\frac{d\sigma}{dxdy}(eN) = \frac{4\pi\alpha^2}{t^2} s \left(\frac{1 + (1-y)^2}{2} \right) \sum_i \lambda_i^2 x\{q_i(x) + \overline{q}_i(x)\} \tag{8.13}$$

where λ_i is the charge of the ith quark in units of the proton electric charge.

8.2.1
Structure Functions

In order to have model independent general expression formulas for the lepton DIS cross section, we define the following variables

$$\begin{aligned}
&l(k) + N(p) \to l'(k') + X(P_F) \\
&k^\mu = (\omega, \boldsymbol{k}), \quad k'^\mu = (\omega', \boldsymbol{k}'), \quad p^\mu = (E, \boldsymbol{p})
\end{aligned}$$

where X represents the whole of hadrons in the final state. In the following, we use expressions of the weak processes because they are more general, but the discussions apply equally to e, μ DIS if one replaces the weak coupling with the electromagnetic coupling and drops the parity violating term W_3 in Eq. (8.16). The scattering amplitude and cross sections are given in Vol. 1, Eqs. (17.102)–(17.104) which are reproduced here.

$$S_{fi} - \delta_{fi} = i(2\pi)^4 \delta(k + p - k' - P_{\mathrm{F}}) \frac{G_{\mathrm{F}}}{\sqrt{2}} j_\mu J_p{}^\mu \tag{8.14a}$$

$$j_\mu = \overline{u}(k')\gamma_\mu(a - b\gamma^5)u(k), \quad J^\mu = \langle X(P_{\mathrm{F}})|J^\mu(0)|N(p)\rangle \tag{8.14b}$$

$a = b = 1$ for the charged current interaction, but we retain a, b for later use.

$$d\sigma = \frac{(G_{\mathrm{F}}/\sqrt{2})^2}{4(k \cdot p)} 2L_{\mu\nu} 4\pi M W^{\mu\nu} \frac{d^3k'}{(2\pi)^3 2\omega'} \tag{8.15a}$$

$$\begin{aligned} L_{\mu\nu} &= \frac{1}{2}\sum_{\text{spin}} [\overline{u}(k')\gamma_\mu(a - b\gamma^5)u(k)] \left[\overline{u}(k')\gamma_\nu(a - b\gamma^5)u(k)\right]^* \\ &= 2\left[(|a|^2 + |b|^2)\left\{k'_\mu k_\nu + k'_\nu k_\mu - (k \cdot k')g_{\mu\nu}\right\} + (|a|^2 - |b|^2)g_{\mu\nu}mm' \right. \\ &\quad \left. - 2i\mathrm{Re}\left\{(ab^*)\varepsilon_{\mu\nu\rho\sigma}k'^\rho k^\sigma\right\}\right] \end{aligned} \tag{8.15b}$$

The lepton mass m, m' can be dropped in the following.

$$\begin{aligned} W^{\mu\nu} &= \frac{(2\pi)^4}{4\pi M} \sum_{\mathrm{F}} \delta^4(p + q - P_{\mathrm{F}}) \left(\frac{1}{2}\right) \sum_{\text{spin}} \langle p|J^{\mp\mu}|P_{\mathrm{F}}\rangle\langle P_{\mathrm{F}}|J^{\pm\nu}|p\rangle \\ &= \left(-g^{\mu\nu} + \frac{q^\mu q^\nu}{q^2}\right) W_1(Q^2, \nu) + \left(p^\mu - \frac{(p \cdot q)q^\mu}{q^2}\right)\left(p^\nu - \frac{(p \cdot q)q^\nu}{q^2}\right) \\ &\quad \times \frac{W_2(Q^2, \nu)}{M^2} - i\epsilon^{\mu\nu\alpha\beta} p_\alpha q_\beta \frac{W_3(Q^2, \nu)}{2M^2} \end{aligned} \tag{8.16}$$

W_3 contributes to parity violating process and is absent for the electron scattering. The spin averaged and summed cross section in LAB frame ($p^\mu = (M, 0, 0, 0)$) is given by

$$\begin{aligned} \frac{d^2\sigma^{\nu,\overline{\nu}}}{d\Omega\, d\omega'} &= \frac{G_{\mathrm{F}}^2}{2\pi^2}\omega'^2 \left[2W_1^{\nu,\overline{\nu}}(Q^2, \nu)\sin^2\frac{\theta}{2} + W_2^{\nu,\overline{\nu}}(Q^2, \nu)\cos^2\frac{\theta}{2} \right. \\ &\quad \left. \mp W_3^{\nu,\overline{\nu}}(Q^2, \nu)\,\frac{\omega + \omega'}{M}\sin^2\frac{\theta}{2}\right] \end{aligned} \tag{8.17}$$

The structure functions W_i^{ν} for the neutrino and $W_i^{\overline{\nu}}(i = 1-3)$ for the antineutrino are different in general, but using the charge symmetry, they can be related. As the charge symmetry operator $\exp[-i\pi I_y]$ changes $p \to -n$, and $J^{\pm} = \overline{N}\tau_{\pm}N \to J^{\mp}$, the following equations hold.

$$W_i^{\nu p} = W_i^{\overline{\nu} n}, \quad W_i^{\overline{\nu} p} = W_i^{\nu n}, \quad i = 1, 2, 3 \tag{8.18}$$

For the isoscalar target ($n_p = n_n$),

$$W_i^{\nu} = (W_i^{\nu p} + W_i^{\nu n})/2 = (W_i^{\overline{\nu} p} + W_i^{\overline{\nu} n})/2 = W^{\overline{\nu}} \tag{8.19}$$

Namely, for the isoscalar target, the difference between $d\sigma^{\nu}$ and $d\sigma^{\overline{\nu}}$ is only the sign of W_3. In the (Bjorken scaling) limit:

$$s \to \infty, \quad Q^2 \to \infty, \quad \nu \to \infty, \quad x = Q^2/2(p.q) = \text{finite} \tag{8.20}$$

where the mass becomes negligible compared with the energy, the Bjorken scaling claims that the structure functions $W_i(Q^2, \nu)(i = 1-3)$ become functions of the ratio $x = Q^2/2(p \cdot q) = Q^2/2M\nu$ only. It is conventional to rewrite them in terms of F_is which are scaling limits of W_is defined by

$$M\,W_1(Q^2, \nu) \to F_1(x), \quad \nu\,W_2(Q^2, \nu) \to F_2(x), \quad \nu\,W_3(Q^2, \nu) \to -F_3(x) \tag{8.21a}$$

$$M\,W_L \equiv \left(1 + \frac{\nu^2}{Q^2}\right) M\,W_2 - M\,W_1 \to \frac{F_2(x) - 2x\,F_1(x)}{2x} \equiv \frac{F_L(x)}{2x} \tag{8.21b}$$

In order to rewrite the cross section in terms of x, y, we use Eq. (8.5) and

$$dx\,dy = \frac{1}{2\pi M}\frac{1-y}{y}d\Omega\,d\omega' \tag{8.22}$$

to obtain

$$\frac{d^2\sigma^{\nu(\overline{\nu})N}}{dx\,dy} = \frac{G_F^2}{2\pi}s\left[2x\,F_1(x)\frac{y^2}{2} + (1-y)F_2(x) \pm y\left(1 - \frac{y}{2}\right)x\,F_3(x)\right] \tag{8.23}$$

where $\pm$ is for $\nu(\overline{\nu})N$ scattering. In the parton model, the interacting constituents have spin 1/2 in which case the Callan–Gross relation holds:

$$F_2(x) = 2x\,F_1(x) \tag{8.24}$$

Then, the cross section is expressed as

$$\frac{d^2\sigma^{\nu}}{dx\,dy} = \frac{G_F^2}{2\pi}s\left[\frac{(F_2 + x\,F_3)}{2} + \frac{(F_2 - x\,F_3)}{2}(1-y)^2\right] \tag{8.25a}$$

$$\frac{d^2\sigma^{\overline{\nu}}}{dx\,dy} = \frac{G_F^2}{2\pi}s\left[\frac{(F_2 + x\,F_3)}{2}(1-y)^2 + \frac{(F_2 - x\,F_3)}{2}\right] \tag{8.25b}$$

Comparing Eq. (8.25) with Eqs. (8.12) and (8.13), we see that the structure functions in the parton model are expressed as

$$F_2^{\nu} = \sum_i x[q_i(x) + \overline{q}_i(x)]$$

$$xF_3 = \sum_i x[q_i(x) - \overline{q}_i(x)]$$

$$F_2^{\rm EM} = \sum_i \lambda_i^2 x[q_i(x) + \overline{q}_i(x)]^{6)}$$

$$F_L = 0 \tag{8.26}$$

8.2.2 Equivalent Photons

Having defined the structure function in its general form, we now want to investigate its properties in more details. We will discuss the electron-deep inelastic scattering because it is the most thoroughly investigated process. Generalization to other cases is straightforward. It is convenient for the following discussions to treat the lepton–parton scattering process as due to absorption of a virtual photon (or $W^{\pm}$) having the virtual mass $q^2 = -Q^2$ and four-momentum q^{μ}. Then, absorption cross sections of transverse and longitudinal polarization of the equivalent photon are given by (Vol. 1, Eq. (17.83))

$$\sigma_{\rm T} = \frac{1}{2}\frac{4\pi^2\alpha}{K}\sum_{\lambda=\pm}\varepsilon_{\mu}(\lambda)\varepsilon_{\nu}(\lambda)^* W^{\mu\nu}(Q^2,\nu) = \frac{4\pi^2\alpha}{KM}M\,W_1(Q^2,\nu) \tag{8.28a}$$

$$\sigma_{\rm L} = \frac{4\pi^2\alpha}{K}\varepsilon_{\mu}(0)\varepsilon_{\nu}(0)^* W^{\mu\nu}(Q^2,\nu) = \frac{4\pi^2\alpha}{KM}M\,W_{\rm L}(Q^2,\nu)^{7)} \tag{8.28b}$$

$$K = \sqrt{\nu^2 + Q^2} \tag{8.28c}$$

where K is a quantity that can also be interpreted as the energy of the incoming photon[8] and M is the mass of the target particle. In terms of the transverse and longitudinal cross sections, the structure functions are expressed as

$$2F_1 = \frac{\sigma_{\rm T}}{\sigma_{\rm B}}, \quad \frac{F_{\rm L}}{x} = \frac{\sigma_{\rm L}}{\sigma_{\rm B}}, \quad \frac{F_2}{x} = \frac{\sigma_{\rm T}+\sigma_{\rm L}}{\sigma_{\rm B}} \tag{8.29}$$

6) For an isoscalar target and to neglect the strange quark components in the approximation,

$$F_2^{\rm EM}(x) = \frac{1}{2}\left[\left(\frac{2}{3}\right)^2 x\left(u(x)+\overline{u}(x)\right) + \left(-\frac{1}{3}\right)^2 x\left(d(x)+\overline{d}(x)\right) + (u,\overline{u} \leftrightarrow d,\overline{d})\right]$$
$$= \frac{5}{18}\left[x\left\{u(x)+d(x)+\overline{u}(x)+\overline{d}(x)\right\}\right] = \frac{5}{18}F_2^{\nu} \tag{8.27}$$

7) The reason to use the scalar polarization $\varepsilon_{\mu}(0)$ instead of longitudinal polarization $\varepsilon_{\mu}(3)$ is to make $(q\cdot\varepsilon(\lambda)) = q^{\mu}\varepsilon_{\mu}(\lambda) = 0$. Its origin is the spacelike Q^2.

8) The definition of the virtual photon energy is somewhat ambiguous. Any variable that becomes $q^0 = |\boldsymbol{q}|$ of the real photon in the limit $Q^2 \to 0$ is acceptable. The energy of the incoming virtual photon $K = \nu$, which some authors adopt, may be appealing intuitively. Our definition of the photon energy was motivated to give the correct incoming flux ($= 4KM$) in the proton rest frame.

σ_B is the Born approximation for the photon absorption cross section.

$$\sigma_B = \frac{8\pi^2\alpha}{\text{Flux}} = \frac{2\pi^2\alpha}{KM}$$
$$\text{Flux} = 4[(q\cdot p)^2 - q^2p^2]^{1/2} = 4KM \tag{8.30}$$

where the "Flux" is the incoming flux factor. In the parton model, $\sigma_L = 0$ and there are no differences between $2F_1 = \sigma_T/\sigma_B$ and $F_2/x = (\sigma_T + \sigma_L)/\sigma_B$. When higher order terms in QCD are included, they have to be treated differently. However, in the LLA approximations, there are no differences.

The structure functions can be reformulated in terms of the virtual photon cross sections using Eq. (8.28) in which physical interpretation is more transparent. As stated at the beginning, the physical image of the structure functions has become clear in the parton model in which they are considered as a function of the single variable x. However, it has become clear that the model is valid at reasonably large ($Q^2 \gtrsim 1\,\text{GeV}/c^2$) but not at too large values of Q^2.

The structure functions F_i's regain Q^2 dependence as Q^2 becomes much larger ($\geq 30\,\text{GeV}/c^2$). It is referred to as the scaling violation. This is the effect of higher-order QCD corrections which we are going to investigate. In the following discussions, we investigate F_2/x as a representative for discussing properties of the structure functions. Then, it is expressed as

$$F(x,Q^2) = \frac{F_2^{\text{EM}}(x,Q^2)}{x} = \frac{\sigma_T + \sigma_L}{\sigma_0}$$
$$= \sum_i \int_0^1 dz \int_0^1 dy\,\delta(x - zy) f_i(y) \frac{\hat{\sigma}_{iT}(z,Q^2) + \hat{\sigma}_{iL}(z,Q^2)}{\hat{\sigma}_0} \tag{8.31}$$

$$\hat{\sigma}_i(z,Q^2) = \hat{\sigma}\left[\gamma^*(q) + q_i(y) \to q_i(z) + X\right] \tag{8.32}$$

where $f_i(x) = q_i(x)$ or $\overline{q}_i(x)$.

We denoted the parton cross section with a hat '^' to distinguish it from one with the nucleon target. The right-hand side of Eq. (8.31) can be separated into the Born term and higher order QCD correction terms. The Born term is given by Eq. (8.30). Taking into account of the parton target instead of the nucleon target, we distinguish parton momentum $p = xP$ from that of the parent nucleon (P). The cross section is given by

$$\sigma_{T\text{Born}} = \frac{8\pi^2\alpha}{2(\hat{s}+Q^2)}\lambda_i^2 z\delta(1-z) \equiv \hat{\sigma}_0\lambda_i^2\delta(1-z) \tag{8.33a}$$

$$\hat{\sigma}_0 = \frac{8\pi^2\alpha}{2(\hat{s}+Q^2)} \tag{8.33b}$$

$$\hat{s} = (q+p)^2\,, \quad z = \frac{Q^2}{2(q\cdot p)} = \frac{x}{y}\,, \quad x = \frac{Q^2}{2(q\cdot P)} \tag{8.33c}$$

Substituting Eq. (8.33) in Eq. (8.31), we obtain the parton model expression for $F(x)$:

$$F(x, Q^2) = \sum_i \int_0^1 dz \int_0^1 dy\,\delta(x - zy)\lambda_i^2 f_i(y)\delta(1-z) = \sum_i \lambda_i^2 f_i(x) \equiv F_0(x) \tag{8.34}$$

which recovers the original parton model expression.[9)]

Problem 8.1

Derive Eq. (8.33). Prove $\sigma_{L_{\text{Born}}} = 0$.

8.3
QCD Corrections

8.3.1
Virtual Compton Scattering

We now calculate the higher order diagrams in Figure 8.3c,d which contribute to $F(x, Q^2)$ in addition to the Born term. For the moment, we assume that the quark only has one flavor for simplicity. One can see that the process is almost identical to the Compton scattering. The difference is two-fold. The photon is virtual, having spacelike mass $q^2 = -Q^2$ and the outgoing photon is replaced with a gluon. The invariant matrix squared for the virtual photon Compton scattering was already given by Vol. 1, Eq. (7.60).

$$\overline{|\mathcal{M}|^2} = 2e^4\left(-\frac{u}{s} - \frac{s}{u} + \frac{2Q^2 t}{su}\right) \tag{8.35}$$

To adapt Eq. (8.35) to $\gamma^* q \to qg$, we define kinematical variables in their CM frame as follows.

$$\gamma^*(q_\gamma) + q(p) \to q(p') + g(q_g) \tag{8.36}$$

where γ^* stands for the virtual photon. We define parton variables as follows. Parton variables are expressed with hats to distinguish them from those of hadronic particles.

$$\hat{s} = (q_\gamma + p)^2 = (q_g + p')^2 = -Q^2 + 2(q_\gamma \cdot p) = 4|\boldsymbol{k}' *|^2 \tag{8.37a}$$

$$\hat{t} = (q_\gamma - p')^2 = (q_g - p)^2 = -2(q_g \cdot p) = -2|\boldsymbol{k}^*||\boldsymbol{k}'^*|(1 - \cos\theta^*) \tag{8.37b}$$

$$\hat{u} = (q_\gamma - q_g)^2 = (p' - p)^2 = -2(p \cdot p') = -2|\boldsymbol{k}||\boldsymbol{k}'^*|(1 + \cos\theta'^*) \tag{8.37c}$$

$$\hat{s} + \hat{t} + \hat{u} + Q^2 = 0 \tag{8.37d}$$

9) To be exact, in the spirit of QCD, one has to specify the reference momentum transfer Q_0^2 at which the distribution function is defined. The parton model assumes that the distribution is independent of Q^2.

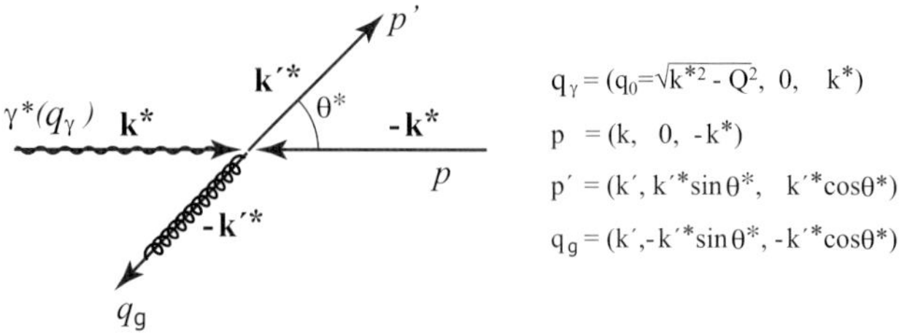

Figure 8.4 Kinematics in the CM of the virtual photon Compton process.

where $\boldsymbol{k}^*, \boldsymbol{k}^{'*}, \theta^*$ are momenta of the virtual photon, the scattered parton and the scattering angle in the γ^*- parton center of mass frame (Figure 8.4). We define a variable z in analogy to the $x = Q^2/2(q \cdot P)$ in Eq. (8.20) which was the fractional momentum of a parton in the hadron.

$$z \equiv \frac{Q^2}{2(q_\gamma \cdot p)} = \frac{Q^2}{\hat{s} + Q^2} \tag{8.38}$$

Namely, z is interpreted as the fractional momentum of the target parton. Or to be more exact, the momentum of the target parton is reduced to fraction z of the original by the initial state gluon radiation before it interacts with the incoming virtual photon. The interpretation will be justified as we go through the following discussions. We further define

$$p_\text{T}{}^2 \equiv (|\boldsymbol{k}^{'*}| \sin\theta^*)^2 = \frac{\hat{s}\hat{t}\hat{u}}{(\hat{s} + Q^2)^2} = -(1-z)\hat{t}\left(1 + \frac{\hat{t}}{\hat{s} + Q^2}\right) \tag{8.39}$$

Some relations among the variables which we will use later are

$$|\boldsymbol{k}^*|^2 = \frac{(\hat{s} + Q^2)^2}{4\hat{s}}, \quad |\boldsymbol{k}^*||\boldsymbol{k}^{'*}| = \frac{Q^2}{4z} \tag{8.40a}$$

$$|\hat{t}_\text{max}| = Q^2/z, \quad |\hat{t}_\text{min}| = 0 \tag{8.40b}$$

$$(p_\text{T}^2)_\text{max} = \frac{\hat{s}}{4} = \frac{Q^2}{4}\frac{1-z}{z} \tag{8.40c}$$

Our desired formula for the matrix element squared can be obtained from Eq. (8.35) by replacing s, t, u with $\hat{s}, \hat{u}, \hat{t}$ (notice $\hat{u} \leftrightarrow \hat{t}$) and correcting the color factor for the gluon. Then, the invariant matrix element squared for the diagrams in Figure 8.3c,d is given by

$$|\overline{\mathcal{M}}(\gamma^* q \to qg)|^2 = 2e^2\lambda_q^2 g_s^2 C_F \left(-\frac{\hat{t}}{\hat{s}} - \frac{\hat{s}}{\hat{t}} + \frac{2Q^2\hat{u}}{\hat{s}\hat{t}}\right) \tag{8.41}$$

where the average for the incoming particles and sum over the outgoing particles for spin, polarization and color degrees of freedom is taken for $|\overline{\mathcal{M}}|^2$. There are eight colors for the gluon and three colors for the quark. Averaging the color degree of freedom for the quark and summing that of the gluon gives the factor 8/3. The coupling constant of the photon e^2 has to be replaced with $g_s^2/2$, giving the color factor of $C_F = 4/3$ (normalization of non-Abelian matrices[10]). Using the cross section expressions (see Eq. (A.12))

$$\left.\frac{d\sigma}{d\Omega}\right|_{\text{CM}} = \frac{|\boldsymbol{k}_f^*|}{|\boldsymbol{k}_i^*|}\frac{|\overline{\mathcal{M}}_{fi}|^2}{64\pi^2 s}, \quad \frac{d\sigma}{dt} = \frac{\pi}{|\boldsymbol{k}_i^*||\boldsymbol{k}_f^*|}\left.\frac{d\sigma}{d\Omega}\right|_{\text{CM}} \tag{8.42}$$

and Eqs. (8.38) and (8.40), the cross section becomes

$$\frac{1}{\hat{\sigma}_0}\frac{d\hat{\sigma}}{d\hat{t}} = \lambda_q^2 \frac{\alpha_s}{2\pi}\frac{C_F}{\hat{t}}\left[\frac{1+z^2}{1-z} + \left(\frac{\hat{t}}{Q^2}\right)\frac{2z^2}{1-z} + \left(\frac{\hat{t}}{Q^2}\right)^2 \frac{z^2}{1-z}\right]. \tag{8.43}$$

We expect $\hat{s} \gtrsim Q^2 \gg \hat{t}, \hat{u}$ in the limit of large incident momentum. In this frame (infinite momentum frame), the first term in $[\cdots]$ gives the dominant contribution which is also referred to as the LLA (leading log approximation, see explanations after Eq. (8.58) and (8.66)). Then, we obtain

$$\begin{aligned} \frac{d\hat{\sigma}(\gamma^* q \to qg)}{\hat{\sigma}_0} &= \lambda_q^2 \frac{\alpha_s}{2\pi} P_{q\leftarrow q}(z) d\ln\hat{t} \\ P_{q\leftarrow q}(z) &= C_F \frac{1+z^2}{1-z} \end{aligned} \tag{8.44}$$

which can be interpreted as the probability for the initial parton to split into the parton with fractional momentum z by emitting a gluon. One sees that the probability is identical to that of the final parton in $e^- e^+ \to q\overline{q}$ reaction to emit a gluon (see Eq. (7.143a)). Again, we have the factorization of energy variables (those including z) and angle variables (those containing $\hat{t}$) with the same splitting function. Indeed, it can be shown (see Appendix K) that the splitting function is universal in the LLA approximation.

To repeat, the splitting function $P_{q\leftarrow q}(z)$ in DIS can be interpreted as (proportional to) the probability of a quark in the gluon cloud having fractional momentum z so long as z is away from the region $z \approx 1$.

The function is singular at $z = 1$ which is an example of the infrared divergence. It is known that the infrared divergence is superficial (see discussions in Sect. 7.3.2 or Vol. 1, Sect. 8.1.3). If we take the virtual gluon exchange diagrams shown in Figure 8.5 into account, the singularity disappears. It is controllable and will be remedied with proper prescription later. Therefore, we will not worry about it for the following discussions.

8.3.2 Factorization

When one tries to integrate Eq. (8.44), one encounters another singularity at $\hat{t} \simeq 0$ (or $p_T \simeq 0$) which is referred to as the collinear or mass singularity. It happens when $m_q = 0$ and $\cos\theta \to 0$ (i.e., $p_T^2 \simeq 0$). Note, the quark is not quite massless,

10) Sometimes, it is quoted that the QCD coupling strength $g_s^2/2$ corresponds to e^2 in QED.

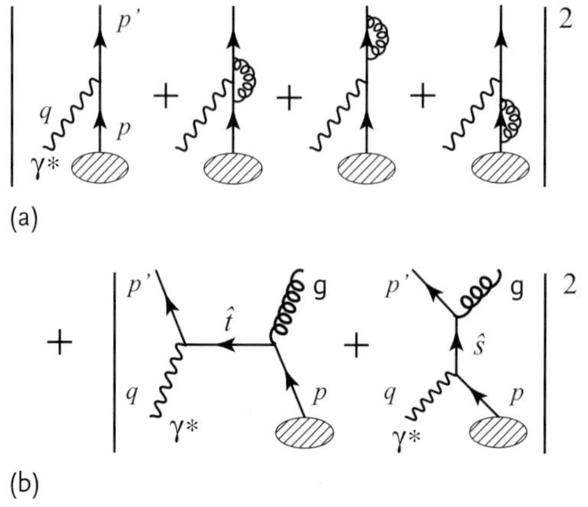

Figure 8.5 Infrared divergence does not appear if we take the virtual gluon exchange together with the real gluon emission into account, order by order. The real gluon emission processes (b) give $O(\alpha_s)$ contribution whose divergent part is canceled by interference term between the Born process ($O(1)$) and the virtual gluon processes ($O(\alpha_s^2)$).

but in QCD, where the energy scale is much higher than a typical light quark mass, appearance of the mass singularity is inevitable.

Its origin is the Coulomb type long range force and we need nonperturbative treatments to handle it correctly. To proceed within the framework of perturbation theories, we take a temporary measure by introducing a small mass "m" which serves as a cut-off to avoid the singularity. Its precise value does not matter as will be clarified in a moment. In the large Q^2 limit, we can use $\ln \hat{t}_{\max} \simeq \ln Q^2$ and obtain

$$\frac{\hat{\sigma}(\gamma^* q \to qg)}{\hat{\sigma}_0} \simeq \lambda_q^2 \frac{\alpha_s}{2\pi} P_{q\leftarrow q}(z) \ln\left(\frac{Q^2}{m^2}\right) \tag{8.45}$$

Substituting Eq. (8.45) in Eq. (8.31), and adding the Born term Eq. (8.33), we obtain

$$\begin{aligned} F(x, Q^2) &= \sum_i \int_0^1 dz \int_0^1 dy\, \delta(x - yz) f_i(y) \lambda_i^2 \\ &\quad \times \left[\delta(1-z) + \left\{\frac{\alpha_s}{2\pi} P_{q\leftarrow q}(z) \ln\left(\frac{Q^2}{m^2}\right) + C_{\text{DIS}}(z)\right\}\right] \\ &= F_0(x) + \int_0^1 dz \int_0^1 dy\, F_0(y) \delta(x - yz) \\ &\quad \times \left[\frac{\alpha_s}{2\pi} P_{q\leftarrow q}(z) \ln\left(\frac{Q^2}{m^2}\right) + C_{\text{DIS}}(z)\right] \end{aligned} \tag{8.46}$$

where $F_0(x) = \sum_i \lambda_i^2 f_i(x)$ is the original parton distribution function supplied by the nucleon, and $C_{\text{DIS}}(z)$ represents the subleading terms in Eq. (8.43). As we can see, the corrected $F(x)$ has $\ln Q^2$ dependence which violates the scaling in the parton model. This is the reason why we have included the Q^2 dependence in the structure function and denoted it by $F(x, Q^2)$ instead of Q^2 independent $F(x)$ in the parton model. Equation (8.46) can be expressed as

$$F(x, Q^2) = F_0(x) + \frac{\alpha_s}{2\pi} \int_x^1 \frac{dy}{y} F_0(y) \left[P_{q\leftarrow q}\left(\frac{x}{y}\right) \ln\left(\frac{Q^2}{m^2}\right) + C_{\text{DIS}}\left(\frac{x}{y}\right)\right] . \tag{8.47}$$

The correction term contains a large logarithm which originates from the collinear configuration of the process. This is a long range effect and needs nonperturbative treatment as we stated before. As it is largely process independent (see the boxed paragraph below), we introduce a new scale μ_{F} which separates the long range part from the short and try to fold it in the original distribution function. We can rewrite Eq. (8.47) as follows.

$$\begin{aligned} F(x, Q^2) &= F_0(x) + \frac{\alpha_s}{2\pi} \int_x^1 \frac{dy}{y} F_0(y) \left[P_{q\leftarrow q}\left(\frac{x}{y}\right) \ln\left(\frac{\mu^2}{m^2}\right) + C'_{\text{DIS}}\left(\frac{x}{y}\right)\right] \\ &+ \frac{\alpha_s}{2\pi} \int_x^1 \frac{dy}{y} F_0(y) \left[P_{q\leftarrow q}\left(\frac{x}{y}\right) \ln\left(\frac{Q^2}{\mu^2}\right) + \left\{ C_{\text{DIS}}\left(\frac{x}{y}\right) - C'_{\text{DIS}}\left(\frac{x}{y}\right)\right\}\right] \end{aligned} \tag{8.48}$$

If one considers $F_0(x)$ as a bare distribution which appears in the original equation, the first line is the first higher order correction to it. In the same spirit as the renormalization procedure to handle the divergent part of radiative corrections (see discussions in Chapter 5), we may consider the first line as the physically observable distribution and replace the whole expression with $O(\alpha_s)$ corrected $F(x, \mu^2)$. Namely, by folding the nonperturbative long range part into the distribution function, we can obtain an expression

$$\begin{aligned} F(x, Q^2) &= F(x, \mu^2) + \frac{\alpha_s}{2\pi} \int_x^1 \frac{dy}{y} F(y, \mu^2) \\ &\times \left[P_{q\leftarrow q}\left(\frac{x}{y}\right) \ln\left(\frac{Q^2}{\mu^2}\right) + C^F_{\text{DIS}}\left(\frac{x}{y}\right)\right] \end{aligned} \tag{8.49}$$

$$C^F_{\text{DIS}}\left(\frac{x}{y}\right) = C_{\text{DIS}}\left(\frac{x}{y}\right) - C'_{\text{DIS}}\left(\frac{x}{y}\right) \tag{8.50}$$

where $F_0(y)$ in the integrand has been rewritten by the redefined distribution function which is valid to order α_s. In deriving Eq. (8.49), we made use of

$$\ln\left(\frac{Q^2}{m^2}\right) = \ln\left(\frac{Q^2}{\mu^2}\right) + \ln\left(\frac{\mu^2}{m^2}\right) \tag{8.51}$$

namely, we separated the high Q^2 part from the low energy part. This is what is commonly referred to as "factorization." Our argument here is correct only in LLA approximation, but it has been proved that the factorization is valid to all orders of the QCD perturbation series [385].

The separation is somewhat arbitrary as it is equivalent to subtraction of infinite term. The term $C^F_{\text{DIS}}(x/y)$ represents this small ambiguity. Further discussion on an adjustment of the small ambiguities will be given in Sect. 8.4.3 within a context of factorization scheme setting. The boundary of long range and short range is set by μ^2 which is referred to as the factorization scale. In the high energy part, the asymptotic freedom can be applied while in the low energy part, nonperturbative approach is necessary. Below the scale μ^2, the collinear gluon has to be considered as a part of the parton distributions in the hadron, inseparable from the hadron structure. The physical observable should not depend on how one sets the factorization scale just like the renormalization scale. In principle, the factorization scale can be determined independently of the renormalization scale which separates the ultraviolet divergence. Though conventionally, they are often set at the same value.

As the left-hand side of Eq. (8.49) should not depend on μ, its derivative with respect to μ^2 should vanish. As a result, we obtain an evolution equation for $F(x,\mu^2)$.

$$\frac{dF(x,\mu^2)}{d\ln\mu^2} = \frac{\alpha_s(\mu^2)}{2\pi}\int_x^1 \frac{dy}{y} F(y,\mu^2) P_{q\leftarrow q}\left(\frac{x}{y}\right) \tag{8.52}$$

where we replaced α_s with μ dependent $\alpha_s(\mu^2)$. Our argument to justify Eq. (8.52) is valid only to $O(\alpha_s)$, but the validity of the equation is proven to all orders of perturbations [385–387].

This is an integrodifferential equation with the initial function given at some reference point $F(x,\mu^2 = Q_0^2)$. It is commonly referred to as the DGLAP (Dokshitzer–Gribov–Altarelli–Parisi) evolution equation (in its simple form) [388–390]. Although the equation does not give any information for x dependence of the distribution function, if it is given at some reference point at $\mu^2 = Q_0^2$, all the μ^2 dependence can be calculated from the DGLAP equation.

The DGLAP evolution equation, at least, in LLA, is derived by working at infinite momentum frame. At finite momenta, correction terms of the form $F(x,Q^2)/Q^2$ come in [see Eq. (8.43)]. From formal arguments, the DGLAP evolution equation is the first term of the operator product expansion (OPE) [391, 392]. The expansion is organized in terms of the operator's twist (mass dimension-spin). At finite values of Q^2, there are higher twist corrections which are suppressed as

$$\frac{\left[\ln\left(Q^2/Q_0^2\right)\right]^{m<n}}{Q^n} \tag{8.53}$$

The twist term has to be worked out in nonperturbative treatments and is generally hard to evaluate accurately.

Factorization: an intuitive picture

Let us digress a bit to obtain an intuitive picture of the factorization in the deep inelastic scattering where one considers the process as an interaction of a virtual photon with mass $-Q^2$. Partons in the hadron undergo multi-interactions long before one of them interacts with the virtual photon. They do so by exchanging gluons and $q\bar{q}$ pairs. The parton distribution which the incoming virtual photon sees is a result of such QCD interactions.

First, we notice that exchange of hard gluons among quarks inside the proton is suppressed by a factor $O(m_N^2/Q^2)$. This can be shown as follows. Consider two-gluon exchange between two quarks in the upper part of Figure 8.6. The loop contribution is proportional to

$$\sim \int_Q^\infty d^4 q \frac{\not{p}_1 - \not{q} + m}{(p_1-q)^2 - m^2} \frac{\not{p}_2 + \not{q} + m}{(p_2+q)^2 - m^2} \left(\frac{1}{q^2}\right)^2 \sim O\left(\frac{m^2}{Q^2}\right) \tag{8.54}$$

In other words, the hard gluon exchange occurs in a time scale $1/Q$ and has disappeared long before the virtual photon comes in.

The typical time scale of mutual interactions is given by the inverse of the binding energy of the nucleon, which is $\sim 1/m_\pi$ times the Lorentz expansion factor $\gamma = E/m_p$, namely, $t_{\text{BE}} \sim E/m_\pi m_p \sim ER^2$ where R is the typical hadron size. If a hard probe hits the proton in a time scale $\sim 1/Q$, short enough so that there is no time for quarks to act cooperatively with other quarks, the struck quark receives no feedback from its partners and would act as a free particle. In this case, there is no time for the partons to rearrange themselves. Their distribution cannot depend on the nature of the incoming probes.

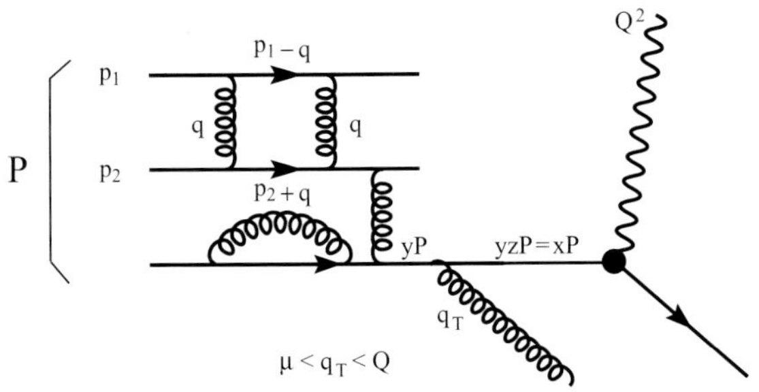

Figure 8.6 Illustration of the factorization.

When Q^2 is low, many of the emitted gluons will have enough time to be reabsorbed in which case their contribution is built into the parton distribution seen by the photon. On the other hand, emitted gluons will have less and less time to be reabsorbed and more gluons will appear as Q^2 becomes large. They contribute to the higher order QCD corrections. Namely, if the parton distribution is already resolved to a scale μ, its scale will be increased to Q (spatially more resolvable)

by emitting more gluons (see Eq. (8.66)). Thus, the low energy part of the QCD correction as represented by emissions and reabsorption of low energy gluons are independent of the hard interaction. It can be separated and folded into the parton distribution with increased μ^2. Thus, the distribution is scale dependent, but independent of the hard process in consideration. This is the reason that the low energy part of the QCD process can be factored out from the high energy process.
Though the parton distribution is not possible to calculate from the first principle, the scale dependent distribution is universal, that is, independent of the hard process in question. Thus, it can be determined experimentally from variety of processes (DIS, Drell–Yan, W, Z production hadron–hadron interactions and so on.) and can be used to predict cross sections of other reactions, for instance, the top, Higgs production at other hadron colliders like LHC. The high energy part of the QCD correction can be separated and folded into the hard process. They can be calculated using the perturbation theories as the asymptotic freedom is at work here. Thus, the factorization enabled the QCD to predict strong interaction processes with reasonable approximation.

8.3.3
Power Expansion of the Evolution Equation

By changing Eq. (8.52) into an integral form

$$F(x, Q^2) = F_0(x, Q_0^2) + \frac{\alpha_s}{2\pi} \int_{Q_0^2}^{Q^2} d\ln Q'^2 \int_x^1 \frac{dy}{y} P_{q\leftarrow q}\left(\frac{x}{y}\right) F(y, Q'^2) \quad (8.55)$$

and by inserting $F(x, Q_0^2)$ into $F(x, Q^2)$ in the integrand successively, one gets a series of gluon emission contribution.

$$F_1(x, Q^2) = \frac{\alpha_s}{2\pi} \int_{Q_0^2}^{Q^2} d\ln Q_1^2 \int_x^1 \frac{dx_1}{x_1} P_{q\leftarrow q}\left(\frac{x}{x_1}\right) F_0(x_1, Q_0^2)$$

$$F_2(x, Q^2) = \left(\frac{\alpha_s}{2\pi}\right)^2 \int_{Q_0^2}^{Q^2} d\ln Q_2^2 \int_{Q_0^2}^{Q_2^2} d\ln Q_1^2$$

$$\times \int_x^1 \frac{dx_2}{x_2} P_{q\leftarrow q}\left(\frac{x}{x_2}\right) \int_{x_2}^1 \frac{dx_1}{x_1} P_{q\leftarrow q}\left(\frac{x_2}{x_1}\right) F_0(x_1, Q_0^2)$$

$$\cdots\cdots \quad (8.56)$$

The nth term can be calculated as

$$F_n(x, Q^2) = \left(\frac{\alpha_s}{2\pi}\right)^n \int_{Q_0^2}^{Q^2} d\ln Q_n^2 \cdots \int_{Q_0^2}^{Q_2^2} d\ln Q_1^2$$

$$\times \int_x^1 \frac{dx_n}{x_n} P_{q\leftarrow q}\left(\frac{x}{x_n}\right) \cdots \int_{x_2}^1 \frac{dx_1}{x_1} P_{q\leftarrow q}\left(\frac{x_2}{x_1}\right) F_0(x_1, Q_0^2) \tag{8.57}$$

For constant α_s, the first line of Eq. (8.57) gives $[(\alpha_s/2\pi)\ln(Q^2/Q_0^2)]^n/n!$. Denoting the second line as $C_n(x)$, we have

$$F(x, Q^2) = F_0 + F_1 + F_2 + \cdots = \sum_n \frac{C_n}{n!}\left(\frac{\alpha_s}{2\pi}\ln\frac{Q^2}{Q_0^2}\right)^n . \tag{8.58a}$$

$$C_n = \int_x^1 \frac{dx_n}{x_n} P_{q\leftarrow q}\left(\frac{x}{x_n}\right) \cdots \int_{x_2}^1 \frac{dx_1}{x_1} P_{q\leftarrow q}\left(\frac{x_2}{x_1}\right) F_0(x_1, Q_0^2) \tag{8.58b}$$

Namely, by substituting the $O(\alpha_s)$ splitting function (Eq. (8.44)) into the DGLAP equation (8.52), one actually collects all the power series in $\alpha_s \ln(Q^2/\mu^2)$.

Normally, C_n is $\sim O(1)$, but in certain circumstances, for instance, in the soft gluon limit ($x \to 0$), C_n is large and it has to be resummed, too. We will come back to this problem in Sect. 8.5.2.

The series expansion may be pictorially represented by diagrams in Figure 8.7d. This shows that the DGLAP equation resums the dominant leading logarithmic terms. This is why the approximation to pick up only the first term in Eq. (8.43) is also referred to as the LLA (Leading Log Approximation) approximation.

The reader is cautioned that for the expansion series to be meaningful, $\alpha_s \ln(Q^2/\mu^2) \lesssim 1$ has to be satisfied. As was discussed in the previous chapter, α_s is a decreasing function of Q^2 (asymptotic freedom) and the formalism we developed is valid only for large Q^2. For the deep inelastic process, the Q^2 value of $\gtrsim (10\,\text{GeV}/c)^2$ may be used where $\alpha_s \lesssim 0.2$ is considered to be safe. Notice also that as can be seen from Eq. (8.56), the DGLAP equation accumulates a special class of virtual processes where the virtuality of successive partons are strongly ordered.

$$Q_0^2 < Q_1^2 < \cdots < Q_n^2 < \cdots < Q^2 \tag{8.59a}$$

$$x_1 > \cdots > x_n > \cdots > x \tag{8.59b}$$

Namely, in adopting the LLA approximation, successive interactions were treated as a stochastic process. The interference effect is included in the subleading terms. The strong ordering is primarily a consequence of working in the infinite momentum frame where every object is evaluated in the Bjorken scaling limit Eq. (8.20). We will show why this is in the following.

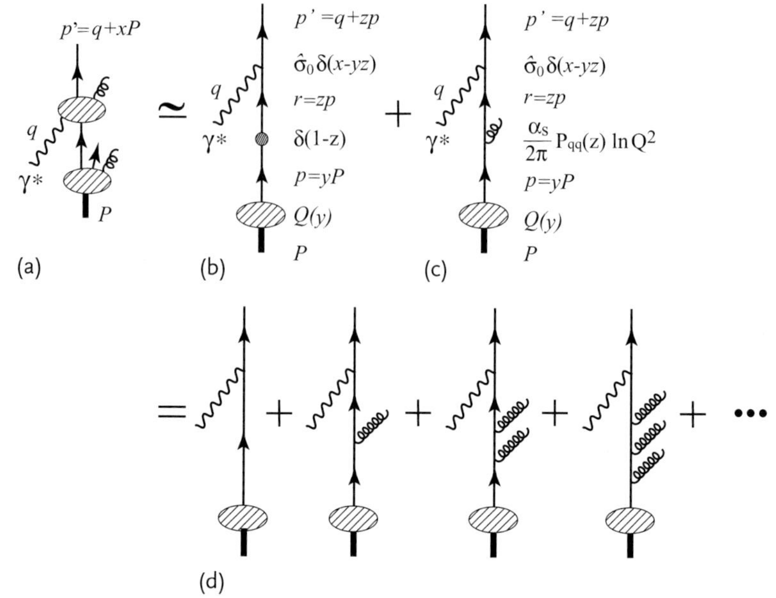

Figure 8.7 QCD corrections to the deep inelastic process: (b) the parton model=Born term. (c) LLA approximation to the Compton like corrections in Figure 8.3. Substitution of LLA (c) into DGLAP equation amounts to adding all diagrams in (d).

8.3.4
Cascade Branching of the Partons

Let us consider the virtuality relations among the partons a, b and c in the infinite momentum frame. There are several equivalent definitions for the fractional momentum z in the infinite momentum frame. Here, we find it convenient to use the light cone variable $p_\pm = E \pm p_z$. We have

$$p^2 \equiv p_\mu p^\mu = p_+ p_- - p_x^2 - p_y^2 = p_+ p_- - p_\mathrm{T}^2, \Rightarrow p_- = \frac{p^2 + p_\mathrm{T}^2}{p_+} \tag{8.60}$$

Suppose the parton "a" branches to the partons "b" and "c" with

$$p_{b+} = z p_{a+}, \quad p_{c+} = (1-z) p_{a+} \tag{8.61}$$

From the momentum conservation,

$$p_{a-} = p_{b-} + p_{c-} \Rightarrow \frac{p_a^2}{p_{a+}} = \frac{p_b^2 + p_\mathrm{T}^2}{z p_{a+}} + \frac{p_c^2 + p_\mathrm{T}^2}{(1-z) p_{a+}} \tag{8.62}$$

which leads to

$$p_a^2 = \frac{p_b^2}{z} + \frac{p_c^2}{(1-z)} + \frac{p_\mathrm{T}^2}{z(1-z)} \tag{8.63}$$

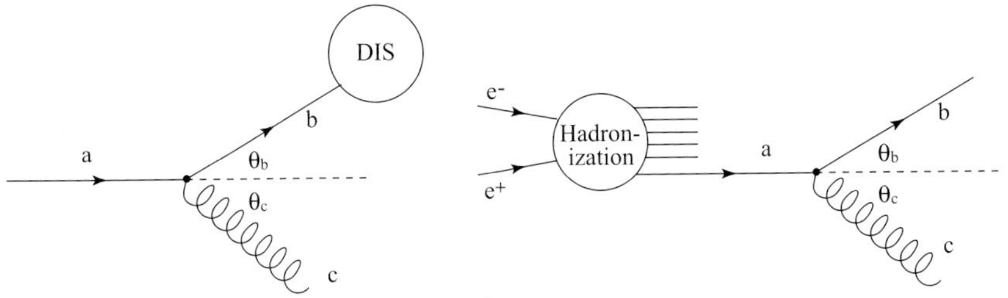

Figure 8.8 (a) In DIS, virtuality of the initial parton increases as it emits another parton. (b) In hadronization, virtuality of the initial parton decreases as it emits another parton: $q(a) \to q(b) + g(c)$.

We consider two cases:

$$\begin{aligned} &\text{Case A;} \quad \text{"}a\text{," "}c\text{" are real, that is,} \quad p_a^2 \simeq p_c^2 \simeq 0 \Rightarrow p_b^2 = -p_T^2/(1-z) < 0 \\ &\text{Case B;} \quad \text{"}b\text{," "}c\text{" are real, that is,} \quad p_b^2 \simeq p_c^2 \simeq 0 \Rightarrow p_a^2 = p_T^2/z(1-z) > 0 \end{aligned}$$

Case A corresponds to the initial parton radiation in DIS (Figure 8.8a). If "a" is also virtual with virtuality $p_a^2 = -Q_0^2$, then

$$Q_1^2 \equiv -p_b^2 = -(p_a - p_c)^2 = -p_a^2 + 2E_a E_c(1 - \cos\theta) = Q_0^2 + Q^2 > Q_0^2 \quad (8.64)$$

The case B corresponds to the final parton radiation in $e^- e^+ \to$ hadrons or the process of the parton's fragmentation (Figure 8.8b). If "a" is also virtual with virtuality $p_a^2 = Q_0^2$, then

$$Q_1^2 \equiv p_b^2 = (p_a - p_c)^2 = p_a^2 - 2E_a E_c(1 - \cos\theta) = Q_0^2 - Q^2 < Q_0^2 \quad (8.65)$$

For successive branching of virtual partons, we have

$$\begin{cases} Q_0^2 < Q_1^2 < \cdots < Q_n^2 & \text{in DIS} \\ Q_0^2 > Q_1^2 > \cdots > Q_n^2 & \text{in hadronization} \end{cases} \quad (8.66)$$

In other words, the virtuality of the incoming parton which eventually interacts with the virtual photon grows as it branches until it reaches Q^2 where it interacts with the photon. Or rather, the photon with the virtuality Q^2 chooses such a configuration of the partons. Now, we have shown that the LLA approximation amounts to the adoption of the infinite momentum frame.

On the other hand, the virtuality of the final state radiation decreases as it branches, eventually reaching $Q^2 \simeq 1\,\text{GeV}^2$ when the branching stops and the parton hadronizes. Details of the cascade phenomena of hadronization will be discussed in Sect. 9.2.

Gluon Emission after the Interaction Equation (8.52) describes Q^2 evolution of the parton distribution. This presupposes that the interaction in the gluon emission is complete before the parton interacts with the incoming photon. However, as can be seen in Figure 8.3d, the QCD amplitude includes gluon emission from the final state quark as well as from the initial state. Equation (8.43) includes both contributions added coherently. Therefore, in our approximation, the final state emission as well as its interference with the initial state emission was ignored. In the treatment of the collinear approximation, we worked in the infinite momentum frame where the process was treated in the time-ordered formalism and only the initial state emission was considered. Namely, we separated the initial gluon emission from the final state emission and treated it as if it is a statistically independent phenomenon. The reason to justify the argument is as follows.

The amplitude for the initial state emission before interacting with the photon is proportional to $1/\hat{t}$, while that of the final state emission is proportional to $1/\hat{s}$ and the latter can be neglected in the LLA approximation. One might argue that the interference term may not still be completely negligible compared to $1/\hat{t}$. However, a careful look clarifies that the final state emission exists to maintain the gauge invariance and this term exists only to cancel the unphysical scalar and longitudinal contribution. If one only considers the transverse polarizations of the physical gluon, for instance, if one adopts the axial gauge, one needs only to consider the gluon emission from the initial state [393] and that the assumption of statistical independence is warranted. This consideration is important as it enables cascade emission of gluons which can be treated with Monte Carlo formalism.

8.4 DGLAP Evolution Equation

8.4.1 Gluon Distribution Function

Thus far, we have only considered effects of the quarks to emit gluons. However, the gluon also exists *ab initio* in the nucleon and emits a quark–antiquark pair. We need to consider the photon being scattered off from those sea quarks. Their corresponding diagrams are depicted in Figure 8.9.

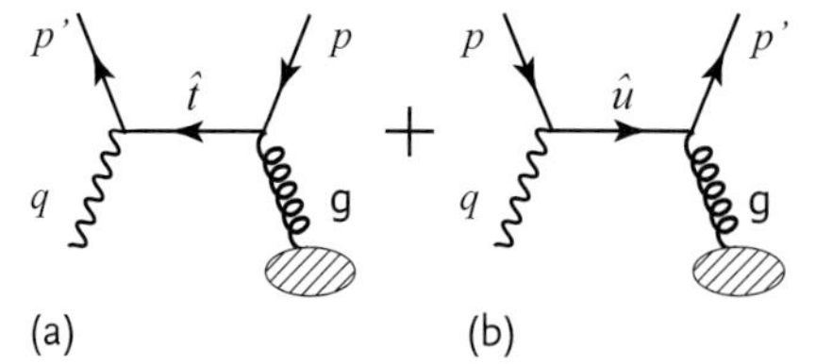

(a) (b)

Figure 8.9 Gluon splitting contributions to DGLAP equations.

The transition amplitude squared corresponding to those diagrams can be derived in the same manner as the QCD Compton process Eq. (8.41).

$$|\overline{\mathcal{M}}(\gamma^* g \to q\overline{q})|^2 = 2e^2\lambda_q^2 g_s^2 T_R\left(-\frac{\hat{u}}{\hat{t}} - \frac{\hat{t}}{\hat{u}} + \frac{2Q^2\hat{s}}{\hat{t}\hat{u}}\right), \quad T_R = \frac{1}{2} \tag{8.67}$$

The equation can be obtained from Eq. (8.35) by crossing and replacing $e^4 \to e^2 g_s^2 T_R = e^2 g_s^2/2$. The gluon color summation gives nothing because the 8 degrees of freedom has to be divided by 8 to take the average. By picking the dominant $1/\hat{t}$ term, the corresponding cross section is given by

$$\frac{\hat{\sigma}(\gamma^* g \to q\overline{q})}{\hat{\sigma}_0} = \frac{\alpha_s}{2\pi}[P_{q\leftarrow g}(z) + P_{\overline{q}\leftarrow g}(z)]\ln\left(\frac{Q^2}{m^2}\right) \tag{8.68a}$$

$$P_{q\leftarrow g}(z) = P_{\overline{q}\leftarrow g}(z) = T_R[z^2 + (1-z)^2] \tag{8.68b}$$

where z is defined by Eq. (8.38). There is no infrared singularity, but the mass singularity exists at $\hat{t} \to 0$ and $\hat{u} \to 0$. Equation (8.68) was obtained by integrating over $\hat{t}$ and $\hat{u}$ at their dominant region respectively. The one integrated over $\hat{t}$ gives the splitting function for the gluon to emit the quark $P_{q\leftarrow g}$ and the other over $\hat{u}$ gives the splitting function to emit the antiquark $P_{\overline{q}\leftarrow g}$.

Generalization of Eq. (8.52) to include the gluon contribution is straightforward. Introducing the variable $\tau \equiv \ln(Q^2/\mu^2)$ and denoting the gluon distribution function in the nucleon by $g(x,\tau)$ and that of the quark by $q(x,\tau)$, the formula to include the gluon splitting effect can be written as

$$\frac{dq(x,\tau)}{d\tau} = \frac{\alpha_s}{2\pi}\int_x^1 \frac{dy}{y}\left[q(y,\tau)P_{q\leftarrow q}\left(\frac{x}{y}\right) + g(y,\tau)P_{q\leftarrow g}\left(\frac{x}{y}\right)\right] \tag{8.69a}$$

Similarly, for the gluon distribution function, we have

$$\frac{dg(x,\tau)}{d\tau} = \frac{\alpha_s}{2\pi}\int_x^1 \frac{dy}{y}\left[q(y,\tau)P_{g\leftarrow q}\left(\frac{x}{y}\right) + g(y,\tau)P_{g\leftarrow g}\left(\frac{x}{y}\right)\right] \tag{8.69b}$$

At this stage, we include different quark flavors as well as antiquarks, and introduce $q_i(x,\tau)$, $\overline{q}_i(x,\tau)$. We define a singlet (Σ) and the nonsinglet (Δ) distribution function as follows.

$$\begin{aligned}\Sigma(x,\tau) &\equiv \sum[q_i(x,\tau) + \overline{q}_i(x,\tau)] \sim \frac{F_2(x,\tau)}{x}\\ \Delta(x,\tau) &\equiv \sum[q_i(x,\tau) - \overline{q}_i(x,\tau)] \sim F_3(x,\tau)\end{aligned} \tag{8.70}$$

The evolution equations for them can be obtained by substituting Eq. (8.70) in Eq. (8.69) and are given by

$$\frac{d\Delta(x,\tau)}{d\tau} = \frac{\alpha_s(\tau)}{2\pi}\int_x^1 \frac{dy}{y}\left[\Delta(y,\tau)P_{q\leftarrow q}\left(\frac{x}{y}\right)\right] \tag{8.71a}$$

$$\frac{d\Sigma(x,\tau)}{d\tau} = \frac{\alpha_s(\tau)}{2\pi}\int_x^1 \frac{dy}{y}\left[\Sigma(y,\tau)P_{q\leftarrow q}\left(\frac{x}{y}\right) + 2n_f g(y,\tau)P_{q\leftarrow g}\left(\frac{x}{y}\right)\right] \tag{8.71b}$$

$$\frac{dg(x,\tau)}{d\tau} = \frac{\alpha_s(\tau)}{2\pi}\int_x^1 \frac{dy}{y}\left[\Sigma(y,\tau)P_{g\leftarrow q}\left(\frac{x}{y}\right) + g(y,\tau)P_{g\leftarrow g}\left(\frac{x}{y}\right)\right] \tag{8.71c}$$

where n_f is the number of flavors. This is the DGLAP equation in its entirety. These equations contain two independent variables, x and Q^2. We repeat that it is not possible to solve them for x with given Q^2 because the x distribution is primarily a low energy phenomenon and nonperturbative treatment is required. However, if a distribution function is given at some initial $Q^2 = Q_0^2$, then their Q^2 evolution can be uniquely determined by the formulae.

To solve the equations, one needs to prepare the splitting functions. There is a general formula to derive the splitting functions in LLA which are given in Appendix K. Using Eq. (K.9), we can derive the following splitting functions:

$$P_{q\leftarrow q}(z) = C_F\frac{1+z^2}{1-z} \tag{8.72a}$$

$$P_{q\leftarrow g}(z) = P_{\bar{q}\leftarrow g}(z) = T_R[z^2 + (1-z)^2] \tag{8.72b}$$

$$P_{g\leftarrow q}(z) = C_F\frac{1+(1-z)^2}{z} \tag{8.72c}$$

$$P_{g\leftarrow g}(z) = 2C_A\left[\frac{1-z}{z} + \frac{z}{1-z} + z(1-z)\right] \tag{8.72d}$$

They are universal in the LLA approximation in the sense that it does not depend on particular processes in question.

The splitting functions have the following properties.

$$\begin{aligned} P_{q\leftarrow q}(z) &= P_{g\leftarrow q}(1-z)\\ P_{q\leftarrow g}(z) &= P_{\bar{q}\leftarrow g}(1-z)\\ P_{g\leftarrow g}(z) &= P_{g\leftarrow g}(1-z) \end{aligned} \tag{8.73}$$

8.4.2
Regularization of the Splitting Function

The splitting functions $P_{q\leftarrow q}$, $P_{g\leftarrow g}$ diverge at $z = 1$. It originates from the soft $\omega \to 0$ gluon emission. We factorized the two divergences, the infrared part which is in the splitting function and the collinear part which is in $\ln(Q^2/\mu^2)$. As we stated before, the infrared divergence is not really a divergence. It does not exist in the inclusive reaction like $e^-e^+ \to$ hadrons (see arguments in Sect. 7.3.2).

$$\sigma_{\text{TOT}}^{ee} = \sigma_0^{ee}\left(1 + \frac{\alpha_s}{\pi} + \cdots\right) \tag{8.74}$$

It is because if virtual gluon contributions are included, the infrared divergence of the gluon emissions is compensated, and is divergence free at each order of perturbation expansion. The compensation is universal and works in the DIS as well. If the total DIS cross section is integrated after including the virtual gluon exchange with regularization, the total cross section is known to be finite.

$$\sigma_{\text{TOT}}^{\text{DIS}} = \sigma_0^{\text{DIS}} + \int_0^1 dz \frac{d\sigma^q}{dz} + \sigma(\text{virtual}) = \sigma_0^{\text{DIS}} \left(1 - \frac{\alpha_s}{\pi} + \cdots\right) \tag{8.75}$$

Notice the sign difference of the correction in the $e\bar{e}$ and DIS reaction. For the $e\bar{e}$ reaction, the final quarks and the gluon are in a singlet state, inducing an attractive force, giving a positive correction. The first order QCD correction to the DIS total cross section, on the other hand, is negative. Equation (8.75) can be rewritten as

$$\int_0^1 dz \left[\frac{1}{\sigma_0}\frac{d\sigma^q}{dz} + \left\{\frac{\sigma(\text{virtual})}{\sigma_0} + \frac{\alpha_s}{\pi}\right\} \delta(1-z)\right] = 0 \tag{8.76}$$

We now introduce a "+" prescription which requires

$$\int_0^1 dz \left[\frac{d\sigma^q}{dz}\right]_+ = 0 \tag{8.77}$$

Then, the gluon emission cross section can be expressed as

$$\frac{1}{\sigma_0}\frac{d\sigma^{\text{DIS}}}{dz} = \frac{1}{\sigma_0}\left[\frac{d\sigma^q}{dz}\right]_+ - \frac{\alpha_s}{\pi}\delta(1-z) \tag{8.78}$$

As the first term on the right-hand side equals the second term in Eq. (8.46), we have

$$\frac{1}{\sigma_0}\left[\frac{d\sigma^q}{dz}\right]_+ = \frac{\alpha_s}{2\pi} P_{q\leftarrow q}(z) \ln\left(\frac{Q^2}{m^2}\right) \tag{8.79a}$$

$$P_{q\leftarrow q}(z) = C_F \left[\frac{1+z^2}{1-z}\right]_+ \tag{8.79b}$$

This redefined splitting function equals that defined in Eq. (8.44) for $z \neq 1$, but differs from it by a condition that its integral vanishes.

$$\int_0^1 dz\, P_{q\leftarrow q}(z) = 0 \tag{8.80}$$

In order to make Eq. (8.80) legitimate, one needs to regularize $P_{q\leftarrow q}$ which can be realized by introducing a "+" function which has the following property.

$$\frac{1}{(z-a)_+} = \frac{1}{z-a}, \quad z \neq a$$

$$\int dz \frac{f(z)}{(z-a)_+} = \int dz \frac{f(z)-f(a)}{z-a} \tag{8.81}$$

This means that the "+" function is not an ordinary function, but is a "distribution" which is meaningful only in the integral. Using Eqs. (8.80) and (8.81), $P_{q\leftarrow q}(z)$ is uniquely specified and is given by

$$P_{q\leftarrow q}(z) = C_F\left[\frac{1+z^2}{1-z}\right]_+ = \frac{4}{3}\left[\frac{1+z^2}{(1-z)_+} + \frac{3}{2}\delta(1-z)\right] \tag{8.82}$$

Originally, the splitting function was defined in Eq. (8.46) which had the form

$$dP(z,\tau)dz = \left[\delta(1-z) + \frac{\alpha_s}{2\pi}P_{j\leftarrow i}(z)\tau\right] \tag{8.83}$$

As the integrated parton flux should be equal to what the initial nucleon supplied, the integration of $dP(z,\tau)$ over z should give 1. Therefore, the requirement of Eq. (8.80) is logical.[11]

Note that Eq. (8.80) means the quark number conservation. This can be seen as follows. Denoting N_v as the number of valence quarks, we write down the quark number and momentum conservation rules. They are

$$\int dz \sum [q_i(z,\tau) - \overline{q}_i(z,\tau)] = N_v \tag{8.84a}$$

$$\int dz \left[\sum z[q_i(z,\tau) + \overline{q}_i(z,\tau)] + zg(z,\tau)\right] = 1 \tag{8.84b}$$

By taking derivatives of the above equations and using Eqs. (8.70) and (8.69), we obtain Eq. (8.80) and

$$\int_0^1 dz z[P_{q\leftarrow q}(z) + P_{g\leftarrow q}(z)] = 0$$

$$\int_0^1 dz z[2n_f P_{q\leftarrow g}(z) + P_{g\leftarrow g}(z)] = 0 \tag{8.85}$$

Problem 8.2

Using Eq. (8.72d), prove the following equation for the regularized splitting function.

$$P_{g\leftarrow g}(z) = 2C_A\left[\frac{1-z}{z} + \frac{z}{(1-z)_+} + z(1-z)\right] + \left(\frac{11}{6}C_A - \frac{2}{3}n_f T_R\right)\delta(1-z) \tag{8.86}$$

11) If one considers $O(\alpha_s)$ correction to the total cross section (Eq. (8.75)), the coefficient of $\delta(1-z)$ is also modified to $1-\alpha_s/\pi$. However, Eq. (8.83) is still valid to $O(\alpha_s^2)$ if one renormalizes by the total cross section.

By introducing the "+" function, Eqs. (8.78) and (8.83) can be determined unambiguously if one specifies the expression for $z \neq 1$ and the total integral. There is no need to introduce redundant parameters either. It does not need the elaborate infrared compensating calculations. It is a concise, legitimate and useful expression. However, one should be aware of its limitation that it cannot be used in the neighborhood $z \to 1$ as the approximation is poor.

8.4.3
Factorization Scheme

We have described the essence of the DGLAP equations. We comment a few technicalities that appear when it is used in practical applications.

DIS Scheme and $\overline{MS}$ Scheme In deriving the divergence free DGLAP equation Eq. (8.49), we left a small ambiguous term denoted by $C^F_{\mathrm{DIS}}(x/y)(= C_{\mathrm{DIS}} - C'_{\mathrm{DIS}})$. C_{DIS} are called coefficient functions and are different in general for F_1, F_2 and F_3. Choice of subtraction constant C'_{DIS} is arbitrary because the factorization prescription contains the subtraction of infinite term and a small shift of the infinity does not alter its validity. The simplest prescription is to subtract the whole nondominant term $C_{\mathrm{DIS}}(x/y)$ in Eq. (8.47) such that the residual coefficient function $C^F_{\mathrm{DIS}}(x/y)$ in Eq. (8.49) vanishes. This subtraction method is referred to as the DIS scheme [394, 395]. In this scheme, the equation for $F_2(x, Q^2)$ takes a simple form. An alternative which is more conventional nowadays is to use $\overline{MS}$ scheme. In this scheme, using the dimensional regularization method, only the divergent term $(1/\epsilon + \ln 4\pi - \gamma_E)$ is subtracted and the nondominant coefficient function $(C_{\mathrm{DIS}}(x))$ are retained. Therefore, the finite terms differ, depending on which factorization scheme is adopted. Besides, the structure function which we have loosely referred to as the quark distribution has to be defined precisely, flavor for flavor, for detailed discussions. We do not discuss the details of the flavor structure, but simply list relations between the two schemes [392, 396–398]. The coefficient functions which appear below are defined according to

$$\begin{aligned}
\underset{k=1,2}{f_k}(x, Q^2) = \int_x^1 \frac{dy}{y}\Bigg[& f_k(y, \mu^2) \\
& \times \left[\delta\left(1-\frac{x}{y}\right) + \frac{\alpha_s}{2\pi}\left\{P_{q\leftarrow q}\left(\frac{x}{y}\right)\ln\frac{Q^2}{\mu^2} + C_k^q\left(\frac{x}{y}\right)\right\}\right] \\
& + g(y, \mu^2)\left[\frac{\alpha_s}{2\pi}\left\{P_{q\leftarrow g}\left(\frac{x}{y}\right)\ln\frac{Q^2}{\mu^2} + C_k^g\left(\frac{x}{y}\right)\right\}\right]\Bigg]
\end{aligned} \tag{8.87a}$$

$$f_3(x,Q^2) = \int_x^1 \frac{dy}{y} f_3(y,\mu^2) \times \left[\delta\left(1-\frac{x}{y}\right) + \frac{\alpha_s}{2\pi}\left\{P_{q\leftarrow q}\left(\frac{x}{y}\right)\ln\frac{Q^2}{\mu^2} + C_3^q\left(\frac{x}{y}\right)\right\}\right] \tag{8.87b}$$

where $f_k(x,Q^2)$ is any of the structure functions:

$$f_k = \left(2F_1, \frac{F_2}{x}, F_3\right) \tag{8.87c}$$

For DIS scheme

$$\begin{aligned} C_1^q(z) &= -C_F z\,, \quad C_1^g(z) = -T_F 4z(1-z) \\ C_2^q(z) &= C_2^g(z) = 0 \\ C_3^q(z) &= -C_F(1+z) \end{aligned} \tag{8.88}$$

For the $\overline{MS}$ scheme,

$$\begin{aligned} C_1^q(z) &= \frac{1}{2}C_2^q(z) - C_F z\,, \quad C_1^g(z) = \frac{1}{2}C_2^g(z) - T_F 4z(1-z) \\ C_2^q(z) &= \frac{C_F}{2}\left[\frac{1+z^2}{1-z}\left\{\ln\left(\frac{1-z}{z}\right) - \frac{3}{4}\right\} + \frac{9+5z}{4}\right]_+ \\ C_2^g(z) &= T_F z\left[(z^2+(1-z)^2)\ln\frac{1-z}{z} - 1 + 8z(1-z)\right] \\ C_3^q(z) &= C_2^q(z) - C_F(1+z) \end{aligned} \tag{8.89}$$

The relation between quark distributions in the two schemes are given by

$$\begin{aligned} q_i^{\text{DIS}}(x,\mu^2) &= q_i^{\overline{MS}}(x,\mu^2) \\ &\quad + \frac{\alpha_S}{2\pi}\int_x^1 \frac{dy}{y}\left[q_i^{\overline{MS}}(y,\mu^2)C_2^q\left(\frac{x}{y}\right) + g^{\overline{MS}}(y,\mu^2)C_2^g\left(\frac{x}{y}\right)\right] \\ g^{\text{DIS}}(x,\mu^2) &= g^{\overline{MS}}(x,\mu^2) \\ &\quad - \frac{\alpha_S}{2\pi}\int_x^1 \frac{dy}{y}\left[\sum_{i=q,\bar{q}} q_i^{\overline{MS}}(y,\mu^2)C_2^q\left(\frac{x}{y}\right) + g^{\overline{MS}}(y,\mu^2)C_2^g\left(\frac{x}{y}\right)\right] \end{aligned} \tag{8.90}$$

One can see that in the DIS scheme, $C_2 = 0$ at all orders by definition, making the functional form of F_2 particularly simple. This is why it was introduced first and why we used it in discussing the DGLAP equation. The choice also simplifies the quark distribution function used in the deep inelastic scattering. However, it contains terms specific to the DIS reaction. On the other hand, in the $\overline{MS}$ scheme, the coefficient functions are relatively complicated, but the parton distribution function is simple and contains no terms related to other hadron reactions. Simplicity of DIS scheme applies only to the deep inelastic scattering. Therefore, $\overline{MS}$ scheme is generally preferred.

Higher Order Contributions The splitting functions we have given so far are those of LLA approximation. When one goes to higher order calculations, they, too, need to be modified.

$$P^{(0)}_{j\leftarrow i}(z) \to P^{(0)}_{j\leftarrow i}(z) + \frac{\alpha_s}{2\pi} P^1_{j\leftarrow i}(z) + \cdots \tag{8.91}$$

Detailed expressions for the higher order splitting functions are given in [397, 399, 400].

8.5 Solutions to the DGLAP Equation

8.5.1 Method of Moments

Anomalous Dimension Solutions to the DGLAP equation can be obtained by method of moments (the Mellin transformation). It is particularly simple for the nonsinglet distribution function. We apply the method to expansion coefficients of the nonsinglet DGLAP equations in Eq. (8.58).

$$F(x, Q^2) = \sum_n \frac{C_n(x)}{n!} \left(\frac{\alpha_s}{2\pi} \ln \frac{Q^2}{Q_0^2} \right)^n \tag{8.92a}$$

$$C_n(x) = \int_x^1 \frac{dx_n}{x_n} P_{q\leftarrow q}\left(\frac{x}{x_n}\right) \cdots \int_{x_2}^1 \frac{dx_1}{x_1} P_{q\leftarrow q}\left(\frac{x_2}{x_1}\right) F_0(x_1, Q_0^2) \tag{8.92b}$$

Taking moments of $C_n(x)$, Eq. (8.92b) is converted to

$$\begin{aligned} \tilde{C}_n(N) &\equiv \int_0^1 \frac{dx}{x} x^N C_n(x) \\ &= \int_0^1 \frac{dx}{x} x^N \int_x^1 \frac{dx_n}{x_n} P_{q\leftarrow q}\left(\frac{x}{x_n}\right) \cdots \int_{x_2}^1 \frac{dx_1}{x_1} P_{q\leftarrow q}\left(\frac{x_2}{x_1}\right) F_0(x_1, Q_0^2) \\ &= (\gamma(N))^n \tilde{F}_0(N, Q_0^2) \end{aligned} \tag{8.93a}$$

$$\gamma(N) = \int_0^1 \frac{dx}{x} x^N P_{q\leftarrow q}(x) \tag{8.93b}$$

$$\tilde{F}_0(N, Q_0^2) = \int_0^1 \frac{dx}{x} x^N F_0(x, Q_0^2) \tag{8.93c}$$

where $\gamma(N)$ is referred to as the anomalous dimension. Therefore, expressed in moments, the equation for the structure function is converted to

$$\begin{aligned}\tilde{F}(N, Q^2) &= \int_0^1 \frac{dx}{x} x^N F(x, Q^2) = \sum_{n=0}^{\infty} \frac{1}{n!} \left(\frac{\alpha_s}{2\pi} \ln \frac{Q^2}{Q_0^2} \gamma(N) \right)^n \tilde{F}_0(N, Q_0^2) \\ &= \exp\left[\frac{\alpha_s}{2\pi} \gamma(N) \ln \frac{Q^2}{Q_0^2} \right] \tilde{F}_0(N, Q_0^2)\end{aligned} \tag{8.94}$$

Then, a formal solution of $F(x, Q^2)$ can be obtained by the inverse Mellin transformation:

$$\begin{aligned}F(x, Q^2) &= \frac{1}{2\pi i} \int_{N_0 - i\infty}^{N_0 + i\infty} dN x^{-N} \tilde{F}(N, Q^2) \\ &= \frac{1}{2\pi i} \int_{N_0 - i\infty}^{N_0 + i\infty} dN x^{-N} \exp\left[\frac{\alpha_s}{2\pi} \gamma(N) \ln \frac{Q^2}{Q_0^2} \right] \tilde{F}_0(N, Q_0^2) \\ &= \frac{1}{2\pi i} \int_{N_0 - i\infty}^{N_0 + i\infty} dN \exp\left[h(N) \right] \tilde{F}_0(N, Q_0^2) \\ h(N) &= N \ln \frac{1}{x} + \frac{\alpha_s}{2\pi} \gamma(N) \ln \frac{Q^2}{Q_0^2}\end{aligned} \tag{8.95}$$

where the integral path $N_0 + iy$ $(-\infty < y < \infty)$ has to be to the right of the singularities in $\tilde{F}_0(N, Q_0^2)$.

Let us investigate properties of the moments of the structure function. We calculated moments of the power expansion series in Eq. (8.93) for the sake of later discussions, but Eq. (8.94) can also be obtained directly by applying the Mellin transformation to the DGLAP equation (8.71a). Defining

$$M(N, \tau) \equiv \int_0^1 \frac{dx}{x} x^N \Delta(x, \tau), \quad \tau = \ln \frac{Q^2}{Q_0^2} \equiv \ln \frac{t}{t_0} \tag{8.96}$$

we obtain

$$\begin{aligned}\frac{dM(N, \tau)}{d\tau} &\equiv \frac{\alpha_s(\tau)}{2\pi} \int_0^1 \frac{dx}{x} x^N \int_x^1 \frac{dy}{y} \Delta(y, \tau) P_{q \leftarrow q}\left(\frac{x}{y} \right) \\ &= \frac{\alpha_s(\tau)}{2\pi} \gamma(N) M(N, \tau)\end{aligned} \tag{8.97}$$

For constant α_s,

$$\ln \frac{M(N,\tau)}{M(N,0)} \equiv \frac{\alpha_s}{2\pi}\gamma(N)\tau \tag{8.98a}$$

$$\to M(N,\tau) = M(N,0)\exp\left[\frac{\alpha_s}{2\pi}\gamma(N)\tau\right] = M(N,0)\left(\frac{t}{t_0}\right)^{\frac{\alpha_s}{2\pi}\gamma(N)} \tag{8.98b}$$

In deriving the last equality, we used $\tau = \ln(t/t_0)$. For t dependent α_s,

$$\begin{aligned}\frac{\alpha_s}{2\pi}\gamma(N)\tau &\to \int \frac{\alpha_s(\tau)}{2\pi}\gamma(N)d\tau \overset{(7.63)}{=\!=} \frac{\gamma(N)}{2\pi}\int_{\alpha_s(0)}^{\alpha_s(\tau)} \frac{\alpha_s}{\beta(\alpha_s)}d\alpha_s \\ &= -\frac{\gamma(N)}{2\pi}\int_{\alpha_s(0)}^{\alpha_s(\tau)} \frac{d\alpha_s}{\alpha_s b_0(1+b_1\alpha_s+\cdots)} \\ &= \frac{\gamma(N)}{2\pi b_0}\ln\left[\frac{\alpha_s(0)}{\alpha_s(\tau)}\right]+\cdots\end{aligned} \tag{8.99}$$

Therefore,

$$\ln\left[\frac{M(N,\tau)}{M(N,0)}\right] = \frac{d_N}{d_K}\ln\left[\frac{M(K,\tau)}{M(K,0)}\right] \tag{8.100a}$$

$$\text{or} \quad M(N,\tau) = M(N,0)\left[\frac{\alpha_s(0)}{\alpha_s(\tau)}\right]^{-d_N} \tag{8.100b}$$

$$d_N = -\frac{\gamma(N)}{2\pi b_0} \tag{8.100c}$$

If one obtains the moments for all Ns, the structure function can be reconstructed by the inverse Mellin transformation. Note that the above solution is obtained in LLA approximation. According to the solution, the ratio of the moments is determined by d_N only, which means that the evolution of the structure function also is determined by the anomalous dimension only. The $\gamma(N)$'s can be calculated by substituting Eq. (8.82) into Eq. (8.93b)

$$\begin{aligned}\gamma(N) &= \int_0^1 \frac{dx}{x}x^N P_{q\leftarrow q}(x) = \frac{4}{3}\left[\frac{3}{2}+\int_0^1 dx\frac{x^{N-1}+x^{N+1}-2}{1-x}\right] \\ &= \frac{4}{3}\left[-\frac{1}{2}+\frac{1}{N(N+1)}-2\sum_{j=2}^{N}\frac{1}{j}\right]\end{aligned} \tag{8.101}$$

Values of $\gamma(N)$s for a few Ns are

$$\begin{aligned}&\gamma(1)=0, \quad \gamma(2)=-1.78, \quad \gamma(3)=-2.78, \quad \gamma(4)=-3.49, \\ &\gamma(5)=-4.04, \quad \gamma(6)=-4.50\end{aligned} \tag{8.102}$$

As the nonsinglet distribution function can be obtained from $F_3(x, Q^2)$ of the neutrino DIS data, it is possible to test the QCD prediction.

Figure 8.10 gives a comparison of the data with the QCD calculations [401]. The slope of the curves agree well with d_N/d_K as predicted in Eq. (8.100a). This also tests the spin of the gluon. If it has spin 0, it gives a different slope which did not agree with data.

It is also possible to compare $M(N,\tau)$ with data directly, but needs an extra information, that is, the value of $\alpha_s(0)$ which compromises the rigor of the prediction somewhat. We show an example of the test in Figure 8.11. Figure 8.11a plots $M(N,\tau)^{-1/d_N}$ as a function of Q^2. As the LLA approximation $\alpha_s(\tau) = 1/[b_0 \ln(Q^2/\Lambda^2)]$ was used, the curves should be straight lines as a function of Q^2 which are reproduced by the data. Since $M_N^{-1/d_N} \propto \ln Q^2 - \ln \Lambda^2$, the intercept of the line at the abscissa should agree with Λ^2 from which the value of Λ can be obtained. However, as can be seen from Figure 8.11b, a modified version of $M(N,\tau)$ which uses the Nachtman moment $\xi = 2x/[1 + \{1 + 4m_p^2 x^2/Q^2\}]$ instead of x [403, 404] gives a better fit with a different value of Λ. If the value of Λ is dependent on the parameter choice, it means it is not so reliable. Incidentally, if one uses a constant α_s instead of a τ dependent function, Eq. (8.98b) tells that the

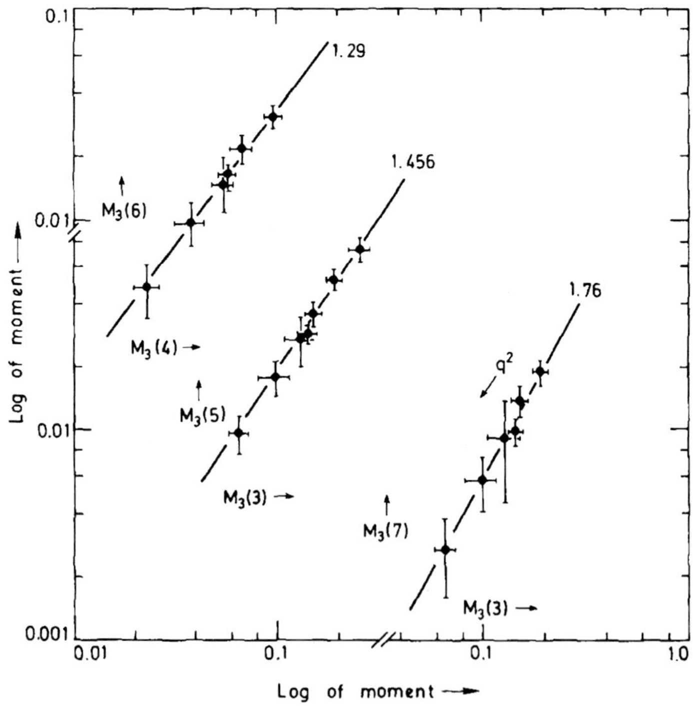

Figure 8.10 Log–log plots of various moments of xF_3 compared with QCD predictions. The slope should give the ratio of the anomalous dimensions d_N/d_K [401].

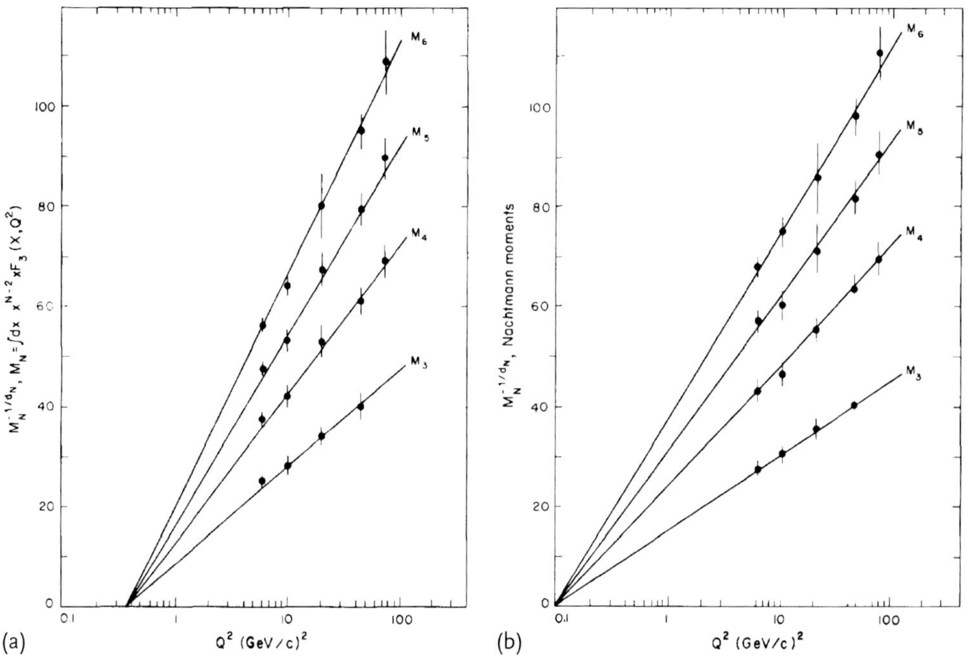

Figure 8.11 Comparisons of QCD prediction with data. The straight lines are fits with a common intercept. (a) ordinary moments: M_N^{-1/d_N} (b) Nachtman moments: $\tilde{M}_N^{-1/d_N}$ for $N = 3, 4, 5$ and 6 [402].

expression for M_N changes to

$$M(N,\tau) = M(N,0)e^{d_N\tau} = M(N,0)\left(\frac{Q^2}{Q_0^2}\right)^{d_N} \tag{8.103}$$

and is not proportional to $\ln Q^2$ but to powers of Q^2 which does not agree with data.

Momentum Distribution at $Q^2 = \infty$ Equations for the moments of the singlet and the gluon distributions can be obtained from Eq. (8.71).

$$\frac{d}{d\tau}\begin{bmatrix}\tilde{\Sigma}_N \\ \tilde{g}_N\end{bmatrix} = \frac{\alpha_s}{2\pi}\begin{bmatrix}\gamma(N)(qq) & 2n_f\gamma(N)(qg) \\ \gamma(N)(gq) & \gamma(N)(gg)\end{bmatrix}\begin{bmatrix}\tilde{\Sigma}_N \\ \tilde{g}_N\end{bmatrix} \tag{8.104a}$$

where $\tilde{\Sigma}_N, \tilde{g}_N, \gamma(N)(qq)\cdots$ mean moments of Σ, G, $P_{q\leftarrow q}, \cdots$. The anomalous dimension $\gamma_b(qq)$, for example, can be obtained similarly as was done in Eq. (8.101):

$$\gamma(N)(qq) = \gamma(N) \tag{8.105a}$$

$$\gamma(N)(qg) = \frac{1}{2}\left[\frac{N^2+N+2}{N(N+1)(N+2)}\right] \tag{8.105b}$$

$$\gamma(N)(gq) = \frac{4}{3}\left[\frac{N^2+N+2}{N(N^2-1)}\right] \tag{8.105c}$$

$$\gamma(N)(gg) = 6\left[-\frac{1}{12} + \frac{1}{N(N-1)} + \frac{1}{(N+1)(N+2)} - \sum_{j=2}^{N}\frac{1}{j} - \frac{N_f}{18}\right] \tag{8.105d}$$

The $N = 2$ moments $(\tilde{\Sigma}_2, \tilde{g}_2)$ represent the average momentum distributions of the quarks and gluons. The calculation gives

$$\gamma(2)(ij) = \begin{bmatrix} -16/9, & n_f/3 \\ 16/9, & -n_f/3 \end{bmatrix} \tag{8.106}$$

which clearly shows the momentum conservation. Solving the secular equation, one obtains two eigenvalues

$$\lambda_+ = -(16 + 3n_f)/9\,, \quad \lambda_- = 0 \tag{8.107}$$

which give the diagonalized anomalous dimension

$$d_{2+} = -\lambda_+/[(33 - 2n_f)/6] > 0\,, \quad d_{2-} = 0 \tag{8.108}$$

According to Eq. (8.100), the moments evolve like

$$\sim \left[\frac{\alpha_s(0)}{\alpha_s(\tau)}\right]^{-d_N} \sim \left[\ln\frac{Q^2}{\Lambda^2}\right]^{-d_N} \tag{8.109}$$

and the solution corresponding to λ_+ vanishes as $Q^2 \to \infty$. The eigen vector corresponding to λ_- gives

$$\begin{bmatrix} \Sigma_2(\infty) \\ G_2(\infty) \end{bmatrix} = \frac{1}{16 + 3n_f}\begin{bmatrix} 3n_f \\ 16 \end{bmatrix}_{n_f=6} = \begin{bmatrix} 9/17 \\ 8/17 \end{bmatrix} \tag{8.110}$$

This means the average momentum of the quark and the gluon at $Q^2 \to \infty$ are $\Sigma_\infty/2n_f = 3/68$ and $8/17$ respectively. This approximately reproduces the fraction of gluon momentum 0.5 in the nucleon observed in DIS data (see Vol. 1, Eq. (17.135)).

Gluon Distribution The gluon does not interact electromagnetically or weakly. Therefore, it does not appear as a directly measurable observable in the DIS. In principle, it is possible to determine its value from the DGLAP equation using DIS scattering data. However, the distribution calculated this way is strongly correlated with the value of Λ. In addition, an initial gluon distribution is required in order to determine the quark distribution at small x. Therefore, it is important to determine the gluon distribution experimentally and independently at least at one fixed Q_0^2 to solve the DGLAP equation. The following processes can be used to determine the gluon distribution.

$$p + p \to \gamma + X \tag{8.111a}$$

$$\gamma + N \to J/\psi + X \tag{8.111b}$$

A main contribution to the direct photon production (not meaning the decay products of π^0, η^0, η', for example) in hadron reactions comes from the QCD Compton process:

$$q + g \to \gamma + q \tag{8.112}$$

Another contribution comes from a pair annihilation:

$$q + \overline{q} \to \gamma + g \tag{8.113}$$

In the pp reaction, the former is dominant. Experimentally, the direct photon includes contributions of decay products from π^0, η^0, for example, as backgrounds. They are soft processes and can be separated by extracting photons at large p_T [405].

The lowest order contribution to the reaction Eq. (8.111b) comes from

$$\gamma + g \to c + \overline{c} \tag{8.114}$$

thus enabling another measurement of the gluon distribution. Figure 8.12 shows a gluon distribution obtained from the J/ψ production [406]. Phenomenologically, the distribution fits well with a function

$$g(x) = 3\frac{(1-x)^5}{x} \tag{8.115}$$

8.5.2
Double Logarithm

Before going to practical and numerical evaluation of the DGLAP equation, we discuss a special yet very practical example which can be solved analytically. Thus far, we concentrated on the large $\ln(Q^2/m^2)$, or collinear region, carefully avoiding the phase space part of $x \simeq 0$, that is, the soft gluon region. As shown in Eq. (8.57), the integration over x gives large $\ln x$ at $x \to 0$ or large $\ln(1-x)$ at $x \to 1$ and the perturbation treatment fails. In the following, we concentrate on the $x \to 0$ region. The small x region is populated by gluons and sea quarks which are created by the gluon. The gluons are produced by cascades. This is due to $1/x$ term in the $P_{g\leftarrow q}(x)$ and $P_{g\leftarrow g}(x)$. We denote the gluon distribution function in the interested region as $g(x)$. In the small x region, the quark contribution can be neglected. Then the DGLAP equation for the gluon distribution function $g(x, \tau)$ becomes

$$\frac{dg(x,\tau)}{d\tau} = \frac{\alpha_s}{2\pi}\int_x^1 \frac{dz}{z} P_{g\leftarrow g}\left(\frac{x}{z}\right) g(z,\tau)\,, \quad \tau = \ln\frac{Q^2}{Q_0^2} \tag{8.116}$$

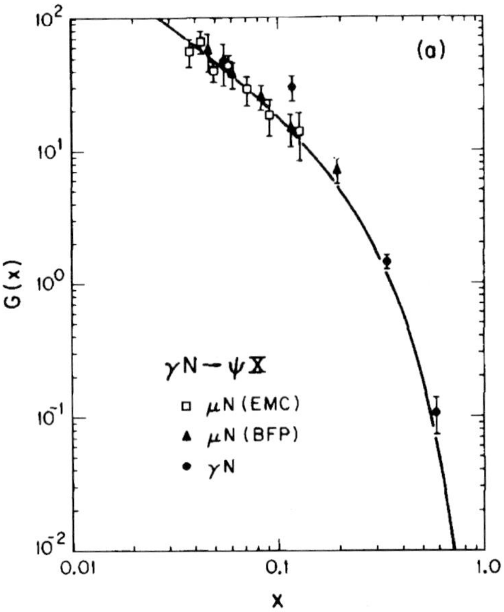

Figure 8.12 Gluon distribution function obtained from $\sigma(\gamma N \to \psi x)$. The line represents the curve $g(x) = 3(1-x)^5/x$ [406].

For a given initial function $g = g_0(x, 0)$, Eq. (8.116) has a solution in the form of the power expansion of τ (see Eq. (8.58))

$$g(x, \tau) = \sum_n \frac{C_n(x)}{n!} \left(\frac{\alpha_s}{2\pi} \ln \frac{Q^2}{Q_0^2} \right)^n \tag{8.117a}$$

$$C_n(x) = \int_x^1 \frac{dx_n}{x_n} P_{g\leftarrow g}\left(\frac{x}{x_n}\right) \cdots \int_{x_2}^1 \frac{dx_1}{x_1} P_{g\leftarrow g}\left(\frac{x_2}{x_1}\right) g_0(x_1, 0) \tag{8.117b}$$

For small $x \to 0$, the gluon splitting function Eq. (8.72d) can be approximated by $P_{g\leftarrow g}(x) \sim 2C_A/x$.

To solve the equation, we need to select the initial function $g_0(x, t_0)$. Since the structure function is related to the photon-proton total cross section by Eq. (8.29), and its total energy by

$$\begin{aligned} s_{\gamma p} &= (q + p_2)^2 = q^2 + 2(p_2 \cdot q) + p_2^2 \\ &= -Q^2 + \frac{Q^2}{x} + m_p^2 \simeq \frac{1-x}{x} Q^2 \xrightarrow{x\to 0} \frac{Q^2}{x} \end{aligned} \tag{8.118}$$

Therefore, $s_{\gamma p} \to \infty$ as $x \to 0$. The high energy behavior of the total cross section is well represented by the Regge theory (see Vol. 1, Sect. 13.7.2) which gives

$$\begin{aligned} &\sigma_{\mathrm{TOT}} \simeq A s_{\gamma p}^{\alpha_P(0)-1} + B s_{\gamma p}^{\alpha_R(0)-1} \\ &\alpha_P(0) \simeq 1\,, \quad \alpha_R(0) \sim 0.5 \end{aligned} \tag{8.119}$$

where $\alpha_P(t)$ and $\alpha_R(t)$ are the Regge trajectories of Pomeron and Reggeon which have quantum numbers of vacuum and ρ mesons respectively. Substituting the Regge expressions in Eq. (8.29), and using $\sigma_B \simeq 2\pi^2\alpha/m_N\nu = (4\pi^2\alpha/Q^2)x$ (see Eq. (8.30)), we have the dominant small x contribution

$$F_2(x) \sim x\frac{\sigma_{TOT}}{\sigma_B} \sim Ax^{1-\alpha_P(0)} + Bx^{1-\alpha_R(0)} \tag{8.120}$$

The first term contributes to flavor singlet sea quarks and gluons and the second term contributes to flavor nonsinglet valence quarks. Since $\alpha_P(0) \simeq 1$ and $F_2(x) \sim xg(x)$, we may try

$$g_0(x, 0) = C/x \quad C = \text{constant} \tag{8.121}$$

For this choice of the initial function and with the approximation $P_{g\leftarrow g}(z) = 2C_A/z$, $C_n(x)$ in Eq. (8.117b) can easily be integrated:

$$C_n(x) = \frac{C}{x}\frac{1}{n!}\left(2C_A \ln\frac{1}{x}\right)^n \tag{8.122}$$

Then, the solution Eq. (8.117a) is expressed as

$$xg(x,\tau) = x\sum_n \frac{C_n(x)}{n!}\left(\frac{\alpha_s}{2\pi}\ln\frac{Q^2}{Q_0^2}\right)^n = C\sum_n \frac{1}{(n!)^2}\left(\frac{C_A\alpha_s}{\pi}\ln\frac{Q^2}{Q_0^2}\ln\frac{1}{x}\right)^n \tag{8.123}$$

When both $\ln Q^2$ and $\ln 1/x$ are large, the former due to collinear configuration and the latter due to soft gluon, the perturbation expansion in terms of α_s is no longer valid. The above expression shows how resummation of the double logarithm can be achieved in special circumstances.

Using the modified Bessel function

$$I_\alpha(z) = i^{-\alpha}J_0(iz) = \sum_{m=0}^{\infty}\frac{1}{m!\Gamma(m+\alpha+1)}\left(\frac{z}{2}\right)^{2m+\alpha} \xrightarrow{\alpha=0,z\to\infty} \sqrt{\frac{1}{2\pi z}}e^z \tag{8.124}$$

we obtain

$$xg(x,\tau) \sim C\exp\left[\sqrt{\frac{4C_A}{\pi}\alpha_s\ln\frac{Q^2}{Q_0^2}\ln\frac{1}{x}}\right] \tag{8.125}$$

For running $\alpha_s(\tau)$, we simply replace $\alpha_s \ln \frac{Q^2}{Q_0^2}$ with $(1/b_0)\ln \frac{\alpha_s(0)}{\alpha_s(\tau)}$ because

$$\alpha_s \ln \frac{Q^2}{Q_0^2} = \alpha_s \int_{Q_0^2}^{Q^2} \frac{dt'}{t'} \to \int_{Q_0^2}^{Q^2} \alpha_s(\tau')d\tau' \stackrel{\text{Eq. (7.65)}}{=\!=} \int_{\alpha_s(0)}^{\alpha_s(\tau)} \frac{\alpha_s}{\beta(\alpha_s)} d\alpha_s$$

$$\simeq -\int_{\alpha_s(0)}^{\alpha_s(\tau)} \frac{1}{b_0 \alpha_s} d\alpha_s = \frac{1}{b_0} \ln \frac{\alpha_s(0)}{\alpha_s(\tau)} = \frac{1}{b_0} \ln \frac{\ln(Q^2/\Lambda^2)}{\ln(Q_0^2/\Lambda^2)}$$

$$b_0 = \frac{1}{4\pi}\left(11 - \frac{2}{3} n_f\right) \tag{8.126}$$

where we used the LLA expression for the β function in the second line. Then, the gluon structure function can be expressed as

$$x g(x,\tau) \sim \exp\left[\sqrt{\frac{4C_A}{\pi b_0} \ln \frac{\ln(Q^2/\Lambda^2)}{\ln(Q_0^2/\Lambda^2)} \ln \frac{1}{x}}\right] \equiv \exp\left[2\sqrt{\eta \xi}\right]$$

$$\eta = \frac{C_A}{\pi b_0} \ln \frac{\ln(Q^2/\Lambda^2)}{\ln(Q_0^2/\Lambda^2)}, \qquad \xi \equiv \ln \frac{1}{x} \tag{8.127}$$

where we introduced η and ξ for later use.

In summary, the gluon distribution function shows a rapid rise as $x \to 0$ which may be parametrized as $x g(x,t) \sim x^{-\lambda}$, consistent with the Regge behavior observed in the high energy hadron production cross section. When the QCD effect is taken into account, the λ increases as Q^2 does.

The distribution at small x which is obtained by ZEUS and H1 at HERA exhibits $x g(x,\tau) \sim x^{-\lambda}$ at small x with λ increasing sharply as a function of Q^2 and can be reproduced quite well with DGLAP based Monte Carlo program which incorporates the NLO contributions [407, 408]. The DGLAP based Monte Carlo method will be discussed shortly. Our analytic expression for the gluon distribution Eq. (8.125) agrees well in the region of interest ($x \lesssim 0.1$) with that obtained from the Monte Carlo program [409–411] (Figure 8.13).

Note, however, that Eq. (8.125) shows that the γp total cross section grows faster than any power of $\ln 1/x$. Since $\ln 1/x \sim \ln s$, it contradicts with the Froissart bound $\ln^2 s$ (Vol. 1, Eq. (13.147)). This means that the DGLAP evolution equation is inadequate as $x \to 0$. The reason can be understood qualitatively as a saturation effect in the growing gluon number density. At some point, the recombination effect counterbalances the growth of the gluon number, or at such small x level, the gluon wave length is so large that the picture of the gluon as particles may no longer be valid. In view of this, an alternative approach which treats the x evolution at fixed Q^2 was developed and is known as the BFKL equation [412–414]. We do not discuss it, but it is a subject of much theoretical interest.

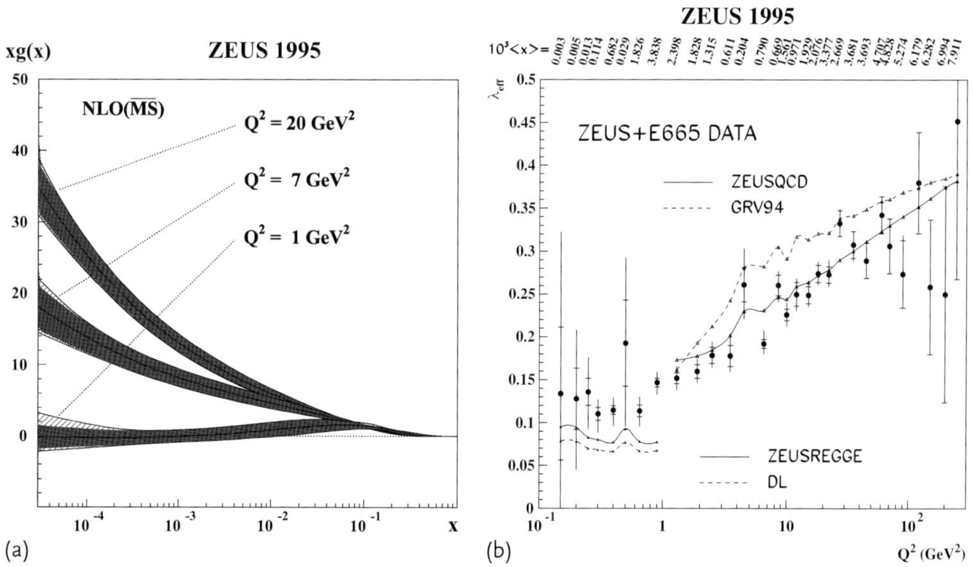

Figure 8.13 (a) Gluon distribution function $xg(x, Q^2)$ at $Q^2 = 1, 7, 20\,\text{GeV}^2$ from the ZEUS QCD fit. (b) Slope $\lambda = d\ln F_2/d\ln(1/x)$ extracted from ZEUS data. The dashed lines (GRV94 and DL) and the solid line below 1 GeV (ZEUSREGGE) are from Regge fits. The solid curve above 1 GeV(ZEUSQCD) is from a NLO Monte Carlo program [408–411].

8.5.3
Monte Carlo Generators

QCD Test Theoretically, the method of moments is clear and tells us how the structure functions evolve. Practically, however, data near $x \sim 0, 1$ are difficult to obtain and it is very hard to extract accurate moments. To circumvent the difficulty, we notice that derivatives $d\ln F_i/d\tau$ at given x only depend on values of $F_i(y)$ at $y > x$, and its value can be extracted from observed data. Experimentally, dominant contributions of sea quarks and gluons come from the region $x \sim 0$ which means they can be neglected in the region $x > x_0$ compared to those from valence quarks on the right-hand side of the DGLAP equation. Then, the evolution equation for the singlet distribution is the same as that of nonsinglet distribution. Practically, one can set $x_0 \approx 0.25 \sim 0.3$. We compare experimental data of $F_2(x)$ as a function of Q^2 and the calculated distributions in Figure 8.14 [415].[12] Figure 8.15 shows the slope $d\ln F_2/d\tau$ at each value of x. It is a function of x and takes different values for different x. It is positive for small $x(\lesssim 0.1)$, showing that F_2 is an increasing function of Q^2 at small x. Whether its values can be reproduced with a single parameter is a stringent QCD test. The data reproduces theoretical values quite well.

12) In fitting the actual data, the second order (NLLA) calculations were used. In this case, one has to use splitting functions which include higher order contributions (see Eq. (8.91)).

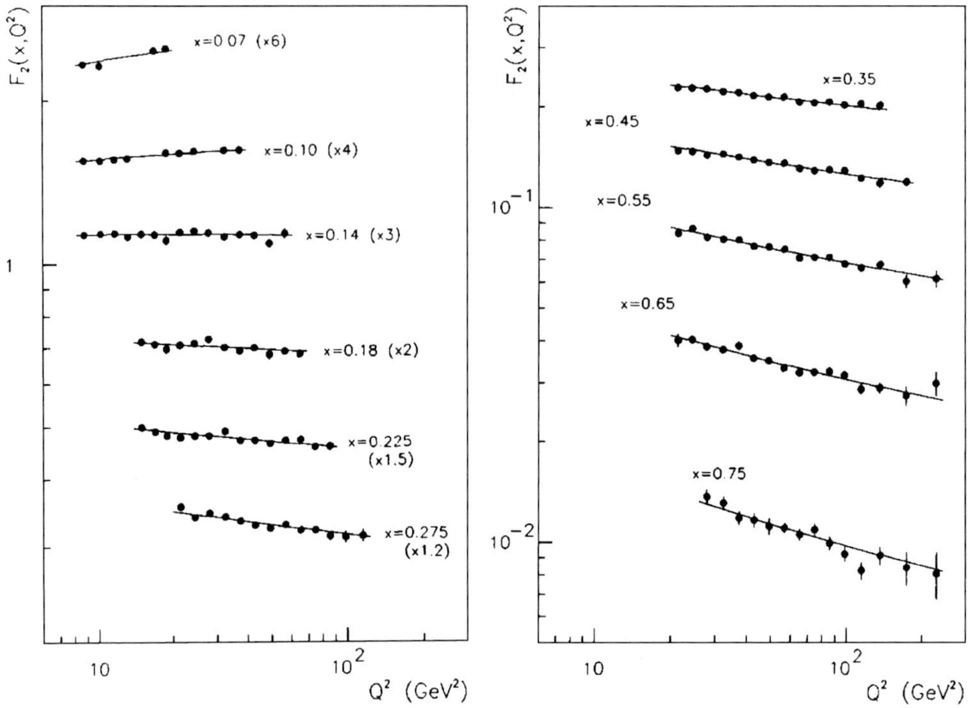

Figure 8.14 The structure function $F_2(x, Q)$: The solid lines represent the singlet + nonsinglet QCD fit discussed in the text. Only statistical errors are shown [415, 416].

It also shows how the slope is dependent on Λ values which gives [417]

$$\Lambda^{(4)}_{\overline{MS}} = 263 \pm 42\,\text{MeV} \tag{8.128}$$

corresponding to $\alpha_s(m_Z^2)$

$$\alpha_s(m_Z^2) = 0.113 \pm 0.003 \tag{8.129}$$

In summary, the ln Q^2 dependence of the structure function is a characteristic feature of the QCD (or more generally, non-Abelian gauge theory), and historically, its validity was established by comparing predictions with the DIS data.

Computer Calculation of the Structure Functions The methods we discussed have been expanded and today, we have several established computer programs that calculate the structure functions at any value of x, Q^2. They start with the given initial

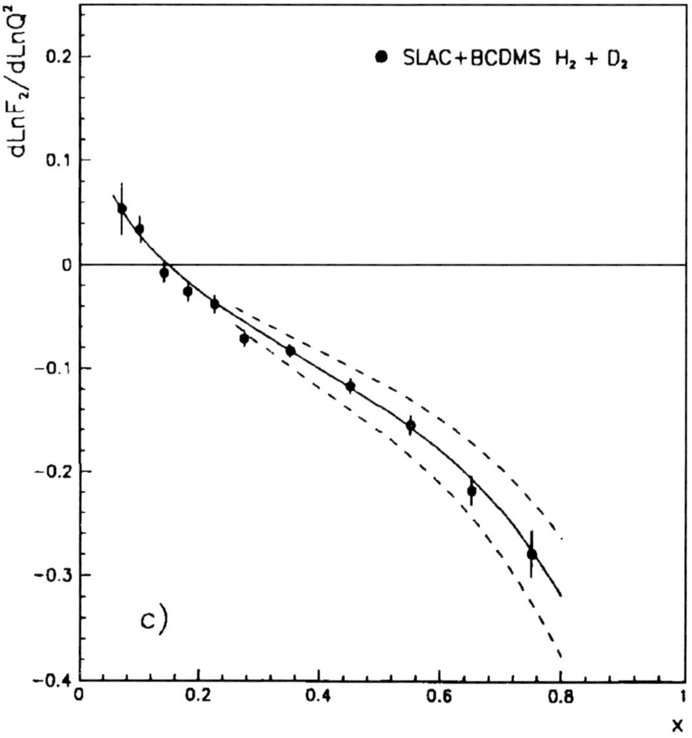

Figure 8.15 $d \ln F_2/d\tau$ at high Q^2. The sensitivity of the high x data to the value of α_s is illustrated here; the solid line is the result of the fit and the dashed lines correspond to $\alpha_s(M_Z^2)$ varied by ± 0.010 (twice the final error) [417].

functions of x at a reference point Q_0 parametrized as, for example,

$$\begin{aligned}
x f_i(x, Q_0) &= A_i x^{B_i}(1-x)^{C_i}(1 + D_i\sqrt{x} + E_i x) \\
A_i &= a_0 + a_1 L + a_2 L^2 \\
B_i &= b_0 + b_1 L + b_2 L^2 \\
C_i &= c_0 + c_1 L + c_2 L^2 \\
L &= \ln\left(\frac{\ln(\mu^2/\Lambda^2)}{\ln(\mu_0^2/\Lambda^2)}\right) > 0
\end{aligned} \tag{8.130a}$$

The exponent B_i is determined to satisfy the Regge trajectories at $x \to 0(Q^2 \to 0, E \to \infty)$. D_is and E_is are included to satisfy different quantum states of Regge trajectories. C_i is determined to reproduce elastic scattering data at $x \to 1$. There are arguments that $C = 2n - 3$ where n is the effective number of partons in the interested region ($C = 3$ for valence quarks, $C = 7$ for sea quarks and $C = 5$ for gluons) [418, 419]. Phenomenologically, they are adjusted in the neighborhood of the given values. For the gluon functions ($f_i = g$), QCD consideration gives $B_g \approx -\alpha_s(12\ln 2/\pi) \approx -0.5$ with a big increase as $x \to 0$, but for $x \geq 0.04$, it is

Table 8.1 Initial functional form at $Q^2 = Q_0^2$.

$$\begin{aligned} xg(x) &= A_g x^{B_g}(1-x)^{C_g}(1 + D_g x^{1/2} + E_g x) \\ x u_v(x) &= A_u x^{B_u}(1-x)^{C_u}(1 + D_u x^{1/2} + E_u x) \\ x d_v(x) &= A_d x^{B_d}(1-x)^{C_d}(1 + D_d x^{1/2} + E_d x) \\ x s(x) &= x[\overline{u}(x) + \overline{d}(x) + \overline{s}(x) + \overline{c}(x)] \\ &= A_s x^{B_s}(1-x)^{C_s}(1 + D_s x^{1/2} + E_s x) \end{aligned}$$

[a] u_v and d_v represent valence quark distribution functions.

Table 8.2 Parameter sets for MRS (A) [18] and H1 [19]. Initial values for MRS (A) are at $Q_0^2 = 4\,\text{GeV}^2$, and HI values at $Q_0^2 = 5\,\text{GeV}^2$. Values denoted by – are determined by boundary conditions.

	A_g	B_g	C_g	D_g	E_g	A_u	B_u	C_u	D_u	E_u
MRS (A)	0.775	−0.3	5.3	0	5.2	–	0.538	3.96	−0.39	5.13
H1	2.24	−0.2	8.52	0	0.0	2.84	0.55	4.19	4.42	−1.40

	A_d	B_d	C_d	D_d	E_d	A_s	B_s	C_s	D_s	E_s
MRS (A)	–	0.33	4.71	5.03	5.56	0.411	−0.3	9.27	−1.25	15.6
H1	1.05	0.55	6.44	−1.16	3.87	0.27	−0.19	1.66	0.16	−1.00

well approximated by

$$x^{B_g}(1 + D_g\sqrt{x} + E_g x) \sim \text{constant} \tag{8.130b}$$

In fact, the data in Figure 8.12 are well fitted by the function $xg(x) = 3(1-x)^5$. The data at HERA extends the available x down to 10^{-4}. In determining the parameters, one adjusts them to reproduce all or most of the available data, faithfully including photoproduction and the Drell–Yan processes. However, the functional form and which data should have more weights differ depending on the group's principle in making programs.

As examples from recent prescriptions, we show functions adopted by MRS (set A) [18] and H1 group [19] in Table 8.1 and their parameter values in Table 8.2.

Figure 8.16 shows calculated functions at $Q^2 = 20$ and $10^4\,\text{GeV}^2$ using up-to-date parameters [15]. An example of theoretical reproducibility using a NLO computer program is given in Figure 8.17.

The qualitative feature of the data can be understood as follows. As Q^2 grows, the virtual photon sees more and more gluons, which in turn creates $q\overline{q}$ pairs. This means the number of quarks in the large x region decreases while that of small x increases. This sort of behavior originates from multiemissions of the gluons. It is a common feature of gauge field theories. However, this amazing data reproducibility together with other evidences of the quark model firmly established the validity of QCD as the right theory of the quarks and gluons.

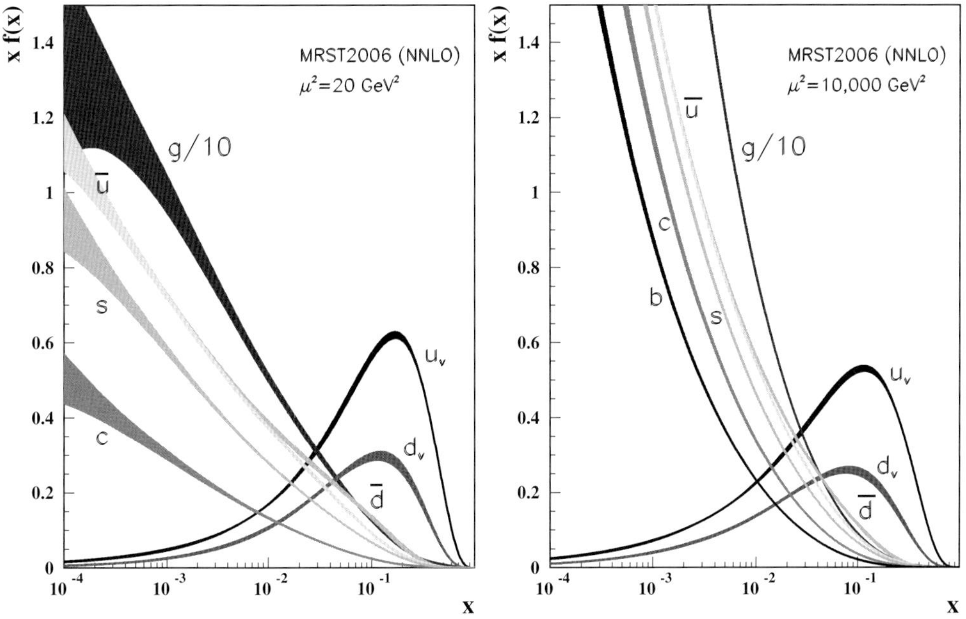

Figure 8.16 Parton distribution functions by MRS (A). Distributions of x times the unpolarized parton distributions $f(x)$ (where $f = u_v, d_v, u, d, s, c, b, g$) and their associated uncertainties using the NNLO MSTW2008 parameterization [15] at a scale $\mu^2 = 20\,\text{GeV}^2$ and $\mu^2 = 10\,000\,\text{GeV}^2$ [3]. For a color version of this figure, please see the color plates at the beginning of the book.

These distributions not only reproduce the deep inelastic scattering data, but also are useful in calculating and estimating the various hadron production cross sections including top quarks, W, Z and Higgs bosons in the TEVATRON and LHC. We shall see later in Chapter 11 that the precise fitting of the parton distribution function (PDF) at small x is very important in the reproducibility of high energy hadron–hadron reactions.

8.6 Drell–Yan Process

The Drell–Yan process (will be referred to as the DY) is a process to produce a heavy muon pair in hadronic reactions

$$p(p_A) + \overline{p}(p_B) \to \mu^-(p_c) + \mu^+(p_d) + X \tag{8.131}$$

and is a good test bench of the validity of the factorization we derived in the DGLAP equation. In the parton model, it is a process where a parton and an antiparton pair annihilates, is converted to a virtual photon creating muon antimuon pair.

$$q(p_1) + \overline{q}(p_2) \to \gamma^*(q) \to \mu^-(p_c) + \mu^+(p_d) \tag{8.132}$$

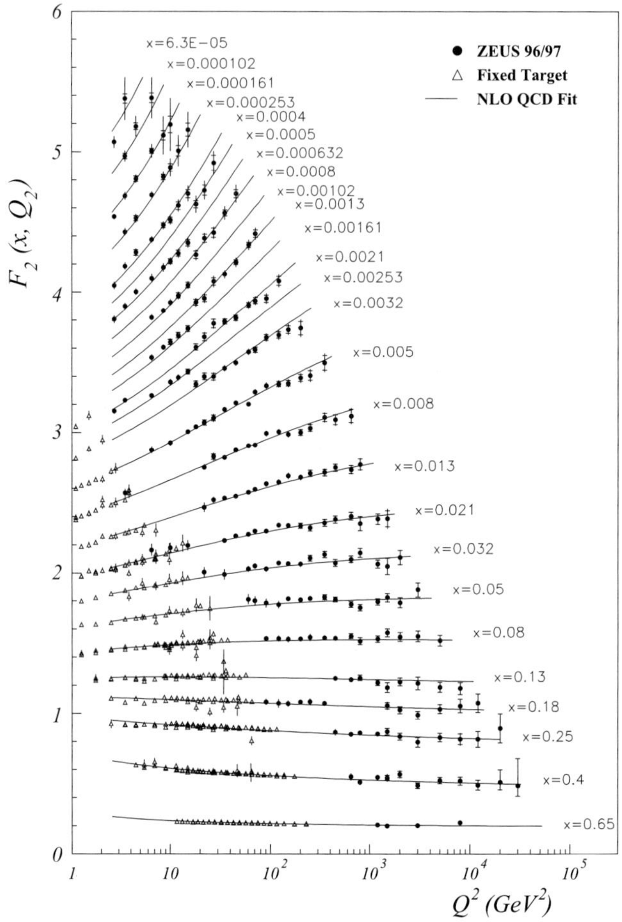

Figure 8.17 $F_2(x, Q^2)$ plotted as a function of Q^2 for different values of x. The QCD fitting to the data in Figure 8.17 uses NLO (Next Leading Order) approximation in which the splitting functions, for example, include order α_s^2 contributions. The data [3]are a compilation of those obtained from $e{-}P$ and $\mu{-}P$ scattering and includes those of ZEUS [407], NMC [423], BCDMS [415] and E665 [424] groups. The lines are NLO QCD fittings.

Since this is kinematically the same reaction as $e^-e^+ \to \mu^-\mu^+$, the cross section at the parton level is given by

$$\hat{\sigma}(q + \overline{q} \to \mu^-\mu^+; Q^2) = \frac{1}{3} \times \frac{1}{3} \times 3 \times \lambda_q^2 \hat{\sigma}(e^-e^+ \to \mu^-\mu^+) = \frac{\lambda_q^2}{3} \frac{4\pi\alpha^2}{3Q^2} \tag{8.133}$$

where $Q^2 = (p_1 + p_2)^2$ is the invariant mass squared of the $q\overline{q}$ system, λ_q the quark charge and α is the fine structure constant. The color factor is explained as follows: the muon pair is color neutral, and hence if the quark 1 has color "R", the antiquark 2 has to have $\overline{R}$. Probability of the quark or antiquark for having R or $\overline{R}$ is 1/3, and hence $1/3 \times 1/3$, and there are three kinds of colors giving the net factor 1/3. Hadronic cross sections for creating a muon pair are given by multiplying the parton cross section with the parton distribution functions $q(x), \overline{q}(x)$ in the hadron.

$$d\sigma(p + \overline{p} \to \mu\overline{\mu} + X; s) = \sum_i \hat{\sigma}(q_i + \overline{q}_i \to \mu\overline{\mu}; Q^2) q_i(x_1)\overline{q}_i(x_2) dx_1 dx_2 \tag{8.134}$$

The four-momenta of the incoming partons are taken to be

$$p_1 = x_1 p_A = \frac{\sqrt{s}}{2}(x_1, 0, 0, x_1) \quad p_2 = x_2 p_B = \frac{\sqrt{s}}{2}(x_2, 0, 0, -x_2) \tag{8.135}$$

where p_A and p_B are evaluated at the center of mass frame of the colliding $p\overline{p}$ (or pp) system and the hadron mass is neglected. The total energy of $q\overline{q}$ system is given by

$$Q^2 = (p_1 + p_2)^2 \simeq 2p_1 \cdot p_2 = 2x_1 p_A \cdot x_2 p_B \simeq x_1 x_2 (p_A + p_B)^2 = x_1 x_2 s \tag{8.136}$$

Using Eq. (8.136), the cross section Eq. (8.134) can be rewritten as

$$\begin{aligned} d\sigma(p + \overline{p} \to \mu\overline{\mu} + X; s) &= \frac{4\pi\alpha^2}{9Q^2} \\ &\times \int dx_1 dx_2 \delta(x_1 x_2 s - Q^2) dQ^2 \sum_i \lambda_i^2 \left[q_i(x_1)\overline{q}_i(x_2) + q_i(x_2)\overline{q}_i(x_1)\right] \end{aligned} \tag{8.137}$$

or concisely

$$Q^2 \frac{d\sigma}{dQ^2}(p + \overline{p} \to \mu\overline{\mu} + X)\bigg|_{Q^2 = \tau s} = \hat{\sigma}_0 \frac{dL}{d\tau}(\tau) \quad \hat{\sigma}_0 = \frac{4\pi\alpha^2}{9s} \tag{8.138a}$$

$$\frac{dL}{d\tau}(\tau) = \int dx_1 dx_2 \delta(x_1 x_2 - \tau) \sum \lambda_i^2 [q_i(x_1)\overline{q}_i(x_2) + q_i(x_2)\overline{q}_i(x_1)] \tag{8.138b}$$

where λ_i is the electric charge of the ith quarks. $dL/d\tau$ is the parton luminosity provided by the incoming proton and antiproton. This is the expression that the parton model gives.

What we are going to discuss here is how Eq. (8.138) is modified when the QCD higher order effect is taken into account. Their Feynman diagrams are depicted in Figure 8.18. One immediately recognizes their similarity with those of the DIS process. In fact, the DY process can be obtained from DIS by crossing. The difference being the DY photon is timelike ($q^2 = Q^2 = \hat{s} > 0$) as compared to the spacelike $q^2 = -Q^2 < 0$ in DIS process. As the spacetime structure is identical except those modified by crossing, the QCD correction due to Figure 8.18e–h produces similar corrections to the parton distribution functions. Just like the DIS case, the same mass singular term of the form $\ln(Q^2/m^2)$ appears with the same coefficient. The difference lies in that in the DY process, there are two partons in the initial state and product of two parton distributions has to be taken. According to calculations, including the virtual gluon contribution of Figure 8.18b–d [397, 425–427], the expression after renormalization is given by

$$\begin{aligned}\frac{dL}{d\tau} = \int dx_1 dx_2 dz \delta(x_1 x_2 z - \tau) \sum \lambda_i^2 \Big[&\{q_i(x_1)\overline{q}_i(x_2) + (1 \leftrightarrow 2)\} \\ &\times \left\{\delta(1-z) + \frac{\alpha_s}{2\pi} 2P_{q\leftarrow q}(z)\left(\ln\frac{Q^2}{\mu^2}\right) + \frac{\alpha_s}{2\pi} DY_q(z)\right\} \\ &+ \{g(x_1)\left(q_i(x_2) + \overline{q}_i(x_2)\right) + (1 \leftrightarrow 2)\} \\ &\times \left\{\frac{\alpha_s}{2\pi} 2P_{q\leftarrow g}(z)\left(\ln\frac{Q^2}{\mu^2}\right) + \frac{\alpha_s}{2\pi} DY_g(z)\right\}\Big]\end{aligned} \tag{8.139}$$

$P_{q\leftarrow q}(z)$, $P_{q\leftarrow g}(z)$ are the same splitting functions which appeared in the DIS process. μ^2 is the factorization scale factor that appears as the mass singularity $\ln(Q^2/m^2)$ is renormalized in the definition of the quark distribution function $q(x) \to q(x, Q^2)$. The factor 2 in front of $P_{q\leftarrow q}(z)$ originates from the fact that there are two initial quarks that can emit a gluon. Terms other than the dominant LLA contributions are contained in the DY function. Note that the dominant LLA terms are the same as in DIS, but subdominant terms including the DY functions differ from corresponding terms in the DIS process. In order to clarify the scale dependence, we fold $\ln(Q^2/\mu^2)$ parts into the distribution functions $q(x)$, and rewrite $q(x) \to q(x, \mu)$, $\alpha_s \to \alpha_s(\mu)$ in Eq. (8.139). Then, we have

$$\begin{aligned}\frac{dL}{d\tau} = &\int dx_1 dx_2 dz \delta(x_1 x_2 z - \tau) \\ &\times \sum \lambda_i^2 \Big[\{q_i(x_1,\mu)\overline{q}_i(x_2,\mu) + (1 \leftrightarrow 2)\}\left\{\delta(1-z) + \frac{\alpha_s}{2\pi} f_q(z)\right\} \\ &+ \{g(x_1,\mu)(q_i(x_2,\mu) + \overline{q}_i(x_2,\mu)) + (1 \leftrightarrow 2)\}\left\{\frac{\alpha_s}{2\pi} f_g(z)\right\}\Big]\end{aligned} \tag{8.140a}$$

$$\begin{aligned}f_q(z) &= DY_q - 2DI_q \\ &= C_F\Big[\delta(1-z)\left(1 + \frac{4\pi^2}{3}\right) - 6 - 4z + \left(\frac{3}{1-z}\right)_+ \\ &\quad + 2(1+z^2)\left(\frac{\ln(1-z)}{1-z}\right)_+\Big]\end{aligned} \tag{8.140b}$$

$$\begin{aligned}f_g(z) &= DY_g - DI_g \\ &= T_R\left[\{z^2 + (1-z)^2\}\ln(1-z) + \frac{3}{2} - 5z + \frac{9}{2}z^2\right]\end{aligned} \tag{8.140c}$$

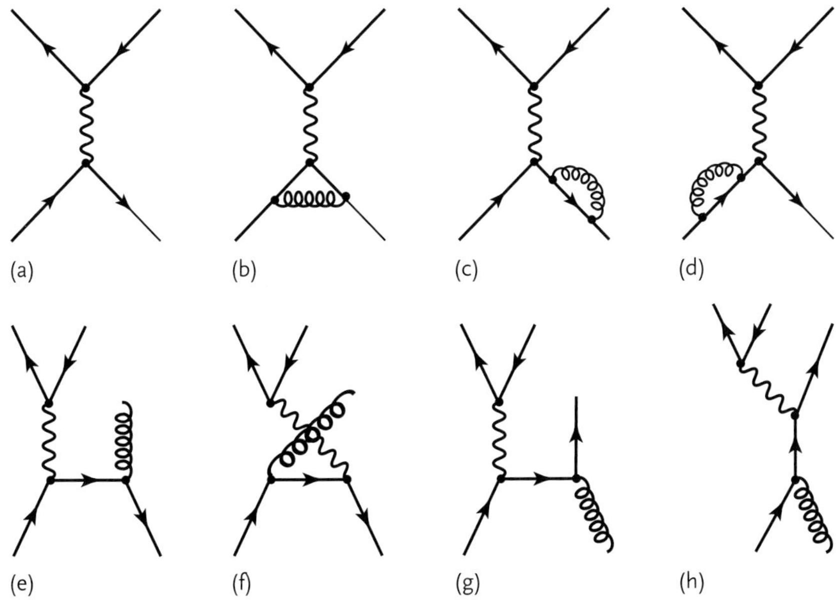

Figure 8.18 Drell–Yan process: $p\overline{p} \rightarrow \mu\overline{\mu} + X$. (a) Tree diagram which is the parton model. (b–d) are virtual effects of order α_s^2. Their interference with the tree process cancels the infrared divergence in the radiative processes (e)–(h) of order α_s.

where DI_is are the functions in DIS process corresponding to DY_is.[13] This shows that the QCD corrections which are soft or collinear can be brought into the scale dependent quark distributions and the hard part can be added to Eq. (8.138a) as corrections to the cross section. The QCD corrected formula of Eq. (8.138) can be expressed as

$$\frac{d\sigma}{d\tau} = \hat{\sigma}\left(q\overline{q} \rightarrow \mu\overline{\mu} : \mu_{\mathrm{F}}^2\right) \frac{dL}{d\tau}\left(\tau, \mu_{\mathrm{F}}^2\right), \quad \tau = \frac{\hat{s}}{s} \tag{8.141a}$$

$$\hat{\sigma}\left(q\overline{q} \rightarrow \mu\overline{\mu} : \mu_{\mathrm{F}}^2\right) = \frac{4\pi\alpha^2}{9\hat{s}}(1 + O(\alpha_s) + \cdots) \tag{8.141b}$$

$$\frac{dL}{d\tau}(\tau, \mu_{\mathrm{F}}^2) = \int dx_1 dx_2 \delta(x_1 x_2 - \tau) \times \sum \lambda_i^2 \left[q_i\left(x_1, \mu_{\mathrm{F}}^2\right)\overline{q}_i(x_2, \mu_{\mathrm{F}}^2) + q_i\left(x_2, \mu_{\mathrm{F}}^2\right)\overline{q}_i\left(x_1, \mu_{\mathrm{F}}^2\right)\right] \tag{8.141c}$$

In other words, we have shown that the factorization works in the Drell–Yan process.

13) Expressions here are calculated in the DIS scheme mentioned in Sect. 8.4.3.

K-Factor One notices that the pair annihilation cross section contains a term proportional to $\delta(1-z)$ in the nondominant parts. This term does not change the parton distribution derived in LLA approximation, but changes the overall normalization by

$$K_{\text{DIS}} = 1 + \frac{\alpha_s}{2\pi} C_F \left(1 + \frac{4\pi^2}{3}\right) \tag{8.142}$$

It is large and positive. Note that the exact value depends on the renormalization scheme and NLLA corrections. For instance, in $\overline{MS}$ scheme, it gives

$$K_{\overline{MS}} = 1 + \frac{\alpha_s}{2\pi} C_F \left(\frac{2}{3}\pi^2 - 8\right) \tag{8.143}$$

which is much smaller. The rest of f_q and the Compton term f_g give an overall negative contribution. Consequently, we combine the first and the second order effect and introduce a "K" factor which is defined by

$$\hat{\sigma}_0 \to \hat{\sigma}(\text{QCD}) = K(s, Q^2)\hat{\sigma}_0(1 + O(\alpha_s) + \cdots) \tag{8.144}$$

It is approximately constant in the range, that is, $K \sim K_{\text{DIS}}$ for $\tau = 0.2 \sim 0.7$. In the low energy fixed target experiments ($Q \approx 20$–30) GeV, the K factor is sizable ($K = 1.5 \sim 2.3$). Experimental data are consistent with this value as, for example, shown in Figure 8.19.

The reason for the large K-factor arises from the soft gluon emission which is analytically continued from the spacelike region ($Q^2 < 0$) to timelike region ($Q^2 > 0$). It was discussed in Sect. 7.3.4 that the soft gluon emission probability in the limit $p_T^2 \to 0$ has double logarithmic dependence ($S(p_T^2) \sim -(\alpha_s/2\pi)C_F \ln^2(p_T^2)$ in Eq. (7.146)). The corresponding term in DIS has negative $\sim q^2 = -Q^2$ which produces extra terms as follows

$$\begin{aligned}
\ln(-Q^2) &= \ln(Q^2) - i\pi \\
\ln^2(-Q^2) &= \ln^2(Q^2) - 2i\pi\ln(Q^2) - \pi^2
\end{aligned} \tag{8.145}$$

The contribution of the last π^2 term to K_{DIS} is $(\alpha_s/2\pi)C_F\pi^2$ which is a large part of K_{DIS}. Since, in the soft limit, $\alpha_s \ln^2 p_T^2$ is $\sim O(1)$, the double logarithmic term has to be resummed and incorporated as an exponential which is referred as the Sudakov form factor [431] (see also discussions in Vol. 1, Sect. 8.1.4). Therefore, this part of the K factor should be exponentiated to give

$$K = \exp\left(\frac{\alpha_s}{2\pi} C_F \pi^2\right)(1 + B\alpha_s + \cdots) = \exp\left(\frac{2\alpha_s}{3}\pi\right)(1 + B\alpha_s + \cdots) \tag{8.146}$$

The exponent stays there if further higher order corrections are introduced. In the fixed target energy region where $\alpha_s \sim 0.3$, it gives a correction factor ~ 2.

In the high energy collider experiments, such as TEVATRON ($\sqrt{s} = 1.8\,\text{TeV}$), where the value of τ is much smaller, the K-factor diminishes because of negative

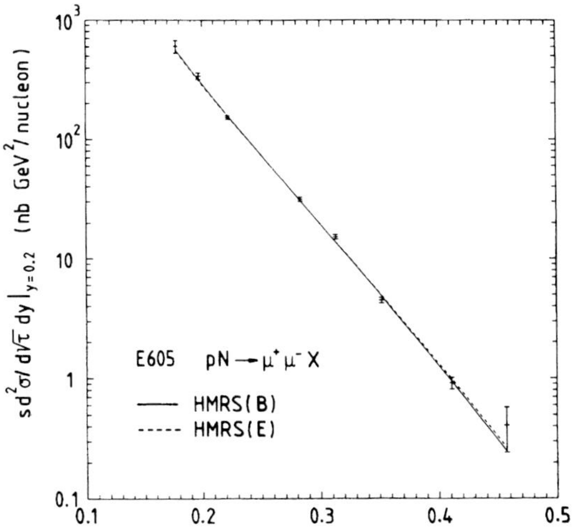

(a)

Experiment	Interaction	Beam Momentum	$K = \sigma_{\text{meas.}}/\sigma_{\text{DY}}$
E288 [Kap 78]	$p\ Pt$	300/400 GeV	~ 1.7
WA39 [Cor 80]	$\pi^{\pm}\ W$	39.5 GeV	~ 2.5
E439 [Smi 81]	$p\ W$	400 GeV	1.6 ± 0.3
NA3 [Bad 83]	$(\bar{p}$ - $p)Pt$	150 GeV	2.3 ± 0.4
	$p\ Pt$	400 GeV	$3.1 \pm 0.5 \pm 0.3$
	$\pi^{\pm}\ Pt$	200 GeV	2.3 ± 0.5
	$\pi^{-}\ Pt$	150 GeV	2.49 ± 0.37
	$\pi^{-}\ Pt$	280 GeV	2.22 ± 0.33
NA10 [Bet 85]	$\pi^{-}\ W$	194 GeV	$\sim 2.77 \pm 0.12$
E326 [Gre 85]	$\pi^{-}\ W$	225 GeV	$2.70 \pm 0.08 \pm 0.40$
E537 [Ana 88]	$\bar{p}\ W$	125 GeV	$2.45 \pm 0.12 \pm 0.20$
E615 [Con 89]	$\pi^{-}\ W$	252 GeV	1.78 ± 0.06

(b)

Figure 8.19 (a) Drell–Yan process fitted with next-to-leading order QCD prediction [16, 428]. (b) Compilation of experimental K-factors [429, 430].

contributions from the quark-gluon scattering ($f_g(z)$ term in Eq. (8.140c)) and the lowest order cross section increases. For W, Z production, it increases by about $\sim 20 \sim 30\%$. Experimentally, it is generally considered as an adjustable normalization factor.

In summary, the factorization also works for the Drell–Yan process. The higher order corrections can be separated into both a soft and hard part, and the soft part can be renormalized as the scale dependent quark distribution and the hard part can be added to the cross section as corrections. It is worth noticing that the distribution $q(x,\mu)$ in the deep inelastic scattering ($\mu^2 = -q^2 = Q^2$) and that in the Drell–Yan process ($\mu^2 = q^2 = Q^2$) have exactly the same form in the LLA approximation.

8.6.1 Factorization in Hadron Scattering

Extension of the treatment applied to the Drell–Yan process to general hadron reactions is straightforward. At the parton level, the cross section of the parton scattering process $a + b \to c + X$ is described as

$$\sigma(AB \to cX) = \int dx_a dx_b \Big[f_A{}^a(x_a) f_B{}^b(x_b) + (A \leftrightarrow B,\ \text{if } a \neq b) \Big] \hat{\sigma}_0(ab \to cX) \tag{8.147}$$

Here, $\hat{\sigma}_0(ab \to cX)$ denotes the parton scattering cross section. It is to be understood that the cross section is averaged over the initial and summed over the final color states. Please note that the parton model is valid only for hard processes ($Q^2 \gg M^2$), where M is a typical hadron mass. Equation (8.147) can be pictorially expressed as Figure 8.20.

The factorization theorem means that the QCD corrected cross section can be expressed by a formula

$$\begin{aligned} \sigma(AB \to cX) &= \int dx_a dx_b \Big[f_A{}^a(x_a,\mu) f_B{}^b(x_b,\mu) + (A \leftrightarrow B,\ \text{if } a \neq b) \Big] \\ &\quad \times \hat{\sigma}(ab \to cX; Q, \mu) \\ \hat{\sigma}(ab \to cX; Q, \mu) &= \hat{\sigma}_0 \Big[1 + \sum_j c_j \alpha_s^j \Big] \end{aligned} \tag{8.148}$$

where $f_A^a(x,\mu)$ includes the soft part of the QCD correction, namely, it is the distribution function evolved to μ á la DGLAP and the hard part is included in c_js. The hard part can be calculated using perturbation theories and it does not depend on the details of the hadron wave function or the type of incoming particle. The factorization is a fundamental property of the theory that makes QCD a reliable calculational tool.

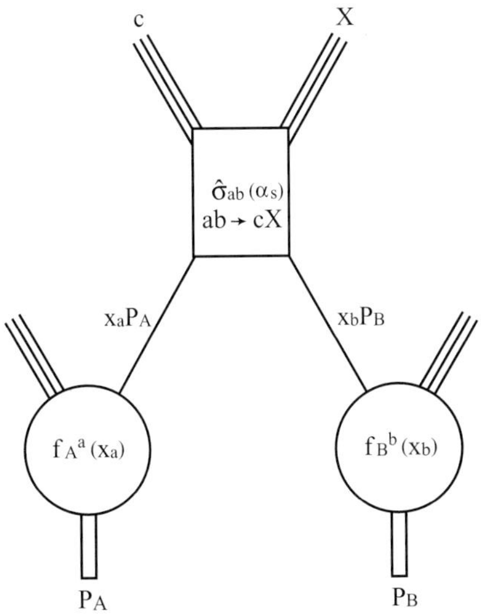

Figure 8.20 Schematic of the parton model description of a hard scattering process. A parton "$a(b)$" in the hadron $A(B)$ having momenta $x_a P_A (x_b P_B)$ reacts to make the $a + b \to c + X$ reaction.

Factorization Scale In separating the collinear part from x dependent part, we had to deal with the mass singularity and introduced the factorization scale $\mu = \mu_F$. This μ_F is different from the renormalization scale $\mu = \mu_r$ which regularizes ultraviolet singularities. It can be set independently regardless of the renormalization scale, provided it does not exceed Q^2, the typical energy scale in consideration. Practically, $\mu_F = \mu_r = Q^2$ is adopted in most cases. The choice of which variable one should adopt for the scale factor is somewhat arbitrary. Conventionally, the invariant mass of the muon pair is used for the Drell–Yan process, the mass for W or heavy quark production, total energy s for the $e^- e^+$ reaction or p_T^2 for the jet production reactions. There is no universal rule as to which variable is to be adopted for setting the scale.

Classical Argument of the Factorizability Intuitively, whether the notion of scale dependent parton distribution developed in the deep inelastic scattering can be extended to the hadron–hadron collisions is not self-evident. When the lepton is the probe, one can imagine that the parton distribution maintains its distribution until the hard process (at large Q^2) occurs. In the case of hadron–hadron scattering, however, the probe being also a parton, it has long range interactions and the emission of soft gluons which modifies the distribution in the hadron A would also affect that of hadron B before the collision occurs. We can justify use of the scale dependent distribution in hadronic collisions using a simple classical argument [432].

The electromagnetic vector potential at $x^\mu = (t; \boldsymbol{r})$ made by a charged particle at $x'^\mu = (t'; \boldsymbol{r}')$ moving with constant velocity $\boldsymbol{v}$ is given by the well-known Liénard–

Wiechert retarded potential

$$A^\mu = (\phi, A) = \frac{q}{4\pi}\left(\frac{1}{r - \boldsymbol{r}\cdot\boldsymbol{v}}, \frac{\boldsymbol{v}}{r - \boldsymbol{r}\cdot\boldsymbol{v}}\right) \tag{8.149a}$$

with the condition

$$t-t' = r = |\boldsymbol{r}-\boldsymbol{r}'(t')| = [(x-x'(t'))^2+(y-y'(t'))^2+(z-z'(t'))^2]^{1/2} \tag{8.149b}$$

which is obtained from

$$\begin{aligned} A^\mu &= \frac{1}{4\pi}\int d^4x' \frac{j^\mu(x')}{|\boldsymbol{x}-\boldsymbol{x}'(t')|}\delta(t-t'-|\boldsymbol{x}-\boldsymbol{x}'(t')|) \\ j^\mu(x) &= q\delta(\boldsymbol{x}-\boldsymbol{r}(t))(1;\boldsymbol{v}), \quad \boldsymbol{r}(t) = \boldsymbol{v}t \end{aligned} \tag{8.150}$$

If the incoming particle is moving along the x-axis ($\boldsymbol{r}(t) = (vt, 0, 0)$) and the target is at $\boldsymbol{x} = (0, y, 0)$, the potential is given by

$$\begin{aligned} \phi(x) &= \frac{q}{4\pi}\frac{\gamma}{\sqrt{\gamma^2(vt-x)^2+y^2+z^2}}, \\ A_x(x) &= \frac{q}{4\pi}\frac{\gamma v}{\sqrt{\gamma^2(vt-x)^2+y^2+z^2}}, \\ A_y(x) &= A_z(x) = 0 \end{aligned} \tag{8.151}$$

where $\gamma = 1/\sqrt{1-v^2} \sim s/m^2$. In the $\gamma \to \infty$ limit, the potential becomes a pure gauge

$$\lim_{\gamma\to\infty} A^\mu \to \partial^\mu q \ln(vt - x) \tag{8.152}$$

and produces no electric or magnetic filed. The electric field, for instance, E_x at finite γ, is given by

$$E_x(t, \boldsymbol{r}) = -\frac{\partial A_x}{\partial t} - \frac{\partial A_t}{\partial x} = -\frac{e\gamma(vt-x)}{[\gamma^2(vt-x)^2+y^2+z^2]^{3/2}} \tag{8.153}$$

This is a pulse force whose strength is $O(1/\gamma^2) \sim m^4/s^2$. Thus, the force experienced by a charge at any fixed time before the collision decreases as m^4/s^2, or conversely, the breakdown of the factorization at order $(m^2/s)^2$ is to be expected.

9
Jets and Fragmentations

9.1
Partons and Jets

We learned that, owing to the asymptotic freedom, perturbation formalism can be applied in solving QCD problems at high Q^2. However, in order to test theoretical predictions, one needs to make a correspondence between partons in QCD calculations and hadrons that are observed experimentally. The first observation of the parton as a jet was made in "$e^- + e^+ \rightarrow q + \overline{q}$" in 1975 [433, 434]. The jet is a phenomenon in which a bundle of hadrons are emitted more or less in the same direction. Since then, the jets have been observed in deep inelastic scatterings and proton-antiproton reactions. Nowadays, the partons with energy over 5 GeV are routinely observed as hadronic jets. In order to make a correspondence between the partons and jets, low energy ($Q^2 < 1\,\text{GeV}^2$) treatment of QCD is required. It is a region where the asymptotic freedom is ineffective and nonperturbative formalism is necessary. Although, in principle, it should be solvable from the fundamental QCD Lagrangian, one has to rely on models at the present stage of QCD development. In this chapter, we clarify characteristics of the jets by studying $e^- + e^+ \rightarrow q + \overline{q} + (g)$ and apply them to analyze hadron–hadron collisions in Chapter 11.

Observation of Jets The jet formation mechanism may be understood as follows. The potential that works between a quark and an antiquark has qualitatively the following form.

$$V(r) = -\frac{4}{3}\frac{\alpha_s}{r} + kr \tag{9.1}$$

The first term is a Coulomb potential similar to that of QED produced by exchange of massless gluons. The factor 4/3 originates from the color degrees of freedom. For distances much shorter than a typical hadron size ($\sim 1/m_\pi$), the first term dominates. The asymptotic freedom ensures the smallness of the coupling constant α_s and allows perturbative treatments in solving problems. The second term, which is referred to as the confining potential, has a nonperturbative origin and is believed to be a result of gluon self-couplings. It produces a force that does not reduce with distance. The quark and antiquark are, in a sense, tied together by

Elementary Particle Physics, Volume 2: Foundations of the Standard Model, First Edition. Yorikiyo Nagashima.

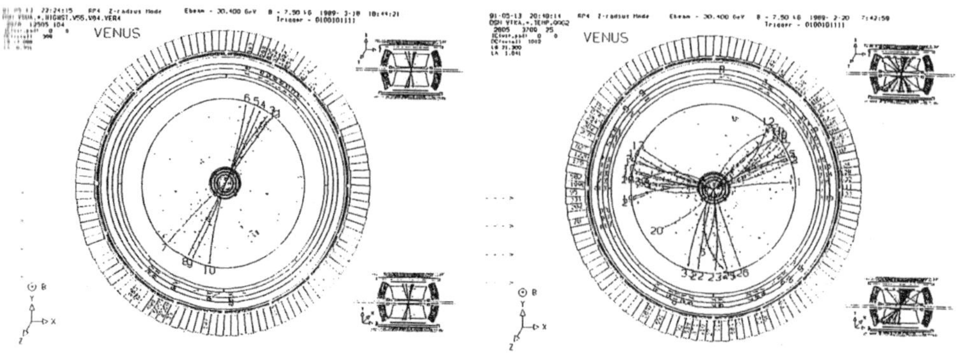

Figure 9.1 Two-jet and three-jet event displays at VENUS/TRISTAN.

a string which requires infinite energy to separate them. This is the reason why isolated quarks are never found, a phenomenon referred to as the confinement.

When a quark pair ($q\overline{q}$) is produced in the center of frame of a e^-e^+ system, they fly apart in opposite directions with equal momentum. Because of the confining potential, a large amount of energy is stored on a line connecting the pair (formation of a string due to color confinement). On the other hand, the vacuum is not tranquil, creating and annihilating pairs of particles and antiparticles. Some of them appear on the string and can make a color singlet with one member of the pair forming a hadron. When the distance between the pair exceeds $\sim 1\,\text{fm}\,(10^{-15}\,\text{m})$, the stored energy exceeds the mass energy of typical hadrons and the creation of an extra hadron is energetically favored over that of stretching the string further. Many hadrons are created in this way on the line between the original $q\overline{q}$ pair. They are dragged by their parent q or $\overline{q}$, and are emitted in concentration in the opposite directions which are observed as jets.

Figure 9.1 shows examples of 2- and 3-jets observed in the VENUS detector at TRISTAN [435, 436]. One observes that many hadrons are emitted in two or three bunches. The collective nature of the jet is more apparent if one observes them not as tracks, but as energy flow. Figure 9.2 shows an event display of a hadronic reaction at much higher energy. In Figure 9.2b, one observes that a large amount of energy is deposited in the electromagnetic calorimeter. That they represent a two body final state in a back-to-back scattering configuration is quite apparent in the so-called Lego plot in Figure 9.2a. Here, a three-dimensional display of the energy deposit is plotted on the η $(= \ln \cot \theta/2) - \phi$ (azimuthal angle around the beam axis) plane.

In order to reconstruct jet phenomena, it is necessary to measure tracks, energy-momenta and possibly to identify the particle species of every component in the jet. Collider detectors capable of observing jets have many features in common, namely, they are of general purpose, hermetic (meaning, they cover an almost full 4π solid angle), uniform response over the entire solid angle, fine modularity and so on (see Sect. 4.3 or Vol. 1, Sect. 12.6 for details). Examples of collider detectors were already shown (see Figure 3.1 for UA1, Figure 4.6 for SLD and Figure 6.20 for BELLE).

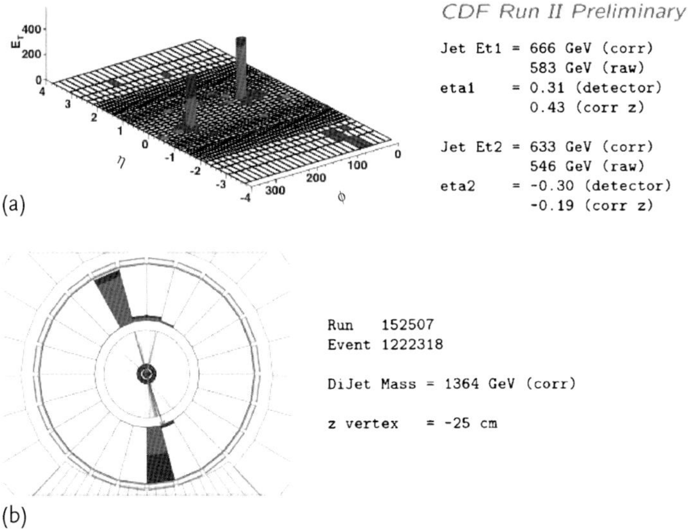

Figure 9.2 Event display of 2-jets in hadronic reactions with mass 1.36 TeV. Energies (666 and 633 GeV) of the two jets are shown as two columns (a) or two colored bars (b). The pink bars correspond to energy collected by the electromagnetic calorimeter and the blue bars to the hadronic calorimeter. Their angle positions ($\eta = \ln \cot \theta/2$ and ϕ clearly indicate that they are emitted in back-to-back configuration). Figure from CDF/FNAL [22]. For a color version of this figure, please see the color plates at the beginning of the book.

9.1.1 Fragmentation Function

Jets are considered to have the following properties.

1. If the jet is an end result of a parton, its form is determined by intrinsic properties of its parent and does not depend on its production mechanism of the parent particle (i.e., $e^- e^+ \to hX, ep \to ehX, p\overline{p} \to hX$).
2. Every hadron in a jet is basically emitted in the same direction. Longitudinal momenta ($p_{\|}$) of the offsprings would increase in proportion to their parent's momentum, while transverse momenta (p_{T}) would be limited by the intrinsic motion of the parent's primordial motion in the hadron (which is $\sim$ 300 MeV) as observed in the transverse momentum distribution of hadronic reactions (see Vol. 1, Fig. 13.30).

From above considerations, it is inferred that the momentum distributions of hadrons in the hadronization of the partons (referred to as fragmentation distribution[1]) obeys the scaling law in the approximation to neglect the particle mass

1) The words fragmentation and hadronization are often used interchangeably, however there are subtle difference between the two. The fragmentation is a hadronization of a parton with high virtuality producing jets. The hadronization is more general, and denotes a process whatever converts partons to hadrons.

and p_T. In other words, the hadron distribution in the quark $\to$ jets is expected to only depend on the scaling variable $z = \frac{p_h}{|p|}$ where momenta of the parton and its offspring hadron are denoted as p and $p_h(h = \pi^\pm, K, p, \overline{p}$ and so on.). We define the fragmentation distribution function denoted as $D_q^h(z)dz$ as the number of particles in a momentum range $z \sim z + dz$. It can be expressed as

$$dN_h = D_q{}^h(z)dz \tag{9.2}$$

Then, the production cross section of a particle species "h" in the e^-e^+ reaction can be written as

$$\frac{d\sigma}{dz}(e^-e^+ \to hX) = \sum_q \sigma(e^-e^+ \to q\overline{q})\left[D_q{}^h(z) + D_{\overline{q}}{}^h(z)\right]$$

$$z = \frac{p_h}{|p|} \approx \frac{E_h}{E_{\text{CM}}} = \frac{2E_h}{\sqrt{s}} \tag{9.3}$$

where $\sqrt{s} = 2E_{\text{CM}}$ is the total energy. This is illustrated pictorially in Figure 9.3a. As

$$\sigma(e\overline{e} \to q\overline{q}) = N_c Q_q^2 \sigma(e\overline{e} \to \mu\overline{\mu}) \tag{9.4}$$

where Q_q, N_c $(= 3)$ are the quark charge and the color degrees of freedom, we can write

$$\frac{1}{\sigma}\frac{d\sigma}{dz}(e^-e^+ \to hX) = \frac{\sum Q_q^2\left[D_q{}^h(z) + D_{\overline{q}}{}^h(z)\right]}{\sum Q_q^2} \tag{9.5}$$

The right-hand side is a function of z only, and does not depend on s. In other words, the scaling should hold. A similar equation can be obtained for the deep inelastic scattering (DIS). Using the fact that the DIS cross section is a function of scaling variables $x = Q^2/2M\nu$ and $y = \nu/E_{\text{in}} = (E_{\text{in}} - E_{\text{out}})/E_{\text{in}}$ (see Eq. (8.13)), we can write

$$\frac{1}{\sigma}\frac{d\sigma(eN \to ehX)}{dz} = \frac{\sum_q Q_q^2 q(x) D_q{}^h(z)}{\sum_q Q_q^2 q(x)} \tag{9.6}$$

and expect the scaling in the fragmentation in the deep inelastic scatterings as well (Figure 9.3b). Here, $q(x)$ denotes the momentum distribution function of the parton in the hadron.

Figure 9.4a shows z distributions produced by e^-e^+ reaction for various values of $\sqrt{s}$ and Figure 9.4b shows those made by various reactions (e^-e^+, νN, pp). The variable z is denoted as x_p in Figure 9.4a and x_{ee} in Figure 9.4b. In Figure 9.4a, $s d\sigma/dx_p$ is plotted because $d\sigma/dx_p$ is proportional to s^{-1} as a function of s. One can see that the distributions satisfy the scaling law that is, they are functions of z only and do not depend on s or particle species confirming our conjecture. The broken scaling at small z (< 0.2) is considered as due to soft gluon emission and the effect of finite mass of the hadrons.

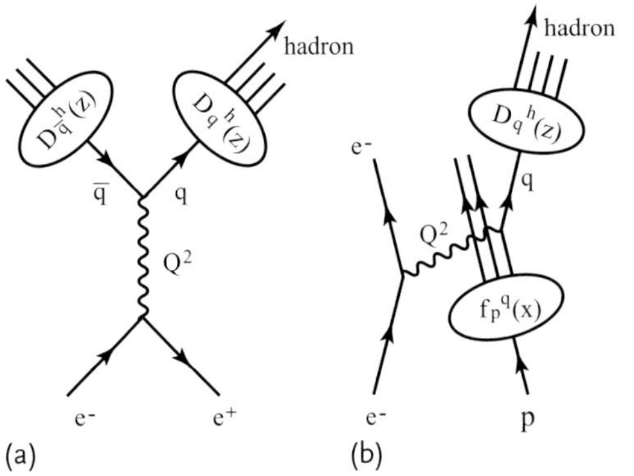

Figure 9.3 Schematic block diagrams of $e^-e^+ \to$ 2-jets and deep inelastic scattering producing a jet.

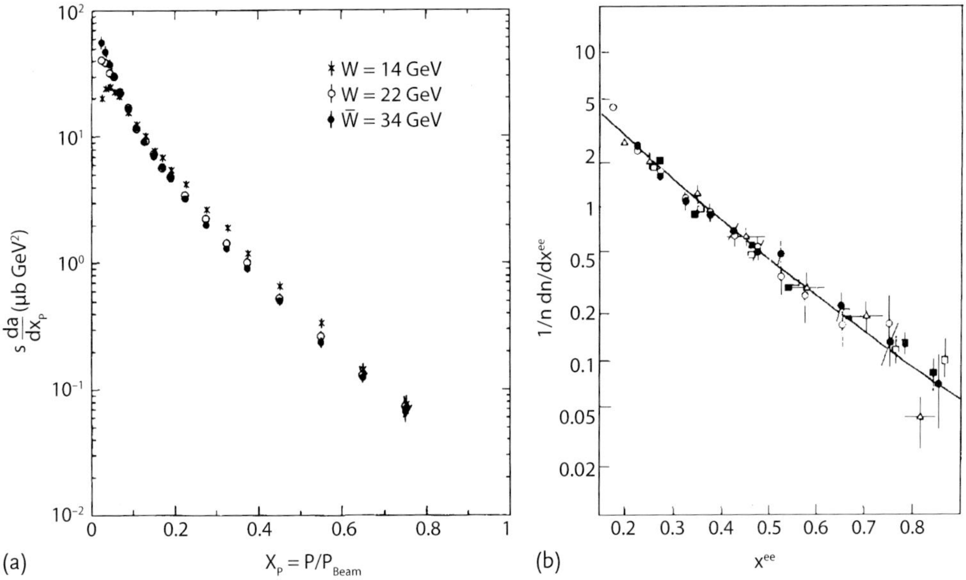

Figure 9.4 Scaling in fragmentation. (a) Data at different energies: $s(d\sigma/dx_p)(e^-e^+ \to h^+X)$, $x_p = 2p/\sqrt{s}$, $\sqrt{s} = 14, 22,$ 34 GeV [437]. (b) Data obtained from different reactions ($\nu - p(\Delta)$, pp (●: CS $p_T < 6$, ○: CS $p_T > 6\,\text{GeV}/c$, ■: CCOR $p_T > 5$, □: CCOR $p_T > 7\,\text{GeV}/c$. They show x dependence of the cross sections independent of the total energy or production reactions. The line denotes data obtained from e^-e^+ reactions [438].

The momentum conservation requires

$$\frac{\sum_h \int dz\, z\, D_q{}^h(z)}{\sum_h \int dz\, D_q{}^h(z)} = 1 \tag{9.7}$$

Isospin independence and charge symmetry require, for instance, that the probability for a u quark for picking up $\bar{d}$ from vacuum ($d\bar{d}$) to make π^+ and that of d picking up $\overline{u}$ to make π^- are the same. Therefore, we expect

$$D_u^{\pi^+} = D_d^{\pi^-} = D_{\overline{u}}^{\pi^-} = D_{\overline{d}}^{\pi^+} \tag{9.8a}$$

$$D_d^{\pi^+} = D_u^{\pi^-} = D_{\overline{d}}^{\pi^-} = D_{\overline{u}}^{\pi^+} \tag{9.8b}$$

$$D_s^{\pi^+} = D_s^{\pi^-} = D_{\overline{s}}^{\pi^-} = D_{\overline{s}}^{\pi^+} \tag{9.8c}$$

$$D_q^{\pi^0} = (D_q^{\pi^+} + D_q^{\pi^-})/2 \tag{9.8d}$$

For the case of neutrino deep inelastic scatterings, the formula corresponding to Eq. (9.6) becomes simple.

$$\frac{1}{\sigma}\frac{d\sigma}{dz}(\nu p \to \mu^- h X) = \frac{d(x) D_d{}^h(z) + (1-y)^2 \overline{u}(x) D_{\overline{u}}{}^h}{d(x) + (1-y)^2 \overline{u}(x)} = D_d{}^h(z) \tag{9.9}$$

where we have neglected the s and c quark components in the nucleon and used Eq. (9.8) in going from the second to the third equality. Similarly,

$$\frac{1}{\sigma}\frac{d\sigma}{dz}(\overline{\nu} p \to \mu^+ h X) = D_u{}^h(z) \tag{9.10}$$

Thus, the fragmentation functions can be directly determined from experiments.

Parton Shower Model for the Fragmentation Function The fragmentation process can be interpreted as a cascade phenomenon of parton splittings. There is a nice parallelism between the parton distribution function in the DIS and the fragmentation function in the jet production. In fact, considering the fact that gluon emission by partons in the final state of $e^- e^+$ reaction has the same splitting function in LLA (see Eq. (7.143a)) as that of the initial parton in DIS (Eq. (8.44)), the same arguments that led to the DGLAP evolution equation show that the fragmentation function obeys the same DGLAP evolution equation with the same splitting function in LLA. The difference lies in the virtuality of the fragmenting objects being timelike as opposed to the spacelike virtuality of the DIS. In the DIS, virtuality of the parton (say, the parton mass) starts from zero (i.e., free parton) and reaches Q^2 when the reaction occurs (see Eq. (8.66) and arguments following it). On the other hand, in the fragmentation, the parton is produced at high virtuality which reduces its value as it splits in cascade. Eventually, at low virtuality ($Q \sim 1\,\text{GeV}$), the partons are converted to hadrons. We shall come back to the parton shower model and other hadronization models later, but we will discuss properties of jets and how to separate them first.

9.1.2 Jet Shape Variables

Thrust and Sphericity In order to separate jets and investigate their characteristics, it is convenient to define a few variables referred to as topology or jet shape variables. In the high energy $e^- e^+$ reactions, light quark pairs ($q\overline{q}$: $q = u, d, s$) would generate a pair of jets. In this case, particles would be produced in back-to-back configuration. On the other hand, if a hard gluon is emitted, 3-jet configuration is expected. Since $q\overline{q}g$ are in a plane, we expect a flat particle distribution. Variables to characterize these properties include "thrust T" and "acoplanarity A" [439, 440] which are defined as

$$T = \max\left[\frac{\sum_i |\boldsymbol{p}_i \cdot \boldsymbol{e}_{\mathrm{T}}|}{\sum_i |\boldsymbol{p}_i|}\right] \tag{9.11a}$$

$$A = 4\min\left[\frac{\sum_i |\boldsymbol{p}_i \cdot \boldsymbol{e}_{\mathrm{A}}|}{\sum_i |\boldsymbol{p}_i|}\right] \tag{9.11b}$$

where $\boldsymbol{p}_i$ is the momentum of the ith particle in an event. $T(A)$ is maximized (minimized) by moving the axis vectors $\boldsymbol{e}_{\mathrm{T}}(\boldsymbol{e}_{\mathrm{A}})$. $\boldsymbol{e}_{\mathrm{T}}$ is referred to as the thrust axis. The thrust T gives a measure of how the particle directions are aligned along the jet (i.e., thrust) axis. When all the particles point to the same direction (pencil-like configuration), $T = 1$, and when their distribution is isotropic (sphere like configuration), $T = 0.5$. Contrary to the thrust, the acoplanarity A gives a degree of swelling and becomes $A = 1(0)$ for perfect isotropic (flat) distribution.

Problem 9.1

Show $T = 0.5$ and $A = 1$ for perfectly isotropic distribution.

After the thrust axis is determined, the major axis $\boldsymbol{e}_2$ perpendicular to the thrust axis in the decay plane is defined in such a way to maximize F_{major}

$$F_{\mathrm{major}} = \max\left[\frac{\sum_i |\boldsymbol{p}_i \cdot \boldsymbol{e}_2|}{\sum_i |\boldsymbol{p}_i|}\right] \tag{9.12}$$

and the third axis (minor axis) $\boldsymbol{e}_3$ perpendicular to both $\boldsymbol{e}_{\mathrm{T}}$ and $\boldsymbol{e}_2$. A quantity F_{minor} is defined as

$$F_{\mathrm{minor}} = \left[\frac{\sum_i |\boldsymbol{p}_i \cdot \boldsymbol{e}_3|}{\sum_i |\boldsymbol{p}_i|}\right] \tag{9.13}$$

Sometimes the "oblateness"

$$O \equiv F_{\mathrm{major}} - F_{\mathrm{minor}} \tag{9.14}$$

is used instead of acoplanarity.

Q-plot and C-parameter Other variables that are frequently used are eigenvalues $Q_j(Q_1 < Q_2 < Q_3)$ of the momentum tensor [441]

$$I^{ij} = \frac{\sum_a p_a^i p_a^j}{\sum_a |\boldsymbol{p}_a|^2} \tag{9.15}$$

and corresponding normalized eigen vectors $(\boldsymbol{r}_1, \boldsymbol{r}_2, \boldsymbol{r}_3)$ which are illustrated in Figure 9.5a. They are no longer used in the actual analyses because of infrared instability, but we discuss it anyway because it is used in early jet analyses and also serves as a nice introduction to conceptual understanding of a more conventional C-parameter. Q_js are normalized according to

$$Q_j = \frac{\sum_a p_{aj}^2}{\sum_a (\boldsymbol{p}_a \cdot \boldsymbol{p}_a)}, \quad Q_1 + Q_2 + Q_3 = 1 \tag{9.16}$$

This is easily seen if expressed in matrix form $\boldsymbol{I}\boldsymbol{r}_j = Q_j \boldsymbol{r}_j, \mathrm{Tr}[\boldsymbol{I}] = \sum_j I^{jj} = 1$. One can define two kinds of transverse momentum with respect to the jet axis, $p_{\mathrm{T\,in}}$ in the reaction plane and $p_{\mathrm{T\,out}}$ perpendicular to it.

$$\langle p_{\mathrm{T\,in}}{}^2 \rangle = Q_2 \frac{\sum_a (\boldsymbol{p}_a \cdot \boldsymbol{p}_a)}{N} \tag{9.17a}$$

$$\langle p_{\mathrm{T\,out}}{}^2 \rangle = Q_1 \frac{\sum_a (\boldsymbol{p}_a \cdot \boldsymbol{p}_a)}{N} \tag{9.17b}$$

where N is the total number of particles in one event.

Historically, a variable "sphericity S" was proposed first [440].

$$S = \frac{3}{2} \min \left[\frac{\sum_a |\boldsymbol{p}_{\mathrm{T}a}|^2}{\sum_a |\boldsymbol{p}_a|^2} \right] \tag{9.18}$$

Minimization is done by moving the axis to define $\boldsymbol{p}_\mathrm{T}$. The sphericity S is related to Q_js by the relation

$$S = \frac{3}{2}(Q_1 + Q_2), \quad 0 \le S \le 1 \tag{9.19}$$

$S = 1$ for isotropic and $S = 0$ for parallel jet configurations. The eigen vector $\boldsymbol{r}_3$ with the eigenvalue Q_3 is referred as the jet or sphericity axis. "Planarity" defined by $A_P \equiv 3Q_1/2(0 < A_P < 0.5)$ is a variable similar to the acoplanarity. Because of Eq. (9.16), a set of (Q_1, Q_2, Q_3) in a jet event represents a point in a shaded triangle ABC in Figure 9.5b. A scatter plot of Q_js in jet events are called the Q-plot. A quick observation of Q-plots, whether they are concentrated close to apexes A or B or C in Figure 9.5c, immediately tells us the jet's topological structure.

At high energies where the asymptotic freedom is effective and when the total energy $\sqrt{s}$ in $e^- e^+ \to q\overline{q}$ reaction is large compared with the mass of the quark q, the $q\overline{q}$ pair is expected to appear as two jets in back-to-back configuration. The

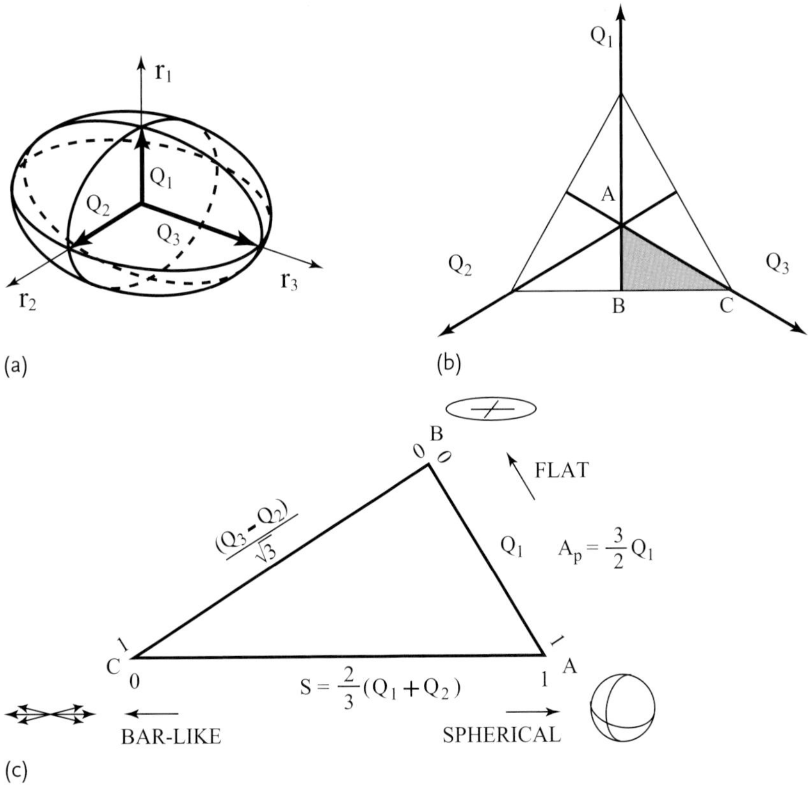

Figure 9.5 Illustration of Q-plot and sphericity. (a) Eigen vectors $\boldsymbol{r}_i$ and corresponding eigenvalues Q_j ($Q_1 \le Q_2 \le Q_3$) of the momentum tensor. (b) As $Q_1 + Q_2 + Q_3 = 1$, a set of (Q_1, Q_2, Q_3) gives a point in a triangle ABC. (c) Q-plot: Choosing two axes $Q_1 \parallel AB$ and $(Q_3 - Q_2)/\sqrt{3} \parallel BC$ are defined. Points near A represent spherical, near B, flat and near C, a pencil-like (2-jet) event shape.

average thrust $\langle T \rangle$ would be close to one. Let us see how the thrust grows with energy. The thrust distribution to $O(\alpha_s)$ perturbation is given by

$$\frac{1}{\sigma_0}\frac{d\sigma}{dT} = C_F\frac{\alpha_s}{2\pi}\left[\frac{2(3T^2 - 3T + 2)}{T(1-T)}\ln\left(\frac{2T-1}{1-T}\right) - \frac{3(3T-2)(2-T)}{1-T}\right] \tag{9.20}$$

Derivation of Eq. (9.20) will be given later (Eq. (10.3a)). Here, we will limit our discussion to its qualitative feature. Figure 9.6a shows the average thrust data

$$\langle 1 - T \rangle = \frac{1}{\sigma_0}\int (1-T)\frac{d\sigma}{dT}dT \tag{9.21}$$

as a function of energy. Above 10 GeV, one can see the sharp increase of T, indicating the fast rise of the jet production rate as the energy increases. The data agrees

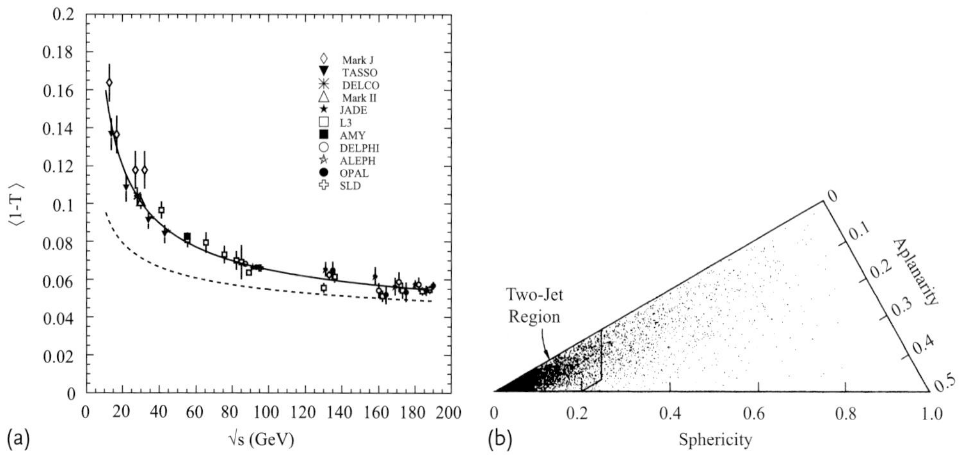

Figure 9.6 (a) The thrust as a function of energy: As the energy increase, a fraction of the jet production rate rises sharply [442–445]. The data agrees well with the Monte Carlo fragmentation models [446]. The two lines are predictions of perturbative QCD with (real line) and without (dashed line) power corrections [447–449]. (b) Q-plot at $\sqrt{s} = 29\,\text{GeV}$. Events are concentrated in the neighborhood of $S \sim 0$ region which represents 2-jet configuration [450, 451].

well with predictions of Monte Carlo calculation using various hadronization models described in the next section. The perturbative QCD calculation Eq. (9.21) gives the dashed line and the real line shows one with power correction [447, 449]. The data agrees with the predictions. Contribution of the power correction will be discussed later in Section 9.4.2.

Figure 9.6b shows an example of Q-plot distribution at $\sqrt{s} = 29\,\text{GeV}$ obtained by HRS group [451]. Comparing with Figure 9.6a, one sees that this is the energy region where the jet structure begins to appear clearly ($\langle 1 - T\rangle \lesssim 0.1$). One can see events are concentrated in the $S \to 0$ region where the 2-jet configuration dominates.

Infrared safe variables: We saw in Section 7.3.2, that $q\overline{q}g$ configuration develops a singularity if the gluon becomes soft or collinear with q or $\overline{q}$ which is an effect of the long distance interaction. If an emitted extra hadron in the fragmentation falls in this configuration, there is a danger of the observable going to diverge. To make a quantitative discussion, let us denote an observable O with n final hadrons by $O_n(p_1, p_2, \cdots, p_n)$ and express the average value of the observable by

$$\langle O\rangle = \sum_n \frac{1}{n!}\int dV_n \frac{d\sigma[n]}{dV_n} O_n(p_1, p_2, \cdots, p_n) \tag{9.22}$$

where dV_n denotes n-body phase space volume element. The "infrared safe" means for $\lambda \to 0$ (soft) or $0 < \lambda < 1$ (collinear)

$$O_{n+1}(p_1, \cdots, p_n = (1-\lambda)p_n, p_{n+1} = \lambda p_n) = O_n(p_1, \cdots, p_n) \tag{9.23}$$

A typical observable that is infrared safe is an inclusive cross section like the total hadronic production cross section by $e^- e^+$ annihilation. We saw for an emission of the gluon, the cross section to $O(\alpha_s)$ is $\sigma(e^- e^+ \to X) = \sigma_0(1+\alpha_s/\pi)$ (Eq. (7.139)) which is free from divergence.

We can show that the thrust is also infrared safe, namely, if a particle splits into two particles with momentum $\boldsymbol{p}_1 + \boldsymbol{p}_2 = \boldsymbol{p}$ where $|\boldsymbol{p}_2| \to 0$ (soft) or $\boldsymbol{p}_2 \parallel \boldsymbol{p}_1$ (collinear).

Proof: We need to show that $T_{n+1} \to T_n$ with

$$T_{n+1} = \max \frac{\sum_{i=1}^{n+1} |\boldsymbol{p}_i \cdot \boldsymbol{n}'|}{\sum_{i=1}^{n+1} |\boldsymbol{p}_i|}, \quad T_n = \max \frac{\sum_{i=1}^{n} |\boldsymbol{p}_i \cdot \boldsymbol{n}|}{\sum_{i=1}^{n} |\boldsymbol{p}_i|} \tag{9.24}$$

for $|\boldsymbol{p}_{n+1}| \to 0$ or $\boldsymbol{p}_n \parallel \boldsymbol{p}_{n+1}$. For a soft gluon limit,

$$\sum_{i=1}^{n+1} |\boldsymbol{p}_i| = \sum_{i=1}^{n} |\boldsymbol{p}_i| + |\boldsymbol{p}_{n+1}| \xrightarrow{|\boldsymbol{p}_{n+1}| \to 0} \sum_{i=1}^{n} |\boldsymbol{p}_i|$$

$$\sum_{i=1}^{n+1} |\boldsymbol{p}_i \cdot \boldsymbol{n}'| = \sum_{i=1}^{n} |\boldsymbol{p}_i \cdot \boldsymbol{n}'| + |\boldsymbol{p}_{n+1} \cdot \boldsymbol{n}'| \xrightarrow{|\boldsymbol{p}_{n+1}| \to 0} \sum_{i=1}^{n} |\boldsymbol{p}_i \cdot \boldsymbol{n}'| \tag{9.25}$$

Optimization for the max value varying $\boldsymbol{n}'$ obviously agrees with that of varying $\boldsymbol{n}$ since they are the same formulas. Therefore, $T_{n+1} \to T_n$ in this limit. For collinear splitting, we have $\boldsymbol{p} = \boldsymbol{p}_i + \boldsymbol{p}_j$, $\boldsymbol{p}_i \parallel \boldsymbol{p}_j$. For parallel vectors,

$$|\boldsymbol{p}_i \cdot \boldsymbol{n}'| + |\boldsymbol{p}_j \cdot \boldsymbol{n}'| = |\boldsymbol{p} \cdot \boldsymbol{n}'|, \quad |\boldsymbol{p}_i| + |\boldsymbol{p}_j| = |\boldsymbol{p}| \tag{9.26}$$

and therefore, the denominator as well as the numerator reduces to those of the n-particle case. □

The sphericity and the momentum tensor played an important role in the early days of jet studies, but they are no longer used because the momentum tensor is in danger of being infrared divergent and the weight of particles with large transverse momentum is overestimated when there are a large number of decay particles thus unsuitable for comparing with theories. One can use a similar, but infrared, safe observables by solving a tensor defined by [452]

$$\theta^{ij} = \frac{1}{\sum_a |\boldsymbol{p}_a|} \sum_a \frac{p_a^i p_a^j}{|\boldsymbol{p}_a|} \tag{9.27}$$

Denoting three eigenvalues of the θ^{ij} as $\lambda_1, \lambda_2, \lambda_3$, an event shape variable "C-parameter" is defined as

$$C = \frac{3}{2} \frac{\sum_{a,b} |\boldsymbol{p}_a||\boldsymbol{p}_b| \sin^2\theta_{ab}}{(\sum_a |\boldsymbol{p}_a|)^2} = 3(\lambda_1\lambda_2 + \lambda_2\lambda_3 + \lambda_3\lambda_1) \tag{9.28}$$

The tensor θ^{ij} and related variables are linear in $|\boldsymbol{p}|$ and infrared safe. C is similar to the sphericity and takes value $1/3(0)$ for spherical (pencil-like) events.

Problem 9.2

Show that the C-parameter is infrared safe, but the sphericity is not.

Other event shape variables include: the heavy jet mass M_H, the jet broadening parameter B and B_W. In order to define those parameters, the jet axis (usually taken as the thrust axis $\boldsymbol{n}_\mathrm{T}$) has to be defined. The whole event sphere which contains all the emitted particles is separated into two hemispheres divided by a plane perpendicular to the thrust axis. Let us call the two hemispheres S_+ and S_-. In each hemisphere, the invariant mass is defined and the heavier of the two is denoted as the (normalized) heavy jet mass ρ_H. Explicitly, they are defined by

$$\rho_H = \frac{1}{E_\mathrm{vis}^2} M_H^2 \,, \quad M_H^2 = \max\left[M_+^2, M_-^2\right] \tag{9.29a}$$

$$M_\pm^2 = \left(\sum_{a \in S_\pm} p_a\right)^2 , \quad E_\mathrm{vis} = \sum_a E_a \tag{9.29b}$$

where p_a are the four-momentum of the particle "a." The jet broadening variables are defined as

$$B_\pm = \frac{\sum_{a \in S_\pm} |\boldsymbol{p}_a \times \boldsymbol{n}_\mathrm{T}|}{2 \sum_a |\boldsymbol{p}_a|} \tag{9.30a}$$

$$B_W = \max[B_+, B_-] \quad \text{wide jet broadening} \tag{9.30b}$$

$$B_\mathrm{T} = B_+ + B_- \quad \text{total jet broadening} \tag{9.30c}$$

9.1.3 Applications of Jet Variables

The jet variables are fundamental in analyzing the structure of jet events and to test the validity of the perturbative QCD. Formalism to express them in terms of the coupling constant and to extract its strength will be discussed later in Section 9.4.2. Here, we take a quick look at their usage.

(1) Spin of the Quark Figure 9.7 shows the angular distribution of the thrust axis. Since the two jets are considered as two hadronized quarks, the angular distribution of the jet axis would reflect that of the two quarks. The experimental data exhibits angular distribution $\sim (1 + \cos^2\theta)$ expected from spin 1/2 parton pair production in $(e^- + e^+ \to q + \overline{q})$. If the spin of the parton is zero, the distribution would have been $\sim \sin^2\theta$.

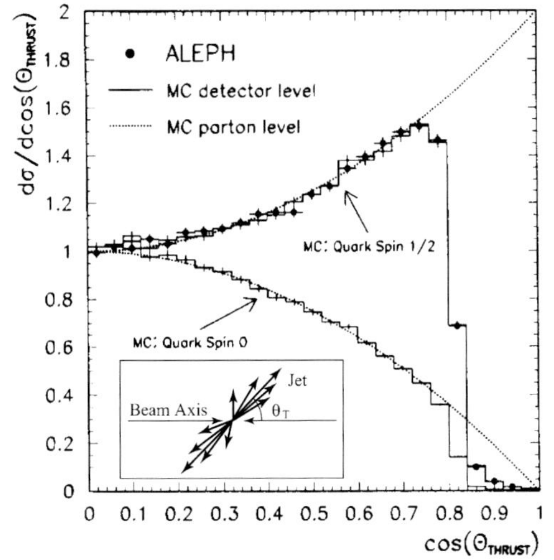

Figure 9.7 The dotted curves show the shape of the theoretical prediction for the primary quarks produced in hadronic Z decays. The ALEPH thrust data (solid points) compared to the fully simulated Monte Carlo (histograms), for both the spin-1 and the spin-0 hypotheses. The errors are purely statistical. The distributions are normalized to one at $\theta = 90$. The inset illustrates a qualitative feature of the thrust axis (direction of the jet) [453].

(2) Separation of 2- and 3-Jets The MARK-J group did not use track information of the jets, but only energy flows [454]. Here, one cannot determine the jet axis on the event by event basis, but has to determine it statistically. Figure 9.8a,c shows that events with small thrust ($T < 0.8$) and large oblateness ($O > 0.1$) which spreads like a disc appear as three-jets. Figure 9.8b,d shows that emission probability of the gluon grows with energy as indicated by the increase of flat events (large oblateness).

(3) New Particle Search The electron collider is most suited for discovering a new particle (Q) with mass M since it can produce any new particle which has coupling with γ or Z through a reaction

$$e^- + e^+ \to (\gamma^*, Z) \to Q + \overline{Q} \tag{9.31}$$

provided the total energy $\sqrt{s}$ exceeds $2M$. Although no new particles were found below or above the Z peak to this day since the discovery of the bottom quark, it is instructive to discuss a method of how to discover one in the e^-e^+ collider. Above and near the threshold of a heavy particle production, ($\sqrt{s} \sim 2M + \epsilon, \epsilon \ll M$), the $Q\overline{Q}$ are produced almost at rest and one expects the fragmented hadrons are distributed isotropically subject to the phase space boundary. This means small thrust and large acoplanarity. Figure 9.9 shows data obtained by VENUS/TRISTAN [436]. It was one of the searches for the top quark in vain at $\sqrt{s} = 60\,\text{GeV}$. Figure 9.9 shows the event shape distribution obtained from TRISTAN. If the top quark was produced, it would have produced events with large acoplanarity and a small thrust

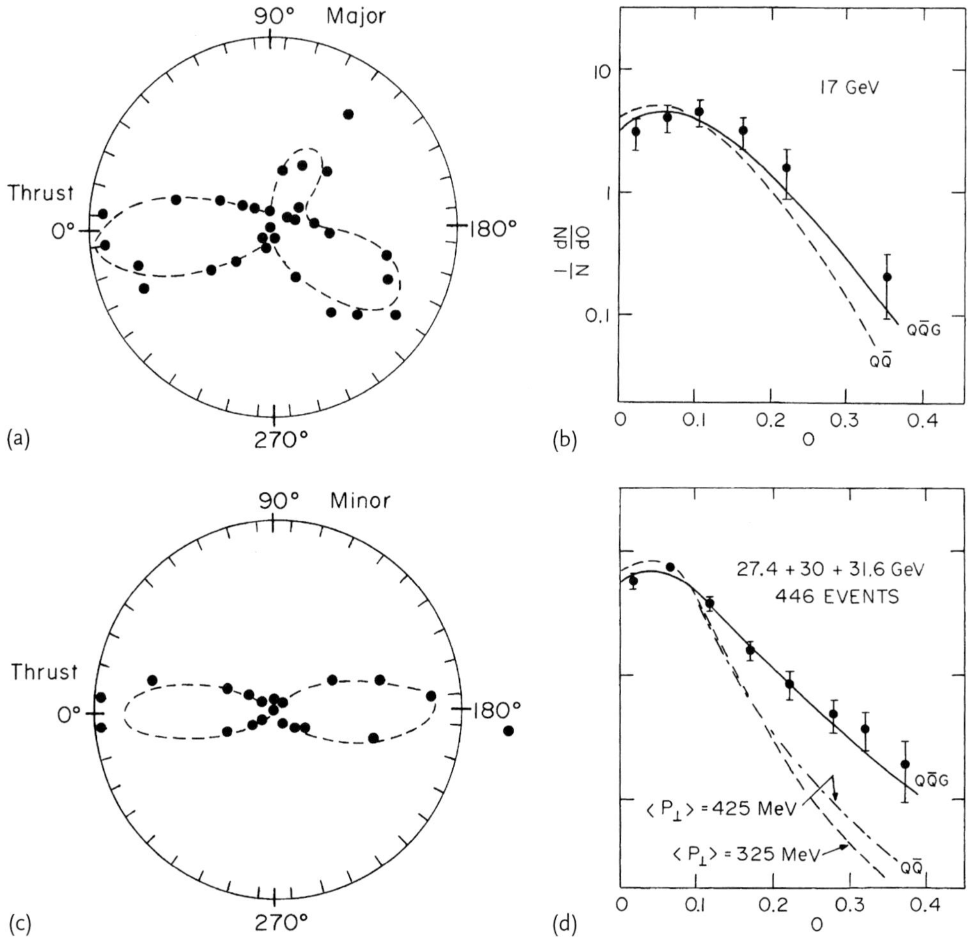

Figure 9.8 Energy flow as observed by the MARK-J group. (a,c) The 2- and 3-jet structure are observed. (b,d) The oblateness 'O' grows as the energy increases showing the growth of the 3-jets [454].

as shown by the dashed lines. The observed distribution can be well fitted with known quarks (u, d, s, c, and b) and it was concluded that the top was not produced.

A similar method was tried at LEP/CERN $\sqrt{s} = 90 \to 180\,\text{GeV}$. Eventually, the top quark was discovered at TEVATRON/Fermilab which has much larger $\sqrt{s} = 1.8\,\text{TeV}$ and its mass was determined to be $\sim 172\,\text{GeV}$. It will be discussed in Section 11.9.3.

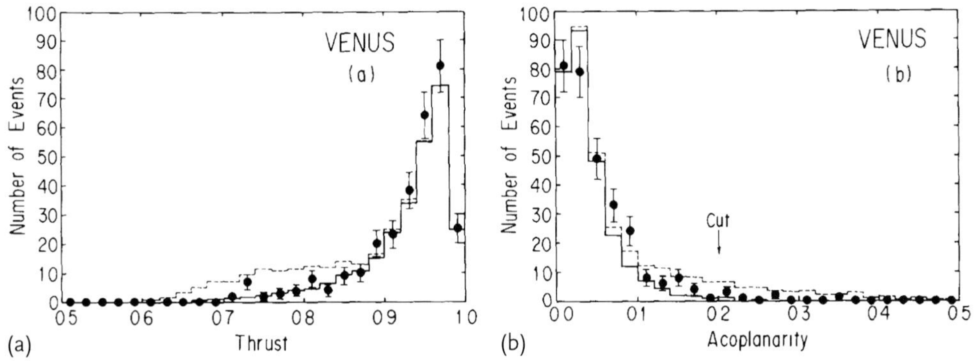

Figure 9.9 Jet event shape distributions of the reaction $e^-e^+ \to q\overline{q}$ at $\sqrt{s} = 60\,\text{GeV}$. Data taken from VENUS/TRISTAN [436]. (a) Thrust and (b) acoplanarity distributions. Histograms show the Monte Carlo calculations including (u, d, s, c, and b). The dashed line is the expectation, including the top quark with mass 25.5 GeV. From the figure, one concludes that no top was produced.

9.1.4
Jet Separation

Method of ϵ, δ We realized that the partons become jets, and that the 2- and 3-jet structures were observed experimentally. However, it is not always true that the separation of jets are well-defined. One can see in the Q-plots in Figure 9.6 that the transition from 2- to 3-jet structures is continuous and what we call 2-jet or 3-jet has to be defined clearly. For instance, 2-jet and 3-jet processes in the e^-e^+ reactions can be, in principle, described by the Feynman diagrams as depicted in Figure 9.10. However, if one of the three final partons is soft or collinear with other partons, they cannot be distinguished from the 2-jet configuration. A quantitative treatment of the jet was first given by Sterman and Weinberg [455]. Defining ε as the fractional energy of a jet included in a cone of half angle δ from the jet axis as described in Figure 9.11, and regarded the 2-jet production cross section as that where the partons are contained in the two cones. Then, calculating the cross section of $e^-e^+ \to q\overline{q} + q\overline{q}g$ to $O(\alpha_s)$ in the limit $\epsilon, \delta \ll 1$, one obtains [398, 455]

$$\sigma_{\text{2-jet}} = \sigma_0 \left[1 - \frac{\alpha_s}{\pi} 4 C_F \left\{ \ln\frac{1}{\delta} \left[\ln\left(\frac{1}{2\epsilon} - 1\right) - \frac{3}{4} \right] + \frac{\pi^2}{12} - \frac{7}{16} \right. \right.$$
$$\left.\left. + O(\epsilon \ln\delta, \delta^2 \ln\epsilon) \right\} \right] \tag{9.32a}$$

$$\sigma_0 = \frac{4\pi\alpha^2}{3s} N_c \sum_q Q_q^2 \tag{9.32b}$$

Here σ_0 is the Born approximation of the $\sigma_{\text{TOT}}(e^-e^+ \to \gamma^* \to \text{hadrons})$. Note, however, for very small values of ε and δ, the above expression becomes negative and cannot be used. This is due to the infrared and collinear divergences. As the

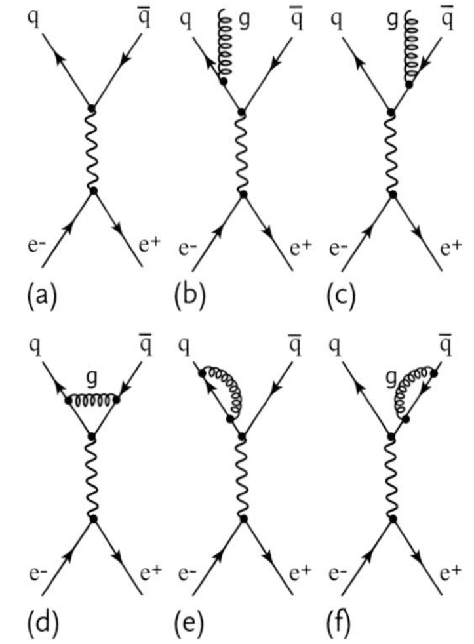

Figure 9.10 Feynman diagrams that contribute to the definitions of 2- and 3-jets. In (b,c), three partons are emitted, but if one of them is soft (small energy) or collinear with other partons, they are not resolvable from diagrams (a,d,e and f).

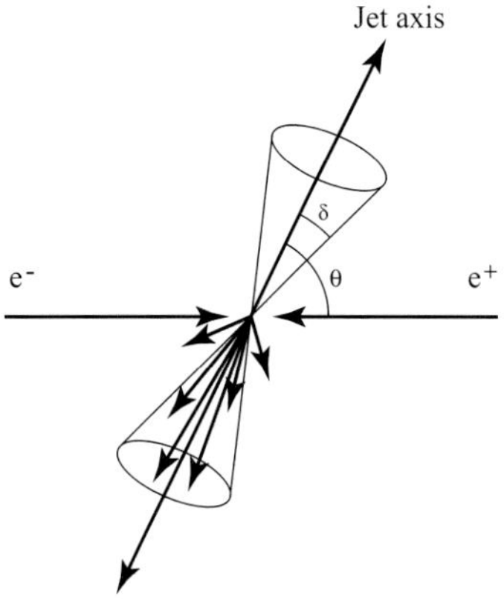

Figure 9.11 Jet resolving parameters ϵ, δ. ε is fractional energy that is included in the half angle δ from the jet axis [455].

total cross section to order $O(\alpha_s)$ is given by (Eq. (7.139))

$$\sigma_{\text{TOT}} = \sigma_0 \left(1 + \frac{\alpha_s(Q^2)}{\pi}\right) \tag{9.33}$$

the difference between Eqs. (9.33) and (9.32) should give the 3-jet production cross section.

$$\sigma_{3\text{-jet}} = \sigma_0 \frac{\alpha_s}{\pi} 4C_F \left\{ \ln\frac{1}{\delta}\left[\ln\left(\frac{1}{2\epsilon} - 1\right) - \frac{3}{4}\right] + \frac{\pi^2}{12} - \frac{7}{16} + O(\epsilon \ln\delta, \delta^2 \ln\epsilon) \right\} \tag{9.34}$$

Equations (9.32) and (9.34) show that if one specifies ε and δ, it is possible to treat the 2- and 3-jet production quantitatively. At this stage, we are not sure if they (parton definition of 2 and 3 jets) can really represent the 2- and 3-jet productions observed experimentally because, in comparing with experiments, we always have to go through the hadronization process. Today, the method of ε and δ is extensively used in separating jets in hadron–hadron reactions.[2)]

Method of y_{cut} The second jet resolving parameter which is commonly used in e^-e^+ reactions is y_{cut}. It is concerned with the invariant mass of the jets. When partons i, j exist, the invariant mass of the parton y_{ij} can be defined as

$$y_{ij} \equiv \frac{(p_i + p_j)^2}{s} \tag{9.35}$$

The divergence can be avoided by requiring $y_{ij} > y_{cut}$. The total cross section does not depend on y_{cut}, but the n-jet ($n = 2, 3 \cdots$) production cross section does.

Referring to the parton variables Eq. (7.127) in $q\overline{q}g$ CM frame, one sees that the requirement of $y_{ij} > y$ is equivalent to that for the fractional energy of the ith parton $1 - x_i > y$. The condition is expressed as

$$2y < x_1, x_2 < 1 - y\,, \quad x_1 + x_2 > 1 + y \tag{9.36}$$

whose boundaries are depicted as contours A, B and C in Figure 9.12. Then fraction of 3-jet events $R_{3\text{-parton}}$ can be obtained by integrating Eq. (7.133a) over the inner area of a triangle bounded by the contours A, B and C. Note that as y is varied, apexes move on the dot-dashed lines which (if extrapolated) connect them to middle points of bases. This is achieved by integrating first along the contours and then sweeping the triangle toward the center of the triangle, namely, integrating over y in the range from y_{cut} to the limit of the phase space $y = 1/3$. The integral along the contours is given by

$$\begin{aligned}
\frac{1}{\sigma_0}\frac{d\sigma_{3\text{-jet}}}{dy} &= \frac{\alpha_s}{2\pi} C_F \int_{2y}^{1-y} dx_1 \left[2\frac{x_1^2 + x_2^2}{(1-x_1)(1-x_2)}\bigg|_{x_2=1-y} \right. \\
&\quad \left. + \frac{x_1^2 + x_2^2}{(1-x_1)(1-x_2)}\bigg|_{x_2=1+y-x_1} \right] \\
&= \frac{\alpha_s}{2\pi} C_F \int_{2y}^{1-y} dx \left[2\frac{(1-y)^2 + 1 - (1-x)(1+x)}{y(1-x)} \right. \\
&\quad \left. + \frac{1}{1-y}\left(\frac{1}{1-x} + \frac{1}{x-y}\right)\{1 + y^2 - 2(1-x)(x-y)\} \right] \\
&= \frac{\alpha_s}{2\pi} C_F \left[2\frac{3y^2 - 3y + 2}{y(1-y)} \ln\frac{1-2y}{y} - 3\frac{(1-3y)(1+y)}{y} \right]
\end{aligned} \tag{9.37}$$

2) Referred to as the cone algorithm which defines a circle with radius R on the rapidity-azimuthal angle (η–ϕ) (see Section 11.4.1 and 11.4.2).

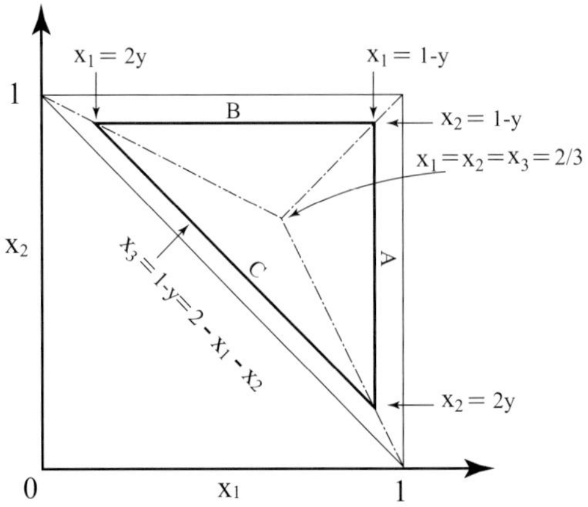

Figure 9.12 Contours of 3-jets with fixed y. The 3-jet production cross section is obtained by integrating over the area inside a triangle which is constrained by $y_{\text{cut}} < y < 1/3$.

where the first term in the right-hand side of the first line is over the contours A and B, and the second is over the contour C. We notice a characteristic logarithmic divergence as $y \to 0$. Integrating Eq. (9.37) from y to the phase space limit $1/3$ gives the total 3-jet production rate $R_3 \equiv \sigma_{\text{3-jet}}/\sigma_{\text{tot}}$.

$$\begin{aligned} R_3 &= C_F \frac{\alpha_s}{2\pi}\left[2\ln^2\frac{y}{1-y} + 3(1-2y)\ln\frac{y}{1-2y}\right. \\ &\quad \left. + 4\text{Li}_2\left(\frac{y}{1-y}\right) - \frac{\pi^2}{3} + \frac{5}{2} - 6y - \frac{9}{2}y^2\right] \\ R_2 &= 1 - R_3 \end{aligned} \tag{9.38}$$

where $\text{Li}_2(x)$ is the dilogarithm function

$$\begin{aligned} \text{Li}_2(x) &= -\int_0^x dy \frac{\ln(1-y)}{y} = -\int_{1-x}^1 dy \frac{\ln y}{1-y} \\ &= \sum_{n=1}^{\infty} \frac{x^2}{n^2} \quad \text{for} \quad |x| < 1 \end{aligned} \tag{9.39}$$

which satisfies

$$\begin{aligned} &\text{Li}_2(0) = 0, \quad \text{Li}_2(1) = \pi^2/6 \\ &\text{Li}_2(x) + \text{Li}_2(1-x) = \pi^2/6 - \ln x \ln(1-x) \end{aligned} \tag{9.40}$$

Just like ϵ, δ, if the value of y becomes too small, R_3 goes negative which means a breakdown of the perturbation expansion. The small y part, however, can be resummed and the divergent terms in Eq. (9.38) can be exponentiated.

Problem 9.3

Derive Eq. (9.38).)

To order α_s^2, the cross sections are given by [456–459]. Relative cross sections can be expressed as

$$R_n \equiv \frac{\sigma_{n\text{-jet}}}{\sigma_{\text{TOT}}} \tag{9.41a}$$

$$R_2 = 1 + C_{21}(y_{\min})\alpha_s(Q^2) + C_{22}(y_{\min}, f)\alpha_s(Q^2)^2 \tag{9.41b}$$

$$R_3 = \qquad C_{31}(y_{\min})\alpha_s(Q^2) + C_{32}(y_{\min}, f)\alpha_s(Q^2)^2 \tag{9.41c}$$

$$R_4 = \qquad\qquad + C_{42}(y_{\min}, f)\alpha_s(Q^2)^2 \tag{9.41d}$$

$f = \mu^2/Q^2$ is a scale parameter for renormalization and is taken usually as $f = 1$ in the e^-e^+ reactions. However, experimentally, a much smaller value is preferred which is considered as effectively taking some part of the higher order effect into account.

The defining formulae for variables given by Eqs. (9.35) and (9.38) are given for the partons. Reconstruction of jets in experiments are carried out as follows. It first picks up a pair of particles i, j and define its momentum p_{ij} and y_{ij} as

$$\begin{aligned} p_{ij} &= p_i + p_j \\ y_{ij} &= 2\min(E_i^2, E_j^2)(1 - \cos\theta_{ij})/E_{\text{vis}}^2 \end{aligned} \tag{9.42}$$ [3)]

This is carried out for every pair of final state particles. Then, the pair with the smallest value of y_{ij} if $y_{ij} < y_{\text{cut}}$ are combined together and replaced by a pseudoparticle with four momentum p_{ij}. Choose the third particle k, and form the second quasiparticle such that the value of y_{ijk} made of ij and k pair with $p_{ijk} = p_{ij} + p_k$ is the smallest among those picked up. The process is repeated for all pairs of objects (particles and/or pseudoparticles) until they have $y_{ijk\cdots} > y_{\text{cut}}$. Whatever objects remain at this stage are called jets.

The same procedure is applied to theoretical partons and $R_{n\text{-parton}}$ can be determined as a function of y_{cut}. In order to compare the theoretical $R_{n\text{-parton}}$ and that of experimental $R_{n\text{-jet}}$, one has to take the hadronization effects into account. Since it requires nonperturbative treatments, one has to depend on models which will

3) This method is called the Durham algorithm [460, 462]. Historically, $y_{ij} = 2E_i E_j(1 - \cos\theta_{ij})$ (JADE algorithm) have also been used frequently. Theoretically, the Durham algorithm is desirable since in this definition, it is possible to resum large double logarithms to all orders [463]. In the analysis of the LEP data, this is the preferred algorithm. There are other variations (E-scheme, P_0-scheme, and so on) which differ in the way the energy and/or momenta of the pair is combined to form those of the pseudo particles.

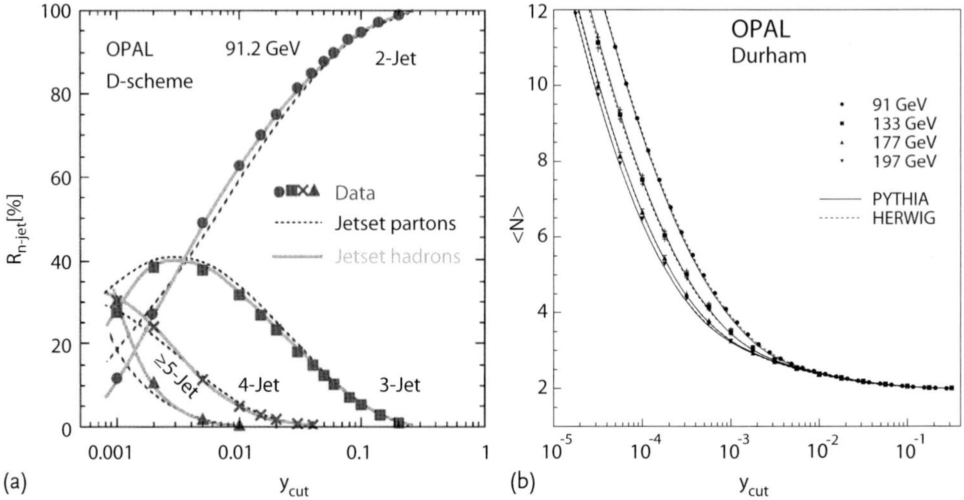

Figure 9.13 Jet multiplicity: (a) 2–3-jets separation with y_{cut} method. Data are compared with a parton level calculation and with fragmentation [445]. (b) Number of jets as a function of y_{cut} and energy [464].

be discussed in the next section. Figure 9.13 shows n-jet production cross sections and how the number of jets increases as y_{cut} is decreased [464]. Figure 9.13 and the detailed Monte Carlo calculations demonstrate that corrections due to hadronization are small for the n-jet separation algorithm which is not necessarily true for other jet variables.

9.2 Parton Shower Model

As the examples in the previous section demonstrate, we need to understand the process of the parton's hadronization in order to make reliable comparisons between theoretical predictions and observed data. At present, there exists no mathematically rigorous method to connect them. There are, however, several models depending on how one approaches the problem. Some start from conceptually very different assumptions, but if they reproduce data well, their basic concept must be connected at somewhere. If we understand characteristic features that survive with precision data, we hope we will eventually reach the correct theory.

Prescriptions for Comparing Theories and Experiments In comparing theoretical predictions with experimental data, there are two classes of prescriptions. In the first prescription, one calculates physical observables (matrix elements) at the par-

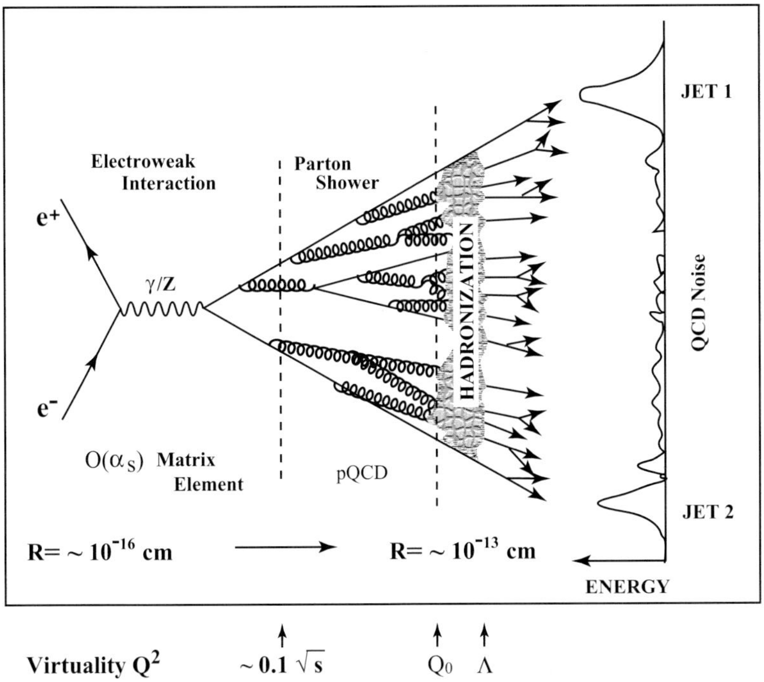

Figure 9.14 Making a correspondence between the perturbative QCD at the parton level and hadrons in the experimental level. The matrix element method applies hadronization models directly to the final partons, that is, at $Q^2 = s$ while the parton shower model develops until the virtuality of the partons reach $\sim 1\,\text{GeV}^2$ and applies the hadronization models in the final stage.

ton level using perturbative QCD (pQCD) and converts the final state partons to hadrons using a hadronization model (referred to as matrix method). In the second prescription, after calculating the matrix elements, the patrons are evolved à la DGLAP formalism, generating cascade showers and eventually hadronizing when the virtuality (a measure of energy scale) reaches the typical hadron mass (method of parton shower). The two methods are shown pictorially in Figure 9.14.

In this drawing, the matrix method corresponds to calculating the matrix element to the left of the dashed line and skips the shower generating process (denoted as pQCD) and goes directly into hadronization. In some special cases, LPHD (local parton hadron duality) is used. LPHD claims that the hadron distribution is directly proportional to that of the parton, provided the virtuality at which hadronization commences is sufficiently small. Its applicability will be discussed in Section 10.3.4.

Here, we will concentrate on the matrix element and the parton shower method. In the parton shower method, one chooses suitable virtuality variables at each step of branching. For the virtuality variable Q, invariant mass of the parent or transverse momentum of the child are often used. When the virtuality reaches a certain value $Q = Q_0$ GeV, the partons are converted to hadrons using a hadronization model. For valid cascade generations, the coupling constant $\alpha_s(Q)$ has to be small, namely, $Q_0 \gg \Lambda$ is required. Usually, $Q_0 \sim 1$ GeV is adopted. In the region $Q \lesssim \Lambda$, a nonperturbative treatment is required.

The matrix method is exact as a function of α_s, at least to the nth order. However, except for $\sigma(e^-e^+ \to \text{hadrons})$, theoretical calculations are limited in most cases to $n \leq 2$, namely, up to four final partons including loop corrections. Moreover, as the hadronization is directly applied to partons with high virtuality (typically, $Q \sim 0.1\sqrt{s}$), it is very model dependent. In the parton shower method, contrary to the matrix element method, there are no restrictions to the number of partons, soft and collinear gluon generation is built-in naturally, and the dependence on the hadronization model is small as it is carried out at low enough virtuality. However, it uses collinear/soft approximation at every step of branches, and inherently difficult to reproduce hard processes (phenomena which contain emission of hard gluons such as energy correlation EEC or large p_T phenomena). Here, the matrix method has an advantage. However, there are approaches to connect them, that is, to take into account of hard process in the parton shower model [465–468].

Sudakov form Factor The DGLAP evolution equation was formulated by successive splitting of the partons. Cascade showers are made possible by repeating the parton splitting as shown in Figure 9.14. If one adopts the invariant mass $\hat{t}_i$ of the parton for the virtuality, it decreases at each step of splitting ($\hat{t}_a > \hat{t}_b, \hat{t}_c$ for $a \to b, c$) (see Eq. (8.66)). According to the evolution equation, probability of a parton splitting for a small change $d\tau = dQ^2/Q^2$ of the evolution variable $\tau = \ln(Q^2/\mu^2)$ is given by

$$\frac{dW_{bc\leftarrow a}}{d\tau} = \int dz \frac{\alpha_s(Q^2)}{2\pi} P_{bc\leftarrow a}(z) \tag{9.43}$$

where $P_{bc\leftarrow a}(z)$ is the splitting function defined by Eq. (7.143b). It is necessary to normalize the whole event generation rate in such a way that the total probability becomes one. Suppose a parton emission occurs at τ, it is necessary that no partons are emitted starting from $\tau = \tau_{\max}$ until τ is reached. As the probability that the splitting does not occur is given by $1 - \delta\tau(dW/d\tau)$, the probability that no partons are emitted between $\tau_{\max} \sim \tau$ is given by

$$
\begin{aligned}
S_a(\tau; \tau_{\max}) &= \lim_{N\to\infty}\left(1 - \delta\tau_1 \frac{d W_{bc\leftarrow a}}{d\tau}(\tau_1)\right)\left(1 - \delta\tau_2 \frac{d W_{bc\leftarrow a}}{d\tau}(\tau_2)\right)\cdots \\
&\quad \cdots\left(1 - \delta\tau_N \frac{d W_{bc\leftarrow a}}{d\tau}(\tau_N)\right) \\
&= \exp\left[\lim_{N\to\infty}\left(-\sum_{i=1} \delta\tau_i \frac{d W_{bc\leftarrow a}}{d\tau}(\tau_i)\right)\right] \\
&= \exp\left[-\int_{\tau}^{\tau_{\max}} d\tau' \frac{d W_{bc\leftarrow a}}{d\tau}(\tau')\right] \\
&= \exp\left[-\int_{\tau}^{\tau_{\max}} d\tau' \int_{z_{\min}}^{z_{\max}} dz \frac{\alpha_s(\tau')}{2\pi} P_{bc\leftarrow a}(z)\right] \\
\delta\tau_i &= \frac{\tau_{\max} - \tau}{N}
\end{aligned}
\tag{9.44}
$$

This is referred to as the Sudakov form factor. Then, the probability that the parton "a" having $\tau_{\max}$ initially splits during $\tau \sim \tau + d\tau$ is given by

$$
S_a(\tau : \tau_{\max}) \frac{d W_{bc\leftarrow a}}{d\tau} d\tau \tag{9.45}
$$

Once the parton "a" splits, then treating "b" and "c" as new parents, one can repeat the same process, thus generating showers in cascade. The procedure stops when τ reaches $\tau_{\min}$. The singularity at $z = 0, 1$ can be avoided by requiring the virtuality to be greater than Q_0. This can be seen from the relation (see Eq. (8.63))

$$
\hat{t}_a = \frac{\hat{t}_b}{z} + \frac{\hat{t}_c}{1-z} + \frac{p_{\mathrm{T}}^2}{z(1-z)} \tag{9.46}
$$

Requiring $\hat{t}_b, \hat{t}_c > \hat{t}_0$, we have $z(1-z) > \hat{t}_0/\hat{t}_a$. Since $Q^2 \simeq \hat{t}$, we have a natural cut-off for z by requiring the virtuality to be greater than Q_0^2.

Problem 9.4

Consider a quark splitting to another quark and a gluon ($a = b = q$, $c = g$). In the LLA approximation ($\alpha_s = [b_0\tau]^{-1}$), show

$$
S_a(\tau) = \left[\frac{\alpha_s(\tau)}{\alpha_s(\tau_{\max})}\right]^{-\gamma_0/b_0}, \quad \gamma_0 = \frac{1}{2\pi}\int_{z_{\min}}^{z_{\max}} dz\, P_{qg\leftarrow q}(z) \tag{9.47}
$$

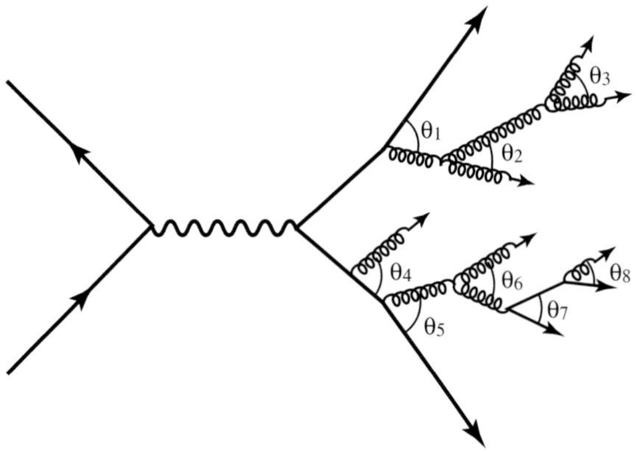

Figure 9.15 Angular ordering of the parton shower model: Cascade showers are created by successive application of parton splitting. The color coherence effect can be taken into account by angular ordering ($\theta_1 > \theta_2 > \theta_3$, $\theta_4 > \theta_5 > \theta_6 > \theta_7 > \theta_8$.

The choice of variables for the virtuality Q^2 differs, depending on which hadronization program one uses. In PYTHIA [469, 470], it is the invariant mass $\hat{t}$, in HERWIG [471–473], it is the angle (θ^2 or $(E\theta)^2$ to be exact) and in ARIADNE [474], it is the transverse momentum p_T^2. They give the same $d\tau = dQ^2/Q^2$, but different formalism, although they are equivalent in the collinear limit. Note that choice of z is also different. Some use fractional momentum and others use light cone variables $(E + p_z)$.

Angular Ordering A practical parton shower model incorporates one feature which has not been discussed so far, namely, the color coherence effect which results from the interference among the partons. The color coherence effect is discussed in detail in Section 10.2. Here, suffice it to say that the interference effect, which is completely discarded in the stochastic treatment, can be taken into account by the angular ordering [471, 475–477]. There, the emission angle of later gluons is constrained to be smaller than the previous one as shown in Figure 9.15. The angular ordering generates the string effect similar to the string model described in the following.

9.3 Hadronization Models

The partons are not physical, as only their remnant hadrons are observed experimentally. In order to convert partons to hadrons, their spacetime structure, flavor and spin, for example, have to be specified. There are three distinct hadronization models, namely, the independent fragmentation model (IF), the (Lund) string fragmentation model (SF) and the cluster model (CL).

9.3.1
Independent Fragmentation Model

The independent fragmentation model [478] is the pioneer of hadronization models and in the early days of jet phenomenon research, it played an important role. The basic tools were expanded and continue to be used in many other models.

a. Fragmentation Function The model assumes scaling for the longitudinal momentum distribution and the energy independent transverse momentum distribution. They are based on the observed scaling (see Figure 9.4) and transverse momentum distributions (see Vol. 1, Fig. 13.30). Neglecting the transverse momentum and assuming the zero mass, the parton q_i of longitudinal momentum $p_{q\parallel}$ is considered to pick up $\overline{q}_{i+1}$ from $q_{i+1}\overline{q}_{i+1}$ in vacuum with probability $f(z)dz$ and constitutes a meson $M(q_i\overline{q}_{i+1})$ with longitudinal momentum $p_{Mz} = zp_{q\parallel}$. The process is described pictorially in Figure 9.16a. The residual parton q_{i+1} which has received the longitudinal momentum $(1-z)p_{q\parallel}$ continues to proceed. When the mass is finite, the light cone variable $E + p_\parallel$ is often used instead of longitudinal momentum. In that case, z is defined by $z = (E_h + p_{h\parallel})/(E_q + p_{q\parallel})$. If the hadronization occurs independent of each other, the process of hadronization can be repeated and the Monte Carlo calculation is possible (Figure 9.16b). It is repeated until the available energy becomes too low to create another hadron. This lower energy cut-off is set conveniently around $\sim O(1\,\text{GeV})$. Assuming one flavor for simplicity, we define the function $f(z)dz$ for the first hadronized meson to leave fractional momentum z to the remaining parton. Then, the total probability $D(z)dz$ of the first parton to completely hadronize having fractional momentum z is given as the sum of the probability $f(1-z)dz$ to become a meson of momentum

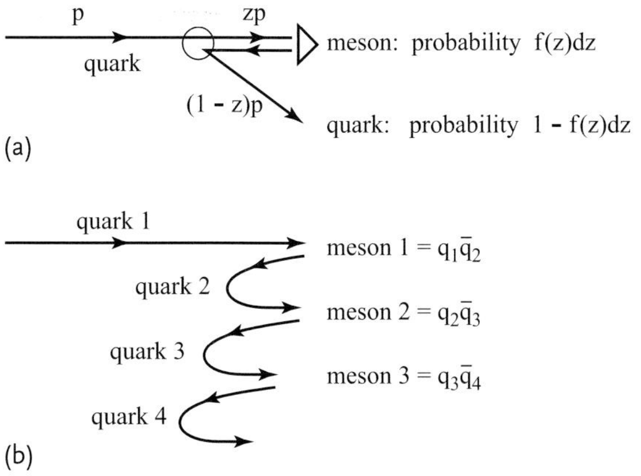

Figure 9.16 Independent fragmentation model. The first quark q_i picks up $q_2\overline{q}_2$ from vacuum and makes a meson 1$= q_1\overline{q}_2$. The remaining q_2 picks up another pair $q_3\overline{q}_3$ and makes the second meson $= q_2\overline{q}_3$ and the process is recursively repeated until the energy is too low to make any more hadrons.

z and the total probability of the residual parton to completely hadronize

$$D(z) = f(1-z) + \int_z^1 f(\eta) d\eta \int_z^1 \delta(z - \eta y) D(y)\, dy \tag{9.48a}$$

$$\int dz\, f(z) = 1 \tag{9.48b}$$

One can solve the equation by taking the Mellin transformation.

$$\tilde{D}(N) = \int_0^1 \frac{dx}{x} x^N D(x), \quad \tilde{f}(N) = \int_0^1 \frac{dx}{x} x^N f(x) \tag{9.49}$$

Then, an integral equation becomes an algebraic relation and the inverse Mellin transformation recovers the solution. A simple form one can adopt for $f(x)$ and hence for $D(x)$ is given by

$$f(z) = (1+\alpha) z^\alpha, \quad \alpha \simeq 0.6 \tag{9.50}$$

$$D(z) = (1+\alpha) \frac{(1-z)^\alpha}{z} \tag{9.51}$$

One can also easily prove that the momentum conservation is guaranteed.

$$\int_0^1 x\, D(x) = 1 \tag{9.52}$$

The function $D(z)$ behaves like $\sim dz/z$ for $z \to 0$ as it should in order to reproduce the infrared behavior of the soft gluon emission. If plotted as a function of the rapidity

$$\eta = \frac{1}{2} \ln \frac{E + p_\parallel}{E - p_\parallel} = \ln \frac{E + p_\parallel}{m_\mathrm{T}} \approx \ln z + \text{const}$$
$$m_\mathrm{T}^2 = p_\mathrm{T}^2 + m^2 \tag{9.53}$$

it is flat in η in the region $\eta \simeq 0$. $1/z$ behavior is also necessary in order to incorporate the experimental law that the mean multiplicity increases logarithmically. The total multiplicity of the hadron induced by the parton is given by

$$n_h = \sum_q \int_{z_\mathrm{min}}^1 dz [D_q^h(z) + D_{\bar{q}}^h(z)] \tag{9.54}$$

Experimentally, the fragmentation function is often parametrized as [479–481]

$$D(x) = N x^{\alpha}(1-x)^{\beta}\left(1+\frac{\gamma}{x}\right) \tag{9.55a}$$

$$N \simeq 0.5 \quad \alpha \simeq -1.5 \quad \beta \simeq 1.0 \quad \gamma \simeq -0.02 \sim -0.1 \quad \text{for} \quad \pi^{\pm} \tag{9.55b}$$

which reproduces the data quite well (Figure 9.17a). Note that the parameters are functions of the scale of reactions as the fragmentation function is subject to evolution à la DGLAP equation. Figure 9.17b–d shows a compilation of the currently available data shown separately for $\pi^{\pm}$, $K^{\pm}$ and $p, \overline{p}$. Note that the diagrams in Figure 9.17 are plotted as a function of x in the Figure 9.17a and $\ln x$ in the Fig-

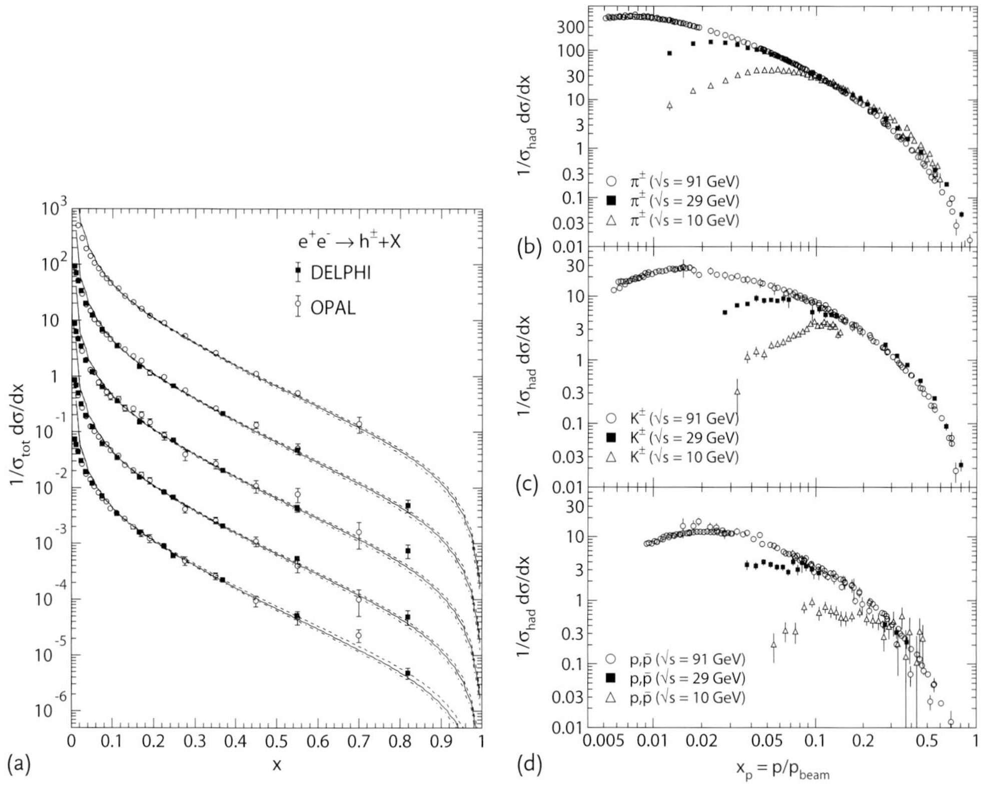

Figure 9.17 Fragmentation distribution function: data and parametrized fit. (a) $(1/\sigma_{\text{tot}})d\sigma/dx$ of $e^+e^- \to h^{\pm} + X$ as a function of scaled momentum x at CM energies $\sqrt{s} = 133$, 161, 172, 183, and 189 GeV (from bottom to top in this order). The NLO predictions for the scale parameter $\xi = \mu/\sqrt{s} = 1/2$ (dashed lines), 1 (solid lines), and 2 (dot-dashed lines) are compared with data from DELPHI (solid boxes) and OPAL (open circles). Each set of curves is rescaled relative to the nearest upper one by a factor of 1/10 [480, 481]. (b–d) $(1/\sigma_{\text{tot}})d\sigma/dx$ of $e^+e^- \to h^{\pm} + X$ for $h = \pi, K, p\overline{p}$ [11]. Scaling violation at small x is considered as due to soft gluon emission and effect of finite mass of hadrons. Notice the abscissa in the left figure is linear while in the right figure, it is logarithmic.

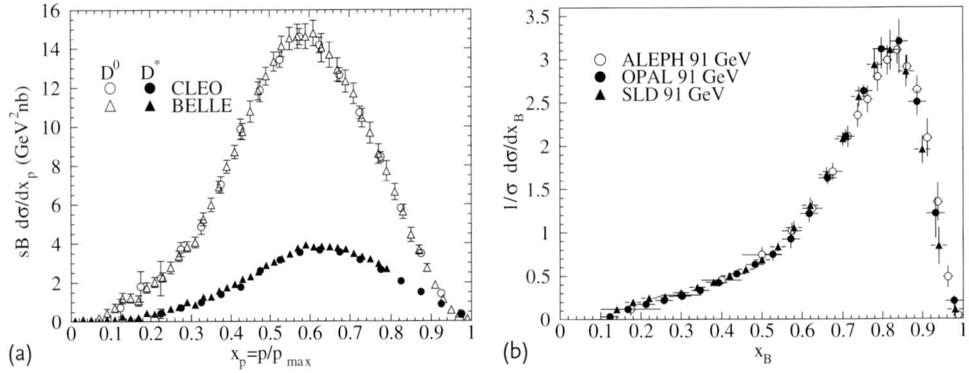

Figure 9.18 Fragmentation distribution function for c- and b-quark. They are reproduced well by the Peterson function (see text) [11, 468].

ure 9.17b–d. One should be aware of how the behavior looks different if plotted in different scales.

For heavy quarks ($q = c, b$), the Peterson function [482]

$$f(z) = \frac{1}{z}\left(1 - \frac{1}{z} - \frac{\epsilon_q}{1-z}\right)^{-2}$$
$$\epsilon_c \sim 0.05, \quad \epsilon_b \sim 0.006 \quad \text{for LLA} \tag{9.56}$$

is popular [482]. By choosing ϵ_q suitably, the function reproduces the data in Figure 9.18 well. The origin of the functional form is considered as follows. Since the heavy quark Q hardly changes its energy if it picks up light quarks ($q\bar{q}$) from vacuum, the amplitude for transformation $Q \to M(Q\bar{q}) + q$ should be inversely proportional to ΔE.

$$\Delta E = \left(m_Q^2 + z^2 P^2\right)^{1/2} + \left[m_q^2 + (1-z)^2 P^2\right]^{1/2} - \left(m_Q^2 + P^2\right)^{1/2}$$
$$\simeq \frac{m_Q^2}{2P}\left[\frac{1}{z} + \frac{\left(m_q^2/m_Q^2\right)}{1-z} - 1\right] \tag{9.57}$$

Experimentally, $\varepsilon_c \sim 0.01$–0.2, $\varepsilon_b \simeq (m_c^2/m_b^2)\varepsilon_c \simeq 0.08$ [11]. This functional form has a peak which moves toward $z \sim 1$ as m_Q becomes large. For the top quark ($m_t \gg m_b$), it is almost $\sim \delta(1-z)$.

b. Gluon The gluon is treated in the same way as the quark [483] or forced to hadronize after being converted to the $q\bar{q}$ pair according to the splitting function $P_{q \leftarrow g}$, $P_{\bar{q} \leftarrow g}$ [484, 485].

c. Transverse Momentum p_T The transverse momentum distribution is independently generated. The chosen function is Gaussian based on the experimental observations.

$$\propto \quad \exp\left(-\frac{p_{\mathrm{T}}^2}{2\sigma^2}\right) \qquad \sigma \sim 0.3\,\mathrm{GeV} \tag{9.58}$$

The p_{T} is generated to conserve locally between q' and $\overline{q}'$.

d. Flavor and Spin The pick up probability of u, d, s, c, and b from vacuum is set to $1 : 1 : 0.3 : 0 : 0$ to reproduce the observed data. This means the heavy quarks c, b do not appear unless they are produced directly in the initial reaction.

The statistical weight between scalar and vector mesons would be $1 : 3$ regarding the naive guess. However, considering that vector mesons are heavier, subsequent loss of the phase space volume, $V/(P + V) \sim 0.6$, is adopted. Tensor mesons with spin larger than two could be treated in a similar manner, but were ignored. Baryons are generated by picking up the $qq\overline{qq}$ pair from vacuum with preassigned probability ($qq/q \sim 0.08$) and forming color neutral qqq or $\overline{qqq}$. Once the flavor and the spin are specified, final state hadrons including resonances are uniquely determined. The resonances are decayed further.

The independent fragmentation model is historically the first formulation of the hadronization phenomena. So far, as one does not regard the fine structure of jets, it remains as a viable model to this day. However, various concepts used in the construction of the model are completely determined phenomenologically with no connection to the QCD principle. It is not Lorentz invariant and the energy-momentum is not conserved because $q\overline{q}$ pairs are produced independently. The conservation of the energy-momentum is restored by rescaling after adding all of them in the CM frame. Finally, there is no distinction between the gluon and the quark. The model cannot reproduce the observed string effect. For these reasons, they are not popular nowadays, but the basic ingredients of the formalism are effectively adopted by other late models.

9.3.2 String Model[4)]

a. Yo-yo The string fragmentation model (SF to be short) takes the string concept inherent in QCD into account and was epoch-making, in that sense, but otherwise conceptually very similar to the independent fragmentation model. In its early days, it was also referred to as the Lund model [197, 489]. When $q\overline{q}$ is produced at $t = x = 0$, they move on the light cone ($t^2 - x^2 = 0$), assuming $m_q = 0$. The SF model assumes that the potential energy exists between the two quarks proportional to the distance between the pair. The concept predates that of lattice QCD prediction [490] except the Coulomb potential is neglected. The energy stored in the string is expressed as $\Phi = \kappa x, \kappa \sim 1\,\mathrm{GeV/fm}$, $1\,\mathrm{fm} = 10^{-15}\,\mathrm{m}$ (see Vol. 1, Sect. 14.5.4). κ is referred to as the string tension. Because of initial kinetic energy, the string is stretched, but decelerated by the string tension and oscillates. The en-

4) [197, 486–489]

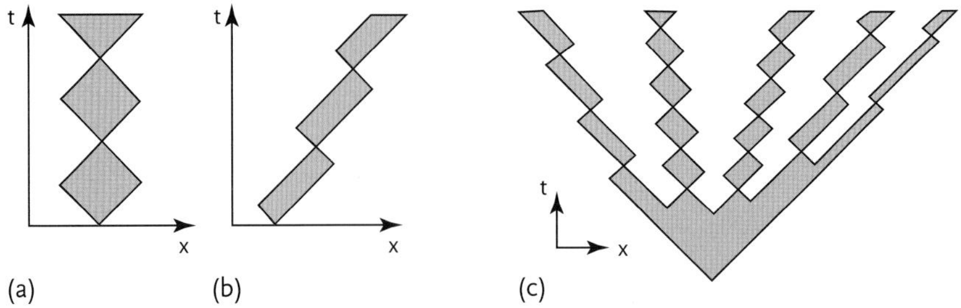

Figure 9.19 Pictorial description of the string model. (a) When the energy of the string is small, it does not split, but oscillates and is referred to as yo-yo. (b) Lorentz boosted yo-yo. (c) When the string is stretched beyond the typical hadron size ($\sim 10^{-15}$ m), the string is split into pieces and many yo-yos are produced.

ergy momentum of partons (with mass m) at both ends of the string are expressed as

$$\kappa x = \sqrt{p_0^2 + m^2} - \sqrt{p^2 + m^2}\,, \quad \kappa t = p_0 - p \tag{9.59}$$

where p_0 denotes the initial momentum at $t = 0$. The trajectory is a hyperbola in general, and straight lines if $m = 0$. The partons change directions, periodically making the trajectory a jig-zag shape which is referred to as yo-yo (Figure 9.19a) When it is Lorentz boosted, the trajectory becomes like that in Figure 9.19b. When the initial energy given to the $q\overline{q}$ pair is large, the stored energy in the stretched string is also large. If a $q'\overline{q}'$ pair pops up on the string from vacuum, and if q' with the original $\overline{q}$ or $\overline{q}'$ with q makes a color neutral pair, the attractive force between the newly formed pair is stronger than that between the original pairs ($q'\overline{q}'$ and $\overline{q}q$), the string is cut. The cutting of the string can happen anywhere and in more than one place (at different time).

b. Fragmentation Function Creation probability of the quark pair ($q'\overline{q}'$) from vacuum can be derived from an analogous known phenomenon in QED. A strong electric field makes the vacuum unstable. An e^-e^+ pair is produced and its creation probability is known to be proportional to [491]

$$\frac{\alpha E^2}{\pi^2} \exp\left(-\frac{\pi m^2}{eE}\right) \tag{9.60}$$

The SF model adopts

$$f(p_\mathrm{T}) \propto \exp\left(-\frac{\pi m_\mathrm{T}^2}{\kappa}\right), \quad m_\mathrm{T}^2 = m^2 + p_\mathrm{T}^2 \tag{9.61a}$$

The reason to use m_T^2 instead of m^2 is because the string is one-dimensional. The functional form can be understood as the tunneling effect because of the energy conservation that the pair is created not at a point, but separated at some distance. The SF model generates $q'\overline{q}'$ pairs at random and independently on the string until

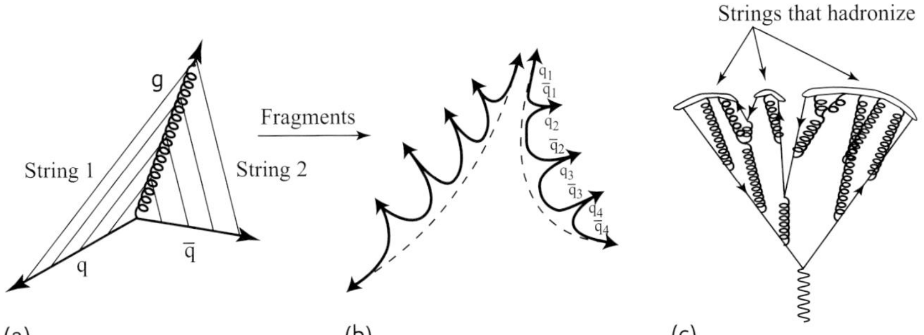

Figure 9.20 (a) In the string model, the gluons are considered as a kink of the string. (b) Hadrons are produced on the string, but those on the opposite side of the gluon direction are suppressed. (c) Strings are stretched between two quarks at both ends with gluons forming kinks in between.

the remaining string has energy too small to produce an extra hadron. Many yo-yos are produced in this way (Figure 9.19c). The functional form of the fragmentation distribution is almost uniquely determined if one requires uniform and symmetric distribution of the pair generation on the string. It is given by

$$D(z) = \frac{(1-z)^a}{z} \exp\left(-b\frac{m_T^2}{z}\right)$$
$$z = \frac{(E+p_{\parallel})_M}{(E+p_{\parallel})_q} \tag{9.62}$$

where suffix M denotes the produced meson and q its parent quark. The constants a, b are adjustable parameters and are approximately $a \simeq 1$, $b \simeq 0.4\,\text{GeV}^{-2}$. Experimental reproducibility of the fragmentation function is as good as any other fitted function like one in Eq. (9.55). The function has more weight on the large side of z for production of heavy quarks and automatically reproduces a similar shape as given by Eq. (9.56).

c. Gluon The gluon is considered as a kink of the string (Figure 9.20). Therefore, hadrons are produced on the two strings with the kink at the apex. No new fragmentation function, and hence no new parameter is introduced for the gluon. However, to avoid the divergence, those of low virtuality $Q < Q_0 = 5 \sim 10\,\text{GeV}$ are treated as $q\bar{q}$ pairs. Thus, fragmentation is not symmetric between the quark and gluon as is the case for the independent fragmentation model.

c. p_T, Flavor and Spin Because of the tunneling effect, the transverse momentum distribution obeys the Gaussian distribution Eq. (9.61a). Because of the mass factor in the exponent, suppression of heavy quarks is automatic. Inserting the quark mass values gives $u, d, s, c \sim 1 : 1 : 0.3 : 10^{-11}$ (i.e., the charm quark production is practically inhibited) and reproduces the experimental values. However, some of the parameters are deliberately put in by hand just like the IF model to improve agreements with experiments. Spin effect and production of baryons are treated similarly as the IF model.

The string model takes into account the confinement effect of QCD, Lorentz invariance, energy-momentum conservation and has very few parameters as to the space time structure (i.e., energy-momentum distribution). It introduces, however, many parameters to the flavor structure of the hadron productions. Reproducibility of the data is very good.

9.3.3
Cluster Model

The simplest method of the hadronization apart from LPHD assumption is the cluster model in which all the virtual gluons are forced to convert to quark–antiquark pairs when the virtuality Q goes down below some prescribed value Q_0. Then, a color neutral pair q and $\overline{q}'$ (referred to as a cluster) which lie mutually close in the phase space is picked up (Figure 9.21a). Then, they are forced to decay to a pair of mesons or baryons having flavors $q\overline{q}'$, $q'\overline{q}$ or $qq'q''$, $\overline{q}, \overline{q}'\overline{q}''$ and spin s_1, s_2 with weight proportional to the phase space factor $p_{\text{CM}} \cdot (2s_1 + a)(2s_2 + 1)$. The decay occurs isotopically. If the cluster mass is too light to have decay products, $q\overline{q}'$ is converted to hadrons with the right quantum number. If the decay products are resonances, further decay processes ensue.

Preconfinement A conspicuous characteristic of the cluster model is its simplicity, weighted only with the phase space volume thus having no adjustable parameters. This was intentional because it was introduced as a supplement to the parton shower generator and today is best known as the hadronization model implemented in the HERWIG event generator [471, 476]. All the essential features are supposedly in the QCD motivated shower model and the hadronization model should play a minor role if the virtuality at the end of the shower development is low enough. The cluster was motivated by a phenomenon referred as the preconfinement demonstrated in the perturbative QCD [493–495]. At the end of cascading branches, the mass and spatial distributions of color singlet clusters of partons spanned by quark–antiquark pairs have a universal distribution and peaks around $\sim 1\,\text{GeV}$ (Figure 9.21b). In the cluster fragmentation, the remaining color singlets at the cut-off scale Q_0 are forcibly split into light $q\overline{q}$ pairs so that in the planar approximation, neighboring quark–antiquark pairs form color singlets. This is pictorially illustrated in Figure 9.21a. The first parton shower model (Webber model [471, 476]) used this very simple cluster model as the hadronization model. Since the preconfinement is universal, the parameters should be able to tune with one experiment at single energy in principle. Parameters that the Webber model used are only Q_0 and $\Lambda_{\overline{MS}}$. Surprisingly, it has survived severe experimental tests to this day. In HERWIG [471–473], the successor of the Webber model, a few more parameters are added to improve the accuracy, resulting in the increase of parameters, but the number is still small compared to other models.

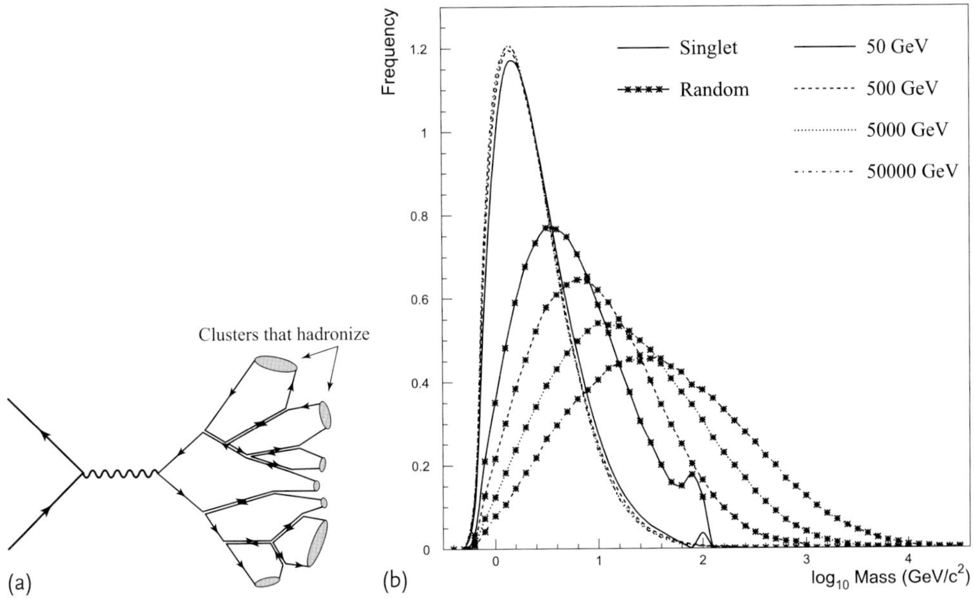

Figure 9.21 Preconfinement and the cluster model. (a) The cluster model: lines are color flow and double lines are gluons. In HERWIG event generators, gluons are forced to convert to $q\overline{q}$ pairs when the virtuality Q reaches $Q_0 \sim 1\,\text{GeV}$, constituting high mass mesons (clusters) which, in turn, decay to resonances and stable hadrons. The shaded areas denote the clusters. Preconfinement tends to combine pairs in the immediate neighborhood to form color neutral clusters. Compare the cluster formation with that of the string model in Figure 9.20. (b) Preconfinement: After showering, the mass spectrum of color singlet is universal and peaks around $\sim 1\,\text{GeV}$ regardless of the initial energies. In the random sample, $q\overline{q}$ pairs from gluon splitting are excluded [492].

9.3.4 Model Tests

Available fragmentation generators use a combination of the matrix element, the parton shower generation and eventual hadronization models. A representative program of the matrix method is the matrix version of JETSET (successor of Lund model) [197, 489]. Those of the parton shower models include the parton shower version of the JETSET (virtuality = invariant mass) [496], HERWIG (virtuality = $E\theta$, cluster model for hadronization) [471–473], ARIADNE (virtuality = p_T and $2 \to 3$ branching (color dipole model) [474], hadronization = SF model) and for hadron reactions MRSA [497, 498] and CTEQ [421, 499]. Every model has its own advantages and disadvantages, and needs to be used with some qualifications. At present, there are no omnipotent programs and it is desirable to use more than one program in comparing with data.

Low Energy vs. High Energy Hadronization models, although constructed on rough arguments in the beginning, have refined themselves with accumulation of

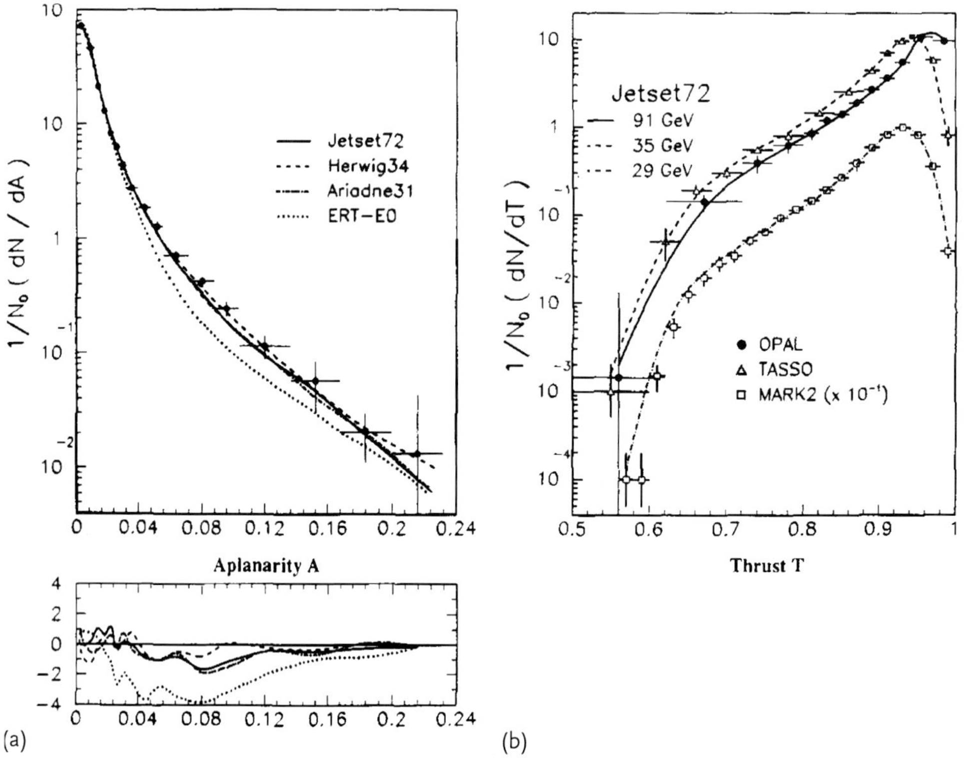

Figure 9.22 Test of hadronization models. (a) Aplanarity: Parameters of various programs were tuned with low energy data at PEP/PETRA and applied to high energy data obtained by OPAL/LEP. ERT-EO is a result of matrix method which bypasses the parton shower generation [461, 501]. (b) Thrust: Parameters are tuned with the data at LEP. Then, the predicted thrust distribution is compared with low energy data at TASSO and MARKII.

data using QCD as a guiding principle. Today, they can reproduce the vast range of data and have predictive powers in testing the theoretical ideas. Comparisons of models for specific observables will be discussed later. Parameters of the models are tuned with various observables ($d\sigma/dx$, $d\sigma/dp_T$, particle multiplicity n_{ch}, and global event shape variables such as thrust, aplanarity and so on). In the early days, data by MARKII/PEP ($\sqrt{s} = 29$ GeV) were used to tune the parameters [500]. However, after LEP and SLC at $\sqrt{s} = 91$ GeV began to produce data [501, 502], consistency between the low and high energy data were examined.

Here, we choose to pick up the thrust T and aplanarity A_p as benchmarks. The global event shape variables very much depend on the jet structure of events and relative weight of n-jets, and the above two variables are suitable for discerning differences of the various distributions. For instance, many jets are produced in a variety of distributions, spherical ($T, A_p \to 0.5$), 3-jets with disc-like distribution ($T \simeq 2/3, A_p \simeq 0$), back-to-back 2-jet production ($T = 1, A_p = 0$), and their difference can be discerned.

Figure 9.22a shows the aplanarity distribution obtained at LEP and compared with the model tuned with low energy data. Figure 9.22b shows the thrust distribution where the parameters are tuned with the OPAL data and tested the model's reproducibility at low energy. One can see that many different models can reproduce data at high energy as well as at low energy. One caveat is the analytical matrix element calculation accompanied with immediate application of hadronization models without going through the parton shower process (ERT of Figure 9.22) has a poor fit at a large value of A_p (a measure of the momentum flow out of the reaction plane). This is considered due to the high virtuality where the fragmentation model is applied ($Q \sim 10\,\text{GeV}$ at the Z resonance region) and the weight of the model compared to perturbative QCD is big.

One concludes from the analysis that the parameters of the hadronization models do not depend on the energy, and that the difference between data of low and high energy regions can be taken into account by the running coupling constant $\alpha_s(Q^2)$. Namely, a single parameter set which takes into account the asymptotic freedom can explain data both at low and high energy in a universal way.

Difference Among the Models The difference between the Independent fragmentation model and the other two is most apparent in the azimuthal angular distribution of 3-jets (Figure 9.23). Although all the models can reproduce the overall shape of various distributions, the IF models fail to reproduce a fine structure at a valley between the most energetic and the second energetic jets. This is the string effect of the gluon which we will discuss in Section 10.2.2.

One sees that both the string and the parton shower model have string effects, but the IF model does not. An intuitive explanation that the hadrons are suppressed on the opposite side of the gluon jet is pictorially depicted in Figure 9.24. It is interesting to note that the basic concept of the string and parton shower model is widely diverse, yet both can reproduce the similar string effect.

Particle Spectra In order to test how precisely the models are tuned, let us see the reproducibility of hadron spectra, although it has no direct connection with jet phenomena. Tables 9.1 and 9.2 shows a comparison of the measured multiplicities of various particles in the Z-decay with three representative models. The underlined numbers show results which are more than three standard deviations from experimental measurements.

The string models (JETSET74 [487], UCLA74[505]) have better reproducibility, but considering the number of parameters, it is no surprise. Even then, there is no predictability as to the tensor mesons. On the other hand, reproducibility of the Webber model is poor at the baryon sector and shows the necessity to increase adjustable parameters. On the whole, however, one may argue that the hadronization models have the fair ability to reproduce not only jet structures, but also particle spectra as well. Note that the two tables are shown to give an idea about the models' overall reproducibility circa 1996. Tuning parameters of the models are continuously updated. Therefore, the reader should refer to the most recent version.

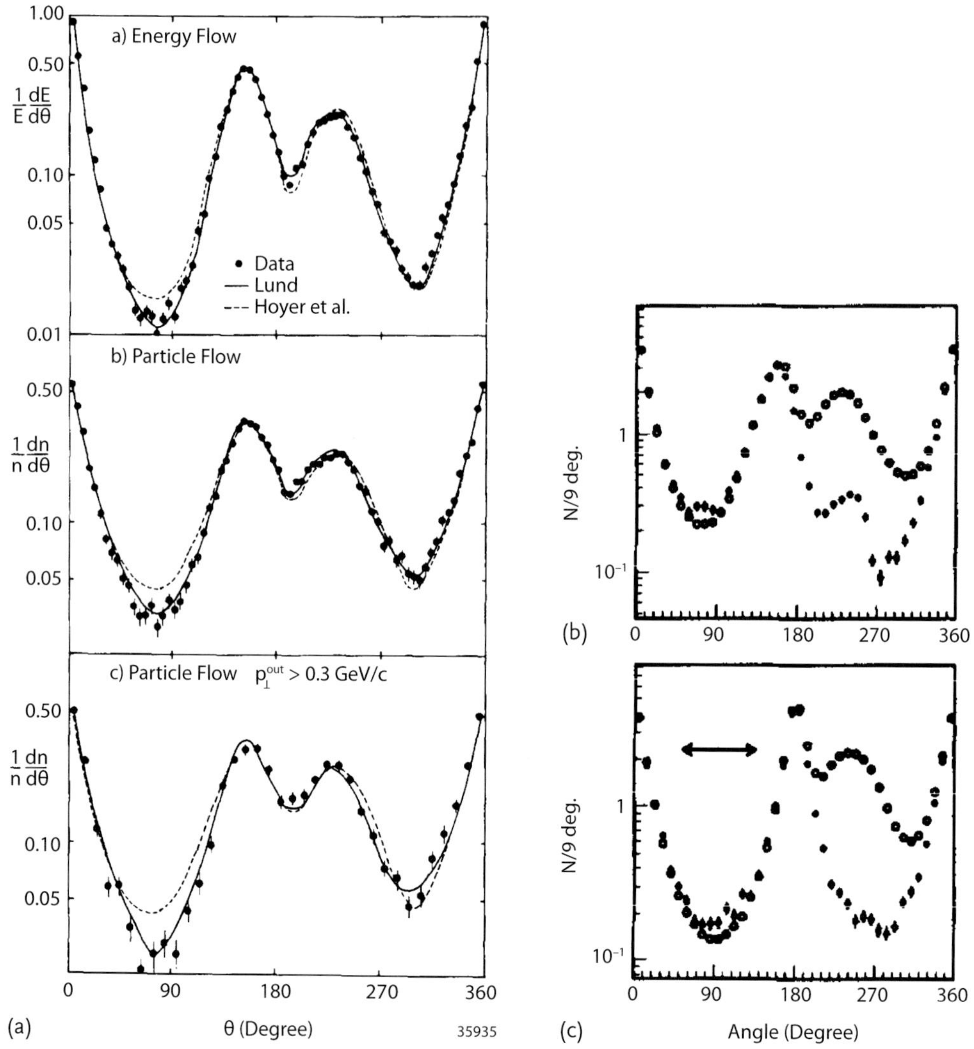

Figure 9.23 Particle flow data of JADE [503] (a) and L3 (b,c) [504]. Particle flows in a valley (at ~ 90° in the figure) between the most and the second most energetic jets are compared with models. The JADE data (a) fits well with SF (real lines) but not with IF (dashed lines) models. The L3 data compares $q\overline{q}g$ (b) and $q\overline{q}\gamma$ (c) in both panels. (b) is a distribution in the laboratory frame and (c) is that of $q\overline{q}$ center of mass frame with γ removed. Notice that the least energetic jet disappeared if γ is removed. Their differences are well reproduced with SF or parton shower models, but not with IF.

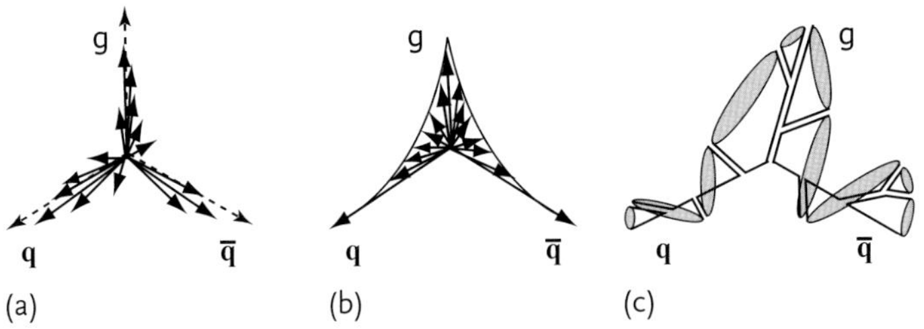

Figure 9.24 A conceptual illustration of the independent fragmentation (IF), string (SF) and parton shower model (PS). (a) The IF model treats the gluon and the quarks equally. (b) The SF model treats the gluon as a kink, and hence less hadrons are produce at the opposite side of the gluon direction. (c) The cluster model: due to preconfinement, after the parton shower development, the cluster tends to be formed among color neutral pairs, nearby in the phase space. The gluon has the ability to produce twice as many pairs (the strength $C_A/C_F = 3/(4/3) \sim 2$), and many clusters are produced on the gluon side.

Table 9.1 Meson yields in Z^0 decay. Experiments: A = Aleph, D = Delphi, L = L3, M = Mark II, O = Opal. Bold letters indicate new data and underlined numbers show disagreement with data by more than 3σ.

Particle	Multiplicity	HERWIG 5.9	JETSET 7.4	UCLA 7.4	Expts
Charged	20.96(18)	20.95	20.95	20.88	ADLMO
$\pi^\pm$	17.06(24)	17.41	16.95	17.04	ADO
π^0	9.43(38)	9.97	9.59	9.61	ADLO
$\boldsymbol{\eta}$	0.99(4)	1.02	1.00	0.78	ALO
$\rho(770)^0$	1.24(10)	1.18	1.50	1.17	AD
$\omega(782)$	1.09(9)	1.17	1.35	1.01	ALO
$\boldsymbol{\eta'(958)}$	0.159(26)	0.097	0.155	0.121	ALO
$f_0\mathbf{(980)}$	0.155(8)	0.111	$\sim$ 0.1	–	ADO
$a_0(980)^\pm$	0.14(6)	0.240	–	–	O
$\phi(1020)$	0.097(7)	0.104	0.194	0.132	ADO
$f_2\mathbf{(1270)}$	0.188(14)	0.186	$\sim$ 0.2	–	ADO
$f_2'(1525)$	0.012(6)	0.021	–	–	D
$K^\pm$	2.26(6)	2.16	2.30	2.24	ADO
K^0	2.074(14)	2.05	2.07	2.06	ADLO
$K^*(892)^\pm$	0.718(44)	0.670	1.10	0.779	ADO
$K^*(892)^0$	0.759(32)	0.676	1.10	0.760	ADO
$K_2^*(1430)^0$	0.084(40)	0.111	–	–	DO
$D^\pm$	0.187(14)	0.276	0.174	0.196	ADO
D^0	0.462(26)	0.506	0.490	0.497	ADO
$D^*(2010)^\pm$	0.181(10)	0.161	0.242	0.227	ADO
$D_S^\pm$	0.131(20)	0.115	0.129	0.130	O
B^*	0.28(3)	0.201	0.260	0.254	D
$B_{u,d}^{**}$	0.118(24)	0.013	–	–	D
J/ψ	0.0054(4)	0.0018	0.0050	0.0050	ADLO
$\psi(3685)$	0.0023(5)	0.0009	0.0019	0.0019	DO
χ_{c1}	0.0086(27)	0.0001	–	–	DL

Table 9.2 Baryon yields in Z^0 decay. Legends as in Table 9.1.

Particle	Multiplicity	HERWIG 5.9	JETSET 7.4	UCLA 7.4	Expts
p	1.04(4)	0.863	1.19	1.09	ADO
Δ^{++}	0.079(15)	0.156	0.189	0.139	D
	0.22(6)	0.156	0.189	0.139	O
Λ	0.399(8)	0.387	0.385	0.382	ADLO
$\Lambda(1520)$	0.0229(25)	–	–	–	DO
$\Sigma^{\pm}$	0.174(16)	0.154	0.140	0.118	DO
Σ^0	0.074(9)	0.068	0.073	0.074	ADO
$\Sigma^{*\pm}$	0.0474(44)	0.111	0.074	0.074	ADO
Ξ^-	0.0265(9)	0.0493	0.0271	0.0220	ADO
$\Xi(1530)^0$	0.0058(10)	0.0205	0.0053	0.0081	ADO
Ω^-	0.0012(2)	0.00556	0.00072	0.0011	ADO
Λ_c^+	0.078(17)	0.0123	0.059	0.026	O

Summary We have investigated two distinct approaches in comparing theoretical predictions at the parton level with observed hadron distributions. In many ways, the parton shower model has advantages over the matrix element method. However, it has a conceptual weakness in treating hard processes where the reverse is true. Recently, there has been some progress in merging the two methods. The above shortage is greatly improved by at least incorporating the first hard gluon emission matching accurately with the matrix method [465, 466] and also by including the next leading order approximation in the shower generation [467, 468].

9.4 Test of the Asymptotic Freedom

The running coupling constant $\alpha_s(Q^2)$ stands out as the single most important fundamental constant of QCD and plays a role of the fine structure content in QED. However, its value and the energy dependence has to be determined by experiments. It is the most important theme of QCD. Here, we discuss various methods to measure it.

9.4.1 Inclusive Reactions

In order to determine α_s or equivalently $\Lambda_{\overline{MS}}$, it is best to use processes that do not depend on hadronization models or jet shape variables. In this sense, inclusive reactions, including the deep inelastic cross sections and the R_γ value which is the ratio of the total hadronic to $\mu\bar{\mu}$ pair production cross sections by e^-e^+ collision, serve the best. QCD corrections to similar observables, the relative Z decay width R_Z and hadronic decay width of the τ lepton R_τ, provide benchmarks at the highest and lowest Q^2 values thus far available. Determination using jet shape variables is complementary and serves to clarify various aspects of the jet structure. Although the QCD correction to the electroweak processes in the inclusive reactions receive the most unambiguous treatment theoretically, there is one caveat, that is, it is a correction to the main electroweak process ($\delta R/R, \delta\Gamma_Z/\Gamma_Z \sim 5\%$), and hence it is not easy to obtain an accurate value of Rs experimentally.

Deep inelastic scattering The method to determine α_s using the deep inelastic scattering data is the first and still one of the most accurate [24]. It shares the same advantage with the R values which is independent of hadronization models. While the QCD contribution in other inclusive reactions are corrections to main electroweak processes and need precision experiments to extract it, the DIS process has the advantage to see the Q^2 evolution directly.

The nonsinglet structure function F_3 offers, in principle, the most precise test of the theory because the Q^2 evolution is independent of the gluon distributions. One way to extract α_s in a similar manner from the DIS data is to use the Gross–Llewellyn–Smith sum rule (see Vol. 1, Sect. 17.8.5 [506]). The QCD correction is derived up to $O(\alpha_s^3)$ [507, 508], estimates of $O(\alpha_s^4)$ [509] and nonperturbative contribution [508] also exist.

$$\int_0^1 dx\, F_3(x) = 3\left(1 - \frac{\alpha_s}{\pi} - 3.5833\left(\frac{\alpha_s}{\pi}\right)^2 - 18.976\left(\frac{\alpha_s}{\pi}\right)^3 + \cdots\right) - \Delta HT \tag{9.63}$$

where ΔHT is the so-called higher twist term which is a nonperturbative correction and is estimated to be $(0.09 \pm 0.045)/Q^2$ [508, 509]. A measurement of the α_s was carried out by the CCFR group [510] and gives

$$\alpha_s(1.73\,\text{GeV}) = 0.28 \pm 0.035(\text{exp}) \pm 0.05(\text{sys}) \pm 0.04(\text{theory}) \tag{9.64}$$

which corresponds to

$$\alpha_s(m_Z) = 0.118 \pm 0.011 \tag{9.65}$$

The comparison of evolving structure functions with data provides another powerful test. Although it is theoretically not as clean as the sum rules because of the need for gluon distribution functions, it is a comprehensive test in the sense that, in principle, quite a large number of measured points can be fitted with a single parameter x. A global fit combining muon and neutrino data and extrapolated to $Q = m_Z$ using NLLA expressions for α_s [Eqs. (7.84a) and (7.79)] gives [511, 512]

$$\begin{aligned} \alpha_s(m_Z) &= 0.117 \pm 0.0045 \\ \Lambda_{\overline{MS}} &= 305 \pm 25 \pm 50\,\text{MeV} \end{aligned} \tag{9.66}$$

Note that the value obtained from DIS data is $\Lambda_{\overline{MS}} = \Lambda_{\overline{MS}}(4)$ and has to be converted to $\Lambda_{\overline{MS}}(5)$ in comparing with those obtained in the e^-e^+ or hadron colliders at large $\sqrt{s}$.

R_γ Inclusive hadron reactions defined by

$$R_\gamma = \frac{\sigma(e^- e^+ \to \text{hadrons})}{\sigma(e^- e^+ \to \mu^- \mu^+)} \tag{9.67}$$

and similar quantities R_Z, R_τ which we will discuss shortly are best suited for determining the coupling constant, as they provide theoretically clean expressions. Before elaborating on α_s power expansions, it is instructive to discuss the general features of the inclusive reactions. Using a similar expression as Eq. (7.118a), R_γ can be decomposed to leptonic and hadronic parts:

$$\begin{aligned} \sigma_{\text{TOT}} &= \frac{1}{2s}\sum_f |\mathcal{M}_{fi}|^2 = \frac{e^4}{2s} L_{\mu\nu} \frac{1}{q^4} H^{\mu\nu}(q^2) \\ q^2 &= s = (k_1 + k_2)^2 \end{aligned} \tag{9.68}$$

where $2s$ is the flux factor and $L_{\mu\nu}$ and $H^{\mu\nu}$ represent the leptonic and hadronic part.

$$\begin{aligned} L^{\mu\nu} &= \frac{1}{4}\sum_{\text{spin}} [\overline{u}(k_1)\gamma_\mu v(k_2)][\overline{u}(k_1)\gamma_\nu v(k_2)]^* \\ &= k_{1\mu} k_{2\nu} + k_{1\nu} k_{2\mu} - (k_1 \cdot k_2) g_{\mu\nu} \end{aligned} \tag{9.69a}$$

$$\begin{aligned} H^{\mu\nu}(q^2) &= \int d^4 x e^{i q\cdot x} \langle \Omega | J^\mu(x) J^\nu(0) | \Omega \rangle \\ &= \sum_f (2\pi)^4 \delta^4(q - P_{\text{F}}) \langle P_{\text{F}} | e J^\nu(0) | \Omega \rangle \langle P_{\text{F}} | e J^\mu(0) | \Omega \rangle^* \\ &\equiv (-q^2 g^{\mu\nu} + q^\mu q^\nu) H(q^2) \end{aligned} \tag{9.69b}$$

The current $J^\mu(x)$ is defined by

$$J^\mu(x) = \overline{q}(x)\gamma^\mu q(x) \tag{9.70}$$

which is the quark current of flavor q. We used translational symmetry to extract

$$\langle f | J^\mu(x) | \Omega \rangle = \langle f | e^{i p_f \cdot x} J^\mu(0) e^{-i p_f \cdot x} | \Omega \rangle = e^{i P_f \cdot x} \langle f | J^\mu(0) | \Omega \rangle \tag{9.71}$$

and the last equality in Eq. (9.69b) is obtained by the Lorentz invariance and current conservation. It also defines the spectral density function $H(q^2)$.

The cross section given by Eq. (9.68) is exact to the order $O(\alpha)$ in the electromagnetic interaction, but exact to all orders in the strong interaction as the $H^{\mu\nu}(q^2)$ contains all the hadronic final states. The $H^{\mu\nu}(q^2)$ is related to the Fourier transform of a two-point function $\Pi^{\mu\nu}(x, y)$ by the unitarity relation

$$H^{\mu\nu}(q^2) = 2\text{Im}\Pi^{\mu\nu}(q^2) \tag{9.72}$$

where $\Pi^{\mu\nu}(q^2)$ is defined by

$$\begin{aligned} \Pi^{\mu\nu}(q^2) &\equiv i \int d^4 x e^{i q\cdot x} \langle \Omega | T[J^\mu(x) J^\nu(0)] | \Omega \rangle \\ &= (-q^2 g^{\mu\nu} + q^\mu q^\nu) \Pi(q^2) \end{aligned} \tag{9.73}$$

This is a special example of more general relations for the two-point correlation function (Appendix I).

Since $L_{\mu\nu}q^\mu = 0$ by current conservation, contraction of $L_{\mu\nu}H^{\mu\nu}$ gives

$$L_{\mu\nu}H^{\mu\nu} = [k_{1\mu}k_{2\nu}+k_{1\nu}k_{2\mu}-(k_1\cdot k_2)g_{\mu\nu}](-q^2g^{\mu\nu}+q^\mu q^\nu)H(q^2) = q^4 H(q^2) \tag{9.74}$$

Therefore, the total cross section can be expressed as

$$\sigma_{\text{TOT}} = \frac{e^4}{2s}H(q^2) = \frac{(4\pi\alpha)^2}{s}\text{Im}\Pi(q^2) \tag{9.75}$$

The $\Pi^{\mu\nu}(q^2)$ is related to the photon propagator (see Eq. (I.34) and (I.39)). Namely, it is the higher order correction to the photon propagator. In one loop approximation, $e^2\Pi^{\mu\nu}(q^2)$ reduces to $\Sigma^{\mu\nu}$ given by Eq. (I.12a).

The two-point function Eq. (9.73) is an analytical function and contains all the necessary information of the strong interaction. Using the unitarity, its imaginary part is expressed in terms of observables and its real part can be obtained by dispersion relations (see Appendix I, Eq. (I.40)). It is suitable for calculating higher order perturbation series and also for organizing the perturbative as well as nonperturbative contribution using the operator product expansion (OPE) method, but that is beyond the scope of this book and we will not elaborate any further.

As to R_γ, R_Z and R_τ, perturbation calculations have been carried out up to next-next-to-leading order (NNLO), namely, to $O(\alpha_s^3)$ (see Eq. (7.89), [358–360] and [513] for comprehensive review).

$$\begin{aligned}
R_\gamma &= \frac{\sigma(e^-e^+\to\text{hadrons})}{\sigma(e^-e^+\to\mu^-\mu^+)} = 12\pi\text{Im}\Pi(s) \\
&= R_0\left(1+\delta_{\text{QCD}}(0)+\delta_m+\delta_{np}\right) \\
&= R_0\left(1+\frac{\alpha_s}{\pi}+1.4092\left(\frac{\alpha_s}{\pi}\right)^2-12.805\left(\frac{\alpha_s}{\pi}\right)^3+O\left(\alpha_s^4\right)\right) \\
R_0 &= N_c\sum_q Q_q^2
\end{aligned} \tag{9.76}$$

where $\delta_{\text{QCD}}(0)$ represents the correction term for massless quarks and is the dominant term. δ_m and δ_{np} represent the mass correction and nonperturbative part (the so-called twist $\sim O(1/m_\tau^4)$) term [514]). For the data at $\sqrt{s} = 10 \sim 40$ GeV, both δ_m and δ_{np} can be neglected. Equation (9.76), applied to data at $\sqrt{s} = 35$ GeV [445, 462], gives

$$\begin{aligned}
\alpha_s(34\,\text{GeV}) &= 0.146 \pm 0.03 \\
\Lambda_{\overline{MS}} &= 223^{+160}_{-109}\,\text{MeV}
\end{aligned} \tag{9.77}$$

R_τ R_τ is the hadronic part of the τ decay, namely,

$$R_\tau = \frac{\Gamma(\tau\to\nu_\tau+\text{hadrons})}{\Gamma(\tau\to e\overline{\nu}_e\nu_\tau)} = \frac{\text{BR}(\tau\to\nu_\tau+\text{hadrons})}{\text{BR}(\tau\to e\overline{\nu}_e\nu_\tau)} \tag{9.78a}$$

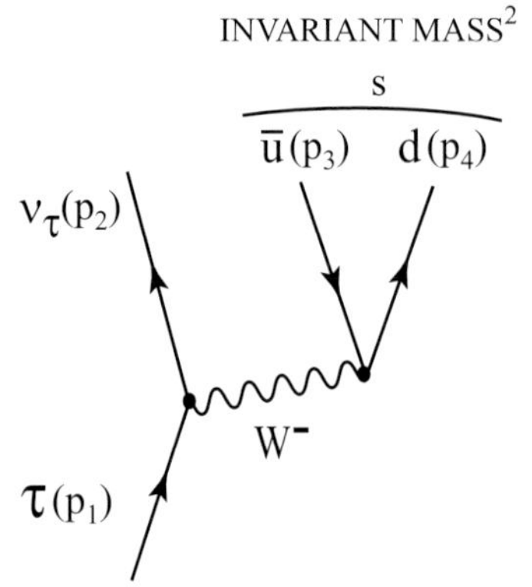

Figure 9.25 The decay diagram for $\tau \to \nu_\tau +$ hadrons) can be obtained from $e^- e^+ \to q\overline{q}$ by changing to $e^- e^+ \to \tau \nu_\tau$, crossing and replacing γ by W. The hadronic invariant mass which corresponds to s in the $e^- e^+$ reaction takes continuous values due to the presence of ν_τ in the final state. To obtain the total hadronic decay width, its contribution has to be integrated over the invariant mass "s."

It is very similar to $R_\gamma(e^- e^+ \to$ hadrons) as it is a crossed version of the latter with γ replaced with W^- (see Figure 9.25). Since $m_\tau < m_c$, a naive quark model prediction gives

$$R_\tau = \frac{\Gamma(\tau^- \to \nu_\tau \overline{u}(d + s))}{\Gamma(\tau^- \to \nu_\tau e^- \overline{\nu}_e)} = N_c \left(|V_{ud}|^2 + |V_{us}|^2\right) = 3 \tag{9.79}$$

The denominator $\Gamma(\tau \to \nu_\tau e^- \overline{\nu}_e)$ can be obtained from the expression for $\mu \to \nu_\mu e^- \overline{\nu}_e$ with suitable replacement of variables ([515], see also Vol. 1, Eq. (15.86)).

$$\Gamma(\tau \to l \overline{\nu}_l \nu_\tau) = \frac{G_F^2 m_\tau^5}{192\pi^3} f\left(\frac{m_l^2}{m_\tau^2}\right) \delta_W^l \delta_\gamma^l \tag{9.80a}$$

$$f(x) = 1 - 8x + 8x^3 - x^4 - 12x^2 \ln x \tag{9.80b}$$

$$\delta_\gamma^l = 1 + \frac{\alpha(m_l)}{2\pi}\left(\frac{25}{4} - \pi^2\right) \tag{9.80c}$$

$$\delta_W^\tau = 1 + \frac{3}{5}\frac{m_\tau^2}{m_W^2} - 2\frac{m_l^2}{m_W^2} \tag{9.80d}$$

$$r_{EW} = \delta_W^\tau \delta_\gamma^\tau = 0.9960 \tag{9.80e}$$

However, because of the existence of ν_τ in the final state, the invariant mass s of $q\overline{q}$ is continuous from m_π^2 to m_τ^2 and has to be integrated. Its value goes into the nonperturbative region and the use of the perturbative QCD has to be justified.

The expression for the hadronic width can be obtained from the formula to calculate $\Gamma(\tau^- \to \nu_\tau e^- \overline{\nu}_e)$ by replacing the charged electron current $j_e^\mu(x) = \overline{e}\gamma^\mu(1 - \gamma^5)\nu_e(x)$ with the hadronic current $J_-^\mu(x) = \overline{d}'(x)\gamma^\mu(1 - \gamma^5)u(x)$ where

d' is the Cabibbo-rotated d quark field. Setting kinematical variables as shown in Figure 9.25, the decay amplitude is given by

$$\begin{aligned} S_{fi} - \delta_{fi} &= -(2\pi)^4 \delta^4(p_1 - p_2 - p_f)\mathcal{M}_{fi} \\ &= -i(2\pi)^4 \delta^4(p_1 - p_2 - p_f) \\ &\quad \times \left[\frac{G_F V_{ud}^*}{\sqrt{2}} \left[\overline{u}(p_2)\gamma_\mu(1-\gamma^5)u(p_1)\right] \langle f | J_-^\mu(0)|0\rangle + (d \to s) \right] \end{aligned} \tag{9.81}$$

where "f" denotes the hadronized final states of $\overline{u}$ and d'. The hadronic decay rate is given by

$$\begin{aligned} \Gamma_\tau &= \frac{1}{2m_\tau}(2\pi)^4 \delta^4(p_1 - p_2 - p_f)|\overline{\mathcal{M}}|^2 dLIPS \\ &= \frac{1}{2m_\tau} \frac{G_F^2 |V_{ud}|^2}{2} L_{\mu\nu}^\tau H^{\mu\nu}(q) \frac{d^3 p_2}{(2\pi)^3 2E_2} + (d \to s) \end{aligned} \tag{9.82a}$$

$$q = p_1 - p_2 \tag{9.82b}$$

$$\begin{aligned} L_{\mu\nu}^\tau &= \frac{1}{2}\sum_{\text{spin}} \left[\overline{u}(p_2)\gamma_\mu(1-\gamma^5)u(p_1)\right]\left[\overline{u}(p_2)\gamma_\mu(1-\gamma^5)u(p_1)\right]^* \\ &= 4[p_{1\mu}p_{2\nu} + p_{1\nu}p_{2\mu} - g_{\mu\nu}(p_1 \cdot p_2)] \\ &\quad + (\text{antisymmetric in } p_1 \text{ and } p_2) \end{aligned} \tag{9.82c}$$

$$\begin{aligned} H^{\mu\nu}(q) &\equiv \int d^4x e^{iq\cdot x}\langle 0|J_+^\mu(x)J_-^\nu(0)|0\rangle \\ &= (2\pi)^4 \sum_f \delta^4(q - p_f)\langle f|J_-^\nu(0)|0\rangle\langle f|J_-^\mu(0)|0\rangle^* \\ &= 2\left[(-q^2 g^{\mu\nu} + q^\mu q^\nu)\text{Im}\Pi^{(1)}(q^2) + q^\mu q^\nu \text{Im}\Pi^{(0)}(q^2)\right] \end{aligned} \tag{9.82d}$$

which defines the spectral functions $\text{Im}\Pi^{(1,0)}(q)$. The superscript ($J = 0, 1$) denotes the angular momentum components in the hadronic rest frame. $\Pi^{\mu\nu}(q^2)$ is given by (see Eq. (I.1))

$$\Pi^{\mu\nu}(q) = \int d^4x e^{iq\cdot x}\Pi^{\mu\nu}(x) = (-g^{\mu\nu}q^2 + q^\mu q^\nu)\Pi^{(1)}(q^2) + q^\mu q^\nu \Pi^{(0)}(q^2) \tag{9.83a}$$

$$\Pi^{\mu\nu}(x) = i\langle 0|T\left[J_+^\mu(x)J_-^\nu(0)\right]|0\rangle \tag{9.83b}$$

Equation (9.83) looks similar to Eq. (9.73), but differs in two points. The $J_\pm^\mu$ is the charged weak current which changes the electric charge and it includes the axial vector A^μ as well as the vector current V^μ. Since the axial current does not conserve, it produces the extra contribution $\Pi^{(0)}$ in the Lorentz decomposition.

The spectral function $\text{Im}\Pi^{(1,0)}$ contains the dynamical information on the invariant mass distribution of the final hadrons. Denoting the total energy of the hadronic system as s and using

$$\begin{aligned}
2(p_1 \cdot p_2) &= p_1^2 + p_2^2 - (p_1 - p_2)^2 = m_\tau^2 - q^2 \equiv m_\tau^2 - s \\
\frac{d^3 p_2}{(2\pi)^3 2E_2} &= -\frac{m_\tau^2}{16\pi^2}\left(1 - \frac{s}{m_\tau^2}\right)\frac{ds}{m_\tau^2}
\end{aligned} \tag{9.84}$$

the hadronic τ width can be written as

$$R_\tau = 12\pi \int_{m_\pi^2}^{m_\tau^2} \frac{ds}{m_\tau^2}\left(1 - \frac{s}{m_\tau}\right)^2 \left[\left(1 + 2\frac{s}{m_\tau^2}\right)\text{Im}\Pi^1(s) + \text{Im}\Pi^0(s)\right] \tag{9.85}$$

General properties of the two-point function are investigated by [516–519]. For an analytic function without poles, the Cauchy's theorem dictates that the contour integral along the loop depicted in Figure 9.26 vanishes. Then, it is written as

$$\begin{aligned}
0 = \oint = &\int_{|s|=m_\tau^2} f(z)ds + \int_{m_\pi^2}^{m_\tau^2} dx\, f(x + i\epsilon) + \int_{m_\tau^2}^{m_\pi^2} dx\, f(x - i\epsilon) \\
\to \quad &2i \int_{m_\pi^2}^{m_\tau^2} dx \text{Im} f(x) + \int_{|s|=m_\tau^2} f(z)ds = 0
\end{aligned} \tag{9.86}$$

The integral Eq. (9.85) can therefore be expressed as a contour integral in the complex s plane.

$$R_\tau = 6\pi i \oint_{|s|=m_\tau^2} \frac{ds}{m_\tau^2}\left(1 - \frac{s}{m_\tau}\right)^2 \left[\left(1 + 2\frac{s}{m_\tau^2}\right)\text{Im}\Pi^1(s) + \text{Im}\Pi^0(s)\right] \tag{9.87}$$

Now, one sees that the integrand is always evaluated at $|s| = m_\tau^2$ which is large enough to justify the use of perturbative QCD. Moreover, it enables us to use the short distance OPE (operator product expansion) formalism to organize the perturbative and nonperturbative contributions which results in [520, 521]

$$R_\tau = N_c(|V_{ud}|^2 + |V_{us}|^2) S_{\text{EW}} \left(1 + \delta'_{\text{EW}} + \delta_{\text{QCD}}(0) + \delta_{m_q} + \delta_{np}\right) \tag{9.88a}$$

$$N_c(|V_{ud}|^2 + |V_{us}|^2) = 3 \tag{9.88b}$$

where S_{EW}, δ'_{EW} are the known electroweak corrections, $\delta_{\text{QCD}}(0)$ is the perturbative QCD contribution with the quark mass set at zero, δ_{m_q} is the quark mass correction and finally δ_{np} is the nonperturbative contribution. They are expressed as [515,

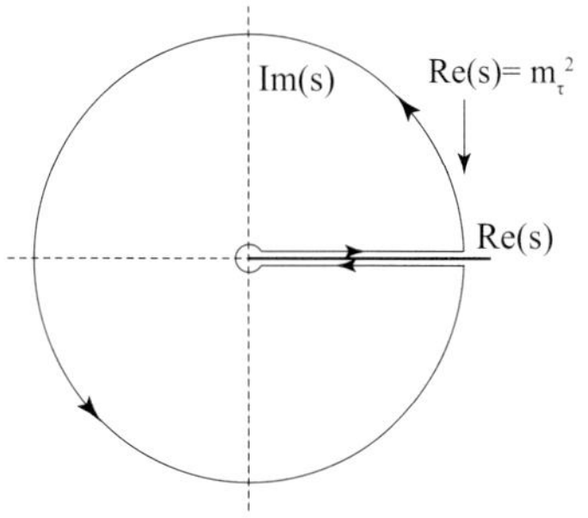

Figure 9.26 Integration contour in the complex s plane used to evaluate R_τ value.

521]

$$S_E = \left(\frac{\alpha(m_b^2)}{\alpha(m_\tau^2)}\right)^{\frac{9}{19}} \left(\frac{\alpha(m_W^2)}{\alpha(m_b^2)}\right)^{\frac{9}{20}} \left(\frac{\alpha(m_Z^2)}{\alpha(m_W^2)}\right)^{\frac{36}{17}} = 1.0194 \quad [219] \tag{9.89a}$$

$$\delta'_{EW} = \frac{5}{12}\frac{\alpha(m_\tau^2)}{\pi} \simeq 0.0010 \quad [143] \tag{9.89b}$$

$$\delta_{QCD}(0) = \left(\frac{\alpha_s}{\pi} + 5.2023\left(\frac{\alpha_s}{\pi}\right)^2 + 26.37\left(\frac{\alpha_s}{\pi}\right)^3 + O(\alpha_s)^4\right) \quad [24, 520, 521] \tag{9.89c}$$

$$\delta_{m_q} = |V_{ud}|^2\delta_{u,d} + |V_{us}|^2\delta_s \quad [375, 520] \tag{9.89d}$$

$$\delta_q = -8\frac{m_q(m_\tau^2)}{m_\tau^2}\left[1 + \frac{16}{3}\left(\frac{\alpha_s(m_\tau^2)}{\pi}\right) + 46.00\left(\frac{\alpha_s(m_\tau^2)}{\pi}\right)^2 + \cdots\right] \quad [521] \tag{9.89e}$$

The nonperturbative contribution can be estimated using OPE (operator product expansion). The leading correction ($O(1/s)$) comes from the quark mass. They are tiny for the up and down quarks ($\delta_{ud} \simeq -0.08\%$), but sizable for the s quark ($\delta_s \simeq -20\%$). However, because of the $|V_{us}|^2$ suppression, the effect on the total ratio R_τ is only $-(1.0 \pm 0.2)\%$ [515].[5)] The next leading power correction turned out to be $O(1/s^3)$ due to the vanishing $1/s^2$ contribution integrated over the circle and furthermore the vector and axial vector contribute in opposite sign. Thus, the theoretical estimate gives a relatively unambiguous and tiny contribution ($\Delta R_\tau \sim 0.005$) from the nonperturbative part [518–521].

Although R_τ can be determined solely from the decay lifetime and the leptonic decay branching ratios,

$$R_\tau = \frac{\Gamma(\tau^- \to \nu_\tau \text{hadrons})}{\Gamma(\tau^- \to \nu_\tau e^- \overline{\nu}_e)} = \frac{1 - B_e - B_\mu}{B_e} = 3.642 \pm 0.012 \quad [521] \tag{9.90}$$

it is instructive to look at the spectral function and how it behaves.

5) Note, however, that the R_τ data is precise enough such that it can be used to determine the mass value of the s-quark (see Section 7.2.5 and [379, 380]).

Spectral Function As the energy is relatively low, the spectral function is expected to exhibit nonperturbative behavior at low or middle energy region if it reaches the asymptotic freedom around $s \approx m_\tau^2$. The nonperturbative effect of the long range force appears as resonances. Experimental data for the spectral functions $\mathrm{Im}\Pi_{V,A}^{J=1,0}$ can be obtained by dividing the normalized invariant mass-squared distribution $(1/N_{V/A})(dN_{V/A}/ds)$ for a given hadronic mass $\sqrt{s}$ by the appropriate kinematic factor. For the nonstrange part of the data, they are given by

$$v_1/a_1 = \frac{m_\tau^2}{6|V_{ud}|^2 S_{\mathrm{EW}}} \frac{B(\tau^- \to V^-/A^- \nu_\tau)}{B(\tau^- \to e^- \overline{\nu}_e \nu_\tau)} \frac{1}{N_{V/A}} \frac{dN_{V/A}}{ds} \times \left[\left(1 - \frac{s}{m_\tau^2}\right)^2 \left(1 + \frac{2s}{m_\tau^2}\right)\right]^{-1} \tag{9.91a}$$

$$a_0 = \frac{m_\tau^2}{6|V_{ud}|^2 S_{\mathrm{EW}}} \frac{B(\tau^- \to \pi^- \nu_\tau)}{B(\tau^- \to e^- \overline{\nu}_e \nu_\tau)} \frac{1}{N_{\mathrm{A}}} \frac{dN_{\mathrm{A}}}{ds} \left(1 - \frac{s}{m_\tau^2}\right)^{-2} \tag{9.91b}$$

where v_1, a_1, a_0 denote vector (V), axial vector (A) and pseudoscalar parts of the spectral function. There is no scalar part for the vector current because of the current conservation. a_0 is almost solely represented by the pion pole which theoretically can be expressed as $a_0 = 4\pi^2 f_\pi^2 \delta(s - m_\pi^2)$ using PCAC (see Eq. (7.42)). The vector and axial vector current parts of nonstrange hadronic decays are separated by the G-parity, that is, the number of pions and also by looking at angular distributions. Experimentally, they were dominated by $1\pi, 2\pi, 3\pi$ modes which, in turn, are dominated by $\pi, \rho(770), a_1(1260)$ and mainly contribute to a_0, v_1, a_1. The rest of the decay modes include $\omega\pi, \eta\pi\pi$ and strange particles ($K\overline{K}, K\overline{K} + n\pi$) which are difficult to separate V from A and consequently are treated with statistical weights. The ρ, a_1 dominance in the vector decay modes is apparent by looking at the $v_1 - a_1$ data (Figure 9.27a). The data on $v_1 + a_1$ (Figure 9.27b) shows that it reaches the asymptotic region around $s \approx m_\tau^2$, although a wavy behavior is seen in the low and middle energy region. The predictions of the parton model with and without QCD correction are indicated by the lines in Figure 9.27. Notice there is no distinction between V and A in the parton model which is valid in the asymptotic region.

Knowing the precise value of R_τ, one can extract that of $\alpha_s(m_\tau^2)$ using the formula Eq. (9.88) and gives the $\alpha_s(Q^2)$ value at the lowest Q value [11, 521–523].

$$\alpha_s(m_\tau^2) = 0.345 \pm 0.004_{\mathrm{exp}} \pm 0.009_{\mathrm{th}} \tag{9.92}$$

Notice that the invariant mass $Q^2 = m_\tau^2$ is in an exquisite position in the boundary of perturbative and nonperturbative regime. The value of $\alpha_s(m_\tau^2)$ is large, and hence the QCD correction is sizable. However, the scale m_τ^2 is so small that when extrapolated to $Q^2 = m_Z^2$, its error is diminished, that is,

$$\begin{aligned} \alpha_s\left(m_Z^2\right) &= 0.1215 \pm 0.0004_{\mathrm{exp}} \pm 0.0010_{\mathrm{th}} \pm 0.0005_{\mathrm{evol}} \\ &= 0.1215 \pm 0.0012 \end{aligned} \tag{9.93}$$

where the error with suffix "evol" originates from the ambiguity of the evolution procedure. This gives the most accurate value at scale $Q = m_Z$ and provides the most critical test of the asymptotic freedom.

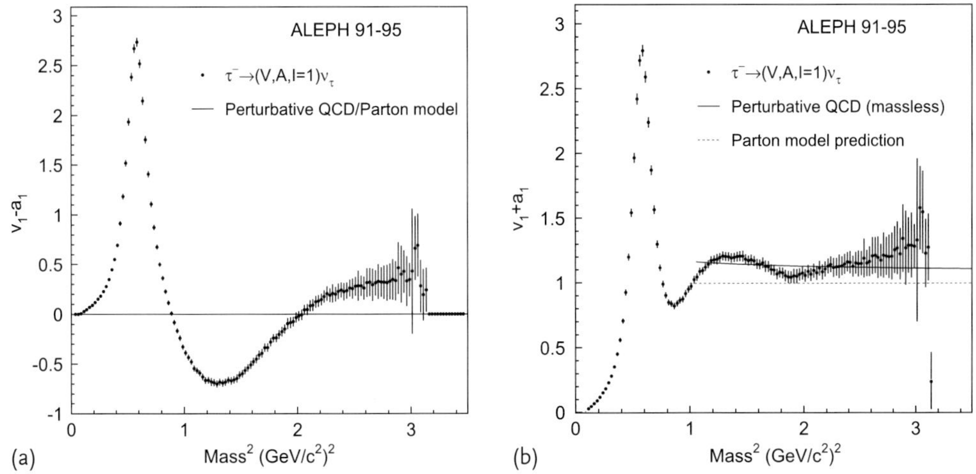

Figure 9.27 (a) The vector minus axial-vector ($v_1 - a_1$) spectral function is shown for data and the parton model. In the parton model as well as in perturbative QCD, the vector and axial-vector contributions are degenerate. (b) The inclusive vector plus axial-vector ($v_1 + a_1$) spectral function and predictions from the parton model and from massless perturbative QCD using $\alpha_s(m_Z^2) = 0.120$ [521, 522].

R_Z QCD correction to R_z is given by [24, 524].

$$\begin{aligned}
R_Z &= \frac{\Gamma(Z \to \text{hadrons})}{\Gamma(Z \to e^- e^+)} = R_{\text{EW}} N_c \left(1 + \delta_{\text{QCD}}(0) + \delta_m\right) \\
&= 19.934 \left(1 + 1.045 \frac{\alpha_s}{\pi} + 0.94 \left(\frac{\alpha_s}{\pi}\right)^2 - 15 \left(\frac{\alpha_s}{\pi}\right)^3 + O(\alpha_s)^4\right) \\
\Gamma(Z \to \text{hadrons}) &= \frac{G_F m_Z^3}{6\pi\sqrt{2}} \sum_q (v_q^2 + a_q^2) \\
R_{\text{EW}} &= \frac{\sum_q (v_q^2 + a_q^2)}{v_e^2 + a_e^2} (1 + \delta_{\text{EW}})
\end{aligned} \tag{9.94}$$

where R_{EW} is the purely electroweak contribution and $\delta_{\text{QCD}}(0)$ is the QCD correction with massless quarks and δ_m is the mass correction [525]. v_q, a_q are the vector and axial vector coupling strength given in Eq. (4.3a). Corrections to R_Z are different from R_γ because Z couples both to the axial vector as well as the axial vector current and the mass correction including that of the top is sizable.

At LEP ($\sqrt{s} = m_Z = 91$ GeV), the value of the coupling constant is measured to be [7, 526, 527]

$$\alpha_s(m_Z^2) = 0.1226 \pm 0.004 \tag{9.95}$$

9.4.2
Jet Event Shapes

Methods to determine the coupling constant using inclusive reactions are relatively free of theoretical ambiguities and superior to others in principle. However, the inclusive reactions do not address details of final states such as those described by the event shape variables. In order to test the validity of the perturbative QCD describing such details, it is necessary to derive values of the coupling constant used in the reconstruction of the event shape variables and check if theoretical predictions reproduce their data using the same coupling constant. Variables to characterize jet properties include jet multiplicity, average jet mass, thrust, energy correlation [528–532], planer triple-energy correlation [533] and so on. Here, we derive values of the coupling constants using jet multiplicity, the energy correlation and discuss power corrections to pQCD applied to generic event shape variables.

Jet Multiplicity Formulas for n-jet production were given by Eqs. (9.38) and (9.41). To compare them with experiments (see Figure 9.13), either method of matrix element or parton shower can be used. OPAL [464, 534] analyzed the differential two-jet rate and the average jet rate using the JADE, Durham and other algorithms. Differential jet rates D_n and mean multiplicity $\langle N \rangle$ are defined as

$$D_n(y_{\text{cut}}) = \frac{d R_n(y_{\text{cut}})}{d y_{\text{cut}}} \tag{9.96}$$

$$\langle N \rangle = \frac{1}{\sigma_{\text{TOT}}} \sum_n n \sigma_n(y_{\text{cut}}) = \frac{1}{N_{\text{TOT}}} \sum_n n N_n(y_{\text{cut}}) \tag{9.97}$$

where R_n, σ_n are (relative) n-jet production cross sections. They fitted their data taken at LEP at $\sqrt{s} = 91$ GeV, 130–136 GeV and 161 GeV to NLLA ($O(\alpha_s^2)$) predictions of the QCD using JETSET [496], PYTHIA [469], and HERWIG [535] Monte Carlo calculations. The combined fitted value gives

$$\begin{aligned} \alpha_s(58.5\,\text{GeV}) = 0.1177 &\pm 0.0006(\text{stat.}) \pm (\text{expt.}) \pm 0.0010(\text{had}) \\ &\pm 0.0032(\text{theo.}) \end{aligned} \tag{9.98}$$

The errors are contributions from statistical, experimental, hadronization and theoretical uncertainties.

Energy Correlation When one measures the hadronic energy produced in the $e^- e^+$ reaction by two calorimeters separated by an angle χ as depicted in Figure 9.28, the energy correlation EEC (energy-energy correlation) and its asymmetry

AEEC (asymmetry of EEC) are defined by [536]

$$\mathrm{EEC}(\chi) = \frac{1}{N_{ev}} \sum_{ev} \sum_{i,j} x_i x_j \frac{1}{\Delta} \int_{\chi-\Delta/2}^{\chi+\Delta/2} d\chi' \delta(\chi_{ij} - \chi')$$

$$x_i = E_i / E_{\mathrm{CM}} \tag{9.99a}$$

$$\mathrm{AEEC}(\chi) = \mathrm{EEC}(\pi - \chi) - \mathrm{EEC}(\chi) \tag{9.99b}$$

where Δ is the angle width ($= \delta\chi$) spanned by the two detectors. Here, i, j are particle numbers in one event, and the sum over all the recorded events are taken for the calculation of EEC. The theoretical expression for EEC is obtained by calculating

$$\frac{1}{\sigma} \frac{d \sum_C^{\mathrm{EEC}}}{d\chi} = \frac{1}{\sigma} \sum_{ij} \int d x_i d x_j \frac{d\sigma}{d x_i d x_j d\chi} x_i x_j \tag{9.100a}$$

$$\frac{1}{\sigma} \int d\chi \frac{d \sum_C^{\mathrm{EEC}}}{d\chi} = 1 \tag{9.100b}$$

Experimentally, EEC is integrated over θ (polar angle of the detector with respect to the beam axis). Integrating the above formula over θ gives [531]

$$\frac{1}{\sigma} \frac{d\Sigma_C^{\mathrm{EEC}}}{d\chi} = \frac{\alpha_s(Q^2)}{\pi} F(\zeta), \quad \zeta = \frac{1 - \cos\chi}{2} \tag{9.101a}$$

$$F(\zeta) = \frac{C_F}{8} \frac{3 - 2\zeta}{\zeta^5(1-\zeta)} \left[2(3 - 6\zeta + 2\zeta^2)\ln(1-\zeta) + 3\zeta(2 - 3\zeta)\right] \tag{9.101b}$$

In the 2-jet configuration, hadrons are mainly emitted in 180° directions, and particles populate at $\chi = 0, \pi$, that is, $\zeta = 0, 1$. The singularity at $\zeta = 1$ in Eq. (9.101) reflects this feature. Soft gluons to be emitted symmetrically are also considered, but in the hard 3-jet configuration, the correlation becomes strong somewhere in between $0 < \chi < \pi$. AEEC is devised to eliminate this correlation and should enhance the 3-jet effect. EEC and AEEC have the following characteristics.

1. They are independent of the fragmentation mechanism.
2. They are infrared safe. R_n discussed before was also infrared safe, but it was due to artificially introduced y_{cut}. The EEC, on the other hand, is inherently an infrared safe variable.
3. They do not need to define the jet axis.
4. The perturbation expansion converges quickly and is stable against $O(\alpha_s^2)$ corrections, 25% for EEC and $\sim$ 12% for AEEC.
5. They are hard processes and the parton shower method is likely less reliable.

Because of the many reasons listed above, the process is best suited for the matrix method. Especially those at higher energy should be reliable because α_s is even smaller and also suffers less from variations due to the fragmentation model.

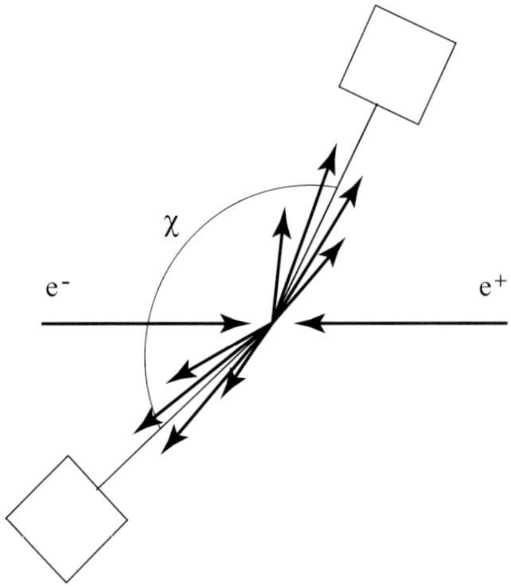

Figure 9.28 (a) EEC and (b) AEEC measure correlation and the asymmetry of energy flow in two directions.

Figure 9.29 shows EEC and AEEC data obtained by the OPAL group [536]. They unfolded their data back to the parton level and compared it with the analytical theory values. Combined analysis of the data obtained by the four LEP groups give [538]

$$\alpha_s(m_Z) = \begin{cases} 0.125^{+0.002}_{-0.003}(\text{Exp}) \pm 0.012(\text{theory}) & \text{EEC} \\ 0.114 \pm 0.005(\text{exp}) \pm 0.004(\text{theory}) & \text{AEEC} \end{cases} \tag{9.102}$$

The high value of $\alpha_s(m_Z)$ obtained from EEC is considered to reflect the larger $O(\alpha_s^2)$ effect as compared to AEEC and gives more credibility to that obtained from AEEC.

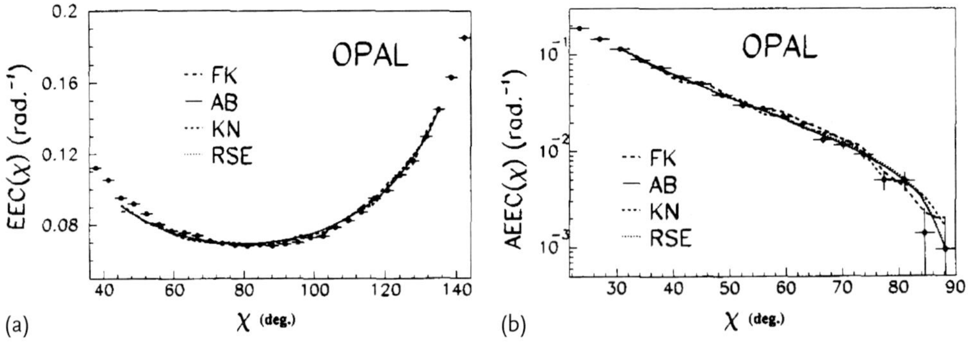

Figure 9.29 (a) EEC and (b) AEEC distributions compared with theoretical predictions [536, 537].

Power Corrections to Event Shape Variables We have seen that values of the mean thrust are reproduced well by Monte Carlo programs, but about a 20% discrepancy exists with the pQCD without power correction (see Figure 9.6). The major difference between the pQCD treatment and the Monte Carlo calculation lies in the treatment of fragmentation or soft gluon emissions. Obviously, nonperturbative contributions cannot be neglected, even at LEP energies if one tries to reproduce the details of the jet structure. Although there are no valid universal treatments for nonperturbative aspects of QCD, there exists a phenomenological ansatz which can relate nonperturbative corrections for different shape variables to a few measurable parameters.

Basic arguments [447, 448] go as follows. We consider a generic event shape variable denoted as y which goes to zero as the jet shape becomes pencil-like.[6] We consider the average of the generic event shape variable y defined by

$$\langle y \rangle = \frac{1}{\sigma_{\rm TOT}} \int_0^{y_{\max}} y \frac{d\sigma}{dy} dy \tag{9.103}$$

and decompose the moment into perturbative and nonperturbative components.

$$\langle y \rangle = \langle y \rangle^{\rm pert} + \langle y \rangle^{\text{non-pert}} \tag{9.104}$$

The pQCD prediction for a physical observable in next-to-leading order (NLO) has the form (see Eq. (7.87))

$$\begin{aligned} \langle y \rangle^{\rm pert} &= A(Q)\alpha_s(\mu_{\rm R}) + \left(B(Q) - A(Q) b_0 \ln \frac{Q^2}{\mu_{\rm R}^2} \right) \alpha_s^2(\mu_{\rm R}) + O(\alpha_s^3) \\ b_0 &= \frac{11 C_A - 2 n_f}{12\pi} \end{aligned} \tag{9.105}$$

where $A(Q)$ and $B(Q)$ are calculable functions. $\mu_{\rm R}$ is the renormalization scale used to extract the higher order corrections. Conventionally, the scale parameter Q is set to $Q = \sqrt{s}$ for $e^- e^+$ reactions and $Q = \sqrt{-q^2}$ in the deep inelastic scatterings.

Let us consider the function $\langle y \rangle^{\rm pert}$. It will have contributions from the gluon radiation spanning a range from very small momenta, either in the infrared region or collinear region, which are essentially nonperturbative. We can formally write the dominant diverging part in a form

$$\langle y \rangle = \int_0^Q dk_{\rm T} f(k_{\rm T}) \quad \text{with} \quad f(k_{\rm T}) = c_{\rm F} \alpha_s(k_{\rm T}) \frac{k_{\rm T}^p}{Q^{p+1}} + O(\alpha_s^2) \quad \text{for } k_{\rm T} \to 0 \tag{9.106}$$

The running $\alpha_s(k_{\rm T})$ can be obtained by effectively resumming a part of the higher order perturbative contributions as was done, for instance, for the double logarithm (see Section 8.5.2).

6) For this reason, $1 - T$ is used instead of T when considering the thrust distribution.

At sufficiently large Q^2, $\langle y \rangle$ would take a purely perturbative form given by Eq. (9.105). Equation (9.106) can now be used to examine the low energy contribution to $\langle y \rangle$ up to a matching scale μ_I which should be somewhere between the perturbative and nonperturbative regime, $\Lambda_{\overline{MS}} \ll \mu_I \ll Q$. Conventionally, $\mu_I = 2\,\text{GeV}$ is used [539]. As the coupling constant in the nonperturbative region is not known, we replace it by an effective coupling constant $\overline{\alpha}_p(\mu_I)$ defined by

$$\int_0^{\mu_I} dk_T \alpha_s(k_T) k_T^p = \frac{\mu_I^{p+1}}{p+1} \overline{\alpha}_p\left(\mu_I^2\right) \tag{9.107}$$

An underlying assumption is that the real physical expression is sufficiently regular for the integral to exist. As such Eq. (9.107) is an effective way to parametrize low energy part of the QCD, including both perturbative and nonperturbative contributions [448, 540, 541]. The parameter $\overline{\alpha}_p$ is scale dependent, but otherwise universal and can be applied to various event shape variables in common. For the thrust or three parton final state like $R_{3\text{-jet}}$ given by Eq. (9.37), pQCD calculation gives the power index $p = 0$ [448] and also determines c_F.

Before adding this contribution to $\langle y \rangle^{\text{non-pert}}$, we have to subtract the perturbative part in the integral to avoid double counting. To second order, it amounts to

$$\alpha_s(\mu_R) + b_0\left(\ln \frac{\mu_R^2}{\mu_I^2} + 2\right) \alpha_s^2(\mu_R) \tag{9.108}$$

Given A and B in Eq. (9.105), the nonperturbative correction can be calculated to give [449, 542, 543]

$$\langle y \rangle^{\text{non-pert}} = a_y \mathcal{P} \tag{9.109}$$

$$\mathcal{P} = \frac{4C_F}{\pi^2} \mathcal{M} \frac{\mu_I}{Q} \left[\overline{\alpha}_0(\mu_I) - \alpha_s(Q) - b_0\left(\ln \frac{Q^2}{\mu_I^2} + \frac{K}{2\pi b_0} + 2\right) \alpha_s^2(Q)\right] \tag{9.110}$$

$$\overline{\alpha}_0(\mu_I) = \frac{1}{\mu_I} \int_0^{\mu_I} \alpha_s(Q) dQ \tag{9.111}$$

$$\mathcal{M} = 1 + \frac{1.575\,C_A - 0.104\,n_f}{4\pi b_0}, \qquad K = C_A\left(\frac{67}{18} - \frac{\pi^2}{6}\right) - \frac{5}{9} n_f \tag{9.112}$$

The factor K takes the $\overline{MS}$ scheme into account and the Milan factor $\mathcal{M}$ accounts for two-loop effects and its value is 1.49 for $n_f = 3$ [544]. a_y is a given constant for each shape variable and is listed in Table 9.3 [542]. The first term containing $\overline{\alpha}_0$ in $a_y \mathcal{P}$ can be obtained by picking up the infrared divergent component of the per-

Table 9.3 Coefficients a_y of power correction $\propto 1/Q$ of event shape variables in the dispersive model [539, 547].

y	$1-T$	C	B_T	B_W	M_H^2/E_{vis}^2
a_y	2	3π	1	1/2	1

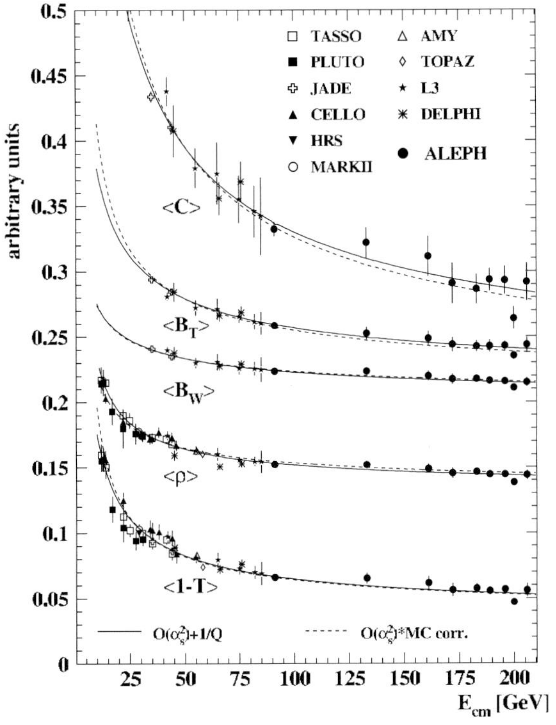

Figure 9.30 CM energy dependence of the mean values of several event shape observables. The solid lines show the result of the fit of combined $O(\alpha_s)$ QCD calculations with power corrections while the dashed lines indicate results obtained with Monte Carlo based corrections [543, 545].

turbative QCD and parametrizing the divergent part using the prescription given in Eq. (9.107) [448]. A simple shift of the variables is expected for $1-T$, M_H^2 and C, whereas for the jet broadening variables B_T and B_W, an additional squeeze of the distribution is expected [542]. This is due to interdependence of nonperturbative and perturbative effects which cannot be neglected for these observables. Equation (9.109) allows us a two-parameter $(\overline{\alpha}_0, \alpha_s)$ fit to the event shape variable data to extract values of the coupling constant $\alpha_s(\mu_R^2)$.

For the event distribution, the effect of the power corrections is to shift the perturbative spectra by the same amount as the correction to the mean [539].

$$\left.\frac{1}{\sigma}\frac{d\sigma}{dy}\right|_{\text{corrected}} = \left.\frac{1}{\sigma}\frac{d\sigma(y-\Delta y)}{dy}\right|_{\text{pert}}$$
$$\Delta y = a_y \mathcal{P} \tag{9.113}$$

Again, the concept of a constant shift Δy does not apply to jet broadening where the shift is B-dependent.

Figure 9.30 shows a few examples of the jet shape variables plotted as a function of the center of mass energy [543, 545]. Both NLO Monte Carlo calculations and the perturbative QCD with power corrections reproduce data well. Fits to the data are used to obtain values of the coupling constant α_s (see Figure 9.31).

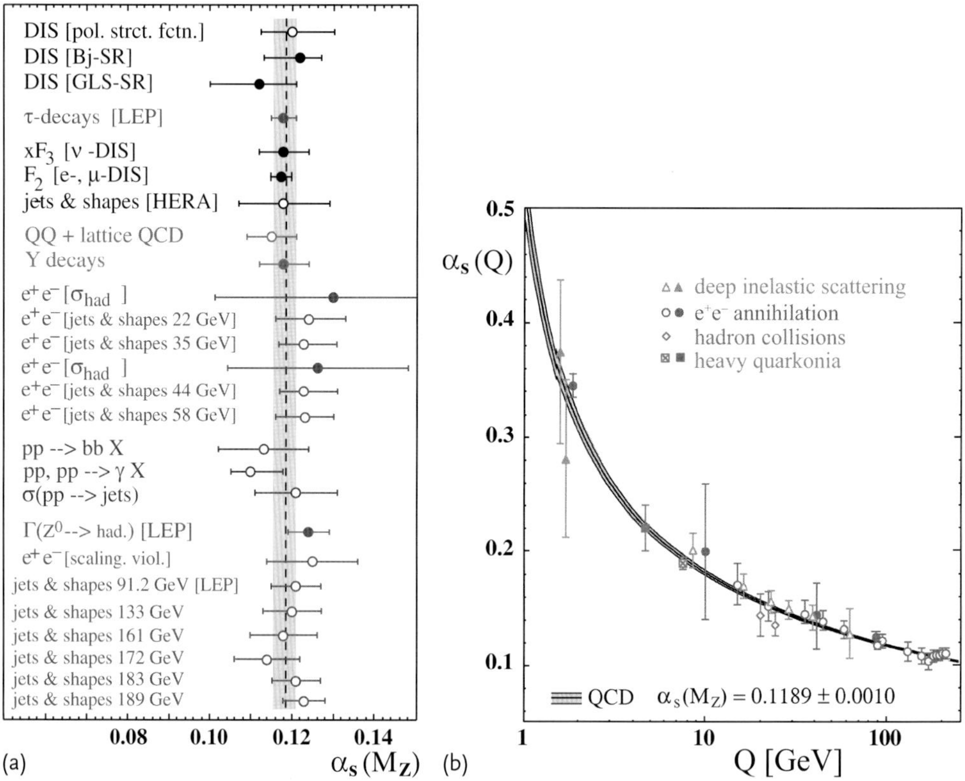

Figure 9.31 Asymptotic freedom and running coupling constant: (a) All $\alpha_s(m_Z^2)$s in the fourth column of Table 9.4 compared with a fourth order theoretical prediction (dashed line) [23, 24]. (b) $\alpha_s(Q^2)$s in the third column of Table 9.4 plotted as a function of Q [24, 25].

9.4.3 Summary of the Running $\alpha_s(Q^2)$

Table 9.4 compares values of $\alpha_s(Q^2)$ obtained from variety of processes at different Q values [24]. When translated to values at $Q = m_Z$, the total uncertainty of $\alpha_s(m_Z^2)$ is 2.6%.

The results for $\alpha_s(Q^2)$, given in the third column of Table 9.4, are presented in Figure 9.31a. As one can see, the strength of the strong coupling obtained at low energies, extrapolated to $Q = m_Z$ and that obtained there agree with high accuracy. This confirms the claimed statement that the whole QCD is described by a single coupling constant with the asymptotic freedom. The vertical dashed

Table 9.4 World summary of measured α_s [24] .

Process	Q (GeV)	$\alpha_s(Q)$	$\alpha_s(M_{Z^0})$	$\Delta\alpha_s(M_{Z^0})$ Expt	$\Delta\alpha_s(M_{Z^0})$ Theory	Theory
DIS (pol. strct. fctn)	0.7–8		$0.120^{+0.010}_{-0.008}$	$^{+0.004}_{-0.005}$	$^{+0.009}_{-0.006}$	NLO
DIS (Bj-SR)	1.58	$0.375^{+0.062}_{-0.081}$	$0.121^{+0.005}_{-0.009}$	–	–	NNLO
DIS (GLS-SR)	1.73	$0.280^{+0.070}_{-0.068}$	$0.112^{+0.009}_{-0.012}$	$^{+0.008}_{-0.010}$	0.005	NNLO
τ-decays	1.78	0.323 ± 0.030	0.1181 ± 0.0031	0.0007	0.0030	NNLO
DIS $[\nu; xF_3]$	5.0	0.214 ± 0.021	0.118 ± 0.006	0.005	0.003	NNLO
DIS $(e/\mu; F_2)$	2.96	0.252 ± 0.011	0.1172 ± 0.0024	0.0017	0.0017	NNLO
DIS $(e - p$; jets)	6–100		0.118 ± 0.011	0.002	0.011	NLO
$Q\overline{Q}$ states	4.1	0.216 ± 0.022	0.115 ± 0.006	0.000	0.006	LGT
⊠ decays	4.75	0.22 ± 0.02	0.118 ± 0.006	–	–	NNLO
e^+e^- (σ_{had})	10.52	0.20 ± 0.06	$0.130^{+0.021}_{-0.029}$	$^{+0.021}_{-0.029}$	0.002	NNLO
e^+e^- (jets and shapes)	22.0	$0.161^{+0.016}_{-0.011}$	$0.124^{+0.009}_{-0.006}$	0.005	$^{+0.008}_{-0.003}$	Resum
e^+e^- (jets and shapes)	35.0	$0.145^{+0.012}_{-0.007}$	$0.123^{+0.008}_{-0.006}$	0.002	$^{+0.008}_{-0.005}$	Resum
e^+e^- (σ_{had})	42.4	0.144 ± 0.029	0.126 ± 0.022	0.022	0.002	NNLO
e^+e^- (jets and shapes)	44.0	$0.139^{+0.011}_{-0.008}$	$0.123^{+0.008}_{-0.006}$	0.003	$^{+0.007}_{-0.005}$	Resum
e^+e^- (jets and shapes)	58.0	0.132 ± 0.008	0.123 ± 0.007	0.003	0.007	Resum
$p\overline{p}$ ⊠ $b\overline{b}X$	20.0	$0.145^{+0.018}_{-0.019}$	0.113 ± 0.011	$^{+0.007}_{-0.006}$	$^{+0.008}_{-0.009}$	NLO
$p\overline{p}, pp$ ⊠ γX	24.3	$0.135^{+0.012}_{-0.008}$	$0.110^{+0.008}_{-0.005}$	0.004	$^{+0.007}_{-0.003}$	NLO
$\sigma(p\overline{p}$ ⊠ jets)	30–500		0.121 ± 0.010	0.008	0.005	NLO
$e^+e^-(\Gamma(Z^0$ ⊠ had.))	91.2	0.124 ± 0.005	0.124 ± 0.005	0.004	$^{+0.003}_{-0.002}$	NNLO
e^+e^- scaling viol.	14–91.2		0.125 ± 0.011	$^{+0.006}_{-0.007}$	0.009	NLO
e^+e^- (jets and shapes)	91.2	0.121 ± 0.006	0.121 ± 0.006	0.001	0.006	Resum
e^+e^- (jets and shapes)	133.0	0.113 ± 0.008	0.120 ± 0.007	0.003	0.006	Resum
e^+e^- (jets and shapes)	161.0	0.109 ± 0.007	0.118 ± 0.008	0.005	0.006	Resum
e^+e^- (jets and shapes)	172.0	0.104 ± 0.007	0.114 ± 0.008	0.005	0.006	Resum
e^+e^- (jets and shapes)	183.0	0.109 ± 0.005	0.121 ± 0.006	0.002	0.005	Resum
e^+e^- (jets and shapes)	189.0	0.110 ± 0.004	0.123 ± 0.005	0.001	0.005	Resum

line in Figure 9.31a is a fit with the four-loop QCD prediction for the running $\alpha_s(Q^2)$ [23] which is the improved version of Eq. (7.84a). Figure 9.31b is a graphical representation of the coupling constant as a function of the energy which clearly shows the validity of the asymptotic freedom.

Further evidences of the asymptotic freedom are:

1. the scale evolution of the structure functions in the deep inelastic scattering as discussed in Chapter 8 and
2. the fragmentation model can reproduce data at high as well as low energies in a unified way by a single coupling constant with the asymptotic freedom. Including the direct measurements in a variety of phenomena described here, we may conclude that the asymptotic freedom has firm experimental establishments.

In summary, all the QCD data at different energies can be described by a single value of α_s at $Q = m_Z$. Here, we quote one given by the Particle Data group in 2008 [11].

$$\begin{aligned} \alpha_s\left(m_Z^2\right) &= 0.119 \pm 0.002^{7)} \\ \Lambda_{\overline{MS}} &= 219 \pm 25\ \text{MeV} \end{aligned} \tag{9.114}$$

7) A more recent world average

$$\alpha_s\left(m_Z^2\right) = 0.1184 \pm 0.0007 \tag{9.115}$$

with astonishing precision of 0.6 % is quoted by [3, 546]

10
Gluons

10.1
Gauge Structure of QCD

The quark model was very successful. When people realized that quarks have three color degrees of freedom, the QCD was naturally proposed as the theory of the strong interaction. The discovery of the asymptotic freedom which has justified the parton model and offered a powerful tool to make quantitative estimates of various phenomena has elevated its status to the Standard Model of the strong interaction. It is a gauge theory based on color $SU(3)$. The outstanding properties of the gauge theory is that the force carrier, the gluon, is a massless vector field and that the non-Abelian nature of QCD requires existence of self-coupling of the gluon. The self-coupling makes the theory nonlinear. This feature does not exist in QED. The gluon is fundamentally different from the photon. The feature exists also in the electroweak theory, but because of its small coupling, its effect is hardly felt. In QCD, however, the peculiar characteristics of the self-coupling manifest themselves in a variety of phenomena. In order to elucidate the non-Abelian nature of the QCD, it is essential to study properties of the gluon.

10.1.1
Spin of the Gluon

If the gluon is the gauge boson of the QCD, its spin has to be one. We have already seen one piece of evidence that the vector gluon is favored against the scalar gluon in comparing moments of the structure function in deep inelastic scattering (see Figure 8.10 and discussions in Section 8.5.1). Here, we solidify the vector nature of the gluon with direct evidence. Experimentally, it can be determined by studying the Dalitz plots in $q\overline{q}g$ CM frame [548]. Energy distribution of the spin 1 gluon in the reaction was already given in Eq. (7.133a). One can obtain a corresponding distribution for the scalar gluon. As an investigative starting point, the spin structure

Elementary Particle Physics, Volume 2: Foundations of the Standard Model, First Edition. Yorikiyo Nagashima.

of the gluon, we list both of them in the following [549, 550].

$$\text{vector gluon} \quad \frac{1}{\Gamma}\frac{d\Gamma}{dx_1 dx_2} = \frac{\alpha_s}{2\pi} C_F \frac{x_1^2 + x_2^2}{(1-x_1)(1-x_2)} \tag{10.1a}$$

$$\text{scalar gluon} \quad = \frac{\tilde{\alpha}_s}{2\pi} C_F \frac{x_3^2}{(1-x_1)(1-x_2)} \tag{10.1b}$$

$$x_i = 2E_i/Q, \quad x_1 + x_2 + x_3 = 2 \tag{10.1c}$$

where $\tilde{\alpha}_s$ is the coupling strength of the scalar gluon. Here x_1, x_2 are fractional, momenta of the quarkquark q and antiquark $\overline{q}$ and $x_3 = 2 - x_1 - x_2$ denotes the fractional momentum of the gluon [see Eq. (7.127)]. Since $\tilde{\alpha}_s$ has no reference to determine its absolute value, a normalized distribution has to be used for comparison. Given the above $e^- e^+ \to q\overline{q}g$ distributions for the different gluon spin, our next task is to find observables that could tell the difference between Eqs. (10.1a) and (10.1b). There are several such observables. Here, we choose two examples, the thrust distribution and the so-called Ellis–Karliner angular distribution.

Method 1: Thrust Distribution: It is possible to calculate the thrust distributions for the spin 1 and 0 gluon [442, 443]. Let us consider a case where the minimum jet mass is required to be "y" in the definition of Eq. (9.35). The jet mass in terms of the parton variables can be obtained using x_1, x_2, x_3. The definition of the thrust Eq. (9.11a) if applied to the partons in $e^- e^+ \to q\overline{q}g$, constrains $T = \max x_i$, which is the fractional energy of the parton "i." Let us consider the region in Figure 9.12 bounded by $x_1 > x_2, x_3$ and $x_1 < 1 - y$. It is the inner triangle bounded by the contour A and two inner sides with the apex at the middle point. In this region, $T = \max x_i = \max x_1 = 1 - y$. Therefore, the contours A, B, and C coincide with $x_1 = T, x_2 = T, x_3 = T$ with T fixed at $1 - y$.

Therefore, the thrust distribution to $O(\alpha_s)$ can be obtained by integrating Eq. (10.1a) along the three contours A, B, and C, given by Eq. (9.37) with $1 - y$ replaced with T.

$$\frac{1}{\sigma_0}\frac{d\sigma}{dT} = \frac{\alpha_s}{2\pi} C_F \int_{2-2T}^{T} dx_1 \left\{ 2 \frac{x_1^2 + x_2^2}{(1-x_1)(1-x_2)}\bigg|_{x_2=T} + \frac{x_1^2 + x_2^2}{(1-x_1)(1-x_2)}\bigg|_{x_2=2-T-x_1} \right\} \tag{10.2}$$

The factor 2 in the first term arises because two contours with $x_2 = T$ and $x_1 = T$ give the same contribution. The integral gives

$$\frac{1}{\sigma_0}\frac{d\sigma}{dT}[q\overline{q}g\ (s=1)] = C_F \frac{\alpha_s}{2\pi}\left[\frac{2(3T^2 - 3T + 2)}{T(1-T)} \ln\left(\frac{2T-1}{1-T}\right) - \frac{3(3T-2)(2-T)}{1-T}\right] \tag{10.3a}$$

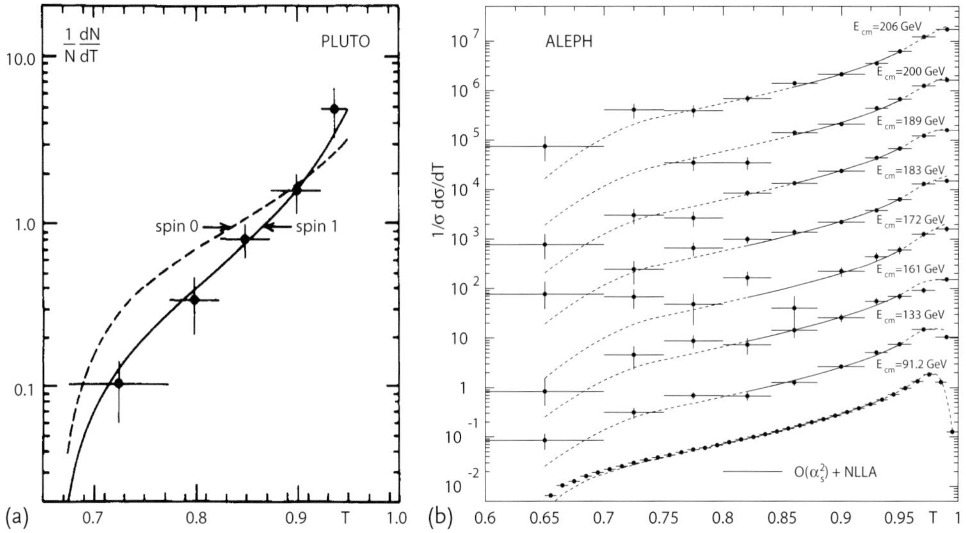

Figure 10.1 Analytical approach to event shape distribution "Thrust." (a) The thrust distribution in $O(\alpha_s)$ is obtained by integrating $d\sigma(ee^+ \to q\overline{q}g)$ along the contour (see text). (a) shows data obtained by PLUTO compared with the vector and scalar gluon predictions [551]. (b) Thrust distributions compared with the perturbative QCD calculation with power correction. It includes only one additional parameter $\alpha_0(\mu_I) = (1/\mu_I)\int_0^{\mu_I} \alpha_s(k)dk$ which stands for the unknown behavior of α_s below the infrared matching scale μ_I [445, 543]. For explanations of the power correction parameter $\alpha_0(\mu_I)$, see Section 9.4.2.

$$\frac{1}{\sigma_0}\frac{d\sigma}{dT}[q\overline{q}\tilde{g}\ (s=0)] = \frac{\tilde{\alpha}_s}{3\pi}\left[2\ln\left(\frac{2T-1}{1-T}\right) + \frac{(4-3T)(3T-2)}{1-T}\right] \tag{10.3b}$$

Figure 10.1a shows an early PLUTO data on thrust distribution compared with the vector and scalar gluon spin curves given in Eqs. (10.3a) and (10.3b). The data confirm the vector gluon. Figure 10.1b shows data taken at LEP. More precise thrust distributions have been obtained, again confirming the vector gluon [445, 543].

Method 2: Ellis–Karliner angular distribution Another method to test the spin of the gluon is the Ellis–Karliner angle plot which is a variation of the Dalitz plot [552, 553]. Referring to Figure 10.2a in the CM frame of $q\overline{q}g$, we have relations among the kinematical variables. In the absence of quark/gluon identification, the three jets are energy ordered.

$$x_1 + x_2 + x_3 = 2\,, \quad x_3 \le x_2 \le x_1$$

$$x_i = \frac{2\sin\theta_i}{\sin\theta_1 + \sin\theta_2 + \sin\theta_3} \qquad (\text{assume} \quad m_q = m_g = 0) \tag{10.4}$$

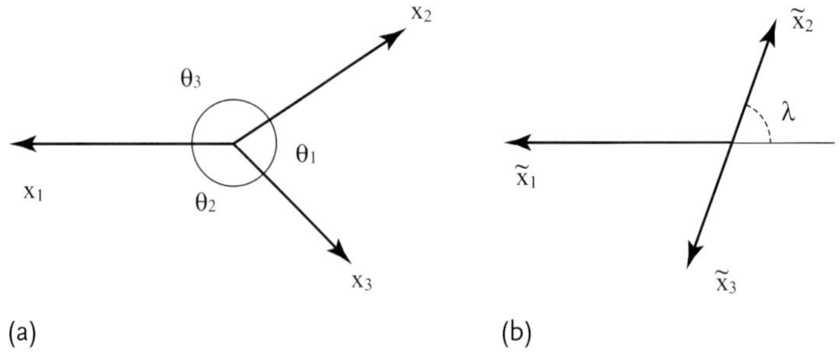

Figure 10.2 Variables to determine the gluon spin. (a) Kinematical variables in the CM frame of $q\bar{q}g$. (b) Energy $\tilde{x}_i$ and the Ellis–Karliner angle λ in the CM frame of parton 2 and 3 [552, 553].

According to Ellis–Karliner's prescription, one moves to the CM frame of the parton 2 and 3, and defines the $E-K$ angle λ as shown in Figure 10.2b. Then,

$$\cos\lambda = \frac{x_2 - x_3}{x_1} = \frac{\sin\theta_2 - \sin\theta_3}{\sin\theta_1} \tag{10.5}$$

To determine the gluon spin by the $E-K$ prescription, it is necessary to separate 3-jet events from 2-jet events. L3 group [554] at LEP ($\sqrt{s} = 91.2$ GeV) analyzed jet events using the JADE algorithm (see footnote 3) in Chapter 9). The jets are numbered with $x_1 > x_2 > x_3$. Since the gluon is emitted by either of the $q\bar{q}$, the least energetic jet is most likely the gluon. A jet resolution parameter $y_{\text{cut}} \geq 0.02$ corresponding to the jet pair mass of 13 GeV or more is used. An important feature of the $\cos\lambda$ distribution is that it is singular at $\cos\lambda = 1$ ($x_1 = 0$) or $x_2 = 1$. The cross section for the scalar gluon does not contain such a singularity. Figure 10.3 shows distributions of x_2 and the Ellis–Karliner angle ($\cos\lambda$) as measured by the L3

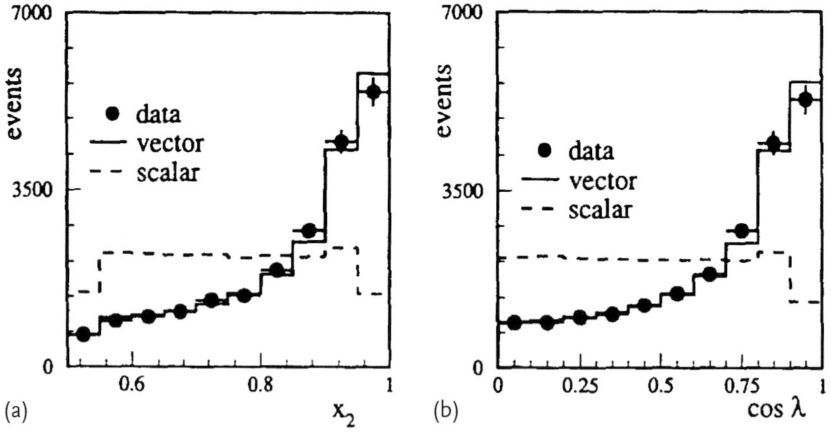

Figure 10.3 Comparison of measured and predicted distributions with 3-jet events for $0.02 < y_{\text{cut}} < 0.05$. The data are corrected for detector effects. The solid and dashed lines show predictions for the vector and scalar gluons, respectively. (a) scaled energy of the second jet, (b) cosine of Ellis–Karliner angle, $\cos\lambda$ [554].

group. Figure 10.3 also shows expectations of the vector and scalar gluon models. The vector hypothesis is overwhelmingly favored and the scalar gluon is incompatible with the data.

10.1.2
Self-Coupling of the Gluon

The non-Abelian nature of the QCD can be considered as proven from evidences of the asymptotic freedom, confinement, and the string effect which will be discussed in Section 10.2.2. It is desirable, however, if one can show direct evidences of the nonlinear coupling of the gluon. If one sets $\alpha_s(U(1)) = C_F\alpha_s(QCD)(C_F = 4/3)$, one cannot distinguish QCD from the Abelian QCD in phenomena, including 2- and 3-jet production cross sections. However, the nonlinear coupling of the gluon can be tested by the 4-jet production in which the Feynman diagrams in Figure 10.4a contributes.

In Figure 10.4c,d, the diagrams are dominant in the collinear region, but their contribution diminishes for the hard gluon emission. Therefore, we need to differentiate Figure 10.3a,b. Behavior of the emitted gluon is contained in the splitting functions (see Eqs. (8.72b) and (8.72d)). When the polarization of the gluon spin is not averaged, they are modified to (see Eq. (K.27) and (K.35))

$$P_{gg\leftarrow g}(z) = 2C_A\left[\frac{(1-z+z^2)^2}{z(1-z)} + z(1-z)\cos 2\phi\right] \quad C_A = 3 \tag{10.6a}$$

$$P_{q\bar{q}\leftarrow g}(z) = n_f T_R\left[z^2 + (1-z)^2 - 2z(1-z)\cos 2\phi\right] \quad T_R = 1/2 \tag{10.6b}$$

where n_f is the number of quark flavors, z is emitted parton's relative momentum, and ϕ is the angle of the polarization relative to the production plane of gg or $q\bar{q}$. While $P_{gg\leftarrow g}(z)$ diverges at $z \sim 0, 1$, $P_{q\bar{q}\leftarrow g}(z)$ does not have such singularity. Therefore, to distinguish their contributions, we only need to choose an observable that can demonstrate this feature.

Notice that the angular correlation of $P_{gg\leftarrow g}$ favors an orientation in which the plane of the emitted pair aligns with the polarization of the emitting gluon. How-

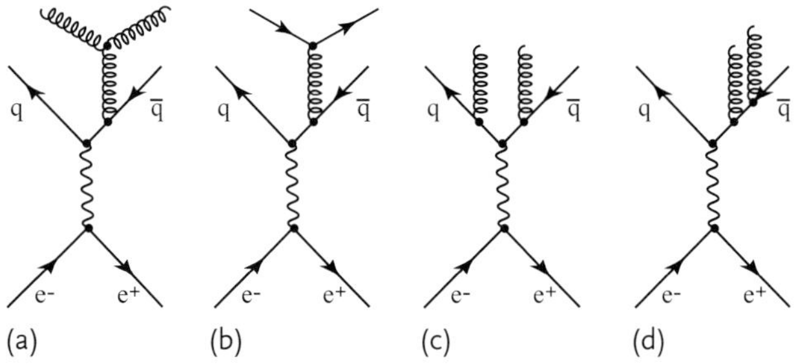

Figure 10.4 Feynman diagrams that contribute to 4-jet topology. (a) is the self-coupling which is characteristic of the QCD and has to be distinguished from diagram (b). The dominant contribution of diagrams (c) and (d) comes from collinear configuration which is small in the hard 4-jet configuration.

ever, the correlation is quite weak. Its coefficient $z(1-z)$ vanishes in the enhanced region $z \to 0, 1$ and reaches its maximum at $z = 1/2$, where it is still only 1/9 of the unpolarized contribution.

The polarization vector lies in a plane made by the first $q\overline{q}$ produced by e^-e^+ annihilation and the virtual gluon. This is because the parity conservation requires the amplitude to be proportional to $\epsilon_g \cdot (\boldsymbol{p}_q - \boldsymbol{p}_{\overline{q}})$. Therefore, the ϕ dependence of Eq. (10.6) can be obtained by looking at the relative orientation of two planes made by the primary $q\overline{q}$ and the emitted gg or secondary $q\overline{q}$.

Experimentally, in order to apply the above considerations, it is necessary to:

1. extract clean 4-jet samples,
2. differentiate between the quark jet and the gluon jet.

For the condition (1), the energy has to be sufficiently high. The condition (2) can be taken into account by naming the four jets 1, 2, 3, and 4 in the descending order of the jet energy and considering 1 and 2 as the quark jets, and 3 and 4 as the gluon jets. At PEP and PETRA ($\sqrt{s} \sim 34\,\text{GeV}$), the energy was too low and the first evidence was obtained at TRISTAN ($\sqrt{s} = 60\,\text{GeV}$) [555]. Here, we take a look at LEP ($\sqrt{s} = 91\,\text{GeV}$). The four jets are separated by the y_{cut} method, requiring $y_{\text{cut}} > 0.01$. With this condition, contribution of the diagram in Figure 10.4b amounts to 34.1% in the Abelian theory, but it is 4.7% in QCD [556]. Variables to distinguish the nonlinear coupling are θ_{BZ} and θ_{NR} as described in Figure 10.5.

They are defined by [557, 558]

$$\cos\theta_{\text{BZ}} = \frac{(\boldsymbol{p}_1 \times \boldsymbol{p}_2) \cdot (\boldsymbol{p}_3 \times \boldsymbol{p}_4)}{|\boldsymbol{p}_1 \times \boldsymbol{p}_2||\boldsymbol{p}_3 \times \boldsymbol{p}_4|} \tag{10.7a}$$

$$\cos\theta_{\text{NR}} = \frac{(\boldsymbol{p}_1 - \boldsymbol{p}_2) \cdot (\boldsymbol{p}_3 - \boldsymbol{p}_4)}{|\boldsymbol{p}_1 - \boldsymbol{p}_2||\boldsymbol{p}_3 - \boldsymbol{p}_4|} \tag{10.7b}$$

Figure 10.6 shows OPAL and L3 data [559, 560]. The experimental data clearly denies the Abelian QCD.

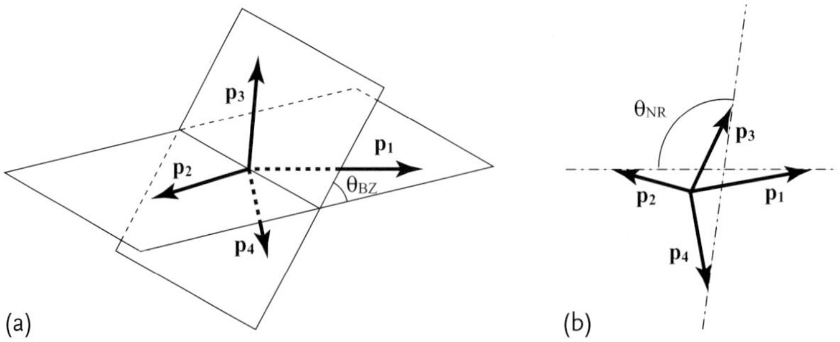

Figure 10.5 Variables used to test the nonlinear coupling of the gluon, θ_{BZ} (a) and θ_{NR} (b). The energy suffix is arranged as $E_1 > E_2 > E_3 > E_4$.

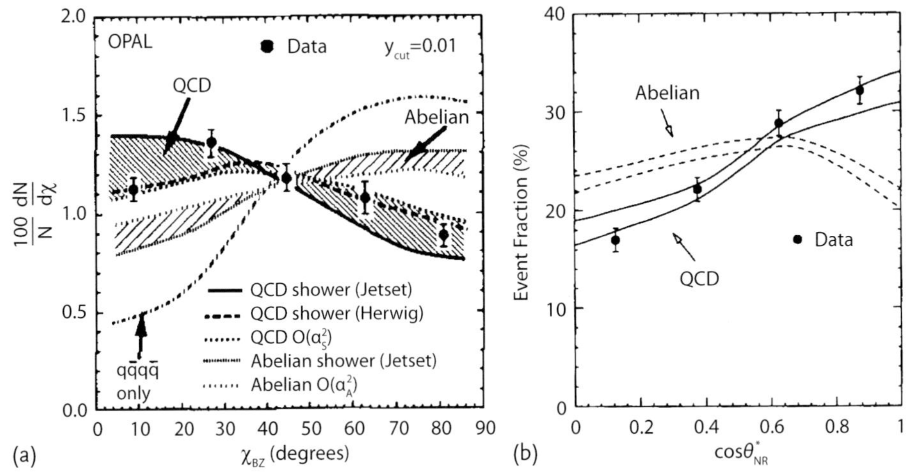

Figure 10.6 Evidence for the nonlinear coupling of the gluon. (a) OPAL data for θ_{BZ} [559] and (b) L3 data for θ_{NR} [560].

10.1.3 Symmetry of QCD

One can test the symmetry structure of the QCD by expressing the cross section in terms of the general group parameters C_F, C_A, T_R. [484, 561–563]

$$d\sigma_{4\text{-jet}} = \left(\frac{\alpha_s}{2\pi}\right)^2 \left(C_F^2\sigma_a + C_F C_A \sigma_b + C_F n_f T_F \sigma_c\right) \tag{10.8}$$

and solving for C_A and C_F. The partial cross section σ_a gets its contribution from two gluon bremsstrahlung, σ_b from the triple gluon coupling and σ_c from the gluon to $q\bar{q}$ branch. The strength of the color factors are listed in Table 10.1. Figure 10.7 shows a two-dimensional plot of the C_A–C_F relation and constraints due to experimental data. One can see excellent agreement with the QCD.

Table 10.1 Color factors in $SU(N)$

Color factor	C_F	C_A	T_R
$SU(N)$	$(N^2-1)/2N$	N	1/2
QCD	4/3	3	1/2
Abelian group	1	0	3
LEP	1.30 ± 0.09	2.89 ± 0.21	

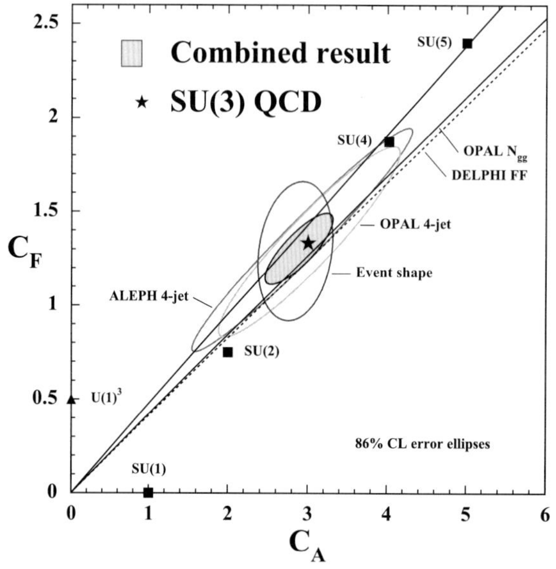

Figure 10.7 Comparison of the coupling strength of $g \rightarrow gg(C_A/C_F)$ versus $g \rightarrow q\overline{q}(T_R/C_F)$ [445, 545].

10.2
Color Coherence

When the wave length of successively emitted gluons is long, they overlap in space. If two gluons have opposite colors, they interfere destructively. In Figure 10.8, if the gluon's wavelength is long (the figure is for QED, but assume the photon is a gluon), it cannot resolve its parent q from neighboring $\overline{q}$ and behaves as if it is emitted from their composite. It is referred to as the color coherence. In the LLA approximation, the effect appears when the emission angle of the second gluon is larger than that of the first $q\overline{q}$ pair.

The angular ordering property of soft emission is common to all gauge theories. In QED, it is known as the Chudakov effect, which has a simple intuitive explanation in a time ordered perturbation theory. Consider that an electron and a positron pair is produced with opening angle θ_{ij} (Figure 10.8). The electron is excited to a virtual state and emits a photon of momentum $\boldsymbol{k}$ with opening angle θ_i and goes back to a physical state with momentum $\boldsymbol{p}$. By the Heisenberg uncertainty principle, the duration of the electron staying in the virtual state is related to the energy difference ΔE between the virtual and the final state. For $|\boldsymbol{k}| \ll |\boldsymbol{p}|$ and $\theta_i = \cos^{-1}(\boldsymbol{k}\cdot\boldsymbol{p}/|\boldsymbol{k}\cdot\boldsymbol{p}|) \to 0$, it is given by

$$\Delta E = \sqrt{\boldsymbol{p}^2 + m^2} + |\boldsymbol{k}| - \sqrt{(\boldsymbol{p}+\boldsymbol{k})^2 + m^2} \sim |\boldsymbol{k}|\theta_i^2 \tag{10.9a}$$

$$\therefore \quad \Delta t \sim \frac{1}{|\boldsymbol{k}|\theta_i^2} \sim \frac{1}{|\boldsymbol{k}_\mathrm{T}|\theta_i} \simeq \frac{\lambda_\mathrm{T}}{\theta_i} \tag{10.9b}$$

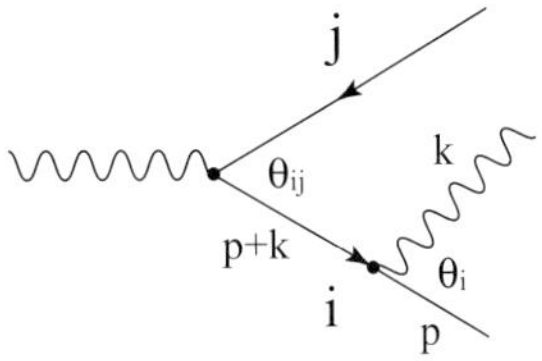

Figure 10.8 The Chudakov effect: a soft photon emitted from an electron (particle *i*) cannot resolve its parent from the companion positron (particle *j*).

where $\lambda_{\rm T}$ is the transverse wave length. During a time interval Δt, the electron–positron separation becomes

$$\Delta d = \Delta t \theta_{ij} = \lambda_{\rm T} \frac{\theta_{ij}}{\theta_i} \tag{10.10}$$

If $\Delta d < \lambda_{\rm T}$, the photon cannot distinguish the positron from the electron and recognize them as an electrically neutral object, hence the emission is suppressed. From Eq. (10.10), we see that the suppression occurs if $\theta_i > \theta_{ij}$. In QCD, the color cancellation, which is formulated below, is a bit more complicated, but similar reasoning is valid [565, 566].

10.2.1 Angular Ordering

The angular ordering is a manifestation of the color coherence [567]. We consider emission of a soft gluon from one of the N external partons. Suppose a gluon with color "a" is emitted from the ith parton with color "b" and converted to color "c," the transition amplitude is modified as

$$\begin{aligned}
&\mathcal{M}_c^a(N+1; k) \\
&= \sum_i \overline{u}(p_i + k) V_c^{aN}(p_i + k) \\
&= \sum_i g_s \left[T_i^a\right]_{cb} \overline{u}(p_i)\, \not{\varepsilon}^*(k) \frac{(\not{p}_i + \not{k} + m)}{(p_i + k)^2 - m^2} V_b^N(p_i + k) \\
&= \sum_i g_s \left[T_i^a\right]_{cb} \overline{u}(p_i) \frac{\left[-(\not{p}_i - m)\, \not{\varepsilon}^*(k) + 2(p_i \cdot \varepsilon^*) + \not{\varepsilon}^* \, \not{k}\right]}{2(p_i \cdot k)} V_b^N(p_i + k) \\
&\simeq \sum_i g_s \left[T_i^a\right]_{cb} \frac{(p_i \cdot \varepsilon^*)}{(p_i \cdot k)} \overline{u}(p_i) V_b^N(p_i) + O(k) \\
&= \sum_i g_s \left[T_i^a\right]_{cb} \frac{(p_i \cdot \varepsilon^*)}{(p_i \cdot k)} \mathcal{M}_b(N; p_i) + O(k)
\end{aligned} \tag{10.11}$$

where we have used $\not{p} = m$, $p^2 = m^2$, $p \gg k$ and neglected the small term in the numerator which is proportional to the gluon momentum. Gluon emissions from internal lines can be neglected because they are deeply off-shell. $\boldsymbol{T}_q = \{t^A\} = \{\lambda^A/2\}$ for the quarks and $\boldsymbol{T}_g = \{[T^a]_{bc}\} = -i f^{abc}$ for the gluons are color matrices we defined in Eqs. (7.4d) and (7.4e). This gives [566].

$$\mathcal{M}_{N+1} = \left[g_s \sum_i \boldsymbol{T}_i{}^a \frac{(\varepsilon^* \cdot p_i)}{(p_i \cdot k)} \right] \mathcal{M}_N \tag{10.12a}$$

$$\begin{aligned} d\sigma_{N+1} &= [|\mathcal{M}_N|^2 \cdots] \sum_{\varepsilon,a} \left[g_s \sum_i \boldsymbol{T}_i{}^a \frac{(\varepsilon^* \cdot p_i)}{(p_i \cdot k)} \right]^2 \frac{d^3k}{(2\pi)^3 2\omega} \\ &= d\sigma_N \frac{d\omega}{\omega} \frac{d\Omega_k}{2\pi} \frac{\alpha_s}{2\pi} \sum_{i,j} C_{ij} W_{ij} \end{aligned} \tag{10.12b}$$

where the ellipsis denotes the flux and phase space factor for N body system, $d\Omega_k$ is the solid angle for the emitted gluon, $C_{ij} = -\boldsymbol{T}_i \cdot \boldsymbol{T}_j$ is the color factor to be computed and the sum over i, j is to be taken for all external lines. Separation of the color factor from the matrix element was made possible because

$$\sum_{a,c} [T_i^a]_{cb} [T_j^a]_{cb'} = [\boldsymbol{T}_i \cdot \boldsymbol{T}_j]_{bb'} = \frac{1}{2} \langle (\boldsymbol{T}_i + \boldsymbol{T}_j) \cdot (\boldsymbol{T}_i + \boldsymbol{T}_j) - \boldsymbol{T}_i \cdot \boldsymbol{T}_i - \boldsymbol{T}_j \cdot \boldsymbol{T}_j \rangle \delta_{bb'} \tag{10.13}$$

is a Casimir operator and proportional to the unit matrix (see Eq. (7.35a) and (7.38a)). As $\sum_\lambda \epsilon_\mu(\lambda) \epsilon_\nu^*(\lambda)$ is given by Eq. (7.15), we are allowed to make the replacement

$$\begin{aligned} &\sum_{ij} \boldsymbol{T}_i \cdot \boldsymbol{T}_j \sum_{\text{pol}} \frac{(\epsilon \cdot p_i)(\epsilon^* \cdot p_j)}{(k \cdot p_i)(k \cdot p_j)} \to -\sum_{ij} \boldsymbol{T}_i \cdot \boldsymbol{T}_j (J_i \cdot J_j) \\ &J_i^\mu = \frac{p_i^\mu}{(k \cdot p_i)} - \frac{n^\mu}{(k \cdot n)} \end{aligned} \tag{10.14}$$

For a color neutral system $\sum_i \boldsymbol{T}_i = 0$, and hence the second term of J^μ does not contribute and we can safely drop n^μ from J^μ. Therefore, the radiation function W_{ij} in Eq. (10.12b) is given by

$$\begin{aligned} W_{ij} &= -\sum_{\text{pol}} \frac{(\varepsilon^* \cdot p_i)(\varepsilon \cdot p_j)}{(p_i \cdot k)(p_j \cdot k)} |\boldsymbol{k}|^2 = \frac{(p_i \cdot p_j)}{(p_i \cdot \hat{k})(p_j \cdot \hat{k})} \\ &= \frac{1 - \beta_i \beta_j \cos\theta_{ij}}{(1 - \beta_i \cos\theta_i)(1 - \beta_j \cos\theta_j)} \qquad \hat{k} = \frac{k}{|\boldsymbol{k}|} \end{aligned} \tag{10.15}$$

$$\begin{aligned} \sum_{ij} C_{ij} W_{ij} &= \sum_{i \neq j} \boldsymbol{T}_i \cdot \boldsymbol{T}_j W_{ij} + \sum_i \boldsymbol{T}_i^2 W_{ii} \\ &= \sum_{i \neq j} \boldsymbol{T}_i \cdot \boldsymbol{T}_j \left[\frac{(p_i \cdot p_j)}{(\hat{k} \cdot p_i)(\hat{k} \cdot p_j)} - \frac{1}{2} \frac{p_i^2}{(\hat{k} \cdot p_i)^2} - \frac{1}{2} \frac{p_j^2}{(\hat{k} \cdot p_j)^2} \right] \end{aligned} \tag{10.16}$$

where we used the identity $\boldsymbol{T}_i = -\sum_{i \neq j} \boldsymbol{T}_j$ for color neutrals (see Problem 10.2) in going to the second line. For massless partons, the second and third term in $[\cdots]$ vanish.

Problem 10.1

Show that Eq. (10.12a) holds if a gluon is emitted from another gluon provided the color matrix is replaced by that of the gluon.

Problem 10.2

Show that $\boldsymbol{T}_q^2 = \boldsymbol{T}_q \cdot \boldsymbol{T}_q = C_F = 4/3,\ \boldsymbol{T}_g^2 = C_A = 3$. Also confirm that $(\boldsymbol{T}_q + \boldsymbol{T}_{\overline{q}})|q_i, \overline{q}_j\rangle = 0$, $(\boldsymbol{T}_q + \boldsymbol{T}_{\overline{q}} + \boldsymbol{T}_g)|q_i, \overline{q}_j, g\rangle = 0$, $(\boldsymbol{T}_{g_1} + \boldsymbol{T}_{g_2} + \boldsymbol{T}_{g_3})|g_a, g_b, g_c\rangle = 0$ for color singlet states.

For simplicity, we consider a case of massless partons emitting the soft gluon, and set $\beta_i = \beta_j = 1$. The radiation function can be arranged as

$$W_{ij} = E_{ij}^{[i]} + E_{ij}^{[j]} \tag{10.17a}$$

$$E_{ij}^{[i]} = \frac{1}{2}\left(W_{ij} + \frac{1}{1-\cos\theta_i} - \frac{1}{1-\cos\theta_j}\right) \tag{10.17b}$$

We shall show that $E_{ij}^{[i]}$ satisfies the relation

$$\int_0^{2\pi} \frac{d\phi_i}{2\pi} E_{ij}^{[i]} = \begin{cases} \frac{1}{1-\cos\theta_i} & \text{if } \theta_i < \theta_{ij} \\ 0 & \text{otherwise} \end{cases} \tag{10.18}$$

Proof of Eq. (10.18): We note that

$$1 - \cos\theta_j = a - b\cos\phi_i \tag{10.19a}$$

$$a = 1 - \cos\theta_{ij}\cos\theta_i\,, \qquad b = \sin\theta_{ij}\sin\theta_i \tag{10.19b}$$

Defining $z = e^{i\phi_i}$ or equivalently $\cos\phi_i = (z^2+1)/2z$, we have

$$I_{ij} \equiv \int \frac{d\phi_i}{2\pi}\frac{1}{1-\cos\theta_j} = -\frac{1}{i\pi b}\oint_{|z|=1}\frac{dz}{(z-z_+)(z-z_-)} \tag{10.20a}$$

$$z_\pm = \frac{a}{b} \pm \sqrt{\frac{a^2}{b^2} - 1} \tag{10.20b}$$

Since $|z_+| > 1, |z_-| < 1$, only the pole at $z = z_-$ contributes and we obtain

$$I_{ij} = \frac{1}{\sqrt{a^2 - b^2}} = \frac{1}{|\cos\theta_i - \cos\theta_{ij}|} \tag{10.21}$$

Hence,

$$\int_0^{2\pi} \frac{d\phi_i}{2\pi} E_{ij}^{[i]} = \frac{1}{(1-\cos\theta_i)}\left[\frac{1 + (\cos\theta_i - \cos\theta_{ij})I_{ij}}{2}\right] \tag{10.22}$$

which proves Eq. (10.18). □

In short, the radiation function which represents the interference contribution from lines i and j can be separated into two independent collinear singularities which has a remarkable property that the later emission is suppressed if its emission angle is larger than the previous emission angle after azimuthal averaging.

This is the angular ordering we described in Figure 9.15. This feature allows us to properly include the color coherence which is an amplitude interference effect in a stochastic Monte Carlo generation. This is the reason why the parton shower generator is capable of producing the color coherence effect by making the emission angle of later gluons smaller relative to previous one.

10.2.2
String Effect

Now, we show that the interference of soft gluons effectively produces the string effect. Let us consider a distribution of the second, but soft gluon after a hard three-jet configuration is formed. The three-jet configuration is assumed to constitute a color singlet.

$$e^- + e^+ \rightarrow q(p_1) + \overline{q}(p_2) + g_1(p_3) + g_2(k) \tag{10.23}$$

where $E_1 \sim E_2 \sim E_3 \gg \omega$ is assumed. The process is a sum of each external line emitting a soft gluon whose emission rate is given by Eq. (10.12). Expressing the color factor explicitly, the emission rate is proportional to

$$\begin{aligned} W(k; q(p_1)\overline{q}(p_2)g(p_3)) &\equiv W(\theta, \phi) \\ &\propto \sum_\lambda \left| \boldsymbol{T}_q \frac{(p_1 \cdot \varepsilon(\lambda)^*)}{(p_1 \cdot \hat{k})} + \boldsymbol{T}_{\overline{q}} \frac{(p_2 \cdot \varepsilon(\lambda)^*)}{(p_2 \cdot \hat{k})} + \boldsymbol{T}_g \frac{(p_3 \cdot \varepsilon^*(\lambda))}{(p_3 \cdot \hat{k})} \right|^2 \\ &= C_A \left[\frac{(p_1 \cdot p_3)}{(p_1 \cdot \hat{k})(p_3 \cdot \hat{k})} + \frac{(p_2 \cdot p_3)}{(p_2 \cdot \hat{k})(p_3 \cdot \hat{k})} \right] + (2C_F - C_A) \frac{(p_1 \cdot p_2)}{(p_1 \cdot \hat{k})(p_2 \cdot \hat{k})} \end{aligned} \tag{10.24}$$

Note that the replacement of $\varepsilon(\lambda)$ with k produces $\boldsymbol{T}_q + \boldsymbol{T}_{\overline{q}} + \boldsymbol{T}_g = 0$ for a color singlet. Therefore, polarization sum $\sum_{\lambda=\pm} \varepsilon_\mu(\lambda)\varepsilon_\nu(\lambda)^*$ for the physical gluon can be replaced with $-g_{\mu\nu}$ (see Problem 7.2). The color factor is obtained as

$$\begin{aligned} 2\boldsymbol{T}_q \cdot \boldsymbol{T}_{\overline{q}} &= (\boldsymbol{T}_q + \boldsymbol{T}_{\overline{q}})^2 - \boldsymbol{T}_q^2 - \boldsymbol{T}_{\overline{q}}^2 = (-\boldsymbol{T}_g)^2 - \boldsymbol{T}_q^2 - \boldsymbol{T}_{\overline{q}}^2 = C_A - 2C_F \\ 2\boldsymbol{T}_g \cdot \boldsymbol{T}_q &= (\boldsymbol{T}_g + \boldsymbol{T}_q)^2 - \boldsymbol{T}_g^2 - \boldsymbol{T}_q^2 = (-\boldsymbol{T}_{\overline{q}})^2 - \boldsymbol{T}_g^2 - \boldsymbol{T}_q^2 = -\boldsymbol{T}_g^2 = -C_A \end{aligned} \tag{10.25}$$

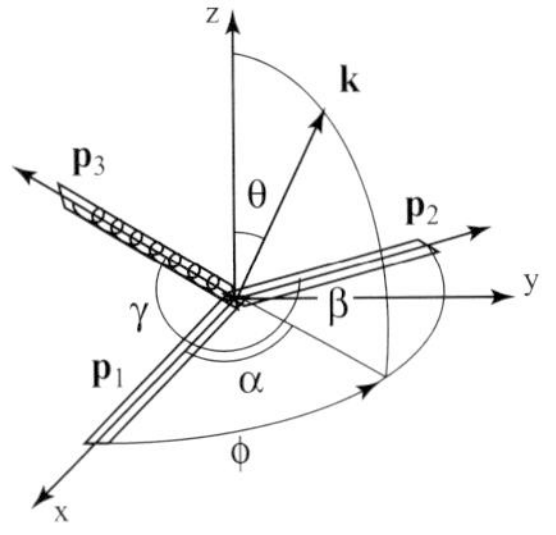

Figure 10.9 Coordinate variables of three-jets. Momenta of $q, \bar{q}, g$ (hard jets) are denoted as $\boldsymbol{p}_1, \boldsymbol{p}_2, \boldsymbol{p}_3$ and that of the soft gluon as $\boldsymbol{k}$.

Projected particle flow distributions in the three-jet plane (x–y plane in Figure 10.9) can be obtained from [568]

$$\begin{aligned}
W(\phi) &= \int d\theta \sin\theta \, W(\theta, \phi) = (2C_F - C_A) a_{12} V(\alpha, \beta) \\
&\quad + C_A[a_{13} V(\alpha, \gamma) + a_{23} V(\beta, \gamma)] \\
V(\alpha, \beta) &= \int \frac{d\cos\theta}{2} \frac{1}{(p_1 \cdot \hat{k})(\hat{p}_2 \cdot \hat{k})} = \frac{2}{\cos\alpha - \cos\beta} \left(\frac{\pi - \alpha}{\sin\alpha} - \frac{\pi - \beta}{\sin\beta} \right) \\
a_{ij} &= (\hat{p}_i \cdot \hat{p}_j) = 1 - \cos\theta_{ij} \\
\alpha &= \phi\,, \quad \beta = \theta_{12} - \alpha\,, \quad \gamma = \theta_{13} + \alpha
\end{aligned} \tag{10.26}$$

Problem 10.3

Derive Eq. (10.26).

To demonstrate the gluon's string effect, we compare its emission rate in $q\bar{q}g$ configuration with that of $q\bar{q}\gamma$. The emission rate of the latter is obtained from Eq. (10.26) by setting $C_A = 0$, $C_F = 1$ and replacing $g_s \to e$. The flow rate of the photon and the gluon is shown in Figure 10.10 [569]. It is shown as a directivity diagram projected on to event planes defined by q, $\bar{q}$ and g momentum vectors. The dashed (solid) line shows the flow rate of the photon (gluon), whose distance from the origin represents the density $\ln[(1/N)dn/d\phi]$. The azimuthal angle ϕ is defined relative to the most energetic quark jet axis. One can see a reduction of soft gluons emitted opposite to the hard gluon, that is, the soft gluons are dragged by the hard gluon. This is due to negative interference of the gluon which is a result of non-Abelian QCD effects as represented by the coefficient of the second term in Eq. (10.24). The reduction rate is $(2C_F - C_A)/2C_F = [2 \cdot (4/3) - 3]/2(4/3) = 1/8$.

One can see that the observed soft hadron angular distribution matches well with calculated soft gluon rate without hadronization model corrections. This is an example known as the "local parton hadron duality (LPHD)." However, those on the hard gluon side are grossly overestimated. This is partly due to the fact that the calculation assumes jet momenta in the asymptotic region, whereas the

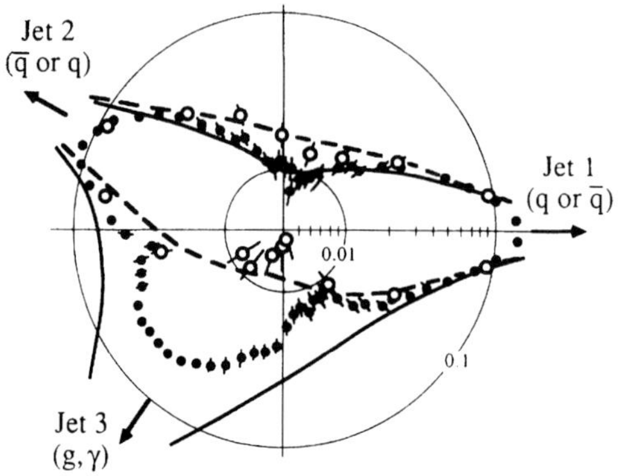

Figure 10.10 String effect due to gluon interference. The particle flow rate is described as a radial distance ($\ln[(1/N)dn/d\phi]$) from the origin as a function of the emission angle. Solid circles are those from $q\overline{q}g$ and open circles are from $q\overline{q}\gamma$ configuration. One sees that the particle flow opposite to the gluon emission is reduced compared to that of γ. Data by TPC/Two gamma collaboration [569].

observation was made at the total center of mass energy $\sqrt{s} = 29\,\text{GeV}$, that is, $E_{\text{jet}} < 10\,\text{GeV}$ and partly due to purity of the jet sample. The three jets assigned to $q, \overline{q}, g$ were chosen as those clusters having small sphericity with $E_1 > E_2 > E_3$ and their identification probability was estimated to be about $60 \sim 70\%$.

10.3
Fragmentation at Small *x*

The angular ordering in the fragmentation process provides a way to take the coherence effect into account in the evolution of the otherwise stochastic DGLAP equation.

10.3.1
DGLAP Equation with Angular Ordering

The equation to govern the evolution of the fragmentation function

$$D_i^h(x, Q^2) = D_i^h(x, Q_0^2) + \frac{\alpha_s(t)}{2\pi} \sum_j \int_x^1 \frac{dz}{z} \int_{Q_0^2}^{Q^2} \frac{dt'}{t'} P_{j\leftarrow i}(z, \alpha_s(t)) D_j^h(x/z, t') \tag{10.27}$$

is valid for $\alpha_s \ln Q^2 \ll 1$. Apart from the timelike argument $Q^2 > 0$ of the fragmentation function, this is the same DGLAP equation which governs the parton distribution function (pdf) inside a hadron. The splitting functions are identical in LLA, but different in NLLA. One should be careful that the off-diagonal elements

$P_{j\leftarrow i}$ and $P_{i\leftarrow j}$ are interchanged. This is because initial partons split first and supply another parton in the fragmentation stepping down on the magnitude of the virtuality while partons supplied by the pdf split in to climb up the virtuality ladder in the development of the parton distribution function.

When x is small ($x \ll 1$), $\alpha_s(Q^2)\ln x$ is large and the perturbation expansion in terms of α_s cannot be used. This is the overlap region of the collinear and soft gluons depicted in Figure 9.12. Here, a different treatment is necessary. We must consider bringing the coherence effect into evolution of the fragmentation by angular ordering. Here, a relevant parameter would be $\theta^2 \simeq t/E^2$. Then, the angular ordering $\theta_b^2, \theta_c^2 < \theta_a^2$ would mean $t_b/E_b^2, t_c/E_c^2 < t_a/E_a^2$, that is, $t_b < z^2 t_a$, $t_c < (1-z)^2 t_a$ (see Eq. (8.61)). The evolution equation would be modified from Eq. (10.27) to

$$\begin{aligned} &D_i^h(x, Q^2) \\ &= D_i^h(x, Q_0^2) + \frac{\alpha_s(t)}{2\pi}\sum_j \int_x^1 \frac{dz}{z} \int_{Q_0^2}^{z^2 Q^2} \frac{dt'}{t'} P_{j\leftarrow i}(z, \alpha_s(t)) D_j^h(x/z, t') \end{aligned} \tag{10.28}$$

or in differential form

$$t\frac{\partial D_i^h(x,t)}{\partial t} = \sum_j \int_x^1 \frac{dz}{z}\frac{\alpha_s(t)}{2\pi} P_{j\leftarrow i}(z, \alpha_s(t)) D_j^h(x/z, z^2 t) \tag{10.29}$$

For a nonsinglet function like the u or d quark fragmentation functions, there is no contribution from the gluon fragmentation and the equation takes a simple form. We can solve the equation by taking moments (see Section 8.5.1)

$$\tilde{D}^h(N, Q^2) = \int_0^1 \frac{dx}{x} x^N D^h(x, x^2 Q^2) \tag{10.30}$$

Then, the integral equation Eq. (10.29) becomes

$$t\frac{\partial \tilde{D}(N,t)}{\partial t} = \frac{\alpha_s}{2\pi}\int_0^1 \frac{dz}{z} z^N P(z)\tilde{D}(N, z^2 t) \tag{10.31}$$

Notice the only modification of the DGLAP equation is the z-dependent scale on the right-hand side and is not important for most values of x. However, for small x, it is crucial because the evolution of the fragmentation function at small x is sensitive to moments near $N = 1$, as can be seen from the formal solution to the DGLAP equation in Eq. (8.95). Notice that the anomalous dimensions are singular at $N = 1$ because of $1/z$ singularity in the splitting function (see Eq. (8.105)).

To solve the equation for fixed α_s, we may use a trial function of the form

$$\tilde{D}^h(N,t) \sim t^{d_N}\,, \quad d_N = \frac{\alpha_s}{2\pi}\gamma(N) \tag{10.32}$$

as suggested by Eq. (8.94). Referring to Eq. (10.31), the function d_N should satisfy

$$d_N(\alpha_s) = \frac{\alpha_s}{2\pi}\int_0^1 \frac{dz}{z} z^{N+2d_N(\alpha_s)} P(z) \tag{10.33}$$

where we made α_s dependence of d_N explicit. If $N-1 >> d_N(\alpha_s) \sim \alpha_s$, the second term $2d_N(\alpha_s)$ in the exponent of Eq. (10.33) may be neglected and we recover the expression Eq. (8.98b). Here, in the region where $x \to 0$, a singular term $\sim 1/x$ in the splitting function P dominates, which produces a pole in $d_N(\alpha_s)$. In the $x \to 0$ region, gluon multiplication would be the most important, that is, contribution from a term in $P_{g\leftarrow g}(z) \to 2C_A/z$ would dominate. Writing the moment of $P_{g\leftarrow g}(x)$ as $\gamma_{gg}(N)$ and the corresponding d_N function as $d_N^{gg}(N)$, Eq. (10.33) gives

$$d_N^{gg}(\alpha_s) = \frac{C_A\alpha_s}{\pi}\frac{1}{N-1+2d_N^{gg}(\alpha_s)} \tag{10.34}$$

hence,

$$d_N^{gg}(\alpha_s) = \frac{1}{4}\left[\sqrt{(N-1)^2 + \frac{8C_A\alpha_s}{\pi}} - (N-1)\right] \tag{10.35a}$$

$$= \sqrt{\frac{C_A\alpha_s}{2\pi}} - \frac{1}{4}(N-1) + \frac{1}{32}\sqrt{\frac{2\pi}{C_A\alpha_s}}(N-1)^2 + \cdots \tag{10.35b}$$

When scale dependence of α_s is taken into account (see Eq. (7.63)),

$$\tilde{D}^h(N,t) \sim t^{d_N^{gg}(\alpha_s)} = \exp\left[d_N^{gg}(\alpha_s)\ln t\right] \to \exp\left[\int^t \frac{dt'}{t'} d_N^{gg}(\alpha_s)\right]$$

$$= \exp\left[\int^{\alpha_s(\tau)} \frac{d_N^{gg}(\alpha_s)}{\beta(\alpha_s)} d\alpha_s\right] \tag{10.36}$$

Substituting the LLA expression for the β function

$$\beta(\alpha_s) = -\alpha_s^2 b_0(1 + b_1\alpha_s + \cdots)$$
$$b_0 = \frac{\beta_0}{4\pi} = \frac{1}{4\pi}\left(11 - \frac{2}{3}n_f\right) \tag{10.37}$$

and Eq. (10.35b) in Eq. (10.36), we find

$$\tilde{D}^h(N,t) \sim \exp\left[\frac{1}{b_0}\sqrt{\frac{2C_A}{\pi\alpha_s}} - \frac{1}{4b_0\alpha_s}(N-1) + \frac{1}{48b_0}\sqrt{\frac{2\pi}{C_A\alpha_s^3}}(N-1)^2 + \cdots\right]_{\alpha_s=\alpha_s(t)} \tag{10.38}$$

This is a result that takes gluon coherence into account and, as will be shown later, reproduces data well. However, before we compare the result with the experiment, let us clarify the origin of peculiar $\sqrt{\alpha_s}$ dependence in the equation.

10.3.2
$\sqrt{\alpha_s}$ Dependence and $N = 1$ Pole in the Anomalous Moment

Equation (10.35) exhibits the behavior of the anomalous dimension $d_N(\alpha_s) \sim \sqrt{\alpha_s}$ in the leading order. This is counterintuitive. From the definition, we expect

$$d_N(\alpha_s) = \frac{\alpha_s}{2\pi}\gamma(N) \overset{\text{Eq. (8.93b)}}{=\!=} \frac{\alpha_s}{2\pi}\int_0^1 \frac{dx}{x} x^N P_{ij}(x) \sim \alpha_s \tag{10.39}$$

The $\sqrt{\alpha_s}$ dependence originates from setting the functional form of the fragmentation function as Eq. (10.32) which makes the equation for γ_{gg} Eq. (10.34) nonlinear. The assumption was $N - 1 \ll 8C_A\alpha_s/\pi$ to obtain d_N^{gg} in Eq. (10.35b) which is expanded in powers of $(N - 1)$. This is not the usual assumption in making power expansion of α_s which is supposed to be the small parameter.

If we expand Eq. (10.35a) with the ordinary assumption $N - 1 \gg 8C_A\alpha_s/\pi$, we will get

$$d_N(\alpha_s) = \frac{\alpha_s}{2\pi}\left(\frac{2C_A}{N-1}\right) - 2\left(\frac{C_A\alpha_s}{\pi}\right)^2 \frac{1}{(N-1)^3} + \cdots \tag{10.40}$$

This has a conventional form expanded in ascending powers of α_s. However, this has a pole at $N = 1$ and has to be resummed to obtain a legitimate result. We will show that resummation of the first $N = 1$ pole in Eq. (10.40) is equivalent to resummation of powers of double logarithm $\alpha_s \ln t \ln 1/x$ without angular ordering as was done in Section 8.5.2.

Let us apply the standard formula given in Eq. (8.93b) to solve the gluon distribution in the small x region. Putting $P_{ij}(x) \to P_{gg}(x) \sim 2C_A/x$, we obtain

$$\gamma(N) = \int_0^1 \frac{dx}{x} x^N P_{gg}(x) \approx \frac{2C_A}{N-1} \tag{10.41}$$

Substituting Eq. (10.41) in Eq. (8.94), we obtain the solution for the moment of the gluon distribution

$$\tilde{g}(N,t) = \tilde{g}_0(N,t_0)\exp\left(\frac{\eta}{N-1}\right) \tag{10.42a}$$

$$\eta \equiv \frac{\alpha_s}{2\pi}2C_A \ln\frac{t}{t_0} \overset{\text{Eq. (8.126)}}{=\!=} \frac{C_A}{\pi b_0}\ln\frac{\ln(t/\Lambda^2)}{\ln(t_0/\Lambda^2)} \tag{10.42b}$$

where the last equality applies to the running α_s. Equation (10.42a) shows that powers of $1/(N-1)$ are indeed resummed. The gluon distribution can be obtained by the inverse Mellin transformation:

$$\begin{aligned}
x g(x,t) &= \frac{1}{2\pi i}\int_{N_0-i\infty}^{N_0+i\infty} dN x^{-(N-1)}\tilde{g}(N,t) \\
&= \frac{1}{2\pi i}\int_{N_0-i\infty}^{N_0+i\infty} dN \tilde{g}_0(N,t_0)\exp[h(N)] \\
h(N) &= (N-1)\xi + \frac{\eta}{N-1} = h(N_s) + \frac{h''(N_s)}{2}(N-N_s)^2 + \cdots \\
\xi &\equiv \ln\frac{1}{x}, \quad h(N_s) = 2\sqrt{\eta\xi}\,, \quad h'(N_s) = 0\,,
\end{aligned} \tag{10.43}$$

In the limit where both η and ξ become asymptotically large, one can expand the integral around the saddle point (referred to as the method of steepest slope) which is determined by $h'(N)|_{N=N_s} = 0$. Then,

$$h(N) = 2\sqrt{\xi\eta} + O(N-N_s)^2 + \cdots\,, \qquad N_s = 1 + \sqrt{\frac{\eta}{\xi}} \tag{10.44}$$

which gives the asymptotic solution

$$x g(x,t) \approx \tilde{g}(N_s, t_0)\exp\left(2\sqrt{\eta\xi}\right) \tag{10.45}$$

This reproduces the resummed double logarithm solution Eq. (8.127) without angular ordering.

10.3.3
Multiplicity Distribution

We now compare our result Eq. (10.38) with experimental data. Multiplicity is given as the $N = 1$ moment of the fragmentation function. In the $e^- e^+$ reaction,

$$\begin{aligned}
\langle n(s)\rangle &= \int_0^1 dx\, D(x,s) = \tilde{D}(1,s) \overset{\text{Eq. (10.38)}}{\sim} \exp\left[\frac{1}{b_0}\sqrt{\frac{2C_A}{\pi\alpha_s(s)}}\right] \\
&= \exp\left[\sqrt{\frac{2C_A}{\pi b_0}}\ln\left(\frac{s}{\Lambda^2}\right)\right](1 + \text{NLLA correction})
\end{aligned} \tag{10.46}$$

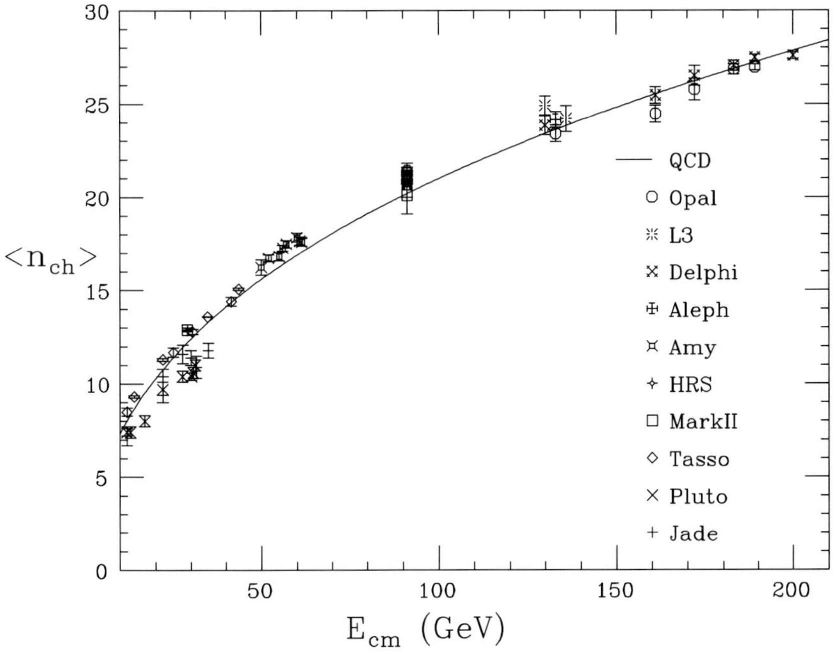

Figure 10.11 Mean multiplicity of charged particles from e^-e^+ reactions [453, 570].

The NLLA correction adds a term of order $\ln \alpha_s(s)$, which gives a multiplicative factor in powers of $\alpha_s(s)$. The multiplicity can be expressed as [571, 572]

$$\langle n(s)\rangle = n_0 \alpha_s^A e^z$$

$$A = \frac{1}{4} + \frac{5n_f}{54\pi b_0} = 0.49(\text{ for } n_f = 5), \quad z = \sqrt{\frac{2C_A}{\pi b_0} \ln\left(\frac{s}{\Lambda^2}\right)} \tag{10.47}$$

n_0 is a normalization factor which cannot be calculated perturbatively. Similar formulas are also obtained by [573, 574]. Figure 10.11 gives the mean multiplicity of hadrons in the final state of the e^-e^+ reaction. Its energy dependence is reproduced precisely by the pQCD. Note, apart from overall normalization which cannot be calculated perturbatively, the formula contains no adjustable parameters if a value of $\alpha_s(s)$ is taken from other experiments [443].

10.3.4
Humpback Distribution: MLLA

The gluon distribution $g(x, t)$ which takes into account the angular ordering can be obtained from inverse Mellin transformations. Denoting $\tilde{D}^h(N, t)$ in Eq. (10.38) as $\tilde{g}(N,t)$, we have

$$
\begin{aligned}
&xg(x,t)\\
&= \frac{1}{2\pi i}\int_{N_0-i\infty}^{N_0+i\infty} dN x^{1-N}\tilde{g}(N,t) \sim \frac{1}{2\pi i}\int_{N_0-i\infty}^{N_0+i\infty} dN x^{1-N} t^{d_N^{gg}(\alpha_s)}\\
&= \frac{1}{2\pi i}\int_{N_0-i\infty}^{N_0+i\infty} dN \exp\left[(N-1)\ln(1/x)\right]\\
&\times \exp\left[\frac{1}{b_0}\sqrt{\frac{2C_A}{\pi\alpha_s}} - \frac{1}{4b_0\alpha_s}(N-1) + \frac{1}{48b_0}\sqrt{\frac{2\pi}{C_A\alpha_s^3}}(N-1)^2 + \cdots\right]
\end{aligned}
\tag{10.48}
$$

$\tilde{g}(N,t)$ is assumed to have no singularity to the left of N_0. We can evaluate the asymptotic value of the integral by the saddle point method. The saddle point z_{sp} is where the slope of the exponent vanishes and if the exponent is given by $f(z)$,

$$
f(z) = A + Bz + \frac{C}{2}z^2 + \cdots = f(z_{\text{sp}}) + \frac{1}{2}f''(z_{\text{sp}})(z-z_{\text{sp}})^2 + \cdots \tag{10.49}
$$

Changing the variable to $z = z_{\text{sp}} + iy$, integration over y produces $(1/\sqrt{2\pi C})$ $\exp[(A - B^2/2C)]$, which is expressed in terms of the original variables by [575]

$$
xg(x,s) = K\alpha_s^{3/4}\exp\left[\frac{1}{b_0}\sqrt{\frac{2C_A}{\pi\alpha_s}}\right]\exp\left[-\frac{(\xi-\xi_p)^2}{2\sigma^2}\right] \tag{10.50a}
$$

$$
\xi = \ln\frac{1}{x}, \quad \xi_p = \frac{1}{4b_0\alpha_s(s)} \sim \frac{1}{4}\ln\frac{s}{\Lambda^2} \tag{10.50b}
$$

$$
\sigma = \left(\frac{1}{24b_0}\sqrt{\frac{2\pi}{C_A\alpha_s^3(s)}}\right)^{1/2} \sim \left[\ln\frac{s}{\Lambda}\right]^{3/4} \tag{10.50c}
$$

This shows the so-called humpback distribution which characterizes reduction of the gluon distribution at large ξ (small x).

The above formula is a simple version of what is commonly referred to as MLLA (modified leading log approximation) [565, 571, 575–580]. MLLA is an analytical expression which takes into account all of the essential ingredients of parton multiplication in NLO (next-to-leading order), that is, of the running coupling constant as well as the splitting function with the angular ordering incorporated. Phenomenologically, skewness of the deformed Gaussian distribution and behaviors in the neighborhood of $x \sim 1$ are taken into account by subleading contributions. The MLLA expression is given by [580–582]

$$
\begin{aligned}
\frac{1}{\sigma}\frac{d\sigma}{d\xi} &= \frac{4N_c}{b}\Gamma(B)K(Y)\times\int_{a_0-i\pi/2}^{a_0+i\pi/2}\frac{d\tau}{\pi}e^{-B\alpha}\left[\frac{C(\alpha,\xi,Y)}{(4N_c/b)\,Y\alpha/\sinh\alpha}\right]^{B/2}\\
&\times I_B(\sqrt{D(\alpha,\xi,Y)})
\end{aligned}
\tag{10.51a}
$$

$$D(a, \xi, Y) = \frac{16N_c}{b} Y \frac{\alpha}{\sinh \alpha} C(\alpha, \xi, Y) \tag{10.51b}$$

$$C(\alpha, \xi, Y) = \cosh \alpha + \left(\frac{2\xi}{Y} - 1\right) \sinh \alpha \tag{10.51c}$$

$$b = (11C_A - 2n_f)/3\,, \quad B = (11C_A + 2n_f/C_A^2)/3b \tag{10.51d}$$

$$Y = \ln \frac{E_{CM}}{2\Lambda_{eff}}\,, \quad \alpha = a_0 + i\tau \tag{10.51e}$$

a_0 is determined by the relation

$$\tanh a_0 = 1 - 2\xi / Y \tag{10.52}$$

and I_B is the modified Bessel function of order B (see Eq. (8.124)). In this formula, the evolution of the partons and the fragmentation are factored out. The formula is obtained at the parton level, but assuming LPHD, it can be directly compared with the observed hadron distribution. All the hadronization effect is contained in the over all normalization factor. By LPHD, $K(Y)$ is considered as a function of Y only. The LPHD (local parton hadron duality) is a simple version of the fragmentation models which claims that at sufficiently low virtuality, the (energy or angular) distribution of final hadrons closely follow that of partons [579]. The above formula only contains one parameter Λ_{eff}, except the normalization factor $K(Y)$. The Λ_{eff} plays essentially a role of cut off parameter hence is expected to grow with particle mass.

The formula reproduces data on particle distributions both inside and intrajets for different production mechanisms (e^-e^+, e^-p, pp) at different energies (Figure 10.12a). Figure 10.12b shows the peak position as a function of particle species

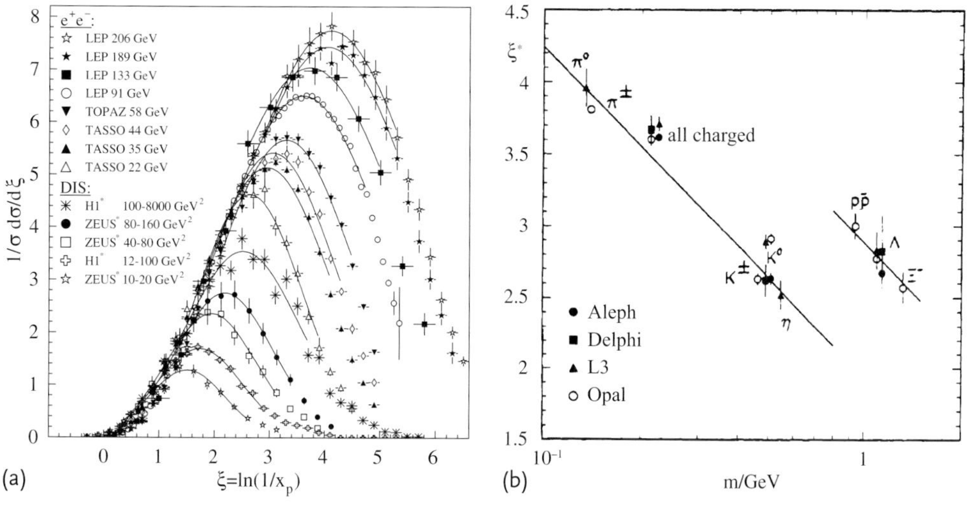

Figure 10.12 (a) MLLA fitting for small x distributions [11]. (b) Peak position ξ^* (ξ_p in the text) plotted at fixed $E_{CM} = m_Z$ as a function of particle mass [526].

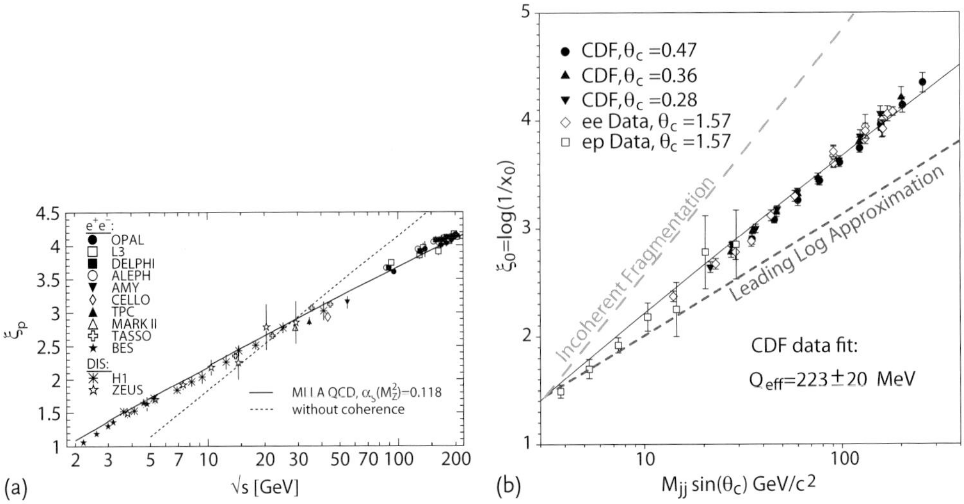

Figure 10.13 Peak positions of hump back distribution at small x. (a) in e^-e^+ and ep reactions and (b) in e^-e^+, ep and $p\overline{p}$ collisions. In hadronic collisions, the virtuality is set as $s = M_{jj} \sin\theta$ where M_{jj} is the invariant mass of di-jet [11, 583].

at fixed E_{CM}. The data shows $\Lambda_{\text{eff}} \sim m/m_0$ with two distinct m_0 for the meson and baryon. The data also confirms the validity of LPHD. Note that also the spectra can also be well reproduced using parton shower/fragmentation Monte Carlo programs with angular ordering taken into account. MLLA is an elaborate analytical expression that can reproduce data in a certain phase space region.

Shift of the peak position in the MLLA is expressed with only one parameter Λ_{eff} and is expressed as

$$\begin{aligned} \xi_p &= \frac{1}{2}Y + \sqrt{cY} - c \\ c &= B\sqrt{\frac{b}{16N_c}} = 0.29(0.35) \quad \text{for} \quad n_f = 3(5) \end{aligned} \tag{10.53}$$

Equation (10.53) predicts the peak position to grow as the logarithm of the energy in CM frame ($\sim \ln E_{\text{CM}}/2$) (Figure 10.13a) with the correct slope [574, 584]. This is a characteristic feature of the QCD which includes the color coherence effect.

An incoherent parton shower model would not produce the factor 1/2 and it is expected to follow the dotted or dashed line in Figure 10.13 which clearly contradicts the data. It is remarkable that the formula also fits to particle distributions in the jets produced in the hadron–hadron reactions (Figure 10.13b) [583]. Here, $\sqrt{s} = m_{jj}\sin\theta$ (m_{jj}: mass of dijet), $x = p/E_{\text{jet}}$, $Y = \ln(E_{\text{jet}}\theta/\Lambda_{\text{eff}})$ was adopted. θ is the cone size that defines a jet in the hadron–hadron interaction.

Intuitive Explanation of the Humpback Distribution The depletion of the observed soft gluon emissions can be understood kinematically as the result of two conflicting tendencies. On one hand, the gluon emission has to satisfy $k_\mathrm{T} \gtrsim 1/R$ to be visible, which restricts its emission at the large angle $\theta > 1/kR$ where $R \sim 1/m_\pi$ is the typical hadron size. On the other hand, the allowed emission angle in the successive cascade decays shrinks because of the angular ordering.

Let us consider the toy model [579, 580] depicted in Figure 10.14a. In the soft region, it is known that the hadron production is flat in the longitudinal phase space $d^3p/E = \pi dz dk_\mathrm{T}^2$ (see Section 11.1), namely, $\rho(k) \equiv dn/d\ln k =$ constant with limited transverse momenta $k\theta = k_\mathrm{T} \sim 1/R$ (shaded region in Figure 10.14b). This is considered as an effect of primordial parton motion in the hadron. At larger transverse momentum (or large virtuality Q), the gluon emission sets in. Denoting the gluon energy and emission angle as ϵ, θ_0, a double logarithm expression for the radiation probability is given as

$$d w_g \propto \alpha_s \frac{d\epsilon}{\epsilon}\frac{d\theta_0}{\theta_0}\theta\left(\epsilon\theta_0 - R^{-1}\right) \tag{10.54}$$

The step function restricts the transverse momentum $p_\mathrm{T} \sim \epsilon\theta_0 > R^{-1}$ to ensure that the gluon formation time (t_form) (see Eq. (10.9)) is much shorter than hadronization time t_had

$$t_\mathrm{form} \approx \frac{\epsilon}{p_\mathrm{T}^2} < t_\mathrm{had} \approx \gamma R \approx \epsilon R^2 \tag{10.55}$$

where $\gamma \simeq \epsilon R$ is the Lorentz contraction factor and the typical hadronization time at rest is taken to be R. The emitted gluon, in turn, will produce a jet whose spec-

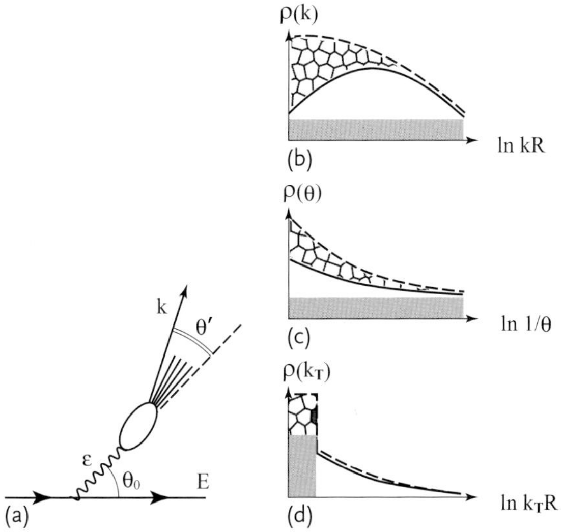

Figure 10.14 (a) First order toy model for illustrating the origin of the humpbacked plateau. (b–d) Effects of color coherence on energy, angle and k_T distributions. Dashed (solid) lines show incoherent (coherent) emission spectra. Mosaic areas show removed regions by turning on the coherence. Shaded areas show primordial hadronic plateau. Adapted from [579, 580].

trum is expected to be flat in the longitudinal phase space

$$R^{-1} < k < \epsilon \tag{10.56}$$

and limited transverse momenta with respect to the gluon, that is, $k\theta' \sim R^{-1}$. This leads to the secondary distribution

$$dn = \frac{dk}{k}\frac{d\theta'}{\theta'}\delta(k\theta' - R^{-1}) \tag{10.57}$$

The δ function can be thought of as a double logarithmic spectrum of the gluon hadronization due to complete bremsstrahlung projected on the domain of the most intensive radiation ($\alpha_s(k\theta')/\pi \sim 1$). By angular ordering, the offspring of the gluon can only be emitted within the opening angle $\theta' < \theta_0$. The restriction applied to Eq. (10.54) results in a constraint on the plateau region:

$$(R\theta_0)^{-1} < k < \epsilon \tag{10.58}$$

which is the major consequence of the coherence. Combining the above considerations, our toy model gives us the following schematic particle multiplicity:

$$\begin{aligned} n = {} & \int^{E} \frac{dk}{k} \int^{1} \frac{d\theta}{\theta}\delta(k\theta - R^{-1}) \\ & + \alpha_s \int^{E} \frac{d\epsilon}{\epsilon} \int^{1} \frac{d\theta_0}{\theta_0}\theta\left(\epsilon\theta_0 - R^{-1}\right) \int^{\epsilon} \frac{dk}{k} \int^{\theta_{\max}} \frac{d\theta'}{\theta'}\delta(k\theta' - R^{-1}) \end{aligned} \tag{10.59}$$

The first term stands for the background quark plateau, and the second one is constructed from the gluon emission (10.54) and fragmentation Eq. (10.57). The parameter $\theta_{\max}$ encodes the difference between the coherent and incoherent emissions:

$$\theta_{\max} = \begin{cases} 1 & \text{incoherent emission} \\ \theta_0 & \text{coherent emission} \end{cases} \tag{10.60}$$

Inserting $\delta(k' - k)$, one obtains a particle energy spectrum $\rho(k) = k\,dn/dk$:

$$\rho^{\text{incoh}} = 1 + \frac{\alpha_s}{2}\left(\ln^2 ER - \ln^2 kR\right) \tag{10.61a}$$

$$\rho^{\text{coh}} = 1 + \alpha_s \ln\frac{E}{k}\ln kR \tag{10.61b}$$

The additional gluon-initiated coherent spectrum exhibits substantial depletion in the soft part of the parton spectrum as compared to the incoherent one (Figure 10.14b). The height of the hump increases with energy and the peak position is at $\sim \sqrt{E}$ or $\xi_p = \ln 1/x_p \sim (1/2)\ln E$. The multiplicity $\int d\ln k(\rho(k) - 1)$ of the incoherent emission appears to be twice as large when compared to the coherent case.

It is noteworthy that the incoherent monotonic spectrum Eq. (10.61a) would also have a peak due to kinematical mass effects ($\rho(k) \propto \sqrt{k^2 - m^2}$ at small energies $k \simeq m$), but placed near the phase space limit $k \sim m \sim R^{-1}$. The peak position without coherence can be reproduced with the independent fragmentation model as shown in Figure 10.13 and has a slope twice as steep.

Problem 10.4

Prove Eqs. (10.61).

Problem 10.5

Show, by inserting a δ function to fix $\theta = \theta_0 + \theta'$ in the multiplicity equation (10.59), that the angular spectrum is given by

$$\begin{aligned} \rho^{\text{incoherent}} &= 1 + \alpha_s \ln^2(E\theta R) \\ \rho^{\text{coherent}} &= 1 + (\alpha_s/2) \ln^2(E\theta R) \end{aligned} \qquad (10.62)$$

10.4 Gluon Fragmentation Function

So far we focused our attention on the small x part of the gluon fragmentation to understand the coherence effect. Here we discuss the overall x dependence of the gluon fragmentation function and identify characteristics peculiar to the gluon as contrasted to the quark fragmentation. We discuss two methods to measure the gluon fragmentation function, one by fitting with the next-to-leading order QCD calculation and the other by positively extracting the gluon in three-jet configuration.

Transverse and Longitudinal Fragmentation Function The unpolarized inclusive single-particle cross section $e^-e^+ \to h + X$ via vector bosons is given by

$$\frac{1}{\sigma_0}\frac{d^2\sigma}{dx\,d\cos\theta} = \frac{3}{8}(1+\cos^2\theta)F_{\mathrm{T}}^h(x) + \frac{3}{4}\sin^2\theta\, F_{\mathrm{L}}^h + \frac{3}{4}\cos\theta\, F_{\mathrm{A}}^h \qquad (10.63)$$

where $x = 2E_h/\sqrt{s}$, $\theta =$ angle in e^-e^+ CM frame and F_{T}^h, F_L^h are fragmentation functions produced by the transverse and longitudinally polarized intermediate states and F_A^h is the parity violating contribution arising from the interference of the vector and axial vector contributions [see Problem 3.2 and Eq. (4.9)]. $\sigma_0 = 4\pi\alpha^2 N_c/3s$ with $N_c = 3$ is the lowest order QED cross section for $e^-e^+ \to \mu^-\mu^+$ times the number of colors N_c. The three contributions can be extracted from the data, either by fitting the angular dependence for each value of x or by weighting each particle in the final state with an appropriate angular factor. We have

$$\left.\frac{d\sigma_P^h}{dx}\right|_{P=\mathrm{T,L,A}} = \int_{-v}^{+v} d\cos\theta\, W_P(\cos\theta; v)\frac{d^2\sigma^h}{dx\,d\cos\theta} \qquad (10.64)$$

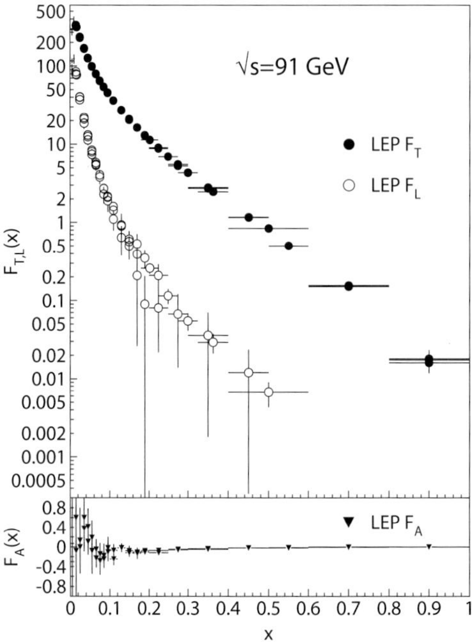

Figure 10.15 Fragmentation functions F_T, F_L and asymmetric F_A. Data points with relative errors greater than 100% are omitted [3].

where $\pm v$ is the maximum (minimum) value of measurable $\cos\theta$ and the weight functions are given by

$$
\begin{aligned}
W_T(u;v) &= \frac{5u^2(3-v^2) - v^2(5-3v^2)}{2v^5} \\
W_L(u;v) &= \frac{v^2(5+3v^2) - 5u^2(3+v^2)}{4v^5} \\
W_A(u;v) &= \frac{2u}{v^3}
\end{aligned}
\tag{10.65}
$$

Figure 10.15 shows extracted F_T and F_L obtained at LEP [3].

From tensor analysis, one can show that Eq. (10.63) is the most general form for the unpolarized inclusive single-particle cross section because for the process

$$
e^-(p_1) + e^+(p_2) \to \gamma^*, Z \to h(k) + X(p_X)
$$

it can be expressed in a form [585]

$$d\sigma = \frac{1}{2s} L_{\mu\nu} \frac{g_{Vi}^2 g_{Vf}^2}{|s - m_V + i m_V \Gamma_V|^2} H^{\mu\nu} \frac{d^3k}{(2\pi)^3 2k^0} \tag{10.66a}$$

$$H^{\mu\nu} = \sum_X (2\pi)^4 \delta(p_1 + p_2 - k - p_X) \langle 0 | J^\mu | k, X \rangle \langle k, X | J^\nu | 0 \rangle \tag{10.66b}$$

$$\frac{d^3k}{(2\pi)^3 2k^0} = \frac{q^2}{32\pi^2} \sqrt{x^2 - \frac{4m_h^2}{q^2}}\, dx\, d\cos\theta \frac{d\phi}{2\pi} ,$$

$$x = \frac{2(q \cdot k)}{q^2} , \quad q^2 = (p_1 + p_2)^2 = s \tag{10.66c}$$

$$L_{\mu\nu} = \frac{1}{4} \sum_{\text{spin}} \left[v(p_2) \gamma_\mu (v - a\gamma^5) u(p_1) \right] \left[v(p_2) \gamma_\mu (v - a\gamma^5) u(p_1) \right]^* \tag{10.66d}$$

Here, g_{Vi}, g_{Vf} are appropriate coupling constants of the vector bosons. The function $H^{\mu\nu}$ has the same structure as the deep inelastic structure functions (Eq. (8.16)) and can be expressed in the form

$$H^{\mu\nu} = A \left(-g^{\mu\nu} + \frac{q^\mu q^\nu}{q^2} \right) + B \left(k^\mu - q^\mu \frac{(q \cdot k)}{q^2} \right) \left(k^\nu - q^\nu \frac{(q \cdot k)}{q^2} \right)$$

$$+ C i \varepsilon^{\mu\nu\rho\sigma} q_\rho k_\sigma + D \frac{q^\mu q^\nu}{q^2} + E(q^\mu k^\nu + k^\mu q^\nu) \tag{10.66e}$$

$$A \propto F_T^h , \quad B \propto \frac{1}{|\boldsymbol{k}|^2} \left(F_L^h - F_T^h \right) , \quad C \propto -\frac{1}{q^0 |\boldsymbol{k}|} F_A^h \tag{10.66f}$$

where $|\boldsymbol{k}|$ is the momentum in the center of mass frame of $e^- e^+$. D and E terms can be neglected because they are multiplied by the lepton mass after contraction with $L_{\mu\nu}$.

In the parton model, only the transverse component contributes. The QCD predictions to $O(\alpha_s)$ for the transverse and longitudinal give ([392], see Eq. (7.139))

$$\sigma_T = \sigma_0 , \quad \sigma_L = \frac{\alpha_s}{\pi} \tag{10.67}$$

Namely, to the first order, the entire QCD correction to the total cross section comes from the longitudinal part. The fragmentation functions can be expressed as a sum of contributions from the different primary partons $i = u, d, \cdots, g$:

$$F_i^h = \sum_i \int_x^1 dz\, C_i(s; z, \alpha_s) D_i \left(\frac{x}{z}, s \right) \tag{10.68}$$

For the leading order where only $q\bar{q}$ pairs are produced, the coefficient functions C_i give $C_i = g_i(s)\delta(1 - z)$ where $g_i(s)$ is the appropriate electroweak coupling (for instance, $g_i \propto Q_i^2$: electric charge of the quarks for $V = \gamma$ (see Eq. (9.5)).

The gluon fragmentation function comes in only as the next leading order. The coefficient functions can be calculated perturbatively and give [392, 396, 397]

$$C_q^{\mathrm{T}}(z) = \delta(1-z) + O(\alpha_s) \tag{10.69a}$$

$$C_g^{\mathrm{T}}(z) = O(\alpha_s) \tag{10.69b}$$

$$C_q^{\mathrm{L}}(z) = C_F \frac{\alpha_s}{2\pi} + O\left(\alpha_s^2\right) \tag{10.69c}$$

$$C_g^{\mathrm{L}}(z) = 4C_F \frac{\alpha_s}{2\pi}\left(\frac{1}{z} - 1\right) + O\left(\alpha_s^2\right) \tag{10.69d}$$

Using Eq. (10.69), the longitudinal fragmentation function can be expressed in terms of the gluon fragmentation function.

$$F_L(x) = C_F \frac{\alpha_s}{2\pi} \int_x^1 \frac{dz}{z}\left[F_{\mathrm{T}}(z) + 4\left(\frac{z}{x} - 1\right) D_g(z)\right] + O\left(\alpha_s^2\right) \tag{10.70}$$

Consequently, the gluon fragmentation function can be extracted to the leading order using experimentally measured F_{T} and F_L. The function D_g is greatest at small values of x. At large x, F_{T} dominates. Since only the integrated values of D_g can be obtained from Eq. (10.70), the conventional method to extract D_g is to parametrize it as [586–588]

$$D_g(x) = P_1 x^{P_2}(1-x)^{P_3} e^{-P_4 \ln^2 x} \tag{10.71}$$

The functional form may be justified by noting that it fits $F_{\mathrm{T}}(x)$ and $F_L(x)$ satisfactorily in the region $x > 0.02$. The exponential term is motivated by the Gaussian spectrum in $1/z$ as given by MLLA, which fits the data well. The DELPHI group obtains [588]

$$P_1 = 0.47 \pm 0.07\,, \quad P_2 = -2.90 \pm 0.02\,, \quad P_3 = 5 \pm 1\,, \quad P_4 = 0.29 \pm 0.01 \tag{10.72}$$

(for fixed value of $\alpha_s^{\mathrm{LO}} = 0.126$).

The method we described to extract the gluon fragmentation function is independent of the jet definition and is potentially more reliable in the small x region. However, it has some limitations because Eq. (10.70) is valid only in the next-to-leading order in QCD and the function had to be parametrized. There is another method in which the gluon jet is selected topologically from the quark jets in three-jet configuration. The two methods are complementary and are in reasonable agreement to each other in the region $x > 0.2$ as is shown in Figure 10.16a. Since we can investigate various properties of the gluon jet by positively identifying it, we will discuss it in some details.

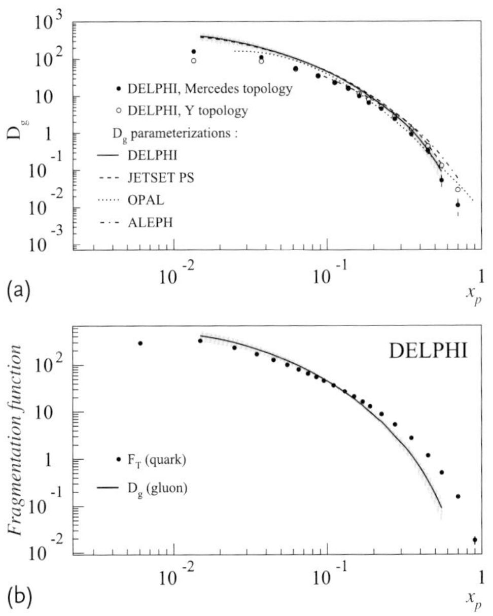

Figure 10.16 (a) Two DELPHI data on the gluon fragmentation function: one extracted from charged particles spectra (full curve, with shaded band showing the uncertainty in D_g) and the other from three-jets sample [589] (open and closed circles use different topology). Also shown are OPAL (dotted curve) and ALEPH (dot-dashed curve) with the same parametrization, compared with the JETSET PS model (dashed curve). (b) Comparison of the gluon fragmentation function $D_g(x)$ with the transverse fragmentation function $F_T(x)$ (as in Figure 10.15). The shaded band shows the range of D_g deviations [588].

10.5 Gluon Jets vs. Quark Jets

As the gluon has larger color charge $C_A = 3 > C_F = 4/3$, its interaction is stronger and therefore will produce more showers. From this consideration alone, one can infer that the gluon jet has larger multiplicity (larger $\langle n_g \rangle$), is softer (smaller $\langle \boldsymbol{p} \rangle$), and more spread (larger $\langle \theta \rangle$). In fact, in the perturbative QCD, $\langle n_g \rangle / \langle n_q \rangle = C_A / C_F = 9/4$ is expected [559]. The gluon is emitted from the quark or antiquark. In the $e^- e^+$ reactions, therefore, the gluon appears as the third jet which, in simple thinking, is considered to have the lowest energy. This is the conventional assumption in experiments and one can obtain properties of the gluon simply by comparing jets of the lowest energy with the first and the second energetic jets.

If the energy is not so high, the shower does not develop well, and the difference may not be so apparent. In fact, early jets at PETRA/DESY ($\sqrt{s} = 34\,\text{GeV}$), the obtained results vary [590–594]. Only at TRISTAN at $\sqrt{s} = 60$ GeV, the difference begin to show up [595, 596]. Even in this case, it was necessary to choose the kinematical variables carefully to differentiate the gluon from the quark. For instance,

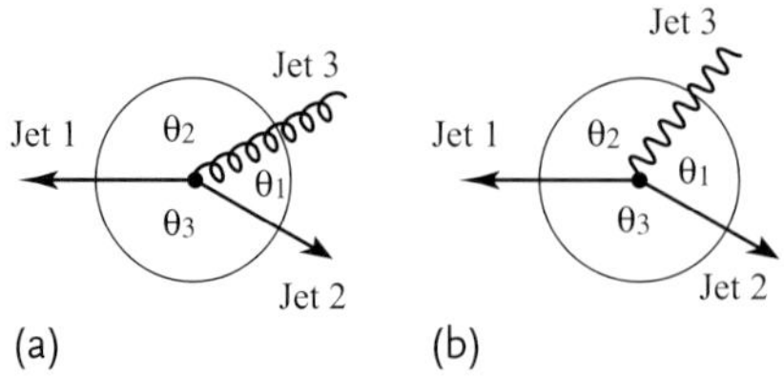

Figure 10.17 Three-jet configuration: (a) Symmetric $q\overline{q}g$ configuration ($\theta_2 \simeq \theta_3$), (b) $q\overline{q}\gamma$.

VENUS group selected three-jet topology consisting of $q\overline{q}g$ and $q\overline{q}\gamma$ and by comparing one to the other, could find the difference in the jet structure [596]. At LEP ($\sqrt{s} = 91$–200 GeV), finding the difference is considerably easier.

The three-jet configuration is often the favorite choice for selecting a good sample of the gluon jets. If one chooses three jet events with symmetric configuration (j_2, j_3, making nearly equal angles relative to j_1 as in Figure 10.17a), they are most likely the $q\overline{q}g$ samples. Purified gluon events can be obtained by asking the quark jets to contain the secondary vertex with short impact parameter (transverse distance) because the jets accompanying the secondary vertices most likely originate from b- or c-quark decays. Since the probability of producing heavy quarks in the jet is negligible, one can assume that doubly tagged events are $c\overline{c}g$ or $b\overline{b}g$ pairs. The energy of the jets are determined from their opening angles using the relation Eq. (10.4).

$$E_j = \frac{\sqrt{s}\sin\theta_j}{\sin\theta_1 + \sin\theta_2 + \sin\theta_3} \tag{10.73}$$

The gluon jet purities are estimated to vary from 95% for low energy jets to 46% for high energy jets for DELPHI events [597]. The light quark-jet samples are taken from $e^- e^+ \to$ 2-jet or from symmetric three-jet events with antitagged bs. It is easy to obtain a pure sample of the light quark jets with high energy because purity of the highest energy in the three-jet events of being the light quark is quite high ($\sim$ 98%). An unbiased sample of low energy light quarks can be obtained by tagging hard isolated photons in $q\overline{q}\gamma$, as illustrated in Figure 10.17b.

(1) The Gluon has Larger Multiplicity than the Quark As the difference between the quark and the gluon jets is more apparent at high energies, we take a look at the data produced at LEP. Figure 10.18 shows multiplicity distributions obtained by the DELPHI group [445, 597].

As stated, the gluon jet is expected to have larger multiplicity compared with the quark jet. Experimentally, however, the multiplicity ratio $\langle n_g\rangle/\langle n_q\rangle$ is much smaller than what is expected from a naive consideration, that is, $C_A/C_F = 9/4$, although the ratio is observed to increase with the scale. This is considered as partly due to the nonperturbative effect, the difference of the valence versus the sea quark, and partly due to color suppression which works on soft gluons, for example. Note, however, the absolute value of the multiplicity cannot be calculated perturbatively.

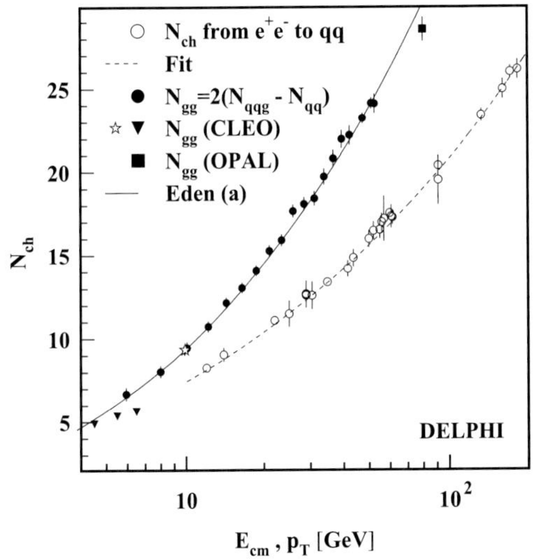

Figure 10.18 (a) Charge multiplicity of gluons and quarks as a function of $\kappa = E_j \sin(\theta/2)$ [445, 597, 598].

Therefore, one expects

$$\begin{aligned} \langle N_q(\kappa)\rangle &= N_0^q + N_{\text{pert}}(\kappa) \\ \langle N_g(\kappa)\rangle &= N_0^g + N_{\text{pert}}(\kappa)\cdot r(\kappa) \\ r(\kappa) &= \frac{C_A}{C_F}(1+\cdots) \end{aligned} \tag{10.74}$$

The first term N_0 represents the nonperturbative contribution and the second term represents the perturbative part. The latter is a function of a pertinent scale variable which we denote as κ. In e^-e^+ reactions, usually $\kappa = \sqrt{s} = 2E_{\text{CM}}$ is chosen. However, at LEP, E_{CM} is fixed and one has to choose another variable that can be applied to the three-jet configuration. The DELPHI group [597, 598] chose $\kappa = E_{\text{jet}}\sin(\theta/2)$ where θ is the angle with respect to the closest neighbor.

N_{pert} is the perturbative prediction for the hadron multiplicity as given in Eq. (10.46) [572–574].

$$N_{\text{pert}} = K\cdot(\alpha_s(\kappa))^b \exp\left(\frac{c}{\sqrt{\alpha_s(\kappa)}}\right)\left[1+O(\sqrt{\alpha_s})\right] \tag{10.75a}$$

$$b = \frac{1}{4}+\frac{2}{3}\frac{n_{\text{f}}}{\beta_0}\left(1-\frac{C_F}{C_A}\right) = \frac{1}{4}+\frac{5n_{\text{f}}}{54\pi b_0} \tag{10.75b}$$

$$c = \frac{1}{b_0}\sqrt{\frac{2C_A}{\pi}} = \frac{\sqrt{32C_A\pi}}{\beta_0}, \quad \beta_0 = 4\pi b_0 = 11-\frac{2}{3}n_{\text{f}}, \quad n_{\text{f}} = 5 \tag{10.75c}$$

From Eq. (10.74), one expects that the slope of the multiplicity distributions is less affected by the nonperturbative effect. At large scale, $N_{\text{gluon}}(\kappa) = C N_{\text{quark}}(\kappa)$.

Therefore, one can extract the perturbative part by taking the derivative and expect that

$$\frac{dN_g/d\kappa}{dN_q/d\kappa} = C \xrightarrow{\kappa\to\infty} C_A/C_F \tag{10.76}$$

By fitting the data with Eq. (10.75), the group obtained

$$\frac{C_A}{C_F} = 2.246 \pm 0.062(\text{stat}) \pm 0.080(\text{syst.}) \pm 0.095(\text{theo.}) \tag{10.77}$$

which is in excellent agreement with the theoretical value $3/(4/3) = 2.25$.

(2) Scale Dependence of the Fragmentation Function Scaling violation of the fragmentation functions is expected because they obey the same DGLAP evolution equations as the structure functions in the deep inelastic scattering. Difference from those of parton distribution functions are that chain of events is reversed, evolution parameters are timelike as contrasted to spacelike q^2 in the DIS with magnitude of the virtuality diminishing as the partons split. Denoting the fragmentation function as $D^h_{g,q}(x_E, s)$, where $s = (2\kappa)^2$, the DGLAP equations in LLA are expressed as [598]

$$\frac{dD^h_g(x_E, s)}{d\ln s} = \frac{\alpha_s(s)}{2\pi}\int_{x_E}^{1}\frac{dz}{z}\left[P_{gg\leftarrow g}(z)\cdot D^h_g\left(\frac{x_E}{z}, s\right) + P_{q\bar{q}\leftarrow g}(z)\cdot D^h_q\left(\frac{x_E}{z}, s\right)\right] \tag{10.78a}$$

$$\frac{dD^h_q(x_E, s)}{d\ln s} = \frac{\alpha_s(s)}{2\pi}\int_{x_E}^{1}\frac{dz}{z}\left[P_{qg\leftarrow q}(z)\cdot D^h_q\left(\frac{x_E}{z}, s\right) + P_{gq\leftarrow q}(z)\cdot D^h_g\left(\frac{x_E}{z}, s\right)\right] \tag{10.78b}$$

Figure 10.19 depicts slopes of fragmentation functions of the gluon and quark with solid lines obtained from Eq. (10.78).

For $x_E \to 1$, one expects a relation

$$r_S(x_E) \equiv \frac{d\ln D^h_g(x_E, s)}{d\ln D^h_q(x_E, s)} \xrightarrow{x_E\to 1} C_A/C_F \tag{10.79}$$

This can be seen from Eq. (10.78). In the limit $x_E \to 1$, terms that have denominator $(1-z)$ become dominant and the second term in Eq. (10.78a) can be neglected. The second term in Eq. (10.78b) can also be neglected as the gluon fragmentation function $D^h_g(z)$ is small for $z > 1/2$. The δ function term in the splitting function only serves to make the singular part disappear. Then, approximating $z = 1$ in the numerator, one gets Eq. (10.79) (see Eq. (8.72a) and (8.72d) for the functional form of the splitting function). By fitting the whole data using Eq. (10.79), one can determine the C_A/C_F ratio more accurately. The DELPHI group obtained a value [598]

$$\frac{C_A}{C_F} = 2.26 \pm 0.09(\text{stat}) \pm 0.06(\text{syst.}) \pm 0.12(\text{clus., scale}) \tag{10.80}$$

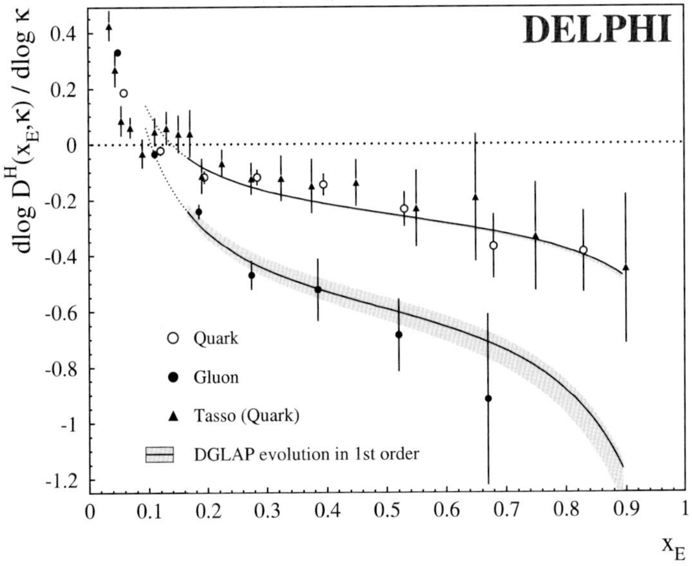

Figure 10.19 The gluon jet is much softer than the quark jet. $d\ln D^h(x_E, \kappa)/d\ln\kappa$ is calculated using the DGLAP equation [598].

which is in agreement with Eq. (10.77).

(3) The Gluon is Broader: Broadness of the gluon was investigated by the OPAL group by comparing their data with that of CDF at $\sqrt{s} = 1.8\,\text{TeV}$ [600]. As the jet selection algorithm is the $\epsilon-\delta$ method in the hadron reaction (see Section 9.1.4), and not the y_{cut} method usually adopted in e^-e^+ reaction, the group applied the same criteria to define the energy content of the jet. Namely, it was the energy

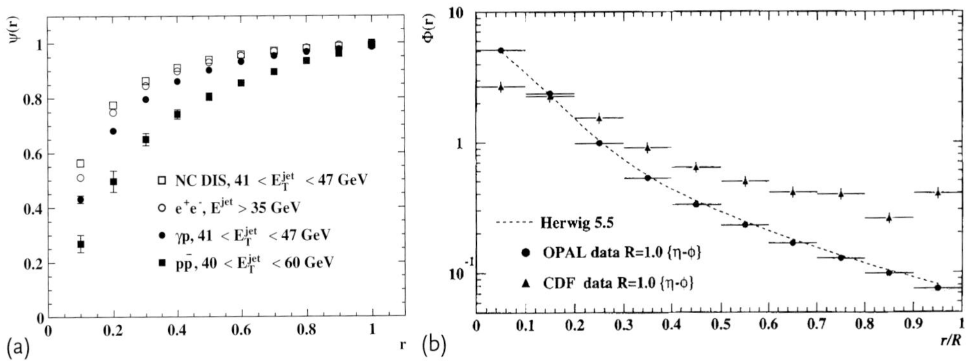

Figure 10.20 Energy profiles of jets: (a) $\psi(r) \equiv E_{\text{jet}}(r)/E_{\text{jet}}(r = R = 0.7)$. $E_{\text{jet}}(r)$ is defined as the total energy in the cone with radius r in $\eta-\phi$ plane [599]. (b) $\Phi(r) = [\psi(r + \Delta r) - \psi(r)]/(\Delta r/R)$. The dashed curves show predictions of the HERWIG model [600, 601].

confined in a cone around the jet axis with radius $R = 0.7$ in the $\eta{-}\phi$ plane (η: rapidity, ϕ the azimuthal angle).

Then, they plotted the energy $E_{\text{jet}}(r)$ contained in the cone radius r as a function of r. In Figure 10.20a, $\psi(r) = E_{\text{jet}}(r)/E_{\text{jet}}(R)$ is defined as the fraction of total energy in the cone with radius r. One sees that the jets produced in the hadron reactions (black squares) which are primarily gluons [see also Figure 11.12] are considerably softer at the core than jets produced in the deep inelastic scattering and in the e^-e^+ reactions. In Figure 10.20b, the differential spectrum $\Phi = [\psi(r+\Delta r) - \psi(r)]/(\Delta r/R)$ is plotted. Here, one sees that the gluon jet is much broader depositing considerable energy at the periphery of the cone.

They show that 90% of the energy is in the $\Delta R \equiv \sqrt{\Delta\eta^2 + \Delta\phi^2} \le 1$ region for the quark jet (OPAL data), while only 75% is included in the same region for the gluon (CDF data). Thus, our conjecture that the gluon is broader is confirmed.

11
Jets in Hadron Reactions

11.1
Introduction

It has been known for a long time that almost all hadronic reactions are so-called soft processes which refer to small angle scatterings and productions. From the QCD point of view, they are processes which are not easily accessible using perturbation theories. They are characterized by two features:

a) Secondary particles distribute more or less uniformly in rapidity η, that is, in logarithm of longitudinal momentum.
b) Cross sections decrease exponentially as a function of the transverse momentum and their average value is typically $\langle p_T \rangle \sim 300$ MeV.

Figure 11.1 shows typical examples of such distributions in the high energy hadron collisions. Considering the above characteristics, convenient variables that characterize hadronic processes are rapidity defined by

$$\eta = \frac{1}{2} \ln \frac{E + p_z}{E - p_z} = \ln \frac{E + p_z}{m_T}, \qquad m_T = \sqrt{p_T^2 + m^2} \tag{11.1}$$

transverse momentum p_T and azimuthal angle ϕ. In terms of these variables, the phase space volume element of a particle is expressed as

$$\frac{d^3 p}{E} = \frac{d\phi d p_T^2 d p_z}{2E} = \frac{1}{2} d\phi d p_T^2 d\eta \tag{11.2}$$

Noting that the phase space along the beam direction becomes large in proportion to the beam energy, a uniform distribution in η means that the total number of secondary particles N_{max} increases in proportion to η_{max}.

$$\eta_{max} = \frac{1}{2} \ln \left. \frac{E + p_z}{E - p_z} \right|_{max} = \ln(E + p_z)|_{max} - \ln m_T \sim \frac{1}{2} \ln s \tag{11.3}$$

Theoretically, the growth with energy has been shown to be limited by $\sim \ln^2 s$, referred to as the Froissart bound [605]. Experimentally, observed multiplicity seems to saturate the bound [3].

Elementary Particle Physics, Volume 2: Foundations of the Standard Model, First Edition. Yorikiyo Nagashima.

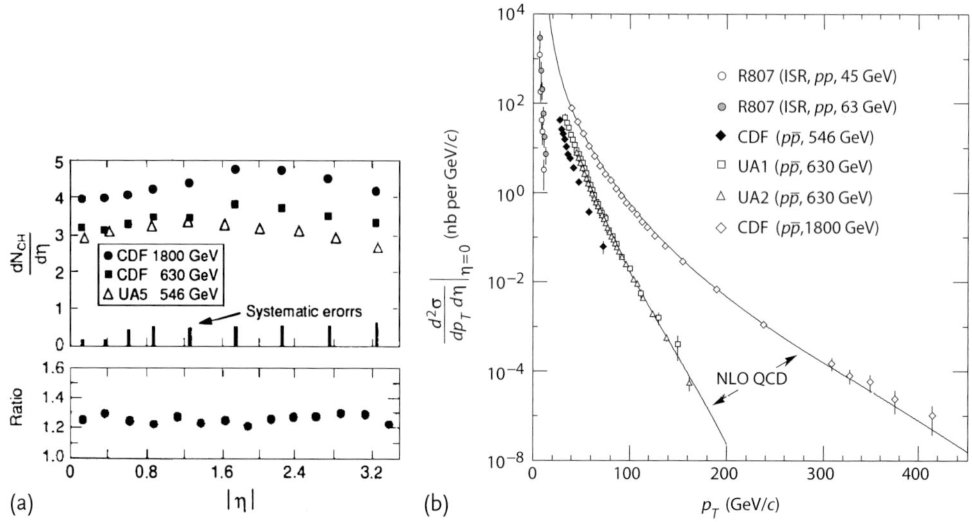

Figure 11.1 Distributions of secondary particles in high energy hadron collisions. (a) Rapidity distribution [602, 603], (b) Transverse momentum distribution [604].

Figure 11.1 shows that at large transverse momentum ($p_{\mathrm{T}} \gtrsim$ a few tens of GeV), the cross section deviates from the exponential decrease. For processes with large transverse momentum, the asymptotic freedom is valid and the perturbative QCD can be applied. This is the main theme of this chapter.

Existence of jet phenomena at large p_{T} was expected theoretically before its observation [606]. Historically, the first jet was observed at ISR/CERN (intersecting storage ring; 24 + 24GeV pp-collider, commissioned in 1972) [607]. However, jets were rare phenomena at the ISR. They were observed in abundance later in $Sp\overline{p}S$/CERN ($\sqrt{s} = 540$ GeV) and Tevatron ($\sqrt{s} = 1.8$ TeV). As the average momentum of the parton is $\sim 1/6$ of the parent proton, which amounts to ~ 4 GeV at ISR, it was not large enough for observation of large p_{T} phenomena.

11.2
Jet Production with Large p_{T}

At the hadron collider $Sp\overline{p}S$ which has $\sqrt{s} = 570$ (later upgraded to 630) GeV, the average momentum per parton exceeds ~ 50 GeV. Many clear and unambiguous jet events are observed. Large angle scattering processes with $p_{\mathrm{T}} \gtrsim 100$ GeV contained jets most of the time, in particular, $2 \to 2$ jet reactions in which two partons scatter elastically, occupied a majority of events. Except for the soft processes we mentioned in the beginning, characteristic of the jet production seemed very similar to $e^- e^+$ reactions.

Figure 11.2 shows data obtained by the UA2 group [602, 603, 608]. When the transverse momentum p_{T} (or equivalently, the transverse energy E_{T}) exceeds

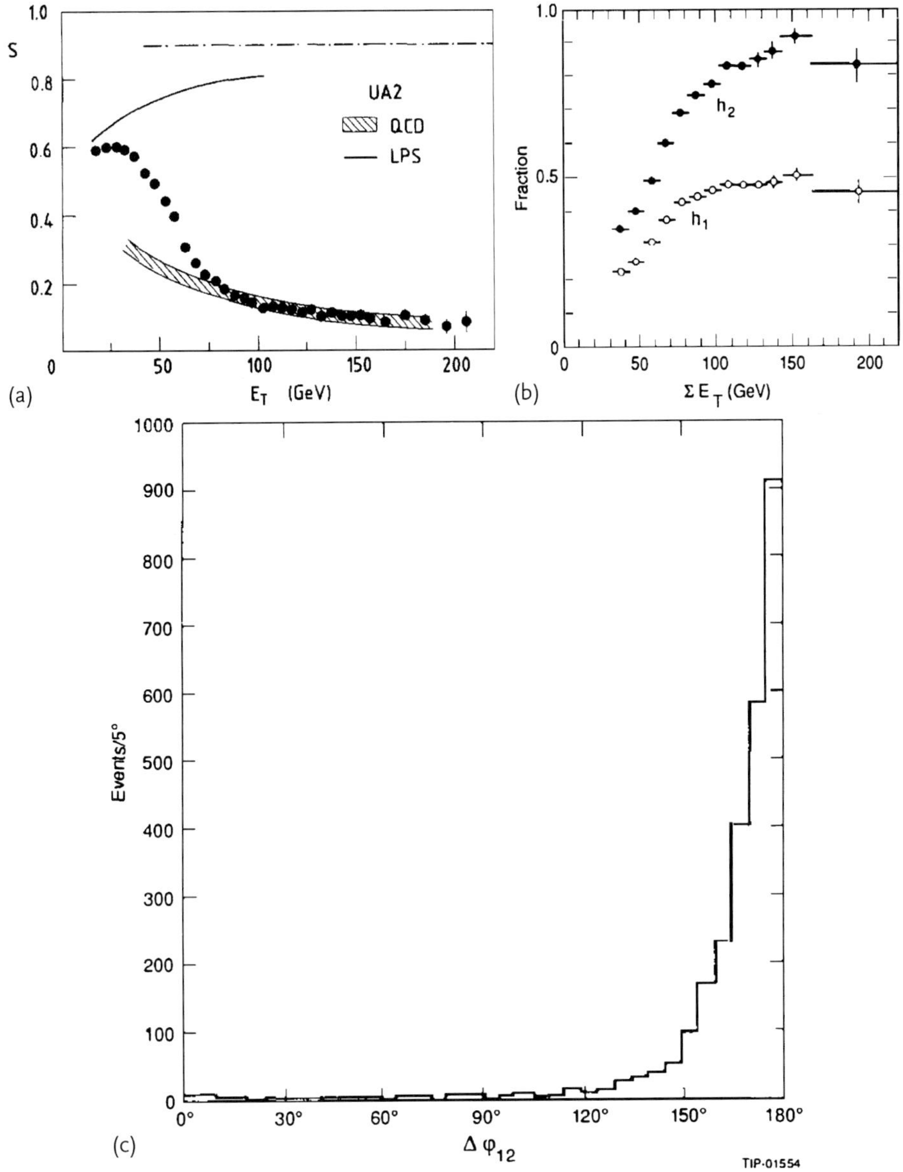

Figure 11.2 (a) Sphericity distributions in the hadronic collision as a function of E_T [602, 608]. At low p_T, the data is close to the LPS(longitudinal phase space program made by UA5 [609] which can reproduce the main features of minimum bias Collider events) (solid line), but at high p_T, the perturbative QCD reproduces data well (shade band). (b) Distributions of $h_1 = E_{T1}/\sum E_T$ and $h_2 = (E_{T1} + E_{T2})/\sum E_T$. At $E_T > 100\,\text{GeV}$, the energy of 2 jets become dominant [110, 591]. (c) A distribution of the azimuthal angle of 2 jets. The figure shows that the 2 jets are emitted back to back [603].

$\sim$ 80 GeV, the sphericity (an index of how strong the jet-like behavior is) becomes very small (Figure 11.2a), and a fraction of the 2-jet events become dominant (Figure 11.2b). Besides, Figure 11.2c shows that the dominant 2 jets are emitted back to back. The data strongly indicates that large p_T events are almost all jet phenomena and, in particular, are dominated by two-body scattering events.

The mechanism of these 2-jet events is illustrated by two diagrams depicted in Figure 11.3. One of the partons in the incoming hadron A reacts with another par-

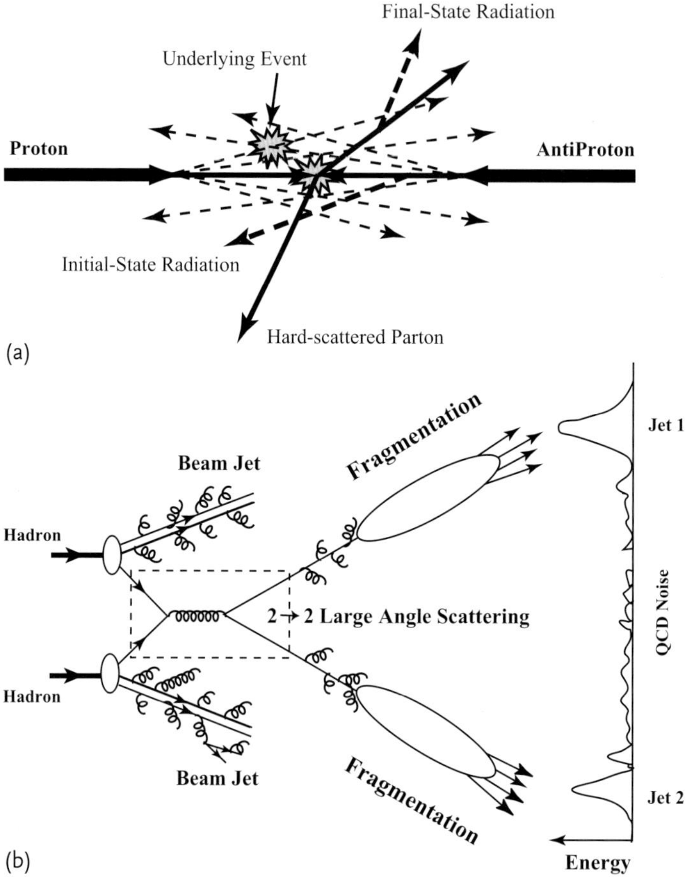

Figure 11.3 Two sketches of 2 → 2 large angle scatterings are illustrated, (a) in configuration space and the other (b) as a pseudo-Feynman diagram. It occupies most of the large p_T jet events. Spectator partons that did not participate in the hard (large p_T) reaction pass through and produce beam jets in the direction of incoming protons. The two hard scattered partons produce jets in their direction depositing large amount of energy in the back to back position, sometimes producing extra jets by initial- or final-state interaction. Soft hadrons which occupy phase space uniformly are beam remnants left over after a parton is knocked out of each other of the initial two incoming hadrons. There are additional semi-hard 2 → 2 parton-parton scatterings (multiple parton interactions). They are generally referred as the underlying events.

ton in the hadron B, is scattered elastically making 2 jets in back to back direction ($2 \to 2$ reaction) possibly producing extra hard jets by initial- or final-state radiations. The remaining partons which did not participate in the scattering (referred as the spectator partons) pass through and produce beam jets. The beam jets mostly go into the beam pipe and majority of them miss the detector. Therefore they are difficult to measure quantitatively. Anyway, they do not contain important physics contents and will not be discussed further. Soft hadrons which are remnants left over a parton is knocked out of each of the initial two hadrons or those produced by semi-hard two-to-two parton-parton scatterings (multiple parton reaction) occupy the phase space uniformly. They are referred as the underlying events. The transverse momentum density (scalar sum of p_T in the unit area on the $\eta - \phi$ plane) of the underlying events is measured to be $d^2 p_T / d\eta d\phi \simeq 0.3$ GeV [610].

A typical collider detector is made to cover as much solid angle as possible surrounding the collision point. Reconstructed particles can be expressed by their energy E, polar angle θ and azimuthal angle ϕ, but as stated in the beginning of the chapter, it is common to use the rapidity (or pseudorapidity $\eta^* = \ln \cot(\theta/2)$ which neglects the mass of the particle). An event is often represented by an energy histogram in the η–ϕ plane which is referred as the Lego plot.[1)]

As stated, a majority of events are soft and their distribution is uniform on the η–ϕ plane. Therefore, if a cluster of energy is concentrated at a particular point, it is legitimate to consider it as a jet. Figure 11.4 shows an example of 2-jet and 3-jet events. One sees a clear $2 \to 2$ reaction at back to back position. A circle at the root of the jets in Figure 11.4b has a radius $R = \sqrt{(\Delta\eta)^2 + (\Delta\phi)^2}$, a variable which is used to define or separate jets. The CDF/D0 group at the Fermilab defines the jet energy as that contained in a circle with $R = 0.7$. One can interpret Figure 11.4b

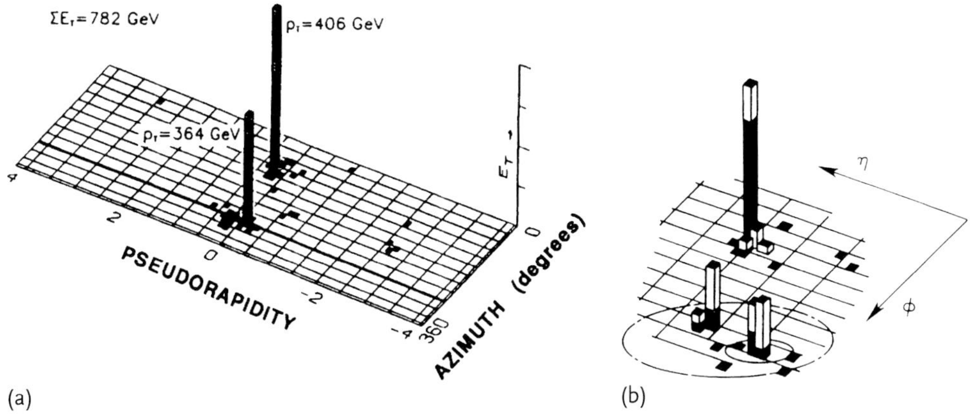

(a) (b)

Figure 11.4 (a) A typical example of 2-jet productions at the Fermilab. One sees two energy clusters stand out in the η–ϕ plane where positions of the two tall towers match those of back to back scattering. (b) Another example of a 2-jet event, but it can be interpreted as a 3-jet event if the jet cone size (circle in the plot) is diminished [611, 612].

1) A plot to express $z = f(x, y)$ where x–y plane is divided in squares of $\Delta x, \Delta y$, and z is represented by a column of height z.

as showing a 3-jet event where a high energy parton has emitted a gluon ($2 \to 3$ reaction), but it can also be interpreted as a $2 \to 2$ reaction if one uses a larger cutoff on R.

11.3 2 → 2 Reaction

11.3.1 Kinematics and Cross Section

Here, we consider details of $2 \to 2$ jet reactions. Figure 11.5 shows some of the Feynman diagrams for two-body elastic scattering by the partons.

Let us consider a reaction $A + B \to c + d$ in which a parton "a" in the hadron A, and a parton "b" in the hadron B collide and become parton "c" and "d" in the final state.

$$a + b \to c + d \tag{11.4}$$

The cross section for the hard (large p_T) process "$ab \to cd$" can be expressed as (see Eq. (8.148))

$$\sigma(AB \to cd) = \int dx_a dx_b \left[f_A{}^a(x_a) f_B{}^b(x_b) + (A \leftrightarrow B, \text{ if } a \neq b) \right] \hat{\sigma}_0(ab \to cd) \tag{11.5}$$

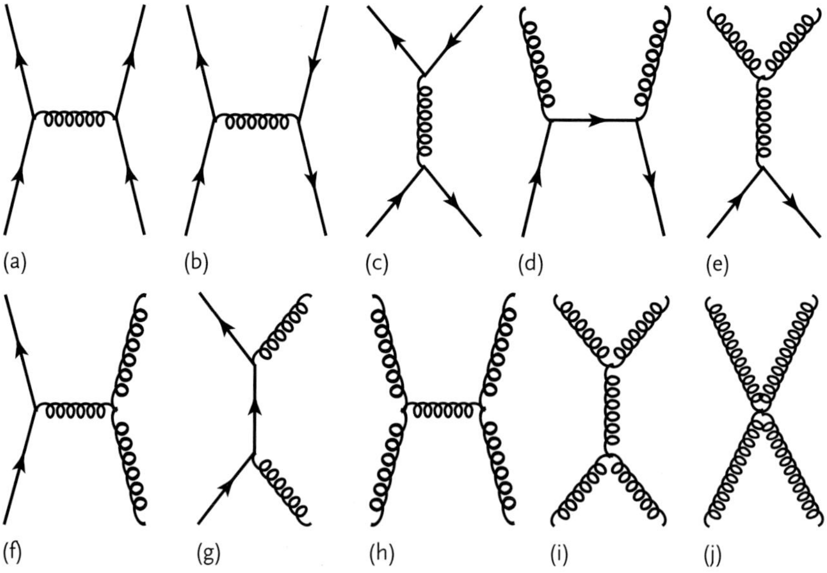

Figure 11.5 $O(\alpha_s^2)$ processes for the $2 \to 2$ reaction. (a) qq scattering, (b,c) $q\overline{q}$ scattering, (d,e) $q\overline{q} \to gg$ annihilation, (f,g) qg scattering, (h–j) gg scattering.

where $f_A^a(x)$ is the parton momentum distribution function (pdf) with flavor a in the hadron A. It is understood that the distribution function is properly evolved to the factorization scale μ^2 (see Eq. (8.49)) relevant for the process á la DGLAP. Its structure and evolution equations were discussed in Section 8.6 for the Drell–Yan process. Here, we treat it as given. Its reliability will be discussed in Section 11.5.2 in this chapter. The cross section $\hat{\sigma}$ for the parton reaction $(ab \to cd)$ is to be understood to take the average of the initial and the sum of the final state spin and color degrees of freedom. The total energy $\hat{s}$ in the parton ab system is defined as

$$\hat{s} = (x_a p_A + x_b p_B)^2 \simeq 2x_a x_b (p_A \cdot p_B) \simeq x_a x_b (p_A + p_B)^2 = x_a x_b s \equiv \tau s \tag{11.6}$$

The cross section for the parton scattering is given by

$$\begin{aligned} d\hat{\sigma}(ab \to cd) &= \frac{|\overline{\mathcal{M}}|^2}{2\hat{s}}(2\pi)^4\delta^4(p_a + p_b - p_c - p_d)\frac{d^3 p_c}{(2\pi)^3 2E_c}\frac{d^3 p_d}{(2\pi)^3 2E_d}\frac{1}{1+\delta_{cd}} \\ &= \frac{|\overline{\mathcal{M}}|^2}{1+\delta_{cd}}\frac{1}{32\pi}\frac{\beta}{\hat{s}}d(\cos\theta^*) \end{aligned} \tag{11.7}$$

where $\beta = p/E$ is the velocity of the produced particles. $\beta = 1$ for massless partons, but is retained for later discussion of the threshold effect for heavy particle productions. The factor δ_{cd} accounts for the statistical factor for identical final particles. The second equation is useful for calculating the total cross section since the process we are considering is a two-body scattering and there is only one independent variable for fixed $\hat{s}$. Table 11.1 gives the matrix elements squared for the Feynman diagrams given in Figure 11.5. One notices that large contributions come from elastic scatterings, in particular, those of gluons. This reflects the fact that the gluon has double colors and the coupling is strong proportionately (i.e., $C_A/C_F = 9/4$).

Denoting $\hat{t} = (p_a - p_c)^2$, $\hat{u} = (p_a - p_d)^2$, $p_T^2 = \hat{u}\hat{t}/\hat{s} = (\hat{s}/4)\sin^2\theta^*$ where variables in the parton center of the mass system are denoted with an asterisk, alternative expressions for the cross section are given by

$$\frac{d\hat{\sigma}}{d\hat{t}} = \frac{1}{16\pi\hat{s}^2}|\overline{\mathcal{M}}|^2\frac{1}{1+\delta_{cd}} \tag{11.8a}$$

$$\frac{d\hat{\sigma}}{dp_T^2} = \frac{d\hat{\sigma}}{d\hat{t}}\frac{1}{\cos\theta^*} = \frac{d\hat{\sigma}}{d\hat{t}}\frac{\hat{s}}{(\hat{t}-\hat{u})} \tag{11.8b}$$

One can express the four-momentum p^μ in terms of p_T, azimuthal angle ϕ and rapidity η,

$$p^\mu = \left(\sqrt{p_T^2 + m^2}\cosh\eta, p_T\cos\phi, p_T\sin\phi, \sqrt{p_T^2 + m^2}\sinh\eta\right) \tag{11.9}$$

Denoting rapidities of the partons as η_c, η_d, the rapidity η of the two-parton system (cd) and the relative rapidity $\Delta\eta$ of c, d in the hadron (A, B) center of mass (CM)

Table 11.1 Invariant matrix squared $|\overline{\mathcal{M}}|^2$ for two-to-two parton subprocesses with massless partons [615]. Color and spin indices are averaged (summed) over initial (final) states .

Process	$\sum \lvert\mathcal{M}\rvert^2/g_s^4$	at $\theta = \pi/2$
$qq' \to qq'$	$\frac{4}{9}\frac{\hat{s}^2+\hat{u}^2}{\hat{t}^2}$	2.22
$q\bar{q}' \to q\bar{q}'$	$\frac{4}{9}\frac{\hat{s}^2+\hat{u}^2}{\hat{t}^2}$	2.22
$qq \to qq$	$\frac{4}{9}\left(\frac{\hat{s}^2+\hat{u}^2}{\hat{t}^2}+\frac{\hat{s}^2+\hat{t}^2}{\hat{u}^2}\right)-\frac{8}{27}\frac{\hat{s}^2}{\hat{u}\hat{t}}$	3.26
$q\bar{q} \to q'\bar{q}'$	$\frac{4}{9}\frac{\hat{t}^2+\hat{u}^2}{\hat{s}^2}$	0.22
$q\bar{q} \to q\bar{q}$	$\frac{4}{9}\left(\frac{\hat{s}^2+\hat{u}^2}{\hat{t}^2}+\frac{\hat{t}^2+\hat{u}^2}{\hat{s}^2}\right)-\frac{8}{27}\frac{\hat{u}^2}{\hat{s}\hat{t}}$	2.59
$q\bar{q} \to gg$	$\frac{32}{27}\frac{\hat{t}^2+\hat{u}^2}{\hat{t}\hat{u}}-\frac{8}{3}\frac{\hat{t}^2+\hat{u}^2}{\hat{s}^2}$	1.04
$gg \to q\bar{q}$	$\frac{1}{6}\frac{\hat{t}^2+\hat{u}^2}{\hat{t}\hat{u}}-\frac{3}{8}\frac{\hat{t}^2+\hat{u}^2}{\hat{s}^2}$	0.15
$gq \to gq$	$-\frac{4}{9}\frac{\hat{s}^2+\hat{u}^2}{\hat{s}\hat{u}}+\frac{\hat{u}^2+\hat{s}^2}{\hat{t}^2}$	6.11
$gg \to gg$	$\frac{9}{2}\left(3-\frac{\hat{t}\hat{u}}{\hat{s}^2}-\frac{\hat{s}\hat{u}}{\hat{t}^2}-\frac{\hat{s}\hat{t}}{\hat{u}^2}\right)$	30.4

frame are given by

$$\eta = (\eta_c + \eta_d)/2\,, \quad \Delta\eta = \eta_c - \eta_d \tag{11.10}$$

As rapidities are additive with respect to the Lorentz boost, those of partons c and d in their CM frame can be obtained from η_c and η_d by subtracting η from them which give $\Delta\eta/2$. Therefore, scattering angle θ^* of the parton c in the parton CM frame is given by

$$\begin{aligned} \cos\theta^* &= \frac{p_z^*}{E^*} = \frac{\sinh(\Delta\eta/2)}{\cosh(\Delta\eta/2)} = \tanh\left(\frac{\Delta\eta}{2}\right) \\ \sin\theta^* &= \frac{1}{\cosh(\Delta\eta/2)} \end{aligned} \tag{11.11}$$

By the energy-momentum conservation, longitudinal momenta x_a, x_b and the transverse momentum x_T

$$x_T = 2m_T/\sqrt{s}\,, \quad m_T \equiv \sqrt{p_T^2 + m^2} \tag{11.12}$$

are related by

$$x_a = x_T e^{\eta}\cosh(\Delta\eta/2)\,, \quad x_b = x_T e^{-\eta}\cosh(\Delta\eta/2) \tag{11.13}$$

where m_{T} can be identified with the transverse momentum p_{T} for light quarks, but the mass m cannot be neglected for heavy quarks. Then, knowing the rapidities η_c, η_d of the produced jets which supposedly reproduce η_c, η_d of the partons, the momenta of the partons a, b in the CM frame of particles A, B

$$p_a^\mu = x_a p_{\mathrm{A}}^\mu, \quad p_b^\mu = x_a p_{\mathrm{B}}^\mu \tag{11.14}$$

can be calculated from the relations

$$\eta = \frac{1}{2}\ln\frac{x_a}{x_b}, \quad x_{\mathrm{T}}^2 = \frac{4m_{\mathrm{T}}^2}{s} = \frac{2x_a x_b}{1+\cosh\Delta\eta} \tag{11.15}$$

The invariant mass M_{jj}^2 of a 2-jet system is given by

$$M_{jj}^2 = \hat{s} = 2m_{\mathrm{T}}^2(1+\cosh\Delta\eta) \tag{11.16}$$

Using the relations $dp_{cz}dp_{dz}/E_c E_d = d\eta_c d\eta_d$,

$$\begin{aligned}
&\delta^4(p_a + p_b - p_c - p_d)\frac{d^3 p_c d^3 p_d}{E_c E_d} dx_a dx_b \\
&= \pi d p_{\mathrm{T}}^2 d\eta_c d\eta_d \delta\left[(x_a - x_b)P_z^* - p_{cz} - p_{dz}\right] \\
&\times \delta\left[(x_a + x_b)E^* - E_c - E_d\right] dx_a dx_b \\
&= \frac{2\pi}{s} dp_{\mathrm{T}}^2 d\eta_c d\eta_d\Big|_{\hat{s}=x_a x_b s},
\end{aligned} \tag{11.17}$$

where E^*, $P_z^*(\simeq E^*)$ denote the hadron energy and momentum in the center of mass frame of incoming hadrons. Equation (11.5) is rewritten as

$$\frac{d^3\sigma}{d\eta_c d\eta_d dp_{\mathrm{T}}^2} = \frac{1}{16\pi s^2}\sum_{ab}\frac{f_{\mathrm{A}}^a(x_a)f_{\mathrm{B}}^b(x_b) + (1-\delta_{ab})(a\leftrightarrow b)}{x_a x_b}\overline{|\mathcal{M}|^2}\frac{1}{1+\delta_{cd}} \tag{11.18}$$

If one uses the relation

$$d\eta_c d\eta_d dp_{\mathrm{T}}^2 = \frac{s}{2} dx_a dx_b d\cos\theta^* \tag{11.19}$$

one gets alternative expressions,

$$\frac{d^3\sigma}{dx_a dx_b d\cos\theta*} = f_{\mathrm{A}}^a(x_a)f_{\mathrm{B}}^b(x_b)\frac{d\hat{\sigma}}{d\cos\theta*} \tag{11.20a}$$

$$\frac{d\hat{\sigma}}{d\cos\theta*} = \frac{\hat{s}}{2}\frac{d\hat{\sigma}}{d\hat{t}} = \sum_{cd}\frac{1}{32\pi\hat{s}}\overline{|\mathcal{M}(ab\to cd)|^2}\frac{1}{1+\delta_{cd}} \tag{11.20b}$$

Furthermore, using Eq. (11.16) and putting $M_{jj}/s = \tau$, the cross section for producing the 2-jet production can be written as

$$\begin{aligned}
\frac{d^2\sigma}{dM_{jj}^2 d\cos\theta*} &= \sum_{ab}\int dx_a dx_b f_{\mathrm{A}}^a(x_a)f_{\mathrm{B}}^b(x_b)\delta(x_a x_b s - M_{jj}^2)\frac{d\hat{\sigma}}{d\cos\theta*} \\
&\equiv \sum_{ab}\frac{\tau}{M_{jj}^2}\frac{dL_{ab}}{d\tau}\cdot\frac{d\hat{\sigma}}{d\cos\theta*}
\end{aligned} \tag{11.21a}$$

where

$$\frac{dL_{ab}}{d\tau} = \frac{1}{1+\delta_{ab}} \int dx_a dx_b \delta(\tau - x_a x_b) \left[f_A^a(x_a) f_B^b(x_b) + (a \leftrightarrow b) \right] \tag{11.21b}$$

is the collision luminosity for the parton a and b in the reaction between the hadron A and B. However, as the producer of the jets is indistinguishable, either the term with $\hat{t} \leftrightarrow \hat{u}$ is added or the integration must be stopped at $\theta = 90°$ unless $c = d$.

11.3.2
Jet Productions Compared with pQCD

We stated that jet phenomena are conspicuous if one picks up events with large p_T. This is the experience of UA1 and UA2 at $Sp\overline{p}S$ collider ($\sqrt{s} \sim 630\,\text{GeV}$). Then, it should be even more so at the Tevatron which has much larger energy ($\sqrt{s} = 1.8\,\text{TeV}$). Figure 11.6 shows a few examples of the cross section data as a function of p_T. This is an inclusive cross section which can be obtained by integrating Eq. (11.18) over η_d and denoting η_c as y. It is amazing that a simple perturbative calculation reproduces the data over nine orders of magnitude in scale dif-

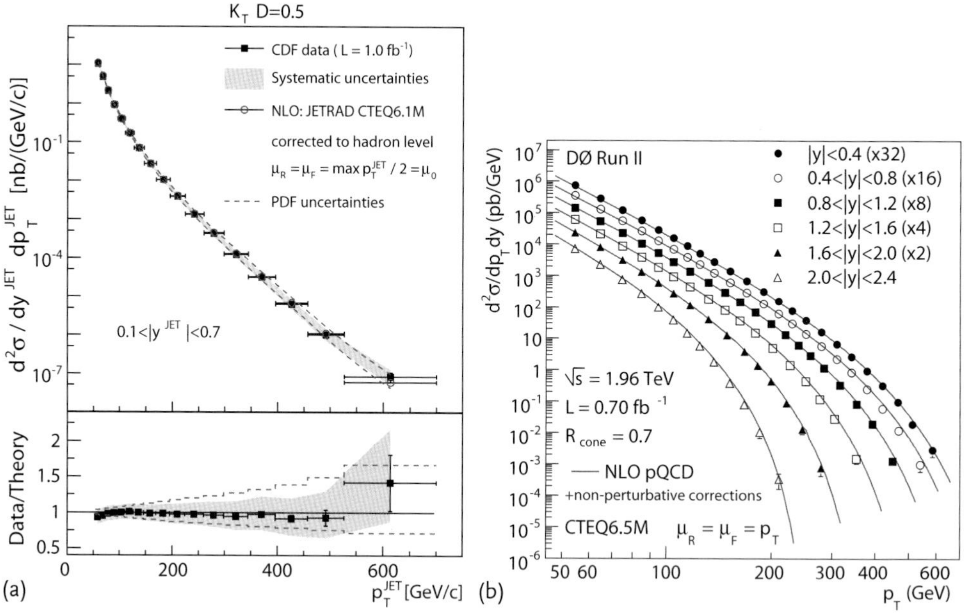

Figure 11.6 $d^2\sigma/dp_T dy$: (a) compares data with the hadronization program JETRAD [26]. The data used k_T algorithm (see Section 11.4.2 $D = 0.5, 1.0$) to define the jet energy. The lower part shows ratios of data/theory [27]; (b) compares data with the perturbative QCD (pQCD) at various values of jet rapidity y (η_c or η_d in the text). The data uses jet energy algorithm with $R = 0.7$ to define jets [28]. For a color version of this figure, please see the color plates at the beginning of the book.

ference. Figure 11.6a compares CDF data with a hadronization generator program (JETRAD which uses CTEQ6.1M for pdf [499]), while Figure 11.6 (right) compares D0 data with the NLO [next leading order $= O(\alpha_s^3(Q^2))$] perturbative QCD (pQCD).

Figure 11.7a shows $d^3\sigma/dM_{jj}d\eta_1 d\eta_2$ and Figure 11.7b, c shows the integrated distribution as a function of the dijet mass M_{jj}. Again, the data are compared with the JETRAD program [26] in Figure 11.7a and with NLO pQCD in Figure 11.7b,c. They also show good agreements.

The transverse energy E_T distribution is sensitive to the x distribution of the parton distribution function. Figure 11.8 shows their relative contribution to the dijet production cross section as a function of E_T based on CTEQ6M parton distribution function [633]. As is shown in the next section, the main contribution to the dijet production process comes from the elastic scatterings and relative contributions among the various processes are more or less independent of the scattering angle. The gg scattering is dominant at small E_T because the gluon distribution is large at small x and the cross section itself is also large. At high E_T, the qq or $/q\bar{q}$ scatterings dominate.

In conclusion, the fact that the 2-jet production cross section can be well described by 2 → 2 parton scatterings depicted in Figure 11.5 is clear evidence that

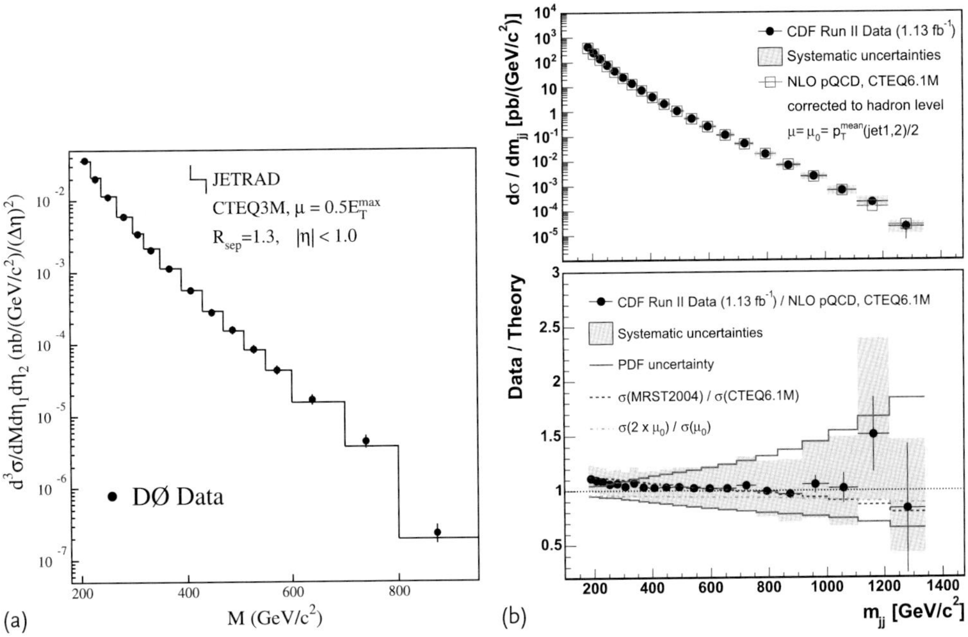

Figure 11.7 (a) D0 data for dijet production cross section $d^3\sigma/dMd\eta_1 d\eta_2$ for $\eta_{\text{jet}} < 1.0$. The histogram represents the JETRAD prediction [29, 30]. CDF data for dijet mass cross section: (b) measured dijet mass spectrum for both jets to have $|y| < 1$ compared to the NLO pQCD prediction obtained using the CTEQ6.1 PDFs. (c) The ratio of the data to the NLO pQCD prediction [31, 32]. For a color version of this figure, please see the color plates at the beginning of the book.

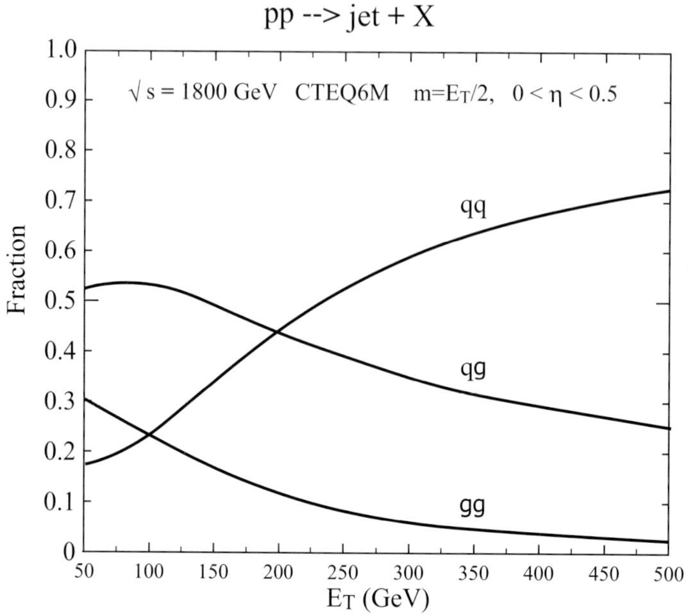

Figure 11.8 Contributions of the various subprocesses to the inclusive jet cross section. This plot was generated with CTEQ6M and $\mu = E_T/2$ [499, 614, 633].

the gluon–gluon couplings exists and provides an indirect evidence for QCD to be a non-Abelian gauge field theory.

Rutherford Scattering in QCD If the $2 \to 2$ jet reactions are elastic scatterings at the parton level, and if QCD is the gauge theory which has a Coulomb type potential at short distance, they should show characteristics of the Rutherford scattering. The angular distributions in Table 11.1 are not all identical, but we notice that dominant contributions at $\theta < \pi/2$ come from $\hat{t}^2 \sim (1 - \cos\theta)^2$. At small $\hat{t}$, the angular distribution of $qq \to qq$, $qg \to qg$, $gg \to gg$ are the same and can be represented by that of gg which gives the dominant contribution as

$$\begin{aligned} &|\mathcal{M}(qq \to qq)|^2 : |\mathcal{M}(qg \to qg)|^2 : |\mathcal{M}(gg \to gg)|^2 \\ &\approx 1 : \frac{C_A}{C_F} : \left(\frac{C_A}{C_F}\right)^2 = 1 : \frac{9}{4} : \left(\frac{9}{4}\right)^2 \end{aligned} \tag{11.22}$$

The ratio was derived from Table 11.1. Then, the cross section for $2 \to 2$ jet scattering has the form [615, 616]

$$\frac{d^3\sigma(AB \to jj)}{dx_a dx_b d\cos\theta*} = F(x_a)F(x_b)\frac{d\hat{\sigma}(gg \to gg)}{d\cos\theta*} \tag{11.23a}$$

$$F(x) = g(x) + (4/9)[q(x) + \overline{q}(x)] \tag{11.23b}$$

$$\frac{d\hat{\sigma}(gg \to gg)}{d\cos\theta^*} = \frac{9}{4}\frac{\pi\alpha_s^2}{x_a x_b s} f(\chi) \tag{11.23c}$$

$$f(\chi) = \chi^2 + \chi + 1 + \chi^{-1} + \chi^{-2} \tag{11.23d}$$

where χ is defined by

$$\chi = \frac{\hat{u}}{\hat{t}} = \frac{1 + \cos\theta^*}{1 - \cos\theta^*} = e^{\Delta\eta} \tag{11.24}$$

Therefore, $d\hat{\sigma}/d\cos\theta^* \sim (\theta^*)^{-4}$ for $\theta^* \sim 0$ and has the same angular dependence as the Rutherford scattering. In terms of χ, the cross section is given by

$$\frac{d\hat{\sigma}}{d\chi} = \frac{(1 - \cos\theta^*)^2}{2}\frac{d\hat{\sigma}}{d\cos\theta^*} \tag{11.25}$$

and is almost a constant at $\chi > 2$. Figure 11.9 shows the angular distributions and confirms the above considerations. The reason that the curve goes upward for $\chi < 1 (\theta^* > \pi/2)$ is due to s channel contribution.

Gluon Distribution in the Hadron By extracting the individual cross section described in Table 11.1, it is possible to separate the parton distribution functions. Thus, the 2-jet cross section data provides a method of extracting the parton distribution function $F(x)$. Figure 11.10 shows $F(x)$ determined from UA1 data [590,

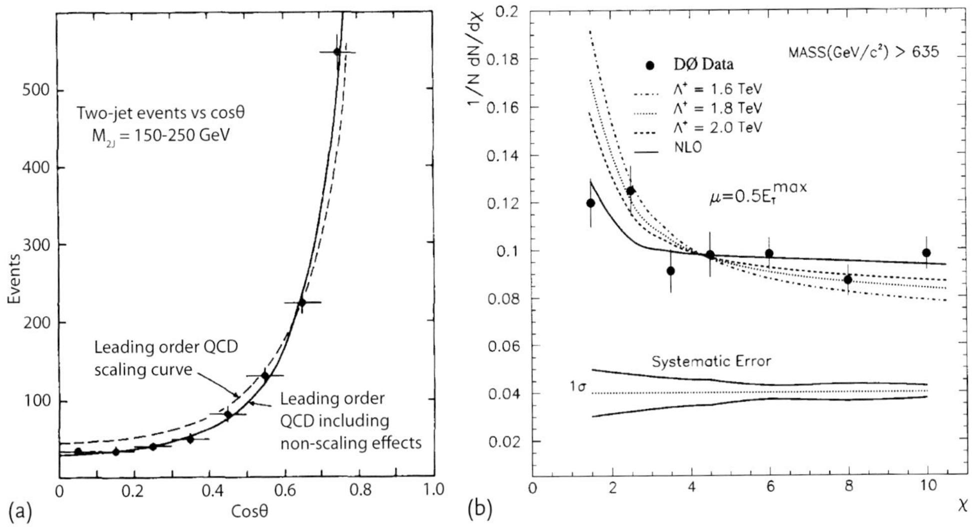

Figure 11.9 (a) $d\sigma/d\cos\theta^*$ (M_{jj} = 150–250 GeV). (b) $d\sigma/d\chi$ (M_{jj} = 200–240 GeV). In (a), the dashed line indicates LO QCD with scaling and the solid line corrected for nonscaling effects. In (b), the solid line indicates the NLO QCD and dashed lines are for various composite models [see Section 11.7] with Λ_c up to 2 TeV [611, 617–620].

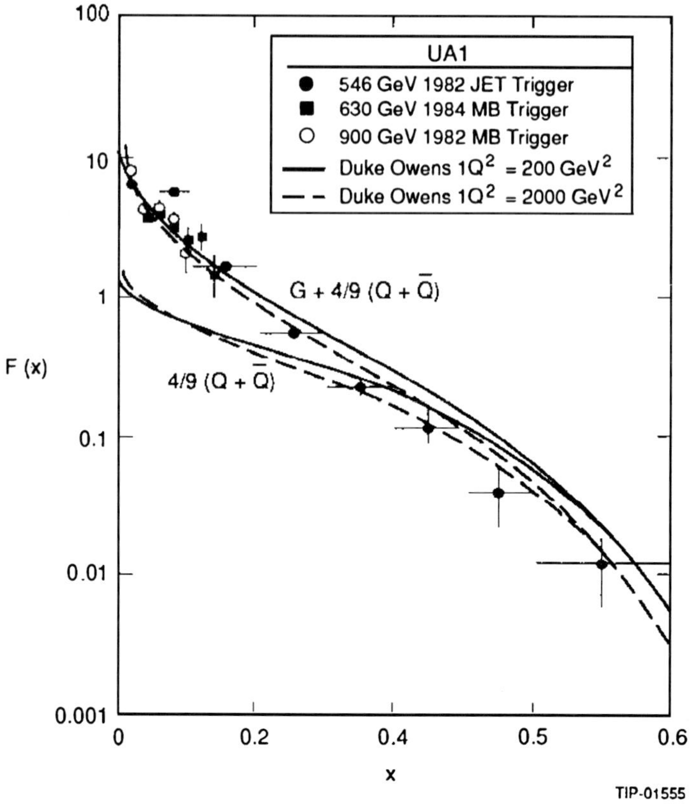

Figure 11.10 The parton distribution function determined from 2 $\to$ 2 jet reactions [590, 603, 621].

603] and compares it with the parton distribution obtained from the deep inelastic scattering data at $Q^2 = 200\,\text{GeV}^2$ (solid line). The dashed line is obtained by evolving the parton distribution function (pdf) at $Q^2 = 200\,\text{GeV}^2$ to $Q^2 = 2000\,\text{GeV}^2$ using the parametrization by [621] (see Eq. (8.130a)). It is a clear evidence that the jet cross section at small x cannot be explained by the qq and $q\bar{q}$ scattering alone but it is in good agreement with the full QCD. According to Figure 11.10, one can extract an almost pure gluon distribution at small x and an almost pure quark distribution at large x. It also shows that the Q^2 evolved structure function from low energy DIS data reproduces that of high Q^2 (see arguments in Section 8.5.3).

11.4
Jet Clustering in Hadronic Reactions

We have seen that the perturbative QCD reproduces overall features of the hadronic jet production processes quite well. However, when one wants to discuss details of the jet structure and compare with theory quantitatively, one has to check the stability of the scale dependence as well as matching of the parton-jet correspondence more quantitatively. The leading order (LO) QCD alone has strong dependence on the scale and has no predictability as to the transverse spread of jets. Since early stage of ISR and $Sp\overline{p}S$ up to Tevatron, a vast amount of data have been accumulated. Continuing efforts on phenomenological improvements guided by the QCD have greatly reduced discrepancies between the QCD predictions at the parton level and measured data on the hadronic level. We shall describe a few of them.

Jet productions in hadronic reactions are considerably different from those in e^-e^+ productions. In the latter, the number of partons are limited and high energy partons are expected to produce clear jets. In the hadronic reactions, the high p_T partons we want to investigate are induced by only one member of participating partons, while others pass through the target hadron with almost no impact. They are referred to as the spectator partons which produce beam jets. Although their energy is high, their received impact is minimum, and thus the asymptotic freedom does not apply. In other words, long range effects of beam jets have to be taken into account.

Their contribution is soft and is considered to make a uniform (flat) distribution in the η–ϕ plane as stated in the beginning of this chapter. The jet energy is conventionally defined as that which is contained within a limited cone centered at the peak of an energy cluster. It is necessary to determine what fraction of the total jet energy is contained in the designated cone. Conventionally, it is estimated using Monte Carlo hadronization models. The estimated optimum fraction differs depending on the model and also on the jet clustering algorithm which determines how to combine energy clusters distributed continuously or discontinuously on the η–ϕ plane.

In early jet formalisms, ambiguities due to inadequate theoretical understanding were large. In handling low energy jets, a major problem was the lack of theoretical understanding of the hadronization models, but for high energy jets, it was largely due to ambiguity in the large tail part of hard jets.

The LO perturbative QCD cannot handle differences originating from the jet expansion, in principle, because in the LO, the jet is a single parton and is not spatially spread by itself and hence does not concern the size of detectors. In NLO, when the gluon emission is included, broadening of the jet can be taken into account.

In the e^-e^+ jet production, rotational symmetry around the interaction point exists and relevant variables are the energy E, polar angle θ, and azimuthal angle ϕ. In the hadronic reactions, the events are boosted along the beam direction and the relevant variables are the transverse energy E_T, (pseudo)rapidity $\eta = -\frac{1}{2}\ln\tan\frac{\theta}{2}$ and the azimuthal angle ϕ.

11.4.1 Cone Algorithm

Because of the difference between e^-e^+ and hadronic reactions, the cone algorithm[2] is conventional in constructing jets in the hadronic reactions as contrasted to the y_{cut} method in the e^-e^+ processes. This is a method to define the opening angle of a cone along the jet axis and fraction of energy contained in it. In the cone algorithm, jet reconstruction goes like this.

1. Preclustering: the energy deposit in calorimeter cells are sorted out by E_T, the transverse energy relative to the beam axis, and a set of seed clusters are formed. Starting with the highest E_T cells, neighboring cells are combined to make energy clusters with $E_\text{T} > E_{\text{T}\,\text{cut}}$ (for instance, $E_{\text{cut}} = 1\,\text{GeV}$).
2. Cone clustering: next, nearby energy clusters within a cone ($\Delta R = \sqrt{\Delta\eta^2 + \Delta\phi^2} < 0.7$ is the convention) are combined to form a protojet. Variables of the protojet are determined by their centroid, that is,

$$E_{\text{Tc}} = \sum_i E_{\text{T}i}\,, \quad \eta_\text{c} = \frac{\sum_i \eta_i E_i}{E_{\text{Tc}}}\,, \quad \phi_\text{c} = \frac{\sum_i \phi_i E_i}{E_{\text{Tc}}} \tag{11.26}$$

   The process is repeated until it stabilizes, that is,

$$\eta_\text{c}(\text{Jet+cluster } i) = \eta_\text{c}(\text{Jet})\,, \quad \phi_\text{c}(\text{Jet+cluster } i) = \phi_\text{c}(\text{Jet}) \tag{11.27}$$

3. Splitting and merging: sometimes, a cluster is shared by two jet cones. If the overlap is large (for instance, the shared energy is more than 50% of either of the jet clusters), they are merged. Otherwise, they are split and shared energy is divided between the two by assigning an individual cell to the nearest centroid.

Figure 11.11a shows dependence of the cross section as a function of R which defines the size of a circle in η–ϕ region [601, 623]. Comparisons with theories [624–626] show that the theoretical dependence is minimum at $R = 0.7$ which is the size both CDF and D0 groups adopted. However, depending on the kind of reactions, for instance, in assessing multijets, other values ($R = 0.4$) are occasionally adopted. Figure 11.11b compares fraction of jet energy in the jet as a function of the size $r \le R$ for data with $O(\alpha_s^3)$ theoretical calculations as well as HERWIG hadronization model [31, 627]. The QCD estimates have some residual scale dependence but the hadronization model reproduces the data well. The LO estimate had about factor 2 ambiguities but NLO is accurate to 20%. A vast amount of experimental data are accumulated at TEVATRON and are compared with NNLO ($O(\alpha_s^4)$) QCD calculations with potentially a few percent accuracy.

Jets become more collimated as the jet transverse momentum increases. This is illustrated in Figure 11.12 where fraction of peripheral energy in the anulus $0.3 < r < R$ is plotted as a function of the jet transverse momentum.

The collimation of the jet is a result of three effects:

2) $\epsilon - \delta$ method is discussed in Section 9.1.4.

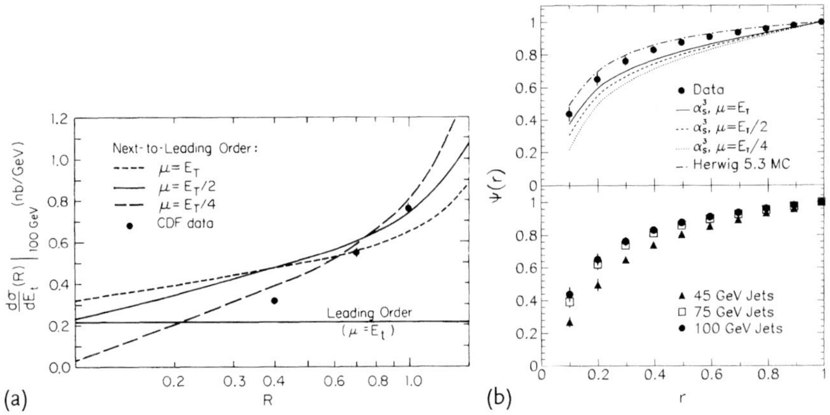

Figure 11.11 (a) The cross section as a function of the cone radius R used in defining a jet with energy 100 GeV. Lines are QCD theoretical predictions. [623] (b) Upper figure: Distribution of the p_{T} fraction of a cone for 100 GeV E_{T} jets and cone size of $R_0 = 1.0$. $\Psi(r)$ is the ratio of p_{T} within a cone of radius r to the p_{T} within a cone of radius $R_0 = 1.0$. Lines are QCD calculations using HMRS B structure functions for $\Lambda_{\mathrm{QCD}} = 122$ MeV and HERWIG. (b) Lower figure: $\Psi(r)$ for 45, 70, and 100 GeV jets [601]. See also Fig. 10.20b.

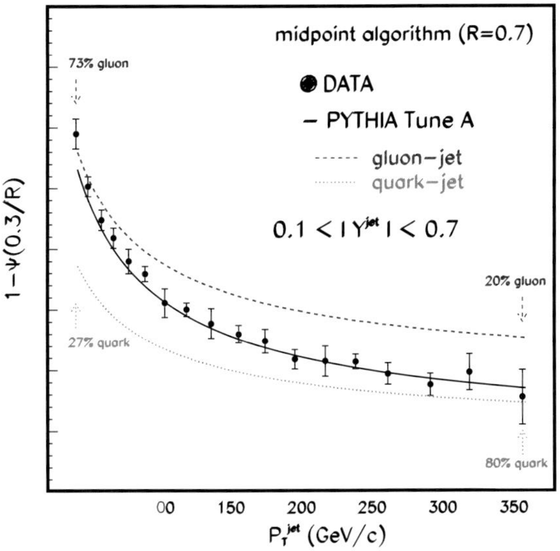

Figure 11.12 The fraction of the transverse momentum in a cone of radius 0.7 that lies in the anulus from 0.3 to 0.7 as a function of the transverse momentum of the jet. Comparison is made to PYTHIA (tune A) [469] and to the separate predictions for quark and gluon jets [628, 629].

1. the probability of a hard gluon to be radiated (the dominat factor in the jet shape) decreases as $\alpha_s(p_{\mathrm{T}}^2)$,
2. non-perturbative power corrections that tend to broaden the jet decreases as $1/p_{\mathrm{T}}$ or $1/p_{\mathrm{T}}^2$,

3. fraction of quark component increases as p_T increases [see Figure 11.8].
4. Figure 11.12 shows that fraction of the quark is 27% at low p_T increasing to 80% at $p_T = 350\,\text{GeV/c}$. Some fraction of fragmented partons end up outside the cone (known as splash-out) which has to be corrected to obtain the jet energy. The numeriacl value of the splash-out energy is roughly constant at 1 GeV/C for a cone of radius 0.7, independent of the jet transverse momentum. This constancy may seem counter intuitive, but is a result of the jet collimation stated above.

11.4.2
k_T Algorithm

The k_T algorithm [630, 631] is a partial recurrence to y_{cut} method in the e^-e^+ reactions. While the cone algorithm combines particles nearby in the configuration space, the k_T algorithm combines those in the momentum space. The formalism goes as follows.

1. For each particle (pseudoparticle) i in the list, define $d_i = p_{Ti}^2$. For each pair (i, j) of momenta $(i \neq j)$, define

$$d_{ij} = \min\left(p_{T,i}^2, p_{T,j}^2\right)\frac{\Delta R_{ij}^2}{D^2} \tag{11.28a}$$

   where

$$\Delta R_{ij} = (\eta_i - \eta_j)^2 + (\phi_i - \phi_j)^2 \tag{11.28b}$$

   D is a free parameter. Conventionally, the parameter is set $D = 1$.
2. Find the minimum of all the d_i and d_{ij} and label it d_{min}.
3. If d_{min} is d_{ij}, remove particles (pseudoparticles) i and j from the list and replace them with a new, merged pseudoparticle p_{ij} given by the recombination scheme. Variables of the recombined pseudoparticles may be defined by the centroid method as given by Eq. (11.26)[3)]
4. If d_{min} is d_i, remove particle (pseudoparticle) i from the list of particles and add it to the list of jets.
5. If any particles remain, go to step 1. The algorithm produces a list of jets, each separated by $\Delta R_{ij} > D$.

Two distinct approaches exist as to when one stops clustering in the algorithm. Catani *et al.*, [630] introduced a parameter d_{cut} to define the scale of the hard process a priori or on an event-by-event basis. When $d_{min} > d_{cut}$ clustering stops, jets with

3) Or, one can use the E recombination scheme which is defined by

$$p_{ij} = p_i + p_j \tag{11.29}$$

$p_T^2 < d_{cut}$ are considered as gbeam jets. Ellis and Soper [631] proposed to keep merging clusters until all remaining jets are separated by $(\eta_i - \eta_j)^2 + (\phi_i - \phi_j)^2 > D^2$.

An advantage of the k_T algorithm is that there is no ambiguity in the assignment of the pseudoparticle to jets and is claimed to be cleaner theoretically [632]. By comparing QCD calculations with $O(\alpha_s^3)$, residual uncertainties are claimed to be smaller compared to the cone algorithm. Its disadvantage is that the algorithm is intuitively less clear, the jet structure is more amorphous which results in diminished resolving power and can be very computer intensive. So far, the difference between the two algorithms has been small if any. It is a technology which needs further watching.

11.5 Reproducibility of the Cross Section

11.5.1 Scale Dependence

In calculating cross sections for the hadronic reactions, one has to convolute the parton cross section by the parton luminosity function which is evolved to the factorization scale μ_F. Note that the parton cross section has its own (renormalization) scale dependence on μ_R as well.

$$\sigma(AB \to cd; Q) = \sum_{ab} \int dx_a dx_b f_A^a(x_a, \mu_F) f_B^b(x_b, \mu_F) \hat{\sigma} \times (ab \to cd; Q, \mu_F, \alpha_s(\mu_R)) \tag{11.30}$$

The origin of the μ_F dependence stems from separation (factorization) of soft gluon emission processes from the hard parton interactions and folding into the scale dependent parton distribution functions. It includes higher order (NLO) corrections to the tree level two-body reaction process (LO). Therefore, consistent treatment of the process can only be obtained by including the NLO contribution to the matrix element as well. It is expected that by adjusting the $\mu(\equiv \mu_R)$ dependence of the matrix element and $(\equiv \mu_F)$ of the distribution function, the overall μ dependence would be minimized. The μ_R need not be the same as the μ_F, but generally they are taken to be equal ($\mu = \mu_F = \mu_R$).

As the physical quantity should not depend on μ, the cross section has to satisfy the relation:

$$\mu \frac{\partial \sigma}{\partial \mu}(AB \to cd) = 0 \tag{11.31}$$

We learned in Section 7.2.3 that if all the powers of the expansion series are added, the statement should be accurate. However, at finite order, residual μ dependence remains and its stability (constancy) is a measure of the higher order contribution. Figure 11.13 shows the scale dependence of jet production cross section

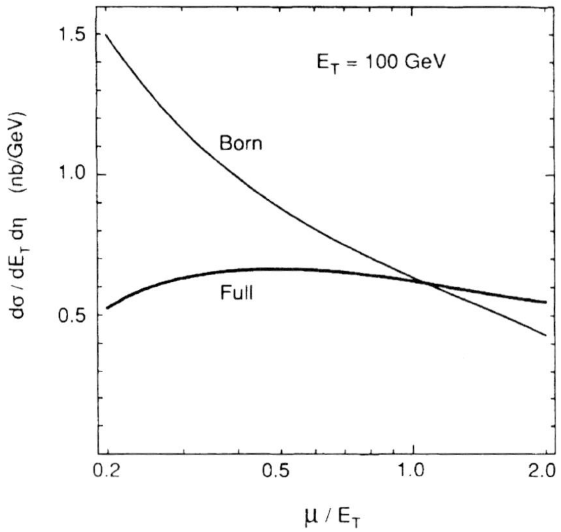

Figure 11.13 Comparison of the scale dependence of the jet production cross section ($p\overline{p} \to$ jets $+ X$) at $E_T = 170\,\text{GeV}$ for the LO and NLO calculations [625, 633].

($p\overline{p} \to$ jets $+ X$). The first order (LO) and the second order (NLO) calculations of $d^2\sigma/d\eta\, d E_T$ are shown [625, 633]. One can see that the scale dependence is considerably improved in the NLO calculation. Today, all the QCD calculations that try to reproduce data quantitatively use at least NLO corrections.

11.5.2
Parton Distribution Function

The scale dependent parton distribution functions (pdf) are determined from various processes as discussed in Section 8.5.3. There are several standard sets of the parton distribution functions and Monte Carlo generators to accommodate them.

When one realizes the amazing reproducibility of the QCD predictions on hadron data, one often forgets that hadron phenomenology is based on lots of assumptions, including the externally given parton distributions.

Figure 11.14 describes a comparison of the inclusive jet production cross sections with several QCD based event generators. Figure 11.14a shows a considerable discrepancy at $E_T > 300\,\text{GeV}$ which aroused a considerable excitement. A possible new phenomenon was suspected for a while, but it calmed down when an improved parton distribution was shown to agree with the data. Figure 11.14b shows more recent data fitted with the new k_T algorithm. This example demonstrates the importance of preparing precise parton distributions.

Figure 11.14 also demonstrates the importance of using at least NLO approximation for calculation. Nowadays, as the understanding of the high energy jet structure deepens, discrepancy between data and theory is mostly less than 20%.

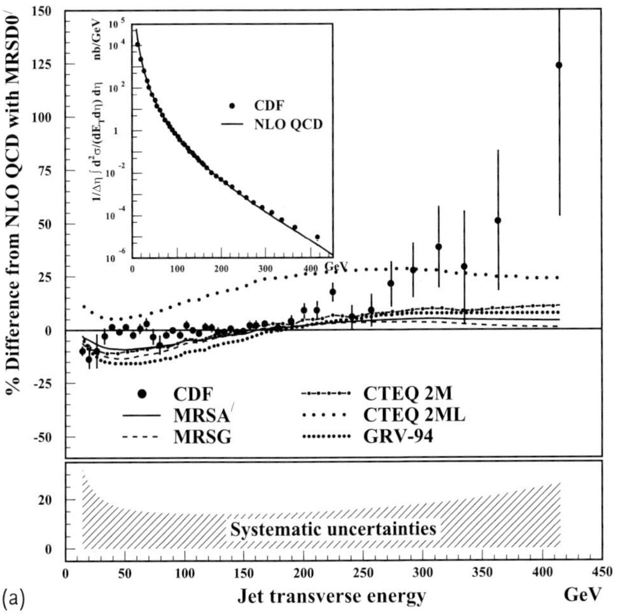

(a)

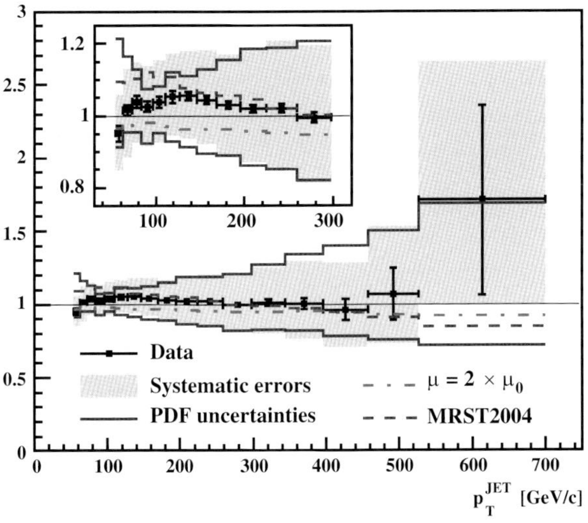

(b)

Figure 11.14 (a) Inclusive jet cross section measured by the CDF compared with predictions from the EKS program with $\mu = E_T^{\text{jet}}/2$, $R_{\text{sep}} = 2.0$ and MRSD0'. The cone algorithm was used. The shaded band represents the quadrature sum of the correlated systematic uncertainties. Additional curves show predictions with other parton distribution functions available at the time. For these predictions, the percentage difference relative to the default theory (MRSDO') is shown. Inset: the cross section data, from which the comparison with the theory is taken [627]. (b) More recent data on the ratio "data/theory" as a function of p_T fitted using the k_T algorithm. The enclosed figure expands the region $p_T < 300\,\text{GeV}/c$. The error bars (shaded band) show the total statistical (systematic) uncertainty on the data. The solid lines indicate the PDF uncertainty on the theoretical prediction. The dashed line presents the ratio of MRST2004 and CTEQ6.1M predictions. The dotted-dashed line shows the ratio of predictions with $2\mu_0$ and μ_0 [31, 634].

11.6
Multijet Productions

A process in which N partons are produced in the final state is, in principle, expected to be observed as N-jet event. Figure 11.15 shows an example of multijet production.

The $N \geq 3$ multijet events can be understood as emission of extra gluons from energetic final partons in $2 \to 2$ scatterings. Emission probability is large when the gluon is soft or in collinear configuration where they are hardly separated from their parent at the hadron level. However, the hard gluon emissions can be separated experimentally and allow clear theoretical treatments. Separation of the individual jet is generally carried out by requiring

$$p_{Ti} > p_{T\min}\,, \quad |\eta_{ij}| < \eta_{\max}\,, \quad R_{ij} > R_{\min} \tag{11.32}$$

where R_{ij} is the distance between the jet i and j in the $\eta-\phi$ plane.

In the case of $N = 3$, adopting $x_i = 2E_i/\sum E_i$, $\sum x_i = 2$, $x_3 > x_4 > x_5$ as variables, we can apply the Dalitz plot to the final jets. Just like the $e\overline{e} \to q\overline{q}g$, the density of the phase space in the $x_3 - x_4$ plane is uniform with triangular boundary determined by $2/3 \leq x_3 \leq 1, 1/2 \leq x_4 \leq 1$ (see Section 7.3.2). The regions for $x_3, x_4 \to 1$, $x_3 + x_4 \to 1$ are dominated by either soft or collinear jets where the 3-jets are hard to separate from 2-jets.

Figure 11.16a shows a 3-jet data by CDF [635, 636]. The cut is applied to eliminate the soft and the collinear region. Distribution of the data is clearly not uniform and agrees with the QCD prediction. Figure 11.16b shows a comparison with a NLO Monte Carlo calculation using CTEQ4M [636]. The agreement is better than 25%.

For $N \geq 4$ data, we show mass distributions of multijet events in Figure 11.17a and angular distributions of the jet with the highest energy in Figure 11.17b [637, 638]. For tree diagrams, QCD calculations are available [639, 640], and reproduce

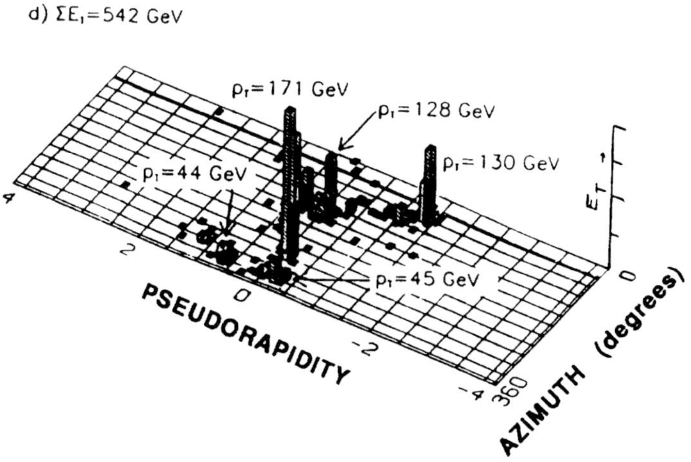

Figure 11.15 Lego plot of a multijet event [611].

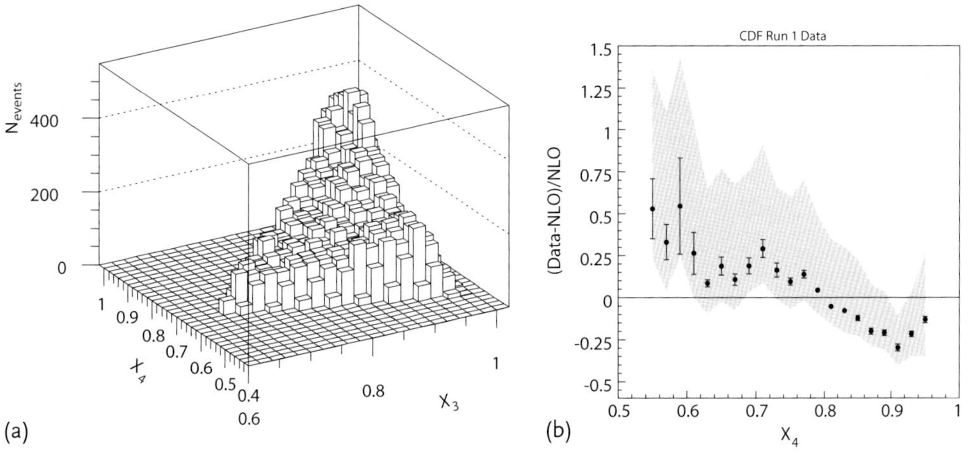

Figure 11.16 (a) shows the 3-jet event distribution in x_3–x_4 plane. (b) Fractional difference between data and theoretical predictions, using the NLO Monte Carlo calculation with CTEQ4M, as a function of x_4, averaged over x_3. The vertical bands show systematic uncertainties [635, 636].

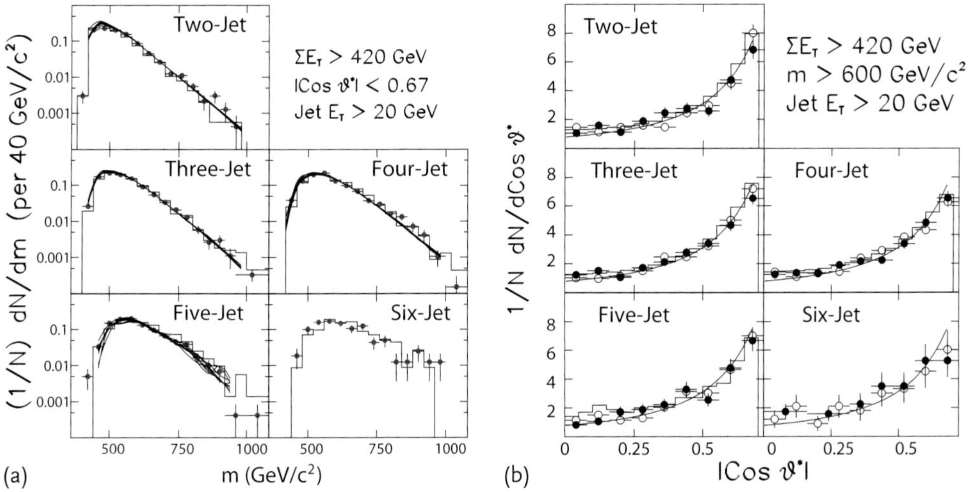

Figure 11.17 (a) Exclusive multijet distributions. (b) Leading-jet angular distributions. The data (solid points) are compared with HERWIG predictions (histogram in (a) and open points in (b)) and NJETS for various parton distribution functions and p_T^2 cuts (solid lines in (a) and histogram in (b)). The curves in (b) show the Rutherford scattering form $(1 - \cos\theta^*)^{-2}$ [637, 638].

experimental features well where they are available. Simulations using the parton shower generators reproduce data well for $N \lesssim 6$, but tend to underestimate as N grows. The fact that angular distributions of the leading-jet agree well with that of the Rutherford scattering and the parton shower models which are based on the

DGLAP evolution equation strongly favors our notion that the multijet productions are $2 \to 2$ reactions followed by gluon emissions.

11.7 Substructure of the Partons?

Having confirmed the validity of the QCD to a high degree of precision, we may launch a hunt for new physics by detecting discrepancy from the Standard Model predictions. Here, we discuss one such example. If the quarks are not elementary particles but structured objects consisting of more fundamental constituents (subquarks), we may expect their effect to appear as deviations from the QCD predictions. If the subquarks exist, their effective interaction is expected to be point-like at scales much larger than their size. The effect of the extended quark may be parametrized as a form factor to the coupling constant[4)]

$$\alpha_s(Q^2) \to \alpha_s(Q^2)F(Q^2), \quad F(Q^2) = \left(1 + Q^2/\Lambda_c^2\right)^{-1} \tag{11.33}$$

Here, Λ_c represents the inverse size of the quark, that is, $\Lambda_c = \infty$ corresponds to a point quark. For the leptons, it is known from $e\overline{e} \to l\overline{l}$ data that Λ_c is not smaller than $\sim 10\,\text{TeV}$ [11]. These effects generally reduce cross sections. One may remember that at low energies ($s \ll m_W^2$), the weak interaction can be approximated by the four-fermion interaction. Therefore, the new effective Lagrangian that originates in the compositeness may be represented by similar four-fermion interactions:

$$\mathcal{L} = \eta \frac{g_{ab}}{2\Lambda_c^2} \sum_{a,b=\text{L,R}} \left(\sum_{i=u,d,\cdots} \overline{q}_{ia}\gamma^\mu q_{ia} \right) \left(\sum_{i=u,d,\cdots} \overline{q}_{ib}\gamma_\mu q_{ib} \right) + \cdots \tag{11.34}$$

where $\eta = \pm 1$ and we assumed the vector type interaction. Generally, there are LL, LR, RR couplings, but for simplicity, we limit our discussion to LL coupling only and further assume $g_{ab}^2/4\pi = g_s^2/4\pi = 1$. If the interaction is added to the Standard Model, the matrix elements listed in Table 11.1 are modified. Since Λ_c is large, contribution from Eq. (11.34) alone is small and we may only consider the interference effect as a correction to the Standard Model. For instance, we have [641, 642]

$$\begin{aligned}
|\mathcal{M}(q\overline{q} \to q\overline{q})|^2 &= \text{QCD} + \frac{8}{9} g_s^2 \frac{4\pi\eta}{\Lambda_c^2} \left(\frac{\hat{u}^2}{\hat{t}} + \frac{\hat{u}^2}{\hat{s}} \right) + O(\Lambda_c^{-4}) \\
|\mathcal{M}(qq \to qq)|^2 &= \text{QCD} + \frac{8}{9} g_s^2 \frac{4\pi\eta}{\Lambda_c^2} \left(\frac{\hat{s}^2}{\hat{t}} + \frac{\hat{s}^2}{\hat{s}} \right) + O(\Lambda_c^{-4}) \\
|\mathcal{M}(q\overline{q} \to q'\overline{q}')|^2 &= |\mathcal{M}(qq' \to qq')|^2 = |\mathcal{M}(q\overline{q}' \to q\overline{q}')|^2 \\
&= \text{QCD} + O(\Lambda_c^{-4})
\end{aligned} \tag{11.35}$$

4) See discussions in Vol. 1, Sect. 17.2. A similar method was utilized to look for possible deviations from QED as discussed in Vol. 1, Sect. 8.2.3.

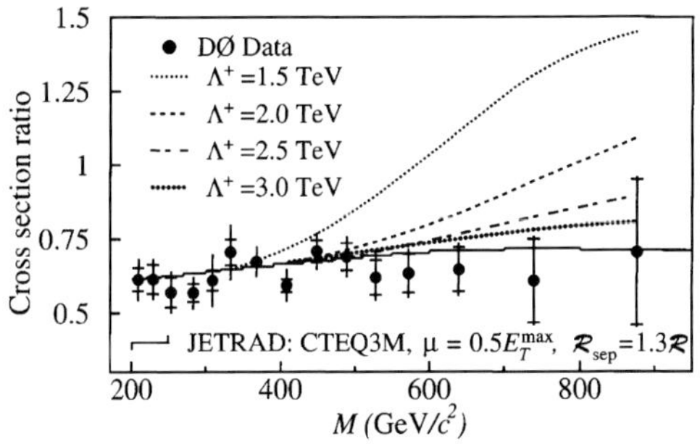

Figure 11.18 Ratio of dijet production cross sections ($d^3\sigma/dM_{jj}\,d\eta_1 d\eta_2$) at $|\eta_{\text{jet}}| < 0.5$ with respect to $0.5 < |\eta_{\text{jet}}| < 1.0$ are plotted. Solid circles show data compared with theories (various lines). The error bars show statistical and systematic uncertainties added in quadrature, and the crossbar shows the size of the statistical error [29–31].

As η can be positive or negative, the cross section does not necessarily increase. Figure 11.18 shows comparison of data on the production cross section of dijets with possible contributions of the quark compositeness [29–31]. One can see that Λ_c is at least as large as $\sim 2\,\text{TeV}$. In the low energy region ($\sqrt{s} \ll \Lambda_c$), the correction is small. However, if $\sqrt{s} \gg \Lambda_c$, there is a possibility that rich structures such as abundance of resonances may appear, just like many hadron resonances in QCD at $\sqrt{s} \gg m_\pi$. Namely, it is one method to search for new physics at a large $\sqrt{s}$, E_T region. It could be one of the main subjects in a new energy frontier hadron collider like LHC.

11.8
Vector Particle Production

Here, we discuss productions of W, Z and hard photons. The Born approximation formula was already given in Chapter 3 primarily to investigate the electroweak properties of the gauge bosons. However, clarification of the production mechanism in the hadron reactions belongs to QCD. Mathematically, the vector boson production is nothing but the Drell–Yan process with a discrete value of Q^2. The processes $p + \overline{p} \to W \to l + \nu$ and $p + \overline{p} \to Z \to l + \overline{l}$ can be treated almost identically with the Drell–Yan process $p + \overline{p} \to \gamma^* \to l\overline{l}$. Therefore, mathematical expressions at the parton level, such as those given in Section 8.6, can be directly applied to the vector production. In the following, we mainly concentrate on the transverse momentum p_T distribution which was not discussed there. We first treat the hard photon production where the perturbative QCD works well. Then, we turn to the W production at relatively low transverse momenta ($p_T \ll m_W$) where the resummation technique is required to compare theory with data. Distributions of the Z can be trivially obtained from that of W by replacing the Born term cross section and with due selection of flavors in convoluting parton distributions.

11.8.1
Direct Photon Production

Production of large p_{T} photons and that of jets in the $p\overline{p}$ reactions are closely connected and are used to determine the gluon distribution in the hadron. Feynman diagrams which contribute mainly to the process are depicted in Figure 11.19. One sees that Figure 11.19c,d can be a good probe for picking up gluon distributions in the hadron. The photon does not fragment and there are no ambiguities associated with jet productions, and hence its direction, energy and number can be well defined. Thus, the process can provide a reference point to calibrate QCD predictions as well as jet analyses. Invariant matrix elements of the process are given in Table 11.2.[5] We express formulas using Q^2 for m_V^2. For the photons, set $Q^2 = 0$. The difference of the coupling constant e_V^2 for $V = \gamma^*,\ W^{\pm}$ and Z can be understood by considering the electroweak Lagrangian.

$$-\mathcal{L}_{\mathrm{EW}} = Q_q e\overline{q}\gamma^{\mu} q A_{\mu} + \frac{g_{\mathrm{W}}}{2\sqrt{2}}\left[\overline{u}_i\gamma^{\mu}(1-\gamma^5)d_j V_{ij} W_{\mu}^{-} + \mathrm{h.c.}\right] + \frac{g_Z}{2}\overline{q}\gamma^{\mu}(v_q - a_q\gamma^5)q Z_{\mu} \tag{11.36}$$

where V_{ij} is the Cabbibo–Kobayashi–Maskawa matrix elements. If the polarization of the W/Z is not observed, the vector and the axial vector part give equal contribution. Therefore, for the case of $V = W^{\pm}$, we have

$$e^2 \to (g_{\mathrm{W}}/(2\sqrt{2})^2)|V_{ij}|^2 \times 2 = \sqrt{2}G_{\mathrm{F}} m_{\mathrm{W}}^2 |V_{ij}|^2 \tag{11.37}$$

We have listed four diagrams in Figure 11.19, but which process is dominant depends on the experimental conditions. For instance, in pp, p-nucleus collisions for not too large p_{T}, the gluon-Compton scattering dominates, but for $p\overline{p}$, collisions with large p_{T}, the $q\overline{q}$ process is dominant.

We define kinematic variables by $q(p_1) + \overline{q}'(p_2) \to V(k) + g(k')$ for Figure 11.19a,b, $q(p_1) + g(p_2) \to V(k) + q(k')$ for Figure 11.19c,d, and treat partons

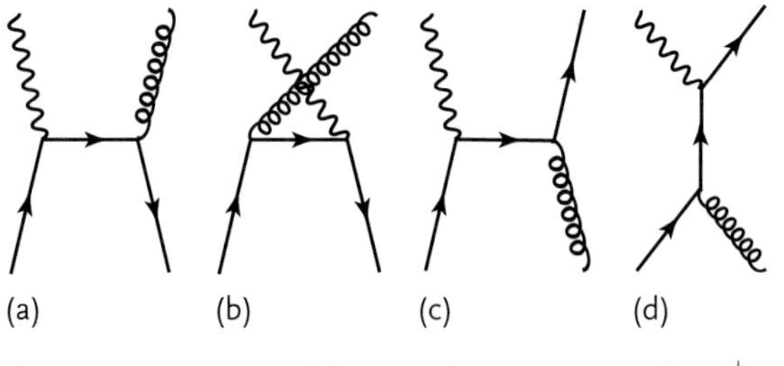

Figure 11.19 Diagrams of the vector boson V ($V = \gamma^*, W^{\pm}, Z$) production in NLO. (a,b) $q\overline{q} \to V+$ gluon. (c,d) QCD Compton process ($g + q \to V + q$).

5) The diagrams in Figure 11.19 can be obtained by crossing from those of the deep inelastic QCD Compton scattering (see Figures 8.3 and 8.9 and their respective matrix elements Eq. (8.35) and Eq. (8.67)).

Table 11.2 Matrix elements for the production of the vector boson [397, 697]. $Q^2 = \hat{s} + \hat{t} + \hat{u} = m_V^2$. For the real photons, set $Q^2 = 0$. $\overline{\sum}$ denotes the average for initial and sum for final states over color and spin degrees of freedom.

Process	$\overline{\lvert\mathcal{M}\rvert^2}$	Coupling strength	
$q\bar{q} \to Vg$	$e_V^2 g_s^2 \dfrac{2C_F}{N} \dfrac{\hat{t}^2 + \hat{u}^2 + 2\hat{s}Q^2}{\hat{t}\hat{u}}$	$e_V^2 = (Q_q e)^2$	for $V = \gamma^*$
		$e_V^2 = g_W^2 \lvert V_{qq'}\rvert^2/4 = \sqrt{2} G_F m_W^2 \lvert V_{qq'}\rvert^2$	for $V = W^\pm$
$gq \to Vq$	$-e_V^2 g_s^2 \dfrac{1}{N} \dfrac{\hat{s}^2 + \hat{u}^2 + 2\hat{t}Q^2}{\hat{s}\hat{u}}$	$e_V^2 = g_Z^2 \left(v_q^2 + a_q^2\right)/4$	
		$= \sqrt{2}\rho G_F m_Z^2 (v_q^2 + a_q^2)$	for $V = Z$

* $g_s^2 = 4\pi\alpha_s$, $N = 3$, $C_F = 4/3$ for $SU(3)$.

as massless. The Mandelstam variables and their expression in the CM frame are defined as follows:

$$
\begin{aligned}
k^\mu &= (\omega, \boldsymbol{k}) \\
\hat{s} &= (p_1 + p_2)^2 = 4|\boldsymbol{p}|^2 \\
\hat{t} &= (p_2 - k')^2 = -2|\boldsymbol{p}^*||\boldsymbol{k}^*|(1 - \cos\theta^*) \\
\hat{u} &= (p_1 - k')^2 = -2|\boldsymbol{p}^*||\boldsymbol{k}^*|(1 - \cos\theta^*)
\end{aligned}
\tag{11.38}
$$

$$
\begin{aligned}
&\hat{s} + \hat{t} + \hat{u} = Q^2 \\
&|\hat{t}|_{\max} = (\hat{s} - Q^2), \quad |\hat{t}|_{\min} = 0 \\
&z = Q^2/\hat{s} = 1 - |\boldsymbol{k}^*|/|\boldsymbol{p}^*| \\
&k_T^2 = |\boldsymbol{k}^*|^2 \sin^2\theta^{*2} = \frac{\hat{t}\hat{u}}{\hat{s}}
\end{aligned}
\tag{11.38b}
$$

Hats ˆ are attached to the parton variables to distinguish them from those of hadrons.

The vector boson production cross section by the partons is obtained by substituting the invariant matrix elements given in Table 11.2 in the following formula.

$$
\frac{d\sigma}{d\hat{t}} = \frac{1}{16\pi}\frac{1}{\hat{s}^2}\overline{|\mathcal{M}|^2}
\tag{11.39}
$$

The cross section by the $p\bar{p}$ collision can be obtained by convoluting the cross section with the parton distributions. We denote the rapidity of the photon as

$$
\eta = \frac{1}{2}\ln\frac{\omega + k_z}{\omega - k_z}
\tag{11.40}
$$

Using the relation

$$dk_x dk_y d\eta = \frac{d^3k}{\omega} = \frac{1}{2\hat{s}} d\hat{t} d\hat{u} d\phi = \frac{\pi}{\hat{s}} d\hat{t} d\hat{u} \tag{11.41}$$

The cross section is expressed as

$$\frac{d\sigma}{dk_x dk_y d\eta} = \sum_{ij} \int dx_1 dx_2 f_i(x_1, \mu_F^2) \overline{f}_j(x_2, \mu_F^2) \left[\frac{\hat{s}}{\pi} \frac{d\sigma_{ij}}{d\hat{t} d\hat{u}} \right]_{\substack{p_1 = x_1 P_1 \\ p_2 = x_2 P_2}} \tag{11.42}$$

where i, j ($f_j = \overline{f}_i$ for the photon and Z) denotes the parton flavor in the hadron, μ_F is the factorization scale which is conventionally set to $\mu_F = Q$ and

$$\frac{d\sigma}{d\hat{t} d\hat{u}} = \frac{d\sigma}{d\hat{t}} \delta(\hat{s} + \hat{t} + \hat{u} - Q^2) \tag{11.43}$$

For the case of the hard photon production where no infrared divergence enters, the integration of Eq. (11.42) is straightforward.

Figure 11.20 compares the data and the QCD calculation. QCDs reproducibility is good.

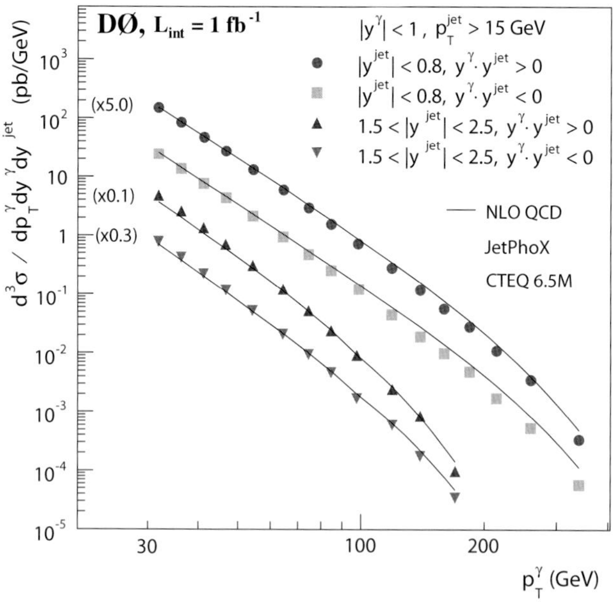

Figure 11.20 The measured differential $p\overline{p} \to \gamma + \text{jet} + X$ cross section as a function of p_T for the four measured rapidity intervals [604, 644–647].

11.8.2
W : p_{T} Distribution

Distribution at Large p_{T}^2 The leading order formula for W boson production in the electroweak process $q + \overline{q}' \to W$ does not have transverse momentum dependence even if one uses μ_{F}^2 (factorization scale) developed parton distributions because only gluons emitted in parallel to the quarks are considered. As long as we neglect the primordial transverse momentum,[6] the W boson can have transverse momentum only when the gluon is emitted at the same time. In order to consider p_{T} distributions, we need to consider higher order QCD effects. They include the gluon emission by the quark–antiquark annihilation and W production by the gluon–quark Compton scattering processes (see Figure 11.19 and replace γ^* with the W).

The W production cross section is obtained from Eq. (11.42) by inserting suitable variables in the above discussion. However, this time, we need to treat the region where the transverse momentum is small, relative to the W mass. To see singularities of the matrix elements, we rewrite it using the fractional momentum z. The matrix element for $q\overline{q}' \to Wg$ can be expressed as

$$\begin{aligned} C_F\left[\frac{\hat{u}}{\hat{t}} + \frac{\hat{t}}{\hat{u}} + \frac{2m_W^2\hat{s}}{\hat{t}\hat{u}}\right] &= \frac{4}{3}\left[\frac{1+z^2}{1-z}\left(\frac{\hat{s}}{-\hat{t}} + \frac{\hat{s}}{-\hat{u}}\right) - 2\right] \\ &= P_{qq}(z)\left(\frac{\hat{s}}{-\hat{t}} + \frac{\hat{s}}{-\hat{u}}\right) - \frac{8}{3} \end{aligned} \tag{11.45}$$

where $P_{qq}(z)$ is the split function given in Eq. (8.72a). We encountered this type of formula before. It has the collinear divergence ($\hat{t} \to 0$ or $\hat{u} \to 0$). It also diverges as $k \to 0$ or $z \to 1$ (infrared divergence). The cross section has singularity as $\sim p_{\mathrm{T}}^2 \to 0$. As we learned, the collinear divergence is folded into the structure function to give the evolving parton distribution function. The infrared divergence is not a real divergence, but can be eliminated by adding the interference effect of the virtual gluon exchange process Figure 11.21b with the Born term Figure 11.21a. Calculation of Eq. (11.42) including $O(\alpha_s)$ diagrams in Figure 11.19 and Figure 11.21 can be carried out by using the dimensional regularization.[7] The

6) The primordial momentum is one which the partons possess intrinsically inside hadrons. From deep inelastic data, we estimate that each quark has $\sim 1/3$ of its parent nucleon momentum on average. Naively, then

$$\begin{aligned} \langle p_{\mathrm{T}}^2\rangle &= \langle p_x^2\rangle + \langle p_y^2\rangle \\ &\sim (2/3)(m_N/3)^2 \\ &\sim (0.26\ \mathrm{GeV}/c)^2 \end{aligned} \tag{11.44}$$

In the $q\overline{q}$ annihilation, two quarks participate, giving $\sqrt{2} \times 0.26\ \mathrm{GeV}/c \sim 0.36\ \mathrm{GeV}/c$. This estimates agrees with the experimental facts that pions produced in the high energy pp collision has the average transverse momentum $\sim 0.35\ \mathrm{GeV}/c$.

7) The formula has the same form as that of Drell–Yan process as the W production can be obtained by setting the invariant mass squared as $Q^2 = m_W^2$.

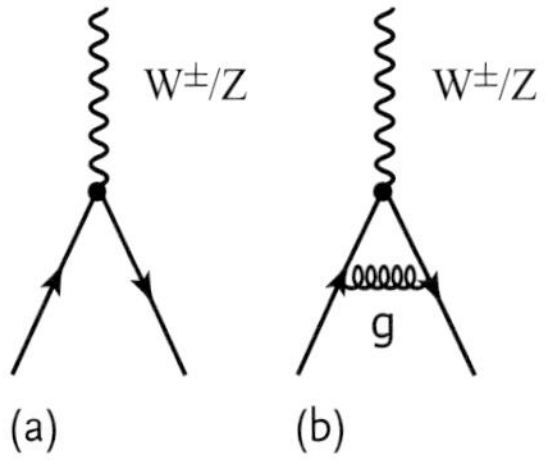

Figure 11.21 (a) Born process of the W/Z boson production. (b) The virtual gluon exchange process that compensates the infrared divergence of the soft gluon emission.

result up to $O(\alpha_s^2)$ can be expressed in the following form [648–652].

$$\frac{1}{\hat{\sigma}_0}\frac{d\sigma}{dp_T^2} = \frac{\alpha_s}{2\pi}\left[A\frac{\ln(m_W^2/p_T^2)}{p_T^2} + B\frac{1}{p_T^2}\right] + C(p_T^2)$$
$$A = A_1 + A_2\frac{\alpha_s}{2\pi} = 2C_F\left[1 + \frac{\alpha_s}{2\pi}\left[\left(\frac{67}{18} - \frac{\pi^2}{6}\right)N_c - \frac{10}{9}T_R\right] + \cdots\right]$$
$$B = B_1 + B_2\frac{\alpha_s}{2\pi} = -3C_F[1 + E\alpha_s + \cdots] \tag{11.46}$$

In this expression, we included up to $O(\alpha_s^2)$ contributions. Here, $\hat{\sigma}_0$ is the lowest order cross section and $C(p_T^2)$ is a regular function of p_T at $p_T = 0$. As we already know, the infrared divergence is cured by including the interference of the virtual gluon process with the Born term. In view of the fact that it contributes only at $p_T = 0$, the cross section can be conveniently reformulated in a form which is regular at $p_T \to 0$ and reproduces $O(\alpha_s)$ result for $p_T \neq 0$.

$$\frac{d\sigma}{dp_T^2} = \hat{\sigma}_0\left[(1 + D)\delta(p_T^2) + C(p_T^2) + \frac{\alpha_s}{\pi}\left\{A\left(\frac{\ln(m_W^2/p_T^2)}{p_T^2}\right)_+ + B\left(\frac{1}{p_T^2}\right)_+\right\}\right] \tag{11.47}$$

The integral of the plus function vanishes (see Eq. (8.77)) and the constant D is determined to reproduce the total cross section. Integrating Eq. (11.47), we have

$$\sigma_{TOT} = \hat{\sigma}_0\left[1 + D + \int_0^{p_{T\max}^2} dx\, C(x)\right] \tag{11.48}$$

which shows how to determine D.

Notice that the A term in Eq. (11.47) is the same as the transverse momentum distribution in the LLA approximation for the process $e^-e^+ \to$ jets in Eq. (7.147). Namely, this form of p_T dependence is generic in QCD because it is caused by the gluon emission.

Equation (11.47) reproduces experimental data well at large transverse momentum ($p_T \gtrsim 30$ GeV) as is shown in Figure 11.22a. However, it does not reproduce the data well at low transverse momentum $p_T \lesssim 20$ GeV as is shown in Figure 11.22b (dashed line denoted as pQCD).

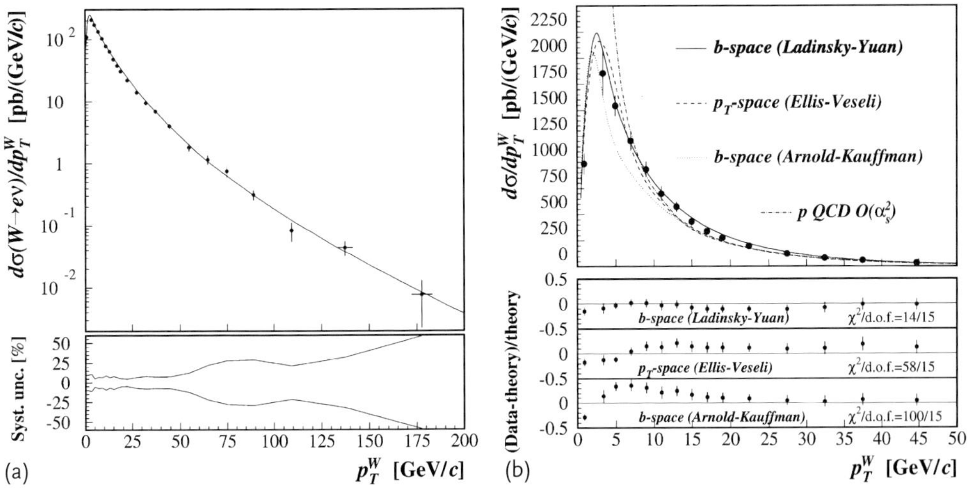

Figure 11.22 (a) Differential cross section for $p + \overline{p} \to W \to e\nu$ production. The dashed line is the theoretical prediction of [653]. Data points only show statistical uncertainties. The fractional systematic uncertainty, shown as the band in the lower plot, does not include an overall 4.4% normalization uncertainty in integrated luminosity. (b) Differential cross section for W boson production compared to three resummation calculations and to the fixed-order calculation. Also shown are the fractional differences (Data-Theory)/Theory between data and the resummed predictions [654].

Distribution at Small p_T^2 The reason that the perturbative calculation does not reproduce the data so well in the small p_T region is expected from the general arguments for the perturbative QCD. Let us take a look at Figure 11.23 which shows the scale dependence of the cross section obtained by the fixed order perturbation [650]. One sees that the $O(\alpha_s^2)$ cross section is fairly stable at large p_T ($\gtrsim 100\,\mathrm{GeV}$), but is not so much at small p_T ($\lesssim 20\,\mathrm{GeV}$). This is because at small p_T, multiemission of soft gluons ($\sim [\ln(m_\mathrm{W}^2/p_\mathrm{T}^2)]^n$) cannot be neglected. Fortunately, the singular terms at $p_\mathrm{T} \to 0$ can be resummed. Their contribution is of the form

$$\frac{d\sigma}{dp_\mathrm{T}^2} \sim \frac{1}{p_\mathrm{T}^2} \sum_{n=0}^{\infty} \alpha_s^n \left[\sum_{m=0}^{2n-1} \left(\ln \frac{m_\mathrm{W}^2}{p_\mathrm{T}^2} \right)^m \right] \tag{11.49}$$

that is, when $\alpha_s \ln(m_\mathrm{W}^2/p_\mathrm{T}^2) \sim 1$, the leading term ($\alpha_s^n L^{2n-1}$) in each bracket can be resummed for $O(\alpha_s)$ calculation. Resummation has to be done in two-dimensional p_T (i.e., $p_x - p_y$) space to take into account the linear sum of p_T properly for the multiple gluon emission. Therefore, the expression in Eq. (11.46) is Fourier transformed to the impact parameter space ($\boldsymbol{b} = (x, y)$), resummed and transformed back to the original p_T space [397, 648, 649, 651, 655–658]. The resultant small p_T

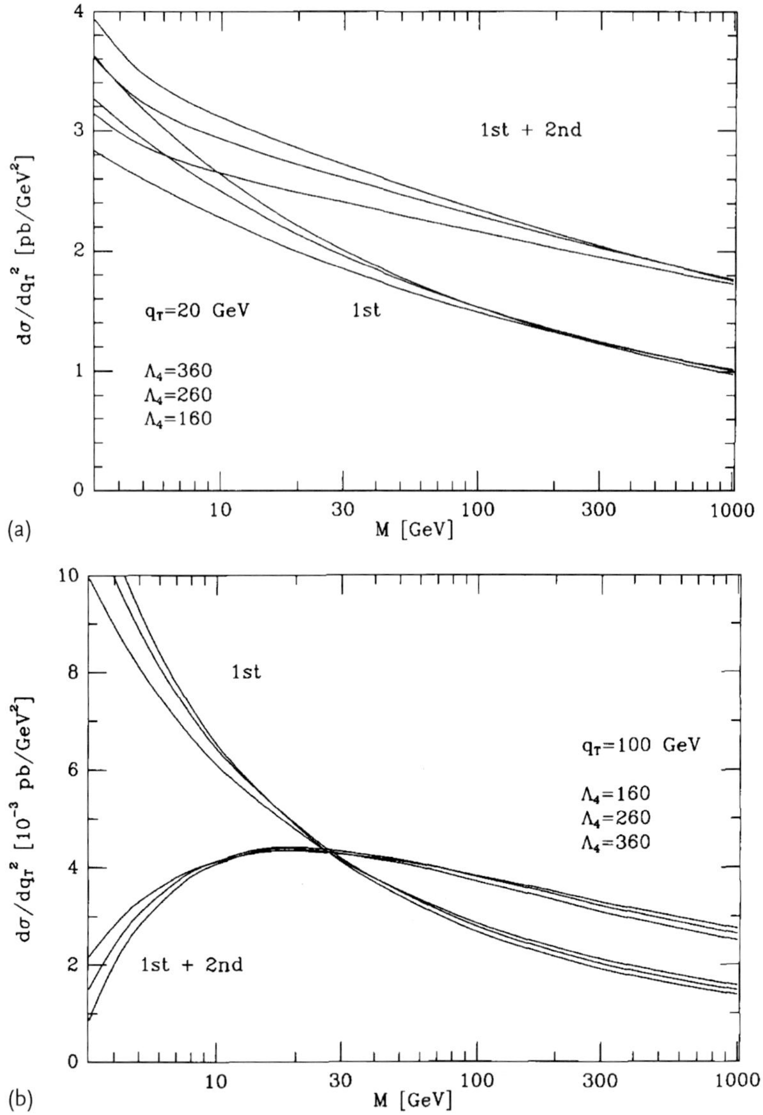

Figure 11.23 The factorization scale ($M = \mu_F$) dependence of $d^2\sigma/dp_T^2(p\overline{p} \to Wg)$ for $p_T = 20\,\text{GeV}$ (a) and 100 GeV (b) at $\sqrt{s} = 1.8\,\text{TeV}$. Different curves correspond to three choices of $\Lambda = 360, 260, 160\,\text{MeV}$. One sees that the $O(\alpha_s^2)$ cross section is fairly stable at large p_T ($\gtrsim 100$) GeV, but unstable at small p_T ($\lesssim 20\,\text{GeV}$) [650, 652].

distribution is given by

$$\frac{d\sigma}{dp_T^2} \simeq \sum_i \hat{\sigma}_0 \frac{1}{2} \int d b\, b\, J_0(b p_T) \exp[-S(b, m_W)]$$
$$\times \int_0^1 dx_1 dx_2 \delta\left(x_1 x_2 - \left(\frac{m_W^2}{s}\right)\right)$$
$$\times \left[q_i\left(x_1, \left(\frac{b_0}{b}\right)^2\right) \overline{q}_i\left(x_2, \left(\frac{b_0}{b}\right)^2\right) + (q \leftrightarrow \overline{q})\right] \quad (11.50)$$

where $b_0 = 2\exp(-\gamma_E)(\gamma_E = 0.5772\ldots$ is the Euler constant). The Born cross section $\hat{\sigma}_0$ is given by[8)]

$$\hat{\sigma}_0 = \begin{cases} \frac{4\pi\alpha^2}{9s} Q_q^2 & V = \gamma^* \\ \frac{\pi}{3s}\sqrt{2} G_F m_W^2 |V_{ij}|^2 & V = W^{\pm} \\ \frac{\pi}{3s}\sqrt{2}\rho G_F m_Z^2 (v_q^2 + a_q^2) & V = Z \end{cases} \quad (11.52)$$

A more complete expression is given in [651, 659]. Note that this expression's factorization scale is set at $\mu_F = b_0/b$. The function $\exp(-S)$ is the Sudakov form factor in b-space where it is given by [655, 657, 660]

$$S(b, Q) = \int_{(b_0/b)^2}^{Q^2} \frac{dq^2}{q^2}\left[A\left(\alpha_s(q^2)\right) \ln \frac{Q^2}{q^2} + B\left(\alpha_s(q^2)\right)\right]$$
$$A(\alpha_s) = \sum_1^{\infty}\left(\frac{\alpha_s}{2\pi}\right)^n A_n, \quad B(\alpha_s) = \sum_1^{\infty}\left(\frac{\alpha_s}{2\pi}\right)^n B_n \quad (11.53)$$

A_n, B_n for $n = 1.2$ are given by Eq. (11.46). Since both the coupling constant and the parton distribution diverge in the low scale region, some cut-off has to be introduced to make the b integral well behaved at small q^2 (large b). This introduces another theoretical uncertainty in addition to setting of the renormalization as well as the factorization scale.

Matching of p_T distribution The whole expression for the entire region of the p_T can be obtained by combining the resummed expression at small p_T and the perturbative expression at large p_T. A simple sum is not adequate because some terms

8) They are Born terms that appear in the parton reactions convoluted with parton distribution functions where a constraint $\delta(x_1 x_2 - m^2/s)$ is imposed to extract the delta function like the spectrum of the vector bosons. As a bare process, the Born cross section contains the delta function (see Eq. (3.16) and (3.18)). For the Drell–Yan process, the cross section $d\sigma/dp_T^2$ is to be interpreted as (see Eq. (8.138a))

$$\frac{d\sigma}{dp_T^2} \to Q^2 \frac{d^2\sigma}{dQ^2 dp_T^2} \quad (11.51)$$

to have a consistent expression with W, Z whose mass is discrete.

are doubly counted. The perturbative expression as is shown in Eq. (11.46) consists of finite terms and those which become singular as $p_T \to 0$. The latter is included in the resummed expression.

One method is to use the resummed expression up to a certain value of $p_T = p_{T\text{match}}$ and use the perturbative expression at $p_T > p_{T\text{match}}$. This method makes the cross section continuous, but its slope changes at $p_T = p_{T\text{match}}$. However, if the matching was carried out at the lower level, that is, with $d^2\sigma/dp_T^2 d\eta$, both the function and its derivative become continuous.

Another method which is suggested in [651] is to subtract terms up to $O(\alpha_s^2)$ from the resummed expression and add to the perturbative one, namely,

$$\frac{d\sigma}{dp_T^2}(\text{total}) = \frac{d\sigma}{dp_T^2}(\text{resum}) + \frac{d\sigma}{dp_T^2}(\text{pert}) - \frac{d\sigma}{dp_T^2}(\text{asym}) \tag{11.54}$$

where $d\sigma/dp_T(\text{asym})$ is the subtraction term consisting of those up to the second order in the power expansion of $d\sigma/dp_T$ (resum). This formula makes the "matching" manifest: at low p_T, the perturbative and the asymptotic terms cancel, leaving the resummed. While at high p_T, the resummed and asymptotic terms cancel to second order, leaving the perturbative. It is important that the perturbative and asymptotic results be evaluated at the same renormalization and factorization scales, μ_R and μ_F, otherwise, some extra terms will be introduced in the expression.

Figure 11.22b shows $D0$ data at the Tevatron in medium to small p_T region versus a few theoretical fittings to the data. The difference of the two curves using the b-space method [651, 653] is due to the different choice of the cut-off factor in the nonperturbative region. Treatment by p_T [661] differs only in higher order terms. One may say agreement with the experiment is very good, considering the uncertainties mentioned above.

11.9 Heavy Quark Production

11.9.1 Cross Sections

Denoting light quarks (u, d, s) as q and heavy quarks (c, b, t) as Q, the processes for the heavy quark production in the lowest order are expressed as in Figure 11.24. Setting variables as depicted in Figure 11.24a,b, the matrix elements are easily calculated. Neglecting the light quark mass and denoting the heavy quark mass as M $(p_1^2 = p_2^2 = 0, p_3^2 = p_4^2 = M^2)$, the spin and color averaged matrix element squared is given by (see, for instance, [662])

$$\overline{|\mathcal{M}(q\bar{q} \to Q\bar{Q})|^2} = \frac{4}{9} g_s^4 \left[\frac{(M^2 - \hat{t})^2 + (M^2 - \hat{u})^2 + 2M^2\hat{s}}{\hat{s}^2}\right] \tag{11.55a}$$

$$\hat{s} = (p_1 + p_2)^2, \quad \hat{t} = (p_1 - p_3)^2, \quad \hat{u} = (p_1 - p_4)^2 \tag{11.55b}$$

Overfix ^ on the variables denote they are parton variables. In order to simplify expressions, we introduce the following variables [663].

$$\tau_{1,2} = \frac{2(p_{1,2} \cdot p_3)}{\hat{s}} = \frac{1 \mp \beta \cos\theta^*}{2}, \quad \rho = \frac{4M^2}{\hat{s}} = 1 - \beta^2 \tag{11.56}$$

Then, Eq. (11.55) becomes

$$\overline{\sum} |\mathcal{M}(q\overline{q} \to Q\overline{Q})|^2 = \frac{C_F}{C_A} g_s^4 \left(\tau_1^2 + \tau_2^2 + \frac{\rho}{2}\right) \tag{11.57a}$$

Similarly,

$$|\overline{\mathcal{M}}(gg \to Q\overline{Q})|^2 = \frac{g_s^4}{8}\left(\frac{C_F}{\tau_1\tau_2} - C_A\right)\left(\tau_1^2 + \tau_2^2 + \rho - \frac{\rho^2}{4\tau_1\tau_2}\right) \tag{11.57b}$$

Referring to Eq. (11.7), the cross section is given by

$$\frac{d\hat{\sigma}}{d\cos\theta^*} = \frac{|\overline{\mathcal{M}}|^2}{1+\delta_{cd}} \frac{1}{32\pi} \frac{\beta}{\hat{s}} \tag{11.58}$$

Substituting Eq. (11.57) in Eq. (11.58), we obtain the total cross sections for the heavy quark production at the parton level.

$$\hat{\sigma}(q\overline{q} \to Q\overline{Q}) = \frac{\alpha_s^2}{M^2}\left[\frac{\pi\beta}{27}\rho(2+\rho)\right] \xrightarrow{\hat{s}\to\infty} \frac{8\pi}{27}\frac{\alpha_s^2}{\hat{s}} \sim \frac{\alpha_s^2}{\hat{s}} \tag{11.59a}$$

$$\begin{aligned}\hat{\sigma}(gg \to Q\overline{Q}) = {} & \frac{\alpha_s^2}{M^2}\frac{\pi\beta}{12}\rho\left[\mathcal{L}(\beta)\left(1+\rho+\frac{\rho^2}{16}\right)\right.\\ & \left.+\left(\frac{3}{4}+\frac{3}{16}\rho+\frac{3}{8}\rho^2\right)\right] \xrightarrow{\hat{s}\to\infty} \frac{\pi}{3}\frac{\alpha_s^2}{\hat{s}}\mathcal{L}(\beta) \sim \frac{\alpha_s^2}{\hat{s}}\mathcal{L}(\beta)\end{aligned} \tag{11.59b}$$

where

$$\mathcal{L}(\beta) = \frac{1}{\beta}\ln\frac{1+\beta}{1-\beta} - 2 \tag{11.59c}$$

The reason to express it in terms of β and $\rho = 1 - \beta^2$ is to make the threshold effect apparent. At high energies $\hat{s} \gg 4M^2$, $\rho = 4M^2/\hat{s}$ goes to 0. Note that the cross section vanishes both at $\beta \to 0$ and $\hat{s} \to \infty$.

The LO cross section for the hadron collision can be obtained by convoluting Eq. (11.59) with the parton distribution functions (Figure 11.25). Referring to Eq. (11.18) and using $\hat{s} = 2M_T^2(1 + \cosh\Delta\eta)$, (11.16)

$$\begin{aligned}\frac{d^3\sigma}{d\eta_Q d\eta_{\overline{Q}} d p_T^2}(P\overline{P} \to Q\overline{Q}) &= \frac{1}{16\pi\hat{s}^2}\sum_{ab} x_1 x_2 f_P^a(x_1) f_{\overline{P}}^b(x_2) |\overline{\mathcal{M}}(ab \to Q\overline{Q})|^2 \\ &= \frac{1}{64\pi M_T^4}\frac{\sum_{ab} x_1 x_2 f_P^a(x_1) f_{\overline{P}}^b(x_2)|\overline{\mathcal{M}}(ab \to Q\overline{Q})|^2}{[1+\cosh(\Delta\eta)]^2}\end{aligned} \tag{11.60}$$

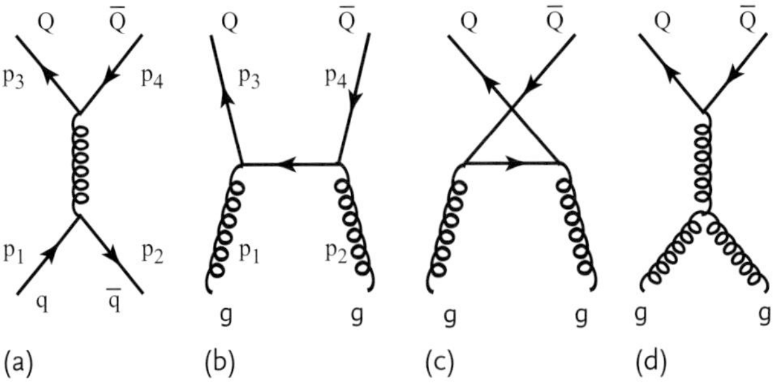

Figure 11.24 Lowest order diagrams for the heavy quark production. (a) Annihilation process by the light quarks. (b–d) gluon fusions.

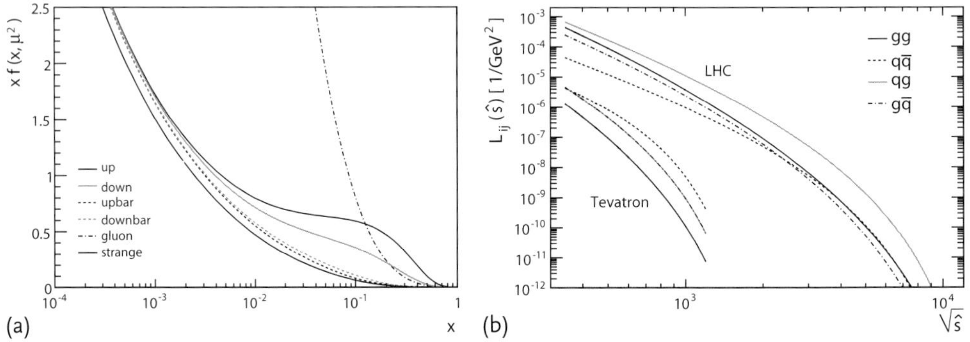

Figure 11.25 (a) The parton distribution function (PDF) of $u, \bar{u}, d, \bar{d}, s$ and gluons inside the proton. The parametrization is CTEQ6.5M [499]. The scale at which the PDFs are evaluated was chosen to be $\mu = 175$ GeV ($\mu^2 = 30\,625$ GeV^2) relevant at Tevatron ($\sqrt{s} = 1.96$ TeV). (b) Parton luminosities for gluon–gluon, quark–antiquark, quark–gluon, and gluon–antiquark interactions at the Tevatron and the LHC [664, 688].

NLO Contributions The next-to-leading-order (NLO) corrections come from two sources of $O(\alpha_s^3)$ diagrams. The tree diagrams accompanied with real gluon emission (some examples are shown in Figure 11.26a–d) and those with virtual emission diagrams (Figure 11.26e–f). The latter is indistinguishable from the tree diagrams, and therefore it has to be added coherently to the tree diagrams. Their interference gives the order $O(\alpha_s^3)$ contribution. Ultraviolet divergences in the virtual diagrams are removed by the renormalization process. Infrared and collinear divergences which appear both in the virtual diagrams and in the integration over the emitted parton cancel each other or can be absorbed in the initial parton densities. The whole calculations are given in [665, 666] (total rate), [667–669] (spectrum) and in [670, 671] (total rate and spectrum).

Incidentally, among the higher order processes, a certain diagram like Figure 11.27a referred to as the flavor excitation exists. In the flavor excitation model, the real (i.e., on the mass-shell) heavy quarks are assumed to be present in the hadron. In this case, the parton line in Figure 11.27a represents the heavy quark.

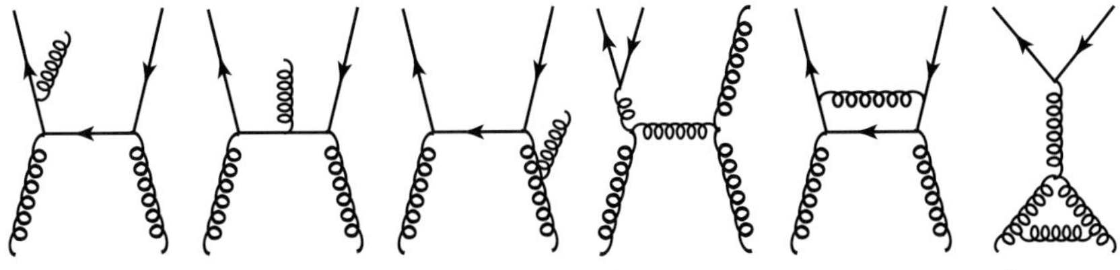

Figure 11.26 Examples of NLO (next-to-leading order) contributions to the heavy quark production. (a–d) Real gluon emission diagrams. (e,f) Virtual gluon emission diagrams. Interference with the LO diagrams Figure 11.24b–d contributes.

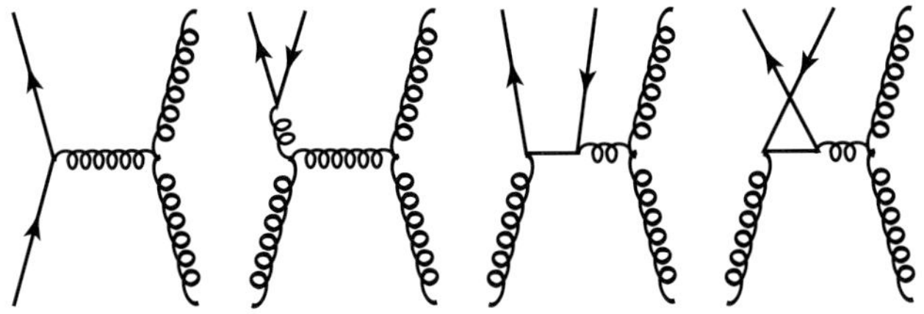

Figure 11.27 (a) Example of flavor excitation diagrams. (b–d) Diagrams that are related to flavor excitation.

Then, the flavor excitation process has a t-channel singularity at $\hat{t} \to 0$ and invalidates the perturbative approach we have discussed. Actually, they are produced by the gluon and become real by scattering with another gluon. However, a careful treatment [672] has shown that the process is already included in the diagrams described in Figure 11.27b–d, and that the largest singularity at $\hat{t} \to 0$ is canceled among them and the rest is absorbed in the initial parton density. Therefore, it was proven that the flavor excitation process should not be included in the perturbative calculation.

Characteristics of the Heavy Quark Production Figure 11.28a shows the total cross section in $p\overline{p}$ collisions for producing the heavy quarks as a function of the hadronic center mass energy $\sqrt{s}$. Notice the different unit used in the case of the top quark relative to others. The suppression is due to the large top quark mass ($m_t = 175$ GeV as compared to $m_c \sim 1.5$, $m_b \sim 5.0$ GeV) and is large at small momentum transfer ($p_T \lesssim M$) as shown in Figure 11.28b. At large p_T, the difference is smaller. The suppression diminishes as the total energy grows.

Looking at the formulae given in Eqs. (11.60) and (11.57), we can grasp some features of the heavy quark productions. Generally, the cross section is large in the region where the denominator of the propagator vanishes. Expressing the denominator as D for each channel contributing to the tree diagrams in Figure 11.24,

$$\hat{s} \text{ channel: } D = \hat{s} = 2(p_1 \cdot p_2) = 2M_T^2(1 + \cosh \Delta\eta) \tag{11.61a}$$

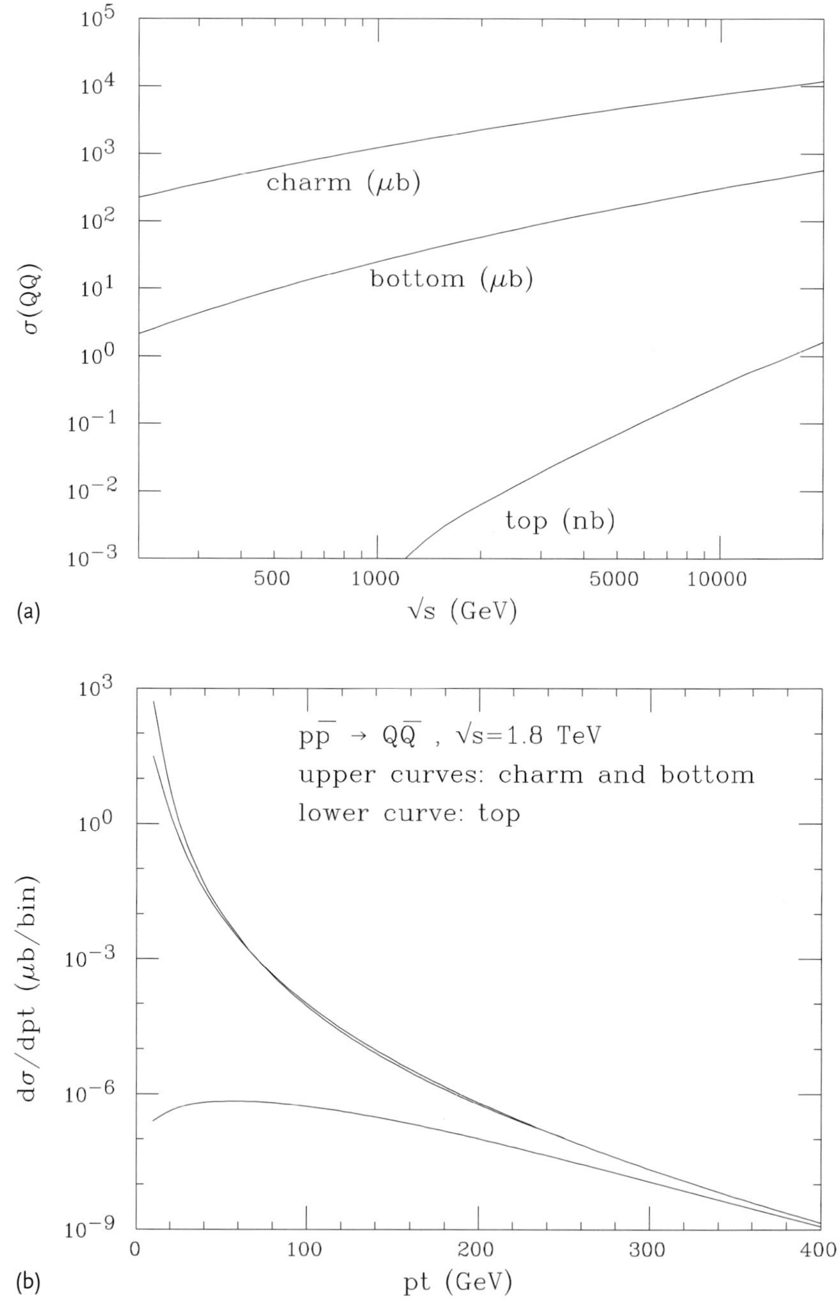

Figure 11.28 (a) $\sigma_{\mathrm{TOT}}(P\overline{P} \to Q\overline{Q})$, (b) $d\sigma/dp_{\mathrm{T}}$ at 1.8 TeV [663]. Note the different unit used for the top production rate. The quark mass is taken to be $m_c = 1.5$, $m_b = 5.0$, $m_t = 175$ GeV.

$$\hat{t} \text{ channel: } D = \hat{t} - M^2 = -2(p_1 \cdot p_3) = -2p_a(E_3 - p_3)$$
$$= -2x_a p(M_T \cosh \eta_3 - M_T \sinh \eta_3) \tag{11.61b}$$

$$\overset{\text{Eq. (11.13)}}{=\!=\!=} -\sqrt{s}x_T e^{\eta} \cosh(\Delta\eta/2)(M_T e^{-\eta} e^{-\Delta\eta/2})$$
$$= -M_T^2(1 + e^{-\Delta\eta}) \tag{11.61c}$$

$$\hat{u} \text{ channel: } D = \hat{u} - M^2 = -2(p_1 \cdot p_4) = -M_T^2(1 + e^{\Delta\eta}) \tag{11.61d}$$

we see that the usual $1/p_T^2 \to \infty$ singularity is screened at $p_T^2 \sim M^2$ by the presence of the heavy mass. Usually, the scale to be applied in evaluating the coupling constant $\alpha_s(\mu^2)$ is taken to be $\mu^2 \sim p_T^2$. Therefore, in a process where the contribution from the small $p_T^2 \to 0$ is dominant, α_s becomes large and makes the perturbative QCD calculation less reliable. In the heavy quark production, however, the mass introduces a natural cut-off and μ is always larger than M. As $m_b, m_t \gg 1\,\text{GeV}$, the validity of the pQCD is warranted. For the charm quark, this is not necessarily true as $m_c \simeq 1.5\,\text{GeV}$. All the p_T^2 dependence is contained in the transverse mass M_T. Since the cross section decreases as M_T^{-4} for large M_T, which becomes $\sim p_T^4$ for large $p_T \gg M$, the contribution to the total cross section comes mainly from the region $p_T^2 \sim M^2$.

For a fixed value of p_T (and hence of M_T), the rate is heavily suppressed when $|\Delta\eta| = |\eta_Q - \eta_{\overline{Q}}|$ becomes large. Therefore, Q and $\overline{Q}$ tend to be produced with the same rapidity. In other words, if a detector catches a b-quark, the probability that the $\overline{b}$-quark is nearby is large. This means by adding a backward spectrometer, the gain in acceptance for the $b\overline{b}$ pair is only two-fold compared to usual four-fold, which would be the case if there is no correlation for the $b\overline{b}$ pair production.

11.9.2
Comparisons with Experiments

In order to compare theoretical expressions with data, we need to take the theoretical ambiguities into account. They include:

1. Scale dependence: ambiguity due to the coupling constant (or equivalently, the uncertainty of $\Lambda_{\overline{MS}}$ in the range $100 \sim 250\,\text{MeV}$). Usually, μ is varied between $\mu_0/2 < \mu_R < 2\mu_0$ where μ_0 is taken to be the quark mass for the heavy quark production. The scale dependence is diminished by including the higher order corrections. Conversely, a large dependence on the scale is a signal that the higher order contribution is sizable. Note that Eq. (11.60) uses the factorization theorem to obtain the hadronic cross section from that of parton, and hence it also depends on the factorization scale μ_F. Usually, $\mu_F = \mu_R$ is chosen, but they are varied independently for fine tuning.
2. Mass dependence: since the quark mass is not directly measurable, it is treated as an adjustable parameter within a range consistent with other experiments.
3. Nonperturbative effect: the formulae we have given in Eq. (11.60) are expressions for $p\overline{p}$ collision, but the final state still consists of partons. To obtain cross

sections for the hadrons in the final state, one has to convolute the above expression with the fragmentation function $f(q \to h_q; z)$ where h_q is the hadron that contains the parton q. Note, the fragmentation effect affects the spectral distribution, but does not change the total cross section. For a recent review of the experimental data, we refer to [673].

Total Cross Section of c and b Let us first discuss the total cross section. In Figure 11.29, the total cross sections for c and b quarks in πN, pN collisions with fixed targets are shown. The scale μ and mass are varied in the range

$$1.2 < m_c < 1.8\,\text{GeV}\,, \quad m_c(\text{default}) = 1.5\,\text{GeV} \tag{11.62a}$$

$$4.5 < m_b < 5 \quad \text{GeV}\,, \quad m_b(\text{default}) = 4.75\,\text{GeV} \tag{11.62b}$$

$$m_c < \mu_\text{R} < 2m_c\,, \quad \mu_\text{F} = 2m_c \quad \text{for} \quad c \tag{11.62c}$$

$$m_b/2 < \mu_\text{R}, \mu_\text{F} < m_b \quad \text{for} \quad b \tag{11.62d}$$

The range of the scale corresponds to the variation of the strong coupling constant $\alpha_s(m_Z) = 0.130, 0.118, 0.105$. Bands in Figure 11.29 denote the scale ambiguities within the range in Eq. (11.62). Note that the parton distribution functions used in converting the parton cross section to that of a hadron are less reliable below $Q^2 = 5\,\text{GeV}^2$. Therefore, the bands for the c quark are to be considered as the lower limit of the scale dependence. Considering the large theoretical uncertainties, we may say the total cross section is in reasonable agreement with the data.

Single Inclusive Distributions (c, b) Spectra of the heavy quark productions contain more information, and hence are more amenable to nonperturbative effects. The effect of the fragmentation is obtained by convoluting the parton cross section with the fragmentation function $D(z)$.

$$\begin{aligned}\frac{d\sigma(h_q)}{dp_\text{T}} &= \int dz\, d\hat{p}_\text{T} \frac{d\sigma}{d\hat{p}_\text{T}} D(q \to h_q; z)\delta(p_\text{T} - z\hat{p}_\text{T}) \\ &= \int \frac{dz}{z}\frac{d\sigma}{d\hat{p}_\text{T}}\left(\hat{p}_\text{T} = \frac{p_\text{T}}{z}\right) D(q \to h_q; z)\,, \quad \hat{p}_\text{T} = p_\text{T}/z\end{aligned} \tag{11.63}$$

A commonly used fragmentation function is that of the Peterson function introduced in Eq. (9.57). At low energies, intrinsic transverse momentum of the parton inside the initial hadrons may also affect the distribution.

c-Quark Distribution We start with fixed target experiments where the total center of mass energy is lower and intrinsic transverse momentum of the incoming partons may be more apparent. We first investigate the p_T^2 distributions obtained in the photon-nucleon collision which is shown in Figure 11.30a. The solid line is the NLO QCD prediction for bare quarks which is significantly harder than the data.

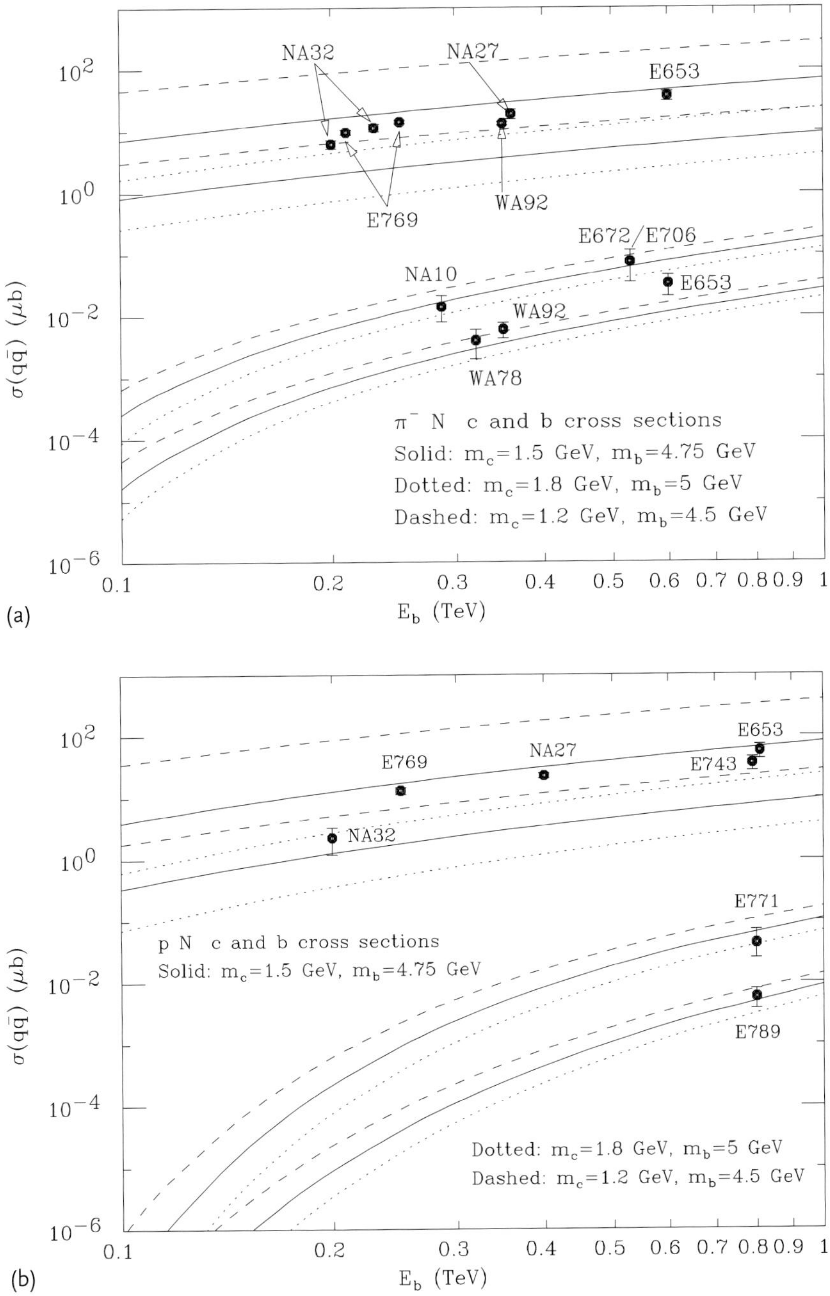

Figure 11.29 Total pair production cross section for c and b quarks in (a) $\pi^- N$ (b) pN collision. Solid lines: $m_c = 1.5$, $m_b = 4.75$ GeV, dotted lines: $m_c = 1.8$, $m_b = 5$ GeV, dashed lines: $m_c = 1.2$, $m_b = 4.5$ GeV. Two lines indicate scale ambiguities as given in Eq. (11.62). Adopted from [663, 671].

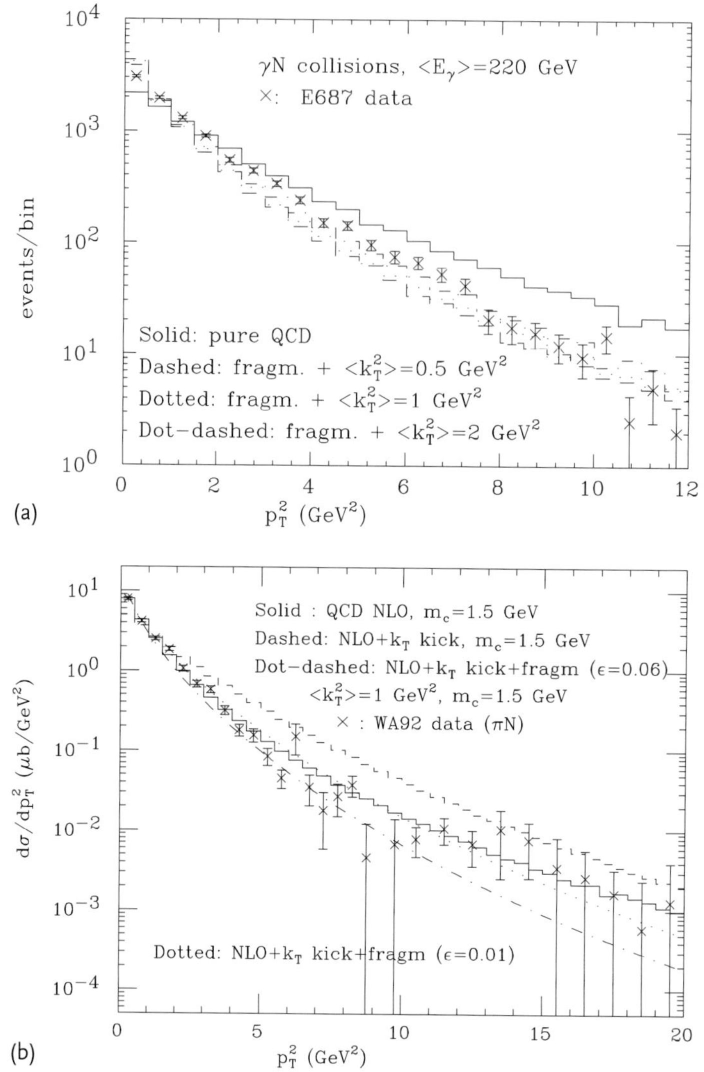

Figure 11.30 (a) $d\sigma/dp_{\mathrm{T}}$ for charm production by γN: experimental data compared to NLO QCD prediction (m_c = 1.5 GeV). Solid: pure QCD. Dashed: fragmentation + $\langle k_{\mathrm{T}}^2\rangle = 0.5\,\mathrm{GeV}^2$. Dotted: fragmentation + $\langle k_{\mathrm{T}}^2\rangle = 1\,\mathrm{GeV}^2$. Dot-dashed: fragmentation + $\langle k_{\mathrm{T}}^2\rangle = 2\,\mathrm{GeV}^2$. (b) Charm production by πN reaction compared with theory. Solid line: NLO QCD $m_c = 1.5$ GeV. Dashed: NLO + k_{T} kick. Dot-dashed: NLO + k_{T} kick + fragmentation ($\epsilon = 0.06$). Dotted line: NLO + k_{T} kick + fragmentation ($\epsilon = 0.01$). ε is the Peterson fragmentation function parameter [663].

Other curves include effects of the fragmentation and intrinsic transverse momentum k_{T} and they agree with the data. Here, $\epsilon = 0.01$ was used for the Peterson function. Notice that the fragmentation effect is essential here and the k_{T} effect is minor.

Figure 11.30b shows data due to hadroproduction. One sees that the data agrees with the solid line which is a pure NLO QCD prediction. However, considering the previous discussion that the fragmentation as well as the k_T effect were necessary to bring the theory near to the experimental data, it is not necessarily a good sign. As one can see, inclusion of k_T kick makes the distribution harder (dashed curve) and further inclusion of the fragmentation effect (dot-dashed line) takes it back to the original position. Here, the k_T kick and the fragmentation almost compensate each other. The intrinsic transverse momentum of the quarks has been known to be about $\sim$ 700 MeV from the study of the Drell–Yan process. Here, $k_T^2 \sim 1\,\text{GeV}^2$ gives good agreement which is reasonable because the gluon is expected to have a broader k_T distribution than the quarks. The reason that the photo-production data is less sensitive to the choice of k_T^2 is because only one of the incident particles has intrinsic distribution. For the fragmentation function, if one adopts a value of $\epsilon_c = 0.01$ for the Peterson distribution function (dotted curve), the agreement is even better.

b-Quark Production Knowing the prescription to adjust the theory to experiments, and with heavier mass which makes the perturbative QCD more reliable, we expect that bottom quark distributions can be reproduced by theory quite well. Figure 11.31a shows the p_T^2 distributions for the b quark (what was actually measured is $\mu\mu$ pair from J/ψ which in turn decayed from the b-quark) compared with FONLL [674, 676] and confirms our expectations.[9)]

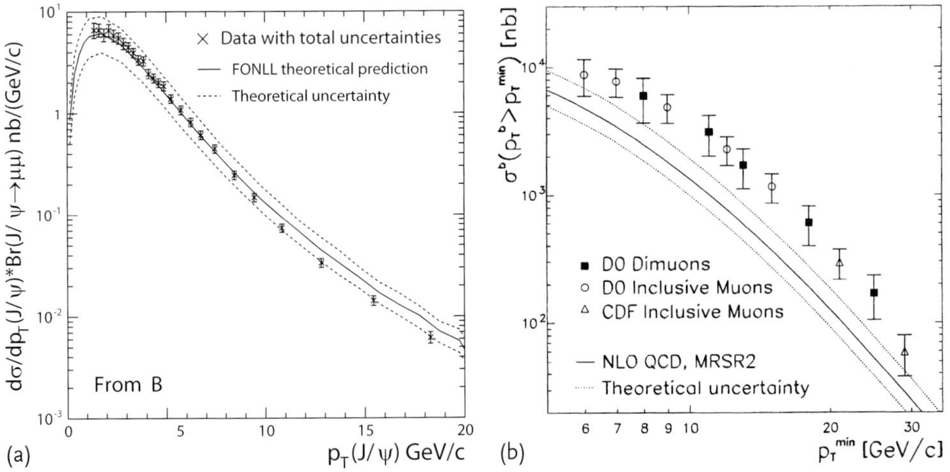

Figure 11.31 (a) J/ψ spectrum from B decays. The theory represents the FONLL. The band shows the ambiguity due to the scale dependence [674–676]. (b) Old fittings were a factor $2 \sim 3$ lower than data [675, 680].

9) FONLL means: NLO ($O(\alpha_s^2) + O(\alpha_s^3)$) calculation + resummation of large $\ln p_T^2/M^2$ [682].

Lessons to be Learned Perhaps it is worth mentioning that there has been a 15 year saga to reach the agreement [675]. Figure 11.31b shows old fittings to data which typically show a factor 2–3 higher than theoretical predictions. The reason the theory underestimated the cross section was mainly due to incorrect use of the fragmentation function [675, 683]. The p_T distribution for heavy quark production varies as $\hat{p}_T^{-N}(N = 4\text{--}5)$ (see Eq. (11.60)) where $\hat{p}_T$ is the transverse momentum of the produced quark. Then, referring to Eq. (11.63), the hadron distribution behaves like

$$\frac{d\sigma}{dp_T} \sim \int \frac{dz}{z}\left(\frac{z}{p_T}\right)^N D(z) \sim D_N \frac{d\sigma}{d\hat{p}_T},$$
$$D_N = \int \frac{dz}{z} z^N D(z) = \langle x_E^{N-1} \rangle \tag{11.64}$$

where $\langle x_E^{N-1} \rangle$ is the average value of the fractional momentum x_E. The experimentally obtained fragmentation function is depicted in Figure 11.32a [211]. The Peterson function is depicted with a real line. Figure 11.32b plots the moment $D_N = \langle x_E^{N-1} \rangle$. Here, we are concerned with the $N = 5$ moment relevant to the heavy quark production. Note that one calculated from the accurate experimental data is significantly larger compared to that obtained from the Peterson function with $\epsilon_b = 0.006$ which is a standard built-in function for many hadronization programs. The Peterson function reproduces the overall shape of the real (i.e., experimental) fragmentation distribution adequately, which is the reason it is used as a standard in many models. However, as we mentioned, the heavy quark production is sensitive to the high moment of the function which is considerably modified if the real or properly parametrized function is not used. Furthermore, the experimental group did not present their hadron data as they are, but those after the b-quark distribution is deconvoluted. To deconvolute the data and convert the hadron distribution to the parton distribution, the fragmentation function is used. Therefore, the published experimental data are not raw data which theorists can use to compare with their calculations. Lessons we have learned suggest that we should remember the following points in interpreting data.

1. The total cross section is often obtained from measurements of a limited phase space. One uses a (theoretical) model to convert the measured data to the total cross section.
2. The experimental data are presented often not as raw data but as deconvoluted parton distributions. In deconvolution, one kind or other model of the fragmentation function is used which necessarily deforms the original distribution.
3. The hadron production is often sensitive to a very limited area of
   a) Parton distribution function: and
   b) Fragmentation function:

In fitting the data with QCD predictions, the major problem was matching the parton level calculation to that of hadrons as was emphasized repeatedly. In obtaining

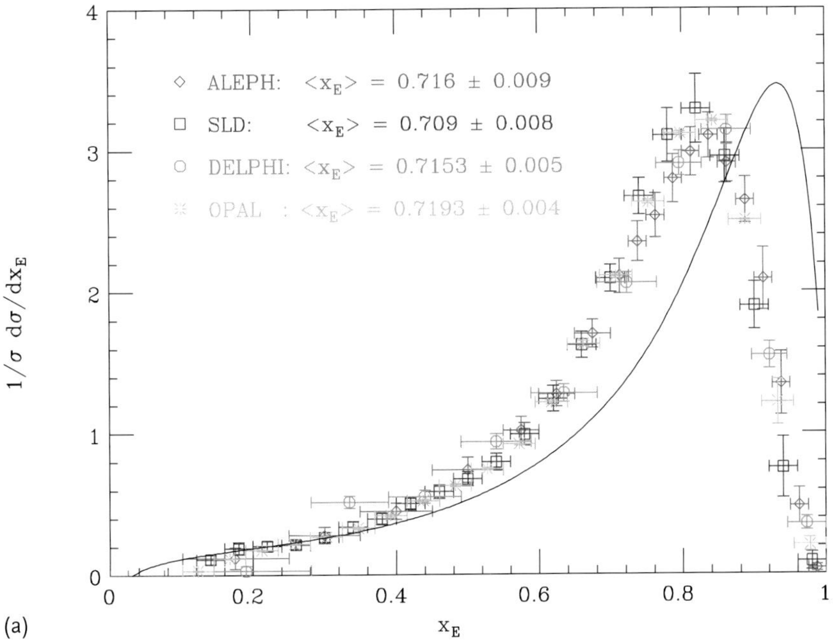

(a)

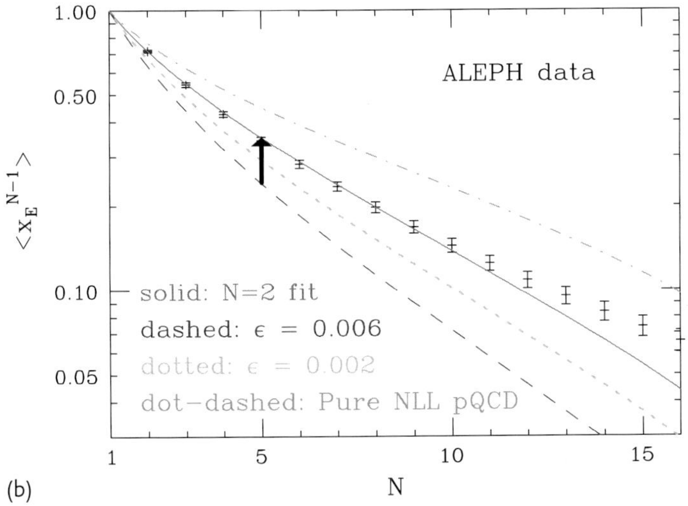

(b)

Figure 11.32 Nonperturbative effect in the heavy quark production. (a) Experimentally obtained bottom fragmentation function $D(z)$ as contrasted to the conventional Peterson function Peterson function (real line) [3]. (b) Moments of the fragmentation function. Points are calculated using the ALEPH [211] data. The solid line uses a parametrized fragmentation function $(\alpha + 1)(\alpha + 2)z^{\alpha}(1 - z)$ with $\alpha = 2M/\Lambda_{\overline{MS}}$ which is good to $N \sim 10$. The dashed line is derived from the Peterson function with the standard value $\epsilon = 0.006$ [683].

the hadron distributions, one uses the parton distribution function for the initial parton and fragmentation function to convert the parton distribution to that of hadrons. The former is mainly obtained from the deep inelastic (DIS) data and the latter from $e\bar{e}$ reactions. The data in hadronic collisions are sensitive to different region of the phase space as opposed to DIS or $e\bar{e}$. The input data does not necessarily have the desired accuracy in the phase space region relevant to the hadronic collision.

11.9.3 Top Quark Production

The top quark was the last of the quarks in the Standard Model and its discovery has been the prime target of the high energy physicists since the discovery of the bottom quark. There were several compelling reasons to believe in the existence of the top quark.

1. GIM mechanism (see Vol. 1, Sect. 15.7.1) for the nonexistence of the flavor changing neutral current requires the b-quark to be a member of an isodoublet, and hence the existence of a partner. The Kobayashi–Maskawa model as the cause of the CP violation required three generations of the quarks.
2. Requirement of no anomaly. For the Standard model to be renormalizable, there should be no triangular anomaly (see Vol. 1, Sect. 19.2.2). In the Standard model, the requirement translates to

$$\sum Q_i = 0 \tag{11.65}$$

   for each generation. The first and second generation satisfy the condition. For the third generation, Eq. (11.65) means a quark with $Q = +2/3$ must exist.
3. Using asymmetry data of $e^- e^+ \to b\bar{b}$ and decay width of Z, the Standard Model can determine the isospin value of the b-quark which clearly showed that $I_{3Lb} = -1/2$, $I_{3Rb} = 0$ (see Figure 2.23). This implies that the b-quark must have a weak isospin partner, that is, the top quark with $I_{3L}(t) = +1/2$ must exist.

However, the mass could not be predicted. TRISTAN ($e\bar{e}$ collider at $\sqrt{s} = 60$ GeV, 1984–1996) did not find it. It was not found in $S\bar{p}pS$ ($\bar{p}p$ collider at $\sqrt{s} = 570 \to 630$ GeV, 1981–1984) either.[10] Then, when the ARGUS group found a large value of the B^0–$\overline{B}^0$ mixing parameter $x_d = \Delta m_d/\Gamma_d \simeq 0.8$ [282], it was realized that the top mass is much heavier than previously expected (see Eq. (6.174)). Therefore, even LEP ($e\bar{e}$ collider at $\sqrt{s} = 91 \to 200$ GeV, 1989–2000) did not have energy high enough to produce it. Finally, it was discovered at Tevatron ($p\bar{p}$ collider at $1.8 \to 1.96$ TeV) [685, 686]. The mass turned out to be $m_t \sim 172$ GeV.

10) Once the UA1 group claimed discovery of the top quark with $30 < m_t < 50$ GeV in the process $W \to t\bar{b}$ followed by $t \to bl\nu$ [684].

Figure 11.33 shows a typical event display of the top production [33]. The event is interpreted as the process

$$
\begin{aligned}
&q+\overline{q}\to t+\overline{t}\\
&\quad t\to W^{+}+b(\text{tagged jet \# 4})\,,\quad W^{+}\to e^{+}+\nu\\
&\quad \overline{t}\to W^{-}+\overline{b}(\text{tagged jet \# 1}),\quad W^{-}\to q\overline{q}'(\text{jet \# 2,\# 3})
\end{aligned}
\tag{11.66}
$$

The two jets from secondary vertices are identified as b-quarks. The Lego plot shows energies of the electron and jets together with that of a neutrino which were calculated from kinematical constraints.

Cross Section Figure 11.34 shows the production data point and theoretical predictions [664, 692]. The NLO correction to LO is small, except near threshold. In Figure 11.34a, theories include part of the NNLO contribution. The data agree with theoretical predictions within errors.

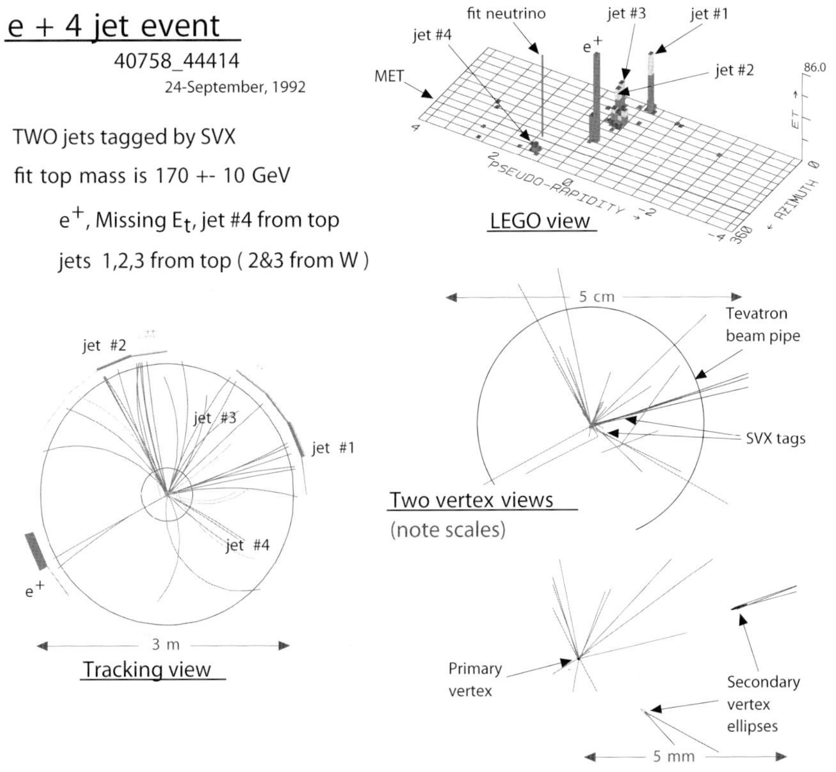

Figure 11.33 Top Events Display. $q\overline{q}\to t\overline{t}\to e^{+}+4$ jets. Two jets from secondary vertices are tagged as *b*-quarks. For a color version of this figure, please see the color plates at the beginning of the book. Figure from CDF/FNAL [CDFwebtop] [33].

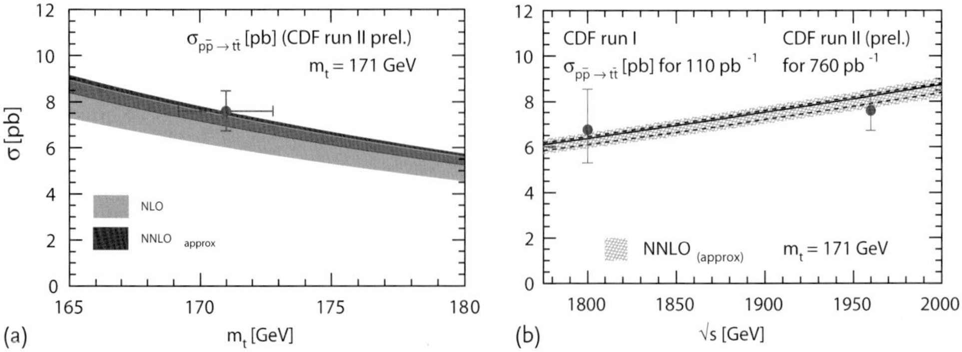

Figure 11.34 Production cross section of $t\overline{t}$ in $p\overline{p}$ collision at Tevatron (1.96 TeV) as a function of (a) m_t [688] and (b) total energy [687]. The data by CDF [689] and DO [690] (not shown) are compared to theoretical predictions [664, 691]. (Figures adopted from [688] and [691].)

Topology of the Top Events Decay modes that can be used to identify the t quark through process $t \to b + W$ are

$$t \to b + W^{+} \tag{11.67a}$$

$$W \to u\overline{d}, u\overline{s}, c\overline{d}, c\overline{s} \quad \text{branching ratio} \sim 2/3 \tag{11.67b}$$

$$W \to e^{+}\nu, \mu^{+}\nu, \tau^{+}\nu \quad \text{branching ratio} \sim 3 \times 1/9 \tag{11.67c}$$

When the b-quark decays, it produces jets or a lepton + jet. The approximate branching ratio of each channel can be obtained by simple quark counting.

$$t\overline{t} \to b\overline{b} + 4\text{ jets} \quad 36/81 \tag{11.68a}$$

$$\to b\overline{b} + 2\text{ jets } + l\nu \quad 36/81 \tag{11.68b}$$

$$\to b\overline{b} + l_1\nu + l_2\nu(l_1 \neq l_2) \quad 6/81 \tag{11.68c}$$

$$\to b\overline{b} + l_1\nu + l_2\nu(l_1 = l_2) \quad 3/81 \tag{11.68d}$$

The mode that does not contain leptons Eq. (11.68a) is difficult to identify because of large backgrounds, but those containing at least one energetic lepton is relatively easy. The decay lepton exhibits characteristics specific to its origin and easy to identify whether it came from t or b, c. Leptons that decayed from the top (via W) has transverse momentum ranging between $m_W/2$ and $m_t/2$ in the t rest frame. After Lorentz boost along the top flight direction, it still retains most of p_T relative to the t-jet axis. Therefore, it has characteristics that (1) it has a large transverse momentum relative to the beam axis and (2) that it is isolated from other jets on the η–ϕ plane.

The "isolated lepton" can be defined as one whose cone has radius $\Delta R = \sqrt{\Delta\phi^2 + \Delta\eta^2}$ on the lepton momentum axis contains little energy associated with a jet (see Figure 11.35a). Leptons that decayed from $b(c)$ has $p_T \sim m_b/3(m_c/3)$ relative to the parent jet axis. In the low energy $e\bar{e}$ reactions investigating the neutral current effect, leptons having large p_T was used to identify $b(c)$ quarks. However, here, b, c-quarks which decayed from the t-quark have large energy and because of the Lorentz boost, one can safely assume that the jet and the decay lepton are emitted in the same direction. Namely, the lepton resides either inside the jet or nearby and is not isolated from the parent jet (Figure 11.35b). Furthermore, the transverse momentum with respect to the beam axis is much smaller than that from the t decay.

In view of the above arguments, a signal for the production of $t\bar{t}$ consists of

1. one or two isolated leptons,
2. existence of a neutrino (missing transverse energy),
3. existence of more than two jets,
4. existence of two b-quarks.

The reason we talk of transverse momentum with respect to the beam axis is due to presence of beam jets. They are produced by spectator partons which did not participate in the scattering. Many of them go through the beam pipe and are not registered by the detector. However, in the collider experiments, the total momentum must vanish and this is also true for the transverse momentum. If the solid angle coverage is (nearly) hermetic, one can assume that the missing energy was carried away by the neutrino provided there is only one missing particle.

The b-quark can be identified using its finite lifetime ($\sim 10^{-12}$ s). Because of the Lorentz boost, it usually decays at a distance from the interaction point which is large enough ($\gtrsim$ 10–1000 μm) to be measured by solid state detectors. Figure 11.33 shows two identified b-quarks by detecting the decay vertices. One can also use the fact that the decay lepton has small transverse momentum relative to the jet axis if the decay vertex is not identified.

Mass of the Top Quark The cleanest signal can be obtained by requiring two leptons. However, the branching ratio is low and the first event was identified by the decay mode containing one lepton + n jets (at least one of them to be identified

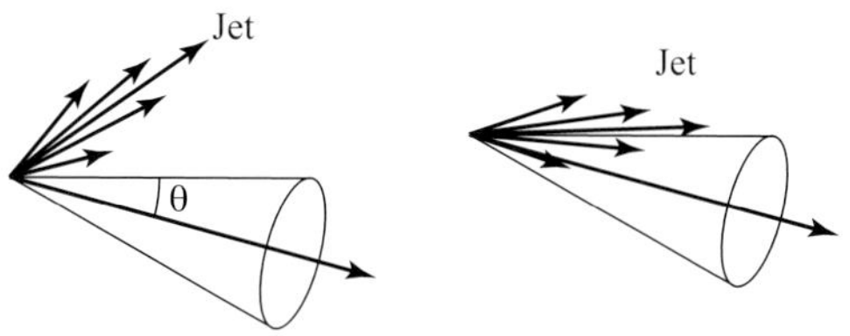

(a) Isolated Lepton (b) Nonisolated Lepton

Figure 11.35 Making a cone of radius $\Delta R = \sqrt{\Delta\phi^2 + \Delta\eta^2}$ or half angle θ with the lepton on its axis, one measures the energy sum $E_T \equiv \sum E_{Ti}$ of particles inside the cone. The lepton is isolated if $E_T \lesssim E_{iso}$ and not if $E_T \gtrsim E_{iso}$

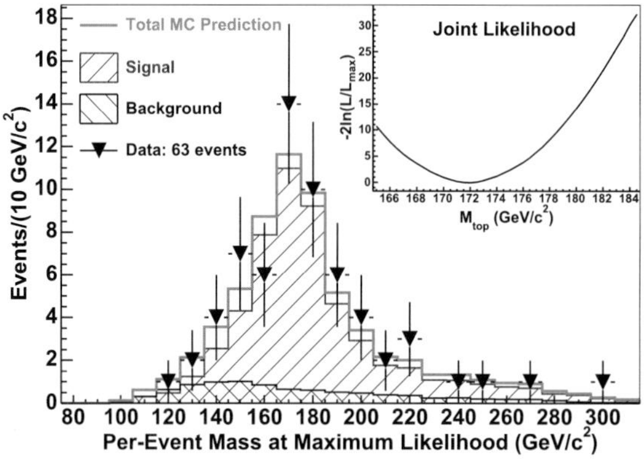

Figure 11.36 The value of the top-quark mass is plotted for each event. Data events (points) are compared to an expected distribution (histogram) comprising simulated $t\bar{t}(m_{\text{top}} = 172.5\,\text{GeV}/c^2)$ and background events. The last bin includes events with masses $> 305\,\text{GeV}/c^2$. The inset shows the joint log likelihood for the 63 events in the figure, before accounting for the presence of background [693].

as the b-quark). Major backgrounds from electroweak process ($W + n$jets) can be reduced by requiring $n \geq 3$. Among the $t\bar{t}$ production processes, "lepton + 4 jets" events ($p\bar{p} \to t\bar{t}$, $t(\bar{t}) \to b(\bar{b})\,W$, $W_1 \to l\nu$, $W_2 \to q\bar{q}'$) can be used to reconstruct the top mass. Figure 11.36 shows a distribution of the top mass [693, 696].

From this distribution, the top quark mass was determined to be $173.2^{+2.6}_{2.4}(\text{stat}) \pm 3.2(\text{sys})$ GeV. Using other decay modes and combining CDF and D0 measurements, the most up-to-date world average gives

$$m_t = 172.0 \pm 1.3\,\text{GeV}/c^2 \quad [3, 688] \tag{11.69}$$

The measured value agrees quite well with an inferred value using the precise electroweak measurements and fitted with theories including the top loop corrections (see Figures 5.8 and 5.9).[11)]

The discovery of the top quark with the predicted mass value is a major triumph of the Standard Model and wiped out any doubt as to the validity of the Standard Model. What remains to be done is to elucidate causes of phenomena that are not explained in the framework of the Standard Model. The flavor structure of the quarks and leptons and dynamic origin of the CP violation remain to be understood. The Higgs mechanism is the underlying base of the Standard Model. It was discovered at LHC with mass 126 GeV [694, 695]. Pursuit of its dynamics will lead to a clue to a new physics beyond the Standard Model.

11) The electroweak radiative corrections include contributions from the top and Higgs loops. For reasonable assumption of the Higgs mass value, for instance, $m_H \sim m_Z - 2m_Z$, the top mass can be indirectly derived

$$m_{\text{top}} = 178.9^{+11.7}_{-8.6}\,\text{GeV}/c^2 \quad [688] \quad (11.71)$$

solely from data on the Z-pole and W production data. The value is rather insensitive to the Higgs mass because of the log dependence.

Appendix A
Gamma Matrix Traces and Cross Sections

Notations and Conventions

Metric tensor:	$g_{\mu\nu} = g^{\mu\nu} = \begin{bmatrix} 1 & 0 & 0 & 0 \\ 0 & -1 & 0 & 0 \\ 0 & 0 & -1 & 0 \\ 0 & 0 & 0 & -1 \end{bmatrix}$	
Contravariant coordinate:	$x^\mu = (x^0, x^1, x^2, x^3) = (t, x, y, z) = (t, \boldsymbol{x})$	(A.1)
Covariant coordinate:	$x_\mu = g_{\mu\nu} x^\nu = (t, -\boldsymbol{x})$	
Scalar product:	$A \cdot B = A_\mu B^\mu = A_\mu g^{\mu\nu} B_\nu = A_0 B_0 - \boldsymbol{A} \cdot \boldsymbol{B}$	
Derivatives:	$\partial^\mu = \frac{\partial}{\partial x_\mu} = \left(\frac{\partial}{\partial t}, -\nabla\right), \quad \partial_\mu = \frac{\partial}{\partial x^\mu} = \left(\frac{\partial}{\partial t}, \nabla\right)$	
Four derivatives:	$\partial_\mu A^\mu = \partial^\mu A_\mu$	

Plane Wave Solution The plane wave solutions of the Dirac equation

$$\begin{aligned} &(\gamma^\mu i\partial_\mu - m)\,\psi(x) = 0 \\ &\psi(x) = \frac{1}{(2\pi)^3} \int \frac{d^3p}{2E} \left[u(p)e^{-ip\cdot x} + \overline{v}(p)e^{ip\cdot x}\right] \end{aligned} \tag{A.2}$$

in the chiral representation (see Eq. (A.7)) are expressed as

$$u_r(p) = \begin{bmatrix} \sqrt{p\cdot\sigma}\,\xi_r \\ \sqrt{p\cdot\overline{\sigma}}\,\xi_r \end{bmatrix} = \sqrt{\frac{E+m}{2}} \begin{bmatrix} \left(1 - \frac{\boldsymbol{\sigma}\cdot\boldsymbol{p}}{E+m}\right)\xi_r \\ \left(1 + \frac{\boldsymbol{\sigma}\cdot\boldsymbol{p}}{E+m}\right)\xi_r \end{bmatrix} \tag{A.3a}$$

$$v_r(p) = \begin{bmatrix} \sqrt{p\cdot\sigma}\,\eta_r \\ -\sqrt{p\cdot\overline{\sigma}}\,\eta_r \end{bmatrix} = \sqrt{\frac{E+m}{2}} \begin{bmatrix} \left(1 - \frac{\boldsymbol{\sigma}\cdot\boldsymbol{p}}{E+m}\right)\eta_r \\ -\left(1 + \frac{\boldsymbol{\sigma}\cdot\boldsymbol{p}}{E+m}\right)\eta_r \end{bmatrix} \tag{A.3b}$$

$$\xi_1 = \begin{bmatrix} 1 \\ 0 \end{bmatrix}, \quad \xi_2 = \begin{bmatrix} 0 \\ 1 \end{bmatrix}, \quad \eta_r = i\sigma_2 \xi_r^* = \begin{bmatrix} 0 & 1 \\ -1 & 0 \end{bmatrix} \xi_r^* \tag{A.3c}$$

Elementary Particle Physics, Volume 2: Foundations of the Standard Model, First Edition. Yorikiyo Nagashima.

They satisfy the following equations.

$$
\begin{aligned}
&\not{p} \equiv p_\mu \gamma^\mu \\
&(\not{p} - m)u(p) = 0, \quad (\not{p} + m)v(p) = 0 \\
&\overline{u}(p)(\not{p} - m) = 0, \quad \overline{v}(p)(\not{p} + m) = 0 \\
&u_s^\dagger(p)u_r(p) = \quad v_s^\dagger(p)v_r(p) = 2E\delta_{rs} \\
&\overline{u}_s(p)u_r(p) = -\overline{v}_s(p)v_r(p) = 2m\delta_{rs} \\
&\overline{u}_s(p)v_r(p) = \quad \overline{v}_s(p)u_r(p) = 0 \\
&\sum_{r=\pm 1/2} u_r\overline{u}_r(p) = \not{p} + m, \qquad \sum_{r=\pm 1/2} v_r\overline{v}_r(p) = \not{p} - m
\end{aligned} \tag{A.4}
$$

Dirac γ Matrices

$$
\begin{aligned}
&\gamma^\mu\gamma^\nu + \gamma^\nu\gamma^\mu = 2g^{\mu\nu} \\
&\gamma^5 = i\gamma^0\gamma^1\gamma^2\gamma^3 = -\frac{i}{4!}\sum \varepsilon_{\mu\nu\rho\sigma}\gamma^\mu\gamma^\nu\gamma^\rho\gamma^\sigma \\
&(\gamma^5)^2 = 1, \quad \gamma^5\gamma^\mu + \gamma^\mu\gamma^5 = 0 \\
&\sigma^{\mu\nu} = \frac{i}{2}[\gamma^\mu, \gamma^\nu], \quad \sigma^{0i} = i\alpha^i, \quad \sigma^{ij} = \varepsilon_{ijk}\sigma^k \\
&\gamma^0(\gamma^\mu)^\dagger\gamma^0 = \gamma^\mu, \quad \gamma^0(\gamma^5)^\dagger\gamma^0 = -\gamma^5 \\
&\gamma^0(\gamma^5\gamma^\mu)^\dagger\gamma^0 = \gamma^5\gamma^\mu, \quad \gamma^0(\sigma^{\mu\nu})^\dagger\gamma^0 = \sigma^{\mu\nu}
\end{aligned} \tag{A.5a}
$$

$$
\begin{aligned}
&\gamma^5\gamma_\sigma = -\frac{i}{3!}\varepsilon_{\mu\nu\rho\sigma}\gamma^\mu\gamma^\nu\gamma^\rho \\
&\gamma^\mu\gamma^\nu\gamma^\rho = g^{\mu\nu}\gamma^\rho - g^{\mu\rho}\gamma^\nu + g^{\nu\rho}\gamma^\mu - i\gamma^5\epsilon^{\mu\nu\rho\sigma}\gamma_\sigma
\end{aligned} \tag{A.5b}
$$

Charge Conjugation Matrices

$$
\begin{aligned}
&C = i\gamma^2\gamma^0, \quad C^T = C^\dagger = -C, \quad CC^\dagger = C^\dagger C = 1, \quad C^2 = -1 \\
&C^{-1}\gamma^\mu C = -\gamma^{\mu T}, \quad C^{-1}\gamma^5 C = \gamma^{5T} \\
&C^{-1}\gamma^5\gamma^\mu C = (\gamma^5\gamma^\mu)^T, \quad C^{-1}\sigma^{\mu\nu}C = -\sigma^{\mu\nu T}
\end{aligned} \tag{A.6}
$$

Chiral (Weyl) Representation

$$
\boldsymbol{\alpha} = \begin{bmatrix} -\boldsymbol{\sigma} & 0 \\ 0 & \boldsymbol{\sigma} \end{bmatrix}, \quad \beta = \gamma^0 = \begin{bmatrix} 0 & 1 \\ 1 & 0 \end{bmatrix}, \quad \boldsymbol{\gamma} = \beta\boldsymbol{\gamma} = \begin{bmatrix} 0 & \boldsymbol{\sigma} \\ -\boldsymbol{\sigma} & 0 \end{bmatrix} \tag{A.7a}
$$

$$
\gamma^5 = \begin{bmatrix} -1 & 0 \\ 0 & 1 \end{bmatrix}, \quad C = \begin{bmatrix} i\sigma^2 & 0 \\ 0 & -i\sigma^2 \end{bmatrix} \tag{A.7b}
$$

Pauli–Dirac Representation

$$\boldsymbol{\alpha} = \begin{bmatrix} 0 & \boldsymbol{\sigma} \\ \boldsymbol{\sigma} & 0 \end{bmatrix}, \quad \beta = \gamma^0 = \begin{bmatrix} 1 & 0 \\ 0 & -1 \end{bmatrix}, \quad \boldsymbol{\gamma} = \beta\boldsymbol{\alpha} = \begin{bmatrix} 0 & \boldsymbol{\sigma} \\ -\boldsymbol{\sigma} & 0 \end{bmatrix} \qquad \text{(A.8a)}$$

$$\gamma^5 = \begin{bmatrix} 0 & 1 \\ 1 & 0 \end{bmatrix}, \quad C = i\gamma^2\gamma^0 = \begin{bmatrix} 0 & -i\sigma^2 \\ -i\sigma^2 & 0 \end{bmatrix} \qquad \text{(A.8b)}$$

Writing wave functions in the Weyl and Pauli–Dirac representations as Φ_W, Φ_D, they are mutually connected by

$$\Phi_D = S\Phi_W, \quad \gamma_D = S\gamma_W S^{-1}, \quad S = \frac{1}{\sqrt{2}}\begin{bmatrix} 1 & 1 \\ -1 & 1 \end{bmatrix} \qquad \text{(A.9)}$$

Traces of the γ Matrices

$$\begin{aligned}
&\gamma_\mu \gamma^\mu = 4 \\
&\gamma_\mu \not{A} \gamma^\mu = -2\not{A} \\
&\gamma_\mu \not{A}\not{B} \gamma^\mu = 4(A \cdot B) \\
&\gamma_\mu \not{A}\not{B}\not{C} \gamma^\mu = -2\not{C}\not{B}\not{A} \\
&\gamma_\mu \not{A}\not{B}\not{C}\not{D} \gamma^\mu = 2[\not{D}\not{A}\not{B}\not{C} + \not{C}\not{B}\not{A}\not{D}]
\end{aligned} \qquad \text{(A.10a)}$$

$$\begin{aligned}
&\text{Tr}[1] = 4, \quad \text{Tr}[\gamma^\mu] = \text{Tr}[\gamma^5] = 0 \\
&\text{Tr}[\gamma^{\mu_1}\gamma^{\mu_2}\cdots\gamma^{\mu_{2n+1}}] = 0 \\
&\text{Tr}[\gamma^{\mu_1}\gamma^{\mu_2}\cdots\gamma^{\mu_n}] = (-1)^n \text{Tr}[\gamma^{\mu_n}\cdots\gamma^{\mu_2}\gamma^{\mu_1}] \\
&\text{Tr}[\gamma^\mu\gamma^\nu] = 4g^{\mu\nu} \\
&\text{Tr}[\gamma^\mu\gamma^\nu\gamma^\rho\gamma^\sigma] = 4[g^{\mu\nu}g^{\rho\sigma} + g^{\mu\sigma}g^{\nu\rho} - g^{\mu\rho}g^{\nu\sigma}] \\
&\text{Tr}[\gamma^5\gamma^\mu\gamma^\nu\gamma^\rho\gamma^\sigma] = -4i\epsilon^{\mu\nu\rho\sigma} = 4i\varepsilon_{\mu\nu\rho\sigma} \\
&\text{Tr}[\gamma^5] = \text{Tr}[\gamma^5\gamma^\mu] = \text{Tr}[\gamma^5\gamma^\mu\gamma^\nu] = \text{Tr}[\gamma^5\gamma^\mu\gamma^\nu\gamma^\rho] = 0
\end{aligned} \qquad \text{(A.10b)}$$

LeviCivita Antisymmetric Tensor

$$\varepsilon^{ijk} = \varepsilon_{ijk} = \begin{cases} +1 & ; ijk = \text{even permutation of 123} \\ -1 & ; ijk = \text{odd permutation of 123} \\ 0 & ; \text{if any two indices are the same} \end{cases} \qquad \text{(A.11a)}$$

$$\varepsilon^{\mu\nu\rho\sigma} = -\varepsilon_{\mu\nu\rho\sigma} = \begin{cases} +1 & ; \mu\nu\rho\sigma = \text{ even permutation of 0123} \\ -1 & ; \mu\nu\rho\sigma = \text{ even permutation of 0123} \\ 0 & ; \text{if any two indices are the same} \end{cases} \qquad \text{(A.11b)}$$

$$
\begin{aligned}
\varepsilon^{\mu\nu\rho\sigma}\varepsilon^{\alpha\beta\gamma\delta} &= -\det[g^{\lambda\lambda'}], \quad \lambda = \mu\nu\rho\sigma, \quad \lambda = \alpha\beta\gamma\delta \\
\varepsilon^{\alpha\nu\rho\sigma}\varepsilon_{\alpha}{}^{\beta\gamma\delta} &= -\det[g^{\lambda\lambda'}], \quad \lambda = \nu\rho\sigma, \quad \lambda = \beta\gamma\delta \\
\varepsilon^{\alpha\beta\rho\sigma}\varepsilon_{\alpha\beta}{}^{\gamma\delta} &= -2(g^{\rho\gamma}g^{\sigma\delta} - g^{\rho\delta}g^{\sigma\gamma}) \\
\varepsilon^{\alpha\beta\gamma\sigma}\varepsilon_{\alpha\beta\gamma}{}^{\delta} &= -6g^{\sigma\delta} \\
\varepsilon^{\alpha\beta\gamma\delta}\varepsilon_{\alpha\beta\gamma\delta} &= -24
\end{aligned} \tag{A.11c}
$$

Cross Section Formula The scattering amplitude and the cross section in the CM frame are given by

$$
\begin{aligned}
\langle f|S - \delta_{fi}|1,2\rangle &= -i(2\pi)^4\delta^4\left(p_1 + p_2 - \sum_f p_f\right)\mathcal{M}_{fi} \\
d\sigma &= \frac{|\mathcal{M}_{fi}|^2}{2s\lambda(1, x_1, x_2)} dLIPS \\
\lambda(1, x, y) &= 1 + x^2 + y^2 - 2x - 2y - 2xy, \quad x_1 = \frac{m_1^2}{s}, \quad x_2 = \frac{m_2^2}{s} \\
dLIPS &= (2\pi)^4\delta^4\left(p_1 + p_2 - \sum_f p_f\right)\prod_f \frac{d^3 p_f}{(2\pi)^3 2E_f} \\
\left.\frac{d\sigma}{d\Omega}\right|_{\mathrm{CM}}(1 + 2 \to 3 + 4) &= \frac{p_3}{p_1}\frac{1}{64\pi^2 s}|\mathcal{M}_{fi}|^2
\end{aligned} \tag{A.12}
$$

Scattering of Polarized Electrons by Polarized Muons

$$
\begin{aligned}
&e_a(p_1) + \mu_b(p_2) \to e_a(p_3) + \mu_b(p_4) \quad a, b = L, R \\
&-i\mathcal{M}_{fi} = [\overline{u}_b(p_4)(-ie\gamma^{\mu})u_b(p_2)]\frac{-ig_{\mu\nu}}{q^2}[\overline{u}_a(p_3)(-ie\gamma^{\nu})u_a(p_1)]
\end{aligned} \tag{A.13}
$$

Spin Sum of $|\mathcal{M}_{fi}|^2$

$$
\begin{aligned}
K^{\mu\nu}(p', p) &= \sum_{r,s=1,2}[\overline{u}_s(p')\gamma^{\mu}(a - b\gamma^5)u_r(p)][\overline{u}_s(p')\gamma^{\nu}(a - b\gamma^5)u_r(p)]^* \\
&= \mathrm{Tr}[\gamma^{\mu}(a - b\gamma^5)(\not{p} + m)(a^* - b^*\gamma^5)\gamma^{\nu}(\not{p}' + m')] \\
&= 4[(|a|^2 + |b|^2)\{p'^{\mu}p^{\nu} + p'^{\nu}p^{\mu} - g^{\mu\nu}(p' \cdot p)\} \\
&\quad + (|a|^2 - |b|^2)g^{\mu\nu}m'm - 2i\mathrm{Re}[ab^*]\varepsilon^{\mu\nu\rho\sigma}p'_{\rho}p_{\sigma}
\end{aligned} \tag{A.14}
$$

For anti-particles, replace m or m' or both $\to -m, -m'$.

LL, LR, RL, RR scattering (L: $a = b = 1/2$, R: $a = -b = 1/2$)

$$
\begin{aligned}
K_L^{\mu\nu}(p_3, p_1) &= 2\left[\{p_3^\mu p_1^\nu + p_3^\nu p_1^\mu - g^{\mu\nu}(p_3 p_1)\} - \varepsilon^{\mu\nu\rho\sigma} p_{3\rho} p_{1\sigma}\right] \\
K_R^{\mu\nu}(p_3, p_1) &= 2\left[\{p_3^\mu p_1^\nu + p_3^\nu p_1^\mu - g^{\mu\nu}(p_3 p_1)\} + \varepsilon^{\mu\nu\rho\sigma} p_{3\rho} p_{1\sigma}\right] \\
K_L^{\mu\nu}(p_3, p_1) K_{L\mu\nu}(p_4, p_2) &= K_R^{\mu\nu}(p_3, p_1) K_{R\mu\nu}(p_4, p_2) \\
&= 16(p_1 \cdot p_2)(p_3 \cdot p_4) = 4(s - m^2 - M^2)^2 \\
&\simeq 4s^2, \quad s \gg m^2, M^2 \\
K_L^{\mu\nu}(p_3, p_1) K_{R\mu\nu}(p_4, p_2) &= K_R^{\mu\nu}(p_3, p_1) K_{L\mu\nu}(p_4, p_2) \\
&= 16(p_1 \cdot p_4)(p_2 \cdot p_3) = 4(u - m^2 - M^2)^2 \\
&\simeq 4u^2, \quad u \gg m^2, M^2
\end{aligned}
\tag{A.15}
$$

$$
d\sigma_{LL} = d\sigma_{RR} = \frac{\alpha^2}{q^4} s, \quad d\sigma_{LR} = d\sigma_{RL} = \frac{\alpha^2}{q^4} \frac{u^2}{s} \tag{A.16}
$$

The cross section for scattering of the unpolarized electron beam by unpolarized muon targets are the average of the initial polarization and sum of final polarization.

$$
\begin{aligned}
d\sigma &= \frac{1}{4}(d\sigma_{LL} + d\sigma_{LR} + d\sigma_{RL} + d\sigma_{RR}) \\
&= \frac{\alpha^2}{q^4} \frac{s^2 + u^2}{2s} = \frac{\alpha^2}{q^4} s \frac{1 + \left(\frac{1+\cos\theta^*}{2}\right)^2}{2}
\end{aligned}
\tag{A.17}
$$

Polarization Sum of the Vector Particle Four polarization vectors $\varepsilon^\mu(\lambda)$, $(\lambda = 0 \sim 3)$ in the rest frame of the particle are defined by

$$
\begin{aligned}
\varepsilon^\mu(0) &= (1, 0, 0, 0): \quad \text{scalar polarization} \\
\varepsilon^\mu(1) &= (0, 1, 0, 0): \quad \text{transverse polarization} \\
\varepsilon^\mu(2) &= (0, 0, 1, 0): \quad \text{transverse polarization} \\
\varepsilon^\mu(3) &= (0, 0, 0, 1): \quad \text{longitudinal polarization}
\end{aligned}
\tag{A.18}
$$

In the coordinate system where the four-momentum of the particle is given by

$$
q^\mu = (\omega, 0, 0, |\boldsymbol{q}|) \tag{A.19}
$$

The four polarization vectors are boosted to

$$
\begin{aligned}
\varepsilon^\mu(s) &= \left(\frac{\omega}{m}, 0, 0, \frac{|\boldsymbol{q}|}{m}\right) \\
\varepsilon^\mu(+) &= -\frac{1}{\sqrt{2}}(0, 1, +i, 0) \\
\varepsilon^\mu(-) &= \frac{1}{\sqrt{2}}(0, 1, -i, 0) \\
\varepsilon^\mu(l) &= \left(\frac{|\boldsymbol{q}|}{m}, 0, 0, \frac{|\omega|}{m}\right)
\end{aligned}
\tag{A.20}
$$

The second and third lines referred to as circular polarizations are alternatives for the transverse components. The polarization vectors satisfy the orthogonality condition

$$\varepsilon^{\mu}(\lambda)\varepsilon_{\mu}(\lambda')^* = g^{\lambda\lambda'} \tag{A.21}$$

and the completeness condition

$$\sum_{\lambda=\text{all}} g^{\lambda\lambda'}\varepsilon^{\mu}(\lambda)\varepsilon^{\nu}(\lambda')^* = g^{\mu\nu} \tag{A.22}$$

Polarization Vectors of a Real/Virtual Particle Polarization vectors of a particle ($\varepsilon^{\mu}(\lambda)$) are orthogonal to the momentum q^{μ} of the particle.

$$q_{\mu}\varepsilon^{\mu}(\lambda) = 0 \tag{A.23}$$

Only the transverse polarizations are allowed for the real photon, and they satisfy

$$\sum_{\lambda=\pm} \varepsilon^{i}(\lambda)\varepsilon^{j}(\lambda) = \delta_{ij} - \frac{q^i q^j}{|\boldsymbol{q}|^2} \tag{A.24}$$

For the case of a massive vector boson,

$$\sum_{\lambda=\pm,l} \varepsilon^{\mu}(\lambda)\varepsilon^{\nu}(\lambda) = -\left(g^{\mu\nu} - \frac{q^{\mu}q^{\nu}}{m^2}\right) \tag{A.25}$$

The requirement of the constraint Eq. (A.23) removes the scalar polarization and the completeness condition becomes Eq. (A.25).

For virtual photons with timelike momentum (i.e., $q^2 > 0$), the above discussion can be extended by simply replacing m by $\sqrt{q^2}$.

For spacelike momentum ($q^2 < 0$), the longitudinal polarization becomes timelike by replacing the mass with $\sqrt{-q^2}$. Accordingly, the metric for the polarization summation has to be changed to keep the completeness condition.

$$\sum_{\lambda=\pm 1, l\to 0} (-1)^{\lambda+1}\varepsilon^{\mu}(\lambda)\varepsilon^{\nu}(\lambda)^* = -\left(g^{\mu\nu} - \frac{q^{\mu}q^{\nu}}{q^2}\right) \tag{A.26}$$

Appendix B
Feynman Rules for the Electroweak Theory

Feynman Rule 1: External Lines: We attach wave functions or polarizations to each incoming or outgoing particles (Figure B.1). Spinor index for fermions are sometimes omitted.

Feynman rules 2: Internal Lines: To each internal line, we attach a propagator depicted in Figure B.2, depending on particle species (Figure B.2). For fermions, the sign of momentum follows that of arrow.

Feynman rule 3: Fermion Gauge Boson Vertices 1: For vertices of the fermion and the gauge bosons, we attach coupling constants and appropriate γ factors (Figure B.3). Charged W bosons couple to left-handed fermions and its strength is given by

$$g_W = \frac{e}{\sin\theta_W} \tag{B.1}$$

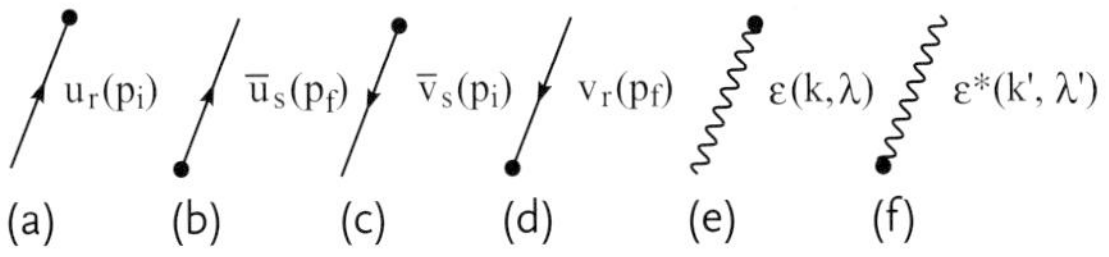

Figure B.1 Feynman rule 1 for external lines.

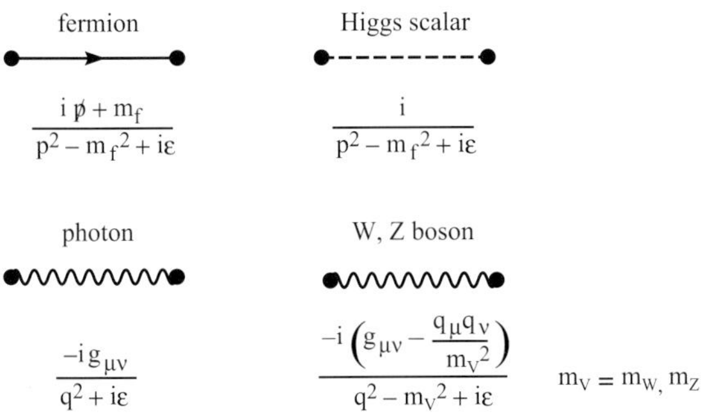

Figure B.2 Feynman rule 2 for external lines.

Elementary Particle Physics, Volume 2: Foundations of the Standard Model, First Edition. Yorikiyo Nagashima.

Cabibbo–Kobayashi–Maskawa matrix elements V_{ji} need to be attached to a quark pair of flavor j and i. The photon couples to the electromagnetic currents with charge $Q_f e$ and is of vector type. The Z boson couples to the neutral current which is a mixture of the left- and right-handed fermions. The coupling constants are product of a common coupling constant

$$g_Z = \frac{e}{\sin\theta_W \cos\theta_W} \tag{B.2}$$

and flavor dependent coupling constants

$$\epsilon_L(f) = I_{3f} - Q_f \sin^2\theta_W\,, \quad \epsilon_R(f) = Q_f \sin^2\theta_W \tag{B.3a}$$

where $f = l = e, \mu, \tau$ or $f = q = u, d, s, c, b, t$. Sometimes, separations according to vector and axial vector couplings are used.

$$g_Z \to g_Z/2\,, \quad v_f = I_3 - 2Q_f \sin^2\theta_W\,, \quad a_f = I_3 \tag{B.4}$$

They are related to the left- and right-handed couplings by

$$v_f = \epsilon_L(f) + \epsilon_R(f)\,, \quad a_f = \epsilon_L(f) - \epsilon_R(f) \tag{B.5}$$

Feynman rule 4: gauge Boson Nonlinear Couplings: Because of the non-Abelian nature of the electroweak theory, the gauge bosons have self-couplings which are shown in Figure B.4. Note there are no $\gamma{-}Z{-}Z$ or $Z{-}Z{-}Z$ couplings. In the figure, all the moments are taken to be inward going.

Feynman rule 5: Higgs Couplings: In Figure B.5, we list vertices where at least one of the particles are the Higgs particles. Notice, the coupling strength is proportional to the mass of the particle.

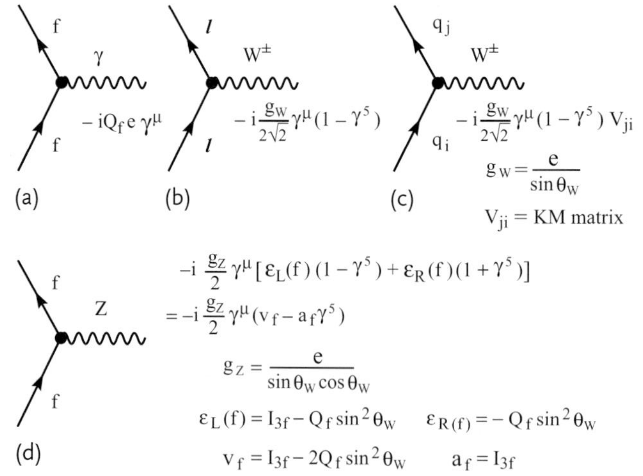

Figure B.3 Feynman rule 3: Vertices of fermions with gauge bosons.

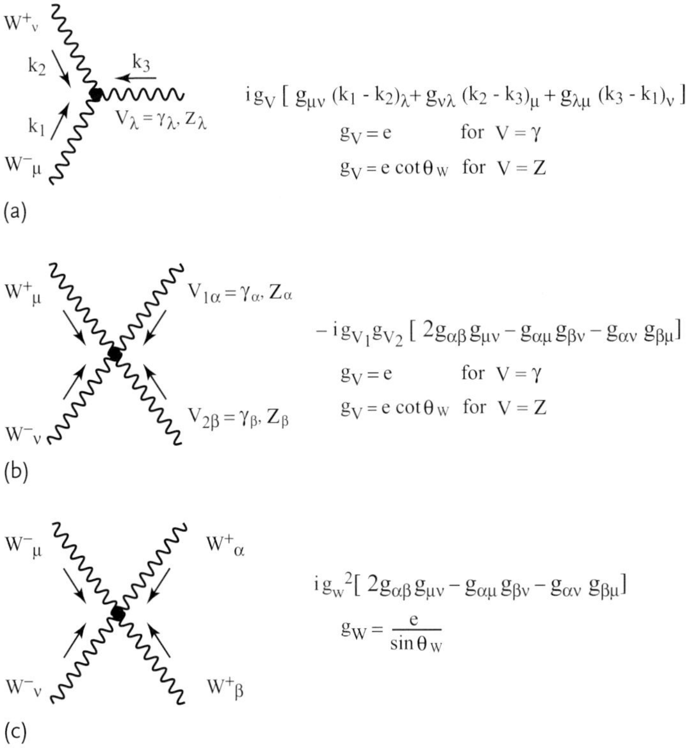

Figure B.4 Feynman rule 4: Nonlinear gauge boson couplings.

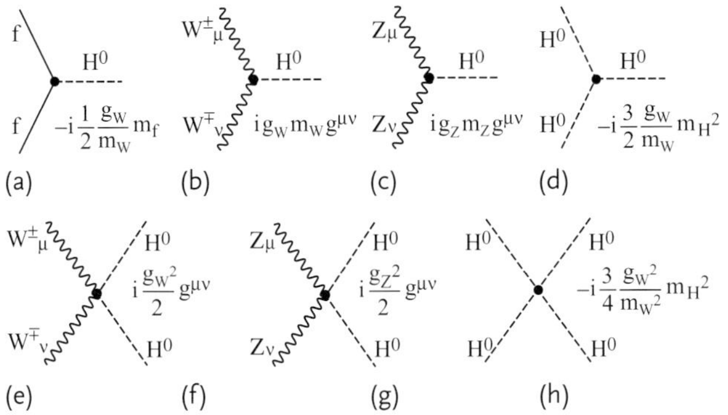

Figure B.5 Feynman rule 5: Vertices of the Higgs.

Feynman rule 6: momentum Assignment and Loops: The momenta of external lines are fixed by the experimental condition. Then, at each vertex, the energy-momentum has to conserve. Assuming all the momenta are defined as inward going, the energy-momentum conservation constrains that sum of energy-momenta of all external lines must vanish. In addition, it fixes all the momenta for tree diagrams which do not contain loops. Each loop leaves one momentum unconstrained and has to be integrated leading to divergent integrals. The integration includes the sum over the spinor index, that is, trace) and polarization depending on the particle species that form the loop. For each closed fermion loop, an extra sign (−) has to be attached. This results from anticommutativity of the fermion fields.

Amplitude for $e^-e^+ \to f\overline{f}$: As an example, we construct the amplitude in $O(\alpha^2)$ process for the reaction $e^-e^+ \to f\overline{f}$ where f is any of the leptons or quarks.

According to the Feynman rules we just described, we attach appropriate functions to every part of the Feynman diagram in Figure B.6.

The S-matrix and the cross section is written as

$$S_{fi} = \delta_{fi} - (2\pi)^4\delta^4(p_1 + p_2 - p_3 - p_4)i\mathcal{M} \tag{B.6a}$$

$$d\sigma = \frac{1}{F}\overline{\sum_{\substack{\text{spin}\\\text{pol}}}}|\mathcal{M}|^2 dLIPS \tag{B.6b}$$

$$dLIPS = (2\pi)^4\delta^4(p_1 + p_2 - p_3 - p_4)\frac{d^3p_3}{(2\pi)^3 2E_3}\frac{d^3p_4}{(2\pi)^3 2E_4} \tag{B.6c}$$

$$F = 4\left[(p_1 \cdot p_2)^2 - m_1^2 m_2^2\right]^{1/2} \simeq 2s \quad \text{for} \quad s \gg m_1^2, m_2^2 \tag{B.6d}$$

where F is the initial flux and $dLIPS$ is the Lorentz invariant phase space of the final state. Using the Feynman diagram, the transition amplitude $\mathcal{M}$ can be written

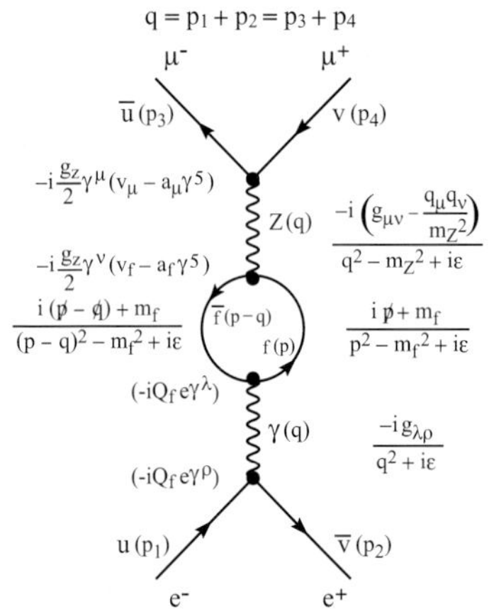

Figure B.6 An example of the Feynman diagram in order $O(\alpha^2)$ for the process $e^-e^+ \to \mu^-\mu^+$. To every element of the Feynman diagram, (wave functions, vertices and propagators), corresponding functions are attached.

as

$$-i\mathcal{M} = \left[\overline{u}(p_3)\left(-i\frac{g_Z}{2}\gamma^\mu\left(v_f - a_f\gamma^5\right)v(p_4)\right]\frac{-i\left(g_{\mu\nu} - q_\mu q_\nu/m_Z^2\right)}{q^2 - m_Z^2 + i\epsilon}$$
$$\times\left(-i\Sigma_{\gamma Z}^{\nu\lambda}(q^2)\right)\frac{-ig_{\lambda\rho}}{q^2 + i\epsilon}\left[\overline{v}(p_2)(-iQ_i e\gamma^\rho)u(p_1)\right] \tag{B.7a}$$

$$-i\Sigma_{\gamma Z}^{\nu\lambda}(q^2) = -\int\frac{d^4p}{(2\pi)^4}\mathrm{Tr}\left[\frac{i(\not{p} - \not{q}) + m_f}{(p-q)^2 - m_f^2 + i\epsilon}\left(-i\frac{g_Z}{2}\gamma^\nu\left(v_f - a_f\gamma^5\right)\right.\right.$$
$$\left.\times\frac{i\not{p} + m_f}{p^2 - m_f^2 + i\epsilon}\left(-iQ_f e\gamma^\lambda\right)\right] \tag{B.7b}$$

where we have separated the fermion loop part of the Feynman diagram because it has to be integrated over the internal momentum and an extra (−) sign has been attached according to the rule (6). $\Sigma_{\gamma Z}^{\nu\lambda}(q^2)$ is a diverging integral and has to be treated with the renormalization prescription which will be discussed in Appendix C.

Appendix C
Radiative Corrections to the Gauge Boson Self-Energy

We calculate the one-loop fermion corrections to $\Delta r, \Delta\alpha$ and $\hat{\Pi}_{\gamma z}(s)$ which appeared in Chapter 5.

Radiative Correction to the Gauge Boson Propagator One-loop correction by a fermion to the gauge boson propagator in the unitary gauge can be expressed as

$$\frac{-i\left(g_{\rho\mu}-(p_\rho p_\mu)/m_{V1}^2\right)}{p^2-m_{V1}^2}\left[-i\Sigma^{\mu\nu}(p^2)\right]\frac{-i\left(g_{\nu\sigma}-(p_\nu p_\sigma)/m_{V2}^2\right)}{p^2-m_{V2}^2} \tag{C.1a}$$

$$\begin{aligned}-i\Sigma^{\mu\nu}(p^2) = -gg'\int\frac{d^4k}{(2\pi)^4}\mathrm{Tr}-\left[\frac{i(\not{k}+m_1)}{k^2-m_1^2}i\gamma^\nu(v'-a'\gamma^5)\right.\\ \left.\times\frac{i(\not{p}+\not{k}+m_1)}{(k+p)^2-m_2^2}i\gamma^\mu(v'-a'\gamma^5)\right]\end{aligned} \tag{C.1b}$$

The variables g, g', v', a, a' are given by

1. $V_1 = V_2 = \gamma$:

$$g = g' = e\,,\quad v = v' = Q_f\,,\quad a = a' = 0\,,\quad m_1 = m_2 = m_f \tag{C.2a}$$

2. $V_1 = \gamma, V_2 = Z$:

$$\begin{aligned}&g = e\,,\quad g' = \frac{g_Z}{2} = \frac{e}{2s_W c_W}\,,\quad v = Q_f\,,\quad v' = I_{3f} - 2Q_f s_W^2\,,\\ &a = 0\,,\quad a' = I_{3f}\,,\quad m_1 = m_2 = m_f\end{aligned} \tag{C.2b}$$

3. $V_1 = V_2 = Z$:

$$\begin{aligned}&g = g' = \frac{g_Z}{2} = \frac{e}{2s_W c_W}\,,\quad v = v' = I_{3f} - 2Q_f s_W^2\,,\quad a = a' = I_{3f}\\ &m_1 = m_2 = m_f\end{aligned} \tag{C.2c}$$

Elementary Particle Physics, Volume 2: Foundations of the Standard Model, First Edition. Yorikiyo Nagashima.

4. $V_1 = V_2 = W^{\pm}$:

$$g = g' = \frac{g_W}{\sqrt{2}} = \frac{e}{\sqrt{2} s_W c_W}, \quad v = v' = a = a' = \frac{1}{2} \tag{C.2d}$$

and $s_W = \sin\theta_W$, $c_W = \cos\theta_W$ are the Weinberg mixing parameters. The trace can be written as

$$\begin{aligned}\mathrm{Tr}[\cdots] = 4\big[&(vv' + aa')\left\{k^\mu(p+k)^\nu + k^\nu(p+k)^\mu - g^{\mu\nu}(k^2 + (p\cdot k)\right\} \\ &+ (vv' - aa')m_1 m_2 g^{\mu\nu}\big]\end{aligned} \tag{C.3}$$

Therefore, $\Sigma^{\mu\nu}$ has a general form

$$\Sigma^{\mu\nu} = g^{\mu\nu}\Sigma_T + p^\mu p^\nu \Sigma_L = \left(g^{\mu\nu} - \frac{p^\mu p^\nu}{p^2}\right)\Sigma_T + \frac{p^\mu p^\nu}{p^2}\left(p^2\Sigma_L + \Sigma_T\right) \tag{C.4}$$

The second term is gauge dependent. In the scattering matrix, the above expression is sandwiched with currents which are conserved (vector current) or partially conserved (axial vector current). In the latter case, the contraction with p^μ gives the fermion mass and for the electron it can be neglected compared to the W mass. Consequently, we only consider Σ_T in the following. Considering

$$k^2 + (p\cdot k) = \frac{1}{2}\left[k^2 - m_1^2 + (p+k)^2 - m_2^2 + (m_1^2 + m_2^2 - p^2)\right] \tag{C.5}$$

we need to calculate the following integrals:

$$-A_1 = -A_0(m_1) = \int_k \frac{1}{(1)}, \quad -A_2 = -A_0(m_2) = \int_k \frac{1}{(2)} \tag{C.6a}$$

$$B_0 = B_0(p^2, m_1, m_2) = \int_k \frac{1}{(1)(2)} \tag{C.6b}$$

$$p^\mu B_1(p^2, m_1 m_2) = \int_k \frac{k^\mu}{(1)(2)} \tag{C.6c}$$

$$p^\mu p^\nu B_{21} - g^{\mu\nu} B_{22} = \int_k \frac{k^\mu k^\nu}{(1)(2)} \tag{C.6d}$$

Here, we used abbreviation $(1) \equiv k^2 - m_1^2 + i\epsilon$, $(2) \equiv (p+k)^2 - m_2^2 + i\epsilon$ and

$$\int_k \equiv \frac{16\pi^2}{i}\mu^{4-D}\int \frac{d^D k}{(2\pi)^D} \tag{C.7}$$

is the integral using dimensional regularization. μ is introduced to keep the physical dimension of the integral. Using the above notation, the Σ_T can be expressed as

$$\begin{aligned}\Sigma_T = \frac{1}{16\pi^2}4gg'\Bigg[&(vv' + aa')\left\{-2B_{22} + \frac{A_0(m_1) + A_0(m_2)}{2} + \frac{p^2 - m_1^2 - m_2^2}{2}B_0\right\} \\ &+ (vv' - aa')\, m_1 m_2 B_0\Bigg]\end{aligned} \tag{C.8}$$

Calculation of the One-Loop Integral As a representative of the integration using the dimensional regularization, we calculate B_0. Counting the power of k, one sees that the integral is logarithmically divergent. In order to make the integral converge, we go from four-dimensional to D dimensional space. As long as $D = 4 - \epsilon < 4$, the integral is convergent. The divergence reappears as $\epsilon \to 0$. As will be shown after integration, the integral is an analytical function in the complex D plane except for few poles at $\mathrm{Re}\, D = n < 4$. Therefore, we are allowed to move to $\mathrm{Re}\, D = n > 4$, carry out the integral formally and take the $\epsilon \to 0$ limit.

$$k^\mu = (k^0; k^1, k^2, k^3) \to (k^0 : k^1, k^2, \ldots, k^{D-1}) \tag{C.9a}$$

$$\frac{d^4 k}{(2\pi)^D} \to \mu^{4-D} \frac{d^D k}{(2\pi)^D} \tag{C.9b}$$

$$k^2 = (k^0)^2 - \sum_i (k^i)^2 \to (k^0)^2 - \sum_i^{D-1} (k^i)^2 \tag{C.9c}$$

$$g_{\mu\nu} g^{\mu\nu} = 4 \to g_{\mu\nu} g^{\mu\nu} = D \tag{C.9d}$$

Firstly, we introduce the following Feynman parameterization.

$$\frac{1}{a_1 a_2 \ldots a_n} = \int dz_1 \cdots dz_n \delta\left(1 - \sum z_i\right) \frac{(n-1)!}{(z_1 a_1 + \cdots + z_n a_n)^n} \tag{C.10}$$

As convergence of the integral is guaranteed, we are allowed to change the integration order.

$$B_0 = \int_k \frac{1}{(1)(2)} = \frac{16\pi^2}{i} \mu^{4-D} \int_0^1 dz \int \frac{d^D k}{(2\pi)^D} \frac{1}{\left[(k + zp)^2 - M^2\right]^2}$$
$$M^2 = -p^2 z(1-z) + m_1^2 (1-z) + m_2^2 z - i\epsilon \tag{C.11}$$

Setting $k' = k + zp$ and introducing $\omega^2 = \sum_{i=1}^{D} (k^{i'})^2 + M^2$, the content in $[\cdots]$ of the denominator can be expressed as

$$[\cdots] = \left(k^{0'} - \omega + i\epsilon\right)\left(k^{0'} + \omega - i\epsilon\right) \tag{C.12}$$

The integral has poles at $\omega - i\epsilon$ and $-\omega + i\epsilon$. We can, therefore, rotate the contour of the integral counterclockwise by 90° (Wick rotation) which is equivalent to changing $k^{0'} \to i k^{0'}$. Rewriting $k^{0'}$ as k^0, we have

$$\left(k^{0'}\right)^2 - \omega^2 \to -(k^0)^2 - \omega^2 \equiv -r^2 \tag{C.13}$$

As r is the Euclidean length in the D dimensional space, we can adopt the polar coordinate system and carry out the $D - 1$ angular variables which gives $D - 1$

dimensional total solid angle Ω_D. Ω_D can be obtained as follows. We calculate

$$\int_{-\infty}^{\infty} dx e^{-x^2} = \sqrt{\pi}$$

$$R \equiv \int_{-\infty}^{\infty} dx_1 \ldots \int_{-\infty}^{\infty} dx_D e^{-x_1^2 \ldots -x_D^2} = \pi^{D/2} \tag{C.14}$$

in the polar coordinate which results in

$$R = \Omega_D \int_0^{\infty} e^{-r^2} r^{D-1} dr \tag{C.15}$$

Using the definition for the Γ function

$$\Gamma(x) = \int_0^{\infty} e^{-t} t^{x-1} dt \tag{C.16}$$

we obtain

$$\Omega_D = \frac{2\pi^{D/2}}{\Gamma(D/2)} \tag{C.17}$$

Using Eq. (C.17), the integration of Eq. (C.11) can be carried out as follows:

$$\int \frac{d^D k}{(r^2+M^2)^2} = i\Omega_D \int_0^{\infty} \frac{r^{D-1} dr}{(r^2+M^2)^2} = i\frac{M^{D-4}\Omega_D}{2} B\left(\frac{D}{2}, 2-\frac{D}{2}\right) \tag{C.18a}$$

$$B(p,q) = \int_0^{\infty} \frac{x^{q-1} dx}{(1+x)^{p+q}} = \int_0^1 t^{p-1}(1-t)^{q-1} dt = \frac{\Gamma(p)\Gamma(q)}{\Gamma(p+q)} \tag{C.18b}$$

Substituting Eqs. (C.18) in Eq. (C.11), we obtain

$$\begin{aligned} B_0 &= \int_0^1 dz \frac{16\pi^2\mu^{4-D} M^{D-4}}{(2\pi)^D} \frac{\pi^{D/2}}{\Gamma(D/2)} \frac{\Gamma(D/2)\Gamma(2-D/2)}{\Gamma(2)} \\ &= \int_0^1 dz \left[\frac{4\pi\mu^2}{M^2}\right]^{2-D/2} \Gamma\left(2-\frac{D}{2}\right) \end{aligned} \tag{C.19}$$

As promised, this is an analytic function of D. At this point, we set $D = 4 - \epsilon$ and use the small ε limit:

$$\Gamma\left(\frac{\epsilon}{2}\right) \approx \frac{2}{\epsilon} - \gamma_E + O(\epsilon)^{1)} \tag{C.20a}$$

1) γ_E is the Euler's constant defined by

$$\gamma_E = \lim_{n\to\infty} \sum_k^n \frac{1}{k} - \log n = 0.577\,215\,66\ldots \tag{C.21}$$

$$x^{\epsilon/2} = e^{(\epsilon/2)\ln x} \approx 1 + \frac{\epsilon}{2}\ln x \tag{C.20b}$$

We finally obtain the expression for B_0:

$$\begin{aligned} B_0(p^2, m_1^2, m_2^2) &= \Delta - \int_0^1 dz \ln(M^2) \\ &= \Delta - \int_0^1 dz \ln\left[-p^2 z(1-z) + m_1^2(1-z) + m_2^2 z - i\epsilon\right] \\ \Delta &\equiv \frac{2}{\epsilon} - \gamma_E + \ln(4\pi) + \ln\mu^2 \end{aligned} \tag{C.22}$$

One realizes that the divergence appears as a pole in the D-plane. The second term is a finite analytical function. Historically, a formalism referred to as the minimum subtraction method which subtracts the divergent term $\epsilon/2$ only was adopted. Nowadays, it is customary to subtract the whole three terms in Δ which is referred to as the modified minimum subtraction because they always appear as a group. It is customary to attach suffix $\overline{MS}$ to the renormalized variables thus obtained. The subtraction methods a là $\overline{MS}$ is equivalent to replace

$$\Delta \to \ln\mu^2 \tag{C.23}$$

$\overline{MS}$ method is the standard renormalization method in QCD where there are no reference point at which observables like $\alpha_s(\mu)$ or mass of the quark $m(\mu)$ can be defined. This is a consequence of the confinement. When observables like α (fine structure constant) or G_μ (Fermi coupling constant in μ decay) can be defined for certain values of the scale μ ($\mu^2 = 0$ or m_W^2), those values can be used as references. They are referred to as an on-shell scheme and are used in the electroweak as well as QED renormalization. In the on-shell scheme, if the $\ln\mu^2$ is retained, it disappears from physical observables and we need not worry about it. In the following, we occasionally use Δ to mean $\ln\mu^2$.

Other integrals can be carried out similarly which are listed below with $s = p^2$. [171]

$$A_0(m) = -m^2\Delta - m^2(1 - \ln m^2) \tag{C.24a}$$

$$B_1 = \frac{1}{2s}\left[-A_0(m_1) + A_0(m_2) - \left(s + m_1^2 - m_2^2\right) B_0\right] \tag{C.24b}$$

$$B_{21} = \frac{1}{3s}\left[-A_0(m_2) - 2\left(s + m_1^2 - m_2^2\right) B_1 - m_1^2 B_0 + \frac{1}{2}\left(\frac{s}{3} - m_1^2 - m_2^2\right)\right] \tag{C.24c}$$

$$B_{22} = \frac{1}{6}\left[A_0(m_2) - \left(s + m_1^2 - m_2^2\right) B_1 - 2m_1^2 B_0 + \left(\frac{s}{3} - m_1^2 - m_2^2\right)\right] \tag{C.24d}$$

B_1, B_{21}, B_{22} can be obtained by contracting Eqs. (C.6c) and (C.6d) with p_μ, $g_{\mu\nu}$. However, in doing so, one has to be careful, keeping in mind that $D \neq 4$, but $D = 4 - \epsilon$ and that B_{22} and so on contains a singular part that behaves like $\sim 1/\epsilon$. Thus,

$$g_{\mu\nu} g^{\mu\nu} = D = 4 - \epsilon \tag{C.25a}$$

$$\lim_{\epsilon \to 0} (\epsilon A_0) = -2m^2 \,, \quad \lim_{\epsilon \to 0} (\epsilon B_0) = 2 \,, \quad \lim_{\epsilon \to 0} (\epsilon B_1) = -1 \tag{C.25b}$$

$$\lim_{\epsilon \to 0} (\epsilon B_{22}) = \frac{\left(s - 3m_1^2 - 3m_2^3\right)}{6} \tag{C.25c}$$

Self energy $\Sigma_{\mathrm{T}}^{\mu\nu}(s)$ Using Eq. (C.22), (C.24a) and (C.24d), the self-energy of the propagator can be expressed as

$$\begin{aligned}
\Sigma_{\mathrm{T}}(s) &= \frac{4gg'}{16\pi^2} \Bigg[\left(v v' + a a'\right) \left\{ -2B_{22} + \frac{A_0(m_1) + A_0(m_2)}{2} + \frac{s - m_1^2 - m_2^2}{2} B_0 \right\} \\
&\quad + \left(v v' - a a'\right) m_1 m_2 B_0 \Bigg] \\
&= \frac{4gg'}{16\pi^2} \frac{1}{3} \Bigg[\left(v v' + a a'\right) \Bigg\{ \left(m_1^2 + m_2^2 - \frac{s}{3} \right) + A_0(m_1) + A_0(m_2) \\
&\quad - \frac{m_1^2 - m_2^2}{2s} \left(A_0(m_1) - A_0(m_2)\right) + \left(s - \frac{m_1^2 + m_2^2}{2} - \frac{\left(m_1^2 - m_2^2\right)^2}{2s} \right) B_0 \Bigg\} \\
&\quad + \left(v v' - a a'\right) 3 m_1 m_2 B_0 \Bigg]
\end{aligned} \tag{C.26}$$

Referring to Eqs. (C.2a)–(C.2d), we can obtain the self-energy of W, Z, for example. Denoting m_1, m_2 as the mass of isospin doublets and using

$$\Delta_j \equiv \frac{2}{\epsilon} - \gamma_{\mathrm{E}} + 4\pi - \ln \frac{m_j^2}{\mu^2} \tag{C.27}$$

we obtain expressions for the self-energy of various gauge bosons [170, 171], namely, $N_c = 3$ for the quark and $N_c = 1$ for the lepton.

$$\begin{aligned}
\Sigma_{WW} &= \frac{N_c \alpha}{4\pi s_W^2} \frac{1}{3} \sum \Bigg[\frac{\Delta_1}{2} \left(s - \frac{5}{2} m_1^2 - \frac{1}{2} m_2^2 \right) + \frac{\Delta_2}{2} \left(s - \frac{1}{2} m_1^2 - \frac{5}{2} m_2^2 \right) \\
&\quad + \left(s - \frac{m_1^2 + m_2^2}{2} \right) \left(1 - \frac{m_1^2 + m_2^2}{m_1^2 - m_2^2} \ln \frac{m_1}{m_2} \right) - \frac{s}{3} \\
&\quad + \left\{ s - \frac{m_1^2 + m_2^2}{2} - \frac{\left(m_1^2 - m_2^2\right)^2}{2s} \right\} F(s, m_1, m_2) \Bigg]
\end{aligned} \tag{C.28a}$$

$$\Sigma_{ZZ}(s) = \frac{N_c\alpha}{4\pi s_W^2 c_W^2}\frac{1}{3}\sum\left[\left(v_f^2 + a_f^2\right)\left\{s\Delta_f + \left(s + 2m_f^2\right)F(s, m_f, m_f) - \frac{s}{3}\right\}\right.$$
$$\left. - \frac{3}{2}m_f^2\left\{\Delta_f + F(s, m_f, m_f)\right\}\right] \quad \text{(C.28b)}$$

$$\Sigma_{\gamma Z}(s) = \frac{N_c\alpha}{4\pi s_W c_W}\frac{2}{3}\sum Q_f v_f\left[\left\{s\Delta_f + (s + 2m_f^2)F(s, m_f, m_f) - \frac{s}{3}\right\}\right] \quad \text{(C.28c)}$$

$$\Sigma_{\gamma\gamma}(s) = \frac{N_c\alpha}{4\pi}\frac{4}{3}\sum Q_f^2\left[\left\{s\Delta_f + (s + 2m_f^2)F(s, m_f, m_f) - \frac{s}{3}\right\}\right] \quad \text{(C.28d)}$$

The function F is derived from B_0 and is defined by

$$B_0(s, m_1, m_2) = B_0(0, m_1 m_2) + F(s, m_1, m_2) \quad \text{(C.29a)}$$

$$B_0(0, m_1, m_2) = \Delta + 1 - \frac{m_1^2 + m_2^2}{m_1^2 - m_2^2}\ln\frac{m_1}{m_2} - \ln m_1 m_2 \quad \text{(C.29b)}$$

$$F(s, m_1, m_2) = -1 + \frac{m_1^2 + m_2^2}{m_1^2 - m_2^2}\ln\frac{m_1}{m_2} \quad \text{(C.29c)}$$

$$-\int_0^1 dz \ln\left[\frac{-sz(1-z) + m_1^2(1-z) + m_2^2 z - i\epsilon}{m_1 m_2}\right] \quad \text{(C.29d)}$$

The function F has the following properties:

$$F(0, m_1, m_2) = 0 \quad \text{(C.30a)}$$

$$F(s, 0, m) = 1 + \left(\frac{m^2}{s} - 1\right)\ln\left(1 - \frac{s}{m^2} - i\epsilon\right) \quad \text{(C.30b)}$$

For $s \ll m_1^2, m_2^2$,

$$F\left(s, m_1^2, m_2^2\right) = \frac{s}{\left(m_1^2 - m_2^2\right)^2}\left(\frac{m_1^2 + m_2^2}{2} - \frac{m_1^2 m_2^2}{m_1^2 - m_2^2}\ln\frac{m_1^2}{m_2^2}\right) + \cdots \quad \text{(C.30c)}$$

For $s \gg m_1^2, m_2^2$,

$$F\left(s, m_1^2, m_2^2\right) = 1 + \frac{m_1^2 + m_2^2}{m_1^2 - m_2^2}\ln\frac{m_1}{m_2} - \ln\frac{|s|}{m_1 m_2} + i\pi\theta(s) + \cdots \quad \text{(C.30d)}$$

For $m_1 = m_2$, it is even more simple and

$$F(s, m, m) = 2 - \beta\ln\frac{\beta + 1}{\beta - 1} \approx \begin{cases} 2 - \ln\dfrac{|s|}{m^2} + i\pi\theta(s) + \cdots & |s| \gg m^2 \\ \dfrac{s}{6m^2} + \dfrac{s^2}{60m^4} + \cdots & |s| \ll m^2 \end{cases}$$

$$\beta = \sqrt{1 - \frac{4m^2}{s}} \quad \text{(C.30e)}$$

$\Delta\alpha$ The gauge invariance guarantees $\Sigma_{\gamma\gamma}(0) = 0$. Inspection of Eq. (C.28c) and (C.28d) show that $\Sigma_{\gamma Z}(0) = 0$ also for the fermion loop. Writing down explicitly,

$$\Sigma_{\gamma\gamma}(0) = \Sigma_{\gamma Z}(0) = 0 \tag{C.31a}$$

$$\Pi_\gamma(s) = \frac{\Sigma_{\gamma\gamma}}{s} = \Pi_\gamma(0) + \hat{\Pi}_\gamma(s) \tag{C.31b}$$

$$\Pi_\gamma(0) = \frac{\alpha}{3\pi}\sum_f Q_f^2 \Delta_f \tag{C.31c}$$

$$\hat{\Pi}_\gamma(s) = \frac{\alpha}{3\pi}\sum_f Q_f^2\left[\left(1 + \frac{2m_f^2}{s}\right)F(s, m_f, m_f) - \frac{1}{3}\right] \tag{C.31d}$$

$$\simeq \begin{cases} \dfrac{\alpha}{3\pi}\sum_f Q_f^2\left[\dfrac{5}{3} - \ln\dfrac{|s|}{m_f^2} + \cdots\right] & |s| \gg m_f^2 \\ \dfrac{\alpha}{3\pi}\sum_f Q_f^2\left[\dfrac{s}{5m_f^2} + \cdots\right] & |s| \ll m_f^2 \end{cases} \tag{C.31e}$$

$$\frac{\Sigma_{\gamma Z}(s)}{s} = \frac{1}{2 s_W c_W}\sum_f \frac{v_f Q_f}{Q_f^2}\Pi_\gamma(s) \tag{C.31f}$$

$\hat{\Pi}_\gamma(s)$ does not contain divergent terms. Reminder: $\sum_f$ includes color for quarks. The above formulas show that $\hat{\Pi}_\gamma(s)$ gets contributions from light fermions ($s \gg m_f^2$). Those from heavy quarks ($s \ll m_f^2$) are negligible which is referred to as decoupling of the heavy quarks. The statement can be summarized as

$$\Delta\alpha = \frac{\alpha(m_Z^2) - \alpha(0)}{\alpha(0)} = -\hat{\Pi}_\gamma(m_Z^2) \approx -\frac{\alpha}{3\pi}\sum_{m_f < m_Z} Q_f^2\left[\frac{5}{3} - \ln\frac{m_Z^2}{m_f^2}\right] \tag{C.32}$$

Δr, $\hat{\Pi}_{\gamma Z(s)}$ For later convenience, we introduce the following variables to clarify the isospin structure of the self-energies. Here, lower index 1,3 means the I_1, I_3 component.

$$\Sigma_{\gamma\gamma} = (Qe)^2\Sigma_{QQ} \tag{C.33a}$$

$$\Sigma_{\gamma Z} = (Qe)g_Z\left(\Sigma_{3Q} - Qs_W^2\Sigma_{QQ}\right) \tag{C.33b}$$

$$\Sigma_{ZZ} = g_Z^2\left(\Sigma_{33} - 2Qs_W^2\Sigma_{3Q} + Q^2 s_W^2\Sigma_{QQ}\right) \tag{C.33c}$$

$$\Sigma_{WW} = g_W^2\Sigma_{11} \tag{C.33d}$$

Then, variables of our interest can be expressed as follows.

$$\Delta\rho = \mathrm{Re}\left[\frac{\Sigma_{ZZ}(0)}{m_Z^2} - \frac{\Sigma_{WW}(0)}{m_W^2} + 2\frac{s_W}{c_W}\frac{\Sigma_{\gamma Z}(0)}{m_Z^2}\right] = \frac{e^2}{m_W^2 s_W^2}\left[\Sigma_{33}(0) - \Sigma_{11}(0)\right] \tag{C.34a}$$

$$\Delta r = -\hat{\Pi}_W(0) = \mathrm{Re}\frac{\Sigma_{WW}(0) - \Sigma_{WW}(m_W^2)}{m_W^2} + \Pi_\gamma(0) - \frac{\delta' s_W^2}{s_W^2} \tag{C.34b}$$

In the following, we omit attaching "Re" unless otherwise noted, but it is meant to be there.

$$\begin{aligned}
\Delta r &= \frac{\Sigma_{WW}(0) - \Sigma_{WW}(m_W^2)}{m_W^2} + \Pi_\gamma(0) \\
&\quad - \frac{c_W^2}{s_W^2}\left[\frac{\Sigma_{ZZ}(m_Z^2)}{m_Z^2} - \frac{\Sigma_{WW}(m_W^2)}{m_W^2} + 2\frac{s_W}{c_W}\frac{\Sigma_{\gamma Z}(0)}{m_Z^2}\right] \\
&= \Delta\alpha - \frac{c_W^2}{s_W^2}\Delta\rho + r_1
\end{aligned} \tag{C.34c}$$

$$\begin{aligned}
r_1 &= \Pi_\gamma(m_Z^2) - \frac{c_W^2}{s_W^2}\Sigma'_{ZZ}(m_Z^2) + \frac{c_W^2 - s_W^2}{s_W^2}\Sigma'_{WW}(m_W^2) \\
&= \frac{c_W^2 - s_W^2}{s_W^2}\epsilon_2 + 2\epsilon_3
\end{aligned} \tag{C.34d}$$

where we have defined

$$\Sigma'_{VV}(s) \equiv \frac{\Sigma_{VV}(s) - \Sigma_{VV}(0)}{s} \tag{C.35}$$

and

$$\Delta\alpha = -\hat{\Pi}_\gamma(m_Z^2) = -\Sigma'(m_Z^2) \tag{C.36a}$$

$$\epsilon_2 = \frac{e^2}{s_W^2}\left[\Sigma'_{11}(m_W^2) - \Sigma'_{33}(m_Z^2)\right] \tag{C.36b}$$

$$\epsilon_3 = \frac{e^2}{s_W^2}\left[\Sigma'_{3Q}(m_W^2) - \Sigma'_{33}(m_Z^2)\right] \tag{C.36c}$$

In obtaining the above results, we used

$$\delta m_W^2 = \Sigma_{WW}(m_W^2)\,, \quad \delta m_Z^2 = \Sigma_{ZZ}(m_Z^2) \tag{C.37a}$$

$$\Pi_\gamma(s) = \frac{\Sigma_{\gamma\gamma}(s)}{s} = \Pi_\gamma(0) + \hat{\Pi}_\gamma(s) \tag{C.37b}$$

and that the coefficient of Σ'_{3Q} is actually QI_3, and sum over doublet members equals to that of Σ'_{33}. Similarly, one can derive

$$\begin{aligned}
\hat{\Pi}_{\gamma Z}(m_Z^2) &= \frac{\hat{\Sigma}_{\gamma Z}(m_Z^2)}{m_Z^2} = \frac{\Sigma_{\gamma Z}(m_Z^2) - \Sigma_{\gamma Z}(0)}{m_Z^2} + \frac{\delta' s_W^2}{c_W s_W} \\
&= \Sigma'_{\gamma Z}(m_Z^2) + \frac{\delta' s_W^2}{c_W s_W} = \frac{c_W}{s_W}\Delta\rho + r_2
\end{aligned} \tag{C.38a}$$

$$r_2 = \Sigma'_{\gamma Z}(m_Z^2) + \frac{c_W}{s_W}\left[\Sigma'_{ZZ}(m_Z^2) - \Sigma'_{WW}(m_W^2)\right] = -\frac{c_W}{s_W}\epsilon_2 - \frac{s_W}{c_W}\epsilon_3 \tag{C.38b}$$

In the following, we will show that $\Delta\alpha$, $\Delta\rho$, ϵ_2, ϵ_3 (hence Δr, $\hat{\Pi}_{\gamma Z}(m_Z^2)$, too) do not contain the divergence and that large contributions are contained in $\Delta\alpha$ and $\Delta\rho$. ϵ_2, ϵ_3 are small numbers in the Standard Model.

$\Delta\rho$ For each generation, Eq. (C.28a) and (C.28b) give

$$\Sigma_{ZZ}(0) = -\frac{N_c\alpha}{16\pi}\frac{1}{s_W^2 c_W^2}\left[2\left(m_1^2\Delta_1 + m_2^2\Delta_2\right)\right] \tag{C.39a}$$

$$\Sigma_{WW}(0) = -\frac{N_c\alpha}{16\pi}\frac{1}{s_W^2}\left[2\left(m_1^2\Delta_1 + m_2^2\Delta_2\right) + \left(m_1^2 + m_2^2\right) - \frac{2m_1^2 m_2^2}{m_1^2 - m_2^2}\ln\frac{m_1^2}{m_2^2}\right] \tag{C.39b}$$

$$\begin{aligned}\therefore \quad \rho_g &= \frac{\Sigma_{ZZ}(0)}{m_Z^2} - \frac{\Sigma_{WW}(0)}{m_W^2} + 2\frac{s_W}{c_W}\frac{\Sigma_{\gamma Z}(0)}{m_Z^2}\\ &= \frac{3\alpha}{16\pi s_W^2 m_W^2}\left[\left(m_1^2 + m_2^2\right) - \frac{2m_1^2 m_2^2}{m_1^2 - m_2^2}\ln\frac{m_1^2}{m_2^2}\right]\end{aligned} \tag{C.39c}$$

Namely, ρ_g for each generation is divergent free. As the mass of the top quark is overwhelmingly large, one obtains an approximate value of

$$\Delta\rho = \sum_g \rho_g \simeq \rho_t = \frac{3\alpha}{16\pi s_W^2}\frac{m_t^2}{m_W^2} \tag{C.40a}$$

ϵ_2 and ϵ_3 Next, we show that ϵ_2 and ϵ_3 are divergence free. Denoting the term in $\Pi_\gamma(0)$ which contains Δ as Sing$[\Pi_\gamma(0)]$, we have for each term

$$\begin{aligned}\text{Sing}\left[\Pi_\gamma\left(m_Z^2\right)\right] &= \text{Sing}\left[\Pi_\gamma(0)\right] = \frac{\alpha}{3\pi}\Delta\left(Q_1^2 + Q_2^2\right)\\ \text{Sing}\left[\Sigma'_{\gamma Z}\left(m_Z^2\right)\right] &= \frac{\alpha}{3\pi}\frac{\Delta}{2s_W c_W}(Q_1 v_1 + Q_2 v_2)\\ \text{Sing}\left[\Sigma'_{ZZ}\left(m_Z^2\right)\right] &= \frac{\alpha}{3\pi}\frac{\Delta}{4s_W^2 c_W^2}\left(v_1^2 + v_2^2 + a_1^2 + a_2^2\right)\\ \text{Sing}\left[\Sigma'_{WW}\left(m_W^2\right)\right] &= \frac{\alpha}{3\pi}\frac{\Delta}{4s_W^2}\end{aligned} \tag{C.41}$$

Substituting $v_f = I_{3f} - 2Q_f s_W^2$, $a_f = I_{3f}$ and

$$\begin{aligned}&Q_1 I_{31} + Q_2 I_{32} = \frac{1}{2}(Q_1 - Q_2) = \frac{1}{2} \quad \rightarrow \quad \sum_f Q_f v_f|_{v_f = I_{3f}} = \frac{1}{2}\\ &\sum_f\left(v_f^2 + a_f^2\right)_{Q_f=0} = 1\end{aligned} \tag{C.42}$$

in Eqs. (C.41), we have

$$\begin{aligned}
\mathrm{Sing}\left[\Sigma'_{33}\left(m_Z^2\right)\right] &= \frac{1}{g_Z^2}\mathrm{Sing}\left[\Sigma'_{ZZ}\left(m_Z^2\right)\right]_{Q_f=0} = \frac{\Delta}{48\pi^2} \\
\mathrm{Sing}\left[\Sigma'_{3Q}\left(m_Z^2\right)\right] &= \frac{1}{e g_Z}\mathrm{Sing}\left[\Sigma'_{\gamma Z}\left(m_Z^2\right)\right]_{v_f=I_{3f}} = \frac{\Delta}{48\pi^2} \\
\mathrm{Sing}\left[\Sigma'_{11}\left(m_W^2\right)\right] &= \frac{1}{g_W^2}\mathrm{Sing}\left[\Sigma'_{WW}\left(m_W^2\right)\right] = \frac{\Delta}{48\pi^2}
\end{aligned} \tag{C.43}$$

Comparing Eq. (C.36b) and (C.36c) with Eqs. (C.43), it follows that the singular terms do not contribute to ϵ_2, ϵ_3.

We next show that dominant terms $\ln(m_Z^2/m_f^2)$ are absent in ϵ_2, ϵ_3. If one neglects m_f^2 compared to $s = m_Z^2$, one notices that function $F(s, m_1^2, m_2^2)$ always appear with Δ_f with the same coefficients. For instance, in Σ_{ZZ} in Eq. (C.28b), it appears as

$$\begin{aligned}
\Delta_f + F\left(m_Z^2, m_f^2, m_f^2\right) &\xrightarrow{m_Z^2 \gg m_f^2} \epsilon - \gamma_E + 4\pi - \ln\frac{m_f^2}{\mu^2} - \ln\frac{m_Z^2}{m_f^2} \\
&= \epsilon - \gamma_E + 4\pi \ln\mu^2 - \ln m_Z^2 = \Delta - \ln m_Z^2
\end{aligned} \tag{C.44}$$

In Σ_{WW} (see Eq. (C.28a)), it appears like

$$\begin{aligned}
&\frac{\Delta_1}{2} + \frac{\Delta_2}{2} + F\left(m_W^2, m_1^2, m_2^2\right) \xrightarrow{m_W^2 \gg m_l^2} \frac{1}{2}\left(\epsilon - \gamma_E + 4\pi - \ln\frac{m_1^2}{\mu^2}\right) \\
&\quad + (1 \leftrightarrow 2) - \ln\frac{m_W^2}{m_1 m_2} = \Delta - \ln m_Z^2 + \ln c_W^2
\end{aligned} \tag{C.45}$$

Therefore, terms including m_Z^2 always appear in combination with Δ. Therefore, for the same reason that ϵ_2, ϵ_3 are divergent free, the large $\ln m_Z^2$ does not exist either. Therefore, there are no contributions from light quarks in them. The dominant contribution to ϵ_2, ϵ_3 comes from the top quark. We write down their contribution and compare with $\Delta\rho$ [236]

$$\Delta\rho = \alpha T = \frac{3 G_F m_t^2}{8\sqrt{2}\pi^2} - \frac{3 G_F m_W^2}{4\sqrt{2}\pi^2}\tan\theta_W^2 \ln\frac{m_H}{m_Z} \tag{C.46a}$$

$$\epsilon_2 = -\frac{\alpha}{4 s_W^2} U = -\frac{G_F m_W^2}{2\sqrt{2}\pi^2}\ln\frac{m_t}{m_Z} \tag{C.46b}$$

$$\epsilon_3 = \frac{\alpha}{4 s_W^2} S = \frac{G_F m_W^2}{12\sqrt{2}\pi^2}\ln\frac{m_H}{m_Z} - \frac{G_F m_W^2}{6\sqrt{2}\pi^2}\ln\frac{m_t}{m_Z} \tag{C.46c}$$

One sees that contribution of the top quark to ϵ_2, ϵ_3 is relatively smaller by factor m_W^2/m_t^2 compared to $\Delta\rho$. The reason to include the Higgs is that it is no longer negligible at this level of approximation. Together with the absence of light quark contributions, this fact makes ϵ_2 ϵ_3 a sensitive probe to unknown new physics.

In summary, the dominant electroweak corrections are those of vacuum polarization to the gauge bosons which can be expressed in terms of $\Delta\alpha$, $\Delta\rho$, ϵ_2, ϵ_3.

Within the Standard Model, the large terms are $\Delta\alpha$ and $\Delta\rho$ which are sufficient for the discussion of up to 1% precision. The light quarks (u, d, s, c, b) contribute to $\Delta\alpha$ and the heavy quark (top) contribute to $\Delta\rho$. In terms of them, the important parameters Δr and $\hat{\Pi}_{\gamma Z}(m_Z^2)$ are expressed as

$$\Delta r \approx \Delta\alpha - \frac{c_W^2}{s_W^2}\Delta\rho + \cdots \tag{C.47a}$$

$$\hat{\Pi}_{\gamma Z}\left(m_Z^2\right) \approx \frac{c_W}{s_W}\Delta\rho + \cdots \tag{C.47b}$$

ϵ_2 and ϵ_3 are small in the Standard Model, however, could be large depending on the structure of new physics. $\Delta\rho$ represents the amount of isospin breaking and the effect is small if the mass difference is small even though the mass itself is large. In this case, the effect can appear in ϵ_3 and could be used for the test of new physics like the technicolor model.

Appendix D
't Hooft's Gauge

In the 't Hooft gauge, the massive vector's propagator, which we will derive shortly, behaves $\sim 1/k^2$ as $k^2 \to \infty$. This is contrasted to that in the unitary (U) gauge which is expressed as

$$i\Delta_F = \frac{-i\left(g_{\mu\nu} - \frac{k_\mu k_\nu}{m^2}\right)}{k^2 - m^2 + i\epsilon} \tag{D.1}$$

The propagator in the U-gauge stays constant as $k^2 \to \infty$ and induces badly diverging loop integrals. Note that, in principle, all the divergences should also disappear in the U-gauge if one calculates carefully. Practically, however, it is hard to keep track of all the divergences.

The Abelian Higgs model is good enough to understand the essence of the 't Hooft gauge.

$$\begin{aligned}
\mathcal{L} &= -\frac{1}{4} F_{\mu\nu} F^{\mu\nu} + (D_\mu\varphi)^\dagger (D^\mu\varphi) - V(\varphi) \\
F_{\mu\nu} &= \partial_\mu A_\nu - \partial_\nu A_\mu \\
D_\mu &= \partial_\mu + i q A_\mu \\
V(\varphi) &= \mu^2(\varphi^\dagger\varphi) + \lambda(\varphi^\dagger\varphi)^2 , \quad \lambda > 0 , \quad \mu^2 > 0
\end{aligned} \tag{D.2}$$

where q is the strength of the coupling. In the U-gauge, the complex field $\varphi = (\phi_1 + i\phi_2)/\sqrt{2}$ is expressed as a product of the modulus and phase field (see Eq. (1.34)). When the symmetry is spontaneously broken (SSB), that is, when μ^2 becomes negative, it is transformed to

$$\varphi = \exp[i\phi_2/v]\frac{\phi_1}{\sqrt{2}} \quad \xrightarrow{\text{SSB}} \quad \frac{v + \phi_1'}{\sqrt{2}} , \quad v = \sqrt{-\frac{\mu^2}{\lambda}} \tag{D.3}$$

Mathematically, it is equivalent to transfer the ϕ_2 away from φ by the gauge transformation. As a result the phase field ϕ_2 is absorbed by A_μ.

$$\varphi \to \varphi' = \exp[-i\phi_2/v]\varphi = \frac{v + \phi_1'}{\sqrt{2}} \tag{D.4a}$$

$$A_\mu \to A_\mu' = A_\mu + \frac{1}{qv}\partial_\mu \phi_2 \tag{D.4b}$$

Elementary Particle Physics, Volume 2: Foundations of the Standard Model, First Edition. Yorikiyo Nagashima.

The Lagrangian after the spontaneous symmetry breakdown becomes

$$\begin{aligned}\mathcal{L}_{\mathrm{SSB}} &= \frac{1}{2}\left(\partial_\mu\phi_1\partial^\mu\phi_1 - m_H^2\phi_1\phi_1\right) - \frac{1}{4}F_{\mu\nu} + \frac{m_{\mathrm{V}}^2}{2}A_\mu A^\mu \\ &\quad - \lambda v\phi_1^3 - \frac{\lambda}{4}\phi_1^4 + \frac{q^2}{2}A_\mu A^\mu\phi_1(2v+\phi_1) \\ m_{\mathrm{V}} &= qv\,, \quad m_H^2 = -2\mu^2\end{aligned} \tag{D.5}$$

where ϕ_1', A_μ' are relabeled as ϕ_1, A_μ. This Lagrangian represents a massive scalar field ϕ_1 (Higgs) with mass $m_H^2 = 2\lambda v^2$, a massive vector field A_μ with mass m_{V} and their interaction (the second line). ϕ_2 has disappeared. One sees that the number of degrees of freedom is maintained, originally 2 for massless vector boson, 2 for complex scalar field and after the symmetry breakdown, 3 for the massive vector field and one for the scalar field.

The 't Hooft gauge uses a linear expression $e^{i\phi_2/v} \to 1 + i\phi_2/v$ for the scalar field and expresses it as

$$\varphi = \frac{1}{\sqrt{2}}(v + \phi_1 + i\phi_2) \tag{D.6}$$

The Lagrangian takes the form

$$\begin{aligned}\mathcal{L}_{\mathrm{SB}} &= \frac{1}{2}\left(\partial_\mu\phi_1\partial^\mu\phi_1 - m_H^2\phi_1\phi_1\right) - \frac{1}{4}F_{\mu\nu}F^{\mu\nu} + \frac{m_{\mathrm{V}}^2}{2}A_\mu A^\mu \\ &\quad + \frac{1}{2}\left(\partial_\mu\phi_2\partial^\mu\phi_2\right) + evA_\mu\partial^\mu\phi_2 \\ &\quad + (\text{higher order terms})\end{aligned} \tag{D.7}$$

where m_H and m_{V} are the same mass as in the U-gauge. The first line represents the same massive scalar field and massive vector boson as those in the U-gauge. The higher order terms which represent interactions are somewhat different and include ϕ_2, but it does not matter for the present discussion. The difference is in the second line, that is, the ϕ_2, the would-be-Goldstone boson, has reappeared. Since the massive vector boson had already taken one degree of freedom from the scalar field, the existence of ϕ_2 is one too many. Obviously, ϕ_2 must be unphysical and should not appear in the final expression for physical processes. It could be removed if desired by a suitable gauge transformation as was done in the U-gauge.

However, an unphysical object has recursively appeared in the gauge theory with reasons, so let us proceed. So far, we have treated the fields as classical. For quantization, we need a gauge fixing term in the Lagrangian. 't Hooft's choice was [39]

$$\mathcal{L}_{\mathrm{gf}} = -\frac{1}{2\xi}\left(\partial_\mu A^\mu - \xi m_{\mathrm{V}}\phi_2\right)^2 \tag{D.8}$$

which cancels the unwanted mixing term in Eq. (D.7) and at the same time, realizes the covariant equation for the vector boson. The Lagrangian is now given by

$$\begin{aligned}\mathcal{L}_{\text{tHooft}} &= -\frac{1}{4}F_{\mu\nu} - \frac{m_{\text{V}}^2}{2}A_\mu A^\mu - \frac{1}{2\xi}\left(\partial_\mu A^\mu\right)^2 \\ &+ \frac{1}{2}\left(\partial_\mu \phi_1 \partial^\mu \phi_1 - m_H^2 \phi_1 \phi_1\right) + \frac{1}{2}\left(\partial_\mu \phi_2 \partial^\mu \phi_2 - \xi\, m_{\text{V}} \phi_2^2\right) \\ &+ \text{interaction terms}\end{aligned} \tag{D.9}$$

One sees that the ϕ_2 obeys the Klein–Gordon equation with mass given by $\xi\, m_{\text{V}}^2$. However, let us take a look at the vector boson first. One can make the equation of motion for the vector field to give

$$\left(\partial_\mu \partial^\mu + m^2\right) A^\nu - \left(1 - \frac{1}{\xi}\right) \partial^\nu \left(\partial_\mu A^\mu\right) = J^\nu \tag{D.10}$$

where $J^\nu = -\delta \mathcal{L}_{\text{int}}/\delta A_\nu$ is the current that couples to the vector field. The coefficients of A^ν in momentum space are expressed as

$$\begin{aligned} g_{\mu\nu}\left(k^2 - m_{\text{V}}^2\right) - \left(1 - \frac{1}{\xi}\right) k_\mu k_\nu &= \left(g_{\mu\nu} - \frac{k_\mu k_\nu}{k^2}\right)\left(k^2 - m_{\text{V}}^2\right) \\ &+ \left(\frac{k_\mu k_\nu}{k^2}\right)\frac{1}{\xi}\left(k^2 - \xi\, m_{\text{V}}^2\right)\end{aligned} \tag{D.11}$$

By making the inverse of Eq. (D.11), one can construct the boson propagator which is expressed as

$$\Delta_{F\mu\nu} = -\left(g_{\mu\nu} - (1-\xi)\frac{k_\mu k \nu}{k^2 - \xi\, m_{\text{V}}^2}\right)\frac{1}{k^2 - m_{\text{V}}^2} \tag{D.12}$$

One sees that $\Delta_{F\mu\nu} \to 1/k^2$ as $k^2 \to \infty$ and assures renormalizability of the theory with the massive vector boson by applying the same logic to prove the renormalizability of the massless gauge boson, that is QED. Since the propagator Eq. (D.12) includes that of U-gauge for a special choice of $\xi \to \infty$, the gauge invariance assures the renormalizability of the spontaneously broken gauge symmetry. Because of this, it is referred to as the renormalizable gauge (R-gauge).

The price to pay is that it has an unwanted pole at $k^2 = \xi\, m_{\text{V}}^2$. Since this is gauge dependent, one doubts if it disappears in the final result. Indeed, it is the unwanted would-be-Goldstone boson ϕ_2 that comes to rescue the situation. As is evident from the Lagrangian in Eq. (D.9), the ϕ_2 is a scalar field which obeys the Klein–Gordon equation with mass $\xi\, m_{\text{V}}^2$. In the perturbation calculation using the Feynman diagrams, the ϕ_2 propagator

$$\frac{i}{k^2 - \xi\, m_{\text{V}}^2} \tag{D.13}$$

has to be included. Cancellation between the two unphysical poles was proved to all orders in perturbation theory by using the Ward identity (BRST symmetry in the non-Abelian theory) [698, 699].

Physical Role of the Would-Be-Goldstone Boson We have seen that the role of the ϕ_2 is exactly to cancel the unphysical pole at $k^2 = \xi m_V^2$. This is a mathematical argument. We want to clarify what it means physically. Its role may be seen by setting $\xi = 1$. Then, the propagator turns into $g_{\mu\nu}/k^2$, the familiar Feynman propagator in QED. Therefore, $\xi = 1$ gauge is often referred as the 't Hooft–Feynman gauge. Because of its simplicity, it is the favored choice for most calculations. There is, however, a crucial difference between the Feynman propagator in QED and the 't Hooft–Feynman propagator in the spontaneously broken gauge theory. In QED, the numerator $g_{\mu\nu} = \sum_\lambda \epsilon_\mu(\lambda)\epsilon_\nu(\lambda)$ is the sum of all four polarization states. However, the longitudinal ($\lambda = L$) and scalar ($\lambda = 0$) parts are unphysical. In Vol. 1, Sect. 5.3.4, we saw that their contribution compensates each other because of the negative norm of the scalar polarization and they do not contribute to physical processes. Here, the vector boson is massive and the longitudinal polarization is physical. We do not want it to be compensated by the scalar polarization. To clarify the role of each component, we rewrite the 't Hooft propagator as

$$\frac{1}{k^2 - m_V^2}\left(g_{\mu\nu} - (1-\xi)\frac{k_\mu k_\nu}{k^2 - \xi m_V^2}\right) = \frac{g_{\mu\nu} - (k_\mu k_\nu)/(m_V^2)}{k^2 - m_V^2} + \frac{(k_\mu k_\nu)/(m_V^2)}{k^2 - \xi m_V^2} \quad \text{(D.14)}$$

As the polarization sum is given by

$$\sum_{\lambda=\text{all}} \epsilon_\mu(\lambda)\epsilon_\nu(\lambda) = g_{\mu\nu} = \underbrace{\left(g_{\mu\nu} - \frac{k_\mu k_\nu}{m_V^2}\right)}_{\text{sum of } \lambda=\pm, L} + \underbrace{\frac{k_\mu k_\nu}{m_V^2}}_{\text{scalar}} \quad \text{(D.15)}$$

one sees that the first part of Eq. (D.14) represents the physical part and is ξ independent. The second part represents the contribution of the scalar polarization which is unphysical and ξ dependent. This is the part that is canceled by the ϕ_2. The would-be-Goldstone boson appears only as an internal line and its role is precisely to cancel the unphysical scalar polarization. It is reminiscent of the ghost in the non-Abelian gauge theory (see Vol. 1, Sect. 11.7.3) whose sole role is to cancel the unphysical part of the self interacting non-Abelian gauge contribution.

In retrospect, every time we encountered the danger of unitarity violation or unrenormalizability of the gauge theory, like scalar and longitudinal polarizations in QED, the ghost in non-Abelian gauge theory, and now the massive gauge boson in the spontaneously broken gauge theory, the bad guy has always been the unphysical polarization. Each time, a different prescription was presented, but it has always been the gauge invariance that guaranteed the ultimate cancellation. This stresses the importance of the formalism to keep manifest gauge invariance.

R-Gauge in the Electroweak Theory Since the R-gauge is realized in the covariant gauge, it needs ghosts when using it in the non-Abelian gauge. As it is beyond our scope to present mathematical and formal details of higher order calculations, we stop here. Instead, we simply list some players that do not appear in the U-gauge. The starting Lagrangian is the same as that used in the U-gauge (Eq. (1.16)) with addition of the gauge fixing and the ghost terms. The parameterization of the

Appendix D 't Hooft's Gauge | 529

symmetry breakdown is the linearized version of the U-gauge, that is,

$$\varphi = \exp\left[i\frac{\boldsymbol{\phi}\cdot\boldsymbol{\tau}}{v}\right]\begin{bmatrix}0\\ \frac{v+H}{\sqrt{2}}\end{bmatrix} \to \left[1 + i\frac{\boldsymbol{\phi}\cdot\boldsymbol{\tau}}{v}\right]\begin{bmatrix}0\\ \frac{v+H}{\sqrt{2}}\end{bmatrix} = \frac{1}{\sqrt{2}}\begin{bmatrix}\phi_2 + i\phi_1\\ v + H - i\phi_3\end{bmatrix} \quad (D.16)$$

fermion: $\dfrac{i\not{p} + m_f}{p^2 - m_f^2 + i\varepsilon}$

Higgs scalar: $\dfrac{i}{q^2 - m_f^2 + i\varepsilon}$

$\gamma, W^{\pm}, Z$: $\dfrac{-i\left(g_{\mu\nu} - (1-\xi)\dfrac{q_\mu q_\nu}{q^2 - \xi m_V^2}\right)}{q^2 - m_V^2 + i\varepsilon}$

$m_V = m_\gamma(=0), m_W, m_Z$

Gauge boson

$\phi^{\pm}$: $\dfrac{i}{q^2 - \xi m_V^2 + i\varepsilon}$

ϕ^0: $\dfrac{i}{q^2 - \xi m_Z^2 + i\varepsilon}$ (Would be Goldstone boson)

Ghosts $\omega^{\pm}$: $\dfrac{-i}{q^2 - \xi m_V^2 + i\varepsilon}$

ω^0: $\dfrac{-i}{q^2 - \xi m_Z^2 + i\varepsilon}$

ω^γ: $\dfrac{-i}{q^2 + i\varepsilon}$

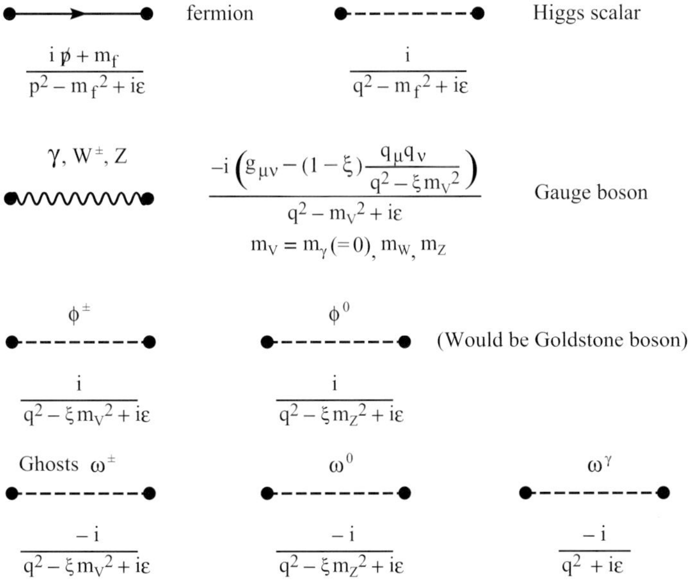

Figure D.1 Propagators in the R-gauge for the fermion, Higgs, the gauge bosons, the would-be-Goldstonebosons, and ghosts.

$f\bar{f}H^0$: $-i\dfrac{1}{2}\dfrac{g_W}{m_W}m_f$

$f_u f_d \phi^{\pm}$: $-i\dfrac{g_W}{2\sqrt{2}\,m_W}[m_d\gamma^\mu(1-\gamma^5) - m_u\gamma^\mu(1+\gamma^5)]V_{KM}$

with $-i\dfrac{1}{2}\dfrac{g_W}{m_W}m_f$

$f_d \bar{f}_d \phi^0$: $\dfrac{1}{2}\dfrac{g_W}{m_W}m_d\gamma^5$

$f_u \bar{f}_u \phi^0$: $-\dfrac{1}{2}\dfrac{g_W}{m_W}m_u\gamma^5$

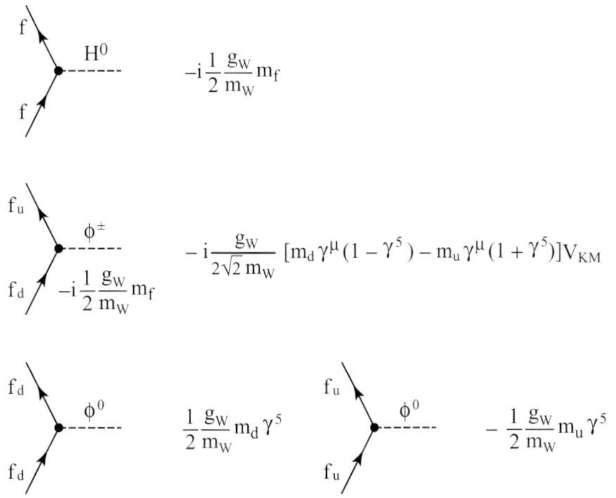

Figure D.2 Interactions of the scalar particles with the fermion field in the R-gauge. Note, the coupling is different for the $f_d \to f_u$, $f_d \to f_d$, and $f_u \to f_u$ where (f_u, f_d) is a weak doublet and V_{KM} is the Kobayashi–Maskawa matrix. $V_{KM} = 1$ for leptons. Note, the ghosts only couple to the gauge bosons.

The 't Hooft gauge fixing terms are

$$\begin{aligned}
\mathcal{L}_{gf} &= -\frac{1}{2\xi}\left[2F_+F_- + F_Z^2 + F_\gamma^2\right] \\
F_\pm &= \partial^\mu W_\mu^\pm - m_W \xi \phi^\pm \\
F_Z &= \partial^\mu Z_\mu - m_Z \xi \phi_3 \\
F_\gamma &= \partial^\mu A_\mu
\end{aligned} \tag{D.17}$$

In addition, four ghosts $\omega_\pm$, ω_Z, ω_γ which couple to $W^\pm$, Z and A are necessary. All the gauge propagators are expressed in the form given by Eq. (D.12). In addition to the Higgs, one needs to include propagators for would-be-Goldstone bosons ($\phi_\pm$, ϕ_3) and the ghosts in the actual calculations. Although we do not use Feynman diagrams in the R-gauge, we illustrate a few of them to have some idea how the calculations are carried out. We show propagators of the scalar particles in Figure D.1, Higgs (H), the would-be-Goldstone-bosons ($\phi^\pm$, ϕ^0), ghosts ($\omega^\pm$, ω^0, ω^γ) and their interactions with the fermion in Figure D.2. Note, the ghosts couple only to gauge bosons. A complete list of the Feynman diagrams in the R-gauge is given in Cheng and Li's textbook [700].

Appendix E
Fierz Transformation

The Fierz transformation is an operation to change the order of the fermion fields in the four-fermion interaction Lagrangian.

$$\Gamma^i \equiv (\overline{\psi}_1 O^i \psi_2)(\overline{\psi}_3 O_i \psi_4) = \sum_j C_{ij} (\overline{\psi}_1 O^j \psi_4)(\overline{\psi}_3 O_j \psi_2) \tag{E.1}$$

where $O^i (i = 1 \sim 16)$ are sixteen 4×4 independent matrices given below.

$$O^i = 1(S), \quad \gamma^\mu(V), \quad \sigma^{\mu\nu}|_{\mu<\nu}(T), \quad \gamma^\mu\gamma^5(A), \quad \gamma^5(P) \tag{E.2a}$$

$$O_i = (O^i)^{-1} = 1(S), \quad \gamma_\mu(V), \quad \sigma_{\mu\nu}|_{\mu<\nu}(T), \quad -\gamma_\mu\gamma^5(A), \quad \gamma^5(P) \tag{E.2b}$$

Note that there are four components in V and A, and six in T, and summation i is meant to include all of them.

To derive C_{ij}, we show first that

$$\mathcal{L} = (\overline{\psi}_1 M \psi_2)(\overline{\psi}_3 N \psi_4) = -\frac{1}{4}\sum_j (\overline{\psi}_1 M O^j N \psi_4)(\overline{\psi}_3 O_j \psi_2) \tag{E.3}$$

Proof: Writing the left-hand side of Eq. (E.3) in the components, it can be written as follows.

$$\mathcal{L} = \sum_{abcd} \overline{\psi}_{1a} M_{ab} \psi_{2b} \overline{\psi}_{3c} N_{cd} \psi_{4d} = -\sum_{abcd} M_{ab} N_{cd} \overline{\psi}_{1a} \psi_{4d} \overline{\psi}_{3c} \psi_{2b} \tag{E.4}$$

Fixing a and d, $M_{ab} N_{cd}$ can be considered as a 4×4 matrix with its elements specified by bc. As any 4×4 matrixes sandwiched by the fermion fields can be expressed as a linear combination of the sixteen O^i's given above, it can be expressed as

$$\begin{aligned} -M_{ab} N_{cd} &= (B^S)_{ad}(1)_{cb} + (B^{V\mu})_{ad}(\gamma_\mu)_{cb} + (B^{T\mu\nu})_{ad}(\sigma_{\mu\nu})_{cb} \\ &\quad + (B^{A\mu})_{ad}(-\gamma_\mu\gamma^5)_{cb} + (B^P)_{ad}(\gamma^5)_{cb} \\ &= \sum_j (B^j)_{ad}(O_j)_{cb} \end{aligned} \tag{E.5}$$

Elementary Particle Physics, Volume 2: Foundations of the Standard Model, First Edition. Yorikiyo Nagashima.

To derive the coefficient $(B^j)_{ad}$, we multiply $(O^j)_{bc}$ on both sides and take traces. Noting $\mathrm{Tr}[O^j O_k] = 4\delta^j_k$, it yields

$$-(M O^j N)_{ad} = 4(B^j)_{ad} \quad \rightarrow \quad (B^j)_{ad} = -\frac{1}{4}(M O^j N)_{ad} \tag{E.6}$$

By substituting Eq. (E.6) in Eq. (E.5), and in turn, Eq. (E.5) in Eq. (E.4), reproduces Eq. (E.3). □

Next, we consider the case where $M = \gamma^A$, $N = \gamma_A$ summed over $A = 1 \sim 4$

$$\begin{aligned}
&\text{l.h.s.}\ (j = S) = -\left(\gamma^\mu 1 \gamma_\mu\right)_{ad} = -4(1)_{ad}\\
&\text{l.h.s.}\ (j = V) = -\left(\gamma^\mu \gamma^\nu \gamma_\mu\right)_{ad} = 2(\gamma^\nu)_{ad}\\
&\text{l.h.s.}\ (j = T) = -\left(\gamma^\mu \sigma^{\rho\sigma} \gamma_\mu\right)_{ad} = -2i(g^{\rho\sigma} - g^{\sigma\rho})_{ad} = 0\\
&\text{l.h.s.}\ (j = A) = -\left(\gamma^\mu \gamma^\nu \gamma^5 \gamma_\mu\right)_{ad} = -2(\gamma^\nu \gamma^5)_{ad}\\
&\text{l.h.s.}\ (j = P) = -\left(\gamma^\mu \gamma^5 \gamma_\mu\right)_{cd} = 4(\gamma^5)_{ad}
\end{aligned} \tag{E.7}$$

Comparing both sides,

$$B^{VS} = -1\,, \quad B^{VV} = 1/2\,, \quad B^{VT} = 0\,, \quad B^{VA} = -1/2\,, \quad B^{VP} = 1 \tag{E.8}$$

which gives

$$\begin{aligned}
&\Gamma^V = \left(\overline{\psi}_1 \gamma^\mu \psi_2\right)\left(\overline{\psi}_3 \gamma_\mu \psi_4\right) = \sum_j C_{Vj}\left(\overline{\psi}_1 O^j \psi_2\right)\left(\overline{\psi}_3 O_j \psi_4\right)\\
&C_{VS} = -1\,, \quad C_{VV} = 1/2\,, \quad C_{VT} = 0\,, \quad C_{VA} = -1/2\,, \quad C_{VP} = 1
\end{aligned} \tag{E.9}$$

Other cases $(M, N) = (O^i, O_i)$, $i = S, T, A, P$ can be calculated similarly. Table E.1 shows the coefficients C_{ij}.

As a special case, we have

$$\begin{aligned}
\psi_{2b}\overline{\psi}_{3c} = \left[\psi_2\overline{\psi}_3\right]_{bc} = -\frac{1}{4}\Big[&\left(\overline{\psi}_3\psi_2\right)1 + \left(\overline{\psi}_3\gamma^\mu\psi_2\right)\gamma_\mu + \left(\overline{\psi}_3\sigma^{\mu\nu}\psi_2\right)_{\mu<\nu}\sigma_{\mu\nu}\\
&- \left(\overline{\psi}_3\gamma^\mu\gamma^5\psi_2\right)\gamma_\mu\gamma^5 + \left(\overline{\psi}_3\gamma^5\psi_2\right)\gamma^5\Big]_{bc}
\end{aligned} \tag{E.10}$$

The parity violating four-Fermi interaction takes another form, namely,

$$\Gamma^{i'} = \left(\overline{\psi}_1 O^i \gamma^5 \psi_2\right)\left(\overline{\psi}_3 O_i \psi_4\right) \tag{E.11}$$

Table E.1 Coefficients of the Fierz transformation.

i\j	S	V	T	A	P
S	−1/4	−1/4	−1/4	−1/4	−1/4
V	−1	1/2	0	−1/2	1
T	−3/2	0	1/2	0	−3/2
A	−1	−1/2	0	1/2	1
P	−1/4	1/4	−1/4	1/4	−1/4

By putting $\psi_2' = \gamma^5 \psi_2$, $\Gamma^{i'}$ can be expressed as

$$\begin{aligned}\Gamma^{i'} &= \sum C'_{ij} \left(\overline{\psi}_{1a}(O^j)_{ab}\psi_{4b}\right)\left(\psi_{3c}\left(O_j\gamma^5\right)_{cd}\psi_{2d}\right) \\ &= \sum M^{i'}_{ac;bd}\overline{\psi}_{1a}\psi_{4d}\overline{\psi}_{3c}\psi_{2b}\end{aligned} \tag{E.12}$$

We can perform a similar calculation for the Eq. (E.12) as we did for $M = O^V$, $N = O_V$ and obtain

$$C'_{ij} = C_{ij} \tag{E.13}$$

From the above table, we can prove

$$\sum_j C_{ij} C_{jk} = \delta_{ik} \tag{E.14}$$

which it should because the repeated transformation takes Γ^i back to where it started. As a special example, the following combination is often used in the neutrino interactions.

$$\left(\overline{\psi}_1\gamma^\mu(1-\gamma^5)\psi_2\right)\left(\overline{\psi}_3\gamma_\mu(1-\gamma^5)\psi_4\right) = \left(\overline{\psi}_1\gamma^\mu(1-\gamma^5)\psi_4\right)\left(\overline{\psi}_3\gamma_\mu(1-\gamma^5)\psi_2\right) \tag{E.15}$$

Appendix F
Collins–Soper Frame

To prove Eqs. (2.86), we use variables that are manifestly covariant under rotation about the z' axis in the CS frame (the rest frame of the lepton pair: see Figure 2.24). The magnitude of the lepton momenta in their rest frame is given by

$$k = |\boldsymbol{k}| = Q/2\,, \quad Q = \sqrt{(k_1 + k_2)} \tag{F.1}$$

where k_1, k_2 are four momenta of the leptons in the CM frame of the two incoming hadrons. Then, without loss of generality, we can choose the four momenta in the CS frame by

$$\begin{aligned} k'_{10} &= k\,, \quad k'_{1x} = k\sin\theta\,, \quad k'_{1z} = k\cos\theta\,, \\ k'_{20} &= k\,, \quad k'_{2x} = -k\sin\theta\,, \quad k'_{1z} = -k\cos\theta \end{aligned} \tag{F.2}$$

We have fixed the x'–z' plane defined by $\boldsymbol{P}_A$, $\boldsymbol{P}_B$, $\boldsymbol{Q}$. Then, the direction of the y' axis is defined by

$$\hat{\boldsymbol{R}_T} \equiv (0, 0, 1, 0) = \frac{\boldsymbol{P}_A \times \boldsymbol{Q}}{|\boldsymbol{P}_A \times \boldsymbol{Q}|} \tag{F.3}$$

By the boost in the $\boldsymbol{Q}_T$ direction with $\beta\gamma = |\boldsymbol{Q}_T|/Q$, $\gamma = \sqrt{Q^2 + \boldsymbol{Q}_T^2}/Q$, they become variables in the $*$ frame. We denote the variables with * in the $*$ frame.

$$\begin{aligned} k^*_{10} &= \gamma k(1 + \beta\sin\theta)\,, \quad k^*_{1x} = \gamma k(\ \sin\theta + \beta)\,, \quad k^*_{1z} = k\cos\theta \\ k^*_{20} &= \gamma k(1 - \beta\sin\theta)\,, \quad k^*_{2x} = \gamma k(-\sin\theta + \beta)\,, \quad k^*_{2z} = -k\cos\theta \end{aligned} \tag{F.4}$$

$$\begin{aligned} k_1^{+*} &= (k^*_{10} + k^*_{1z})/\sqrt{2} = k[\cos\theta + \gamma(1 + \beta\sin\theta)]/\sqrt{2} \\ k_2^{-*} &= (k^*_{20} - k^*_{2z})/\sqrt{2} = k[\cos\theta + \gamma(1 - \beta\sin\theta)]/\sqrt{2} \\ k_1^{-*} &= (k^*_{10} - k^*_{1z})/\sqrt{2} = k[-\cos\theta + \gamma(1 + \beta\sin\theta)]/\sqrt{2} \\ k_2^{+*} &= (k^*_{20} + k^*_{2z})/\sqrt{2} = k[-\cos\theta + \gamma(1 - \beta\sin\theta)]/\sqrt{2} \end{aligned} \tag{F.5}$$

Elementary Particle Physics, Volume 2: Foundations of the Standard Model, First Edition. Yorikiyo Nagashima.

$$
\begin{aligned}
k_1^{+*}k_2^{-*} - k_1^{-*}k_2^{+*} &= (k^2/2)[\{(\cos\theta + \gamma)^2 - \beta^2\gamma^2\sin^2\theta\} \\
&\quad - \{(-\cos\theta + \gamma)^2 - \beta^2\gamma^2\sin^2\theta\}] \\
&= 2k^2\gamma\cos\theta = 2\left(\frac{Q}{2}\right)^2 \frac{\sqrt{Q^2 + Q_\mathrm{T}^2}}{Q}\cos\theta \\
\therefore \quad \cos\theta &= \frac{2}{Q\sqrt{Q^2 + \boldsymbol{Q}_\mathrm{T}^2}}(k_1^+k_2^- - k_1^-k_2^+)^* \\
&= \frac{2}{Q\sqrt{Q^2 + \boldsymbol{Q}_\mathrm{T}^2}}(k_1^+k_2^- - k_1^-k_2^+) \qquad \text{(F.6)}
\end{aligned}
$$

where z boost invariance was used in deriving the last equality.

$$
\begin{aligned}
\Delta_x^* &= k_{1x}^* - k_{2x}^* = \gamma k(\sin\theta + \beta) - \gamma k(-\sin\theta + \beta) = 2\gamma k\sin\theta \\
\boldsymbol{\Delta}_\mathrm{T}^{2*} &= \Delta_x^{2*} = 4k^2\gamma^2\sin^2\theta = \gamma^2 Q^2\sin\theta^2 \\
(\boldsymbol{\Delta}_\mathrm{T}^* \cdot \boldsymbol{Q}_\mathrm{T}^*)^2 &= (\Delta_x^*)^2(Q_\mathrm{T}^*)^2 = (Q\gamma\sin\theta)^2(Q_\mathrm{T})^2 = Q^2Q_\mathrm{T}^2\gamma^2\sin^2\theta \\
\frac{(\boldsymbol{\Delta}_t^* \cdot \boldsymbol{Q}_\mathrm{T})^2}{Q^2(Q^2 + Q_\mathrm{T}^2)} &= \beta^2\gamma^2\sin^2\theta \\
\therefore \quad \frac{\boldsymbol{\Delta}_\mathrm{T}^{2*}}{Q^2} - \frac{(\boldsymbol{\Delta}_\mathrm{T}^* \cdot \boldsymbol{Q}_\mathrm{T}^*)^2}{Q^2(Q^2 + Q_\mathrm{T}^2)} &= \gamma^2\sin^2\theta - \beta^2\gamma^2\sin^2\theta = \sin^2\theta \qquad \text{(F.7)}
\end{aligned}
$$

Thus far, we have only used covariant variables under rotation around the z' axis. An explicit azimuthal angle dependence of the momentum of the produced lepton pair can be derived using the $\hat{\boldsymbol{R}}_\mathrm{T}$ vector defined in Eq. (F.3).

$$
\begin{aligned}
\boldsymbol{\Delta}_\mathrm{T}^* \cdot \hat{\boldsymbol{R}}_\mathrm{T} &= \Delta_{Ty}^* = Q\sin\theta\sin\phi \\
\boldsymbol{\Delta}_\mathrm{T}^* \cdot \hat{\boldsymbol{Q}}_\mathrm{T}^* &= \Delta_{Tx}^* = \gamma Q\sin\theta\cos\phi \\
\therefore \quad \frac{(Q^2 + Q_\mathrm{T}^2)^{1/2}}{Q}\frac{\boldsymbol{\Delta}_\mathrm{T}^* \cdot \hat{\boldsymbol{R}}_\mathrm{T}}{\boldsymbol{\Delta}_\mathrm{T}^* \cdot \hat{\boldsymbol{Q}}_\mathrm{T}^*} &= \gamma\frac{Q\sin\theta\sin\phi}{\gamma Q\sin\theta\cos\phi} = \tan\phi \qquad \text{(F.8)}
\end{aligned}
$$

Notice that the right-hand side of Eqs. (F.6), (F.7) and (F.8) are all manifestly invariant under z boosts, so that the equations apply equally well using laboratory frame variables as well as those in center of mass frame of the two incoming hadrons.

Appendix G
Multipole Expansion of the Vertex Function

We relate observables in the relativistic vertex function given in Eq. (3.58) to magnetic dipole and electric quadrupole moments. Here, we mainly follow arguments of [140]. The vertex function we treat here is a transition matrix element of electric current between initial and final states of a vector boson with momentum p, spin s and helicity h.

$$g_{VWW}\Gamma^{\mu}_{h',h} \equiv g_{VWW}\langle p's'h'|J^{\mu}(0)|psh\rangle \tag{G.1a}$$

$$\begin{aligned}\Gamma^{\mu}_{h',h} = i(\epsilon^*_{h'})_\beta V^{\mu\alpha\beta}(\epsilon_h)_\alpha \overset{(p^2=p'^2=m^2)}{=\!=} -\Big[G_1^V(p+p')^\mu(\epsilon^*_{h'}\cdot\epsilon_h) \\ + G_2^V\{\epsilon^{*\mu}_{h'}(q\cdot\epsilon_h) - \epsilon_h{}^\mu(q\cdot\epsilon^*_{h'})\} + G_3^V(p+p')^\mu(q\cdot\epsilon^*_{h'})(q\cdot\epsilon_h)/m^2\Big]\end{aligned} \tag{G.1b}$$

$$G_1^V(q^2) = 1+\frac{1}{2}\lambda_V\frac{q^2}{m^2}, \quad G_2^V(q^2) = 1+\kappa_V+\lambda_V, \quad G_3^V(q^2) = -\lambda_V \tag{G.1c}$$

$$q^\mu = (p-p')^\mu \tag{G.1d}$$

Breit Frame We work in the so-called brick wall or Breit frame where the momenta of the incoming and outgoing particles are expressed as

$$\begin{aligned} p^\mu &= (E,0,0,p), \quad E = \sqrt{p^2+m^2} \\ p'^\mu &= (E',0,0,-p), \quad E' = \sqrt{p^2+m'^2}\end{aligned} \tag{G.2}$$
[1]

1) The sign of the momenta defined here in the Breit frame is opposite to the convention used in this book elsewhere, but adopted to be consistent with treatments in [140, 701].

Elementary Particle Physics, Volume 2: Foundations of the Standard Model, First Edition. Yorikiyo Nagashima.

We choose the momenta $\boldsymbol{p}, \boldsymbol{p}'$ to be in the $x-z$ plane. The polarization vectors in the Breit frame are expressed as follows:

$$\begin{aligned} q &= p - p' = (0,0,0,2p) \to q^2 = -4p^2 \\ \epsilon_+ &= -2^{-1/2}(0,1,+i,0) \\ \epsilon_- &= \ \ 2^{-1/2}(0,1,-i,0) \\ \epsilon_0 &= (p/m, 0, 0, E/m) \\ \epsilon'_+ &= \epsilon_- = \ \ 2^{-1/2}(0,1,-i,0) \\ \epsilon'_- &= \epsilon_+ = -2^{-1/2}(0,1,+i,0) \\ \epsilon'_0 &= (p/m, 0, 0, -E/m) \end{aligned} \tag{G.3}$$

Then, after straightforward calculations, we obtain

$$\Gamma^-_{0+} = -\frac{2pE}{m} G_2 \tag{G.4a}$$

$$\Gamma^0_{00} = -2E\left[G_1 + \frac{2p^2}{m^2}\left(G_1 - G_2 - \frac{2E^2}{m^2} G_3\right)\right] \tag{G.4b}$$

$$\Gamma^0_{-+} = 2E G_1 \tag{G.4c}$$

Other $\Gamma^\mu_{h'h}$ can be calculated similarly, but we do not need them.

Magnetic Moment The magnetic moment $\boldsymbol{\mu}$ is defined classically by its energy in the magnetic field $\boldsymbol{B}$.

$$E = -\boldsymbol{\mu} \cdot \boldsymbol{B} \tag{G.5}$$

Quantum mechanically, the static magnetic energy is given by the expectation value of the interaction Hamiltonian. For the electromagnetic interaction,

$$H_{\text{mag}} = e \int d^3x\, J_\mu A^\mu = -e \int d^3x\, \boldsymbol{J}(0,\boldsymbol{x}) \cdot \boldsymbol{A}(0,\boldsymbol{x}) \tag{G.6}$$

where we set $t = 0$ and $A^0(0,\boldsymbol{x}) = 0$. Let the electromagnetic potential have a form

$$\boldsymbol{A} = C \frac{\boldsymbol{e}_x - i\boldsymbol{e}_y}{\sqrt{2}} e^{-i\boldsymbol{q}\cdot\boldsymbol{x}}, \quad \boldsymbol{q} = 2\boldsymbol{p} = 2p\boldsymbol{e}_z \tag{G.7}$$

Then, the magnetic field is given by

$$\boldsymbol{B} = \nabla \times \boldsymbol{A} = -i\boldsymbol{q} \times \boldsymbol{A} \to \boldsymbol{A} = \frac{1}{i|\boldsymbol{q}|}(B_y, -B_x, 0) \tag{G.8}$$

Substituting Eq. (G.8) into Eq. (G.6) and equating with Eq. (G.5), we obtain

$$\mu_- \equiv \frac{1}{\sqrt{2}}(\mu_x - i\mu_y) = \lim_{|q|\to 0} \frac{1}{|\boldsymbol{q}|} \int d^3x\, e J_-(0,\boldsymbol{x}) e^{-i\boldsymbol{q}\cdot\boldsymbol{x}} \tag{G.9}$$

Applying the Wigner–Eckart theorem, we have

$$\begin{aligned}\mu &\equiv \langle s s_z = s|\mu_z|s s_z = s\rangle = \langle ss|10ss\rangle\langle s||\mu||s\rangle \\ \langle s s'_z|\mu_-|s s_z\rangle &= \langle s s'_z|1-1 s s_z\rangle\langle s||\mu||s\rangle \\ &= \lim_{|q|\to 0}\frac{1}{|\boldsymbol{q}|}\int d^3x\langle s s'_z|eJ_-(0,\boldsymbol{x})|s s_z\rangle e^{-i\boldsymbol{q}\cdot\boldsymbol{r}}\end{aligned}\tag{G.10}$$

The right-hand side of Eq. (G.10) is related to the vertex function $\Gamma^-_{h'h}$ through Eq. (G.1a). The relativistic vertex function has been expressed as the helicity amplitude in the Breit frame. In this frame, the direction of the final state momentum is along the negative z axis, and hence $h = s_z$, $h' = -s'_z$. We find

$$\begin{aligned}\Gamma^{\mu}_{h'h}/2E &= \langle -\boldsymbol{p}h'|J^{\mu}(0)|\boldsymbol{p}h\rangle/2E \\ &\xrightarrow{|q|/m\to 0} (-1)^{s'+s'_z}\int d^3x\langle s's'_z|J^{\mu}|s s_z\rangle e^{-i\boldsymbol{q}\cdot\boldsymbol{x}}\end{aligned}\tag{G.11}$$

The factor $2E$ is the relativistic normalization factor. The factor $(-1)^{s'+s'_z}$ arises from rotation of the final quantization axis by 180°. By denoting the rotation operator as $R_y(\theta)$, we have

$$\begin{aligned}|s's'_z\rangle &= R_y(\pi)|s'h'\rangle = \sum_{\lambda}|s'\lambda\rangle\langle s'\lambda|R_y(\pi)|s'h'\rangle = \sum_{\lambda} d^{s'}_{\lambda h'}(\pi)|s'\lambda\rangle \\ &= (-1)^{s'-h'}|s'-h'\rangle = (-1)^{s'+s'_z}|s'-h'\rangle\end{aligned}\tag{G.12}$$

where $d^{s}_{h'h}(\theta)$ is the rotation matrix (see Vol. 1, Appendix E). From Eq. (G.10), we obtain

$$\mu = \frac{\langle ss|10ss\rangle}{\langle s s'_z|1-1 s s_z\rangle}\lim_{|q|\to 0}\frac{1}{|\boldsymbol{q}|}\int d^3x\langle s s'_z|eJ_-(0,\boldsymbol{x})|s s_z\rangle e^{-i\boldsymbol{q}\cdot\boldsymbol{r}}\tag{G.13}$$

Substituting Eq. (G.11) in the above expression and using Eq. (G.4a), we finally obtain

$$\begin{aligned}\mu &= \frac{\langle ss|10ss\rangle}{\langle s0|1-1s+\rangle}\lim_{|q|\to 0}\frac{1}{|\boldsymbol{q}|}(-1)\frac{e}{2E}\Gamma^-_{0+} \\ &= \frac{e}{2m}G_2 \overset{\text{Eq. (G.1c)}}{=\!=\!=} \frac{e}{2m}(1+\kappa_\gamma+\lambda_\gamma)\end{aligned}\tag{G.14}$$

Quadrupole Moment In the nonrelativistic treatment, the quadrupole moment Q is defined as

$$\begin{aligned}Q &= \int d^3x\langle s s_z = s|(3z^2-r^2)\rho(\boldsymbol{x})|s s_z = s\rangle \\ &= 2\int d^3x\langle ss|r^2\rho(\boldsymbol{x})P_2(\cos\theta)|ss\rangle\end{aligned}\tag{G.15}$$

where $P_2(\cos\theta)$ is the second order Legendre polynomial and $\rho(\boldsymbol{x}) = eJ_0(0, \boldsymbol{x})$ is the charge density operator normalized as

$$e = \int d^3x \langle ss|\rho(\boldsymbol{x})|ss\rangle \tag{G.16}$$

The expression Eq. (G.15) is contained in the right-hand side of Eq. (G.11). $\rho(\boldsymbol{x}) = eJ_0(0, \boldsymbol{x})$ and the exponential factor can be expanded as functions of spherical Bessel functions and Legendre polynomials of $\cos\theta$ (see Eq. (G.25)). Therefore, our task is to expand the left-hand side of Eq. (G.11) in terms of multipoles.

Multipole Expansion A general formula for multipole expansion of the relativistic vertex function was worked out in [701]. The matrix element of the electromagnetic current between $|\boldsymbol{p}'s', \lambda'\rangle$ and $|\boldsymbol{p}s, \lambda\rangle$ where $ss', \lambda\lambda'$ are spin and helicity of the initial and final states, takes an especially simple form in the Breit frame under the assumption of P and T invariance. As the electric quadrupole moment originates from nonuniform charge distribution, we are interested in the time component of the current ($\rho = eJ_0$) which behaves as a scalar under rotational operation. The corresponding vertex function is expressed as[2)]

$$\Gamma^0_{h'h} = \langle \boldsymbol{p}'s'h'|J^0(0)|\boldsymbol{p}sh\rangle = (-1)^{2s'} \sum_{J=0} \begin{pmatrix} s' & J & s \\ h' & 0 & h \end{pmatrix} Q_J(s', s) \tag{G.17}$$

Q_J is a real function if T invariant and depends on J, s', s but not on h', h, as the result of the Wigner–Eckart theorem. The amplitude vanishes if $h' + h \neq 0$ because of the angular momentum (J_3) conservation. Only terms with J = even or odd, depending on the relative parity of the initial and final states contribute in the summation of Eq. (G.17) because of the parity and rotational invariance through 180° on the y axis. The $3j$ symbol is related to the Clebsch–Gordan coefficients by

$$\begin{pmatrix} j_1 & j_2 & j_3 \\ m_1 & m_2 & m_3 \end{pmatrix} = \frac{(-1)^{j_1-j_2-m_3}}{\sqrt{2j_3+1}} \langle j_1 m_1 j_2 m_2 | j_3 - m_3\rangle \tag{G.18}$$

It vanishes unless

$$\begin{aligned} &m_1 + m_2 + m_3 = 0 \\ \text{and} \quad &|j_i - j_j| \le j_k \le j_i + j_j \quad \text{(triangular relation)} \end{aligned} \tag{G.19}$$

For the γWW triple gauge coupling (TGC) vertex function, we set $m = m', s' = s = 1$ and Eq. (G.17) is simplified to

$$\Gamma^0_{h'h} = \begin{pmatrix} 1 & 0 & 1 \\ h' & 0 & h \end{pmatrix} Q_0 + \begin{pmatrix} 1 & 2 & 1 \\ h' & 0 & h \end{pmatrix} Q_2 \tag{G.20}$$

Terms with $J > 2$ do not contribute because of the triangular relation. To relate the quadrupole moment Q with the vertex function, we need to solve Q_0 and Q_2.

2) The vertex functions corresponding to space components $J_\pm = \mp(J_1 \pm iJ_2)$ have a similar expression [701]. For $m = m'$, J_3 vanishes in the Breit frame because of the current conservation.

Therefore, we substitute Γ^0_{00} and Γ^0_{-+} given in Eq. (G.4) into the left-hand side of Eq. (G.20). Using

$$\begin{pmatrix} 1 & 0 & 1 \\ 0 & 0 & 0 \end{pmatrix} = -\frac{1}{\sqrt{3}}, \quad \begin{pmatrix} 1 & 2 & 1 \\ 0 & 0 & 0 \end{pmatrix} = \sqrt{\frac{2}{15}}$$
$$\begin{pmatrix} 1 & 0 & 1 \\ -1 & 0 & 1 \end{pmatrix} = \frac{1}{\sqrt{3}}, \quad \begin{pmatrix} 1 & 2 & 1 \\ -1 & 0 & 1 \end{pmatrix} = \sqrt{\frac{1}{30}} \tag{G.21}$$

and substituting Eqs. (G.4b) and (G.4c) into Eq. (G.20), we obtain

$$Q_0 = \frac{1}{\sqrt{3}}\left(-\Gamma^0_{00} + 2\Gamma^0_{-+}\right) = 2\sqrt{3}E\left[G_1 + \frac{2p^2}{3m^2}\left\{G_1 - G_2 - \frac{2E^2}{m^2}G_3\right\}\right] \tag{G.22a}$$

$$Q_2 = \sqrt{\frac{10}{3}}\left(\Gamma^0_{00} + \Gamma^0_{-+}\right) = -\sqrt{\frac{10}{3}}\frac{4Ep^2}{m^2}\left[G_1 - G_2 - \frac{2E^2}{m^2}G_3\right] \tag{G.22b}$$

Substituting Eq. (G.1c) into Eq. (G.22) and considering the relativistic normalization factor $2E$, Eq. (G.22) are in the nonrelativistic limit:

$$\frac{eQ_0}{2E} \to \sqrt{3}e\left[1 + \lambda\frac{2p^2}{m^2} + \frac{2p^2}{3m^2}\left\{\left(1 + \lambda\frac{2p^2}{m^2}\right) - (1 + \kappa + \lambda) + 2\lambda\right\}\right] \tag{G.23a}$$

$$\frac{eQ_2}{2E} \to -\sqrt{\frac{40}{3}}\frac{ep^2}{m^2}\left[1 + \lambda\frac{2p^2}{m^2} - (1 + \kappa + \lambda) + 2\lambda\right] \tag{G.23b}$$

We also expand the right-hand side of Eq. (G.11) in the form identical to Eq. (G.20).

$$(-1)^{1+s'_z}\int d^3x\langle 1s'_z|J^0(0,\boldsymbol{x})|1s_z\rangle e^{-i\boldsymbol{q}\cdot\boldsymbol{x}}$$
$$= \begin{pmatrix} 1 & 0 & 1 \\ h' & 0 & h \end{pmatrix} Q_0^{\mathrm{NR}} + \begin{pmatrix} 1 & 2 & 1 \\ h' & 0 & h \end{pmatrix} Q_2^{\mathrm{NR}} \tag{G.24}$$

Q_0^{NR} and Q_2^{NR} can be obtained by substituting

$$e^{-i\boldsymbol{q}\cdot\boldsymbol{r}} = \sum_{J=0}^{\infty}(-i)^J(2J+1)j_J(|\boldsymbol{q}|r)P_J(\cos\theta) \tag{G.25}$$

into the left-hand side of Eq. (G.24). For $s_z = s'_z = 1$ $(h = -h' = 1)$,

$$Q_J^{\mathrm{NR}}(q^2) = \frac{(2J+1)(-i)^J}{\begin{pmatrix} 1 & J & 1 \\ -1 & 0 & 1 \end{pmatrix}}\int d^3x\langle 11|\rho(\boldsymbol{x})P_J(\cos\theta)|11\rangle j_J(|\boldsymbol{q}|r) \tag{G.26}$$

Expanding the spherical Bessel functions up to $|\boldsymbol{q}|^2$,

$$j_0(|\boldsymbol{q}|r) \simeq 1 - |\boldsymbol{q}|^2r^2/6 = 1 - 2p^2r^2/3\,, \quad j_2(|\boldsymbol{q}|r) \simeq |\boldsymbol{q}|^2r^2/15 = 4p^2r^2/15 \tag{G.27}$$

and using Eqs. (G.16) and (G.15), we have

$$Q_0^{\mathrm{NR}} = \sqrt{3}e\left\{1 + O\left(\frac{p^2}{m^2}\right)\right\}, \quad Q_2^{\mathrm{NR}} = -\sqrt{\frac{40}{3}}Qp^2\left\{1 + O\left(\frac{p^2}{m^2}\right)\right\} \tag{G.28}$$

Identifying Eq. (G.23b) with Q_2^{NR} in the limit $p^2 \to 0$, we finally obtain

$$Q = -\frac{e}{m^2}(\kappa - \lambda) \tag{G.29}$$

Appendix H
SU(N)

Generators of the Group A set of all the linear transformations that keep the length $L = \sqrt{\sum_{i=1}^{N} |u_i|^2}$ in N-dimensional coordinate space of N complex variables $u_i (i = 1 \sim N)$ makes a group and is called $U(N)$ [702–705]. With further restriction of the determinant being 1, it is called $SU(N)$, special unitary group. The number of independent variables is N^2 for $U(N)$ and $N^2 - 1$ for $SU(N)$. Any representation matrix of $SU(3)$ group operating on N-dimensional vectors can be expressed as $n \times n$ unitary matrices, which is a function of $N^2 - 1$ independent continuous variables θ_A and traceless hermitian matrices F^A

$$U = \exp\left(i \sum_{A=1}^{N^2-1} \theta_A F^A\right), \quad F^{A\dagger} = F^A, \quad \mathrm{Tr}[F^A] = 0 \tag{H.1}$$

The traceless condition makes the determinant of the matrix equal to one. It can be proved as follows. Writing $U = e^{iQ}$, where Q is a hermitian matrix, it can be diagonalized by using some unitary matrix S. Then,

$$\begin{aligned}\det U &= \det[S^{-1} U S] = \det[S^{-1} \exp(iQ) S] = \det\left[\sum S \frac{(iQ)^n}{n!} S^{-1}\right] \\ &= \det[\exp(i S Q S^{-1})] = \det[e^{i\mathrm{Diag}[Q]}] = \prod_i e^{i\eta_i} \\ &= \exp\left(i \sum_i \eta_i\right) = e^{i\mathrm{Tr}[Q]} \quad (= 1 \text{ if } \mathrm{Tr}[Q] = 0)\end{aligned} \tag{H.2}$$

where η_is are eigenvalues of the diagonalized matrix Q. $N^2 - 1$ F^A satisfy commutation relations called Lie algebra.

$$[F^A, F^B] = i f^{ABC} F^C \tag{H.3}$$

F^As are called generators and f^{ABC} structure constants of the $SU(3)$ group representations. Since the whole representation matrices are analytic functions of θ_A's smoothly connected to the unit matrix in the limit of $\theta_A \to 0$, the Lie algebra

Elementary Particle Physics, Volume 2: Foundations of the Standard Model, First Edition. Yorikiyo Nagashima.

Eq. (H.3) completely determines the local structure of the group.[1] f^{ABC} can be made totally antisymmetric in the indices ABC. Determination of the generator is not unique because different $F^{A'}$ can be made from any independent linear combination of F^A.[2]

Let us write ψ as representing the N-dimensional ($n \geq N$) contravariant vector (with upper index) on which the $n \times n$ $SU(N)$ matrices act. Then, ψ is transformed to ψ' by U.

$$\psi = \begin{bmatrix} \xi^1 \\ \xi^2 \\ \vdots \\ \xi^n \end{bmatrix}, \quad \psi' = \begin{bmatrix} \xi^{1'} \\ \xi^{2'} \\ \vdots \\ \xi^{n'} \end{bmatrix}, \quad n \geq N \tag{H.5a}$$

$$\psi \quad \rightarrow \quad \psi' = U\psi \tag{H.5b}$$

$$\xi^{a'} = U^a{}_b \xi^b, \quad U = e^{-i\theta_i F_i} \tag{H.5c}$$

where simultaneous appearance of indices mean sums over the indices up to n. The collection of n matrices with $n = N$ is called a fundamental representation and adjoint (or regular) representation if $n = N^2 - 1$.

Gell–Mann Matrix It is customary to adopt Pauli matrices for $SU(2)$ ($F_i = \sigma_i/2$) and Gell–Mann's λ_i matrices for $SU(3)$ ($F_i = \lambda_i/2$) as generators of the fundamental representations. The vectors in the fundamental representation can be expressed as

$$\xi^1 = \begin{bmatrix} 1 \\ 0 \\ 0 \end{bmatrix}, \quad \xi^2 = \begin{bmatrix} 0 \\ 1 \\ 0 \end{bmatrix}, \quad \xi^3 = \begin{bmatrix} 0 \\ 0 \\ 1 \end{bmatrix} \tag{H.6}$$

Since $\xi^1 \leftrightarrow \xi^2$ [= $SU(2)$ transformation] is a part of $SU(3)$ transformation, we define λ_i which operates only on ξ^1 and ξ^2.

$$\lambda_i = \begin{bmatrix} & \tau_i & 0 \\ & & 0 \\ 0 & 0 & 0 \end{bmatrix}, \quad i = 1 \sim 3 \tag{H.7a}$$

1) An example of global structure is "connectedness." For example, $SU(2)$ and $O(3)$ has the same structure constants, but $SU(2)$ is simply connected and $O(3)$ is doubly connected and there is one to two correspondences between the two.

2) For instance, instead of angular momentum J_i, we can use $J_\pm = J_x \pm i J_y$ which satisfies

$$[J_+, J_-] = 2J_3, \quad [J_3, J_\pm] = \pm J_\pm \tag{H.4}$$

Then, λ_is are isospin operators. In a similar manner, we define $SU(2)$ operators "V spin" that acts only on $\xi^1 \leftrightarrow \xi^3$, and "$U$ spin" that only acts on $\xi^2 \leftrightarrow \xi^3$.

$$\lambda_4 = \begin{bmatrix} 0 & 0 & 1 \\ 0 & 0 & 0 \\ 1 & 0 & 0 \end{bmatrix}, \quad \lambda_5 = \begin{bmatrix} 0 & 0 & -i \\ 0 & 0 & 0 \\ i & 0 & 0 \end{bmatrix}, \quad V_3 = \begin{bmatrix} 1/2 & 0 & 0 \\ 0 & 0 & 0 \\ 0 & 0 & -1/2 \end{bmatrix}$$
$$\lambda_6 = \begin{bmatrix} 0 & 0 & 0 \\ 0 & 0 & 1 \\ 0 & 1 & 0 \end{bmatrix}, \quad \lambda_7 = \begin{bmatrix} 0 & 0 & 0 \\ 0 & 0 & -i \\ 0 & i & 0 \end{bmatrix}, \quad U_3 = \begin{bmatrix} 0 & 0 & 0 \\ 0 & 1/2 & 0 \\ 0 & 0 & -1/2 \end{bmatrix} \tag{H.7b}$$

$(\lambda_1, \lambda_2, \lambda_3)$, $(\lambda_4, \lambda_5, V_3)$ and $(\lambda_6, \lambda_7, U_3)$ are operators of subgroups $SU(2)_I$, $SU(2)_V$ and $SU(2)_U$. Note, however,

$$I_3 - V_3 + U_3 = 0 \tag{H.7c}$$

and are not independent. We define

$$\lambda_8 \equiv \frac{2}{\sqrt{3}}(U_3 + V_3) = \frac{1}{\sqrt{3}} \begin{bmatrix} 1 & 0 & 0 \\ 0 & 1 & 0 \\ 0 & 0 & -2 \end{bmatrix} \tag{H.7d}$$

The eight matrices $\lambda_i, (i = 1 \sim 8)$ are called the Gell–Mann matrices and constitute generators of the $SU(3)$ in the fundamental representation. The number of diagonal matrices is two, which is the rank of $SU(3)$. The eight λ_i satisfy the following commutation relations

$$\left[\frac{\lambda_I}{2}, \frac{\lambda_j}{2}\right] = i f_{ijk} \frac{\lambda_k}{2} \tag{H.8a}$$

$$\left\{\frac{\lambda_I}{2}, \frac{\lambda_j}{2}\right\} = \frac{1}{3}\delta_{ij} + i d_{ijk} \frac{\lambda_k}{2} \tag{H.8b}$$

f_{ijk}, d_{ijk} are totally antisymmetric or symmetric with respect to $i\,j\,k$ permutations and have values listed in Table H.1

Table H.1 Structure constants of $SU(3)$.

$f_{123} = 1$
$f_{147} = f_{246} = f_{257} = f_{345} = f_{516} = f_{637} = \frac{1}{2}$
$f_{458} = f_{678} = \frac{\sqrt{3}}{2}$
$d_{118} = d_{228} = d_{338} = -d_{888} = \frac{1}{\sqrt{3}}$
$d_{146} = d_{157} = d_{256} = d_{344} = d_{355} = \frac{1}{2}$
$d_{247} = d_{366} = d_{377} = -\frac{1}{2}$
$d_{448} = d_{558} = d_{668} = d_{778} = -\frac{1}{2\sqrt{3}}$

Adjoint Representation The representation matrices with $n = N^2 - 1$ can be expressed in terms of the structure constants.

$$U(N^2-1) = e^{-\sum_A \theta_A F^A} : \quad [F^A]_{BC} = -i f^{ABC}, \quad (A, B, C = 1 \sim N^2-1)^{3)} \tag{H.9}$$

Proof: Insert $A = [F^A]$, $B = [F^B]$, $C = [F^C]$ in the Jacobi identities

$$[[A, B], C] + [[B, C], A] + [[C, A], B] = 0 \tag{H.10}$$

use Eq. (H.3) and $f^{ABC} = -f^{BAC}$, then we obtain

$$f^{ABD} f^{CDE} + f^{BCD} f^{ADE} + f^{CAD} f^{BDE} = 0 \tag{H.11}$$

Then insertion of Eq. (H.9) into the above equality leads to

$$[F^A, F^C] = i f^{ACD} F^D \tag{H.12}$$

Therefore, F^A defined in Eq. (H.9) is an adjoint representation of the $SU(N)$. □

In the quantum field theory, F^As are field operators acting on state vectors (ψ). If the Lagrangian or the equation of motion is invariant under $SU(N)$ transformation, F^A is a conserved observable. The number of matrices that can be diagonalized simultaneously is called the rank of the group. The rank of $SU(N)$ is $N-1$. Matrices that are made of F^As and commute with any F^B are called Casimir operators.Particles which can be represented by the state vectors are specified and classified by the eigenvalues of the Casimir operators and F^Bs that can be diagonalized simultaneously.

Representation matrix of the complex conjugate vector ψ^* is given by U^*. Since its generator $-F^{k*}$ satisfies the same commutation relation Eq. (H.3), ψ^* is also a representation vector and is called a conjugate representation. If ψ represents a particle, then ψ^* represents its antiparticle. Therefore, the quantum numbers of antiparticles are minus those of particles. If we define a covariant vector (with lower indices) by

$$\psi^\dagger = (\psi^*)^T = (\xi^{1*}, \xi^{2*}, \cdots, \xi^{N*}) \equiv (\xi_1, \xi_2, \cdots, \xi_N) \tag{H.13}$$

It transforms as

$$\xi_a \to \xi_a{}' = \sum U^{a*}{}_b \xi^{b*} = \sum U_{ab}{}^* \xi_b = \sum \xi_b U_{ba}{}^\dagger \equiv (U^\dagger)^b{}_a \xi_b \tag{H.14a}$$

$$\text{or} \quad \psi^\dagger \to \psi^{\dagger\prime} = \psi^\dagger U^\dagger \tag{H.14b}$$

3) We use indices i, j, k when they run from 1 to n referring to representations in general, a, b, c ($a, b, c = 1 \sim 3$) to fundamental ($n = N$) representations and A, B, C to adjoint ($A, B, C = 1 \sim N^2 - 1$) representations.

If a representation matrix $U(A)$ of any element A can be expressed as

$$U(A) = \left[\begin{array}{c|c|c} U_1(A) & 0 & 0 \\ \hline 0 & U_2(A) & 0 \\ \hline 0 & 0 & \ddots \end{array}\right] \tag{H.15}$$

It is said reducible and written as

$$U = U_1 \oplus U_2 \oplus \cdots \oplus U_k \tag{H.16}$$

Casimir operator An important quantity to characterize N-dimensional representation of the $SU(N)$ group is the eigenvalue $C_2(n)$ of a quadratic Casimir operator:

$$[F^2(n)]_{jk(j,k=1\sim n)} \equiv [\boldsymbol{F} \cdot \boldsymbol{F}]_{jk} = \sum_{A=1}^{N^2-1} [F^A(n) F^{A\dagger}(n)]_{jk} = C_2(n)\delta_{jk} \tag{H.17}$$

The operator $F^2(n)$ is expressed by a unit matrix because it commutes with any group elements. It can be shown as follows. Making a commutator with one of the generators $F^j(n)$, one gets

$$\begin{aligned} [F^2, F^B] &= \sum_i [F^A F^{A\dagger}, F^B] = \sum_A \{F^A[F^A, F^B] + [F^A, F^B]F^A\} \\ &= \sum_A i f^{ABC}(F^A F^C + F^C F^A) = 0 \end{aligned} \tag{H.18}$$

where we used the fact that $[F^A]$ is a hermitian matrix and f^{ABC} is totally antisymmetric in A, B, C. Since the Casimir operator commutes with any of the generators, it commutes with any group elements.

Direct Product When there are two representations $U(A)$, $U(B)$ of dimensionality n_A, n_B which have elements $A = \{A_j\}$, $B = \{B_k\}$, products of basic elements ξ_j, η_k of each representation $\xi_j \eta_k$ span $n_A \times n_B$ dimensional space. The representation in this space $U(AB)$ is expressed as

$$U(AB)(\xi_j \eta_k) = \xi_j{}' \eta_k{}' = (U(A)\xi_j)(U(B)\eta_k) \tag{H.19}$$

and is called a direct product representation. It is written as

$$U(AB) = U(A) \otimes U(B) \tag{H.20}$$

In general, the direct product is reducible.

Example H.1

Each representation of the angular momentum $J = J_a$ and $J = J_b$ has $2J_a + 1$ and $2J_b + 1$ elements. The representation of the direct product $J_a \otimes J_b$ has $(2J_a + 1)(2J_b + 1)$ elements. It is reducible and decomposed to $J = J_1 \oplus J_2 \oplus \cdots \oplus J_n$ where $J_1 = J_a + J_b$, $J_2 = J_a + J_b - 1, \ldots, J_n = |J_a - J_b|$.

Some Useful Formulas A symmetric commutator of the generators in the fundamental representations (denoted as t^A) corresponding to Eq. (H.3) is given by

$$\{t^A, t^B\} = \frac{1}{N}\delta_{AB} + d^{ABC}t^C \tag{H.21}$$

From Eqs. (H.3) and (H.21), one obtains

$$t^A t^B = \frac{1}{2}\left[\frac{1}{N}\delta_{AB} + (d^{ABC} + i f^{ABC})t^C\right] \tag{H.22}$$

which is consistent with the normalization condition

$$\mathrm{Tr}[t^A t^B] = \sum_{c,d=1\sim N} [t^A]_{cd}[t^B]_{dc} = \frac{1}{2}\delta_{AB} \tag{H.23}$$

The constant d^{ABC} is totally symmetric in A, B, C and can be determined by

$$d^{ABC} = 2\mathrm{Tr}[\{t^A, t^B\}t^C] \tag{H.24}$$

From Eq. (H.22), one can calculate

$$\begin{aligned} \mathrm{Tr}[t^A t^B t^C] &= \frac{1}{4}(d^{ABC} + i f^{ABC}) \\ \mathrm{Tr}[t^A t^B t^C t^D] &= \frac{1}{4N}\delta_{AB}\delta_{CD} + \frac{1}{8}(d^{ABE} + i f^{ABE})(d^{CDE} + i f^{CDE}) \end{aligned} \tag{H.25}$$

Similar relations hold for the adjoint representations (denoted as T^A).

$$\begin{aligned} [T^A]_{BC} &= -i f^{ABC} \quad (A, B, C = 1 \sim N^2 - 1) \\ \mathrm{Tr}[T^A T^B T^C] &= \frac{N}{2} i f^{ABC} \\ \mathrm{Tr}[T^A T^B T^C T^D] &= \delta_{AB}\delta_{CD} + \delta_{AD}\delta_{BC} \\ &\quad + \frac{N}{4}(d^{ABE}d^{CDE} - d^{ACE}d^{BDE} + d^{ADE}d^{BCE}) \end{aligned} \tag{H.26}$$

Some Definitions of Terminology Let $F^A(n)$ be the n-dimensional representation matrix and we define $T_R(n)$ and $C_2(n)$ by

$$\mathrm{Tr}[F^A(n)F^B(n)] = T_R(n)\delta_{ab} \tag{H.27a}$$

$$\boldsymbol{F}\cdot\boldsymbol{F} = C_2(n)\mathbf{1} \tag{H.27b}$$

By convention, the normalization of the $SU(N)$ matrix is chosen to be $T_\mathrm{F} \equiv T_R(n = N) = 1/2$. With this choice, $SU(N)$ matrices satisfy relations

$$T_\mathrm{F} \equiv T_R(N_\mathrm{F}) = \frac{1}{2}, \quad T_R(N_\mathrm{A}) = N \tag{H.28a}$$

$$C_F \equiv C_2(N_\mathrm{F}) = \frac{N^2-1}{2N}, \quad C_A \equiv C_2(N_\mathrm{A}) = N \tag{H.28b}$$

where suffix F denotes fundamental ($n = N_\mathrm{F} = N$) and A denotes adjoint ($n = N_\mathrm{A} = N^2 - 1$).

Appendix I
Unitarity Relation

We prove the following unitarity theorems:

Theorem I.1

$$\mathrm{Im}\int d^4x\, e^{iq\cdot(x-y)}[i\langle\Omega|T[\phi(x)\phi(y)]|\Omega\rangle]$$
$$=\frac{1}{2}\int d^4x\, e^{iq\cdot(x-y)}\langle\Omega|\phi(x)\phi(y)|\Omega\rangle \tag{I.1}$$

where $\phi(x)$ is any field but here, for simplicity, we take it as a scalar field having mass m and production threshold $2m$. For the function $\Pi(q^2)$ defined by

$$\Pi(q^2)\equiv\int d^4x\, e^{iq\cdot(x-y)}i\langle\Omega|T[\phi(x)\phi(y)]|\Omega\rangle \tag{I.2}$$

we have

Theorem I.2

$$\Pi(q^2)=\frac{1}{\pi}\int dt\frac{\mathrm{Im}\Pi(t)}{t-q^2}=\frac{1}{m^2-q^2}+\frac{1}{\pi}\int_{4m^2}^{\infty}dt\frac{\mathrm{Im}\Pi(t)}{t-q^2} \tag{I.3}$$

Mathematical Preparation Let us refresh our memory of the basics of complex analytical functions: A complex analytic function is continuously differentiable and satisfies the Cauchy–Riemann condition

$$f(z)=f(x+iy)=u(x,y)+iv(x,y)$$
$$\frac{\partial u}{\partial x}=\frac{\partial v}{\partial y},\quad \frac{\partial u}{\partial y}=-\frac{\partial v}{\partial x} \tag{I.4}$$

Elementary Particle Physics, Volume 2: Foundations of the Standard Model, First Edition. Yorikiyo Nagashima.

Theorem I.3

The analytic function can be expressed as

$$f(z) = \frac{1}{2\pi i} \oint dy \frac{f(y)}{y - z} \tag{I.5}$$

Theorem I.4

The analytic function is uniquely defined if information on singular points (poles with their residues and branch cuts with discontinuity across the cut) and behavior at infinity are given.

Theorem I.5

A real analytic function which is real on the real axis (i. e., $f(x)^* = f(x) = f(x^*)$) satisfies

$$f(z^*) = [f(z)]^* \tag{I.6}$$

which is referred to as the Schwarz reflection principle.

Proof: As the function $f(z^*)^*$ is related to $f(z)$ by

$$f(z^*)^* = u(x, -y) - iv(x, -y) = u'(x, y) + iv'(x, y) \tag{I.7}$$

It is easily verified that u', v' satisfies the Cauchy–Riemann condition, which means $F(z) = f(z^*)^*$ is also an analytic function. Since $F(z) = f(z)$ on the real axis, by Theorem I.4, they are equal at all z which proves Eq. (I.6). □

Theorem I.6

A real analytic function that vanishes at infinity and has singularities only on the real axis satisfies the dispersion relation

$$f(z) = \frac{1}{\pi} \int_{x_{\min}}^{\infty} dx \frac{\operatorname{Im} f(z + i\epsilon)}{x - z} \tag{I.8}$$

Proof: Consider a contour in Figure I.1 and use Eq. (I.5). If the function vanishes as $|z| \to \infty$,[1] the integral on the circle vanishes and the contour integral can be

1) † $f(z)$ is assumed to vanish faster than $1/|z|$ at infinity. If it does not, one needs to consider the subtracted dispersion relation.

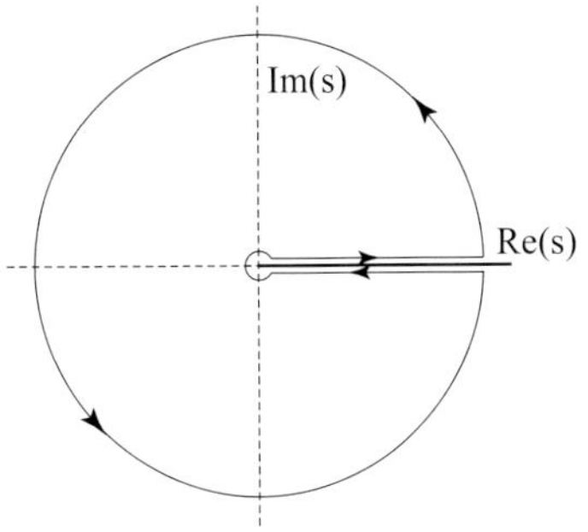

Figure I.1 Integration contour in the complex s plane, where the radius of the circle is at infinity.

expressed as

$$f(z) = \frac{1}{2\pi i}\left[\int_{+\infty}^{x_{\min}} dx \frac{f(x-i\epsilon)}{x-z} + \int_{x_{\min}}^{\infty} dx \frac{f(x+i\epsilon)}{x-z}\right]$$

$$= \frac{1}{2\pi i}\int_{x_{\min}}^{\infty} dx \frac{f(x+i\epsilon) - f(x+i\epsilon)^*}{x-z}$$

$$= \frac{1}{\pi}\int_{x_{\min}}^{\infty} dx \frac{\operatorname{Im} f(x+i\epsilon)}{x-z} \tag{I.9}$$

□

Analytic Properties of the Two-Point Function We shall investigate analytic properties of the following two-point functions and their Fourier transform which are defined by

$$\Pi(x-y) = i\langle \Omega | T[\phi(x)\phi(y)] | \Omega \rangle, \tag{I.10a}$$

$$H(x-y) = \langle \Omega | \phi(x)\phi(y) | \Omega \rangle \tag{I.10b}$$

$$\Pi(q^2) = \int d^4x\, e^{iq\cdot x}\, \Pi(x) \tag{I.10c}$$

$$H(q^2) = \int d^4x\, e^{iq\cdot x}\, H(x) \tag{I.10d}$$

First, we will prove the following properties of the function $\Pi(q^2)$.

Statement (1): The $\Pi(q^2)$ is an analytic function of q^2, with singularities only on the real q^2 axis. It has a pole at $q^2 = m^2$ and branch cut for $q^2 > (2m)^2$, where m is the mass of one particle state that the field in consideration can make from the vacuum state and $(2m)^2$ is the threshold for continuum channel.

Singularities

Singularity in perturbation theory Here, we give an example to get an idea of where singularities are located in the perturbation theory. Let us consider the electromagnetic field and put $\phi = A^\mu$. To the lowest order, the photon two-point function (i.e., propagator) is expressed as (see Vol. 1, Chapt. 8)

$$\langle \Omega | T[A^\mu(x)A^\nu(y)] | \Omega \rangle = i D_F^{\mu\nu}(x-y) = \int \frac{d^4q}{(2\pi)^4} e^{-iq(x-y)} \frac{-ig^{\mu\nu}}{q^2 + i\varepsilon} \tag{I.11}$$

If $A^\mu(x)$ is the interacting field in the Heisenberg picture, the left-hand side of Eq. (I.11) should express the full propagator, including all the radiative corrections. To $O(\alpha)$ in the interaction representation, it was calculated in Eqs. (C.1) and (C.31d).

$$D_F^{\mu\nu}(q^2) = \frac{g^{\mu\nu}}{iq^2} + \frac{1}{iq^2}(-i\Sigma_{\gamma\gamma}^{\mu\nu})\frac{1}{iq^2} + O(\alpha^2) \tag{I.12a}$$

$$\begin{aligned} -i\Sigma_{\gamma\gamma}^{\mu\nu}(q^2) &= -\int \frac{d^4k}{(2\pi)^4} \mathrm{Tr} \\ &\quad \times \left[(-ie\gamma^\mu) \frac{i(\not{k} + m)}{k^2 - m^2 + i\varepsilon} (-ie\gamma^\nu) \frac{i(\not{k} - \not{q} + m)}{(k-q)^2 - m^2 + i\varepsilon} \right] \\ &\equiv (q^2 g^{\mu\nu} - q^\mu q^\nu)\Pi_\gamma(q^2) \end{aligned} \tag{I.12b}$$

$$\begin{aligned} \Pi_\gamma(q^2) &= \Pi_\gamma(0) + \hat{\Pi}_\gamma(q^2) \\ &= \frac{\alpha}{3\pi}\left[\Delta - \left(1 + \frac{2m^2}{q^2}\right) \int_0^1 dz \ln\left\{1 - \frac{q^2}{m^2} z(1-z)\right\} - \frac{1}{3} \right] \end{aligned} \tag{I.12c}$$

$$\Delta = \frac{2}{\epsilon} - \gamma_E + \ln(4\pi\mu^2) \tag{I.12d}$$

where $\Pi_\gamma(0)$ is the divergent part which is to be subtracted. Here, m is the mass of the fermion that makes the loop. The Π_γ function has branch cut for negative argument of the logarithmic function which begins

$$q^2 > \frac{m^2}{z(1-z)} > 4m^2 \tag{I.13}$$

So, at least, the statement seems right in the renormalized perturbation theory.

Spectral representation To obtain a general expression for the two-point function Eq. (I.10a), we use the completeness relation.

$$\begin{aligned}
\mathbf{1} = |\Omega\rangle\langle\Omega| &+ \int \frac{d^3p}{(2\pi)^3 2E_p}|p\rangle\langle p| \\
&+ \int \frac{d^3p_1}{(2\pi)^3 2E_{p_1}} \int \frac{d^3p_2}{(2\pi)^3 E_{p_2}}|p_1, p_2\rangle\langle p_1, p_2| + \cdots \\
&+ \int \frac{d^3p_1}{(2\pi)^3 2E_{p_1}} \cdots \int \frac{d^3p_n}{(2\pi)^3 2E_{p_n}}|p_1, \cdots, p_n\rangle\langle p_1, \cdots, p_n| + \cdots
\end{aligned}$$

which can be combined to make it a compact expression in a form valid for one particle intermediate state

$$\equiv |\Omega\rangle\langle\Omega| + \sum_X \int \frac{d^3p_X}{(2\pi)^3 2E_X}|X_p\rangle\langle X_p| \tag{I.14}$$

where for $X = n$-particle state, $(E_X, \boldsymbol{p}_X)$ and M_X stand for

$$\boldsymbol{p}_X = \boldsymbol{p}_1 + \cdots + \boldsymbol{p}_n, \quad E_X = \sqrt{\boldsymbol{p}_X^2 + M_X^2}, \quad M_X^2 = (p_1 + \cdots + p_n)^2 \tag{I.15}$$

and $\sum_X$ sums all the rest of the freedom, including other momenta and the number of particles. Substituting Eq. (I.14) into Eq. (I.10a), we obtain

$$\begin{aligned}
&\langle\Omega|T[\phi(x)\phi(y)]|\Omega\rangle \\
&= \sum_X \int \frac{d^3p_X}{(2\pi)^3 2E_X}\left[\theta(x-y)\langle\Omega|\phi(x)|X\rangle\langle X|\phi(y)|\Omega\rangle\right. \\
&\left.+\ \theta(y-x)\langle\Omega|\phi(y)|X\rangle\langle X|\phi(x)|\Omega\rangle\right] \\
&= \sum_X \int \frac{d^3p_X}{(2\pi)^3 2E_X} \\
&\times \left[\theta(x-y)e^{-ip_X\cdot(x-y)} + \theta(y-x)e^{ip_X\cdot(x-y)}\right]|\langle\Omega|\phi(0)|X\rangle|^2
\end{aligned} \tag{I.16}$$

where we dropped the first term in Eq. (I.14) because under normal circumstances, the vacuum expectation value of the field operator vanishes and we used the translational invariance

$$\langle p_f|\phi(x)|p_i\rangle = \langle f|e^{i\hat{p}\cdot x}\phi(0)e^{-i\hat{p}\cdot x}|i\rangle = e^{-i(p_i-p_f)\cdot x}\langle f|\phi(0)|i\rangle \tag{I.17}$$

in going to the last line of Eq. (I.16). Lorentz invariance of $|\Omega\rangle$ and $\phi(0)$

$$U|\Omega\rangle = |\Omega\rangle, \quad U\phi(0)U^{-1} = \phi(0) \tag{I.18}$$

allows us to move the state $|X\rangle$ to its rest frame and we denote it as $|X_0\rangle$. Using

$$\theta(x) \equiv \theta(x^0) = \frac{1}{2\pi i}\int_{-\infty}^{\infty} dz \frac{e^{izx_0}}{z - i\epsilon} \tag{I.19}$$

$$
\begin{aligned}
&\langle\Omega|T[\phi(x)\phi(y)]|\Omega\rangle \\
&= \sum_X \int \frac{d^3p_X}{(2\pi)^3 2E_X} \frac{1}{2\pi i} \int dz \\
&\quad \times \left[\frac{e^{-i(E_X - z)t' + i p\cdot x'}}{z - i\epsilon} + \frac{e^{-i(z - E_X)t' - i p\cdot x'}}{z - i\epsilon} \right]_{x' = x - y} |\langle\Omega|\phi(0)|X_0\rangle|^2 \\
&= \sum_X \frac{i}{(2\pi)^4} \int \frac{d^3p_X}{2E_X} \int dp^0 \\
&\quad \times \left[\frac{1}{p^0 - E_X + i\epsilon} - \frac{1}{p^0 + E_X - i\epsilon} \right] e^{-i p_0 t' + i p\cdot x'} |\langle\Omega|\phi(0)|X_0\rangle|^2 \\
&= \sum_X \int \frac{d^4p}{(2\pi)^4} \frac{i}{p^2 - M_X^2 + i\epsilon} e^{-i p\cdot(x-y)} |\langle\Omega|\phi(0)|X_0\rangle|^2 \\
&= \sum_X \Delta_F(x - y; M_X^2) |\langle\Omega|\phi(0)|X_0\rangle|^2
\end{aligned} \tag{I.20}
$$

One notices that the two-point function is expressed in terms of the Feynman propagator $\Delta_F(x - y)$, but with continuous mass M_X and integrated over M_X^2. Conventionally, it is expressed as

$$
i\langle\Omega|T[\phi(x)\phi(y)]|\Omega\rangle = \int_0^\infty \frac{dM^2}{2\pi} \rho(M^2) i\Delta_F(x - y; M^2) \tag{I.21}
$$

where $\rho(M^2)$ is a real positive spectral density function defined by

$$
\rho(M^2) = \sum_X (2\pi)\delta(M^2 - M_X^2) |\langle\Omega|\phi(0)|X_0\rangle|^2 \tag{I.22}
$$

For X = one particle state, $|\langle\Omega|\phi(0)|X_0\rangle|^2 = \text{constant} \equiv Z$. For X = two or more particles, M_X^2 takes a continuous value larger than $(2m)^2$. Therefore, the spectral density function typically has a functional shape as depicted in Figure I.2. Thus, the Fourier transform of the two-point function is expressed as

$$
\begin{aligned}
\int d^4x e^{iq\cdot x} i\langle\Omega|T[\phi(x)\phi(0)]|\Omega\rangle &= \int_0^\infty \frac{dM^2}{2\pi} \rho(M^2) \frac{-1}{q^2 - M^2 + i\epsilon} \\
&= \frac{-Z}{q^2 - m^2 + i\epsilon} + \int_{4m^2}^\infty \frac{dM^2}{2\pi} \rho(M^2) \frac{-1}{q^2 - M^2 + i\epsilon}
\end{aligned} \tag{I.23}
$$

Equation (I.23) demonstrates the analytic structure of the two-point function as a function of complex variable q^2. It has a few poles corresponding to one-particle (and possible bound states) and branch cuts on the real axis starting at $(2m)^2$. This is just as we expected from the consideration of the perturbative treatment of the photon propagator. Thus, we have given the promised proof of the Statement (1).

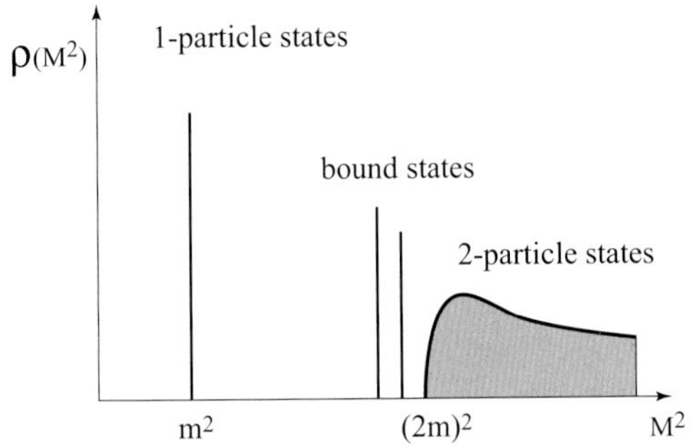

Figure I.2 A typical spectrum of the spectral function $\rho(M^2)$ of an interacting field. The one particle state gives a delta function at m^2 and many particle states have a continuous spectrum above the threshold $(2m)^2$. If there are bound states, they produce delta functions as well.

Pole Mass Notice that m^2 appearing as a pole is the true mass that is observable and not a bare mass that appears in the Lagrangian. $Z = |\langle \Omega|\phi(0)|n=1\rangle|^2$ is the wave function (field strength) renormalization and represents a probability that the state $|n\rangle$ is in a single particle state. If $\phi(x)$ is the free field, then $Z=1$. However, for the interacting field, $Z \neq 0$ and logically should be less than one because the probability to transform to other n ($\neq 1$) particle virtual states is finite. In practice, Z is infinite due to diverging loop integrals. Thus, if we can calculate the spectral function, we can evaluate the two-point function and know the exact mass of the particle. When mass is determined this way, it is referred to as the pole mass.

Spectral Function We have shown that the function $\int d^4x e^{iq\cdot(x-y)} i\langle T\phi(x)\phi(y)\rangle_0$ is a real analytic function of $z = q^2$ with singularities only on the real z-axis. Therefore, it can be expressed as Eq. (I.8). Comparison of Eq. (I.23) with Eq. (I.8) shows that the spectral function can also be expressed as

$$\rho(q^2) = 2\mathrm{Im}\left[\int d^4x e^{iq\cdot x} i\langle \Omega|T[\phi(x)\phi(0)]|\Omega\rangle\right] \tag{I.24}$$

Next, we expand Eq. (I.10d) similarly by inserting intermediate states.

$$\begin{aligned} H(q^2) &= \int d^4x e^{iq\cdot x}\langle \Omega|\phi(x)\phi(0)|\Omega\rangle \\ &= \int d^4x e^{iq\cdot x}\sum_n \langle \Omega|\phi(x)|n\rangle\langle n|\phi(0)|\Omega\rangle \\ &= \int d^4x e^{i(q-p_n)\cdot x}\sum_n \langle \Omega|\phi(0)|n\rangle\langle n|\phi(0)|\Omega\rangle \\ &= (2\pi)^4 \sum_n \delta^4(q-p_n)|\langle \Omega|\phi(0)|p_n\rangle|^2 \\ &= (2\pi)^4 \sum_X \int \frac{d^3p_n}{(2\pi)^3 2E_n}\delta^4(q-p_n)|\langle \Omega|\phi(0)|X_0\rangle|^2 \end{aligned} \tag{I.25a}$$

In obtaining the last equality, we separated integration over momentum from $\sum_n$ and denoted the rest of summation as $\sum_X$ and used Lorentz invariance to change the state $|p_n\rangle$ to $\boldsymbol{p}_n = 0$ which is the state $|X_0\rangle$. The expression can further be rewritten as

$$= \sum_X 2\pi \frac{\delta(q^0 - E_X)}{2E_X} |\langle \Omega | \phi(0) | X_0 \rangle|^2 = \sum_X 2\pi \delta(q^2 - M_X^2) |\langle \Omega | \phi(0) | X_0 \rangle|^2$$
$$= \rho(q^2) \tag{I.25b}$$

Combining Eq. (I.24) and Eq. (I.25b), we have

$$\int d^4x e^{iq\cdot(x-y)} \langle \Omega | \phi(x)\phi(y) | \Omega \rangle = 2\mathrm{Im}\left[\int d^4x e^{iq\cdot(x-y)} i \langle \Omega | T\phi(x)\phi(y) | \Omega \rangle\right] \tag{I.26}$$

which is the promised proof of the Theorem I.1. The relation Eq. (I.26) is referred to as the unitarity relation.

Lorentz Structure of the Two-Point Function If the field in consideration is not a scalar, the two-point function has Lorentz structure consistent with that of the field product. For instance, the Fourier transform of the two-point function for the vector field has a tensor form

$$\Pi^{\mu\nu}(q^2) = (q^2 g^{\mu\nu} - q^\mu q^\nu)\Pi_{\mathrm{T}}(q^2) + q^\mu q^\nu \Pi_{\mathrm{L}}(q^2) \tag{I.27}$$

For the photon, $\Pi_{\mathrm{L}}(q^2)$ can be dropped because of the gauge invariance as we saw in Eq. (I.12b).

Because of the analytic structure of the spectral function, each of $(\Pi_{\mathrm{T}}, \Pi_{\mathrm{L}})$ can be expressed in the form given by Eq. (I.8). If the analytic function represents a transition amplitude, its imaginary part in the forward direction is related to the total cross section by unitarity. We will see it for the case of $e^-e^+ \to$ hadrons in the following example.

Example: the Photon Propagator and the $\sigma_{\mathrm{TOT}}(e^-e^+ \to$ hadrons) Let us consider, as an example, the two-point function made by the electromagnetic current.

$$\Pi^{\mu\nu}(q^2) = i \int d^4x e^{iq\cdot x} \langle \Omega | T[J^\mu(x) J^\nu(0)] | \Omega \rangle \tag{I.28a}$$

$$H^{\mu\nu}(s) = \int d^4x e^{iq\cdot x} \langle \Omega | J^\mu(x) J^\nu(0) | \Omega \rangle$$
$$= \sum_f (2\pi)^4 \delta^4(p_1 + p_2 - P_f) \langle P_f | J^\mu(0) | \Omega \rangle^* \langle P_f | J^\nu(0) | \Omega \rangle \tag{I.28b}$$

Here, eJ^μ represents the electromagnetic current. From Lorentz invariance and current conservation, $H^{\mu\nu}$, $\Pi^{\mu\nu}(q^2)$ can be expressed as

$$H^{\mu\nu}(q^2) = (-q^2 g^{\mu\nu} + q^\mu q^\nu) H_\gamma(q^2)^{2)}$$
$$\Pi^{\mu\nu}(q^2) = (-q^2 g^{\mu\nu} + q^\mu q^\nu) \Pi_\gamma(q^2) \tag{I.29}$$

where we attached suffix "γ" to remind one of the special process we are considering.

$H^{\mu\nu}(q^2)$ is related to the total hadronic cross section $\sigma_{\text{TOT}}(e^- e^+ \to \text{hadrons})$ by

$$\sigma_{\text{TOT}} = \frac{1}{2s}\overline{\sum_f} |\mathcal{M}_{fi}|^2 = \frac{e^4}{2s} L_{\mu\nu} \frac{1}{q^4} H^{\mu\nu}(q^2) = \frac{e^4}{2s} H_\gamma(q^2) \tag{I.30}$$

where

$$\begin{aligned} S_{fi} - \delta_{fi} &= -(2\pi)^4 i \delta^4(k_1 + k_2 - p_X)\mathcal{M}_{fi} \\ &= \overline{v}(k_2)(-ie\gamma_\mu)u(k_1)\frac{-ig^{\mu\nu}}{q^2}\langle f|eJ_\nu(0)|0\rangle \end{aligned} \tag{I.31a}$$

$$\begin{aligned} L_{\mu\nu} &= \frac{1}{4}\sum_{\text{spin}}[\overline{v}(k_2)\gamma_\mu u(k_1)][\overline{v}(k_2)\gamma_\nu u(k_1)]^* \\ &= k_{1\mu}k_{2\nu} + k_{1\nu}k_{2\mu} - (q^2/2)g_{\mu\nu}, \quad s = q^2 = (k_1 + k_2)^2 \end{aligned} \tag{I.31b}$$

The cross section given by Eq. (I.30) is exact to order $O(\alpha)$ in the electromagnetic interaction, but exact to all orders in the strong interaction as the $H^{\mu\nu}(q^2)$ contains all the hadronic final states.

The $\Pi^{\mu\nu}$ appears in higher order correction terms to the photon propagator. To see it, we consider the $e^- e^+ \to e^- e^+$ scattering amplitude which contains the higher process depicted as a black blob in Figure I.3. The scattering amplitude can be expressed as (see Vol. 1, Eq. (6.34))

$$\begin{aligned} S_{fi} - \delta_{fi} &= -i(2\pi)^4 \delta^4(k_1 + k_2 - k_1' - k_2')\mathcal{M}_{fi} \\ &= \frac{1}{4!}\langle k_1' k_2' | \, T\left(i\int d^4x\,(\mathcal{L}_e + \mathcal{L}_h)\right)^4 |k_1 k_2\rangle \end{aligned} \tag{I.32a}$$

$$\mathcal{L}_e = e j^\mu(x) A_\mu(x), \quad \mathcal{L}_h = e J^\mu(x) A_\mu(x) \tag{I.32b}$$

where j^μ is the electron current. It is reduced to

$$\begin{aligned} S_{fi} - \delta_{fi} = e^4 \int & d^4z\, d^4x\, d^4y\, d^4w \langle k_1' k_2' | j^\rho(z)|0\rangle\langle 0|T[A_\rho(z)A_\mu(x)]|0\rangle \\ & \times \langle 0|T[J^\mu(x)J^\nu(y)]|0\rangle\langle 0|T[A_\nu(y)A_\sigma(w)]|0\rangle\langle 0|j^\sigma(w)|k_1, k_2\rangle \end{aligned} \tag{I.32c}$$

2) For convenience, we adopted a different sign for the Lorentz structure of the function having positive $q^2 > 0$.

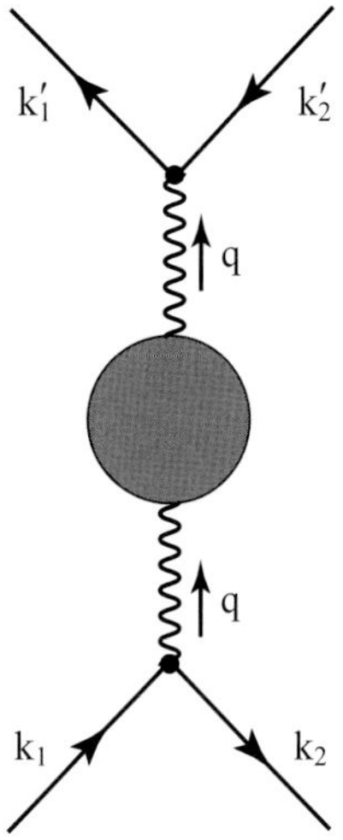

Figure I.3 Feynman diagram for forward scattering of two electrons with higher order corrections to the photon self energy depicted as a black blob.

Replacing the photon propagator with its lowest order expression, we obtain

$$\therefore \quad -i\mathcal{M}_{fi} = \overline{u}(k_1')(e\gamma^{\rho})v(k_2')\left[\frac{-ig_{\rho\mu}}{q^2}\{\cdots\}\frac{-ig_{\nu\sigma}}{q^2}\right]\overline{v}(k_2)(e\gamma^{\sigma})u(k_1) \tag{I.33}$$

$$\{\cdots\} = \left(-ie^2\int d^4(x-y)e^{iq\cdot(x-y)}i\langle\Omega|T[J^{\mu}(x)J^{\nu}(y)]|\Omega\rangle\right) = ie^2\Pi^{\mu\nu}(q^2) \tag{I.34}$$

If one replaces the content of $[\cdots]$ with the photon propagator, Eq. (I.33) is just the Born amplitude, which means that the $\Pi^{\mu\nu}(q^2)$ represents higher order corrections to the photon propagator. Indeed, by comparing Eq. (I.33) with the photon propagator Eq. (I.12a), we see that $e^2\Pi^{\mu\nu}$ agrees with $\Sigma_{\gamma\gamma}^{\mu\nu}$ at $O(\alpha)$.

To justify our assertion that Eq. (I.26) is the unitarity relation, we relate $\Pi^{\mu\nu}$ to the forward scattering amplitude. Setting the final state in Eq. (I.33) equal to the initial state, we have

$$\begin{aligned} i\mathcal{M}_{ii} &= -\overline{u}(k_1')(e\gamma^{\rho})v(k_2')\left[\frac{-ig_{\rho\mu}}{q^2}\left(-ie^2\Pi^{\mu\nu}(q^2)\right)\frac{-ig_{\nu\sigma}}{q^2}\right] \\ &\quad\times \overline{v}(k_2)(e\gamma^{\sigma})u(k_1)\Big|_{k_1',k_2'=k_1,k_2} \\ &= ie^4 L_{\mu\nu}\frac{1}{q^4}\Pi^{\mu\nu}(q^2) = ie^4\Pi_{\gamma}(q^2) \end{aligned} \tag{I.35}$$

where spin average was taken in deriving the penultimate equality.

The unitarity relation is given by (see Vol. 1, Eqs. (9.52) and (9.56))

$$-2\mathrm{Im}\mathcal{M}_{ii} = \sum_n |\mathcal{M}_{ni}|^2 \tag{I.36}$$

which relates the total cross section to the imaginary part of the forward scattering amplitude. Substituting Eq. (I.30) and Eq. (I.35) in Eq. (I.36), we obtain

$$H_\gamma(q^2) = 2\mathrm{Im}[\Pi_\gamma] = \frac{2s}{e^4}\sigma_{\mathrm{TOT}} \tag{I.37}$$

$$\therefore \quad H^{\mu\nu}(q^2) = 2\mathrm{Im}[\Pi^{\mu\nu}(q^2)] \tag{I.38}$$

Thus, we rederived Eq. (I.1) from the unitarity of the scattering amplitude for $\phi = A^\mu$.

Since $\Pi_\gamma(q^2)$ is a real analytic function, we can use the dispersion relation to express it in terms of its imaginary part, and hence the total cross section. Using R_γ, the ratio of the total hadronic cross section to that of $\sigma(e^- e^+ \to \mu^- \mu^+) = 4\pi\alpha^2/(3s)$, we have

$$\mathrm{Im}\Pi_\gamma(q^2) = \frac{s}{e^4}\sigma_{\mathrm{TOT}}(e^- e^+ \to \text{hadrons}) = \frac{R_\gamma}{12\pi} \tag{I.39}$$

As $R_\gamma \to$ is constant for $s \to \infty$, we use the once subtracted dispersion relation to ensure convergence of the integral which gives

$$\mathrm{Re}[\Pi(q^2) - \Pi(0)] = \mathrm{Re}[\hat{\Pi}(q^2)] = \frac{q^2}{12\pi^2}\int_{4m_\pi^2}^{\infty} ds \frac{R_\gamma}{s(s-q^2)} \tag{I.40}$$

The relation is used to derive an accurate value of $\Delta\alpha$ in Chapter 5.

Appendix J
σ Model and the Chiral Perturbation Theory

Linear σ Model The linear σ-model[1] [707] is a model of the PCAC (partially conserved axial current hypothesis), illustrates many features of the spontaneous symmetry break down, reproduces low energy $\pi-N$ dynamics and serves as a useful material in many applications. The Lagrangian density is given by

$$\mathcal{L}_0 = \overline{N}\left[i\gamma^\mu\partial_\mu - g_{\pi NN}(\sigma + i\boldsymbol{\tau}\cdot\boldsymbol{\pi}\gamma^5)\right]N + \frac{1}{2}(\partial_\mu\sigma)^2 + \frac{1}{2}(\partial_\mu\boldsymbol{\pi})^2 - V(\sigma,\boldsymbol{\pi}) \tag{J.1a}$$

$$V(\sigma,\boldsymbol{\pi}) = \frac{\lambda}{4}(\sigma^2 + \boldsymbol{\pi}\cdot\boldsymbol{\pi} - v^2)^2 \tag{J.1b}$$

where $\boldsymbol{\tau} = (\tau_1, \tau_2, \tau_3)$ are the Pauli matrices representing the isospin rotation and $N = (\psi_p, \psi_n)^T$ is a isospin doublet consisting of the proton and the neutron which are assumed to be massless. σ is a hypothetical scalar field and $\boldsymbol{\pi} = (\pi_1, \pi_2, \pi_3)$ are the isospin 1 pseudoscalar fields which can be identified with the pion. $V(\sigma,\boldsymbol{\pi})$ represents the potential of the $\sigma, \boldsymbol{\pi}$ fields and λ is assumed to be positive to ensure stability of the vacuum. v is some positive constant. The Lagrangian $\mathcal{L}_0$ is symmetric under the isospin rotation

$$\begin{aligned}&\boldsymbol{\pi} \to \boldsymbol{\pi}' = \boldsymbol{\pi} + \delta\boldsymbol{\alpha}\times\boldsymbol{\pi}\,, \quad \sigma \to \sigma' = \sigma\\ &N \to N' = [1 - i(\boldsymbol{\tau}\cdot\delta\boldsymbol{\alpha}/2)]N\,, \quad \overline{N} \to \overline{N}' = \overline{N}[1 + i(\boldsymbol{\tau}\cdot\delta\boldsymbol{\alpha})/2]\end{aligned} \tag{J.2}$$

and the chiral transformation

$$\boldsymbol{\pi} \to \boldsymbol{\pi}' = \boldsymbol{\pi}^i + \delta\boldsymbol{\alpha}^i\sigma\,, \quad \sigma \to \sigma' = \sigma - \delta\boldsymbol{\alpha}\cdot\boldsymbol{\pi} \tag{J.3a}$$

$$N \to N' = [1 - i(\boldsymbol{\tau}\cdot\delta\boldsymbol{\alpha}/2)\gamma^5]N\,, \quad \overline{N} \to \overline{N}' = \overline{N}[1 + i(\boldsymbol{\tau}\cdot\delta\boldsymbol{\alpha}/2)\gamma^5] \tag{J.3b}$$

Using the Noether's theorem, one can derive conserved vector and axial vector currents corresponding to each symmetry from the Lagrangian $\mathcal{L}_0$

$$\begin{aligned}\boldsymbol{V}_\mu &= \overline{N}\gamma_\mu\frac{\boldsymbol{\tau}}{2}N + \boldsymbol{\pi}\times(\partial_\mu\boldsymbol{\pi})\\ \boldsymbol{A}_\mu &= \overline{N}\gamma_\mu\gamma^5\frac{\boldsymbol{\tau}}{2}N - (\partial_\mu\sigma)\boldsymbol{\pi} + (\partial_\mu\boldsymbol{\pi})\sigma\end{aligned} \tag{J.4a}$$

1) [354, 706]

Integral of their 0th component defines isospin and axial charge operators.

$$I^a = \int d^3x V_0^a(x) \quad Q_5^a = \int d^3x A_0^a(x) \quad (a = 1-3) \tag{J.4b}$$

Problem J.1

Derive Eqs. (J.4a).

Let us introduce a four component vector in the isopin space

$$\boldsymbol{\Sigma} = (\Sigma^0, \ldots, \Sigma^3) = (\sigma, \boldsymbol{\pi}) \tag{J.5}$$

Then, the potential energy part of the Lagrangian's can be expressed as

$$\begin{aligned} V(\Sigma) &\equiv \frac{\lambda}{4}(\sigma^2 + \boldsymbol{\pi}\cdot\boldsymbol{\pi} - v^2)^2 = \frac{\lambda}{4}(\boldsymbol{\Sigma}\cdot\boldsymbol{\Sigma} - v^2)^2 \\ &= \frac{\lambda}{4}|\boldsymbol{\Sigma}|^4 + \frac{\mu^2}{2}|\boldsymbol{\Sigma}|^2 + V(0) \\ \mu^2 &= -\lambda v^2, \quad V(0) = \frac{\lambda v^4}{4} \end{aligned} \tag{J.6}$$

One realizes that the potential has the same form as that of the Higgs potential Eq. (1.27).

Particle Spectrum in the Wigner Mode Depending on the sign of the μ^2, we have two kinds of vacuum, the Wigner mode ($\mu^2 > 0$) and the Nambu–Goldstone mode ($\mu^2 = -\lambda v^2 < 0$). In the Wigner mode, the potential minimum is at $|\boldsymbol{\Sigma}| = 0$, that is, the vacuum expectation value of the fields vanishes.

$$\langle\sigma\rangle = \langle\boldsymbol{\pi}\rangle = 0 \tag{J.7}$$

Then, the Lagrangian is expressed as

$$\begin{aligned} \mathcal{L}_0 &= \overline{N}[i\gamma^\mu\partial_\mu - g_{\pi NN}(\sigma + i\boldsymbol{\tau}\cdot\boldsymbol{\pi}\gamma^5)]N \\ &\quad + \frac{1}{2}(\partial_\mu\sigma)^2 + \frac{1}{2}(\partial_\mu\boldsymbol{\pi})^2 - \frac{1}{2}m_\Sigma^2(\sigma^2 + \boldsymbol{\pi}\cdot\boldsymbol{\pi}) - \frac{\lambda}{4}(\sigma^2 + \boldsymbol{\pi}^2)^2 \\ m_\Sigma^2 &= \mu^2 > 0 \end{aligned}$$

Neglecting the last term which represents the quartic interaction, this Lagrangian, when quantized, describes a massless nucleon field (the mass term $m\overline{N}N$ is not chiral invariant), and massive mesons with

$$m_\pi^2 = m_\sigma^2 = m_\Sigma^2 = \mu^2 \tag{J.8}$$

One sees that the scalar and pseudoscalar mesons are degenerate. This is a consequence of the chiral symmetry implemented in the Wigner mode. The theory is left–right symmetric, and hence the particle spectrum should be parity degenerate. The above picture is not consistent with our real world where the nucleons have mass and there is no scalar partner to the pion.

Particle Spectrum in the Nambu-Goldstone Mode In the Nambu–Goldstone mode, $\mu^2 = -\lambda v^2 < 0$ and the potential has a ring of minima satisfying $\boldsymbol{\Sigma} \cdot \boldsymbol{\Sigma} = v^2$. We choose the vacuum at $\sigma = v$. Then, the vacuum expectation value of the scalar field no longer vanishes, but rather has a finite value. Then, we need to introduce a new field to describe quantum excitation from the new vacuum

$$\sigma = v + \sigma' \tag{J.9}$$

and rewrite the Lagrangian in terms of the new field. Then, we rename σ' back as σ. The result is

$$\begin{aligned}\mathcal{L}_0 = {} & \overline{N}[i\gamma^\mu \partial_\mu - m_N]N - g_{\pi NN}\overline{N}[(\sigma + i\boldsymbol{\tau} \cdot \boldsymbol{\pi}\gamma^5)]N \\ & + \frac{1}{2}\left[(\partial_\mu \sigma)^2 - m_\sigma^2 \sigma^2\right] + \frac{1}{2}(\partial_\mu \boldsymbol{\pi})^2 \\ & - \frac{\lambda}{4}(\sigma^2 + \boldsymbol{\pi} \cdot \boldsymbol{\pi})^2 - \lambda v(\sigma^3 + \sigma \boldsymbol{\pi} \cdot \boldsymbol{\pi})\end{aligned} \tag{J.10a}$$

$$m_N = g_{\pi NN} v\,, \quad m_\sigma^2 = 2\lambda v^2 \tag{J.10b}$$

Picking up terms quadratic in the fields, we can obtain the particle spectrum. The pions are massless because they are the Goldstone bosons associated with the spontaneous break down of the chiral symmetry $SU(2)_L \times SU(2)_R$. The particle spectrum now consists of a nucleon isodoublet with mass $m_N = g_{\pi NN} v$, an isosinglet of a scalar meson of mass $m_\sigma = \sqrt{2\lambda v^2}$ and an isotriplet massless π-meson.

Actually, the pions get mass because of the explicit mass term of the quark generated by the weak interaction, which in the σ model, can be incorporated by adding an external perturbation term to the Lagrangian. This is discussed in the next section. Here, we simply assume that such a small perturbation exists.

Things look more realistic now. The parity degeneracy has disappeared, and the pion has light mass indeed. The existence of the scalar meson is somewhat problematic, but at least its mass is independent of the pion and the nucleon. The model fits to the reality if we assume that the mass of the scalar is heavy and unobservable.

We shall next show an outcome of the PCAC which provides yet more strong evidence for the validity of the σ model.

PCAC: Partially Conserved Axial Vector Current Hypothesis In a loose sense, PCAC means that the axial vector current is almost conserved. In the language of the field theory, it means that it is broken but recovers in the limit of massless pions. Within the σ model, it is realized by adding a small perturbation to the Lagrangian.

$$\begin{aligned}\mathcal{L}_\sigma &= \mathcal{L}_0 + \mathcal{L}' \\ \mathcal{L}' &= \varepsilon \sigma\end{aligned} \tag{J.11}$$

The linear term tilts the potential and removes the vacuum degeneracy around the circle $|\boldsymbol{\Sigma}|^2 = v^2/2$. The new vacuum moves to

$$v \to v' = v + \frac{\varepsilon}{2\lambda v^2} + O(\varepsilon^2) \tag{J.12}$$

and the resulting mass spectrum is

$$m_N = g_{\pi NN} v' \simeq g_{\pi NN} \left(v + \frac{\varepsilon}{2\lambda v^2} + \cdots \right) \tag{J.13a}$$

$$m_\pi^2 = \frac{\varepsilon}{v'} \simeq \frac{\varepsilon}{v} + \cdots \tag{J.13b}$$

$$m^2 = \lambda(3v'^2 - v^2) \simeq \lambda \left(2v^2 - \frac{3\varepsilon}{\lambda v^2} + \cdots \right) \tag{J.13c}$$

The value of ε can be determined by evaluating the divergence of the axial current. The axial vector current no longer conserves, but satisfies

$$\partial_\mu A_a^\mu = \varepsilon \pi_a \tag{J.14}$$

The equation is known as the PCAC relation. Taking the expectation value of the divergence of the axial current A_a^μ between the vacuum and one pion state, one gets

$$\langle 0 | \partial_\mu A_a^\mu | \pi^a \rangle = \varepsilon \langle 0 | \pi^a | \pi^a \rangle \tag{J.15}$$

The left-hand side appears in the expression for the pion decay width (see Vol. 1, Eqs. (15.44) and (15.47)) and fixes the constant ε in terms of the pion decay constant F_π and mass.

$$\varepsilon = F_\pi m_\pi^2, \quad F_\pi = 94\text{MeV} \tag{J.16}$$

Comparing the above expression with Eq. (J.13b), we obtain

$$v = F_\pi \tag{J.17a}$$

$$m_N = g_{\pi NN} F_\pi \tag{J.17b}$$

The second equality is known as the Golberger–Treiman relation ($m_N g_A = g_{\pi NN} F_\pi$) [708] in its tree version. Historically, it played a very important role by relating the weak axial vector coupling constant $g_A = 1.24$ with the strong interaction coupling constant $g_{\pi NN}^2/4\pi = 14.6$. Inclusion of the higher order terms which can be expressed in terms of the $\pi - N$ total cross section, PCAC reproduces the value of g_A accurately [709, 710]. This agreement supports our main assumption that the pions are Goldstone bosons associated with spontaneous breakdown of the chiral symmetry.

Representations of the σ Model In order to embark on a path to the effective Lagrangian approach, we rewrite the linear σ model Eq. (J.1) as follows:

$$\begin{aligned} \mathcal{L}_0 = {} & \overline{N}_L i \partial\!\!\!/ N_L + \overline{N}_R i \partial\!\!\!/ N_R - g_{\pi NN} \left(\overline{N}_L \boldsymbol{\Sigma} N_R + \overline{N}_R \boldsymbol{\Sigma}^\dagger N_L \right) \\ & + \frac{1}{4} \text{Tr} \left[\partial_\mu \boldsymbol{\Sigma} \partial^\mu \boldsymbol{\Sigma}^\dagger \right] - \frac{\lambda}{16} \left(\text{Tr}[\boldsymbol{\Sigma} \boldsymbol{\Sigma}^\dagger] - 2v^2 \right)^2 \end{aligned} \tag{J.18}$$

where we have introduced $N_{L(R)} = \frac{1}{2}(1 \mp \gamma^5)N, \boldsymbol{\Sigma} = \sigma + i\boldsymbol{\tau} \cdot \boldsymbol{\pi}$. The model is invariant under the $SU(2)_L \times SU(2)_R$ global transformations

$$N_L \to N'_L = e^{-i\boldsymbol{\tau}\cdot\boldsymbol{\alpha}_L/2} N_L \equiv L N_L \tag{J.19a}$$

$$N_R \to N'_R = e^{-i\boldsymbol{\tau}\cdot\boldsymbol{\alpha}_R/2} N_R \equiv R N_R \tag{J.19b}$$

$$\boldsymbol{\Sigma} \to \boldsymbol{\Sigma}' = L\boldsymbol{\Sigma} R^\dagger \tag{J.19c}$$

After symmetry breaking ($\boldsymbol{\Sigma} \to \boldsymbol{\Sigma} + v\mathbf{1}$), the Lagrangian reads

$$\begin{aligned}\mathcal{L}_0 &= \overline{N}(i\not{\partial} - m_N)N - g_{\pi NN}\left(\overline{N}_L \boldsymbol{\Sigma} N_R + \overline{N}_R \boldsymbol{\Sigma}^\dagger N_L\right)\\ &\quad + \frac{1}{4}\mathrm{Tr}\left[\partial_\mu \boldsymbol{\Sigma} \partial^\mu \boldsymbol{\Sigma}^\dagger\right] - \frac{\lambda}{16}\left(\mathrm{Tr}\left[\boldsymbol{\Sigma}\boldsymbol{\Sigma}^\dagger\right] + v\mathrm{Tr}\left[\boldsymbol{\Sigma} + \boldsymbol{\Sigma}^\dagger\right]\right)^2\end{aligned} \tag{J.20}$$

This is the same expression as Eq. (J.10) in somewhat different notation. One sees that the Lagrangian is invariant under the transformation Eq. (J.19) provided $\boldsymbol{\alpha}_L = \boldsymbol{\alpha}_R = \boldsymbol{\alpha}$ or $L = R = U$, namely,

$$N_L \to N'_L = U N_L\,, \quad N_R \to N'_R = U N_R\,, \quad \boldsymbol{\Sigma} \to \boldsymbol{\Sigma}' = U\boldsymbol{\Sigma} U^\dagger \tag{J.21}$$

which is denoted as the $SU(2)_V$ symmetry where subscript "V" stands for vector. One sees that the Lagrangian after symmetry breakdown still satisfies $SU(2)_V$ symmetry. We say that the symmetry $SU(2)_L \times SU(2)_R$ has been broken to $SU(2)_V$.

Nonlinear σ Model [2] We now build and use the chiral perturbation theory (ChPT). ChPT is an effective quantum theory of fields to treat the pseudoscalar fields representing the lightest mesons based on QCD. We start with two flavors ($N_f = 2$). In this case, the theory is built on u and d quarks, and deals with the interaction of the three pions (π^+, π^0, π^-). We can expand later to three flavors ($N_f = 3$). Our starting point is Eq. (J.18) minus the nucleon terms.

$$\mathcal{L}_0 = \frac{1}{2}(\partial_\mu \sigma)^2 + \frac{1}{2}(\partial_\mu \boldsymbol{\pi} \cdot \partial^\mu \boldsymbol{\pi}) - \frac{\lambda}{4}(\sigma^2 + \boldsymbol{\pi}\cdot\boldsymbol{\pi} - v^2)^2 \tag{J.22a}$$

$$= \frac{1}{4}\mathrm{Tr}\left[(\partial_\mu \boldsymbol{\Sigma}) \cdot (\partial^\mu \boldsymbol{\Sigma})^\dagger\right] - \frac{\lambda}{16}\left(\mathrm{Tr}[\boldsymbol{\Sigma}\cdot\boldsymbol{\Sigma}^\dagger] - 2v^2\right)^2 \tag{J.22b}$$

Our aim is to eliminate the unwanted σ field or at least make its contribution negligibly small. Usually, this is done by making the mass of the σ field much larger ($m_\sigma^2 = 2\lambda v^2 \to \infty$) than the energy scale we are interested in. However, looking at Eq. (J.10a), we notice that the coupling strength of the σ to the pion field is proportional to v and λ. The coupling becomes strong in the large mass limit and the σ field is inseparable. This difficulty can be avoided by using a nonlinear representation for the fields after spontaneous symmetry breaking:

$$\boldsymbol{\Sigma} = \sigma + i\boldsymbol{\tau}\cdot\boldsymbol{\pi} \to (v + S)U \equiv (v + S)e^{i\boldsymbol{\tau}\cdot\boldsymbol{\pi}/v} \tag{J.23}$$

2) [711]

where S is the redefined scalar field as the excitation from the new vacuum. This special exponential form for the nonlinear representation is just one form of an infinite number of possibilities and a convenient one for the following discussions. One is allowed to use any one of them since on-shell transition amplitudes do not depend on how they are represented. This is discussed in [712, 713]. The resulting Lagrangian reads

$$\begin{aligned}\mathcal{L}_{nl\sigma} &= \frac{1}{2}\left[(\partial_\mu S)^2 - m_S^2 S^2\right] - \lambda v S^3 - \frac{\lambda}{4} S^4 + \frac{(v+S)^2}{4}\mathrm{Tr}\left[\partial_\mu U \partial^\mu U^\dagger\right]\\ m_S^2 &= 2\lambda v^2\end{aligned} \tag{J.24}$$

The Lagrangian still describes a massive scalar meson and three massless pions. However, we can eliminate the v dependent interaction term by shifting the field.

To see this, consider a scalar field ϕ obeying the Klein–Gordon equation.

$$[\partial_\mu \partial^\mu - m^2]\phi(x) = 0 \tag{J.25}$$

In the path integral approach, the transition amplitude is derived as a functional derivative of the generating functional (see Vol. 1, Chapt. 11)

$$\begin{aligned}Z[J] &= \int D\phi \exp\left[i\int d^4x(\mathcal{L}_{KG} + J\phi)\right]\\ &= \int D\phi \exp\left[i\int d^4x\left\{\frac{1}{2}\left(\partial_\mu\phi(x)\partial^\mu\phi(x) - m^2\phi^2(x)\right) + J\phi\right\}\right]\end{aligned} \tag{J.26}$$

which is an integral of the action over the field variable at every space-time point. The J in the above expression is an external source term introduced to perform the functional derivatives. Since the field is an integration variable in the path integral approach, we are free to shift it in any way we want. Therefore, we choose a special shift of to the field in the expression for the generating functional as follows:

$$\phi(x) \to \phi(x) - i\int d^4y \Delta_F(x-y)J(y) \tag{J.27}$$

Using Eq. (J.26), integration by parts leads to an alternative expression (see Vol. 1, Eqs. (11.10) and (11.17)):

$$\begin{aligned}Z[J] &= \int D\phi \exp\left[-\frac{i}{2}\int d^4x\phi(x)\left(\partial_\mu\partial^\mu + m^2\right)\phi(x)\right.\\ &\quad \left.+ \int d^4x d^4y\, J(x)\Delta_F(x-y)J(y)\right]\\ &= Z(0)\exp\left[-\frac{i}{2}\int d^4x d^4y\, J(x)\Delta_F(x-y)J(y)\right]\\ Z(0) &= \int D\phi \exp\left[-\frac{i}{2}\int d^4x\phi\left(\partial_\mu\partial^\mu + m^2\right)\phi(x)\right]\end{aligned} \tag{J.28}$$

Terms that contain the field ϕ explicitly are completely factorized and were integrated away to give a constant $Z(0)$ which can be dropped since it just gives a normalization constant.[3] In the second line of the above equations, one sees that the field ϕ has disappeared from the entire expression, as it only appears as the propagator in the internal lines. If the field ϕ is massive, the propagator in the momentum space can be expanded in inverse powers of the mass.

$$\Delta_F(k) = \frac{1}{k^2 - m_\phi^2} = -\frac{1}{m_\phi^2}\left[1 + \left(\frac{k^2}{m_\phi^2}\right) + \left(\frac{k^2}{m_\phi^2}\right)^2 + \cdots\right] \tag{J.29}$$

In the limit $m_\phi^2 \to \infty$,

$$\int d^4 y \Delta_F(x - y) \to -\frac{1}{m_\phi^2} \tag{J.30}$$

and

$$Z[J] \to Z_{\text{eff}}[J] = \exp\left[\frac{i}{2m_\phi^2}\int d^4 x\, J(x) J(x)\right] \tag{J.31}$$

The effective Lagrangian is now of JJ coupling type which is analogous to the Fermi interaction in the large mass limit of the W boson.

This is just one example of a general statement called the decoupling theorem [371, 714]. It states that if the remaining low energy theory is renormalizable, the effects of the heavy particle appear either as a renormalization of the coupling constants in the theory or else are suppressed by powers of the heavy particle mass. Notice that renormalizability is an important condition. For instance, effects of the heavy top quark mass in the radiative corrections to the electroweak theory were sizable, as they were proportional to m_t^2 or $\ln m_t^2$ (see, for example, an expression for $\Delta\rho$ Eq. (5.52)). In other words, the top was not decoupled. This is because the electroweak theory is based on $SU(2)_\text{L}$ symmetry, and full members of a doublet (t, b) are required for the renormalizability. One just cannot separate the heavy particle alone in order to apply the decoupling theorem.

Going back to the exponential representation of the σ model, the interaction of the scalar field with the pion appears as $[(v + S)^2/4]\text{Tr}[\partial_\mu U \partial^\mu U^\dagger]$. It has the form

$$(v^2 + 2v S + S^2)\frac{1}{4}\text{Tr}\left[\partial_\mu U \partial^\mu U^\dagger\right] \tag{J.32}$$

The first term is independent of S. The second term has the same form as ϕJ in Eq. (J.26). By neglecting the S^2 interaction for now, we make the following identifications:

$$\phi \to S\,, \quad J \to \frac{v}{2}\text{Tr}\left[\partial_\mu U \partial^\mu U^\dagger\right] \tag{J.33}$$

3) Because of this, the process of removing heavy fields from the Lagrangian is often called "integrating out" the field.

to obtain $Z_{\text{eff}}[J]$ and an effective Lagrangian from Eq. (J.31),

$$Z_{\text{eff}}[J] = \exp\left[i \int d^4x \mathcal{L}_{\text{eff}}\right] \tag{J.34a}$$

$$\mathcal{L}_{\text{eff}} = \frac{v^2}{4}\text{Tr}[\partial_\mu U \partial^\mu U^\dagger] + \frac{v^2}{8m_S^2}\left(\text{Tr}[\partial_\mu U \partial^\mu U^\dagger]\right)^2 + \cdots \tag{J.34b}$$

The dots in Eq. (J.34b) refer to other terms created by integration of high order S terms. Inclusion of the higher order terms in the expansion of the propagator brings in an infinite series of the parameter p_S^2/m_S^2, or equivalently, derivatives of the pion field. Inclusion of $S^2\text{Tr}[\partial_\mu U \partial^\mu U^\dagger]$ brings in loop diagrams which either renormalize the coefficients or are included in the higher order terms.

One sees that Eq. (J.34b) is an expansion series in powers of the pion momentum scaled by the vacuum expectation value v or the scalar mass m_S. Thus, we have succeeded in deriving an effective Lagrangian valid at the low energy with the scalar field decoupled at least to order v^2/m_S^2.

The argument can be applied to a general class of scenarios in the spontaneous symmetry breaking environment. If we adopt the chiral invariance symmetry and use $v = F_\pi$ as in the linear σ model, we have the Chiral Perturbation Theory. Equation (J.34b) is the result of adopting the σ model. If the real pion undergoes other interactions which respect the same $SU_{\text{L}}(2) \times SU(2)_{\text{R}}$ chiral symmetry, they will introduce other terms, including two-derivative, four-derivative, and six-derivative terms, but respecting the same symmetry, for instance, of the form

$$\text{Tr}\left[\partial_\mu U \partial_\nu U^\dagger\right]\text{Tr}\left[\partial^\mu U \partial^\nu U^\dagger\right], \quad \left(\text{Tr}\left[\partial_\mu U \partial^\mu U^\dagger\right]\right)^2, \ldots \tag{J.35}$$

There is only one two-derivative and no derivative-free terms in a list because in the exponential representation of the pion field, the symmetry operation translates to a shift of pion field ($\boldsymbol{\pi} \to \boldsymbol{\pi} +$ constant) and the invariant $\text{Tr}[U U^\dagger]$ is a constant. Thus, the general Lagrangian can be organized by dimensionality of the operators,

$$\mathcal{L} = \mathcal{L}_2 + \mathcal{L}_4 + \mathcal{L}_6 + \cdots \tag{J.36}$$

there are no dimensionality of odd numbers because of the parity invariance. Coefficients of these terms are generally not known, and must be determined phenomenologically.

Symmetry Breaking The symmetry of $SU(2)_{\text{L}} \times SU(2)_{\text{R}}$ can be broken by tilting the potential slightly, for example, by adding a term similar to $\varepsilon\sigma$ in the linear sigma model. It can be expressed as

$$\mathcal{L}_\varepsilon = \varepsilon\sigma \to \frac{\varepsilon}{4}\text{Tr}[\boldsymbol{\Sigma} + \boldsymbol{\Sigma}^\dagger] \tag{J.37}$$

and its extension to the exponential representation as

$$\mathcal{L}_\varepsilon = \frac{\varepsilon}{4}(v + S)\mathrm{Tr}[U + U^\dagger] = \frac{\varepsilon}{4}(v + S)\mathrm{Tr}\left[2 - \left(\frac{\boldsymbol{\tau}\cdot\boldsymbol{\pi}}{v}\right)^2 + \cdots\right]$$

$$= \varepsilon(v + S) - \frac{\varepsilon}{2v}\boldsymbol{\pi}\cdot\boldsymbol{\pi} + \cdots$$

$$\equiv \varepsilon(v + S) - \frac{m_\pi^2}{2}\boldsymbol{\pi}\cdot\boldsymbol{\pi} + \cdots \tag{J.38a}$$

$$\rightarrow \varepsilon = v m_\pi^2 \tag{J.38b}$$

Equation (J.38b) reproduces the same result as Eq. (J.13b). Then, the lowest order expression for the Lagrangian is obtained by setting $S = 0$,

$$\mathcal{L}_2 = \frac{v^2}{4}\mathrm{Tr}\left[\partial_\mu U \partial^\mu U^\dagger\right] + \frac{m_\pi^2}{4}v^2\mathrm{Tr}\left[U + U^\dagger\right] \tag{J.39}$$

This is the starting point of the chiral perturbation theory. Higher order terms will contain terms like

$$\left(m_\pi^2\mathrm{Tr}[U + U^\dagger]\right)^2, \quad m_\pi^2\mathrm{Tr}[U + U^\dagger]\mathrm{Tr}\left[\partial_\mu U \partial^\mu U^\dagger\right] \tag{J.40}$$

The nonlinear σ model can also be applied to the Higgs sector. To consider the Standard Model as the low energy effective theory of the strongly interacting Higgs, one must also consider it as a possible extension of the Standard Model.

Appendix K
Splitting Function

General Formula for Deriving the Splitting Functions We saw that the splitting function[1] which appeared in the final state of "$e\bar{e} \to$ hadrons" is identical to that which appeared in the deep inelastic scattering. In fact, the universality applies to any splitting function in the LLA approximation. Here, we derive a general formula to calculate the splitting functions [391]. First, we consider a process $a + d \to c + X$ in Figure K.1a and decompose it as $a \to c + b, b + d \to c + X$. The particle "$a$" splits into "$b + c$" (or radiates c) before interacting with particle d. Thus, the particle "a" can be considered as a provider of incident flux "b" which interacts with d.

We shall show that in the infinite momentum frame, the cross section of $a + d \to c + X$ can be expressed as

$$\begin{aligned} d\sigma_{N+1} &= d\sigma(a + d \to c + X) \equiv dP(z, Q^2)dz d\sigma_N(b + d \to X) \\ &\equiv \frac{\alpha_s}{2\pi} P_{b\leftarrow a}(z)dz \ln Q^2 d\sigma_N \end{aligned} \tag{K.1}$$

where z is a fraction of energy that the parton "b" carries, that is, $E_b = zE_a$ and $Q^2 = |(p_a - p_c)^2|$ is its invariant mass. We have defined $dP(z, Q^2)dz$ as the flux provided by the parton "a." The second line in Eq. (K.1) defines the splitting function.

We consider in the infinite momentum frame ($p \gg m$) so that for all practical purposes, on shell particles can be considered as massless. Here, the incoming particle "a" branches to "b" and "c" and we assume the virtuality of the emitted parton is much larger than that of particles "a" and "c."

$$t = p_b^2 = (p_a - p_c)^2 \gg p_a^2, p_c^2 \tag{K.2}$$

In the three prong vertex, at least one of them is off mass-shell (here it is b) having the invariant mass squared $Q^2 = |t|$, but this is also a small number (or to be more exact, we limit our consideration in such a region) compared to its momentum, that is, $p \gg Q \gtrsim p_{\rm T}$, and thus it can also be considered approximately on mass-shell. This means that we are working in a (old style) time ordered frame in which energy is not conserved. We define variables in the infinite momentum frame as

1) [391, 397]

Elementary Particle Physics, Volume 2: Foundations of the Standard Model, First Edition. Yorikiyo Nagashima.

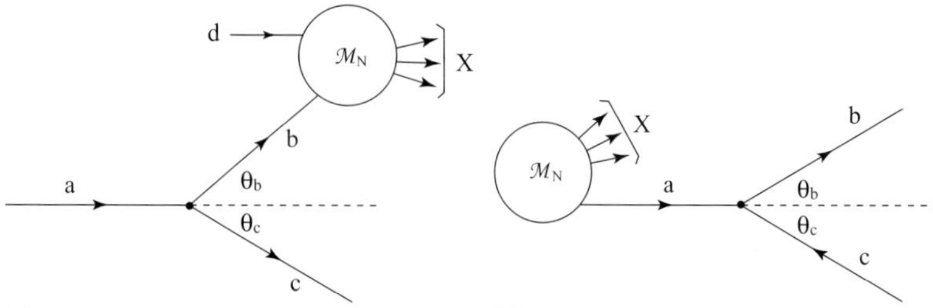

Figure K.1 Diagrams for processes that emit an extra parton (denoted as c) in the initial and final state.

follows.

$$
\begin{aligned}
p_a &= (E_a, 0, p_{za}) \simeq (E_a, 0, E_a) \\
p_b &= (E_b, p_T, p_{zb}) \simeq \left(z E_a + \frac{p_T^2}{2z E_a}, p_T, z E_a \right) \\
p_c &= (E_c, -p_T, p_{zc}) \simeq \left((1-z) E_a + \frac{p_T^2}{2(1-z) E_a}, -p_T, (1-z) E_a \right) .
\end{aligned} \tag{K.3}
$$

Other variables that are useful can be defined as well.

$$E_b = z E_a , \quad E_c = (1-z) E_a \tag{K.4a}$$

$$\theta = \theta_{bc} = \theta_b + \theta_c = \frac{\theta_b}{1-z} = \frac{\theta_c}{z} \tag{K.4b}$$

$$
\begin{aligned}
\hat{t} &= -Q^2 = (p_a - p_c)^2 = -2 E_a E_c (1 - \cos\theta_c) \\
&\simeq -E_a E_c \theta_c^2 = -z^2 (1-z)(E_a \theta)^2
\end{aligned} \tag{K.4c}
$$

$$p_T = E_b \theta_b = E_c \theta_c = z(1-z) E_a \theta \tag{K.4d}$$

where the variable $\hat{t}$ is used instead of t to remind one that the particle in consideration is a parton. The invariant mass of "b" is expressed as

$$Q^2 = p_T^2/(1-z) \tag{K.4e}$$

Next, expressing the invariant matrix element as

$$\mathcal{M}_{N+1} \equiv \mathcal{M}(ad \to cX) = \frac{\mathcal{M}(a \to bc)\mathcal{M}(bd \to X)}{\hat{t}} = \frac{g_s V(a \to bc)\mathcal{M}_N}{\hat{t}} \tag{K.5}$$

The cross section for $ad \to cX$ can be written as

$$
\begin{aligned}
d\sigma_{N+1} &\equiv d\sigma(ad \to cX) \\
&= \frac{1}{F_{ad}}(2\pi)^4\delta^4(p_a + p_d - p_c - p_X)|\mathcal{M}(ad \to cX)|^2\rho_X \frac{d^3 p_c}{(2\pi)^2 2E_c} \\
&= \frac{F_{bd}}{F_{ad}} \frac{g_s^2|\overline{\sum} V(a \to bc)|^2}{\hat{t}^2} \frac{d^3 p_c}{(2\pi)^3 2E_c} \\
&\quad \times \frac{(2\pi)^4\delta^4(p_b + p_d - p_X)}{F_{bd}}|\overline{\mathcal{M}}_N|^2\rho_X \\
&= \frac{F_{bd}}{F_{ad}} \frac{g_s^2\overline{\sum}|V(a \to bc)|^2}{\hat{t}^2} \frac{d^3 p_c}{(2\pi)^3 2E_c} d\sigma_N
\end{aligned}
\tag{K.6a}
$$

$$
F_{ad} = 4\left[(p_a \cdot p_d)^2 - p_a^2 p_d^2\right]^{1/2} \simeq 4(p_a \cdot p_d), \quad F_{bd} \simeq 4(p_b \cdot p_d) \tag{K.6b}
$$

where F_{ad}, F_{bd} are incident fluxes and ρ_X is the phase space for the final state "X." $\overline{\sum}$ and $\overline{\mathcal{M}}$ mean to take spin (and all other degrees of freedom) average in the initial states and sum over final states. By doing so, we treated the two processes, $a \to c + b$ and $b + d \to X$, as independent, considering the particle in the internal line is almost physical which is valid in the collinear approximation. Substituting the relations

$$
\frac{F_{bd}}{F_{ad}} = \frac{E_b}{E_a} = z, \quad Q^2 \equiv -\hat{t} = \frac{p_T^2}{1-z}, \quad \frac{d^3 p_c}{E_c} = \frac{dz\,dp_T^2 d\phi}{2(1-z)} = \frac{dz\,dQ^2 d\phi}{2} \tag{K.7}
$$

in Eq. (K.6), we obtain

$$
\begin{aligned}
d\sigma_{N+1} &= z 4\pi\alpha_s \frac{|\overline{\sum} V(a \to bc)|^2}{Q^2 p_T^2/(1-z)} \frac{d\phi\,dz\,dQ^2}{4(2\pi)^3} d\sigma_N \\
&= \frac{\alpha_s}{2\pi}\left[\frac{z(1-z)}{2} \frac{|\overline{\sum}|V(a \to bc)|^2}{p_T^2} \frac{d\phi}{2\pi}\right] dz\, d\ln Q^2\, d\sigma_N
\end{aligned}
\tag{K.8}
$$

$d\phi/2\pi$ can be replaced with one as polarization states are summed, but retained for future use. Comparing Eq. (K.8) with Eq. (K.1), we obtain the expression for the splitting function.

$$
P_{b \leftarrow a}(z) = \frac{z(1-z)}{2} \frac{\overline{\sum}|V(a \to bc)|^2}{p_T^2} \tag{K.9}
$$

Splitting Function for $q \to q$ + Gluon We have produced a formula to calculate the splitting function in general. Now, we derive one for the gluon emission by a quark. Note, we have to sum only transverse polarizations because the partons are treated as almost physical. Denoting the color matrix for the quark as t^A and using

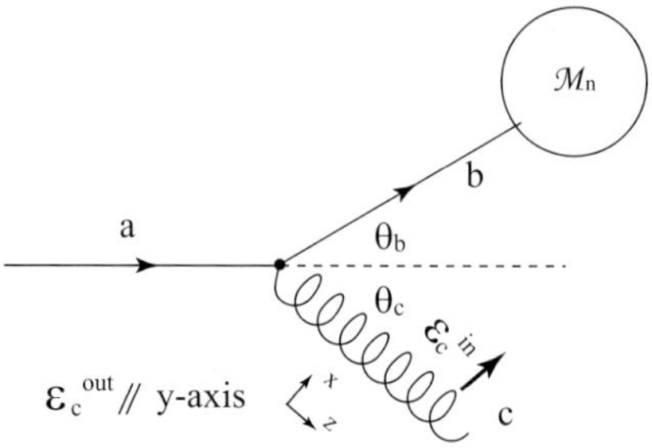

Figure K.2 $q(a) \to q(b) + G(c)$

Eq. (K.3),

$$\overline{\sum}|V(a \to bc)|^2 = \frac{1}{2 \times 3} \sum_{\text{spin, pol, color}} |\overline{u}_s(p_b)[t^A]_{sr}\gamma^\mu u_r(p_a)\epsilon_\mu(p_c)|^2$$

$$= \frac{1}{2} \frac{\sum_A \text{Tr}[t^A t^A]}{3} \sum_{\text{pol}} \left[\text{Tr}\left\{\not{p}_b \not{\epsilon} \not{p}_a \not{\epsilon}^*\right\}\right]$$

$$= 2C_F \sum_{\lambda=\pm} \left[2(p_a \cdot \epsilon_\lambda)(p_b \cdot \epsilon_\lambda^*) - (\epsilon_\lambda \cdot \epsilon_\lambda^*)(p_a \cdot p_b)\right]$$

$$= 4C_F \left[(\boldsymbol{p}_a \cdot \boldsymbol{p}_b) - \frac{(\boldsymbol{p}_a \cdot \boldsymbol{p}_c)(\boldsymbol{p}_b \cdot \boldsymbol{p}_c)}{(\boldsymbol{p}_c \cdot \boldsymbol{p}_c)} + (p_a \cdot p_b)\right]$$

$$= 4C_F E_a E_b (1 - \cos\theta_{ac} \cos\theta_{bc})$$

$$\simeq 2C_F \left[\frac{1+z^2}{z(1-z)^2}\right] p_T^2 \tag{K.10}$$

$$\therefore \quad P_{b \leftarrow a}(z) = C_F \frac{1+z^2}{1-z} \tag{K.11}$$

where C_F is the color factor which is the average of initial quark color and sum of final quark and gluon color.

$$C_F = \frac{1}{3} \sum_A \text{Tr}[t^A t^A] = \frac{4}{3} \tag{K.12}$$

Equation (K.11) reproduces the LLA result Eq. (8.44).

Splitting Function with Azimuthal Angle Dependence The Dirac plane wave functions are expressed as (see Vol. 1, Eq. (4.63))

$$u_r(\boldsymbol{p}) = \begin{bmatrix} \sqrt{p \cdot \sigma}\xi_r \\ \sqrt{p \cdot \overline{\sigma}}\xi_r \end{bmatrix} = \sqrt{\frac{E+m}{2}} \begin{bmatrix} \left(1 - \frac{\sigma \cdot p}{E+m}\right)\xi_r \\ \left(1 + \frac{\sigma \cdot p}{E+m}\right)\xi_r \end{bmatrix}$$

$$\xrightarrow{m=0} \sqrt{2E} \begin{bmatrix} \frac{1-\sigma\cdot\hat{p}}{2}\xi_r \\ \frac{1+\sigma\cdot\hat{p}}{2}\xi_r \end{bmatrix} \tag{K.13a}$$

$$v_r(\boldsymbol{p}) = \begin{bmatrix} \sqrt{p \cdot \sigma}\eta_r \\ -\sqrt{p \cdot \overline{\sigma}}\eta_r \end{bmatrix}$$

$$\xrightarrow{m=0} \sqrt{2E} \begin{bmatrix} \frac{1-\sigma\cdot\hat{p}}{2}\eta_r \\ -\frac{1+\sigma\cdot\hat{p}}{2}\eta_r \end{bmatrix} \tag{K.13b}$$

where $\eta_r = i\sigma_2\xi^*$. In the infinite momentum frame, $p \gg m$ and $p_T \simeq p\theta$. Putting $\boldsymbol{p} \simeq (p\theta, 0, p)$,

$$u_+(\boldsymbol{p}) = \sqrt{2E} \begin{bmatrix} 0 \\ -\theta/2 \\ 1 \\ \theta/2 \end{bmatrix}, \quad u_-(\boldsymbol{p}) = \sqrt{2E} \begin{bmatrix} -\theta/2 \\ 1 \\ \theta/2 \\ 0 \end{bmatrix}$$

$$v_+(\boldsymbol{p}) = \sqrt{2E} \begin{bmatrix} \theta/2 \\ -1 \\ \theta/2 \\ 0 \end{bmatrix}, \quad v_-(\boldsymbol{p}) = \sqrt{2E} \begin{bmatrix} 0 \\ -\theta/2 \\ -1 \\ -\theta/2 \end{bmatrix} \tag{K.14}$$

Putting $\boldsymbol{p}_1 = (p_1\theta_1, 0, p_1)$, $\boldsymbol{p}_2 = (p_2\theta_2, 0, p_2)$

$$\begin{aligned}
\overline{u}_+(p_2)\, \not{\epsilon}'_x u_+(p_1) &= \overline{u}_-(p_2)\, \not{\epsilon}'_x u_-(p_1) = -u_+^\dagger(p_2)\alpha_x u_+(p_1) \\
&= -\sqrt{E_1 E_2}(\theta_2 + \theta_1) \\
\overline{u}_+(p_2)\, \not{\epsilon}'_x u_-(p_1) &= \overline{u}_-(p_2)\, \not{\epsilon}'_x u_+(p_1) = 0 \\
\overline{u}_+(p_2)\, \not{\epsilon}'_y u_+(p_1) &= -\overline{u}_-(p_2)\, \not{\epsilon}'_y u_-(p_1) = -u_+^\dagger(p_2)\alpha_y u_+(p_1) \\
&= -i\sqrt{E_1 E_2}(\theta_2 - \theta_1) \\
\overline{u}_+(p_2)\, \not{\epsilon}'_y u_-(p_1) &= \overline{u}_-(p_2)\, \not{\epsilon}'_y u_+(p_1) = 0
\end{aligned} \tag{K.15}$$

The above expression is generic except the z-axis was taken along the gluon direction. Now, we define x, y axis as in Figure K.2. Then, we have to replace $\theta_1 \to \theta_c$ and $\theta_2 \to \theta_b + \theta_c$. Other variables are defined in Eq. (K.4). Substituting them in Eq. (K.15),

$$\overline{u}_\pm(p_b)\, \not{\epsilon}_x u_\pm(p_a) = -\sqrt{E_a E_b}(\theta_b + 2\theta_c) = -E_a\theta\sqrt{z}(1+z) \tag{K.16a}$$

$$\overline{u}_\pm(p_b)\, \not{\epsilon}_y u_\pm(p_a) = \pm i\sqrt{E_a E_b}\theta_b = \pm i E_a\theta\sqrt{z}(1-z) \tag{K.16b}$$

Referring to Eq. (K.9),

$$
\begin{aligned}
P_{qG\leftarrow q} &= \frac{z(1-z)}{2p_{\mathrm{T}}^2}\overline{\sum}|V(a\to bc)|^2 = \frac{z(1-z)}{2p_{\mathrm{T}}^2}\overline{\sum}|-i\bar{u}(p_b)t^A\,\not{\epsilon}\, v(p_a)|^2\\
&= \frac{C_F}{2z(1-z)(E_a\theta)^2}\frac{1}{2}\left[\underbrace{\left|E_a\theta\sqrt{z}(1+z)\right|^2}_{\epsilon=\epsilon_x}+\underbrace{\left|\pm i E_a\theta\sqrt{z}(1-z)\right|^2}_{\epsilon=\epsilon_y}\right]\times 2\\
&= C_F\frac{1+z^2}{1-z}
\end{aligned}
\tag{K.17}
$$

where the last factor 2 comes from spin sum because there are two combinations that give the same contribution as shown in Eq. (K.15). In general, two polarization vectors lie in x–y plane. Denoting the azimuthal angle of the polarization ϵ_x as ϕ, we can consider correlation between the polarization of the gluon and decay plane. The function becomes

$$
\begin{aligned}
d\Gamma(\phi) &\propto \left|\cos\phi\,\mathcal{M}(\epsilon_a=\epsilon_x)+\sin\phi\,\mathcal{M}(\epsilon_a=\epsilon_y)\right|^2\\
&\propto \frac{\cos^2\phi(1+z)^2+\sin^2\phi(1-z)^2}{1-z}\\
&= \left[\frac{1+z^2}{1-z}+\frac{2z}{1-z}\cos 2\phi\right]
\end{aligned}
\tag{K.18}
$$

$$
\therefore\quad P_{qG\leftarrow q} = C_F\left[\frac{1+z^2}{1-z}+\frac{2z}{1-z}\cos 2\phi\right] \tag{K.19}
$$

Branching of Partons in the Final State One can show that the same formula holds for the branching of a parton in the final state (see Figure K.1b) as is expected from the identical form of the splitting function in the $e^-e^+\to q\bar{q}g$ to that of deep inelastic scattering. However, kinematical states of the participating partons are different. As it has important applications later for the final state jet production process, we comment on the similarity and difference of the splitting in the initial and final state.

In the infinite momentum frame as is described in Eq. (K.3), kinematic variables shown in Figure K.1b satisfy the following equalities. This time, the parton "a" is virtual and has large time like Q^2.

$$
E_b = zE_a\,,\quad E_c=(1-z)E_a\,,\quad \theta=\frac{\theta_b}{1-z}=\frac{\theta_c}{z} \tag{K.20a}
$$

$$
Q^2=\hat{t}=(p_b+p_c)^2=2E_bE_c(1-\cos\theta)\simeq E_bE_c\theta^2=z(1-z)(E_a\theta)^2 \tag{K.20b}
$$

$$
p_{\mathrm{T}}\simeq E_b\theta_b=E_c\theta_c=z(1-z)E_a\theta \tag{K.20c}
$$

which leads to

$$
Q^2=p_{\mathrm{T}}^2/z(1-z) \tag{K.20d}
$$

Note the difference between $Q^2 = |\hat{t}|$ in Eq. (K.4e) $[Q^2 = p_T^2/(1-z)]$ and that in Eq. (K.20d) $[Q^2 = p_T^2/z(1-z)]$. However, the $X+1$ final state phase space is also modified as

$$\begin{aligned}\rho_{X+1} &= \rho_{X-a}\frac{d^3 p_b}{(2\pi)^3 2E_b}\frac{d^3 p_c}{(2\pi)^3 2E_c} = \rho_{X-a}\frac{d^3 p_a}{(2\pi)^3 2E_a}\left(\frac{E_a}{E_b}\right)\frac{d^3 p_c}{(2\pi)^3 2E_c}\\ &= \rho_X \frac{1}{z}\frac{d^3 p}{(2\pi)^2 E_c}\end{aligned} \tag{K.21}$$

where ρ_{X-a} is the phase space for X final state minus "a" and we used $d^3 p_b = d^3 p_a$ for fixed p_c. The extra $1/z$ factor in the phase space cancels the difference in Q^2. Thus, Eq. (K.8) expressed in terms of the cross section is valid for the final state branching as well. The difference lies in the virtuality of the partons. In the initial state radiation, parton "b" is virtual. However, in the final state radiation, parton "a" is virtual. This makes an essential difference when we consider cascade branching. See arguments in Section 8.3.4 of the main text.

$G \to q\bar{q}$ The Dirac plane wave functions in the infinite momentum frame are expressed as

$$\begin{aligned}\overline{u}_+(p_2)\, \not{\epsilon}'_x v_-(p_1) &= \overline{u}_-(p_2)\, \not{\epsilon}'_x v_+(p_1) = -u_-^\dagger(p_2)\alpha_x v_+(p_1)\\ &= -\sqrt{E_1 E_2}(\theta_2+\theta_1)\\ \overline{u}_+(p_2)\, \not{\epsilon}'_x v_+(p_1) &= \overline{u}_-(p_2)\, \not{\epsilon}'_x v_-(p_1) = 0\\ \overline{u}_+(p_2)\, \not{\epsilon}'_y v_-(p_1) &= -\overline{u}_-(p_2)\, \not{\epsilon}'_y v_+(p_1) = u_-^\dagger(p_2)\alpha_y v_+(p_1)\\ &= i\sqrt{E_1 E_2}(\theta_2-\theta_1)\\ \overline{u}_+(p_2)\, \not{\epsilon}'_y v_+(p_1) &= \overline{u}_-(p_2)\, \not{\epsilon}'_y v_-(p_1) = 0\end{aligned} \tag{K.22}$$

We define variables as in Figure K.3 and Eq. (K.23). In this case, $\theta_2 = \theta_b, \theta_1 = -\theta_c$.

$$E_b = zE_a\,, \quad E_c = (1-z)E_a\,, \quad \theta = \frac{\theta_b}{1-z} = \frac{\theta_c}{z} \tag{K.23a}$$

$$\hat{t} = p_a^2 = (p_b+p_c)^2 = 2E_bE_c(1-\cos\theta) \simeq E_bE_c\theta^2 = z(1-z)(E_a\theta)^2 \tag{K.23b}$$

$$p_T \simeq E_b\theta_b = E_c\theta_c z(1-z)E_a\theta = \sqrt{z(1-z)\hat{t}} \tag{K.23c}$$

Substituting Eqs. (K.23) in Eq. (K.22),

$$\overline{u}_\pm(p_b)\, \not{\epsilon}_x v_\mp(p_c) = -\sqrt{E_bE_c}(\theta_b-\theta_c) = -E_a\theta\sqrt{z(1-z)}(1-2z) \tag{K.24a}$$

$$\overline{u}_\pm(p_b)\, \not{\epsilon}_y v_\mp(p_c) = \pm i\sqrt{E_bE_c}(\theta_b+\theta_c) = \pm iE_a\theta\sqrt{z(1-z)} \tag{K.24b}$$

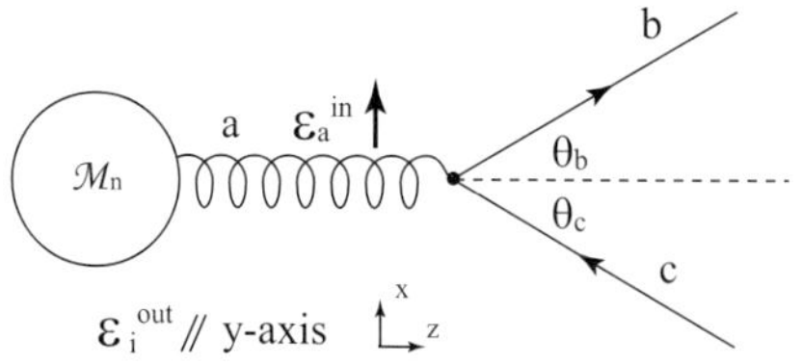

Figure K.3 $G(a) \rightarrow q(b) + \overline{q}(c)$.

$$
\begin{aligned}
P_{G\rightarrow q\overline{q}} &= \frac{z(1-z)}{2p_T^2}\overline{\sum}|V(a\rightarrow bc)|^2 = \frac{z(1-z)}{2p_T^2}\overline{\sum}|\overline{u}(p_b)t^A \not{\epsilon} v(p_c)|^2 \\
&= \frac{T_R}{2z(1-z)(E_a\theta)^2} \\
&\quad \times \left[\underbrace{\left|-E_a\theta\sqrt{z(1-z)}(1-2z)\right|^2}_{\epsilon=\epsilon_x} + \underbrace{\left|\pm i E_a\theta\sqrt{z(1-z)}\right|^2}_{\epsilon=\epsilon_y}\right]\times 2 \\
&= 2T_R[z^2+(1-z)^2]
\end{aligned}
\tag{K.25}
$$

where the last factor 2 comes from spin sum and T_R is the color factor which is the average of gluon colors and the sum of quark colors.

$$
T_R = \frac{1}{8}\mathrm{Tr}[t^A t^A] = \frac{1}{2} \tag{K.26}
$$

The polarization vector lies in $x-y$ plane. Denoting the azimuthal angle of the polarization as ϕ, the correlation function between the polarization of the gluon and decay plane becomes

$$
\begin{aligned}
d\Gamma(\phi) &\propto \sum\left|\cos\phi\,\mathcal{M}(\epsilon_a=\epsilon_x)+\sin\phi\,\mathcal{M}(\epsilon_a=\epsilon_y)\right|^2 \\
&\propto \cos^2\phi(1-2z)^2+\sin^2\phi \\
&= z^2+(1-z)^2-2z(1-z)\cos 2\phi
\end{aligned}
\tag{K.27}
$$

Note, this has negative angular correlation as $\phi = 0 \rightarrow \pi/2$ and the $q\overline{q}$ pair tends to be produced in a plane perpendicular to the polarization.

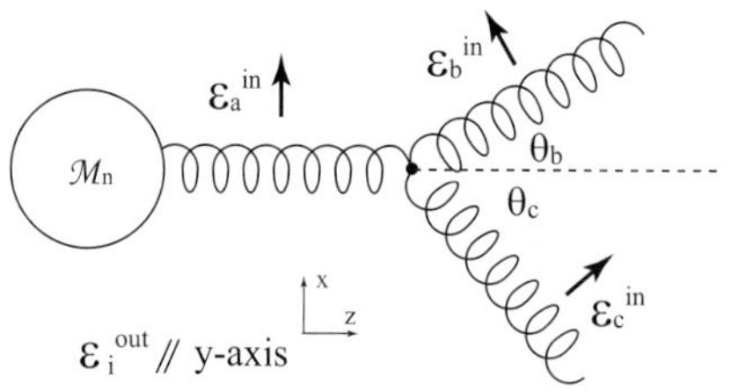

Figure K.4 $G \rightarrow GG$

G → GG We define variables as given in Figure K.4.

$$E_b = zE_a\,, \quad E_c = (1-z)E_a \tag{K.28a}$$

$$\theta = \theta_b + \theta_c = \frac{\theta_b}{1-z} = \frac{\theta_c}{z} \tag{K.28b}$$

$$t = (p_b + p_c)^2 = 2E_b E_c(1-\cos\theta) \simeq z(1-z)(E_a\theta)^2 \tag{K.28c}$$

$$p_T \simeq E_b\theta_b = E_c\theta_c = z(1-z)E_a\theta \tag{K.28d}$$

The matrix element of the 3-gluon vertex is expressed as

$$\begin{aligned}
&\langle 0|\int d^4x\mathcal{H}_{\text{int}}|abc\rangle \equiv (2\pi)^4\delta^4(p_a+p_b+p_c)gV(G\to GG)\\
&V(G\to GG) = if^{ABC}\epsilon_a^\mu\epsilon_b^\nu\epsilon_c^\rho[g_{\mu\nu}(p_a-p_b)_\rho + g_{\nu\rho}(p_b-p_c)_\mu + g_{\rho\mu}(p_c-p_a)_\nu]
\end{aligned} \tag{K.29}$$

where ϵ_i^α represents the polarization vector of the gluon i and all momenta are defined as incoming so that $p_a + p_b + p_c = 0$. Since $(\epsilon_i \cdot p_i = 0)$, we can rewrite Eq. (K.29)

$$V_{G\to GG} = -2if^{ABC}[(\varepsilon_a\cdot\varepsilon_b)(\varepsilon_c\cdot p_b) - (\varepsilon_b\cdot\varepsilon_c)(\varepsilon_a\cdot p_b) - (\varepsilon_c\cdot\varepsilon_a)(\varepsilon_b\cdot p_c)] \tag{K.30}$$

Since all the gluons are almost on mass-shell, their polarization vectors can be taken as transverse. We take one direction in the plane of decay, that is, in $x-z$ plane and the other normal to the plane, that is, in y direction. Then,

$$\begin{aligned}
&\varepsilon_{i,\text{in}}\cdot\varepsilon_{j,\text{in}} = \varepsilon_{i,y}\cdot\varepsilon_{j,y} = -1 + O(\theta^2)\\
&\varepsilon_{i,x}\cdot\varepsilon_{j,y} = \varepsilon_{i,y}\cdot p_j = 0
\end{aligned} \tag{K.31}$$

Using this convention, we have

$$\begin{aligned}
&\varepsilon_{c,\text{in}}\cdot p_b = -E_b\theta = -zE_a\theta\\
&\varepsilon_{a,\text{in}}\cdot p_b = -E_b\theta_b = -z(1-z)E_a\theta\\
&\varepsilon_{b,\text{in}}\cdot p_c = +E_c\theta = (1-z)E_a\theta
\end{aligned} \tag{K.32}$$

Substituting Eq. (K.32) in Eq. (K.30), we have for $\varepsilon_i = \varepsilon_{i,\text{in}} i = a, b, c$

$$\begin{aligned}
\sum_{b,c}\frac{|V_{G\to GG}|^2}{z(1-z)} &= 4|f^{ABC}|^2(E_a\theta)^2\left[\frac{z}{1-z} + z(1-z) + \frac{1-z}{z}\right]\\
&\equiv 4|f^{ABC}|^2(E_a\theta)^2 S(z;\varepsilon_a,\varepsilon_b,\varepsilon_c)
\end{aligned} \tag{K.33}$$

For other polarization combinations, we refer to Table K.1. Taking the average of the initial and the sum of final state polarization, we have

$$\begin{aligned}
P_{G\to GG}(z) &= \frac{z(1-z)}{2p_T^2}\overline{\sum}|V(G\to GG)|^2\\
&= 2C_A\left[\frac{z}{1-z} + z(1-z) + \frac{1-z}{z}\right]
\end{aligned} \tag{K.34}$$

Table K.1 Polarization dependence of the branching $G \to GG$.

ε_a	ε_b	ε_c	$S(z; \varepsilon_a, \varepsilon_b, \varepsilon_c)$
in	in	in	$(1-z)/z + z/(1-z) + z(1-z)$
in	out	out	$z(1-z)$
out	in	out	$(1-z)/z$
out	out	in	$z/(1-z)$
other combinations			0

where $C_A = \mathrm{Tr}[T^C T^C] = \sum f^{ABC} f^{ABC} = 3$. The correlation function $d\Gamma(\phi)$ can be obtained from

$$\begin{aligned} d\Gamma(\phi) &\propto \sum_{\varepsilon_b,\varepsilon_c} |\cos\phi(\varepsilon_{a,\mathrm{in}}, \varepsilon_b, \varepsilon_c) + \sin\phi(\varepsilon_{a,\mathrm{out}}, \varepsilon_b, \varepsilon_c)|^2 \\ &= \cos^2\phi\left[\frac{1-z}{z} + \frac{z}{1-z} + 2z(1-z)\right] + \sin^2\phi\left[\frac{1-z}{z} + \frac{z}{1-z}\right] \\ &= \frac{(1-z+z^2)^2}{z(1-z)} + z(1-z)\cos 2\phi \end{aligned} \quad (K.35)$$

The angular orrelation favors an orientation in which the plane of the emitted pair aligns with the polarization of the emitting gluon. However, the correlation is quite weak. Its coefficient $z(1-z)$ vanishes in the enhanced region $z \to 0, 1$ and reaches its maximum at $z = 1/2$, where it is still only one-ninth of the unpolarized contribution.

Appendix L
Answers to the Problems

Chapter 7: QCD

Solution to Problem 7.6:

$$
\begin{aligned}
P^{(2)} &= c_0 + c_1\alpha_s + [c_2 - c_1 b_0\tau]\alpha_s^2 + \cdots\,, \\
\frac{d\alpha_s}{d\tau} &= -\beta(\tau) = b_0\alpha_s^2(1 + b_1\alpha_s + \cdots) \\
\frac{dP^{(2)}}{d\tau} &= -c_1 b_0\alpha_s^2 + \frac{d\alpha_s}{d\tau}\left[c_1 + 2(c_2 - c_1 b_0\tau)\alpha_s\right] + \cdots \\
&= \left[c_1 b_0 b_1 + 2b_0(c_2 - c_1 b_0\tau)\right]\alpha_s^3 + O\left(\alpha_s^4\right) \\
\therefore \quad \tau &= \ln\left(\frac{s}{\mu^2}\right) = \frac{2c_2 + c_1 b_1}{2c_1 b_0}\,.
\end{aligned}
$$

Solution to Problem 7.7 Using abbreviations

$$
\frac{\partial P}{\partial\alpha_s} = P_1\,,\quad \frac{\partial P}{\partial m} = P_2\,,\quad \alpha_s(\tau) = \overline{\alpha}_s\,,\quad \beta(\tau) = \overline{\beta}\,,\quad \gamma(\tau) = \overline{\gamma}
$$

$$
\begin{aligned}
&\left[\mu^2\frac{\partial}{\partial\mu^2} + \beta\frac{\partial}{\partial\alpha_s} + \gamma_m m\frac{\partial}{\partial m}\right] P\left(0, \overline{\alpha}, \overline{m}, Q^2\right) \\
&= -\left[\frac{\partial\overline{\alpha}}{\partial\tau}P_1 + \frac{\partial\overline{m}}{\partial\tau}P_2\right] + \beta\left[\frac{\partial\overline{\alpha}}{\partial\alpha_s}P_1 + \frac{\partial\overline{m}}{\partial\alpha_s}P_2\right] + \gamma_m m\frac{\partial\overline{m}}{\partial m}P_2 \qquad \text{(L.1)}
\end{aligned}
$$

which vanishes if one makes use of relations

$$
\begin{aligned}
&\frac{d\overline{\alpha}}{d\tau} = \overline{\beta}\,,\quad \frac{d\overline{\alpha}}{d\alpha_s} = \frac{\overline{\beta}}{\beta} \quad \text{(see Eq. (7.66))}\,, \\
&\frac{d\overline{m}}{d\tau} = \overline{\gamma m}\,,\quad \frac{d\overline{m}}{d\alpha_s} = \frac{\partial\overline{m}}{\partial\alpha_s} + \frac{\partial\overline{\alpha}}{\partial\alpha_s}\frac{\partial\overline{m}}{\partial\overline{\alpha}} = \left(-\frac{\gamma}{\beta} + \frac{\overline{\beta}}{\beta}\frac{\overline{\gamma}}{\overline{\beta}}\right)\overline{m}\,,\quad \frac{d\overline{m}}{dm} = \frac{\overline{m}}{m}
\end{aligned}
\qquad \text{(L.2)}
$$

where equations in the second line can be derived from Eq. (7.96).

Elementary Particle Physics, Volume 2: Foundations of the Standard Model, First Edition. Yorikiyo Nagashima.

Solution to Problem 7.8 Define the pion and eta before and after mixing by

$$\phi_3 = \frac{1}{\sqrt{2}}\left(u\bar{u} - d\bar{d}\right), \quad \phi_8 = \frac{1}{\sqrt{6}}\left(u\bar{u} + d\bar{d} - 2s\bar{s}\right)$$

$$\begin{bmatrix} \pi_0 \\ \eta_8 \end{bmatrix} \equiv U \begin{bmatrix} \phi_3 \\ \phi_8 \end{bmatrix} = \begin{bmatrix} \cos\theta & \sin\theta \\ -\sin\theta & \cos\theta \end{bmatrix} \begin{bmatrix} \phi_3 \\ \phi_8 \end{bmatrix} \tag{L.3}$$

Then, the mass matrix is given by

$$\begin{aligned} M^2 &= U^{-1}\text{diag}(m_{\pi_0}^2, m_{\eta_8}^2)U \equiv \begin{bmatrix} M_3 & M_{38} \\ M_{83} & M_8 \end{bmatrix} \\ &= \begin{bmatrix} \cos^2\theta\, m_{\pi_0}^2 + \sin^2\theta\, m_{\eta_8}^2 & \sin\theta\cos\theta\left(m_{\pi_0}^2 - m_{\eta_8}^2\right) \\ \sin\theta\cos\theta\left(m_{\pi_0}^2 - m_{\eta_8}^2\right) & \sin^2\theta\, m_{\pi_0}^2 + \cos^2\theta\, m_{\eta_8}^2 \end{bmatrix} \end{aligned} \tag{L.4a}$$

From Eq. (7.57a), one can also obtain masses of π^0 and η_8 in terms of the quark masses.

$$\langle \phi_i | \mathcal{L}_m | \phi_j \rangle|_{i,j=3,8} \sim \begin{bmatrix} B(m_u + m_d) & \frac{B}{\sqrt{3}}(m_u - m_d) \\ \frac{B}{\sqrt{3}}(m_u - m_d) & \frac{B}{3}(m_u + m_d + 4m_s) \end{bmatrix} \tag{L.4b}$$

where B is given by Eq. (7.114). By equating Eq. (L.4a) and Eq. (L.4b), one gets

$$\begin{aligned} \tan 2\theta &= \frac{2M_{38}}{M_3 - M_8} = \frac{\sqrt{3}(m_d - m_u)}{2m_s - m_u - m_d} \\ M_3 + M_8 &= \frac{4B}{3}(m_s + m_u + m_d), \quad M_8 - M_3 = \frac{2B}{3}(2m_s - m_u - m_d) \\ M_{38} &= \frac{B}{\sqrt{3}}(m_u - m_d) \end{aligned} \tag{L.5}$$

Solving the secular equation

$$(M_3 - \lambda)(M_8 - \lambda) - M_{38}^2 = 0 \tag{L.6}$$

one gets

$$\begin{aligned} \lambda &= \frac{M_3 + M_8}{2} \pm \sqrt{\left(\frac{M_3 + M_8}{2}\right)^2 - M_3 M_8 + M_{38}^2} \\ &= \begin{cases} \frac{B}{3}(m_u + m_d + 4m_s) + \varepsilon + O(\varepsilon^2) \\ B(m_u + m_d) - \varepsilon + O(\varepsilon^2) \end{cases} \\ \varepsilon &= \frac{B}{2}\frac{(m_d - m_u)^2}{(2m_s - m_u - m_d)} \end{aligned} \tag{L.7}$$

□

Chapter 8: Deep Inelastic Scattering

Solution to Problem 8.1: Denoting four momenta of the virtual photon, target quark and recoiled quark as q, p_1 and p_2

$$\hat{\sigma}_B = \frac{1}{4(q\cdot p_1)}\frac{1}{2}\sum |e\lambda_q \overline{u}(p_2)\ \not{\varepsilon} u(p_1)|^2 (2\pi)^4 \delta^4(q+p_1-p_2)\frac{d^3 p_2}{(2\pi)^3 2E_2} \tag{L.8}$$

Using

$$\text{flux} = 2\left[(q+p_1) - q^2 - m_q^2\right] \simeq 2\left[\hat{s} + Q^2\right]$$

$$\sum |\overline{u}(p_2)\ \not{\varepsilon} u(p_1)|^2 = 4\sum_\lambda \left[2(\varepsilon\cdot p_1)(\varepsilon^*\cdot p_2) - (\varepsilon\cdot\varepsilon^*)(p_1\cdot p_2)\right] = 4Q^2$$

$$\begin{aligned}\frac{\delta(q^0+E_1-E_2)}{2E_2} &= \delta\left[(p_2^0)^2 - (q^0+p_1^0)^2\right] \overset{p_2=q+p_1}{=\!=\!=} \delta\left[p_2^2 - (q+p_1)^2\right]\\ &= \delta\left[Q^2 - 2(q\cdot p_1)\right] = \frac{1}{2m\nu}\delta(z-1) \quad z = \frac{Q^2}{2(q\cdot p_1)}\end{aligned} \tag{L.9}$$

$$\hat{\sigma}_B = \frac{8\pi^2\alpha\lambda_q^2}{2(\hat{s}+Q^2)} z\delta(1-z) = \lambda_q^2\hat{\sigma}_0\delta(1-z) \tag{L.10}$$

Solution to Problem 8.2: Substitute Eq. (8.72b) and (8.72d) into Eq. (8.85).

Chapter 9: Jets

Solution to Problem 9.2: For the sphericity, consider a case where a vector $\boldsymbol{p}$ parallel to the sphericity axis splits to two collinear vectors $\boldsymbol{p}/2 + \boldsymbol{p}/2$. The sum of transverse momenta does not change for this splitting. However, the term in the denominator changes from $\boldsymbol{p}^2 \to 2(\boldsymbol{p}/2)^2 = \boldsymbol{p}^2/2$. Therefore, the sphericity S diminishes and is not stable.

The C-parameter is defined as the eigenvalues of a tensor operator:

$$C^{ij} = \frac{1}{\sum_a |\boldsymbol{p}_a|}\sum_a \frac{p_a^i p_a^j}{|\boldsymbol{p}_a|} \equiv \frac{N}{D} \tag{L.11}$$

Splitting of any $\boldsymbol{p}$ to a sum of collinear momenta is expressed as

$$\boldsymbol{p} = \sum f_b \boldsymbol{p}\,, \quad \text{with} \quad \sum_b f_b = 1 \tag{L.12}$$

Since the splitting is collinear, D is not changed. As for N, one finds

$$\frac{p_a^i p_a^j}{|\boldsymbol{p}_a|} = \sum_b \frac{f_b^2 p_a^i p_a^j}{f_b|\boldsymbol{p}_a|} = \frac{p_a^i p_a^j}{|\boldsymbol{p}_a|}\sum_b f_b = \frac{p_a^i p_a^j}{|\boldsymbol{p}_a|} \tag{L.13}$$

namely, N does not change either.

Solution to Problem 9.3

$$
\begin{aligned}
R_3 &= C_F \frac{\alpha_s}{2\pi}\left[\int_{2y}^{1-y} dx_1 \int_{1+y-x_1}^{1-y} dx_2 \frac{x_1^2 + x_2^3}{(1-x_1)(1-x_2)}\right] \\
\left[\cdots\right] &= \int_{2y}^{1-y} dx_1 \int_{1+y-x_1}^{1-y} dx_2 \left[\frac{2}{(1-x_1)(1-x_2)} - \frac{1+x_1}{1-x_2} - \frac{1+x_2}{1-x_1}\right] \\
&= \int_{2y}^{1-y} dx_1 \frac{2}{1-x_1} \ln\frac{x_1 - y}{y} - \left[\int_{2y}^{1-y} dx_1 (1+x_1) \ln\frac{x_1-y}{y} + (1 \leftrightarrow 2)\right] \\
&\equiv A + [B + C]
\end{aligned}
\tag{L.14}
$$

Using $\int dz \ln z = z\ln z - z$, $\int dz z \ln z = (z^2/2)(\ln z - 1/2)$, it is straightforward to obtain

$$
B + C = 2B = 3(1-2y)\ln\frac{y}{1-2y} + \frac{5}{2} - 6y - \frac{9}{2}y^2
\tag{L.15}
$$

For A, changing the variable from x_1 to $z = (1-x_1)/(1-y)$

$$
\begin{aligned}
A &= -2 \int_{1-y/(1-y)}^{y/(1-y)} dz \frac{\ln\left(\frac{1-y}{y}(1-z)\right)}{z} \\
&= -2\ln\frac{1-y}{y} \int_{1-y/(1-y)}^{y/(1-y)} \frac{dz}{1-z} - 2 \int_{1-y/(1-y)}^{y/(1-y)} dz \frac{\ln(1-z)}{z} dz \\
&= 2\ln\frac{1-y}{y} \ln\frac{1-2y}{y} + D
\end{aligned}
\tag{L.16}
$$

$$
\begin{aligned}
D &= 2\left[-\int_0^{y/(1-y)} + \int_0^{1-y/(1-y)}\right] \frac{\ln(1-z)}{z} dz \\
&= 2\left[\mathrm{Li}_2\left(\frac{y}{1-y}\right) - \mathrm{Li}_2\left(1 - \frac{y}{1-y}\right)\right]
\end{aligned}
\tag{L.17}
$$

Using

$$
\mathrm{Li}_2(1-x) + \mathrm{Li}_2(x) = \frac{\pi^2}{6} - \ln x \ln(1-x)
\tag{L.18}
$$

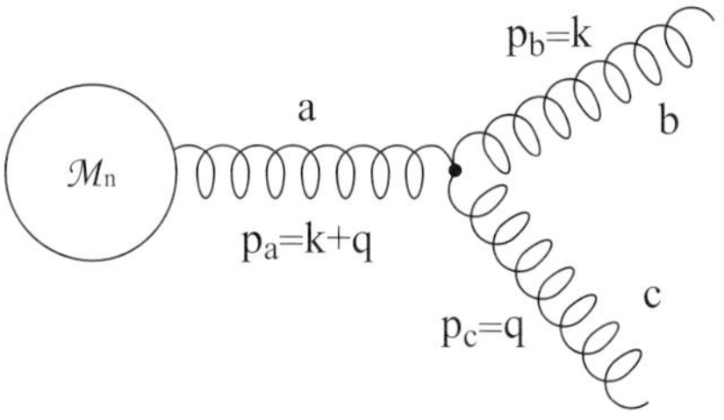

Figure L.1 Soft gluon emission from a gluon

we have

$$\begin{aligned} D &= 4\text{Li}_2\left(\frac{y}{1-y}\right) - \frac{\pi^2}{3} + 2\ln\frac{y}{1-y}\ln\frac{1-2y}{1-y} \\ &= 4\text{Li}_2\left(\frac{y}{1-y}\right) - \frac{\pi^2}{3} + 2\ln^2\frac{y}{1-y} + 2\ln\frac{y}{1-y}\ln\frac{1-2y}{y} \\ \therefore\quad A &= 4\text{Li}_2\left(\frac{y}{1-y}\right) - \frac{\pi^2}{3} + 2\ln^2\frac{y}{1-y} \end{aligned} \tag{L.19}$$

Combining $(A + B + C)$, one gets the answer. □

Chapter 10: Gluons

Solution to Problem 10.1 The amplitude for emission of extra gluon from a gluon in the final state (Figure L.1) is expressed as

$$\begin{aligned} \mathcal{M}_{N+1}(p_b; p_c) &\equiv M_{N+1}(p_{b_\mu}; p_{c_\nu})\varepsilon^*_{b_\mu}\varepsilon^{\nu*}_c = M_N(p_{a_\sigma}, a)\frac{N^{\sigma\rho}}{p_a^2}T^{abc}_{\rho\mu\nu}\varepsilon^{\mu*}_b\varepsilon^{\nu*}_c \\ N^{\sigma\rho} &= \sum_{\lambda=\pm}\varepsilon^\sigma_a(\lambda)^*\varepsilon^\rho_a(\lambda) \\ T^{abc}_{\rho\mu\nu} &= g_s f^{abc}\left[g_{\rho\mu}(p_a - p_b)_\nu + g_{\mu\nu}(p_b - p_c)_\rho + g_{\nu\rho}(p_c - p_a)_\mu\right] \end{aligned} \tag{L.20}$$

Using $p_i \cdot \varepsilon_i = 0$ and inserting $p_a = (q + k)$, $p_b = -k$, $p_c = -q$, we obtain

$$\begin{aligned} \varepsilon^\rho_a T^{abc}_{\rho\mu\nu}\varepsilon^{\mu*}_b\varepsilon^{\nu*}_c &= 2g_s f^{abc}\left[(\varepsilon_a\cdot\varepsilon^*_b)(p_b\cdot\varepsilon^*_c) - (\varepsilon^*_b\cdot\varepsilon^*_c)(p_b\cdot\varepsilon_a)\right. \\ &\quad \left.- (\varepsilon_c\cdot\varepsilon_a)(p_c\cdot\varepsilon_b)\right] \\ &= 2g_s f^{abc}\left[(\varepsilon_a\cdot\varepsilon^*_b)(k\cdot\varepsilon^*_c) - (\varepsilon^*_b\cdot\varepsilon^*_c)(k\cdot\varepsilon_a)\right. \\ &\quad \left.- (\varepsilon_c\cdot\varepsilon_a)(q\cdot\varepsilon_b)\right] \\ &= -2g_s f^{abc}(\varepsilon_a\cdot\varepsilon_c)(q\cdot\varepsilon_b) + O(k) \end{aligned} \tag{L.21}$$

Substituting Eq. (L.21) in Eq. (L.20), we obtain

$$\begin{aligned}\mathcal{M}_{N+1}(p_b; p_c) &= -\sum_{\lambda} M_N(p_{a_\sigma}, a)\varepsilon_a^{\sigma}(\lambda)^* \left(\varepsilon_a(\lambda)\cdot\varepsilon_c(\lambda')\right)\left[g_s f^{abc}\frac{(q\cdot\varepsilon_b)}{(q\cdot k)}\right] \\ &\simeq M_N(p_{c_\nu})\varepsilon_c^{\nu *}\left[g_s f^{abc}\frac{(q\cdot\varepsilon_b)}{(q\cdot k)}\right] = \mathcal{M}_N\left[g_s \boldsymbol{T}_g\frac{(q\cdot\varepsilon_b)}{(q\cdot k)}\right]\end{aligned} \tag{L.22}$$

where the second last equality is obtained by approximating $p_a = q + k \simeq q = p_c$.

□

Solution to Problem 10.3 In the coordinate frame of Figure 10.9,

$$\begin{aligned}\hat{\boldsymbol{k}} &= (\sin\theta\cos\alpha, \sin\theta\sin\alpha, \cos\theta)\,, \\ \hat{\boldsymbol{p}}_1 &= (1,0,0)\,, \quad \hat{\boldsymbol{p}}_2 = (\cos(\alpha+\beta), \sin(\alpha+\beta), 0) \\ \left(\hat{p}_1\cdot\hat{k}\right) &= 1-\sin\theta\cos\alpha\,, \quad \left(\hat{p}_2\cdot\hat{k}\right) = 1-\sin\theta\cos\beta \\ V(\alpha,\beta) &= \int\frac{d\cos\theta}{2}\frac{1}{(\hat{k}\cdot\hat{p}_1)(\hat{k}\cdot\hat{p}_2)} \\ &= \frac{1}{\cos\alpha-\cos\beta}\int\frac{d\cos\theta}{2}\left(\frac{\cos\alpha}{1-\sin\theta\cos\alpha}-\frac{\cos\beta}{1-\sin\theta\cos\beta}\right)\end{aligned} \tag{L.23}$$

The integrand containing the first term in $(\cdots)$ can be written as

$$\int_0^1\frac{d\cos\theta}{2}\frac{\cos\alpha}{1-\sin\theta\cos\alpha} = \int_0^{\pi/2} d\theta\left(\frac{1}{1-\sin\theta\cos\alpha}-1\right) \tag{L.24}$$

Using $t = \tan\theta/2$, $d\theta = 2dt/(1+t^2)$, $\sin\theta = 2t/(1+t^2)$,

$$\begin{aligned}\int_0^{\pi/2}\frac{d\theta}{1-\sin\theta\cos\alpha} &= 2\int_0^{\infty}\frac{dt}{1+t^2-2t\cos\alpha} = 2\int_0^{\infty}\frac{dt}{(t-\cos\alpha)^2+\sin^2\alpha} \\ &= \frac{2}{\sin\alpha}\left[\tan^{-1}\frac{t-\cos\alpha}{\sin\alpha}\right]_0^{\infty} = \frac{2}{\sin\alpha}[\pi-\alpha]\end{aligned} \tag{L.25}$$

$$\therefore \quad V(\alpha,\beta) = \frac{2}{\cos\alpha-\cos\beta}\left[\frac{\pi-\alpha}{\sin\alpha}-\frac{\pi-\beta}{\sin\beta}\right] \tag{L.26}$$

□

Solution to Problem 10.4 A:

$$\int^{E}\frac{d\epsilon}{\epsilon}\int^{1}\frac{d\theta_0}{\theta_0}\vartheta(\epsilon\theta_0-1/R) = \int_{1/R}^{E}\frac{d\epsilon}{\epsilon}\int_{1/\epsilon R}^{1}\frac{d\theta_0}{\theta_0} \tag{L.27}$$

For the incoherent emission: $\theta_{\max} = 1/R$.

$$\int^{\epsilon} \frac{dk}{k} \int^{\theta'_{\max}} \frac{d\theta'}{\theta'} \delta(k\theta' - 1/R) = \int_{1/R}^{\epsilon} \frac{dk}{k} \tag{L.28}$$

As $\int d\theta_0$ and $\int dk$ is independent, we carry out integration over θ_0 which gives $\ln \epsilon R$. Then, the exchanging integration over ε and k, n becomes

$$\begin{aligned} n &\sim \int_{1/R}^{E} \frac{d\epsilon}{\epsilon} \int_{1/R}^{\epsilon} \frac{dk}{k} \ln \epsilon R = \int_{1/R}^{E} \frac{dk}{k} \int_{k}^{E} \frac{d\epsilon}{\epsilon} \ln \epsilon R \\ &= \int d \ln k \frac{1}{2} \left[\ln^2 ER - \ln^2 kR \right] \\ \therefore \quad \frac{dn}{d \ln k} &= \frac{\alpha_s}{2} \left[\ln^2 ER - \ln^2 kR \right] \end{aligned} \tag{L.29}$$

For the coherent emission, $\theta_{\max} = \theta_0$,

$$n = \int_{1/R}^{E} \frac{d\epsilon}{\epsilon} \int_{1/\epsilon R}^{1} \frac{d\theta_0}{\theta_0} \int_{1/\theta_0 R}^{\epsilon} \frac{dk}{k} \tag{L.30}$$

Exchanging the integration order,

$$n = \int_{1/R}^{E} \frac{d\epsilon}{\epsilon} \int_{1/R}^{\epsilon} \frac{dk}{k} \int_{1/kR}^{1} \frac{d\theta_0}{\theta_0} = \int_{1/R}^{E} \frac{d\epsilon}{\epsilon} \int_{1/R}^{\epsilon} \frac{dk}{k} \ln(kR) \tag{L.31}$$

changing the integration order again

$$\begin{aligned} n &\sim \int_{1/R}^{E} \frac{dk}{k} \int_{k}^{E} \frac{d\epsilon}{\epsilon} \ln(kR) = \int_{1/R}^{E} \frac{dk}{k} \ln \frac{E}{k} \ln kR \\ \therefore \quad \frac{dn}{d \ln k} &= \alpha_s \ln \frac{E}{k} \ln kR \end{aligned} \tag{L.32}$$

□

Appendix

Solution to Problem J.1: The Noether's current is expressed as

$$J^{\mu} = \frac{\delta \mathcal{L}}{\delta(\partial_{\mu} \phi_i)} \delta \phi_i \tag{L.33}$$

where ϕ_i is any part of the field that appears in the Lagrangian. J^μ can be extracted by removing the common infinitesimal parameter $\delta\alpha_i$. For the isospin transformation,

$$\frac{\delta\mathcal{L}}{\delta(\partial_\mu N)} = \overline{N}i\gamma^\mu\,, \quad \frac{\delta\mathcal{L}}{\delta(\partial_\mu\pi_i)} = \partial_\mu\pi_i\,, \quad \frac{\delta\mathcal{L}}{\delta(\partial_\mu\sigma)} = \partial_\mu\sigma$$
$$\delta N = -i(\tau_i/2)N\delta\alpha_i\,, \quad \delta\pi_i = (\delta\boldsymbol{\alpha}\times\boldsymbol{\pi})_i\,, \quad \delta\sigma = 0 \tag{L.34}$$

Using

$$\partial_\mu\pi_i(\delta\boldsymbol{\alpha}\times\boldsymbol{\pi})_i = \partial_\mu\boldsymbol{\pi}\cdot\delta\boldsymbol{\alpha}\times\boldsymbol{\pi} = \boldsymbol{\pi}\times\partial_\mu\boldsymbol{\pi}\cdot\delta\boldsymbol{\alpha} = \left(\boldsymbol{\pi}\times\partial_\mu\boldsymbol{\pi}\right)_i\delta\alpha_i \tag{L.35}$$

we have

$$\begin{aligned} J^\mu &= \frac{\delta\mathcal{L}}{\delta(\partial_\mu N)}\delta N + \frac{\delta\mathcal{L}}{\delta(\partial_\mu\pi_i)}\delta\pi_i + \frac{\delta\mathcal{L}}{\delta(\partial_\mu\sigma)}\delta\sigma \\ &= \overline{N}i\gamma^\mu\left[-i(\tau_i/2)N\delta\alpha_i\right] + \partial_\mu\pi_i(\delta\boldsymbol{\alpha}\times\boldsymbol{\pi})_i \\ &= \left[\overline{N}\gamma^\mu\frac{\tau_i}{2}N + \left(\boldsymbol{\pi}\times\partial_\mu\boldsymbol{\pi}\right)_i\right]\delta\alpha_i \\ \therefore\quad J_i^\mu &= \left[\overline{N}\gamma^\mu\frac{\tau_i}{2}N + \left(\boldsymbol{\pi}\times\partial_\mu\boldsymbol{\pi}\right)_i\right] \end{aligned} \tag{L.36}$$

For the chiral transformation,

$$\delta N = -i\gamma^5(\tau_i/2)N\delta\alpha_i\,, \quad \delta\pi_i = \sigma\delta\alpha_i\,, \quad \delta\sigma = -\pi_i\delta\alpha_i \tag{L.37}$$

$$\begin{aligned} J^\mu &= \overline{N}i\gamma^\mu\left[-i\gamma^5(\tau_i/2)N\delta\alpha_i\right] + \partial_\mu\pi_i(\sigma\delta\alpha_i) + \partial_\mu\sigma[-\pi_i\delta\alpha_i] \\ &= \left[\overline{N}\gamma^\mu\gamma^5\frac{\tau_i}{2}N + \sigma\partial_\mu\pi_i - \partial_\mu\sigma\pi_i\right]\delta\alpha_i \\ \therefore\quad J_i^\mu &= \overline{N}\gamma^\mu\gamma^5\frac{\tau_i}{2}N + \sigma\partial_\mu\pi_i - \pi_i\partial_\mu\sigma \end{aligned} \tag{L.38}$$

References

1 Dydack, F. (1992) *Nucl. Phys.*, A546:85c–106c.
2 Grunewald, M.W. (2003) *Nucl. Phys. B (Proc. Suppl.)*, 117:280–297.
3 Nakamura, K. *et al.* (Particle Data Group) (2010) *J. Phys. G: Nuclear and Particle Physics*, 37:075021.
4 Young, R.D. *et al.* (2007) *Phys. Rev. Lett.*, 99:122003.
5 Mnich, J. (1996) *Phys. Rep.*, 271:181–266.
6 Nagashima, Y. (2005) *Butsuri*, 60:171–179.
7 LEP Electroweak working group. (2006) Precision electroweak measurements on the z resonance. *Phys. Rep.*, 427:257–454.
8 Langacker, P. (1995) *Test of the standard model and searches for new physics.* in P. Langacker (ed) *Precision Tests of the Standard Electroweak Model*, World Scientific, Singapore, 883–950.
9 Langacker, P. (2005) *Test of the electroweak theory.* Fermilab Academic Lectures. http://boudin.fnal.gov/AcLec/AcLecLang.html, last access: 20 September 2010.
10 Abazov, V.M. *et al.* (D0) (2009) *Phys. Rev. Lett.*, 103:141801.
11 Amsler, C. *et al.* (Particle Data Group) (2008) *Phys. Lett. B*, 667:1.
12 Lin, S.W. *et al.* (Belle) (2008) *Nature*, 452:332–336.
13 BelleEvent:Belle/KEKB (2012) http://belle.kek.jp/belle/events/, last access: 25 August 2012.
14 Abe, K. *et al.* (Belle) (2005) *Phys. Rev. D*, 71:072003.
15 Martin, A.D. *et al.* (2009) *Eur. Phys. J. C*, 63:189–285.
16 Harriman, P.N., Martin, A.D., Stirling, W.J., and Roberts, R.G. (1990) *Phys. Rev. D*, 42:798–810.
17 Kwiecinski, J., Martin, A.D., Stirling, J., and Roverts, R.G. (1990) *Phys. Rev. D*, 42:3645–3659.
18 Martin, A.D., Stirling, W.J, and Roberts, R.G. (1994) *Phys. Rev. D*, 50:6734–6753.
19 Aid, S. *et al.* (H1) (1995) *Nucl. Phys.*, B470:3–38.
20 Martin, A.D., Stirling, W.J, and Roberts, R.G. (1995) *Phys. Rev. D*, 51:4756–4762.
21 Heinemann, B. (2010) *Physics at hadron colliders.* CERN, Summer Student Lectures.
22 CDF/FNAL (2012) CDFwebjet www-cdf.fnal.gov/physics/talks_transp/2003/cipanp_qcd_mesropian.pdf and www.fnal.gov/pub/tevatron/milestones/interactive-timeline.html, last access: 24 August 2012.
23 Chetyrkin, K.G., Kniehl, B.A., and Steinhauser, M. (1997) *Phys. Rev. Lett.*, 79:2184–2187.
24 Bethke, S. (2000) *J. Phys. G*, 26:R27–R66.
25 Bethke, S. (2007) *Prog. Part. Nucl. Phys.*, 58:351–386.
26 Giele, W.T., Glover, E.W.N., and Kosower, D.A. (1993) *Nucl. Phys. B*, 403:633–667.
27 Abulencia, A. *et al.* (CDF) (2007) *Phys. Rev. D*, 75:092006–19.
28 Abazov, V.M. *et al.* (2008) *Phys. Rev. Lett.*, 101:062001.
29 Abbott, B. *et al.* (1999) *Phys. Rev. Lett.*, 82:062001.
30 Abazov, V.M. *et al.* (D0) (2001) *Phys. Rev. D*, 64:032003.

Elementary Particle Physics, Volume 2: Foundations of the Standard Model, First Edition. Yorikiyo Nagashima.

31 Blazey, G.C. and Flaugher, B.L. (1999) *Annu. Rev. Nucl. Part. Sci.*, 49:633–685.
32 Aaltonen, T. *et al.* (CDF) (2009) *Phys. Rev. D*, 79:112002.
33 CDF/FNAL (2012) CDFwebtop, www-cdf.fnal.gov/events/pic/topev.gif, last access: 25 August 2012 and www.fnal.gov/pub/tevatron/milestones/interactive-timeline.html, last access: 24 August 2012.
34 Glashow, S.L. (1961) *Nucl. Phys. B*, 22:579.
35 Weinberg, S. (1967) A model of leptons. *Phys. Rev. Lett.*, 19:1264.
36 Salam, A. (1968) Elementary Particle Theory, in N. Svaratholm (ed) *Proc. 8th Nobel Symp.*. Almquist and Forlag, Stockholm.
37 Higgs, P.W. (1964) *Phys. Rev. Lett.*, 13:508.
38 't Hooft, G. (1971) *Nucl. Phys. B*, 33:173.
39 't Hooft, G. (1971) *Nucl. Phys. B*, 35:167.
40 Greenberg, O.W. (1964) *Phys. Rev. Lett.*, 13:598.
41 Han, M.Y. and Nambu, Y. (1965) *Phys. Rev.*, 139:B1006.
42 Gross, D.J. and Wilczek, F. (1973) *Phys. Rev. Lett.*, 30:1343–1346.
43 Gross, D.J. and Wilczek, F. (1973) *Phys. Rev. D*, 8:3633.
44 Politzer, D. (1973) *Phys. Rev. Lett.*, 30:1346–1349.
45 Nambu, Y. and Jona-Lasinio, G. (1961) *Phys. Rev.*, 122:345–358.
46 Nambu, Y. and Jona-Lasinio, G. (1961) *Phys. Rev.*, 124:246–254.
47 Georgi, H. and Glashow, S. (1972) *Phys. Rev. Lett.*, 28:1494–1497.
48 Llewellyn Smith, C.H. (1973) *Phys. Lett. B*, 46:233–236.
49 Cornwall, J.M., Levin, D.N., and Tiktopoulos, G. (1973) *Phys. Rev. Lett.*, 30:1268–1270.
50 Cornwall, J.M., Levin, D.N., and Tiktopoulos, G. (1974) *Phys. Rev. D*, 10:1145–1167.
51 Cornwall, J.M., Levin, D.N., and Tiktopoulos, G. (1974) *Phys. Rev. Lett.*, 32:498–501.
52 Reines, F. and Cowan Jr., C.L. (1956) *Phys. Rev.*, 92:8301.
53 Hasert, J. *et al.* (1973) *Phys. Lett. B*, 46:121.
54 Hasert, J. *et al.* (1973) *Phys. Lett. B*, 46:138.
55 Vilain, P. *et al.* (CHARM) (1994) *Phys. Lett. B*, 335:246–252.
56 Ahrens, L. *et al.* (E734) (1990) *Phys. Rev. D*, 41:3297–3316.
57 Amaldi, U. *et al.* (1987) *Phys. Rev. D*, 36:1385–1407.
58 Prescott, C.Y. *et al.* (1978) *Phys. Lett. B*, 77:347–352.
59 Prescott, C.Y. *et al.* (1979) *Phys. Lett. B*, 84:524–528.
60 LlewellynSmith, C.H. (1983) *Nucl. Phys. B*, 228:205–215.
61 Perrier, F. (1995) *The measurements of electroweak parameters from deep inelastic neutrino scattering.* in P. Langacker (ed) *Precision Tests of the Standard Electroweak Model*, World Scientific, Singapore, 385–490.
62 Holder, M. *et al.* (CDHS) (1977) *Phys. Lett. B*, 72:254–260.
63 Paschos, E.A. and Wolfenstein, L. (1973) *Phys. Rev. D*, 7:91–95.
64 Zeller, M. *et al.* (NuTeV) (2002) *Phys. Rev. Lett.*, 88:091802.
65 Masterson, B.P. and Wieman, C.E. (1995) *Atomic parity nonconservation experiments.* in P. Langacker (ed) *Precision Tests of the Standard Electroweak Model*, World Scientific, Singapore, 545–576.
66 Blundell, S.A., Johnson, W.R., and Sapirstein, J. (1995) The theory of atomic parity violation. In P. Langacker (ed) *Precision Tests of the Standard Electroweak Model*, World Scientific, Singapore, 577–598.
67 Macpherson, M.J.D. *et al.* (1987) *Europhys. Lett.*, 4:811–816.
68 Macpherson, M.J.D. *et al.* (1991) *Phys. Rev. Lett.*, 67:2784–2787.
69 Wolfenden, T., Baird, P., and Sanders, P. (1991) *Europhys. Lett.*, 15:731-736.
70 Bouchiat, M.A. *et al.* (1986) *J. Phys. (Paris)*, 47:1709.
71 Noecker, M.C., Materson, B.P., and Wieman, C.E. (1988) *Phys. Rev. Lett.*, 61:310–313.
72 Gilbert, S.L. *et al.* (1985) *Phys. Rev. Lett.*, 55:2680–2683.
73 Drell, P.S. and Commins, E. (1984) *Phys. Rev. Lett.*, 53:968–971.

74 Baltay, C. (1978) *Proc. 19th Int. Conf. High Energy Physics*, Tokyo.

75 Klein, M. and Riemann, T. (1984) *Z. Phys. C*, 24:151–155.

76 Aaron, F. D. *et al.* (H1/Zeus) (2010) *JHEP*, 01:109.

77 Engelen, J. and Kooijman, P. (1998) Prog. Par. Nucl. Phys., **41**, 1–47.

78 Callan, C.G. and Gross, D. (1969) *Phys. Rev. Lett.*, 22:156.

79 Abe, K. *et al.* (VENUS) (1990) *Z. Phys. C*, 48:13–21.

80 Braunschweig, W. *et al.* (TASSO) (1989) *Z. Phys. C*, 44:365–378.

81 Sagawa, H. *et al.* (AMY) (1989) *Phys. Rev. Lett.*, 63:2341–2345.

82 Abe, K. *et al.* (VENUS) (1993) *Phys. Lett. B*, 313:288–298.

83 Braunschweig, W. *et al.* (TASSO) (1983) *Phys. Lett. B*, 126:493–498.

84 Bean, A. *et al.* (CLEO) (1987) *Phys. Rev. D*, 35:3533–3536.

85 Arbajar, G. *et al.* (UA1) (1986). CERN-PPE/91-54.

86 Gronberg, J.B. (1991) *B quark physics:recent resuls from ua1.* in J. Tran Thanh Van (ed) *91 high energy hadronic interactions: proceedings of the XXVIth Rencontre de Moriond*, Editions Frontiers, 23–32.

87 Schaile, D. and Zerwas, P.M. (1992) *Phys. Rev. D*, 45:3262–3265.

88 Elsen, E. *et al.* (JADE) (1990) *Z. Phys. C*, 46:349–359.

89 Behrend, H.J. *et al.* (CELLO) (1990) *Z. Phys. C*, 47:333–342.

90 Shimonaka, A. *et al.* (TOPAZ) (1991) *Phys. Lett. B*, 268:457–464.

91 Rosner, J.L. (1996) *Phys. Rev. D*, 54:1078–1082.

92 Carena, M. *et al.* (2004) *Phys. Rev. D*, 70:093009.

93 Davoudiasl, H., Hewett, J.L., and Rizzo, T.G. (2000) *Phys. Rev. Lett.*, 84:2080–2083.

94 Aaltonen, T. *et al.* (CDF) (2007) *Phys. Rev. Lett.*, 99:171802.

95 Abazov, V.M. *et al.* (D0) (2008) *Phys. Rev. D*, 77:091802.

96 Collins, J.C. and Soper, D.E. (1977) *Phys. Rev. D*, 16:2219–2225.

97 Acosta, D. *et al.* (CDF) (2005) *Phys. Rev. D*, 71:052002.

98 Abazov, V.M. *et al.* (D0) (2008) *Phys. Rev. D*, 77:011106(R).

99 Abazov, V.M. *et al.* (D0) (2008) *Phys. Rev. Lett.*, 101:191801.

100 Hobbs, J.D., Neubauer, M.S., and Willenbrock, S. (2010) *Test of the standard electroweak model at the energy frontier.* arXiv:1003.5733v1.

101 Rubbia, C., McIntyre, P., and Cline, D. (1977) *Proc. Int. Neutrino Conf.*, Aachen, 683.

102 van de Meer, S. (1972) CERN ISR-PO/72-31.

103 Cole, F.T. and Mills, F.E. (1981) *Ann. Rev. Nucl. Part. Sci.*, **31**, 295–335.

104 Rubbia, C. (1993) The discovery of *W* and *Z* bosons. *Phys. Rep.*, 239:215–284.

105 Denegri, D. (2004) *Phys. Rep.*, 403-404:107–145.

106 Astbury, A. *et al. A 4pi solid angle detector for the sps used as a proton-antiproton collider at a center of mass energy of 540 GeV.* Proposal CERN/SPSC/78-06.

107 Arnison, G. *et al.* (UA1) (1983) *Phys. Lett. B*, 122:103–116.

108 Bagnaia, P. *et al.* (UA2) (1983) *Phys. Lett. B*, 129:130–140.

109 Alitti, J.A. *et al.* (UA2) (1992) *Phys. Lett. B*, 276:354–364.

110 Arnison, G. *et al.* (UA1) (1983) *Phys. Lett. B*, 129:273–282.

111 Abe, F. *et al.* (CDF) (1991) *Phys. Rev. D*, 44:29–52.

112 Abulencia, A. *et al.* (CDF) (2007) *J. Phys. G*, 34:2457–2544.

113 Hamberg, R., van Neerven, W.L., and Matsuura, T. (1991) *Nucl. Phys. B*, 359:343–405. Erratum ibid. (2002) 644:403.

114 Sutton, P.J. *et al.* (1992) *Phys. Rev. D*, 45:2349–2359.

115 Rijken, P.J. and van Neerven, W.L. (1995) *Phys. Rev. D*, 51:44–63.

116 van Neerven, W.L. and Zijlstra, E.B. (1992) *Nucl. Phys. B*, 382:11–62. Erratum, ibid. (2004) 680:513.

117 Harlander, R.V. (2002) *Phys. Rev. Lett.*, 88:201801.

118 Denegri, D. (2004) *Phys. Rep.*, 403–404:107–145.

119 Albajar, C. *et al.* (UA1) (1989) *Z. Phys. C*, 44:15–61.

120 Abbott, B. *et al.* (D0) (1998) *Phys. Rev. Lett.*, 80:3008–3013.
121 Abbott, B. *et al.* (D0) (1998) *Phys. Rev. D*, 58:092003.
122 Glenzinski, D.A. (2000) *Annu. Rev. Nucl. Part. Sci.*, 50:207–248.
123 Mele, S. (2004) *Phys. Rep.*, 403-404:255–270.
124 Abazov, V.M. *et al.* (D0) (2009) *Phys. Rev. Lett.*, 103:231802.
125 Affolder, T. *et al.* (CDF) (2000) *Phys. Rev. Lett.*, 85:3347–3352.
126 Aaltonen, T. *et al.* (CDF) (2008) *Phys. Rev. Lett.*, 100:071801.
127 Abbott, B. *et al.* (D0) (2000) *Phys. Rev. D*, 61:072001.
128 Barnett, R.M. *et al.* (Particle Data Group) (1996) *Phys. Rev. D*, 54:1.
129 Rosner, J.L. and Worah, M.P. (1994) *Phys. Rev. D*, 49:1363–1369.
130 Sushkov, O.P., Flambaum, V.V., and Khriplovich, I.B. (1975) *Sov. J. Nucl. Phys.*, 20:537.
131 Alles, W., Boyer, Ch., and Buras, A.J. (1977) *Nucl. Phys. B*, 119:125–140.
132 Brown, R.W. and Mikaelian, K.O. (1979) *Phys. Rev. D*, 19:922–934.
133 The LEP Electroweak Working Group and the SLD Heavy Flavour Group. (1997). CERN-PPE/97-154.
134 The LEP Electroweak Working Group the SLD Electroweak, and Heavy Flavour Group. (2003). arXiv:hep-ex/0312023v2.
135 Denner, A. *et al.* (2000) *Nucl. Phys. B*, 587:67–117.
136 Jadach, S. *et al.* (2001) *Comput. Phys. Commun.*, 140:432–474.
137 Hagiwara, K. *et al.* (1987) *Nucl. Phys. B*, 282:253–307.
138 Gounaris, G.J. (1996) CERN 96-01, **1**, 525.
139 Bilenky, M. *et al.* (1993) *Nucl. Phys. B*, 409:22–68.
140 Kim, K.J. and Tsai, Y.S. (1973) *Phys. Rev. D*, 7:3710–3721.
141 Abbiendi, G. *et al.* (OPAL) (2004) *Eur. Phys. J. C*, 33:463–476.
142 Heister, A. *et al.* (ALEPH) (2001) *Eur. Phys. J. C*, 21:423–441.
143 Braaten, E. and Li, S. (1990) *Phys. Rev. D*, 42:3888–3891.
144 Abreu, P. *et al.* (DELPHI) (2001) *Phys. Lett. B*, 502:9–23.
145 Achard, P. *et al.* (L3) (2004) *Phys. Lett. B*, 586:151–166.
146 Alitti, J. *et al.* (UA2) (1990) *Z. Phys. C*, 47:523–531.
147 Abe, F. *et al.* (CDF) (1992) *Phys. Rev. Lett*, 69:28–32.
148 Abe, F. *et al.* (CDF) (1996) *Phys. Rev. Lett.*, 76:3070–3075.
149 Abbott, V.M. *et al.* (D0) (2007) *Phys. Rev. D*, 61:072001.
150 Martin, A.D., Roberts, R.G., and Stirling, W.J. (1993) *Phys. Review D*, 47:867–882.
151 Martin, A.D., Roberts, R.G., and Stirling, W.J. (1993) *Phys. Lett. B*, 306:145–150.
152 Orito, S. and Takeshita, T. (1991) The Test of Electroweak Theory by LEP: The first results. (in Japanese) *Butsuri*, 46:643.
153 Berends, F.A., Burgers, G.J.H., and van Neerven, W.L. (1987) *Phys. Lett. B*, 185:395–398.
154 Berends, F.A. and W.L. Van Neerven. (1988) *Nucl. Phys. B*, 297:429–478. Erratum, ibid 304 (1988) 921.
155 Berends, F.A. (1989) *Z* line shape. In G. Altarelli, R. Kleiss, and C. Verzegnassi, editors, *Z Physics at LEP*, **1**, CERN 89–127.
156 Yennie, D., Frautschi, S., and Suura, H. (1961) *Ann. Phys.*, 13:379.
157 Cahn, R.N. (1987) *Phys. Rev. D*, 36:2666.
158 Jadach, S. and Ward, B.F.L. (1988) *Phys. Rev. D*, 38:2897–2903.
159 Beenakker, W., Berends, F.A., and van der Marck, S.C. (1990) *Z. Phys. C*, 46:687–691.
160 Greco, M. *et al.* (1975) *Phys. Lett. B*, 56:367–372.
161 Greco, M. *et al.* (1975) *Nucl. Phys. B*, 101:234–262.
162 Ross, D.A. and Veltman, M.J.G. (1975) *Nucl. Phys. B*, 95:135–147.
163 Veltman, M. (1977) *Nucl. Phys. B*, 123:89–99.
164 Altarelli, G. (1989) in *Proc. Int. Symp. on Lepton and Photon Interactions at High Energies, Stanford* 286. World Scientific, Singapore.
165 Stuart, R.G. (1991) *Phys. Lett. B*, 262:113–119.
166 Sirlin, A. (1991) *Phys. Rev. Lett.*, 67:2127–2130.

167 Leike, A. and Riemann, T. (1991) *Phys. Lett. B*, 273:513–518.

168 Beenakker, W., Berends, F.A., and van der Marck, S.C. (1991) *Nucl. Phys. B*, 349:323–368.

169 Willenbrock, S. and Valencia, G. (1991) *Phys. Lett. B*, 259:373–376.

170 Hollik, W.F.L. (1990) *Fortschr. Phys.*, 38:165–260.

171 Jegerlehner, F. *Renormalizing the standard model.* in Cvetic, M. and Langacker, P. (eds) (1991) *Testing the standard model: Proceedings of the Theoretical Advanced Study Institute in Elementary Particle Physics (TASI)*, **1**, World Scientific, Singapore, 476–590.

172 Montagna, G. *et al.* (1993) *Nucl. Phys. B*, 401:3–66.

173 Montagna, G. *et al.* (1999) *Comput. Phys. Commun.*, 117:278–289.

174 Bardin, D. *et al.* (1989) *Z. Phys. C*, 493–502.

175 Bardin, D.Yu. *et al.* (2001) *Comput. Phys. Commun.*, 133:229–395.

176 Arbuzov, A.B. (2001) *JHEP*, 07:043.

177 Steinberger, J. (1991) *Phys. Rep.* 203:346–381

178 ALEPH (1995) http://aleph.web.cern.ch/aleph/dali/, last access: 27 August 2012.

179 Bederede, D. *et al.* (ALEPH) (1995) *Nucl. Instrum. Methods A*, 365:117–134.

180 Alvsvaag, S.J. *et al.* (DELPHI) (1995) *Nucl. Phys. B (Proc. Suppl.)*, 44:116–121.

181 Brock, I.C. *et al.* (L3) (1996) *Nucl. Instrum. Methods A*, 381:236–266.

182 Abreu, P. *et al.* (DELPHI) (1996) *Nucl. Instr. Methods A*, 378:57–100.

183 Abbiendi, G. *et al.* (OPAL) (2000) *Eur. Phys. J. C*, 14:373–425.

184 Ward, B.F.L. *et al.* (1997) *Acta Phys. Pol. B*, 28:925.

185 Brandt, D. *et al.* (2000) *Rep. Prog. Phys.*, 63:939–1000.

186 LEP Energy Working Group. (1993) *Phys. Lett. B*, 307:187–193.

187 Assmann, R. *et al.* (1995) *Z. Phys. C*, 66:567–582.

188 Bravin, E. *et al.* (1998) *Nucl. Instr. Methods*, A417:9–15.

189 Assmann, R. *et al.* (LEP Energy Working Group.) (1999) *Eur. Phys. J. C*, 6:187–223.

190 Arnaudon, L. *et al.* (1995) *Nucl. Instrm. Methods A*, 357:249–252.

191 Sokolov, A.A. and Ternov, I.M. (1964) *Sov. Phys. Dokl.*, 8:1203.

192 Assmann, R. *et al.* (1995) *Nucl. Phys. B Proc. Suppl.*, 109:17–31.

193 Arnaudon, L. *et al.* (1995) *Z. Phys. C*, 66:45–62.

194 Barate, R. *et al.* (ALEPH) (2000) *Eur. Phys. J. C*, 16:597–611.

195 Ackerstaff, K. *et al.* (OPAL) (1994) *Eur. Phys. J. C*, 1:439–459.

196 Sjostrand, T. (1986) *Comput. Phys. Commun.*, 39:347–407.

197 Sjostrand, T. and Bengtsson, M. (1987) *Comput. Phys. Commun.*, 43:367–379.

198 Behnke, T. and Charlton, D.G. (1995) *Phys. Scr.*, 52:133–157.

199 Abbiendi, G. *et al.* (OPAL) (1999) *Eur. Phys. J. C*, 8:217–239.

200 Abe, K. *et al.* (SLD) (2002) *Phys. Rev. D*, 65:092006.

201 Moenig, K. (1998) *Rep. Prog. Phys.*, 61:999–1043.

202 Buskulic, D. *et al.* (ALEPH Z) (1995) *Nucl. Instrum. Methods A*, 360:481–506.

203 Schramm, D.N. and Turner, M.S. (1998) *Rev. Mod. Phys.*, 70:303–318.

204 Abe, K. *et al.* (SLD) (1994) *Phys. Rev. Lett.*, 73:25–29.

205 Abe, K. *et al.* (SLD) (2001) *Phys. Rev. Lett.*, 86:1162–1166.

206 Maruyama, T. *et al.* (1991) *Phys. Rev. Lett.*, 66:2376–2379.

207 Maruyama, T. *et al.* (1992) *Phys. Rev. B*, 46:4261–4264.

208 Calloway, D. (1994) SLAC-PUB-6423.

209 Abreu, P. *et al.* (DELPHI) (2000) *Eur. Phys. J. C*, 16:371–405.

210 Heister, A. *et al.* (ALEPH) (2002) *Eur. Phys. J. C*, 24:177–191.

211 Heister, A. *et al.* (ALEPH) (2001) *Phys. Lett. B*, 512:30–48.

212 Buskulic, D. *et al.* (ALEPH) (1994) *Phys. Lett. B*, 335:99–108.

213 Heister, A. *et al.* (ALEPH) (2001) *Eur. Phys. J. C*, 22:201–215.

214 Bigi, I.I. and Sanda, A.I. (1981) *Nucl. Phys. B*, 193:85–108.

215 Ross, D.A. and Taylor, J.C. (1973) *Nucl. Phys. B*, 51:125–144.

216 Sirlin, A. (1980) *Phys. Rev. D*, 22:971–981.

217 Aoki, K. *et al.* (1982) *Prog. Theor. Phys. Suppl.*, 73:1–226.

218 Bohm, M. and Denner, A. (1992) Radiative corrections in the electroweak standard model. in M.A. Perez and R. Huerta (eds) *Proceedings of the Workshop on High Energy Phenomenology: Mexico City, 1–12 July, 1991*, World Scientific, Singapore.

219 Marciano, W.J. and Sirlin, A. (1988) *Phys. Rev. Lett.*, 61:1815–1818.

220 Altarelli, G. and Barbieri, R. (1991) *Phys. Lett. B*, 253:161–167.

221 Burkhardt, H. *et al.* (1989) *Z. Phys. C*, 43:497–501.

222 Hagiwara, K. *et al.* (2007) *Phys. Lett. B*, 649:173–179.

223 Jegerlehner, F. (1991) Physics of precision experiments with z's. *Prog. Part. Nucl. Phys.*, 27:1–76. PSI-PR-91-16.

224 Akhundov, A.A. *et al.* (1986) *Nucl. Phys. B*, 276:1.

225 Beenakker, W. and Hollik, W. (1988) *Z. Phys. C*, 40:141–148.

226 Marciano, W.J. (1979) *Phys. Rev. D*, 20:274–288.

227 Sirlin, A. (1989) *Phys. Lett. B*, 232:123–126.

228 Fanchiotti, S. and Sirlin, A. (1990) *Phys. Rev. D*, 41:319–321.

229 Degrassi, G., Fanchotti, S., and Sirlin, A. (1991) *Nucl. Phys. B*, 351:49–69.

230 Marciano, W. (1995) *Radiative corrections to neutral current processes.* in P. Langacker (ed) *Precision Tests of the Standard Electroweak Model*, World Scientific, Singapore, 170–200.

231 Hollik, W. (1995) *Renormalization of the standard model.* in P. Langacker (ed) *Precision Tests of the Standard Electroweak Model*, World Scientific, Singapore, 37–116.

232 Erler, J. and Ramsey-Musolf, M.J. (2005) *Phys. Rev. D*, 72:073003.

233 Caso, C. *et al.* (Particle Data Group) (1998) *Eur. Phys. J. C*, 3:1.

234 Hollik, W. (1995) Predictions for e^+e^- processes. in P. Langacker (ed) *Precision Tests of the Standard Electroweak Model*, World Scientific, Singapore, 117–169 *Adv. Ser. Dir. High Ener. Phys.*, **14**.

235 Peskin, M.E. and Takeuchi, T. (1990) *Phys. Rev. Lett.*, 65:964–968.

236 Altarelli, G., Barbieri, R., and Caravaglios, F. (1993) *Nucl. Phys. B*, 405:3.

237 Marciano, W.J. and Rosner, J.L. (1990) *Phys. Rev. Lett.*, 65:2963–2966.

238 Erler, J. and Langacker, P. (2010) *Phys. Rev. Lett.*, 105:031801.

239 Ellis, J., Fogli, G.L., and Lisi, E. (1994) CERN-TH 7448/94.

240 Peskin, M.E. (2001) *Phys. Rev.*, D64:093003.

241 Peskin, M.E. and Takeuchi, T. (1992) *Phys. Rev. D*, 46:381–409.

242 Hagiwara, K., Haidt, D., and Matsumoto, D. (1998) *Eur. Phys. J. C*, 2:95–122.

243 Antonelli, M. *et al.* (2010) Flavor physics in the quark sector. *Phys. Rep.*, 494:197–414.

244 Wolfenstein, L. (1984) *Phys. Rev. Lett.*, 51:1945–1947.

245 Gilman, F.J. and Nir, Y. (1990) *Annu. Rev. Nucl. Part. Sci.*, 40:213–38.

246 Buras, A.J., Matthias, J., and Weisz, P.H. (1990) *Nucl. Phys. B*, 347:491–536.

247 Hardy, J.C. and Towner, I.S. (2009) *Phys. Rev. C*, 79:055502.

248 Boyle, P.A. *et al.* (2008) *Phys. Rev. Lett.*, 100:141601.

249 Antonelli, M. *et al.* (2008) arXiv:0801.1817 [hep-ph].

250 Ambrosino, F. *et al.* (KLOE) (2006) *Phys. Lett. B*, 632:76–80.

251 Kayis-Topaksu, A. *et al.* (CHORUS) (2005) *Phys. Lett. B*, 626:24–34.

252 Abramowicz, H. *et al.* (CDHS) (1982) *Z. Phys. C*, 15:19–31.

253 Rabinowitz, S.A. *et al.* (CCFR) (1993) *Phys. Rev. Lett.*, 70:134–137.

254 Bazarko, A.O. *et al.* (CCFR) (1995) *Z. Phys. C*, 65:189–198.

255 Vilain, P. *et al.* (CHARM) (1999) *Eur. Phys. J. C*, 11:19–34.

256 Eidelman, S. *et al.* (Particle Data Group). (2004) *Phys. Lett. B*, 592:1.

257 Aubin, C. *et al.* (MILC) (2005) *Phys. Rev. Lett.*, 94:011601.

258 Widhalm, L. *et al.* (Belle) (2006) *Phys. Rev. Lett.*, 97:061804.

259 Aubert, B. *et al.* (BABAR) (2007) *Phys. Rev. D*, 76:052005.

260 Besson, D. *et al.* (CLEO) (2009) *Phys. Rev. D*, 80:032005.

261 Neubert, M. (1994) Heavy-quark symmetry. *Phys. Rep.*, 245:259–395.

262 Hoecker, A. and Ligeti, Z. (2006) *CP violation and the CKM matrix*. arXiv:hep-ph/0605217.

263 Isgur, N. and Wise, M.B. (1989) *Phys. Lett. B*, 232:113–117. ibid. (1990) 237, 527–530.

264 Caprini, I., Lellouch, L., and Neubert, M. (1998) *Nucl. Phys. B*, 530:153.

265 Okamoto, M. *et al.* (2005) *Nucl. Phys. (Proc. Suppl.) B*, 140:461–463.

266 Bartlet, J. *et al.* (CLEO) (1999) *Phys. Rev. Lett.*, 82:3746–3750.

267 Abe, A. *et al.* (Belle) (2002) *Phys. Lett. B*, 526:247.

268 Aubert, B. *et al.* (BABAR) (2009) *Phys. Rev. D*, 79:012002.

269 Henderson, S. *et al.* (CLEO) (1992) *Phys. Rev. D*, 45:2212–2231.

270 Aubert, B. *et al.* (BABAR) (2006) *Phys. Rev. D*, 73:012006.

271 Limosani, A. *et al.* (Belle) (2005) *Phys. Lett. B*, 621:28–40.

272 Aubert, B. *et al.* (BABAR) (2005) *Phys. Rev. Lett.*, 95:111801.

273 Becirevic, D. and Kaidalov, A.B. (2000) *Phys. Lett. B*, 478:417–423.

274 Gulez, E. *et al.* (HPQCD) (2005) *Phys. Rev. D*, 73:074502.

275 Ball, P. and Zwicky, R. (2005) *Phys. Rev. D*, 71:014015.

276 Scola, D. and Isgur, N. (1995) *Phys. Rev. D*, 52:2783–2812.

277 Ha, H. *et al.* (Belle) (2011) *Phys. Rev. D*, 83:071101(R).

278 del Amo Sanchez, P. *et al.* BABAR: (2011) *Phys. Rev. D*, 83:032007.

279 Laiho, J. and van de Water, R.S. (2010) *Phys. Rev. D*, 81:034503.

280 Jarlskog, C. (1985) *Phys. Rev. Lett.*, 55:1039–1042.

281 Jarlskog, C. (1989) *Introduction to CP violation*. in C. Jarlskog (ed) *CP Violation*, World Scientific, Singapore, 3–40

282 Albrecht, H. *et al.* (ARGUS) (1987) *Phys. Lett. B*, 192:245–252.

283 Artuso, M. *et al.* (CLEO) (1989) *Phys. Rev. Lett.*, 62:2233–2236.

284 Besson, D. and Skwarnicki, T. (1993) *Annu. Rev. Nucl. Part. Sci.*, 43:333–378.

285 Besson, D. *et al.* (CUSB) (1985) *Phys. Rev. Lett.*, 54:381–384.

286 Albrecht, H. *et al.* (ARGUS) (1991) *Phys. Lett. B*, 255:297–304.

287 Abulencia, A. *et al.* (CDF) (2006) *Phys. Rev. Lett.*, 97:242003.

288 Abdallah, P. *et al.* (DELPHI) (2003) *Eur. Phys. J. C*, 28:155–173.

289 Aubert, B. *et al.* (Babar) (2004) *Phys. Rev. D*, 70:012007.

290 Nir, Y. (1999) *CP violation in and beyond the standard model*. hep-ph/9911321. Lectures given in the XXVII SLAC Summer Institute on Particle Physics CP Violation In and Beyond the Standard Model 7–16 July 1999.

291 Burdman, G. and Shipsey, I. (2003) *Annu. Rev. Nucl. Part. Sci.*, 53:431–499.

292 Blaylock, G., Seiden, A., and Nir, Y. (1995) *Phys. Lett. B*, 355:555–560.

293 Bergmann, S. *et al.* (2000) *Phys. Lett. B*, 486:418–425.

294 Zhang, L.M. *et al.* (Belle) (2006) *Phys. Rev. Lett.*, 96:151801.

295 Aubert, B. *et al.* (BABAR) (2007) *Phys. Rev. Lett.*, 98:211802.

296 Aaltonen, T. *et al.* (CDF) (2008) *Phys. Rev. Lett.*, 100:121802.

297 Staric, M. *et al.* (Belle) (2007) *Phys. Rev. Lett.*, 98:211803.

298 Aubert, B. *et al.* (BABAR) (2009) *Phys. Rev. D*, 80:071103.

299 Inami, T. and Lim, C.S. (1981) *Prog. Theor. Phys.*, 65:297–314. Errata, ibid. (1981) 65:1772.

300 Herrlich, S. and Nierste, U. (1994) *Nucl. Phys. B*, 419:292–322.

301 Herrlich, S. and Nierste, U. (1996) *Nucl. Phys. B*, 476:27–88.

302 Buras, A.J. (1999) *Lecture at 14th Lake Loise Winter Institute*. arXiv:hep-ph/9905437.

303 Follana, E. *et al.* (2008) *Phys. Rev. Lett.*, 100:062002.

304 Donoghue, J.F. *et al.* (1986) *Phys. Lett. B*, 179:361–366.

305 Buras, A.J. and Gerard, J.-M. (1987) *Phys. Lett. B*, 192:156–162.

306 Lusignoli, M. (1989) *Nucl. Phys. B*, 325:33–61.

307 Gaillard, M.K. and Lee, B.W. (1974) *Phys. Rev. D*, 10:897–916.

308 Hagelin, J.S. (1981) *Nucl. Phys. B*, 193:123–149.

309 Chau, L.L. (1983) *Phys. Rep.*, 95:1–94.

310 Lenz, A. and Nierste, U. (2007) *JHEP*, 06:072.

311 Aubert, B. *et al.* (BABAR) (2007) *Phys. Rev. Lett.*, 99:021603.

312 Carter, A.B. and Sanda, A.I. (1980) *Phys. Rev. Lett.*, 45:952–954.

313 Carter, A.B. and Sanda, A.I. (1981) *Phys. Rev. D*, 23:1567–1579.

314 Gronau, M. and London, D. (1991) *Phys. Lett. B*, 253:483–488.

315 Gronau, M. and Wyler, D. (1991) *Phys. Lett. B*, 265:172–176.

316 Atwood, D., Dunietz, I., and Soni, A. (1997) *Phys. Rev. Lett.*, 78:3257–3260.

317 Atwood, D., Dunietz, I., and Soni, A. (2001) *Phys. Rev. D*, 63:036005.

318 Poluektov, A. *et al.* (Belle) (2003) *Phys. Rev. D*, 70:072003.

319 Giri, A. *et al.* (2003) *Phys. Rev. D*, 68:054018.

320 Poluektov, A. *et al.* (Belle) (2006) *Phys. Rev. D*, 73:112009.

321 Poluektov, A. *et al.* (Belle) (2010) *Phys. Rev. D*, 81:112002.

322 Aubert, B. *et al.* (BABAR) (2008) *Phys. Rev. D*, 78:034023.

323 Bigi, I.I. and Sanda, A.I. (1987) *Nucl. Phys. B*, 281:41–71.

324 Einstein, A., Podolsky, B., and Rosen, N. (1935) *Phys. Rev.*, 47:777.

325 Bell, J.S. (1966) *Rev. Mod. Phys.*, 38:447–452.

326 Aspect, A., Grangier, P., and Roger, G. (1982) *Phys. Rev. Lett.*, 49:91–94.

327 Aspect, A., Dalibard, J., and Roger, G. (1982) *Phys. Rev. Lett.*, 49:1804–1807.

328 Hocker, A. *et al.* (2001) *Eur. Phys. J. C*, 21:225–259.

329 Aaltonen, T. *et al.* (CDF) (2008) *Phys. Rev. Lett.*, 100:161802.

330 Abazov, V.M. *et al.* (D0) (2008) *Phys. Rev. Lett.*, 101:241801.

331 London, D. and Peccei, R.D. (1989) *Phys. Lett. B*, 223:257–261.

332 Gronau, M. (1989) *Phys. Rev. Lett.*, 63:1451–1454.

333 Gronau, B. (1989) *Phys. Lett. B*, 229:260–284.

334 Gronau, M. and London, D. (1990) *Phys. Rev. Lett.*, 65:3381–3384.

335 Aubert, B. *et al.* (BABAR) (2002) *Nucl. Instrum. Methods A*, 479:1–116.

336 Abashian, A. *et al.* (Belle) (2002) *Nucl. Instrum. Methods A*, 479:117–232.

337 Iijima, T. *et al.* (Belle) (2000) *Nucl. Instrum. Methods A*, 453:321–325.

338 Adam, I. *et al.* (2005) (BABAR) *Nucl. Instrum. Methods A*, 538:281–357.

339 Buchalla, G. and Buras, A.J. (1994) *Nucl. Phys. B*, 412:106–142.

340 Buras, A.J. and Uhlih, S. (2008) *Rev. Mod. Phys.*, 80:965–1007.

341 Buras, A.J. (1998) *Weak Hamiltonian, CP violation and rare decays.* in F. David and R. Gupta (eds) *Probing the Standard Model of Particle Interactions*, Elsevier. arXiv:hep-ph/9806471.

342 Brod, J. and Gorbahn, M. (2008) *Phys. Rev. D*, 78:034006.

343 Artamonov, A.V. *et al.* (2008) *Phys. Rev. Lett.*, 101:191802.

344 Artamonov, A.V. *et al.* (2009) *Phys. Rev. D*, 79:092004.

345 Littenberg, L.S. (1989) *Phys. Rev. D*, 39:3322–3324.

346 Ahn, J.K. *et al.* (2008) *Phys. Rev. Lett.*, 100:201802.

347 Bigi, I.I. and Sanda, A.I. (1989) On spontaneous CP violation triggered by scalar bosons. in C. Jarlskog (ed) *CP Violation*,World Scientific, Singapore, 362–383.

348 ’t Hooft, G. (1976) *Phys. Rev. Lett.*, 37:8–11.

349 ’t Hooft, G. (1976) *Phys. Rev. D*, 14:3432–3450.

350 Peccei, R. (1989) *The strong CP problem.* in C. Jarlskog (ed) CP Violation, *Adv. Ser. Direct. High Energy Phys.*, **3**, World Scientific, Singapore, 503–551.

351 Mohapatra, R.N. (1989) *CP violation and left–right symmetry.* in C. Jarlskog (ed) *CP Violation*, World Scientific, Singapore, 384–435.

352 Sakharov, A.D. (1967) *Pizma Zh. Exp. Teor. Fiz.*, 5:32.

353 Yundrian, F.J. (1999) *The theory of Quark and Gluon Interactions.* Springer-Verlag, Berlin, Heidelberg, New York.

354 Pich, A. (1995) *Rep. Prog. Phys.*, 58:563–609.

355 Pich, A. (1995). arXiv:hep-ph/0001118v1.

356 Muta, T. (1987) *Foundations of Quantum Chromodynamics*. World Scientific, Singapore.
357 Field, R.D. (1989) *Applications of Perturbative QCD*. Addison-Wesley.
358 Gorishny, S.G., Kataev, A.L., and Larin, S.A. (1991) *Phys. Lett. B*, 259:144–150.
359 Surguladze, L.R. and Samuel, M.A. (1991) *Phys. Rev. Lett.*, 66:560–563.
360 Surguladze, L.R. and Samuel, M.A. (1991) *Phys. Rev. Lett.*, 66:2416–2416.
361 Dissertori, G. *et al.* (2008) *JHEP*, 02:040.
362 Soper, D.E. (1996) Basics of QCD perturbation theory. *XXIV SLAC Summer Institute on Particle Physics*, Stanford, August 1996.
hep-ph/9702203.
363 Surguladze, L.R. and Samuale, M.A. (1996) *Rev. Mod. Phys.*, 68:259–303.
364 Surguladze, L.R. and Samuel, M.A. (1993) *Phys. Lett. B*, 309:157–162.
365 Grunberg, G. (1995) *Phys. Lett. B*, 95:70–74.
366 Stevenson, P.M. (1981) *Phys. Rev. D*, 23:2916–2944.
367 Brodsky, S.J., Lepage, G.P., and Mackenzie, P.B. (1983) *Phys. Rev. D*, 28:228–235.
368 Ingelman, G. and Rathsman, J. (1994) *Z. Phys. C*, 63:589–600.
369 Weinberg, S. (1973) *Phys. Rev. D*, 8:3497.
370 Symanzik, K. (1973) *Commun. Math. Phys.*, 34:7–36.
371 Appelquist, T. and Carazzone, J. (1975) *Phys. Rev. D*, 11:2856–2851.
372 Bernreuther, W. and Wetzel, W. (1982) *Nucl. Phys. B*, 197:228–236.
373 Bernreuther, W. (1983) *Ann. Phys.*, 151:127–162.
374 Marciano, W.J. (1984) *Phys. Rev. D*, 29:580–582.
375 Dashen, R. (1969) *Phys. Rev.*, 183:1245.
376 Leutwyler, H. (1996) *Phys. Lett.*, B 378:313.
377 Prades, J. (1998) *Nucl. Phys. B (Proc. Suppl.)*, 64:253.
378 Bijnens, J. *et al.* (1995) *Phys. Lett. B*, 348:226.
379 Pich, A. and Prades, J. (1999) *JHEP*, 10:004.
380 Narison, S. (1999) *Phys. Lett. B*, 466:345.
381 Becirevic, D. *et al.* (1998) *Phys. Lett. B*, 444:401.
382 Gimenez, V. *et al.* (1999) *Nucl. Phys. B*, 540:472.
383 Aoki, S. *et al.* (1999) *Phys. Rev. Lett.*, 82:1373.
384 Friedman, J.R. and Kendall, H.W. (1972) Deep inelastic electron scattering. *Annu. Rev. Nucl. Sci.*, 22:203.
385 Collins, J.C. and Soper, D.E. (1987) The theorems of perturbative QCD. *Annu. Rev. Nucl. Part. Sci.*, 37:383–409.
386 Georgi, H. and Politzer, H.D. (1974) *Phys. Rev. D*, 9:416–420.
387 Gross, D.J. and Wilczek, F. (1974) *Phys. Rev. D*, 9:980–993.
388 Gribov, G.N. and Lipatov, L.N. (1972) *Sov. J. Nucl. Phys.*, 15:438, 675.
389 Dokshitzer, Yu.L. (1977) *Sov. Phys. JETP*, 46:641.
390 Altarelli, G. and Parisi, G. (1977) *Nucl. Phys. B*, 126:298–318.
391 Altarelli, G. (1982) *Phys. Rep.*, 81:1–129.
392 Nason, P. and Webber, B.R. (1994) *Nucl. Phys. B*, 421:473–517. Erratum ibid. (1996) 480:755.
393 Lipatov, L.N. (1975) *Soviet J. Phys.*, 20:94.
394 Altarelli, G., Ellis, R.K., and Martinelli, G. (1978) *Nucl. Phys. B*, 143:521–545.
395 Altarelli, G., Ellis, R.K., and Martinelli, G. (1978) *Nucl. Phys. B*, 146:544(e).
396 Altarelli, G., Ellis, R.K., Martinelli, G., and Pi, S.-Y. (1979) *Nucl. Phys. B*, 160:301–329.
397 Ellis, R.K., Stirling, W.J., and Webber, B.R. (1996) *QCD and Collider Physics*, Cambridge University Press.
398 Dissertori, G., Knowles, I., and Schmelling, M. (2003) *Quantum Chromodynamics: High Energy Experiments and Theory*, Oxford Science Publications.
399 Curci, G., Furmanski, W., and Petronzio, R. (1980) *Nucl. Phys. B*, 175:27–92.
400 Furmanski, W. and Petronzio, R. (1980) *Phys. Lett. B*, 97:437–442.
401 Bosetti, P. C. *et al.* (BEBC) (1978) *Nucl. Phys. B*, 142:1–28.
402 de Groot, J.G.H. *et al.* (CDHS) (1979) *Phys. Lett. B*, 82:292–296.
403 Nachtman, O. (1973) *Nucl. Phys. B*, 63:237–247.
404 Nachtman, O. (1974) *Nucl. Phys. B*, 78:455–467.
405 Bonesini, M. *et al.* (WA70) (1988) *Z. Phys. C*, 38:371–382.

406 Barger, V., Keung, W.Y., and Phillips, R.J.N. (1980) *Phys. Lett. B*, 91:253–258.

407 Chekanov, S. *et al.* (ZEUS) (2001) *Eur. Phys. J. C*, 21:443–471.

408 Breitweg, J. *et al.* (ZEUS) (1999) *Eur. Phys. J. C*, 7:609–630.

409 Ball, R.D. and Forte, S. (1999) *Phys. Lett. B*, 335:77–86.

410 Ball, R.D. and Forte, S. (1999) *Phys. Lett. B*, 336:77–79.

411 Ball, R.D. and Landshoff, P.V. (2000) *J. Phys. G*, 26:672–682.

412 Kuraev, E.A., Lipatov, L.N., and Fadin, V.S. (1975) *Phys. Lett. B*, 60:50.

413 Kuraev, E.A., Lipatov, L.N., and Fadin, V.S. (1977) *Sov. Phys. JETP*, 45:199.

414 Balitsky, Ya.Ya. and Lipatov, L.N. (1978) *Sov. J. Nucl. Phys.*, 28:822.

415 Benvenuti, A.C. *et al.* (BCDMS) (1989) *Phys. Lett. B*, 223:485.

416 Benvenuti, A.C. *et al.* (BCDMS) (1989) *Phys. Lett. B*, 223:490.

417 Virchaux, M. and Milsztajn, A. (1992) *Phy. Lett. B*, 274:221–229.

418 Farrar, G.R. (1974) *Nucl. Phys. B*, 77:429–442.

419 Gunion, J.F. (1974) *Phys. Rev. D*, 10:242–250.

420 Botts, J. *et al.* (1993) *Phys. Lett. B*, 304:159–166.

421 Lai, H.L. *et al.* (1994) *Phys. Rev. D*, 51:4763–4782.

422 Gluck, M. *et al.* (1995) *Z. Phys. C*, 67:433–447.

423 Arneodo, M. *et al.* (NMC) (1997) *Nucl. Phys. B*, 483:3–43.

424 Adams, M.R. *et al.* (E665) (1996) *Phys. Rev. D*, 54:3006.

425 Kubar-Andre, J. and Paige, F.E. (1979) *Phys. Rev. D*, 19:221.

426 Kubar, J., le Bellac, M., Meunier, J.L., and Plaut, G. (1980) *Nucl. Phys. B*, 175:251–275.

427 Altarelli, G., Ellis, R.K., and Martinelli, G. (1979) *Nucl. Phys. B*, 157:461–497.

428 Brown, C.N. *et al.* (1989) *Phys. Rev. Lett.*, 63:2637–2640.

429 Webb, J.C. (2003) Ph.D. thesis, hep-ex/0301031v1.

430 Grosso-Pilcher, C. and Schochet, M.J. (1986) High mass dilepton production in hadron collisions. *Annu. Rev. Nucl. Part. Sci.*, 36:1–28.

431 Sudakov, V. (1956) *Sov. Phys. JETP*, 3:65.

432 Basu, R., Ramalho, A.J., and Sterman, G. (1984) *Nucl. Phys. B*, 244:221–246.

433 Schwitters, R.F. *et al.* (1975) *Phys. Rev. Lett.*, 35:1320–1322.

434 Hanson, G. *et al.* (1975) *Phys. Rev. Lett.*, 35:1609.

435 Abe, K. *et al.* (VENUS) (1987) *J. Phys. Soc. Japan*, 56:3763.

436 Yoshida, H. *et al.* (VENUS) (1987) *Phys. Lett. B*, 198:570–576.

437 Brandelik, R. *et al.* (TASSO) (1982) *Phys. Lett. B*, 114:65–70.

438 Darriulat, P. (1980) *Ann. Rev. Nucl. Part. Sci.*, 30:159–210.

439 Farhi, E. (1977) *Phys. Rev. Lett.*, 39:1587–1588.

440 Bjorken, J.D. and Brodsky, S. (1970) *Phys. Rev. D*, 1:1416–1420.

441 Wu, S.L. and Zobernig, G. (1979) *Z. Phys.*, 2:107–110.

442 de Rujula, A., Ellis, J., Floratos, E.G., and Gaillard, M.K. (1978) *Nucl. Phys. B*, 138:387–429.

443 Kunszt, Z., Nason, P., Marchesini, G., and Webber, B.R. *Z physics at LEP 1*, CERN 89-08, **1**, 373. ETH-PT/89-39.

444 Duchesneau, D. (1998) in *Proc. 29th Int. Conf. High Energy Phys.*, Vancouver.

445 Bethke, S. (2004) *Phys. Rep.*, 403-404:203–220.

446 Abbiendi, G. *et al.* (OPAL) (2005) *Eur. Phys. J. C*, 40:287–316.

447 Webber, B.R. (1995) arXiv:hep-ph/9510283.

448 Dokshitzer, Yu.L. and Webber, B.R. (1995) *Phys. Lett. B*, 352:451–455.

449 Fernandez, P.A.M., Bethke, S., and Biebel, S., and Kluth, O. (2001) *Eur. Phys. J. C*, 22:1–15.

450 Althoff, M. *et al.* (TASSO) (1984) *Z. Phys. C*, 22:307–340.

451 Bender, D. *et al.* (HRS) (1985) *Phys. Rev.*, D31:1–16.

452 Ellis, R.K., Ross, D.A., and Terrano, A.E. (1981) *Nucl. Phys. B*, 178:421–456.

453 Barate, R. *et al.* (ALEPH) (1998) *Phys. Rep.*, 294:1–165.

454 Barber, D.P. *et al.* (MARKJ) (1979) *Phys. Rev. Lett.*, 43:830–833.

455 Sterman, G. and Weinberg, S. (1977) *Phys. Rev. Lett.*, 39:1436–1439.

456 Kunszt, Z. (1981) *Phys. Lett. B*, 99:429–432.

457 Fabricius, K., Kramer, G., Schierholz, G., and Schmitt, I. (1982) *Z. Phys. C*, 11:315–333.

458 Gutbrod, F, Kramer, G., and Schierholz, G. (1984) *Z. Phys. C*, 21:235–241.

459 Lampe, B. and Kramer, G. (1986) *Prog. Theor. Phys.*, 76:1340–1347.

460 Bethke, S., Kunszt, Z., Soper, D.E., and Stirling, W.J. (1992) *Nucl. Phys. B*, 370:310–334.

461 Bethke, S. (1991) *J. Phys. G*, 17:1455–1480.

462 Bethke, S. and Pilcher, J.E. (1992) *Ann. Rev. Nucl. Part. Sci.*, 42:251–289.

463 Catani, S. *et al.* (1991) *Phys. Lett. B*, 269:432–438.

464 Abbiendi, G. *et al.* (OPAL) (2006) *Eur. Phys. J. C*, 45:547–568.

465 Corcella, G. and Seymour, M. (2000) *Nucl. Phys. B*, 565:227–244.

466 Catani, S. *et al.* (2001) *JHEP*, 11:063.

467 Frixione, S. and Webber, B.R. (2002) *JHEP*, 06:029.

468 Frixione, S., Nason, P., and Webber, B.R. (2003) *JHEP*, 08:007.

469 Sjostland, T. (2001) *Comput. Phys. Commun.*, 135:238–259.

470 Sjostrand, T., Mrenna, S., and Skands, P. (2006) *JHEP*, 05:250.

471 Marchesini, G. and Webber, B.R. (1984) *Nucl. Phys. B*, 238:1–29.

472 Marchesini, G. and Webber, B.R. (1988) *Nucl. Phys. B*, 310:461–526.

473 Marchesini, G., Webber, B.R., Knowles, I.G., Seymour, M.H., and Stanco, L. (1992) *Comput. Phys. Commun.*, 67:465–508.

474 Lonnblad, L. (1992) *Comput. Phys. Commun.*, 71:15–31.

475 Bassetto, A., Ciafaloni, M., and Marchesini, G. (1983) Jet structure and infrared sensitive quantities in perturbative QCD. *Phys. Rep.*, 100:201–272.

476 Webber, B.R. (1984) *Nucl. Phys. B*, 238:492–528.

477 Gribov, G.N., Levin, E.M., and Ryskin, M.G. (1983) *Phys. Rep.*, 100:1–150.

478 Field, R.D. and Feynman, R.P. (1978) *Nucl. Phys. B*, 136:1–76.

479 Binnewies, J., Kniehl, B.A., and Kramer, G. (1995) *Phys. Rev. D*, 52:4947–4960.

480 Kniehl, B.A., Kramer, G., and Potter, B. (2000) *Nucl. Phys. B*, 582:514–536.

481 Kniehl, B.A., Kramer, G., and Potter, B. (2001) *Nucl. Phys. B*, 597:337–369.

482 Peterson, C., Schlatter, D., Schmitt, I., and Zerwas, P.M. (1982) *Phys. Rev. D*, 27:105–111.

483 Hoyer, P. *et al.* (1979) *Nucl. Phys. B*, 161:349.

484 Ali, A. *et al.* (1980) *Nucl. Phys. B*, 167:454–478.

485 Ali, A. *et al.* (1980) *Phys. Lett. B*, 93:155–160.

486 Artru, X. and Mennessier, G. (1974) *Nucl. Phys. B*, 70:93–115.

487 Andersson, B. *et al.* (1983) *Phys. Rep.*, 97:31–145.

488 Artru, X. (1983) *Phys. Rep.*, 97:147–171.

489 Bengtsson, H.U. and Sjostrand, T. (1987) *Comput. Phys. Commun.*, 46:43–82.

490 Bali, G.S. (2001) QCD forces and heavy quark bound states. *Phys. Rep.*, 343:1–136. hep-ph/0001 312.

491 Brezin, E., Itzykson, G., and Sjostrand, T. (1970) *Phys. Rev. D*, 2:1191.

492 Knowles, I.G. and Lafferty, G.D. (1997) Hadronization in Z0 decay. *J. Phys. G*, 23:731–789.

493 Amati, D. and Veneziano, G. (1979) *Phys. Lett. B*, 83, 87–92.

494 Bassetto, A., Ciafaloni, M., and Marchesini, G. (1980) *Nucl. Phys. B*, 163:477–518.

495 Marchesini, G. (1981) *Nucl. Phys. B*, 181:335–346.

496 Sjostrand, T. (1994) *Comput. Phys. Commun.*, 82:74–89.

497 Martin, A.D., Roberts, R.G., and Stirling, W.J. (1984) *Phys. Rev. D*, 50:6734–6752.

498 Martin, A.D., Roberts, R.G., and Stirling, W.J. (2002) *Eur. Phys. J. C*, 23:73.

499 Pumplin, J. *et al.* (2002) *JHEP*, 07:012.

500 Abrams, G.S. *et al.* (1989) *Phys. Rev. Lett.*, 63:1558–1561.

501 Akrawy, M.Z. *et al.* (OPAL) (1990) *Z. Phys. C*, 47:505–521.

502 Aarnio, P. *et al.* (DELPHI) (1990) *Phys. Lett. B*, 240:271–282.

503 Bartel, U. *et al.* (JADE) (1984) *Phys. Lett. B*, 134:275–280.

504 Acciarri, M. *et al.* (L3) (1995) *Phys. Lett. B*, 345:74–84.

505 Chun, S. and Buchanan, C. (1998) *Phys. Rep.*, 292:239–317.

506 Gross, D.J. and Llewellyn-Smith, C.H. (1969) *Nucl. Phys. B*, 14:337.

507 Larin, S.A. and Vermaseren, J.A.M. (1991) *Phys. Lett. B*, 259:345–352.

508 Chyla, J. and Kataev, A. (1992) *Phys. Lett. B*, 297:385–397.

509 Braun, V.M and Kolesnichenko, A.V. (1987) *Nucl. Phys. B*, 283:723–748.

510 Kim, J.H. *et al.* (CCFR) (1998) *Phys. Rev. Lett.*, 81:3595–3598.

511 Altarelli, G. (1994) *The Development of Perturbative QCD, 308, (7) QCD and Experiment.* World Scientific, Singapore.

512 Hincliffe, I. and Manohar, A. (2000) The QCD coupling constant. *Annu. Rev. Nucl. Part. Sci.*, 50:643–678.

513 Chetyrkin, K.G., Kuhn, J.H., and Kwiatkowski, A. (1996) *Phys. Rep.*, 277:189–281.

514 Chetyrkin, K.G., Kuhn, J.H., and Steinhauser, M. (1996) *Nucl. Phys. B*, 482:213–240.

515 Pich, A. (1988) Tau physics. In A.J. Buras and M. Lindner (eds), *Heavy Flavors II*, World Scientific, Singapore, 451–494.

516 Braaten, E. (1988) *Phys. Rev. Lett.*, 60:1606.

517 Braaten, E. (1989) *Phys. Rev. D*, 39:1458–1460.

518 Diberder, F.L. and Pich, A. (1992) *Phys.Lett. B*, 286:147–152.

519 Diberder, F.L. and Pich, A. (1992) *Phys.Lett. B*, 289:165–175.

520 Braaten, E., Narison, S., and Pich, A. (1992) *Nucl. Phys. B*, 373:581–612.

521 Davier, M., Hocker, A., and Zhang, Z. (2006) *Rev. Mod. Phys.*, 78:1043–1109.

522 Barate, R. *et al.* (ALEPH) (1998) *Eur. Phys. J. C*, 4:409–431.

523 Ackerstaff, K. *et al.* (OPAL) (1999) *Eur. Phys. J. C*, 7:571–593.

524 Tournefier, E. (1998) hep-ex/9810042v1: Talk given at the 10th International Seminar [QUARKS'98] Suzdal, Russia, May 18–24.

525 Hebbeker, T., Martinez, M., Passarino, G., and Quast, G. (1994) *Phys. Lett. B*, 331:165–170.

526 Schmelling, M. (1995) *Physica Scripta*, 51:683–713.

527 Abe, K. *et al.* (SLD) (1995) *Phys. Rev. D*, 51:962–984.

528 Basham, C.L. *et al.* (1978) *Phys. Rev. Lett.*, 41:1585–1588.

529 Basham, C.L. *et al.* (1978) *Phys. Rev. D*, 17:2298–2306.

530 Basham, C.L. *et al.* (1979) *Phys. Rev. D*, 19:2018–2045.

531 Richards, D.G., Stirling, W.J., and Ellis, S.D. (1983) *Nucl. Phys. B*, 229:317–346.

532 Falck, N.K. and Kramer, G. (1989) *Z. Phys. C*, 42:459–470.

533 Csikor, F., Pocsik, G., and Toth, A. (1985) *Phys. Rev. D*, 31:1025–1032.

534 Acton, P.D. *et al.* (OPAL) (1992) *Z. Phys. C*, 55:1–24.

535 Corcella, G. (2001) *JHEP*, 0101:010.

536 Akrawy, M.Z. *et al.* (OPAL) (1990) *Phys. Lett. B*, 252:159–169.

537 Abreu, P. *et al.* (DELPHI) (1992) *Z. Phys. C*, 54:55–73.

538 Glover, E.W.N. and Sutton, M.R. (1995) *Phys. Lett. B*, 342:375–380.

539 Dokshitzer, Yu.L., Lucenti, A., Marchesini, G., and Salam, G.P. (1998) *JHEP*, 05:003.

540 Dokshitzer, Yu.L., Marchesini, G., and Webber, B.R. (1996) *Nucl. Phys. B*, 469:93–142.

541 Dokshitzer, Yu.L. and Webber, B.R. (1997) *Phys. Lett. B*, 404:321–327.

542 Abdallah, P. *et al.* (DELPHI) (2003) *Eur. Phys. J. C*, 29:285–312.

543 Heister, A. *et al.* (ALEPH) (2004) *Eur. Phys. J. C*, 35:457–486.

544 Dokshitzer, Yu.L. (1999) arXiv:hep-ph/9911299.

545 Kluth, S. (2006) *Rep. Prog. Phys.*, 69:1771–1846.

546 Bethke, S. (2009) *Eur. Phys. J. C*, 64:689–703.

547 Dokshitzer, Yu.L., Lucenti,A., Marchesini, G., and Salam, G.P. (1998) *Nucl. Phys. B*, 511:396–418.

548 Ellis, J., Gaillard, M.K., and Ross, G.G. (1976) *Nucl. Phys. B*, 111:253–271.

549 Laermann, E., Streng, K.H., and Zerwas, P.M. (1980) *Z. Phys. C*, 3:289–298.

550 Alexander, G. *et al.* (OPAL) (1991) *Z. Phys. C*, 52:543–550.

551 Wiik, B.W. (1981) *Phys. Scr.*, 23:895–913.

552 Ellis, J. and Karliner, I. (1979) *Nucl. Phys. B*, 148:141–147.

553 Ellis, J., Karliner, I., and Stirling, W.J. (1989) *Phys. Lett. B*, 217:363–368.

554 Adeva, B. *et al.* (L3) (1991) *Phys. Lett. B*, 263:551.

555 Park, I.H. *et al.* (AMY) (1989) *Phys. Rev. Lett.*, 62:1713–1716.

556 Bethke, S., Ricker, A., and Zerwas, P.M. (1991) *Z. Phys. C*, 49:59–72.

557 Bengtsson, M. and Zerwas, P.M. (1988) *Phys. Lett. B*, 208:306–308.

558 Nachtman, O. and Reiter, A. (1982) *Z. Phys. C*, 16:45–54.

559 Akrawy, M.Z. *et al.* (OPAL) (1991) *Z. Phys. C*, 49:49–57.

560 Adeva, B. *et al.* (L3) (1990) *Phys. Lett. B*, 248:227–234.

561 Ali, A. *et al.* (1979) *Phys. Lett. B*, 82:285–288.

562 Nagy, Z. and Trócsányi, Z. (1997) *Phys. Rev. Lett.*, 79:3604–3607.

563 Nagy, Z. and Trócsányi, Z. (1998) *Phys. Rev. D*, 59:014020.

564 Kluth, S. (2003) *Proc. 10th Int. QCD Conf., Montpellier, France, 2–9 July 2003* hep-ex/0309070.

565 Dokshitzer, Yu.L., Khoze, V.A., Troyan, S.I., and Mueller, A.H. (1988) QCD coherence in high-energy physics. *Rev. Mod. Phys.*, 60:373–388.

566 Marchesini, G. *et al.* (1990) *Nucl. Phys. B*, 330:261–283.

567 Mueller, A.H. (1981) *Phys. Lett. B*, 104:161–164.

568 Azimov, Ya.I., Dokshitzer, Yu.L., Khoze, V.A., and Troyan, S.I. (1985) *Phys. Lett. B*, 165:147.

569 Aihara, H. *et al.* (1986) *Phys. Rev. Lett.*, 57:945–948.

570 Biebel, O. (2001) *Phys. Rep.*, 340:165–289.

571 Mueller, A.H. (1984) *Nucl. Phys. B*, 241:141–154.

572 Webber, B.R. (1984) *Phys. Lett. B*, 143:501–504.

573 Dokshitzer, Yu.L., Khoze, V.A., and Troyan, S.I. (1992) *Int. J. Mod. Phys. A*, 7:1875.

574 Khoze, V.A. and Ochs, W. (1997) *Int. J. Mod. Phys. A*, 12:2949.

575 Fong, C.P. and Webber, B.R. (1991) *Nucl. Phys. B*, 355:54–81.

576 Mueller, A.H. (1983) *Nucl. Phys. B*, 213:85–108. Erratum ibid. (1984) 241:141.

577 Mueller, A.H. (1983) *Nucl. Phys. B*, 228:351–364.

578 Azimov, Ya.I., Dokshitzer, Yu.L., Khoze, V.A., and Troyan, S.I. (1985) *Z. Phys. C*, 27:65–72.

579 Azimov, Ya.I., Dokshitzer, Yu.L., Khoze, V.A., and Troyan, S.I. (1986) *Z. Phys. C*, 31:213.

580 Dokshitzer, Yu.L., Khoze, V.A., Mueller, A.H., and Troyan, S.I. (1991) *Basics of Perturbative QCD*. Editions Frontiers, France.

581 Dokshitzer, Yu.L. and Troyan, S. (1984) Proc. XIX Winter School of the LNPI. Leningrad preprint, LNPI-922.

582 Akrawy, M.Z. *et al.* (OPAL) (1990) *Phys. Lett. B*, 247:617–628.

583 Safanov, A.N. (2000) *Nucl. Phys. (Proc. Supple.) B*, 86:55–64. hep-ex/0007037v2.

584 Khoze, V.A., Lupia, S., and Ochs, W. (1996) *Phys. Lett. B*, 386:451–457.

585 Mele, B. and Nason, P. (1991) *Nucl. Phys. B*, 361:626–644.

586 Akers, R. *et al.* (OPAL) (1995) *Z. Phys. C*, 68:203–213.

587 Busklic, D.D. *et al.* (ALEPH) (1995) *Phys. Lett. B*, 357:487–499. Erratum ibid. (1995) 364, 247.

588 Abreu, P. *et al.* (DELPHI) (1999) *Eur. Phys. J. C*, 6:19–33.

589 Abreu, P. *et al.* (DELPHI) (1996) *Z. Phys. C*, 70:179–195.

590 Arnison, G. *et al.* (UA1) (1986) *Nucl. Phys. B*, 276:253–271.

591 Banner, M. *et al.* (UA2) (1984) *Phys. Lett. B*, 138:430–440.

592 Derrick, M. *et al.* (1985) *Phys. Lett. B*, 165:449–453.

593 Petersen, A. *et al.* (MKII) (1985) *Phys. Rev. Lett.*, 55:1954–1957.

594 Braunschweig, W. *et al.* (TASSO) (1989) *Z. Phys. C*, 45:1–10.

595 Kim, Y.K. *et al.* (AMY) (1989) *Phys. Rev. Lett.*, 63:1772–1775.

596 Takaki, H. *et al.* (VENUS) (1993) *Phys. Rev. Lett.*, 71:38–41.

597 Abreu, P. *et al.* (DELPHI) (1999) *Phys. Lett. B*, 449:383–400.

598 Abreu, P. *et al.* (DELPHI) (2000) *Eur. Phys. J. C*, 13:573–589.
599 Butterworth, J.M. and Wing, M. (2005) *Rep. Prog. Phys.*, 68:2773–2828.
600 Akers, M.Z. *et al.* (OPAL) (1994) *Z. Phys. C*, 63:197–211.
601 Abe, F. *et al.* (CDF) (1993) *Phys. Rev. Lett*, 70:713–717.
602 Ansari, R. *et al.* (UA2) (1987) *Z. Phys. C*, 36:175–187.
603 Shapiro, M.D. and Siegrist, J.L. (1991) Hadron collider physics. *Annu. Rev. Nucl. Part. Sci.*, 41:97–132.
604 Groom, D.E. *et al.*, (Particle Data Group) (2000) *The European Physical Journal C*, 15:1.
605 Froissart, M. (1961) *Phys. Rev.*, 123:1053.
606 Berman, S.M., Bjorken, J.D., and Kogut, J.B. (1971) *Phys. Rev. D*, 4:3388–3418.
607 Akesson, T. *et al.* (1982) *Phys. Lett. B*, 118:185, 193.
608 Appel, J.A. *et al.* (UA2) (1985) *Phys. Lett. B*, 165:441–448.
609 Alner, G.J. *et al.* (UA5) (1987) *Nucl. Phys. B*, 291:445.
610 Aaltonen, T. *et al.* (CDF) (2010) *Phys. Rev.*, D82:034001.
611 Abe, F. *et al.* (CDF) (1992) *Phys. Rev. D*, 45:2249–2263.
612 Huth, J.E. Proc. *DPF91. Vancouver, Aug., 1991.*
613 Lai, H.L. *et al.* (1996) *Phys. Rev. D*, 55:1280–1296.
614 Affolder, T. *et al.* (CDF) (2001) *Phys. Rev. D*, 64:032001.
615 Combridge, B.L., Kripfganz, J., and Ranft, J. (1977) *Phys. Lett. B*, 70:234–238.
616 Combridge, B.L. and Maxwell, C.J. (1984) *Nucl. Phys. B*, 239:429–458.
617 Arnison, G. *et al.* (UA1) (1985) *Phys. Lett. B*, 158:494–504.
618 Arnison, G. *et al.* (UA1) (1986) *Phys. Lett. B*, 177:244–250.
619 Bagnaia, P. *et al.* (UA2) (1984) *Phys. Lett. B*, 144:283–290.
620 Abbott, B. *et al.* (D0) (1998) *Phys. Rev. Lett.*, 80:666–671.
621 Duke, D.W. and Owens, J.F. (1984) *Phys. Rev. D*, 84:49–54.
622 Terrón, J. (Madrid) (2002) ZEUS Weekly, Aug. 19.
623 Abe, F. *et al.* (CDF) (1992) *Phys. Rev. Lett*, 68:1104–1108.
624 Ellis, S.D., Kunszt, Z., and Soper, D.E. (1989) *Phys. Rev. D*, 40:2188–2222.
625 Ellis, S.D., Kunszt, Z., and Soper, D.E. (1990) *Phys. Rev. Lett.*, 64:2121–2124.
626 Ellis, S.D., Kunszt, Z., and Soper, D.E. (1992) *Phys. Rev. Lett.*, 69:3615.
627 Abe, F. *et al.* (CDF) (1996) *Phys. Rev. Lett.*, 77:438–443.
628 Campbell, J.M., Huston, J.W., and Stirling, W. (2007) *Reg. Prog. Phys.*, 70:89–193.
629 Acosta, D. *et al.* (CDF) (1005) *Phys. Rev. D.*, 71:112002.
630 Catani, S. *et al.* (1993) *Nucl. Phys. B*, 406:187–224.
631 Ellis, S.D. and Soper, D.E. (1993) *Phys. Rev. D*, 48:3160–3166.
632 Nagy, Z. (2002) *Phys. Rev. Lett.*, 88:122003.
633 Stump, D. *et al.* (2003) *JHEP*, 10:046.
634 Abulencia, A. *et al.* (CDF) (2006) *Phys. Rev. Lett.*, 96:122001.
635 Abe, F. *et al.* (CDF) (1992) *Phys. Rev. D*, 45:1448–1458.
636 Acosta, D. *et al.* (CDF) (2005) *Phys. Rev. D*, 71:032002.
637 Abe, F. *et al.* (CDF) (1995) *Phys. Rev. Lett.*, 74:850–854.
638 Abazov, V.M. *et al.* (D0) (2003) *Phys. Rev. D*, 67:052001.
639 Kunszt, Z. and Stirling, W.J. (1986) *Phys. Lett. B*, 171:307–312.
640 Berends, F.A. and Kuijf, H. (1991) *Nucl. Phys. B*, 353:59–86.
641 Eichten, E.J., Lane, K.D., and Peskin, M.E. (1983) *Phys. Rev. Lett.*, 50:811–814.
642 Eichten, E.J., Hincliffe, I., Lane, K.D., and Quigg, C. (1984) *Rev. Mod. Phys.*, 56:579–704.
643 Abazov, V.M. *et al.* (D0) (2009) *Phys. Rev. Lett.*, 103:191803.
644 Abazov, V.M. *et al.* (D0) (2008) *Phys. Lett. B*, 666:435.
645 Abe, F. *et al.* (CDF) (1994) *Phys. Rev. Lett*, 73:2662–2667.
646 Acosta, D. *et al.* (CDF) (2002) *Phys. Rev. D*, 65:112003.
647 Aaltonen, T. *et al.* (CDF) (2009) *Phys. Rev. D*, 80:111106.

648 Altarelli, G., Ellis, R.K., Greco, M., and Martinelli, G. (1984) *Nucl. Phys. B*, 246:12–44.

649 Altarelli, G., Ellis, R.K., and Martinelli, G. (1985) *Z. Phys. C*, 27:617–632.

650 Arnold, P.B. and Reno, M.H. (1989) *Nucl. Phys. B*, 319:37–71. ibid. (1990) 330, 284 (erratum)

651 Arnold, P.B., Kaufman, R.P. (1991) *Nucl. Phys. B*, 349:381–413.

652 Gonsalves, R.J., Pawlowski, J., and Wai, C.-F. (1989) *Phys. Rev. D*, 40:2245–1989.

653 Ladinsky, G.A. and Yuan, C.-P. (1994) *Phys. Rev. D*, 50:R4239–R4243.

654 Abazov, V.M. *et al.* (D0) (2001) *Phys. Lett. B*, 513:292–300.

655 Collins, J.C. and Soper, D.E. *Nucl. Phys. B*, 193:381–443, 1981; ibid. B197:446-476,1982; ibid. B213:543, 1983 (errata).

656 Collins, J.C., Soper, D.E., and Sterman, G. (1985) *Nucl. Phys. B*, 250:199–224.

657 Davies, C.T.H. and Stirling, W.J. (1984) *Nucl. Phys. B*, 244:337–348.

658 Davies, C.T.H., Webber, B.R., and Stirling, W.J. (1985) *Nucl. Phys. B*, 256:413–433.

659 Balázs, C. and Yuan, C.-P. (1997) *Phys. Rev. D*, 56:5558–5583.

660 Kodaira, J. and Trentadue, L. (1982) *Phys.Lett. B*, 112:66–70.

661 Ellis, K. and Veseli, S. (1998) *Nucl. Phys. B*, 511:649–669.

662 Barger, V.D. and Phillips, R.J.N. (1997) *Collider Physics, Updated Edition*. Addison-Wesley.

663 Mangano, M.L. (1997) *Two lectures on heavy quark production in hadronic collisions*. hep-ph/9711337.

664 Moch, S. and Uwer, P. (2008) *Phys. Rev. D*, 78:034003.

665 Nason, P., Dawson, S., and Ellis, R.K. (1988) *Nucl. Phys. B*, 303:607–633.

666 Beenacker, W., Kuijf, H., and van Neerven, W.L. (1989) *Phys. Rev. D*, 40:54–82.

667 Nason, P., Dawson, S., and Ellis, R.K. (1989) *Nucl. Phys. B*, 327:49–92.

668 Nason, P., Dawson, S., and Ellis, R.K. (1990) Erratum. *Nucl. Phys. B*, 335:260.

669 Beenacker, W. *et al.* (1991) *Nucl. Phys. B*, 351:507–560.

670 Mangano, M.L. (1992) *Nucl. Phys. B*, 373:295–345.

671 Frixione, S., Mangano, M.L., Nason, P., and Ridolfi, G. (1994) *Nucl. Phys. B*, 431:453–483.

672 Collins, J.C., Soper, D.E., and Sterman, G. (1986) *Nucl. Phys. B*, 263:37–60.

673 Lourenco, C. and Wohri, H.K. (2006) *Phys. Rep.*, 433:127–180.

674 Acosta, D. *et al.* (CDF) (2005) *Phys. Rev. D*, 71:032001.

675 Cacciari, M. (2004) *Rise and fall of the bottom quark production excess*. hep-ph/0407187v1.

676 Cacciari, M. *et al.* (2004) *JHEP*, 07:033.

677 Abe, F. *et al.* (CDF) (1993) *Phys. Rev. Lett.*, 71:500–504.

678 Abe, F. *et al.* (CDF) (1993) *Phys. Rev. Lett.*, 71:2396–2400.

679 Abe, F. *et al.* (CDF) (1995) *Phys. Rev. Lett.*, 75:1451–1455.

680 Abbott, B. *et al.* (D0) (2000) *Phys. Lett. B*, 487:264–272.

681 Albrow, M.H. *et al.* (CDF) (2002) *Phys. Rev. D*, 65:052005.

682 Cacciari, M., Greco, M., and Nason, P. (1998) *JHEP*, 05:007.

683 Cacciari, M. and Nason, P. (2002) *Phys. Rev. Lett.*, 89:122003.

684 Arnison, G. *et al.* (UA1) (1984) *Phys. Lett. B*, 147:493–506.

685 Abe, F. *et al.* (CDF) (1994) *Phys. Rev. Lett*, 73:225–231.

686 Albrow, M.H. *et al.* (CDF) (2002) *Phys. Rev. D*, 50:2966–3026.

687 Azzi, P. (2003) arXiv:hep-ex/0312 052v1.

688 Incandela, J.R. *et al.* (2009) *Prog. Part. Nucl. Phys.*, 63:239–292. arXiv:hep-ex/0904.2499v33.

689 Lister, A. (2008) *hep-ex/0810.3350v2*. CDF Conf. Note 9448.

690 Abazov, V.M. *et al.* (2008) *Phys. Rev. Lett.*, 100:192004.

691 Moch, S. and Uwer, P. (2008) *Nucl. Phys. Proc. Suppl. B*, 183:75–80.

692 Cacciari, M. *et al.* (2004) *JHEP*, 04:068.

693 Abulencia, A. *et al.* (CDF) (2006) *Phys. Rev. Lett.*, 96:022004.

694 Aad, G. *et al.* (ATLAS) (2012) *Phys. Rev. Lett.*, 108:111803.

695 Aaltonen, T. *et al.* (CDF/D0) (2012) *Phys. Rev. Lett.*, 109:071804.

696 Aaltonen, T. *et al.* (CDF) (2011) *Phys. Rev. D*, 84:071105(R).

697 Ellis, R.K. and Stirling, W.J. (1990) Fermilab-Conf-90/164-T.

698 Lee, B.W. and Jinn-Justin, J. (1978) *Phys. Rev. D*, 7:1049–1056.

699 Fujikawa, K., Lee, B.W., and Sanda, A.I. (1972) *Phys. Rev. D*, 6:2923–2943.

700 Cheng, T.P. and Li, L.F. (1984) *Gauge Theory of Elementary Particle Physics.* Clarendon Press, Oxford.

701 Durand III, L., DeCelles, P.C., and Marr, R.B. (1962) *Phys. Rev.*, 126:1882–1898.

702 Carruthers, P.A. (1966) *Introduction to Unitary Symmetry.* John Wiley & Sons, Ltd, New York.

703 Gasiorowicz, S. (1966) *Elementary Particle Physics.* John Wiley & Sons, Ltd, New York.

704 Close, F.E. (1979) *An Introduction to Quarks and Leptons.* Academic Press.

705 Georgi, H. (1999) *Lie Algebras in Particle Physics.* Perseus Books.

706 Georgi, H. (1984) *Weak Interactions and Modern Particele Theory.* Addison-Wesley.

707 Gell-Mann, M. and Levy, M. (1960) *Nuovo Cim.*, 16:705.

708 Goldberger, M.L. and Treiman, S.B. (1958) *Phys. Rev.*, 110:1178.

709 Adler, S. (1965) *Phys. Rev.*, 140:736.

710 Weisberger, W.I. (1966) *Phys. Rev.*, 143:1302.

711 Weise, W. (1989) *Nucl. Phys. A*, 497:7–22.

712 Coleman, S., Wess, J., and Zumino, B. (1969) *Phys. Rev.*, 177:2239–2247.

713 Callan, C.G. *et al.* (1969) *Phys. Rev.*, 177:2247–2250.

714 Ovrut, B.A. and Schnitzer, H.J. (1980) *Phys. Rev. D*, 22:2518–2533.

Index

Elementary Particle Physics, Volume 2: Foundations of the Standard Model, First Edition. Yorikiyo Nagashima.